THEORETICAL SOLID STATE PHYSICS

Volume 1: Perfect Lattices in Equilibrium

WILLIAM JONES
Lecturer in Physics,
The University of Sheffield

NORMAN H. MARCH
Coulson Professor of Theoretical Chemistry,
University of Oxford
(formerly Professor of
Theoretical Solid State Physics,
Imperial College of Science and Technology,
University of London)

Dover Publications, Inc., New York

Published in Canada by General Publishing Company, Ltd., 30 Lesmill Road, Don Mills, Toronto, Ontario.
Published in the United Kingdom by Constable and Company, Ltd., 10 Orange Street, London WC2H 7EG.

This Dover edition, first published in 1985, is an unabridged and unaltered republication of the work first published by Wiley-Interscience, a division of John Wiley & Sons Ltd., London, in 1973.

Manufactured in the United States of America
Dover Publications, Inc., 31 East 2nd Street, Mineola, N.Y. 11501

Library of Congress Cataloging in Publication Data

Jones, William, 1940–
Theoretical solid state physics.

Reprint. Originally published: London ; New York : Wiley-Interscience, 1973.
Bibliography: p.
Contents: v. 1. Perfect lattices in equilibrium—v. 2. Non-equilibrium and disorder.
1. Solid state physics—Collected works. I. March, Norman H. (Norman Henry), 1927– . II. Title.
[QC176.A1J572 1985] 530.4′1 85-10165
ISBN 0-486-65015-4 (vol. 1)
ISBN 0-486-65016-2 (vol. 2)

Preface

This book has been written with professional theoretical physicists in mind and with a view to the training of postgraduate students in theoretical physics. To this end a fairly wide selection of problems has been included in both Volumes.

The two Volumes of the work are concerned with very different aspects of the subject. Volume I represents an attempt to deal with the fundamental theory of the equilibrium properties of perfect crystalline solids. On the other hand, Volume II deals in detail with non-equilibrium properties and also includes a discussion of defects and disordered systems. This latter field is in a state of very rapid development at the time of writing but we felt it essential to include such an account in a work of the present scale, in order to preserve some balance in our treatment.

The writing began in 1966 and we quickly recognized the Herculean task on which we had embarked. It seemed to us, nonetheless, that what was needed was a fairly detailed treatment of the theory, not by-passing difficult many-body theory, but rather laying down foundations where it seemed clear that further structures could be erected.

We also felt that, whenever limitations of space permitted, we should not leave the theory at these foundations but should carry the discussion as far as current papers on the subject. We are conscious still, in spite of the vastness of this work, of the many areas we have either omitted, or to which we have obviously not done justice.

The theory of solids has, in places, now reached such a high level of sophistication that no theory group that exists at present can claim first-hand knowledge of the whole subject. We can only add that many friends and colleagues who are more expert than us in particular areas have given us the benefit of their criticism and advice. We need hardly say that errors and misunderstandings that remain, as they surely will in an undertaking of this scale, are entirely our own responsibility.

Specifically, in our own laboratory especial thanks are due to Drs J. C. Stoddart and J. W. Tucker for their very generous help. Other friends who have made extremely helpful comments are Drs R. D. Lowde, D. J. Morgan, C. Norris, T. M. Rice, P. Schofield, G. L. Sewell, L. J. Sham, and Professors

G. Rickayzen and W. H. Young. Many other folk have helped us too, and we express here our gratitude to them.

Finally, Miss P. A. Traviss (now Mrs A. Hooper) gave us enormous help with the preparation of the manuscript and our grateful thanks are due to her for her expert typing and for her tolerance and cheerfulness throughout the whole exercise.

Contents

THEORETICAL SOLID STATE PHYSICS

CHAPTER 1

Band Theory and Crystal Symmetry

1.1 Introduction

Two aspects of the problem of electrons moving in crystalline solids will be dealt with in this chapter. The more basic of the two is that which deals with the effect of crystal symmetry on the wave functions. While much of the theory of crystal symmetry is general and can be applied to the many-electron wave functions of the whole crystal, in practice the one-electron approximation is so useful that we shall present the discussion in this framework.

In the Hartree one-electron picture, the electrons move in a local potential, whereas in the Hartree–Fock scheme the potential is non-local or energy dependent (see Appendix 1.1). In principle band theory includes both these cases, but in practice it most often has to do with a common local potential in which all the electrons are assumed to move. It is then conceptually nearer to the Hartree picture than the Hartree–Fock, though in the Hartree method a chosen electron moves in a slightly different potential from its fellow electrons, in any finite system.

The second aspect of the problem is naturally, then, a discussion of the tools by which we can solve the periodic potential problem. This is dealt with at length in sections 1.8–1.11. Some preliminary justification for band theory is considered immediately below, while the group theory is dealt with in sections 1.4–1.7 and in Appendix 1.3.

1.2 Basis of one-electron local potential theory

As indicated above, Hartree introduced the fruitful concept in atoms that, instead of working with the total wave function of the system, one should allocate to each electron its own personal wave function and energy level. In the calculation of a single-particle wave function, account has to be taken of the charge clouds of the other Z-1 electrons, together, of course, with the dominant Coulombic attraction of the nucleus of charge Ze. For many aspects of atomic theory, this approximation turns out to give very accurate agreement with experiment.

However, when we turn to crystals, and in particular to metals where the density of conduction electrons is very high ($\sim 10^{23}/\text{cm}^3$), we might expect the effects of electron–electron interactions to vitiate the simple Hartree picture. In spite of this, the approximation of single-particle levels and wave functions is again supported by a very wide variety of experiments.

The way towards the resolution of this difficulty was pointed out by Bohm and Pines (1951, 1952, 1953). They argued that the long-range Coulomb interaction could lead to collective behaviour of the conduction electrons in a metal, but that, once these organized oscillations were accounted for, the effective interaction was both weak and relatively short ranged compared with a Coulomb field. We shall discuss the collective behaviour of an electron gas in detail in the next chapter, but until that point it will be useful to bear in mind that these collective, or plasma, oscillations are not excited in a metal under normal circumstances. Thus, for many purposes, an independent electron description will again suffice.

1.2.1 Relation between electron density and local potential

It is also of considerable significance that besides the above physical argument for a one-electron picture, there are aspects of the full many-body problem that can be described, in principle exactly, in terms of a local one-body potential.

As an introduction to the real many-body problem, let us consider first the elementary results for a free Fermi gas of electrons at absolute zero. Then, momentum space is filled out to the Fermi momentum p_f and is entirely unoccupied outside a sphere of radius p_f. If ρ_0 is the electron density, then it follows that, since we can divide the phase space into cells of volume h^3 and put two electrons with opposed spins into each cell,

$$N = \frac{2}{h^3}\cdot\frac{4\pi p_f^3}{3}\mathscr{V} \tag{1.2.1}$$

or

$$\rho_0 = \frac{8\pi}{3h^3}p_f^3, \tag{1.2.2}$$

where N is the total number of electrons in the volume $\mathscr{V}$ of the metal.

Let us suppose that we "switch on" to the Fermi gas a potential energy $V(\mathbf{r})$ which is slowly varying in space. Then the Fermi gas density ρ becomes inhomogeneous, and if E_f is the Fermi energy we may write down the classical energy equation for the fastest electron:

$$E_f = \frac{p_f^2(\mathbf{r})}{2m} + V(\mathbf{r}), \tag{1.2.3}$$

where E_f must of course be independent of $\mathbf{r}$, for otherwise the electrons could redistribute themselves to lower the total energy. Using equation (1.2.2) for the local density $\rho(\mathbf{r})$, i.e.

$$\rho(\mathbf{r}) = \frac{8\pi}{3h^3} p_f^3(\mathbf{r}) \tag{1.2.4}$$

and substituting for the maximum momentum $p_f(\mathbf{r})$ from equation (1.2.3) we find

$$\rho(\mathbf{r}) = \frac{8\pi}{3h^3}(2m)^{\frac{3}{2}} [E_f - V(\mathbf{r})]^{\frac{3}{2}}. \tag{1.2.5}$$

This relation between the density of particles $\rho(\mathbf{r})$ and the common potential $V(\mathbf{r})$ in which they are assumed to move is only exact when V is constant in space, but is the basis of the valuable Thomas–Fermi approximation for slow variations in $V(\mathbf{r})$. This means that the potential should vary by only a fraction of its value over a characteristic electron wavelength, which, in the present case, will be the de Broglie wavelength of an electron at the Fermi energy. Unfortunately, this latter condition is not well satisfied in atoms, molecules or solids and we must transcend the approximation (1.2.5).

An exact perturbative generalization of equation (1.2.5) was given by March and Murray (1960, 1961) using the density matrix defined later and the desired expansion in powers of the potential $V(\mathbf{r})$ is set out in Appendix 1.2. When the spatial variation of the potential is slow, then the perturbation series can be summed, and equation (1.2.5) is regained. The conclusion is that the density $\rho(\mathbf{r})$ given by

$$\rho(\mathbf{r}) = \sum_i^{E_f} \psi_i^*(\mathbf{r})\,\psi_i(\mathbf{r}), \tag{1.2.6}$$

with one-electron wave functions $\psi_i(\mathbf{r})$ being generated by the one-particle Schrödinger equation with potential energy $V(\mathbf{r})$,

$$-\frac{\hbar^2}{2m}\nabla^2\psi_i + V(\mathbf{r})\,\psi_i = E_i\,\psi_i \tag{1.2.7}$$

can be obtained as a *functional* of the potential $V(\mathbf{r})$. The relationship between ρ and V is, in general, non-local; that is, the density ρ at point $\mathbf{r}$ is determined by the values of V everywhere, in contrast to the approximation of equation (1.2.5).

More important for our purposes is that the density $\rho(\mathbf{r})$ in a crystal is, in principle (see Chapter 5), accessible experimentally by measuring the absolute intensities for X-ray or electron scattering at the Bragg reflexions. Knowing $\rho(\mathbf{r})$, the functional relation between density and potential referred to above may be inverted, and we can in principle obtain a unique local potential $V(\mathbf{r})$ from experiment, which will yield the exact particle density appropriate to the many-body system of interacting electrons moving in the lattice potential.

Of course this, of itself, will not be helpful unless the potential $V(\mathbf{r})$ can be used to calculate other physical properties. However, we shall see in Chapter 3 that it is highly relevant to a calculation of lattice vibrational properties as this can be reduced, to a good approximation, to the calculation of a ground-state electron density. A much more difficult question is whether such a local, energy-independent, potential $V(\mathbf{r})$ will generate a useful approximation to the low-lying single-particle states. While a precise answer is lacking as yet, the available evidence suggests that this is the case, for example, in a normal metal, where there are excited states which are separated from the ground state by an infinitesimal energy (Sham and Kohn, 1966). The problem of an energy gap, as in the superconducting state (cf. Chapter 8), needs different treatment.

1.2.2 *Total energy of ground state*

While the excited states will remain a prime objective for calculation from such an "experimentally" determined local potential, it can be shown rigorously that the many-body ground state energy is determined uniquely by the density $\rho(\mathbf{r})$ (Hohenberg and Kohn, 1964).

Let us again approach the matter via one-particle theory, where the essential results are known. Thus, if we consider the kinetic energy T consistent with the (ρ, V) relation (1.2.5), we obtain

$$T = \frac{3h^2}{10m}\left(\frac{3}{8\pi}\right)^{\frac{2}{3}} \int d\mathbf{r}\, \rho^{\frac{5}{3}}(\mathbf{r}), \tag{1.2.8}$$

since the kinetic energy/particle in a free-electron gas is simply $\frac{3}{5}E_f$ and hence, from (1.2.2), and (1.2.3) with V set equal to zero, the kinetic energy/unit volume is proportional to $\rho_0^{\frac{5}{3}}$. Equation (1.2.8) merely uses this free-electron relation locally, exactly as we did in passing from (1.2.2) to (1.2.4). The electron–electron interaction energy is readily seen to be

$$V_{ee} = \frac{e^2}{2}\int\int \frac{\rho(\mathbf{r})\,\rho(\mathbf{r}')}{|\mathbf{r}-\mathbf{r}'|}\, d\mathbf{r}\, d\mathbf{r}' \tag{1.2.9}$$

and is again a functional of the density $\rho(\mathbf{r})$. The interaction energy of the electrons with the potential is of course

$$V_{\text{ep}} = \int d\mathbf{r}\, \rho(\mathbf{r})\, [V_{\text{ext}}(\mathbf{r}) + V_{\text{ee}}(\mathbf{r})], \tag{1.2.10}$$

where the external potential $V_{\text{ext}}(\mathbf{r})$ is, for example, in an atom of atomic number Z simply $-Ze^2/r$, and $V_{\text{ee}}(\mathbf{r})$ is the potential energy contribution from the electron cloud.

Equations (1.2.8)–(1.2.10) show that, in Thomas–Fermi theory, the total energy of the ground state can be obtained from a knowledge of the particle density $\rho(\mathbf{r})$, together with the external potential, which clearly must be known before $\rho(\mathbf{r})$ can be calculated.

Furthermore, if we vary the total energy $E = T - V_{\text{ep}} + V_{\text{en}}$ with respect to the particle density ρ, subject only to the requirement that the integral $\int \rho(\mathbf{r})\, d\mathbf{r}$ giving the total number of electrons remains constant, the Euler equation of this variational problem (cf. Appendix 1.2) turns out to be simply equation (1.2.5), E_{f} playing the role of a Lagrange multiplier.‡

Just as the (ρ, V) relation (1.2.5) can be generalized to avoid the slowly varying potential approximation, as set out in Appendix 1.2, so the variational principle consistent with this new (ρ, V) relation can also be found, as shown in the same appendix.

However, whereas these results are still restricted to the one-electron approximation, as mentioned above Hohenberg and Kohn (1964) have shown that, even in the full many-body problem, the total energy of the ground state may be formally calculated from a knowledge of the density $\rho(\mathbf{r})$ and the external potential $V_{\text{ext}}(\mathbf{r})$. This result is of sufficient importance for us to construct the proof here and now from the variational principle for the many-body wave function.

The result hinges on the proof that there exists but one external potential which gives the charge density $\rho(\mathbf{r})$. Suppose there are two external potentials $V(\mathbf{r})$ and $V_1(\mathbf{r})$ with corresponding normalized ground-state wave functions $\Psi_0(\mathbf{r})$ and $\Psi_1(\mathbf{r})$ and energies E_0 and E_1. By the variational principle

$$E_0 = \langle \Psi_0 | H | \Psi_0 \rangle < \langle \Psi_1 | H | \Psi_1 \rangle;$$
$$\langle \Psi | H | \Psi_1 \rangle = \langle \Psi_1 | H_1 | \Psi_1 \rangle + \langle \Psi_1 | V - V_1 | \Psi_1 \rangle, \tag{1.2.11}$$

where H and H_1 are the many-body Hamiltonians for the two external potentials. Let us now suppose that $V(\mathbf{r})$ and $V_1(\mathbf{r})$ give the same particle density. Then this equation becomes

$$E_0 < \langle \Psi_1 | H_1 | \Psi_1 \rangle + \langle \Psi_0 | V - V_1 | \Psi_0 \rangle \tag{1.2.12}$$

since the last term involves only the density and external potentials. Hence

$$E_0 < \langle \Psi_0 | H_1 | \Psi_0 \rangle + \langle \Psi_0 | V - V_1 | \Psi_0 \rangle, \tag{1.2.13}$$

‡ Nuclear–nuclear interactions, independent of ρ, are not shown.

using again the variational principle for H_1. But the right-hand side of this inequality is just E_0, and an obvious contradiction arises. Hence the external potential uniquely determines the charge density.

Since the total energy is determined uniquely by the external potential, the above result leads to the conclusion that the total energy is uniquely defined by the particle density. Unfortunately, a first principles calculation of this functional relation $E_0[\rho]$ presents enormous difficulties in the presence of interactions. Nevertheless, a formal relation between ρ and V can be obtained by writing

$$E_0 = T_0[\rho]+V[\rho], \tag{1.2.14}$$

where $T_0[\rho]$ is the kinetic energy of the system of *non*-interacting particles which has the electron density $\rho(\mathbf{r})$.‡ We stress that $V[\rho]$ includes not only potential energy but some kinetic energy resulting from correlations. Let us now assume the first-order change in $V[\rho]$ (when ρ is varied in a manner arbitrary apart from the requirement of particle conservation) is

$$\delta V[\rho] = \int V(\mathbf{r})\,\delta\rho(\mathbf{r})\,d\mathbf{r} \tag{1.2.15}$$

with

$$\int \delta\rho(\mathbf{r})\,d\mathbf{r} = 0. \tag{1.2.16}$$

Then in such a density variation we find

$$\delta E_0 = \delta T_0[\rho]+\int V(\mathbf{r})\,\delta\rho(\mathbf{r})\,d\mathbf{r} = 0. \tag{1.2.17}$$

The total energy change is zero by the variation principle. But because of the definition of $T_0[\rho]$ this equation also expresses the variational principle for a set of non-interacting particles moving in an external potential $V(\mathbf{r})$. Equation (1.2.17), therefore, constitutes a definition of the local potential reproducing the exact density, provided equation (1.2.15) holds.

Dirac (1930), Slater (1951) and many other workers have considered the introduction of exchange, or a combination of exchange and correlation effects, into the Thomas–Fermi theory. Unfortunately, we do not know the correlation energy exactly, except for very high densities, even in the simplest case of a uniform Fermi gas (Chapter 2). It is therefore unlikely that we shall know $E_0[\rho]$ exactly over the next few decades. However, progress in approximating to this is reported in section 2.11.

Though we are dealing in crystals with a many-body system of electrons and ions which is exceedingly complex, our purpose here has been to show that, while no rigorous prescription exists as yet for a first principles

‡ The (ρ, V) relation of Appendix 1.2 can be inverted to yield a one-body potential $V(\mathbf{r})$ for any given $\rho(\mathbf{r})$.

calculation, some justification can be given for seeking a one-body local potential on which to found the theory of electron states. It should be emphasized again that this is different from a Hartree potential, and takes some account of exchange and correlation effects. Furthermore, in a metallic crystal, we referred to the qualitative argument of Bohm and Pines that, once the plasma or collective oscillations are accounted for, the residual electron–electron interaction is relatively weak and short ranged. Though these two arguments are totally different in character they both point to the same conclusion; namely that one-body potential theory is likely to be of considerable utility in treating electrons in crystals.

It is then obvious on physical grounds that such a potential will have the symmetry of the crystal lattice and we shall turn immediately to discuss the consequences of this property of the potential. We shall thereby be led to a group-theoretical classification of the one-electron wave functions. However, much of the ensuing discussion will be relevant not only to electron states in crystals, but also to the properties of lattice vibrations discussed fully in Chapter 3. We begin with a description of the direct and reciprocal lattices.

1.3 Direct and reciprocal lattices

1.3.1 Bravais and direct lattices

For every lattice we may find three primitive translations $\mathbf{R}_1, \mathbf{R}_2, \mathbf{R}_3$ such that any vector from one position in the crystal to another with exactly similar physical environment is given by

$$\mathbf{R} = l\mathbf{R}_1 + m\mathbf{R}_2 + n\mathbf{R}_3, \tag{1.3.1}$$

where l, m and n are integers. We immediately see that $\mathbf{R}_1$, $\mathbf{R}_2$ and $\mathbf{R}_3$ are the three smallest non-zero lattice vectors. The vectors $\mathbf{R}$ given by equation (1.3.1) define a set of points which is known as the *Bravais lattice* of the crystal.

The *unit cell* is the smallest volume for which translation of all interior points through the vectors $\mathbf{R}$ enables all other points in the crystal to be reached. The unit cell then contains just one Bravais lattice point, and we can construct a unit cell about each lattice point. There is arbitrariness in the positions of the boundaries of the unit cell relative to the lattice point it contains. Generally, however, we shall find it most convenient to construct the boundaries as planes bisecting lines to nearest neighbours, next-nearest neighbours and so on. A little thought will convince the reader that the volume so enclosed satisfies the above definition of the unit cell.

In many important cases, there is only one atom per unit cell. Then, we find it convenient to fix the points of the Bravais lattice on the atomic sites. The *direct lattice*, defined by the atomic positions, is then identical with the

Bravais lattice. Examples with one atom per unit cell are face-centred-cubic (fcc) and body-centred-cubic (bcc) crystals and their unit cells are shown in Figures 1.1 and 1.2. On the other hand, hexagonal close-packed (hcp)

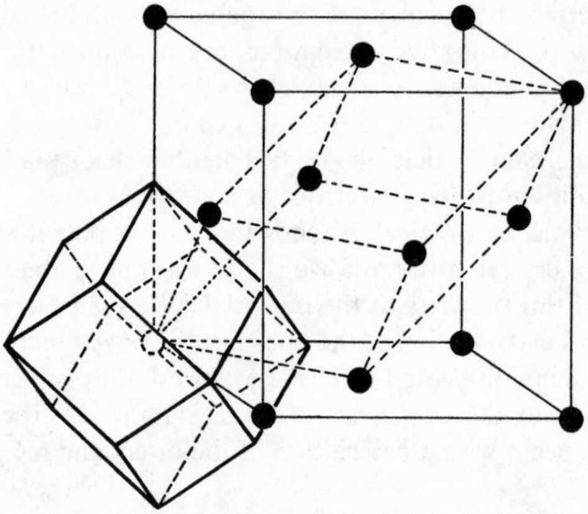

FIGURE 1.1. Unit cell for face-centred cubic (fcc) crystals. Choice of unit cell is not unique. We show, in particular, parallelepiped unit cell of Bravais lattice and Wigner–Seitz choice of unit cell, this being dodecahedron shown.

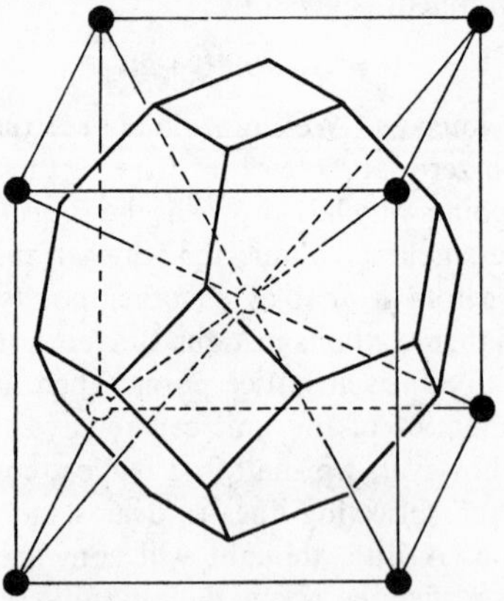

FIGURE 1.2. Unit cell for body-centred cubic (bcc) crystals. Cell shown is that symmetrically placed about atom at body centre.

and diamond lattices have two atoms per unit cell, Figures 1.3 and 1.4 showing their respective structures. It is then convenient to position the Bravais lattice so that the atoms occupy the most symmetrical sites possible in the unit cell.

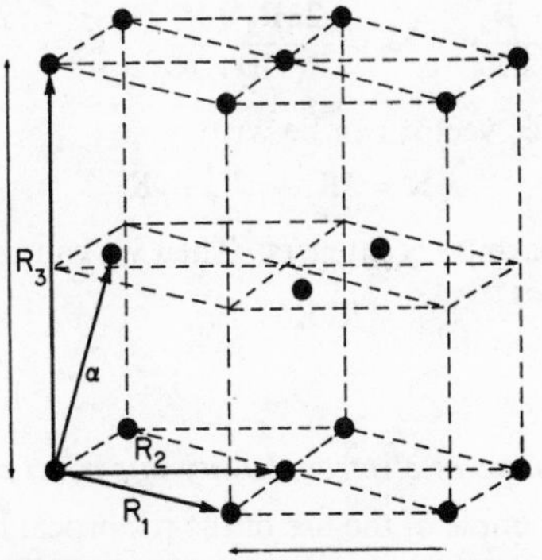

FIGURE 1.3. Hexagonal close-packed (hcp) structure. Three basis vectors are labelled $\mathbf{R}_1$, $\mathbf{R}_2$ and $\mathbf{R}_3$. Fourth vector $\boldsymbol{\alpha}$ is related to basis vectors by

$$\boldsymbol{\alpha} = \tfrac{1}{4}(2\mathbf{R}_2 - \mathbf{R}_1) + \tfrac{1}{2}\mathbf{R}_3$$

and is often used in describing hcp structure. There are two atoms per unit cell.

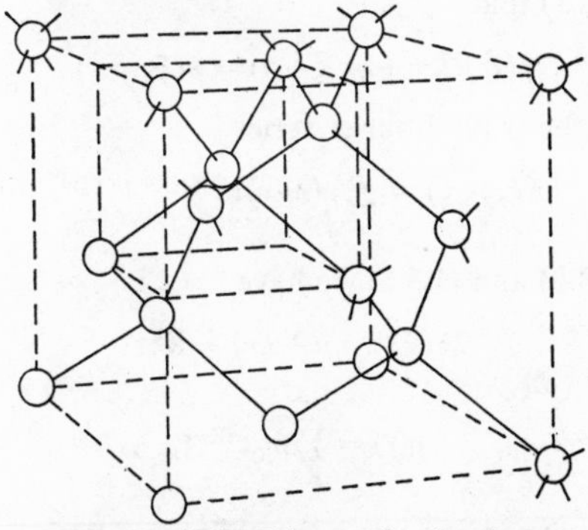

FIGURE 1.4. Diamond structure. Semiconducting Ge and Si are important examples. The tetrahedral bonding should be noted. Atomic polyhedron for this case is given by Kimball (1935).

1.3.2 Reciprocal lattice

We now introduce a second Bravais lattice, namely the *reciprocal lattice,* which has primitive translations $\mathbf{K}_1, \mathbf{K}_2, \mathbf{K}_3$ defined by

$$\mathbf{K}_1 = \frac{2\pi \mathbf{R}_2 \times \mathbf{R}_3}{\mathbf{R}_1 . \mathbf{R}_2 \times \mathbf{R}_3}, \quad \mathbf{K}_2 = \frac{2\pi \mathbf{R}_3 \times \mathbf{R}_1}{\mathbf{R}_1 . \mathbf{R}_2 \times \mathbf{R}_3}, \quad \mathbf{K}_3 = \frac{2\pi \mathbf{R}_1 \times \mathbf{R}_2}{\mathbf{R}_1 . \mathbf{R}_2 \times \mathbf{R}_3}. \tag{1.3.2}$$

Every reciprocal lattice vector can be written as

$$\mathbf{K} = \lambda \mathbf{K}_1 + \mu \mathbf{K}_2 + \nu \mathbf{K}_3 \tag{1.3.3}$$

where λ, μ and ν are arbitrary integers. Then we can see that

$$\mathbf{R}.\mathbf{K} = 2\pi j, \tag{1.3.4}$$

where j is some integer.

1.3.3 Fourier components of electron density in perfect crystal

As an important example of the use of the reciprocal lattice, let us consider the representation of the periodic electron density $\rho(\mathbf{r})$, which we shall deal with in more detail in Chapter 5 but which, as already emphasized above, plays a central role in the many-body problem of the electrons in a solid. Periodicity implies that

$$\rho(\mathbf{r}+\mathbf{R}) = \rho(\mathbf{r}), \tag{1.3.5}$$

or writing

$$\mathbf{r} = \xi \mathbf{R}_1 + \eta \mathbf{R}_2 + \zeta \mathbf{R}_3, \tag{1.3.6}$$

it follows from (1.3.1) that

$$\rho(\xi+l, \eta+m, \zeta+n) = \rho(\xi, \eta, \zeta). \tag{1.3.7}$$

We therefore have the triple Fourier series

$$\rho(\xi, \eta, \zeta) = \sum_{\lambda\mu\nu} \rho(\lambda\mu\nu)\, e^{2\pi i(\xi\lambda+\eta\mu+\zeta\nu)}. \tag{1.3.8}$$

However, from (1.3.3) and (1.3.6) we have

$$2\pi(\xi\lambda+\eta\mu+\zeta\nu) = \mathbf{K}.\mathbf{r} \tag{1.3.9}$$

and hence

$$\rho(\mathbf{r}) = \sum_{\mathbf{K}} \rho_{\mathbf{K}}\, e^{i\mathbf{K}.\mathbf{r}}, \tag{1.3.10}$$

which is the desired expansion using the reciprocal lattice.

Often, as we shall see especially in Chapter 5, the expansion (1.3.10) is most useful when inverted to the form

$$\rho_{\mathbf{K}} = \Omega^{-1} \int_{\Omega} \rho(\mathbf{r})\, e^{-i\mathbf{K}.\mathbf{r}}\, d\mathbf{r}, \tag{1.3.11}$$

which we shall now derive, Ω being the volume of the unit cell. Let **K** and **G** be any two reciprocal lattice vectors. We start from the Fourier representation of the Dirac δ-function:

$$\delta(\mathbf{K}-\mathbf{G}) = \frac{1}{8\pi^3}\int e^{i(\mathbf{K}-\mathbf{G})\,.\,\mathbf{r}}\,d\mathbf{r}, \tag{1.3.12}$$

which is qualitatively obvious since a completely localized function must have constant Fourier components. If we employ equation (1.3.4), we can reduce equation (1.3.12) to a sum of integrations over the unit cell, namely

$$\begin{aligned}\delta(\mathbf{K}-\mathbf{G}) &= \frac{1}{8\pi^3}\sum_{\mathbf{R}}\int_{\Omega} e^{i(\mathbf{K}-\mathbf{G})\,.\,(\mathbf{r}-\mathbf{R})}\,d\mathbf{r}\\ &= \frac{1}{8\pi^3}\sum_{\mathbf{R}}\int_{\Omega} e^{i(\mathbf{K}-\mathbf{G})\,.\,\mathbf{r}}\,d\mathbf{r}.\end{aligned} \tag{1.3.13}$$

It follows that

$$\int_{\Omega} e^{i(\mathbf{K}-\mathbf{G})\,.\,\mathbf{r}}\,d\mathbf{r} = \Omega\delta(\mathbf{K},\mathbf{G}) \tag{1.3.14}$$

where the generalized Kronecker delta is defined by

$$\left.\begin{aligned}\delta(\mathbf{K},\mathbf{G}) &= 1 \quad \text{if } \mathbf{K}=\mathbf{G},\\ &= 0 \quad \text{otherwise.}\end{aligned}\right\} \tag{1.3.15}$$

Equation (1.3.14) implies the result (1.3.11), when used with (1.3.10).

While we began with an expansion of the charge density $\rho(\mathbf{r})$, the above arguments obviously apply to any function which is periodic with the period of the direct lattice. Similarly, for any function $\sigma(\mathbf{k})$ which has the periodicity of the reciprocal lattice,

$$\sigma(\mathbf{k}) = \sum_{\mathbf{R}} \sigma_{\mathbf{R}}\, e^{i\mathbf{k}.\mathbf{R}}. \tag{1.3.16}$$

Furthermore, if **R** and **S** are two direct lattice vectors, then the analogue of equation (1.3.14) is evidently

$$\int_{\Omega_B} e^{i(\mathbf{R}-\mathbf{S}).\mathbf{k}}\,d\mathbf{k} = \Omega_B\,\delta(\mathbf{R},\mathbf{S}), \tag{1.3.17}$$

where Ω_B denotes the volume of the unit cell of the reciprocal lattice, which is termed the *Brillouin zone* (BZ). As we shall see later, this is the basic region inside which we must discuss the dispersion relation of electron waves.

Two useful expansions, which are easily seen to be formally true, are

$$\Omega \sum_{\mathbf{R}} \delta(\mathbf{r}+\mathbf{R}) = \sum_{\mathbf{K}} e^{i\mathbf{K}.\mathbf{r}} \tag{1.3.18}$$

and

$$\Omega_B \sum_{\mathbf{K}} \delta(\mathbf{k}+\mathbf{K}) = \sum_{\mathbf{R}} e^{i\mathbf{k}.\mathbf{R}}. \tag{1.3.19}$$

As an example of equations (1.3.18) and (1.3.19), we can calculate the product of the volumes of the unit cells in the direct and reciprocal lattices. Starting from equation (1.3.12) we may write

$$\begin{aligned}\sum_{\mathbf{R}} \delta(\mathbf{r}+\mathbf{R}) &= \frac{1}{8\pi^3} \sum_{\mathbf{R}} \int e^{i\mathbf{k}.(\mathbf{r}+\mathbf{R})}\, d\mathbf{k} \\ &= \frac{\Omega_B}{8\pi^3} \sum_{\mathbf{K}} \int e^{i\mathbf{k}.\mathbf{r}}\, \delta(\mathbf{k}+\mathbf{K})\, d\mathbf{k} \\ &= \frac{\Omega_B}{8\pi^3} \sum_{\mathbf{K}} e^{i\mathbf{K}.\mathbf{r}} \\ &= \frac{\Omega_B \Omega}{8\pi^3} \sum_{\mathbf{R}} \delta(\mathbf{r}+\mathbf{R}),\end{aligned} \tag{1.3.20}$$

which yields

$$\Omega\Omega_B = 8\pi^3. \tag{1.3.21}$$

This relation may also, of course, be obtained geometrically.

1.4 Translation operators and Bloch's theorem

When we wish to deal with a crystal of finite size, the question of boundary conditions arises. If the crystal is very large, we expect the precise form of these not to affect the physical description of properties over the bulk of the crystal. We may then choose conditions which are most simple mathematically. These are the "cyclic" or "periodic" boundary conditions, where the value of any wave function is the same at equivalent points on opposite sides of the crystal. If N is the number of cells, counting along the direction of each primitive translation, $N\mathbf{R}_i$ $(i = 1, 2, 3)$ brings us to a point not only physically but also mathematically identical to the original point. Let us introduce the translation operator $T_{\mathbf{R}}$ such that

$$T_{\mathbf{R}} f(\mathbf{r}) = f(\mathbf{r}+\mathbf{R}), \tag{1.4.1}$$

where f is an arbitrary function. Then, with these periodic boundary conditions, repeated application of $T_{\mathbf{R}_i}$ to the wave function will clearly yield

$$(T_{\mathbf{R}_i})^N \psi(\mathbf{r}) = \psi(\mathbf{r}) \quad (i = 1, 2, 3). \tag{1.4.2}$$

It follows that

$$(T_{\mathbf{R}_i})^N = 1, \tag{1.4.3}$$

so that every eigenvalue of $T_{\mathbf{R}}$ is a root of unity, of the form $e^{i\alpha}$. Since the translation operators commute among themselves and with the Hamiltonian H (H being unchanged by translation through a lattice vector) we can find a complete set of simultaneous eigenfunctions of H and $T_{\mathbf{R}}$ for all $\mathbf{R}$ (cf. Dirac, 1947) such that, if $\mathbf{R}$ is given by (1.3.1), then

$$T_{\mathbf{R}}\psi = (T_{\mathbf{R}_1})^l (T_{\mathbf{R}_2})^m (T_{\mathbf{R}_3})^n \psi = e^{i(l\alpha_1+m\alpha_2+n\alpha_3)}\psi, \tag{1.4.4}$$

ψ of course being an eigenfunction of H, (1.4.4) following from (1.4.1) and (1.3.1). If we now define $\mathbf{k}_1$, $\mathbf{k}_2$ and $\mathbf{k}_3$ such that

$$\alpha_i = \frac{\mathbf{k}_i . \mathbf{R}_i}{2\pi} \quad (i = 1, 2, 3) \tag{1.4.5}$$

then it follows that

$$T_{\mathbf{R}}\psi = e^{i\mathbf{k}.\mathbf{R}}\psi, \tag{1.4.6}$$

where

$$\mathbf{k} = k_1 \mathbf{K}_1 + k_2 \mathbf{K}_2 + k_3 \mathbf{K}_3. \tag{1.4.7}$$

The vector $\mathbf{k}$ may be used to label the eigenfunction; we have thus shown that a complete set of eigenfunctions $\psi_{\mathbf{k}}$ may be constructed such that

$$H\psi_{\mathbf{k}} = E(\mathbf{k})\psi_{\mathbf{k}} \tag{1.4.8}$$

and

$$\psi_{\mathbf{k}}(\mathbf{r}+\mathbf{R}) = e^{i\mathbf{k}.\mathbf{R}}\psi_{\mathbf{k}}(\mathbf{r}). \tag{1.4.9}$$

Equation (1.4.9) expresses *Bloch's theorem.* An obvious physical consequence of (1.4.9) is that the charge density $\psi_{\mathbf{k}}^*(\mathbf{r})\psi_{\mathbf{k}}(\mathbf{r})$ associated with a Bloch electron of wave vector $\mathbf{k}$ is periodic with the period of the lattice.

An important point to notice is that since, from (1.3.4),

$$e^{i(\mathbf{k}+\mathbf{K}).\mathbf{R}} = e^{i\mathbf{k}.\mathbf{R}} \tag{1.4.10}$$

for every reciprocal lattice vector $\mathbf{K}$, we could equally well specify the wave function by the vector $\mathbf{k}+\mathbf{K}$. In this book, we shall almost always classify the wave functions by the shortest wave vectors, namely those lying within the BZ. This is equivalent to saying that we may specify all the eigenvalues $e^{i\mathbf{k}.\mathbf{R}}$ of the translation operators $T_{\mathbf{R}}$ by restricting $\mathbf{k}$ in this way.

To find the number of allowed values of $\mathbf{k}$ within the BZ, we note that we obtain all possible values of $e^{i\alpha}$ if

$$\alpha_i = \frac{2\pi l_i}{N} \quad (l_i = 1, \ldots, N). \tag{1.4.11}$$

We immediately obtain the number of allowed values as N^3, the number of cells, and we can also see that the density of allowed points is constant over the BZ, the value being $1/(8\pi^3)$ per volume of crystal.

Since the wave functions form a complete set, there must be many orthogonal wave functions labelled by the same **k**, and we distinguish them by an index i: $\psi_{\mathbf{k}}^i$. For later purposes, it may be noted that, for any two functions ψ and ϕ which satisfy Bloch's theorem,

$$\int d\mathbf{r}\, \psi_{\mathbf{k}}^{i*}\, \phi_{\mathbf{k}'}^{j} = 0, \quad \mathbf{k} \neq \mathbf{k}', \tag{1.4.12}$$

whether they are eigenfunctions of H or not.

1.4.1 *Existence of energy bands*

We have just introduced as labels the indices i. In this section we shall demonstrate that the many-valued function $E_i(\mathbf{k})$ consists of branches which are analytic, in the limit of an infinite crystal, except possibly at points of degeneracy. Each of the analytic branches then defines an energy band, i.e. a quasi-continuous set of energy levels and it is to these branches that the indices refer.

The proof of the analyticity of the branches of $E(\mathbf{k})$ goes as follows. Consider the function $e^{-i\boldsymbol{\kappa}.\mathbf{r}}\,\psi_{\mathbf{k}+\boldsymbol{\kappa}}^p(\mathbf{r})$, where we use p as a label denoting the pth branch. Here $\psi_{\mathbf{k}+\boldsymbol{\kappa}}^p(\mathbf{r})$ is, as usual, an eigenfunction of H, with eigenvalue $E_p(\mathbf{k}+\boldsymbol{\kappa})$. The translation operator $T_{\mathbf{R}}$ acting on this function yields

$$T_{\mathbf{R}}\{e^{-i\boldsymbol{\kappa}.\mathbf{r}}\,\psi_{\mathbf{k}+\boldsymbol{\kappa}}^p(\mathbf{r})\} = e^{i\mathbf{k}.\mathbf{R}}\{e^{-i\boldsymbol{\kappa}.\mathbf{r}}\,\psi_{\mathbf{k}+\boldsymbol{\kappa}}^p(\mathbf{r})\}. \tag{1.4.13}$$

This shows that the function considered is an eigenfunction of $T_{\mathbf{R}}$, with eigenvalue $e^{i\mathbf{k}.\mathbf{R}}$, and we can therefore expand $e^{-i\boldsymbol{\kappa}.\mathbf{r}}\,\psi_{\mathbf{k}+\boldsymbol{\kappa}}^p(\mathbf{r})$ solely in terms of the eigenfunctions $\psi_{\mathbf{k}}^i(\mathbf{r})$ of H:

$$\psi_{\mathbf{k}+\boldsymbol{\kappa}}^p(\mathbf{r}) = \sum_n c_{pn}\,\psi_{\mathbf{k}}^n(\mathbf{r})\, e^{i\boldsymbol{\kappa}.\mathbf{r}}. \tag{1.4.14}$$

We may now obtain the determinantal equation for the energy by substituting (1.4.14) into the Schrödinger equation (1.2.7), and writing down the condition that the simultaneous equations for c_{pn} have non-trivial solutions, namely

$$\det|H_{mn} - E| = 0, \tag{1.4.15}$$

where

$$H_{mn} = \int \psi_{\mathbf{k}}^{m*}\, e^{-i\boldsymbol{\kappa}.\mathbf{r}} \left\{-\frac{\hbar^2\nabla^2}{2m} + V(\mathbf{r})\right\} \psi_{\mathbf{k}}^n\, e^{i\boldsymbol{\kappa}.\mathbf{r}}\, d\mathbf{r}.$$

Straightforward use of the properties of the wave functions then yields

$$H_{mn} = \left\{E_n(\mathbf{k})+\frac{\hbar^2 k^2}{2m}\right\}\delta_{mn}+\frac{\hbar^2 \boldsymbol{\kappa}}{2m}\cdot\int_\Omega \psi_{\mathbf{k}}^{m*}\nabla\psi_{\mathbf{k}}^{n}\,d\mathbf{r}. \tag{1.4.16}$$

The elements of this determinant are all analytic in $\boldsymbol{\kappa}$, and so we expect the roots to be analytic in $\boldsymbol{\kappa}$ also. However, some care is needed, as we shall see.

The determinant (1.4.15) is of infinite order. We shall investigate a finite determinant, and assume the results hold true for the infinite case. Let $f(\kappa, E)$ be a determinant analytic in κ, and we shall restrict ourselves to one variable, the extension to many variables being obvious. If the roots are $a_1(\kappa)$, $a_2(\kappa)$, etc., then we may write

$$f(\kappa, E) = [a_1(\kappa)-E]\,[a_2(\kappa)-E]\,[a_3(\kappa)-E]\ldots \tag{1.4.17}$$

and continuity of the a_i in κ follows immediately. To examine the behaviour of the derivatives at the zeros of $f(\kappa, E)$, and in particular when $a_1(\kappa) = E$, we have

$$\frac{\partial f}{\partial \kappa} = f'(\kappa, E) = a_1'(\kappa)\,(a_2-E)\,(a_3-E)\ldots, \tag{1.4.18}$$

$$f''(\kappa, E) = a_1''(\kappa)\,(a_2-E)\,(a_3-E)\ldots+a_1'\,a_2'(a_3-E)\,(a_4-E)\ldots+\ldots. \tag{1.4.19}$$

From equation (1.4.18) and the analytic properties of $f(\kappa)$, $a_1'(\kappa)$ is uniquely defined, and obviously $a_2'(\kappa)$, etc. similarly exist. Thus, since a_1', a_2', etc. exist, it follows from equation (1.4.19) that $a_1''(\kappa)$ is defined and so on. However, this conclusion does *not* hold if two roots are *equal*. For example, if $a_1 = a_2$, equation (1.4.18) no longer defines $a_1'(\kappa)$ uniquely.

So far then, we conclude that $E(\mathbf{k})$ is analytic in $\mathbf{k}$, except possibly at $\mathbf{k}$ values where essential degeneracy‡ occurs. Examples we shall discuss later will reveal such departures from analyticity.

One other consequence of the continuity of $E(\mathbf{k})$ within a band is that it can be expanded in a Fourier series using the direct lattice. The periodicity of $E(\mathbf{k})$ in the reciprocal lattice follows since

$$\psi_{\mathbf{k}+\mathbf{K}}(\mathbf{r}) = \psi_{\mathbf{k}}(\mathbf{r}) \tag{1.4.20}$$

for all reciprocal lattice vectors $\mathbf{K}$. Thus we have

$$E(\mathbf{k}) = \sum_{\mathbf{R}} E_{\mathbf{R}}\, e^{i\mathbf{k}.\mathbf{R}}. \tag{1.4.21}$$

In addition to showing that $E(\mathbf{k})$ has continuous branches, which, as we shall see by direct examples, are generally separated by a forbidden energy gap, it is worth noting that $\psi_{\mathbf{k}}$ within a band is also continuous in $\mathbf{k}$.

‡ That is degeneracy implied by symmetry, as is discussed more fully in section 1.5.3 below.

1.5 Rotational invariance and group theory

Apart from its translational symmetry, the Bravais lattice will be invariant under certain proper and improper rotations. (An improper rotation consists of a proper rotation plus a reflexion in some symmetry plane.) For example, cubic Bravais lattices are invariant under 48 operations. These consist of all transformations by which the vector (x, y, z) goes into all vectors obtained by all possible permutations of the three coordinates x, y, z with all possible signs. When the crystal lattice is the Bravais lattice, the crystal lattice also has this symmetry. For other lattices, some of the Bravais lattice symmetry may not be present in the crystal, but additional symmetry elements may appear; for example, glide planes and screw axes (for precise definitions, see Appendix 1.3).

The exploitation of this symmetry to obtain general information about the electronic states of the crystal is accomplished by use of the apparatus of group theory, the concepts and language of which are reviewed in what follows. A much more detailed discussion can be found in the general Appendix on Group Theory in Volume II.

1.5.1 Groups of unitary transformations

The rotational operations just referred to always leave the length of any vector in the lattice unchanged. We therefore begin our discussion with such transformations of vectors. Instead of using the 48 operations of the full cubic group, let us develop briefly the notion of a group by considering, for simplicity, just five unitary transformations‡ a, b, c, d and e such that

$$\left.\begin{aligned} a\mathbf{r} &= a(x,y,z) = (z,x,y), \quad & b\mathbf{r} &= b(x,y,z) = (y,z,x), \\ c\mathbf{r} &= c(x,y,z) = (x,z,y), \quad & d\mathbf{r} &= d(x,y,z) = (z,y,x) \\ \text{and} \quad e\mathbf{r} &= e(x,y,z) = (y,x,z). \end{aligned}\right\} \tag{1.5.1}$$

Now we notice that any two successive transformations give us either the original vector $\mathbf{r}$, or the same result as a single application of one of the other transformations, e.g.

$$(ab)\,\mathbf{r} = a(y,z,x) = (x,y,z) \tag{1.5.2}$$

and

$$(ac)\,\mathbf{r} = a(x,z,y) = (y,x,z) = e\mathbf{r}. \tag{1.5.3}$$

From equation (1.5.2) we can say that $ab = 1$ and similarly, from equation (1.5.3), $ac = e$.

‡ In the present context, these will leave the length of any vector in the lattice the same. In general terms, a unitary transformation is one whose adjoint is also the inverse.

We can thus construct a multiplication table (see Table 1.1) for the elements $\{abcde1\}$, where we have included the identity transformation 1, which leaves $\mathbf{r}$ unchanged. The following comments are called for:

TABLE 1.1. Multiplication table for example of group of order 6.

	a	b	c	d	e	1
a	b	1	e	c	d	a
b	1	a	d	e	c	b
c	d	e	1	a	b	c
d	e	c	b	1	a	d
e	c	d	a	b	1	e
1	a	b	c	d	e	1

(i) The transformation labelling the column is applied to the vector $\mathbf{r}$ first. This, however, is a convention not always adopted in other texts.

(ii) Any row and column of the table includes each element once.

(iii) The associative law $(ab)c = a(bc)$, etc. holds.

The existence of a multiplication table satisfying (ii) and (iii) is a necessary condition for its constituent elements to be said to form a *group*. We leave the reader to verify that $\{abcde1\}$ satisfies the formal definition:

A set $G = \{\alpha, \beta, \gamma, \ldots\}$ is said to form a group when:

(i) if α, β are members of G, so is $\alpha\beta = \gamma$,

(ii) if α, β, γ are any members of G, $(\alpha\beta)\gamma = \alpha(\beta\gamma)$,

(iii) there exists a unique element 1, the identity element, such that if α is any member of G, $\alpha 1 = 1\alpha = \alpha$.

(iv) for every member of G, α, there is a second unique member β such that $\alpha\beta = \beta\alpha = 1$. We put $\beta = \alpha^{-1}$, and term it the inverse of α.

Since we shall occasionally take the group whose multiplication table is shown in Table 1.1, to exemplify concepts we introduce, it will be denoted by the symbol g_6. We first use it to illustrate the concept of subgroup. If we take the elements $a, b, 1$, we can form the multiplication table (Table 1.2). We see that the elements $a, b, 1$ form a group of themselves. It is a *subgroup* of the group g_6.

TABLE 1.2. Multiplication table of subgroup of order three.

	a	b	1
a	b	1	a
b	1	a	b
1	a	b	1

1.5.2 *Unitary transformations leaving Hamiltonian invariant*

The unitary transformations we are concerned with leave the Hamiltonian H invariant. As we shall see shortly, we are led via the symmetry properties to a classification of the degenerate states of H. If we define operators which when acting on the wave functions $\psi(\mathbf{r})$ have the effect of a unitary transformation of the above type on the argument $\mathbf{r}$, then these operators have the important property that they commute with the Hamiltonian H, though not necessarily with one another. Thus, these operators acting on $\psi(\mathbf{r})$ generate new wave functions which correspond to the same eigenvalue as $\psi(\mathbf{r})$.

Our first task, then, is to set up the symmetry operators described above. As an example, to illustrate how this can be done, we shall define the six operators $\{ABCDEI\}$ corresponding to the elements of the group g_6 such that for any function $f(\mathbf{r})$

$$Af(\mathbf{r}) = f(a^\dagger \mathbf{r}), \quad \text{etc.,} \tag{1.5.4}$$

where $a^\dagger$ is the adjoint of a. Since, in our case, a is unitary, $a^\dagger = a^{-1}$. It should also be noted that $a^\dagger c^\dagger = c^\dagger a^\dagger$.

The set $G_6 = \{ABCDEI\}$ forms a group with the same multiplication table as the set $\{abcde1\}$. For example, let us see the result of the operation AC on $\psi(\mathbf{r})$. Defining $\phi(\mathbf{r})$ through

$$\phi(\mathbf{r}) = C\psi(\mathbf{r}) = \psi(c^\dagger \mathbf{r}), \tag{1.5.5}$$

then

$$AC\psi(\mathbf{r}) = A\phi(\mathbf{r}) = \phi(a^\dagger \mathbf{r}). \tag{1.5.6}$$

If we introduce $\mathbf{r}' = a^\dagger \mathbf{r}$, then we evidently have, from equation (1.5.5),

$$\phi(\mathbf{r}') = \psi(c^\dagger \mathbf{r}') = \psi(c^\dagger a^\dagger \mathbf{r}) = \psi(e^\dagger \mathbf{r}), \tag{1.5.7}$$

the last step following from Table 1.1. From equations (1.5.5) and (1.5.6), we obtain immediately $AC = E$, corresponding to $ac = e$. This is an example of an *isomorphism*. When two groups have the same multiplication table and the same number of distinct elements, they are said to be isomorphic.

It is worth while to note in passing that we might have adopted the definitions $A\psi(\mathbf{r}) = \psi(a\mathbf{r})$, $B\psi(\mathbf{r}) = \psi(b\mathbf{r})$, etc. This would have led to equivalent multiplication tables for the two groups, that is again to isomorphic groups, but not to the same correspondence between A's and a's. In fact, A would then have corresponded to $a^\dagger = b$, etc.

1.5.3 *Invariant manifolds and irreducible representations*

A convenient picture to use is one in which wave functions are represented as vectors in a suitable space. The operators $\{AB\ldots\}$ may then be thought of as inducing unitary transformations on these vectors. To make these ideas clear, we will consider the results of the application of $\{AB...\}$ to the three plane-wave functions

$$\psi_1 = \frac{1}{(2\pi)^{\frac{3}{2}}} e^{i(x-y+z)}, \quad \psi_2 = \frac{1}{(2\pi)^{\frac{3}{2}}} e^{i(y-z+x)}, \quad \psi_3 = \frac{1}{(2\pi)^{\frac{3}{2}}} e^{i(z-x+y)}. \tag{1.5.8}$$

These are orthonormal over the cube of side 2π:

$$\langle \psi_i^* \psi_j \rangle = \int_0^{2\pi} dx\, dy\, dz\, \psi_i^* \psi_j = \delta_{ij} \quad (i,j = 1,2,3). \tag{1.5.9}$$

Using geometrical language, we say that the three independent functions span a three-dimensional *manifold* M, and any function ϕ which is a linear combination of ψ_1, ψ_2, ψ_3 is said to lie in the manifold.

Using the properties of G_6 discussed above, the first row in Table 1.3 below is readily obtained. The multiplication table may be used to find the other two rows. We get no new functions by operating on ψ_1, ψ_2, ψ_3 with members of the group G_6 and if we operate on any function ϕ lying in M, the resulting function also lies in M. For example, if $\phi = c_1\psi_1 + c_2\psi_2 + c_3\psi_3$, then $A\phi = c_3\psi_1 + c_1\psi_2 + c_2\psi_3$.

TABLE 1.3. Operation of G_6 on functions (1.5.8)

$A\psi_1 = \psi_2$	$B\psi_1 = \psi_3$	$C\psi_1 = \psi_2$	$D\psi_1 = \psi_1$	$E\psi_1 = \psi_3$	$I\psi_1 = \psi_1$
$A\psi_2 = \psi_3$	$B\psi_2 = \psi_1$	$C\psi_2 = \psi_1$	$D\psi_2 = \psi_3$	$E\psi_2 = \psi_2$	$I\psi_2 = \psi_2$
$A\psi_3 = \psi_1$	$B\psi_3 = \psi_2$	$C\psi_3 = \psi_3$	$D\psi_3 = \psi_2$	$E\psi_3 = \psi_1$	$I\psi_3 = \psi_3$

We therefore call M an *invariant* manifold of the group G_6. Consider now the functions

$$\chi_1 = \frac{1}{\sqrt{3}}(\psi_1 + \psi_2 + \psi_3), \quad \chi_2 = \frac{1}{\sqrt{2}}(\psi_1 - \psi_2), \quad \chi_3 = \frac{1}{\sqrt{6}}(\psi_1 + \psi_2 - 2\psi_3). \tag{1.5.10}$$

These vectors are also orthonormal, and any vector lying in M may be expanded in terms of them. Thus χ_1, χ_2, χ_3 span the manifold, and may be used to define it instead of ψ_1, ψ_2, ψ_3.

We note at this point that for *any* of the operators of G_6, say A, $A\chi_1 = \chi_1$. Thus χ_1 forms by itself an invariant manifold while χ_2 and χ_3 can be shown to define a two-dimensional invariant manifold.

It is important at this stage to generalize the concepts introduced above, and in particular to extend the geometrical language to cases where we have more than three independent functions. Thus, let us assume we take the set of all independent degenerate eigenfunctions, $\psi_1, \ldots, \psi_n$, corresponding to eigenvalue E. The unitary transformations leaving the Hamiltonian invariant, discussed in the previous section, are such that a symmetry operator A commutes with the Hamiltonian, that is

$$AH = HA. \tag{1.5.11}$$

Any linear combination ϕ of the above degenerate eigenfunctions, namely

$$\phi = \sum_i c_i \psi_i \tag{1.5.12}$$

is such that

$$HA\phi = AH\phi = A\epsilon\phi, \tag{1.5.13}$$

from (1.5.11). This means that $A\phi$ is also an eigenfunction of H with eigenvalue ϵ and is therefore a linear combination of the same form (1.5.12). The set ψ_i is said to span an *n-dimensional invariant manifold.*

If we think of the wave functions ψ_i as basis vectors, then all the degenerate eigenfunctions ϕ can be thought of as vectors in different directions in the same space (manifold).

A further important idea that we must introduce here concerns the possibility that such an invariant manifold can be decomposed into two or more sub-manifolds, which are also invariant. If so, the original manifold is said to be reducible.

In this case, the sub-manifolds are spanned by linear combinations of eigenvectors, all of which correspond to the same energy and so it follows that the eigenfunctions of H can be chosen such that each lies wholly within an *irreducible* manifold of G. Hence, if we can predict the possible dimensions of these irreducible manifolds, we can predict the degeneracies implied by the symmetry of the Hamiltonian. Such degeneracy is said to be *essential,* whereas degeneracy which is dependent on the detailed form of the Hamiltonian (i.e. additional to its symmetry) is termed *accidental.*

It is, in fact, possible to obtain the dimensions of the irreducible manifolds by working in a matrix representation of the operators. This also gives the transformation properties of the eigenvectors spanning the irreducible manifold.

1.5.4 Matrix representation

Given any invariant manifold spanned by orthonormal functions $\psi_1, \ldots, \psi_n$, we may form matrices corresponding to the operators $A, B, \ldots$ of any group:

$$(A) = \begin{pmatrix} A_{11} & A_{12} & \cdots & A_{1n} \\ \cdot & \cdot & \cdot & \cdot & \cdot \\ A_{n1} & & & A_{nn} \end{pmatrix}, \tag{1.5.14}$$

where $A_{ij} = \langle\psi_i A\psi_j\rangle$. This set of matrices satisfies the definition of a group, having the same multiplication table as the group of operators. Suppose, for example, $AC = E$, when the matrix element of AC between ψ_i and ψ_j is given by

$$(AC)_{ik} = \sum_j A_{ij} C_{jk} = \sum_j \langle\psi_i A\psi_j\rangle\langle\psi_j C\psi_k\rangle = \langle\psi_i AC\psi_k\rangle$$
$$= \langle\psi_i E\psi_k\rangle = E_{ik}, \tag{1.5.15}$$

where we have used the completeness relation to sum over j. This relation is true because the ψ's span the manifold. We now have a *matrix representation* of the group of operators, and we say that the functions $\psi_1, ..., \psi_n$ *realize* the representation.

If a manifold M is reducible to the symbolic form

$$M = M_1 + M_2 + M_3 + ..., \tag{1.5.16}$$

where $M_1, M_2, ...$ are irreducible, then the matrix representation (A) can be expressed in the form

$$(A) = \begin{pmatrix} A_1 & & 0 \\ & A_2 & \\ 0 & & A_3 \end{pmatrix}. \tag{1.5.17}$$

Here A_1 is the matrix realized by the functions spanning M_1, etc. A matrix that may be transformed into the form (1.5.17) by a unitary transformation is called *reducible*. Otherwise it is called *irreducible*.

We see that the essential step is now to find the order of the irreducible matrices, which gives the degeneracy. The sum of the diagonal elements of each matrix, that is the trace of the matrix, is termed the *character* in this context. These characters uniquely specify the irreducible representation and procedures for finding these exist. Here, we are concerned with how to use the tables of characters and not with how they are derived. For the reader interested in their derivation, reference can be made to the general Appendix on Group Theory in Volume II (referred to subsequently as GAII). In that appendix, it is also shown that the converse of the result that we can pass uniquely from irreducible manifolds to irreducible matrices holds. The conclusion is that the eigenfunctions of H, when suitably constructed, realize all possible matrix representations of a symmetry group; that is of a group of operators which commute with the Hamiltonian.

Classes. For future purposes, it is useful here to define *classes*. Let us suppose that there exists an operator X of the group such that $B = X^{-1}AX$, where A and B are also members of the group. Then A and B are said to belong to the same class. We may note that B is obtained from A through a unitary transformation. Under such transformations, the trace of a matrix remains invariant. Hence the characters of the elements of a given class are all the same.

1.5.5 Point group symmetry properties

The full set of coordinate transformations, that is of all proper and improper rotations, which leaves the crystal invariant, is termed the point group of the crystal.

We defined above a group of coordinate transformations. If a is a member of this group, and $\psi_{\mathbf{k}}(\mathbf{r})$ is an eigenfunction of H with eigenvalue $E(\mathbf{k})$, we can define another function $\phi(\mathbf{r})$ such that

$$\phi(\mathbf{r}) = \psi_{\mathbf{k}}(a\mathbf{r}). \tag{1.5.18}$$

Then, the operator $A^\dagger$, corresponding to a through (1.5.4), is clearly such that $\phi(\mathbf{r}) = A^\dagger \psi_{\mathbf{k}}(\mathbf{r})$ is another eigenfunction of the Hamiltonian with the same eigenvalue. We still need to know what $\mathbf{k}$ vector labels $\phi(\mathbf{r})$. To ascertain this, we add a direct lattice vector $\mathbf{R}$ to the argument of ϕ, yielding $\psi_{\mathbf{k}}(a\mathbf{r}+a\mathbf{R})$. From Bloch's theorem (1.4.9) we find

$$\psi_{\mathbf{k}}(a\mathbf{r}+a\mathbf{R}) = e^{i\mathbf{k}.a\mathbf{R}}\psi_{\mathbf{k}}(a\mathbf{r}). \tag{1.5.19}$$

In terms of ϕ, equation (1.5.19) becomes

$$\phi(\mathbf{r}+\mathbf{R}) = e^{ia^\dagger\mathbf{k}.\mathbf{R}}\phi(\mathbf{r}), \tag{1.5.20}$$

since $(a^\dagger\mathbf{k}).\mathbf{R} = \mathbf{k}.(a\mathbf{R})$. Comparing equation (1.5.20) with Bloch's theorem (1.4.9), we see that $\phi(\mathbf{r})$ is labelled by the vector $a^\dagger\mathbf{k}$. Thus we may write $\phi(\mathbf{r}) = \psi_{a^\dagger\mathbf{k}}(\mathbf{r})$, and hence from equation (1.5.18)

$$\psi_{a^\dagger\mathbf{k}}(\mathbf{r}) = \psi_{\mathbf{k}}(a\mathbf{r}). \tag{1.5.21}$$

It is also evident that

$$E(a^\dagger\mathbf{k}) = E(\mathbf{k}). \tag{1.5.22}$$

1.5.6 Inversion symmetry of dispersion relation

If the crystal has an inversion centre, that is, if the periodic potential $V(\mathbf{r}) = V(-\mathbf{r})$, then we can choose $a^\dagger$ such that $a^\dagger\mathbf{k} = -\mathbf{k}$ and hence from (1.5.18) we have

$$E(-\mathbf{k}) = E(\mathbf{k}). \tag{1.5.23}$$

In fact, this result is true whether or not an inversion centre exists for the crystal in question, as we shall now prove.

Because of Bloch's theorem (1.4.9), it will be useful to define a function $u_{\mathbf{k}}(\mathbf{r})$ by

$$u_{\mathbf{k}}(\mathbf{r}) = e^{-i\mathbf{k}.\mathbf{r}}\psi_{\mathbf{k}}(\mathbf{r}). \tag{1.5.24}$$

Now we replace **r** by **r**+**R** and clearly the right-hand side of (1.5.24) remains unchanged. Thus $u_{\mathbf{k}}(\mathbf{r})$ is a periodic function, with the period of the lattice.

Writing the Hamiltonian in atomic units‡ as

$$H = -\frac{\nabla^2}{2} + V(\mathbf{r}) \tag{1.5.25}$$

we easily obtain the equation

$$\tfrac{1}{2}(k^2 - \nabla^2)\, u_{\mathbf{k}} - i\mathbf{k}.\nabla u_{\mathbf{k}} + V(\mathbf{r})\, u_{\mathbf{k}} = E(\mathbf{k})\, u_{\mathbf{k}}. \tag{1.5.26}$$

Taking the complex conjugate of this equation, we find

$$\tfrac{1}{2}(k^2 - \nabla^2)\, u_{\mathbf{k}}^* + i\mathbf{k}.\nabla u_{\mathbf{k}}^* + V(\mathbf{r})\, u_{\mathbf{k}}^* = E(\mathbf{k})\, u_{\mathbf{k}}^*. \tag{1.5.27}$$

But this is also the equation of $u_{-\mathbf{k}}$, showing that this latter quantity is the same as $u_{\mathbf{k}}^*$ to within a phase factor. With appropriate choice of the phase, we may set

$$\psi_{-\mathbf{k}}(\mathbf{r}) = \psi_{\mathbf{k}}^*(\mathbf{r}) \tag{1.5.28}$$

from which it follows that

$$E(\mathbf{k}) = E(-\mathbf{k}). \tag{1.5.29}$$

This is a special case of Wigner time-reversal symmetry (see Appendix GAII and section 1.11 below).

1.5.7 Essential degeneracy: touching of bands

Returning now to equation (1.5.21), we have implied that the ψ's on each side of the equation belong to the same band, since we have introduced no band indices. While this is most frequently true, there are important exceptions, which may occur when the point **k** is one of high symmetry. By a point of high symmetry, we mean

$$a\mathbf{k} = \mathbf{k} + \mathbf{K}, \tag{1.5.30}$$

where a is other than the identity element. It is possible for **K** to be zero but it is not necessarily so, because **k** and **k**+**K** are equivalent points, as implied by equation (1.4.20).

We now introduce band indices into equation (1.5.21) to give

$$\psi_{a\mathbf{k}}^i(\mathbf{r}) = \psi_{\mathbf{k}}^j(a^\dagger \mathbf{r}), \tag{1.5.31}$$

where the adjoints have been transposed, as is easily seen to be valid. Using equation (1.5.30), we find immediately

$$\psi_{a\mathbf{k}}^i(\mathbf{r}) = \psi_{\mathbf{k}+\mathbf{K}}^i(\mathbf{r}) = \psi_{\mathbf{k}}^i(\mathbf{r}). \tag{1.5.32}$$

‡ That is $\hbar = e = m = 1$. These units are usually convenient, though in subsequent sections it will sometimes be helpful to retain the fundamental constants explicitly.

But from equation (1.5.31)

$$\psi_{\mathbf{k}}^{j}(a^{\dagger}\mathbf{r}) = A\psi_{\mathbf{k}}^{j}(\mathbf{r}) = \psi_{\mathbf{k}}^{i}(\mathbf{r}). \tag{1.5.33}$$

Now if $\psi_{\mathbf{k}}^{i}(\mathbf{r})$ therefore differs from $\psi_{\mathbf{k}}^{j}(\mathbf{r})$ by more than a phase factor $e^{i\theta}$, then **k** is a point of essential degeneracy, which means the band indices i and j are not equal. Thus we have different bands actually touching at this point **k**. It will be readily seen that the set of operators a obeying (1.5.30) form a group $G_{\mathbf{k}}$, and to obtain the possible essential degeneracies at the point of high symmetry **k**, we follow the procedure already outlined, finding the irreducible representations of $G_{\mathbf{k}}$. To show that the functions $\psi_{\mathbf{k}}$ realize these irreducible representations, we combine $G_{\mathbf{k}}$ with the translation group T.

One then obtains, as irreducible representations of $G_{\mathbf{k}} \times T$, matrices of the form $e^{i\mathbf{k}.\mathbf{R}}(A)$ where (A) is an irreducible matrix representation of A. We have already seen that the eigenfunctions of H realize all representations of a symmetry group of H, and hence we can obtain functions $\psi_{\mathbf{k}}$ realizing (A) without change of **k**. It may be remarked that $\psi_{\mathbf{k}}$ does not in general lie in an irreducible manifold of the point group.

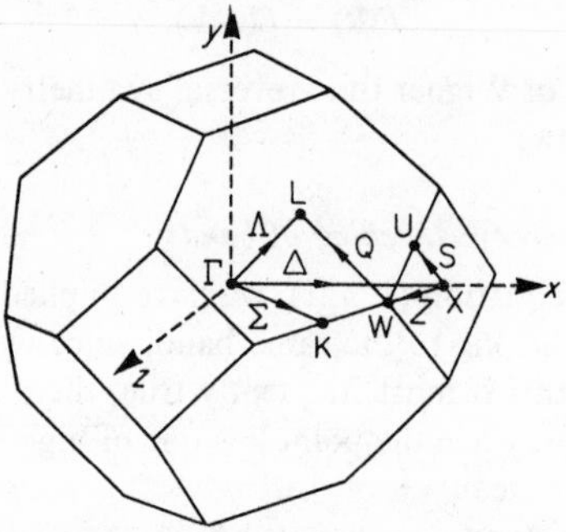

FIGURE 1.5. Brillouin zone for face-centred cubic (fcc) lattice, with symmetry points labelled. Point Γ is at centre of zone. This is unit cell of body-centred cubic crystal (cf. Figure 1.2).

Example of character table for point W of face-centred cubic crystal. As an example of some of the group-theoretical concepts we have introduced, we show a table of the characters associated with the symmetry operators of the point W for the BZ of a face-centred cubic crystal (see Figure 1.5). This can be obtained by the methods of Appendix GAII and we show in the notation developed there the eight operations of the group below Table 1.4.

The five irreducible representations W_1, W_2, etc. are labelled in the Bouckaert, Wigner and Smoluchowski (1936) notation. We show in brackets an alternative notation which immediately tells us the character of the wave function in an orbital angular momentum representation, s, p, d, etc. In this latter notation, which indicates the lowest-order spherical harmonic appearing in the expansion of the wave function, different states with the same l value are distinguished by superscripts 1, 2, etc.

TABLE 1.4. Character table for point W
$\mathbf{k} = (\pi/2a)(\pm 2, \pm 1, 0)$, where $2a$ is the edge of the basic cube

W	E	C_4^2	$2JC_4$	$2C_2$	$2JC_4^2$
$W_1\,(W_s)$	1	1	1	1	1
$W_2\,(W_f)$	1	1	1	-1	-1
$W_1'\,(W_d)$	1	1	-1	1	-1
$W_2'\,(W_p^1)$	1	1	-1	-1	1
$W_3\,(W_p^2)$	2	-2	0	0	0

$$E\begin{pmatrix}x\\y\\z\end{pmatrix}=\begin{pmatrix}x\\y\\z\end{pmatrix} \qquad C_4^2\begin{pmatrix}x\\y\\z\end{pmatrix}=\begin{pmatrix}-x\\y\\-z\end{pmatrix}$$

$$JC_4\begin{pmatrix}x\\y\\z\end{pmatrix}=\begin{pmatrix}\pm z\\-y\\\pm x\end{pmatrix} \qquad C_2\begin{pmatrix}x\\y\\z\end{pmatrix}=\begin{pmatrix}\pm z\\-y\\\pm x\end{pmatrix}$$

$$JC_4^2\begin{pmatrix}x\\y\\z\end{pmatrix}=\begin{pmatrix}\pm x\\y\\\pm z\end{pmatrix}$$

Here J is the inversion operator, while C_2 and C_4 are rotations about two- and four-fold axes respectively (see Appendix GAII).

However, we can note here:

(i) Though there are eight operations, there are only five columns shown. This is because there are five classes, and, as we saw, elements of the same class have the same characters.

(ii) The number of irreducible representations, $W_1, \ldots, W_3$ equals the number of classes.

(iii) The character of the identity element E is simply equal to the order of the irreducible representation in question. Hence W_3 is seen to be doubly degenerate; an example of essential degeneracy.

We shall come back to this character table in section 1.11.1 in connection with the energy bands of the NaCl lattice.

1.6 Surfaces of constant energy; Fermi surface

Since, for a given band, $E(\mathbf{k})$ is continuous in the BZ, we may draw surfaces of constant energy analogous to equipotential surfaces in electrostatics. These surfaces will be tangential to $\nabla_{\mathbf{k}} E(\mathbf{k})$ at every point. Since the electrons will occupy the lowest available states, one of these surfaces will be such that all states described by **k**-vectors within it are occupied, and **k**-vectors outside it label unoccupied states. This particular surface is, of course, the Fermi surface (FS). We shall now show that this surface has the symmetry of the crystal. This means that the operations of the point group leave it invariant. The starting point of the proof is equation (1.5.22) or, equivalently,

$$E(a\mathbf{k}) = E(\mathbf{k}), \tag{1.6.1}$$

for any element a of the point group. Equation (1.6.1) does not immediately prove that the FS has the symmetry of the crystal, for the E's on opposite sides of the equation could refer to different bands. This is, in fact, not the case.

There is no unique set of Bloch functions for any given band, but one particular way in which they may be constructed is to determine the wave functions in a segment S of the BZ (the segment being such that operations on it with all the elements of the point group just makes up the BZ). One may then construct all other Bloch wave functions by operating with all the elements of the point group on all the wave functions with **k**-vectors lying in the segment. The required continuity of $\psi_{\mathbf{k}}$ and $E(\mathbf{k})$ in **k** is assured, since the operator a giving an adjoining segment will leave **k**-values on the surface separating the segment unchanged. We shall see later that these properties ensure the continuity of the momentum eigenfunction (see section 1.9).

1.6.1 Contact with zone boundary

Let us now investigate the situation when the FS contacts the BZ boundaries. Suppose we have faces of a BZ parallel to a reflexion plane, which we assume perpendicularly bisects the reciprocal lattice vector **K** joining opposite faces (see Figure 1.6). If $\mathbf{k}'$ is the reflexion of a point **k** lying on a face, we may write

$$E(\mathbf{k}') = E(\mathbf{k}) \tag{1.6.2}$$

and

$$\mathbf{K}.\nabla_{\mathbf{k}'} E(\mathbf{k}') = -\mathbf{K}.\nabla_{\mathbf{k}} E(\mathbf{k}). \tag{1.6.3}$$

Now **k** and $\mathbf{k}'$ are equivalent points:

$$\mathbf{k}' = \mathbf{k}+\mathbf{K}, \tag{1.6.4}$$

and also

$$E(\mathbf{k}+\mathbf{K}) = E(\mathbf{k}), \tag{1.6.5}$$

so that

$$[\nabla_{\mathbf{k}} E(\mathbf{k})]_{\mathbf{k}=\mathbf{k}'} = \nabla_{\mathbf{k}} E(\mathbf{k}). \tag{1.6.6}$$

Comparing this with equation (1.6.3), we see that

$$\mathbf{K}.\nabla_{\mathbf{k}} E(\mathbf{k}) = 0. \tag{1.6.7}$$

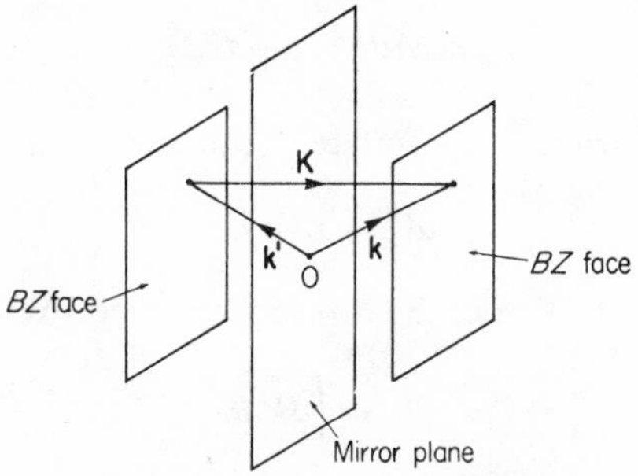

FIGURE 1.6. Reflexion plane perpendicularly bisecting reciprocal lattice vector **K** joining opposite faces of BZ.

This implies that the normal gradient of $E(\mathbf{k})$ is zero at the face.

Thus we see that if a face of a BZ is parallel to a mirror plane, the FS, which is tangential to $\nabla E(\mathbf{k})$, will cut the surface of the BZ normally. Thus the FS cuts normally the surface of the BZ of a body-centred cubic lattice, and will also cut normally the square faces of the BZ of a face-centred cubic lattice. However (see Figure 1.5), this is not necessarily true for the hexagonal faces of the BZ of this lattice, for there is no appropriate mirror plane.

1.7 Density of states

While we require for many purposes the dispersion relation $E(\mathbf{k})$ for the band, a less detailed description is afforded by the density of electron states $n(E)\,dE$. Precisely, this gives the number of states per unit volume of the crystal in energy around E, in the increment of energy dE. For free electrons, it follows trivially by differentiating equation (1.2.5) with $V = 0$ that $n(E) \propto E^{\frac{1}{2}}$.

In the absence of spin–orbit coupling, there is twofold degeneracy for every point in the BZ. Then we need only a factor of 2 in the occupation of states, and we can otherwise forget the spin. This we do in this section. However, the generalization of band theory to include spin–orbit interaction is effected in section 1.11.9 by starting from the Dirac equation. The most important effect of the spin–orbit coupling is to remove the spin degeneracy, when there is no inversion centre.

Our first task is to write $n(E)$ in terms of the dispersion relation. To do so, we choose two constant energy surfaces in **k** space, labelled by energies E and $E+dE$ respectively. If $dV_{\mathbf{k}}$ is the volume of **k** space thereby enclosed, then, since there are $1/8\pi^3$ allowed values of **k** per unit volume of the crystal, as follows from equations (1.4.5) and (1.4.11),

$$n(E)\,dE = \frac{1}{8\pi^3}\,dV_{\mathbf{k}}. \tag{1.7.1}$$

The area dS is a normal to the surface of constant energy; we then have

$$dV_{\mathbf{k}} = \int \Delta\mathbf{k}\,.\,d\mathbf{S}, \tag{1.7.2}$$

where $\Delta\mathbf{k}$ is the separation of the two surfaces. Since $\Delta\mathbf{k}$ and $d\mathbf{S}$ are in the same direction,

$$dV_{\mathbf{k}} = \int \Delta k\,dS, \tag{1.7.3}$$

where the integration is over the constant energy surface $E(\mathbf{k})$. However

$$dE = \nabla_{\mathbf{k}} E\,.\,\Delta\mathbf{k} = |\nabla_{\mathbf{k}} E|\,\Delta k, \tag{1.7.4}$$

and hence

$$n(E) = \frac{1}{8\pi^3}\int \frac{dS}{|\nabla_{\mathbf{k}} E|}. \tag{1.7.5}$$

It will be seen that if $\nabla_{\mathbf{k}} E$ is zero, or discontinuously changes sign, the latter alternative occurring only at points of degeneracy, we might expect some sort of singularity in $n(E)$ to occur. Points in the BZ for which $\nabla_{\mathbf{k}} E = 0$ will be referred to as *analytical critical points,* and points for which $\nabla_{\mathbf{k}} E$ discontinuously changes sign as *non-analytical critical points.* At critical points E has a maximum, minimum or saddle-point. We shall show below that information concerning these critical points can be extracted from topological considerations. A fuller discussion is given in Appendix 1.6.

1.7.1 Effect of critical points

To find the effect of the critical points on the density of states, we choose principal axes k_1, k_2, k_3 with origin at the critical point, so that in its neighbourhood we have the expansion

$$E = E_0 + \alpha_1 k_1^2 + \alpha_2 k_2^2 + \alpha_3 k_3^2. \tag{1.7.6}$$

We classify the points as types 0, 1, 2 and 3, the type-number being the number of negative α's. We see that type 0 is a minimum, types 1 and 2 saddle-points and type 3 a maximum.

We evaluate the integral (1.7.5) in the neighbourhood of the critical point by changing the scale of k_2, if necessary, to make $\alpha_1 = \alpha_2$, and by introducing a further transformation to cylindrical coordinates:

$$r = (k_1^2+k_2^2)^{\frac{1}{2}}, \quad \theta = \arctan\left(\frac{k_2}{k_1}\right). \tag{1.7.7}$$

Then

$$E = E_0+\alpha_1 r^2+\alpha_3 k_3^2 \tag{1.7.8}$$

in the neighbourhood of the critical point. In this neighbourhood, the constant energy surface S is a figure of revolution when considered in terms of our new coordinates r and k_3, as in Figure 1.7, with elements related to that of

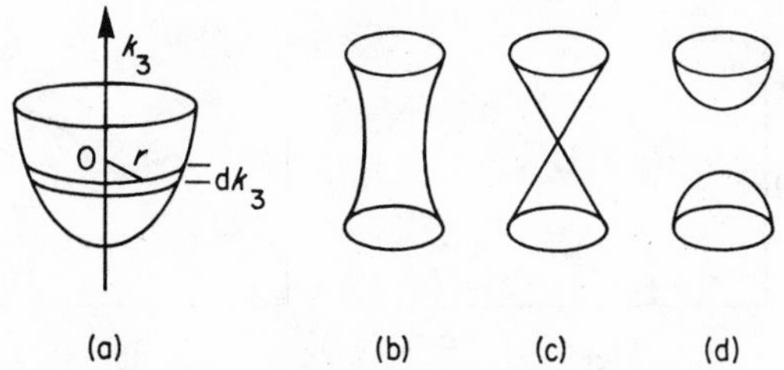

FIGURE 1.7. Showing constant energy surfaces S as figures of revolution near critical points. (a) Showing coordinates introduced in equation (1.7.8); (b), (c) and (d), showing how, near a saddle point, S passes from a one-sheeted hyperboloid to a hyperboloid of two sheets; (c) shows S at the critical point itself.

the original surface by $dS = \lambda\, dS'$, where λ is the scaling factor introduced to make $\alpha_1 = \alpha_2$.

Considering an element of surface between k_3 and k_3+dk_3, its area is

$$dS = \frac{2\pi r\, dk_3}{\cos\gamma}, \tag{1.7.9}$$

where γ is the angle of the normal relative to the (k_1, k_2) plane. But since this normal is in the direction of $\nabla_{\mathbf{k}} E(\mathbf{k})$ it follows that

$$\cos\gamma = \frac{\partial E}{\partial r}\bigg/|\nabla_{\mathbf{k}} E|. \tag{1.7.10}$$

Hence

$$\int \frac{dS}{|\nabla_{\mathbf{k}} E|} = \frac{2\pi\lambda}{|\alpha_1|} \int dk_3. \tag{1.7.11}$$

The contribution to the density of states is thus proportional to the extent of the energy surface in the k_3 direction. We now proceed to evaluate (1.7.11) for the four types of critical point.

Type 0. $\alpha_1 = \alpha_2 > 0$, $\alpha_3 > 0$. We have then

$$-\left(\frac{E-E_0}{\alpha_3}\right)^{\frac{1}{2}} \leqslant k_3 \leqslant \left(\frac{E-E_0}{\alpha_3}\right)^{\frac{1}{2}}$$

and hence the contribution to the density of states from the region about the critical point is proportional to $(E-E_0)^{\frac{1}{2}}$, increasing from zero with infinite slope [Figures 1.8(a) and (b)]. One recognizes here the basic origin of the free-electron density of states $E^{\frac{1}{2}}$, at small E.

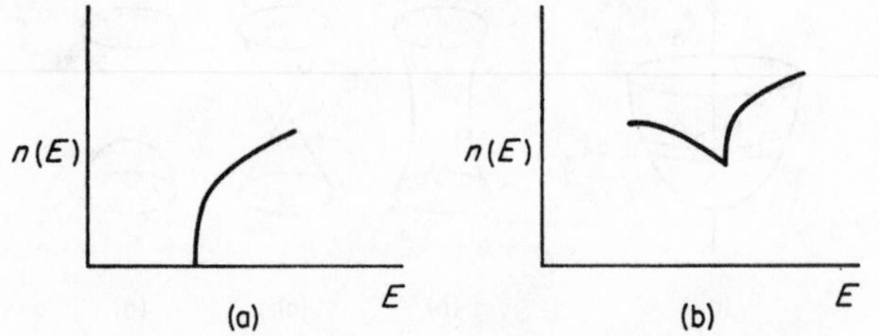

FIGURE 1.8. Forms of density of states near critical points. (a) At absolute minimum. (b) Other minima. Contribution from region about critical point is now superposed on non-zero density of states.

Type 1. $\alpha_1 = \alpha_2 > 0$, $\alpha_3 < 0$. The surface is a hyperboloid of two sheets, and is shown in Figure 1.7. For $E > E_0$, we must remember that (1.7.11) is only valid near the critical point and then the k_3 integration is bounded and of no interest. Thus for $E > E_0$ there is no pathological behaviour. However, when $E < E_0$ there is region of values of k_3 which are not allowed. From (1.7.8) this region is

$$\left(-\left[\frac{E_0-E}{|\alpha_3|}\right]^{\frac{1}{2}}, \left[\frac{E_0-E}{|\alpha_3|}\right]^{\frac{1}{2}}\right).$$

Thus, the contribution of this critical point to the density of states is again proportional to $(E_0-E)^{\frac{1}{2}}$, and $n(E)$ has infinite slope in the limit as E approaches E_0 from below [Figure 1.8(c)].

Type 2. Analogous reasoning to that used above for type 1 shows that here $n(E)$ has finite slope for $E < E_0$, and leaves E_0 with infinite slope [Figure 1.8(d)].

Type 3. The maximum behaves like the minimum, but with reversed E scale as shown in Figures 1.8(e) and (f).

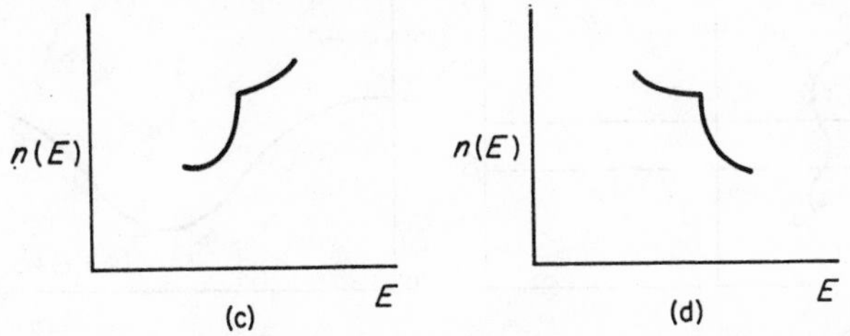

FIGURE 1.8 (c) Type 1 saddle-point. (d) Type 2 saddle-point.

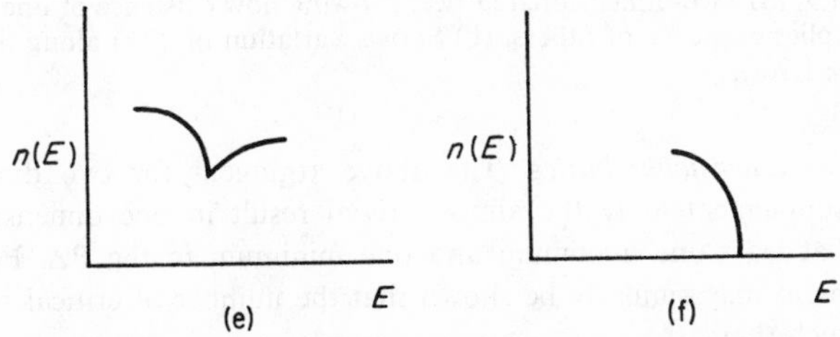

FIGURE 1.8 (e) General maximum. (f) Absolute maximum.

1.7.2 Minimal set of critical points

(a) *Two-dimensional lattice*. The existence of a critical point in a periodic function implies the existence of others as we can see from the following example of a two-dimensional lattice [Figure 1.9(a)]. Because of the periodicity, there must be at least one maximum in the function $E(\mathbf{k})$ in the BZ. This maximum is at point a in Figure 1.9(a) and a corresponding maximum b in an adjacent cell (BZ) of the reciprocal lattice is also shown. Let us now consider the variation of $E(\mathbf{k})$ along any line $(\alpha\beta)$ [see Figure 1.9(a)]. Because of the periodicity condition, the values of $E(\mathbf{k})$ along this line will have a maximum at some point. The locus of these maxima is shown by the solid line joining a to b. Somewhere along this line, one of the maxima must be smaller than any of the others [Figure 1.9(b)]. At this point, therefore, $E(\mathbf{k})$ must have a saddle-point. Similarly, $E(\mathbf{k})$ must have at least one minimum, and an analogous argument shows that a second saddle-point exists.

We have thus shown from simple topological considerations that a function periodic in two dimensions must, in general, have a maximum and two saddle-points.

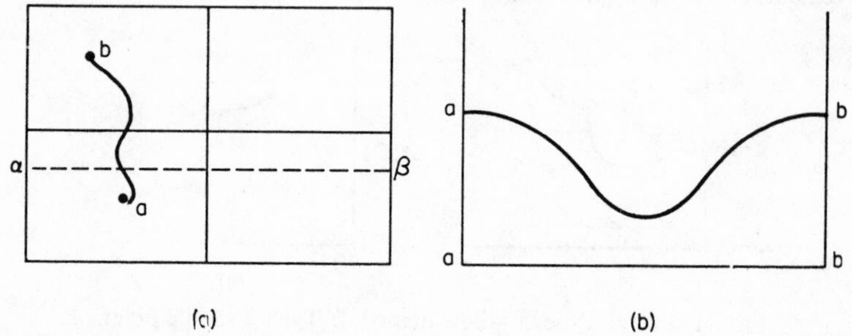

FIGURE 1.9. (a) Two-dimensional lattice, showing how existence of one critical point implies existence of others. (b) Shows variation of $E(\mathbf{k})$ along line (ab) in Figure 1.9(a).

(b) *Three-dimensional lattice.* The above argument for two dimensions may be supplemented by the almost trivial result in one dimension that there are at least one maximum and one minimum in the BZ. For three dimensions, it may similarly be shown that the number of critical points of type i is such that

$$N^i \geqslant \frac{3!}{i!(3-i)!}. \tag{1.7.12}$$

It is important to notice that N^i here refers to the number of critical points at inequivalent positions in the BZ. Thus, critical points at $\mathbf{k}$ and $\mathbf{k}+\mathbf{K}$, $\mathbf{K}$ being a reciprocal lattice vector, are not distinguished in calculating N^i.

More useful results for N^i than (1.7.12) can be extracted, however, from the more sophisticated topological argument set out in Appendix 1.6. These are known as the Morse relations which take the form

$$N^0 \geqslant 1, \quad N^1 - N^0 \geqslant 2, \quad N^2 - N^1 + N^0 \geqslant 1, \quad N^3 - N^2 + N^1 - N^0 = 0. \tag{1.7.13}$$

It is readily verified that these relations imply $N^0 \geqslant 1$, $N^1 \geqslant 3$, $N^2 \geqslant 3$, $N^3 \geqslant 1$, which are precisely the results given by (1.7.12). However, the Morse relations enable us to go further than (1.7.12). For these relations (1.7.13) connect numbers of critical points of different types; having found numbers of critical points by symmetry considerations, we can soon test whether these numbers satisfy (1.7.13). If they do not, we must seek further critical points. The smallest set of critical points consistent with symmetry and topological requirements [the latter requirements being embodied in (1.7.13)], is called the *minimal set*. We shall now go on to illustrate the procedure of finding the minimal set and the resulting structure of the density of states by considering the fcc lattice.

1.7.3 Density of states in face-centred cubic lattice

The BZ for the fcc lattice is shown in Figure 1.5. The point Γ at the centre of the zone is taken as a minimum in $E(\mathbf{k})$. At the corners W (there are three points, inequivalent in the sense that one cannot be reached from another by adding a reciprocal lattice vector, with the same energy), we assume maxima. This is the simplest assumption conforming with a free-electron-like level ordering. Then $N^2 - N^1 = 2$ from the fourth of the Morse relations, since we have assumed N^3, the number of maxima, is 3, and N^0, the number of minima, is 1. Furthermore, as shown in section 1.6.1, the energy surfaces are normal to the square faces. By reference to the four-fold symmetry axis‡ about X, we can deduce that we have saddle-points of type 1. There are 6 inequivalent points and hence 6 saddle-points.

Although there is not a mirror plane parallel to the hexagonal faces, the argument of section 1.6.1 is applicable to the special point L, so that we have further saddle-points of type 1, with four inequivalent points. Hence $N^1 = 10$ and therefore $N^2 = 12$.

If we now examine another high symmetry point K, it is readily verified that there must always be saddle-points there. There are 12 inequivalent points K, and it can be shown that from each point K, we can reach two points U, again of high symmetry and at which saddle-points must occur, by translation through a reciprocal lattice vector. It should be noted that there are 24 points U, which are equivalent in pairs. The conclusion is that the 36 points K plus U represent 12 inequivalent points, which is just the number of saddle-points we have to allocate. This is the number referred to in the Morse relations. It is obvious from our argument that, if $E(\mathbf{k})$ is regarded as a periodic function in the reciprocal lattice, there are 36 saddle-points of this type. With the information thus available, the density of states has the minimal structure of Figure 1.10. We want to stress finally that the critical points in the density of states are a direct consequence of perfect periodicity. Even the introduction of a phonon (discussed in Chapter 3) into the lattice strictly smudges out the critical points. More importantly the

‡ This simply means that rotation by $\pi/2$ about this axis brings the BZ back into itself.

introduction of disorder, as discussed in Chapter 10, will remove these singularities.

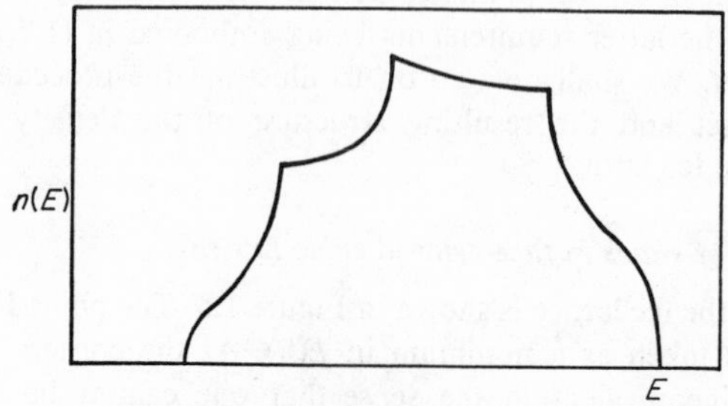

FIGURE 1.10. Schematic form of density of states for face-centred cubic lattice. Effects of critical points are shown as kinks in $n(E)$ as function of E which are frequently referred to as van Hove singularities. Reading from left, critical points are, in the labelling of Figure 1·8, (a), (c), (c), (d) and (f). (b) and (e), which are contributions from local minimum and local maximum, respectively, do not appear in smallest allowable set of singularities shown in this figure.

1.8 Localized Wannier functions

The previous discussion has placed the emphasis on a description in which the wave functions of Bloch form reflect very directly the periodicity of the crystal potential. We shall now show that an alternative description in terms of localized orbitals, the Wannier functions, is also perfectly proper.

We have already seen that $\psi_{\mathbf{k}}^{i}(\mathbf{r})$ is continuous in $\mathbf{k}$ and periodic in the reciprocal lattice. From equation (1.5.26) for the periodic part $u_{\mathbf{k}}(\mathbf{r})$ of the Bloch wave, it is straightforward to show that we may classify a Bloch function by $\mathbf{k}$ or $\mathbf{k}+\mathbf{K}$. Equivalently, $\psi_{\mathbf{k}}(\mathbf{r})$ may be taken as periodic in the reciprocal lattice (see equation 1.4.20) and we can therefore write down the Fourier series

$$\psi_{\mathbf{k}}^{i}(\mathbf{r}) = \sum_{\mathbf{R}} w_{\mathbf{R}}^{i}(\mathbf{r})\, e^{i\mathbf{k}.\mathbf{R}}, \tag{1.8.1}$$

where

$$w_{\mathbf{R}}^{i}(\mathbf{r}) = \frac{1}{\Omega_{\mathrm{B}}}\int_{\Omega_{\mathrm{B}}} d\mathbf{k}\, \psi_{\mathbf{k}}^{i}(\mathbf{r})\, e^{-i\mathbf{k}.\mathbf{R}}$$

$$= \frac{1}{\Omega_{\mathrm{B}}}\int_{\Omega_{\mathrm{B}}} d\mathbf{k}\, \psi_{\mathbf{k}}^{i}(\mathbf{r}-\mathbf{R}), \tag{1.8.2}$$

the last step following from Bloch's theorem. Thus, equation (1.8.1) may be written

$$\psi_i(\mathbf{r}) = \sum_{\mathbf{R}} w^i(\mathbf{r}-\mathbf{R})\, e^{i\mathbf{k}.\mathbf{R}}, \tag{1.8.3}$$

where

$$w^i(\mathbf{r}) = \frac{1}{\Omega_B}\int_{\Omega_B} \psi_{\mathbf{k}}^i(\mathbf{r})\, d\mathbf{k}. \tag{1.8.4}$$

$w^i(\mathbf{r})$ is the Wannier function for the ith band and is a single function implicit in which is a complete specification of the band. The Wannier function for a given band is not unique, for we may multiply any $\psi_{\mathbf{k}}$ in the integral (1.8.4) by a phase factor.

1.8.1 Symmetry of Wannier function

In any given case, however, we will wish to form the Wannier function showing the maximum symmetry possible; another point of obvious interest is whether or not the Wannier function is real. Let us consider these two questions; first, the one of symmetry.

The transformation properties of the wave function for $\mathbf{k} = 0$ are obviously connected with those of the Wannier function: we have, for any operator α of the point group, provided no essential degeneracy exists at $\mathbf{k} = 0$,

$$\psi_0(\alpha\mathbf{r}) = \chi(\alpha)\,\psi_0(\mathbf{r}) \tag{1.8.5}$$

(where $|\chi(\alpha)| = 1$) and we shall look at the situation where $w(\mathbf{r})$ has the same transformation properties as ψ_0. This situation will occur in *simple bands*, which we define as those for which we can write for any $\mathbf{k}$,

$$\psi_{\mathbf{k}}(\alpha\mathbf{r}) = \chi(\alpha)\,\psi_{\alpha^\dagger\mathbf{k}}(\mathbf{r}) \tag{1.8.6}$$

corresponding to equation (1.8.5). For a general $\mathbf{k}$ in the BZ we can force this condition in any energy band, but not at high symmetry points. It is easy to show that in symmorphic lattices‡ non-degenerate bands which are also simple always exist, but we cannot say that *all* non-degenerate bands have this character. (This question is discussed further in Appendix GAII.)

With equation (1.8.6) holding, the symmetry of the Wannier function is almost immediately evident. We have

$$\Omega_B\, w(\alpha\mathbf{r}) = \int_{\Omega_B} \psi_{\mathbf{k}}(\alpha\mathbf{r})\, d\mathbf{k} = \chi(\alpha)\int_{\Omega_B} \psi_{\alpha^\dagger\mathbf{k}}(\mathbf{r})\, d\mathbf{k}$$

$$= \chi(\alpha)\int_{\Omega_B} \psi_{\mathbf{k}}(\mathbf{r})\, d\mathbf{k} = \chi(\alpha)\,\Omega_B\, w(\mathbf{r}). \tag{1.8.7}$$

‡ That is, lattices with no symmetry operations involving non-primitive translations (screw axes or glide planes).

In other words, for simple bands, the Wannier function can be constructed to display the symmetry of the wave function at $\mathbf{k} = 0$.

Let us now consider whether the Wannier function is complex by splitting it into its real and imaginary parts

$$w(\mathbf{r}) = w_{\mathrm{Re}}(\mathbf{r}) + w_{\mathrm{I}}(\mathbf{r}). \tag{1.8.8}$$

If equation (1.8.7) is true, it is true of w_{Re} and w_{I} also. Now

$$\psi_{-\mathbf{k}}(\mathbf{r}) = \sum_{\mathbf{R}} w_{\mathrm{Re}}(\mathbf{r}-\mathbf{R})\, \mathrm{e}^{-i\mathbf{k}.\mathbf{R}} + \sum_{\mathbf{R}} w_{\mathrm{I}}(\mathbf{r}-\mathbf{R})\, \mathrm{e}^{-i\mathbf{k}.\mathbf{R}}, \tag{1.8.9}$$

whereas

$$\psi_{\mathbf{k}}^*(\mathbf{r}) = \sum_{\mathbf{R}} w_{\mathrm{Re}}(\mathbf{r}-\mathbf{R})\, \mathrm{e}^{-i\mathbf{k}.\mathbf{R}} - \sum_{\mathbf{R}} w_{\mathrm{I}}(\mathbf{r}-\mathbf{R})\, e^{-i\mathbf{k}.\mathbf{R}} \tag{1.8.10}$$

so that

$$\psi_{-\mathbf{k}}(\mathbf{r}) + \psi_{\mathbf{k}}^*(\mathbf{r}) = \sum_{\mathbf{R}} w_{\mathrm{Re}}(\mathbf{r}-\mathbf{R})\, \mathrm{e}^{-i\mathbf{k}.\mathbf{R}}. \tag{1.8.11}$$

In a non-degenerate band, $\psi_{\mathbf{k}}^*$ and $\psi_{-\mathbf{k}}$ can differ only by a phase-factor (cf. equation 1.5.28), so that unless $\psi_{-\mathbf{k}}(\mathbf{r}) = -\psi_{\mathbf{k}}^*(\mathbf{r})$, the Wannier function can be taken real, and, moreover, will display the transformation properties of equation (1.8.7) if the band is simple. If, on the other hand,

$$\psi_{-\mathbf{k}}(\mathbf{r}) + \psi_{\mathbf{k}}^*(\mathbf{r}) = 0,$$

the Wannier function is entirely imaginary, and, again, displays the transformation properties of equation (1.8.7) if the band is simple. Of course, we can always force the condition $\psi_{\mathbf{k}}^*(\mathbf{r}) = \psi_{-\mathbf{k}}(\mathbf{r})$ if we do not wish to force other properties. However, lack of continuity in $\mathbf{k}$ of $\psi_{\mathbf{k}}$, for example, may cause $w(\mathbf{r})$ to lose its localized property. We can, again, examine the properties of $\psi_0(\mathbf{r})$ to settle the question. If $\psi_0(\mathbf{r})$ is chosen to be purely real, the Wannier function cannot be purely imaginary, and vice versa.

Let us suppose that, at least for one direction of $\mathbf{k}$, there exists an operator β such that $\beta\mathbf{k} = -\mathbf{k}$. (This can be taken as the inversion if the inversion is in the point group.) Now, if we take equations (1.8.5) and (1.8.6) to hold for this operator, for this direction of $\mathbf{k}$,

$$\lim_{\mathbf{k}\to 0} \psi_{-\mathbf{k}}(\mathbf{r}) = \chi(\beta)\, \psi_0(\mathbf{r}). \tag{1.8.12}$$

Thus, if we are to take $\psi_{\mathbf{k}}^*(r) = \psi_{-\mathbf{k}}(\mathbf{r})$,

$$\psi_{\mathbf{k}}^*(\mathbf{r}) = \chi(\beta)\, \psi_0(\mathbf{r}) \tag{1.8.13}$$

so that for the Wannier function to be entirely real, we must have $\chi(\beta) = 1$. On the other hand, for the Wannier function to be entirely imaginary, $\chi(\beta) = -1$. $\chi(\beta)$ must, in fact, be integral if time-reversal degeneracy (see section 1.5.6) does not exist at $\mathbf{k} = 0$, for then $\psi_0^*(\mathbf{r})$ and $\psi_0(\mathbf{r})$ can only differ by a phase-factor, so that ψ_0 can be chosen to be entirely real or entirely imaginary, as we please. In summary, therefore, for a non-degenerate band the Wannier function can be chosen to be purely real or purely imaginary. But in a simple band, where we can take $w(\mathbf{r})$ to have the symmetry of $\psi_0(\mathbf{r})$, it must be purely real, if $\chi(\beta) = 1$, and purely imaginary if $\chi(\beta) = -1$, in order that it has this symmetry.

1.8.2 Completeness relation

Let us normalize the Bloch wave functions such that

$$\int_{\text{all space}} \psi_{\mathbf{k}_1}^{i*}(\mathbf{r})\,\psi_{\mathbf{k}_2}^{j}(\mathbf{r})\,d\mathbf{r} = \Omega_B\,\delta(\mathbf{k}_1-\mathbf{k}_2)\,\delta_{ij}. \tag{1.8.14}$$

Then it follows from equation (1.8.4) that

$$\begin{aligned}\int w^*(\mathbf{r}-\mathbf{R})\,w^j(\mathbf{r}-\mathbf{S})\,d\mathbf{r} &= \frac{1}{\Omega_B^2}\int_{\Omega_B} \psi_{\mathbf{k}_1}^{i*}(\mathbf{r}-\mathbf{R})\,\psi_{\mathbf{k}_2}^{i}(\mathbf{r}-\mathbf{S})\,d\mathbf{k}_1\,d\mathbf{k}_2\\ &= \frac{1}{\Omega_B^2}\int d\mathbf{r}\int_{\Omega_B} \psi_{\mathbf{k}_1}^{i*}(\mathbf{r})\,\psi_{\mathbf{k}_2}^{j}(\mathbf{r})\,e^{i\mathbf{k}_1.\mathbf{R}}\,e^{-i\mathbf{k}_2.\mathbf{S}}\,d\mathbf{k}_1\,d\mathbf{k}_2\\ &= \frac{\delta_{ij}}{\Omega_B}\int_{\Omega_B} \delta(\mathbf{k}_1-\mathbf{k}_2)\,e^{i\mathbf{k}_1.\mathbf{R}}\,e^{-i\mathbf{k}_2.\mathbf{S}}\,d\mathbf{k}_1\,d\mathbf{k}_2 = \delta_{ij}\,\delta(\mathbf{R},\mathbf{S}).\end{aligned} \tag{1.8.15}$$

Thus we have shown that the consequence of equation (1.8.14) is

$$\int_{\text{all space}} w^{i*}(\mathbf{r}-\mathbf{R})\,w^j(\mathbf{r}-\mathbf{S})\,d\mathbf{r} = \delta_{ij}\,\delta(\mathbf{R},\mathbf{S}). \tag{1.8.16}$$

The Wannier functions evidently form a complete orthonormal set, and so we may write down the completeness relation

$$\sum_{\mathbf{R},i} w^{i*}(\mathbf{r}'-\mathbf{R})\,w^i(\mathbf{r}-\mathbf{R}) = \delta(\mathbf{r}'-\mathbf{r}). \tag{1.8.17}$$

It is worth enquiring at this point what the Schrödinger equation for the Bloch waves implies for the Wannier functions. The wave equation

$$H\psi_{\mathbf{k}}(\mathbf{r}) = E(\mathbf{k})\,\psi_{\mathbf{k}}(\mathbf{r}) \tag{1.8.18}$$

becomes, when we employ equation (1.8.1),

$$\sum_{\mathbf{R}} Hw(\mathbf{r}-\mathbf{R})\,e^{i\mathbf{k}.\mathbf{R}} = \sum_{\mathbf{R}} w(\mathbf{r}-\mathbf{R})\,E(\mathbf{k})\,e^{i\mathbf{k}.\mathbf{R}}. \tag{1.8.19}$$

Substituting the expansion (1.4.21) into the left-hand side and using equation (1.3.14), we obtain

$$Hw(\mathbf{r}) = \sum_{\mathbf{R}} E_{\mathbf{R}}\, w(\mathbf{r}-\mathbf{R}). \tag{1.8.20}$$

Multiplying both sides by $w^*(\mathbf{r}-\mathbf{S})$ and integrating over the whole of space we find that

$$E_{\mathbf{S}} = \int w^*(\mathbf{r}-\mathbf{S})\, Hw(\mathbf{r})\, d\mathbf{r}. \tag{1.8.21}$$

Unfortunately equation (1.8.20) is not unique to the Wannier function. Let us take any function $\phi(\mathbf{r})$ such that

$$\phi(\mathbf{r}) = \sum_{\mathbf{R}} \xi_{\mathbf{R}}\, w(\mathbf{r}-\mathbf{R}) \tag{1.8.22}$$

which obviously reduces to $\psi_{\mathbf{k}}(\mathbf{r})$ with the choice $e^{i\mathbf{k}.\mathbf{R}}$ for $\xi_{\mathbf{R}}$. But from equation (1.8.22) we find immediately

$$H\phi(\mathbf{r}) = \sum_{\mathbf{R},\mathbf{S}} E_{\mathbf{R}}\, \xi_{\mathbf{S}}\, w(\mathbf{r}-\mathbf{R}-\mathbf{S})$$

$$= \sum_{\mathbf{R}} E_{\mathbf{R}}\, \phi(\mathbf{r}-\mathbf{R}), \tag{1.8.23}$$

which is formally identical with the original form (1.8.20). The solution of equation (1.8.20) must clearly be found subject to the orthonormality relations (1.8.16). This is a difficult problem and in working with a single function containing all information about the band we have to pay a high price.

1.8.3 Relation to atomic orbitals

We shall now briefly comment on the significance of the Wannier function. For electrons tightly bound to the nuclei on the lattice sites, e.g. the electrons of the core, we require that the wave function for the electron shall behave as a free atom orbital ϕ in each cell. The function must also satisfy Bloch's theorem, and we see by inspection that the requirements are satisfied by the function

$$\psi_{\mathbf{k}}(\mathbf{r}) = \sum_{\mathbf{R}} \phi(\mathbf{r}-\mathbf{R})\, e^{i\mathbf{k}.\mathbf{R}}. \tag{1.8.24}$$

If the approximation is good, the band will be very narrow and its mean energy $E(0)$ will be accurately given by the energy of the electron in the free atom. Equation (1.8.24) provides the interpretation of the Wannier function as a perturbed atomic orbital. However, the perturbation may often be so large in real crystals as to vitiate the usefulness of this interpretation.

1.9 Momentum eigenfunction

Using Bloch's theorem

$$\psi_{\mathbf{k}}(\mathbf{r}) = e^{i\mathbf{k}.\mathbf{r}}\, u_{\mathbf{k}}(\mathbf{r}) \tag{1.9.1}$$

and expanding the periodic function $u_{\mathbf{k}}(\mathbf{r})$ in the Fourier series

$$u_{\mathbf{k}}(\mathbf{r}) = \sum_{\mathbf{K}} v_{\mathbf{K}}(\mathbf{k})\, e^{i\mathbf{K}.\mathbf{r}} \tag{1.9.2}$$

we find

$$\psi_{\mathbf{k}}(\mathbf{r}) = \sum_{\mathbf{K}} v_{\mathbf{K}}(\mathbf{k})\, e^{i(\mathbf{k}+\mathbf{K}).\mathbf{r}}. \tag{1.9.3}$$

The physical interpretation of this equation is evidently that a Bloch electron with wave vector $\mathbf{k}$ can have momentum $\mathbf{k}+\mathbf{K}$, where $\mathbf{K}$ is any reciprocal lattice vector. The probability of finding momentum $\mathbf{k}+\mathbf{K}$ is simply $|v_{\mathbf{K}}(\mathbf{k})|^2$ apart from a normalization factor [see equation (1.9.6) below].

While, with a definite wave vector $\mathbf{k}$, there are discrete values of the momentum (though an infinite number), by allowing $\mathbf{k}$ to take on all the values within a band we built up a continuous function $v_{\mathbf{K}}(\mathbf{k})$, the *momentum eigenfunction* of the band. In fact, as we now show, v depends only on momentum $\mathbf{k}+\mathbf{K}$.

Using equation (1.3.14), we can immediately solve equation (1.9.3) for $v_{\mathbf{K}}(\mathbf{k})$ and we find

$$v_{\mathbf{K}}(\mathbf{k}) = \frac{1}{\Omega}\int_{\Omega} \psi_{\mathbf{k}}(\mathbf{r})\, e^{-i(\mathbf{k}+\mathbf{K}).\mathbf{r}}\, d\mathbf{r}. \tag{1.9.4}$$

But as we have seen, $\psi_{\mathbf{k}}(\mathbf{r}) = \psi_{\mathbf{k}+\mathbf{K}}(\mathbf{r})$ and thus

$$v_{\mathbf{K}}(\mathbf{k}) = \frac{1}{\Omega}\int_{\Omega} \psi_{\mathbf{k}+\mathbf{K}}(\mathbf{r})\, e^{-i(\mathbf{k}+\mathbf{K}).\mathbf{r}}\, d\mathbf{r} = v(\mathbf{k}+\mathbf{K}). \tag{1.9.5}$$

Hence $v_{\mathbf{K}}(\mathbf{k})$ is solely a function of the momentum $\mathbf{k}+\mathbf{K}$ and this is the desired result.

We can clearly equally well write the Bloch wave function as

$$\psi_{\mathbf{k}}(\mathbf{r}) = \sum_{\mathbf{K}} v(\mathbf{k}-\mathbf{K})\, e^{i(\mathbf{k}-\mathbf{K}).\mathbf{r}}. \tag{1.9.6}$$

We now utilize the Fourier series expansion of the crystal potential, which, by analogy with (1.3.10), we can write as

$$V(\mathbf{r}) = \sum_{\mathbf{K}} V_{\mathbf{K}}\, e^{i\mathbf{K}.\mathbf{r}}. \tag{1.9.7}$$

Substituting (1.9.6) and (1.9.7) into the Schrödinger equation, and equating coefficients of the different plane waves, we obtain almost immediately the set of simultaneous equations

$$\left[\frac{|\mathbf{k}-\mathbf{K}|^2}{2}-E(\mathbf{k})\right]v(\mathbf{k}-\mathbf{K})+\sum_{\mathbf{K}'}V_{\mathbf{K}'-\mathbf{K}}v(\mathbf{k}-\mathbf{K}')=0 \tag{1.9.8}$$

for each value of **K**. In principle, we might choose **k** and find all the coefficients $v(\mathbf{k}-\mathbf{K})$ by solution of the secular equation.

1.9.1 Free electrons

The form (1.9.8), in general, is not convenient for practical calculations, but we shall find it useful to begin the discussion with the "empty lattice", as an example to which we shall return later. Then in this case $V(\mathbf{r}) = 0$, and hence $V_{\mathbf{K}} = 0$ for all **K**, when (1.9.8) becomes

$$\left\{\frac{|\mathbf{k}-\mathbf{K}|^2}{2}-E(\mathbf{k})\right\}v(\mathbf{k}-\mathbf{K})=0 \quad \text{for all } \mathbf{K}. \tag{1.9.9}$$

A solution is evidently given by

$$E(k)=\frac{k^2}{2}, \quad v(\mathbf{k})=\frac{1}{\Omega^{\frac{1}{2}}}, \quad v(\mathbf{k}-\mathbf{K})=0, \quad \mathbf{K}\neq 0 \tag{1.9.10}$$

this solution satisfying the normalization condition (1.8.14). This is the elementary free-electron solution, but it should be noted that further solutions are given by

$$E(\mathbf{k})=\frac{|\mathbf{k}-\mathbf{K}|^2}{2}, \quad v(\mathbf{k}-\mathbf{K})=\frac{1}{\Omega^{\frac{1}{2}}}, \quad v(\mathbf{k}-\mathbf{K}')=0, \quad \mathbf{K}'\neq\mathbf{K}. \tag{1.9.11}$$

We shall come back to the "empty lattice" solution (1.9.11) when we deal with the band structure of Ge.

If we use the form (1.9.10), we have that $E(\mathbf{k})$ is a single-valued function of **k**, which goes on increasing outside the BZ. This is the so-called *extended zone scheme*.

If we used the solutions given by (1.9.11) then, in one dimension, we would have solutions as shown in Figure 1.11. We are really plotting here, as we can see from the figure, the same solution, but using a wave vector $\mathbf{k}-\mathbf{K}$ rather than **K**.

The most generally useful scheme is, as we have already indicated, that in which we define the wave vector of the state by always restricting **k** to lie in the (first) BZ. Then $E(\mathbf{k})$, as we have emphasized already, is multivalued, and we must label $E(\mathbf{k})$ with a band index i. In the empty lattice case, we then find trivially

$$E_i(\mathbf{k}) = \frac{|\mathbf{k}-\mathbf{K}_i|^2}{2} \tag{1.9.12}$$

with corresponding wave functions

$$\psi_{\mathbf{k}i}(\mathbf{r}) = e^{i(\mathbf{k}-\mathbf{K}_i)\cdot\mathbf{r}}. \tag{1.9.13}$$

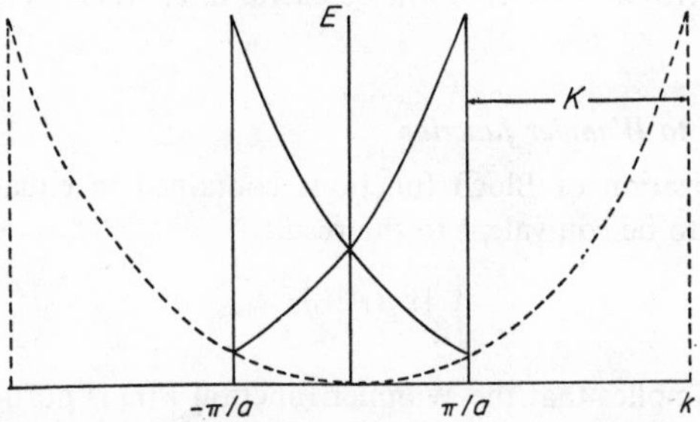

FIGURE 1.11. Free-electron $E(\mathbf{k})$ relation. Dashed curve outside $|\mathbf{k}| = \pi/a$ represents extended zone scheme. Solid curves give energy in reduced zone scheme. $K = 2\pi/a$ is primitive translation of one-dimensional reciprocal lattice.

1.9.2 *Nearly free electron theory*

The method of nearly free electrons, which for reasons that will become apparent later, is highly relevant in discussing the Fermi surfaces of simple metals, essentially solves the equation (1.9.8) by low-order perturbation theory.

From equations (1.9.8), we have immediately

$$v(\mathbf{k}-\mathbf{K}) = \sum_{\mathbf{K}'} \frac{V_{\mathbf{K}'-\mathbf{K}}\, v(\mathbf{k}-\mathbf{K}')}{E(\mathbf{k}) - \dfrac{|\mathbf{k}-\mathbf{K}|^2}{2}}. \tag{1.9.14}$$

Now if $\psi_{\mathbf{k}}$ resembles a plane wave, $v(\mathbf{k}) \sim \Omega^{-\frac{1}{2}}$ for $\mathbf{k}$ vectors within the BZ, the other $v(\mathbf{k}-\mathbf{K})$'s being small. Ignoring the latter on the right-hand side of (1.9.14), and approximating $E(\mathbf{k})$ by $k^2/2$, we obtain

$$v(\mathbf{k}-\mathbf{K}) = \frac{V_{-\mathbf{K}}/\Omega^{\frac{1}{2}}}{\dfrac{k^2}{2} - \dfrac{|\mathbf{k}-\mathbf{K}|^2}{2}}. \tag{1.9.15}$$

This gives us the wave function to first order in perturbation theory when inserted in equation (1.9.6). From this first-order wave function, we obtain the second-order perturbation energy as

$$E(\mathbf{k}) = \frac{k^2}{2} + V_0 + \sum_{\mathbf{K}} \frac{|V_{\mathbf{K}}|^2}{\dfrac{k^2}{2} - \dfrac{(\mathbf{k}+\mathbf{K})^2}{2}}. \tag{1.9.16}$$

Equations (1.9.15) and (1.9.16) will be useful later when we discuss pseudopotentials.

1.9.3 Relation to Wannier function

The normalization of Bloch functions contained in equation (1.8.14) is readily shown to be equivalent to the result

$$\int_{\Omega} |\psi_{\mathbf{k}}(\mathbf{r})|^2 \, d\mathbf{r} = 1, \tag{1.9.17}$$

which in turn implies that the Wannier function $w(\mathbf{r})$ is normalized to unity over all space. Furthermore, it follows from the definition (1.9.6) of v and equations (1.9.17) and (1.3.14) that

$$\sum_{\mathbf{K}} |v(\mathbf{k}+\mathbf{K})|^2 = \Omega^{-1}. \tag{1.9.18}$$

Then

$$\begin{aligned}
\int_{\text{all space}} w(\mathbf{r})\, e^{-i\mathbf{k}.\mathbf{r}}\, d\mathbf{r} &= \frac{1}{\Omega_B} \int_{\Omega_B} d\mathbf{k}' \int d\mathbf{r}\, \psi_{\mathbf{k}'}(\mathbf{r})\, e^{-i\mathbf{k}.\mathbf{r}} \\
&= \frac{1}{\Omega_B} \int_{\Omega_B} d\mathbf{k}' \sum_{\mathbf{R}} \int_{\Omega} d\mathbf{r}\, \psi_{\mathbf{k}'}(\mathbf{r}-\mathbf{R})\, e^{-i\mathbf{k}.(\mathbf{r}-\mathbf{R})} \\
&= \frac{1}{\Omega_B} \int_{\Omega_B} d\mathbf{k}' \int_{\Omega} d\mathbf{r}\, \psi_{\mathbf{k}'}(\mathbf{r})\, e^{-i\mathbf{k}.\mathbf{r}} \sum_{\mathbf{R}} e^{i(\mathbf{k}-\mathbf{k}').\mathbf{R}} \\
&= \int_{\Omega_B} d\mathbf{k}' \int_{\Omega} d\mathbf{r}\, \psi_{\mathbf{k}'}(\mathbf{r}) \sum_{\mathbf{K}} \delta(\mathbf{k}-\mathbf{k}'+\mathbf{K})\, e^{-i\mathbf{k}.\mathbf{r}},
\end{aligned} \tag{1.9.19}$$

the last line following when we make use of equation (1.3.19). Completing the integration over $\mathbf{k}'$, we find

$$\begin{aligned}
\int_{\text{all space}} d\mathbf{r}\, w(\mathbf{r})\, e^{-i\mathbf{k}.\mathbf{r}} &= \int_{\Omega} d\mathbf{r}\, \psi_{\mathbf{k}}(\mathbf{r})\, e^{-i\mathbf{k}.\mathbf{r}} \\
&= \Omega v(\mathbf{k}),
\end{aligned} \tag{1.9.20}$$

where we have used the relation (1.9.4). Thus, we have the basic relation between the Wannier function $w(\mathbf{r})$ and the momentum eigenfunction $v(\mathbf{k})$:

$$v(\mathbf{k}) = \frac{1}{\Omega}\int_{\text{all space}} d\mathbf{r}\, w(\mathbf{r})\, e^{-i\mathbf{k}.\mathbf{r}} \tag{1.9.21}$$

and

$$w(\mathbf{r}) = \frac{1}{\Omega_{\mathrm{B}}}\int_{\text{all space}} d\mathbf{k}\, v(\mathbf{k})\, e^{i\mathbf{k}.\mathbf{r}}. \tag{1.9.22}$$

Equation (1.9.18) shows that a separate discussion of the symmetry of the momentum eigenfunction is unnecessary, and for this we can refer back to section 1.8. There is also a completeness relation for the momentum eigenfunction which we can derive from the corresponding result (1.8.17) for $w(\mathbf{r})$. We first multiply both sides of this equation by $e^{i\mathbf{k}.(\mathbf{r}-\mathbf{r}')}\, e^{i(\mathbf{K}-\mathbf{G}).\mathbf{r}}$. Integrating over the unit cell with respect to $\mathbf{r}'$ and over all space with respect to $\mathbf{r}$, we obtain, using equation (1.9.18),

$$\Omega \sum_i v_i^*(\mathbf{k}+\mathbf{K})\, v_i(\mathbf{k}+\mathbf{G}) = \delta(\mathbf{K},\mathbf{G}), \tag{1.9.23}$$

which is the required completeness relation.

The above tools, particularly the Wannier function, will be used a good deal throughout the book.

Though of less general applicability, we wish to say a little at this stage about a representation closely related to the momentum eigenfunction. Whereas in the definition of the momentum eigenfunction, we analyse a Bloch wave with wave vector $\mathbf{k}$ into its momentum components $\mathbf{p} = \mathbf{k}+\mathbf{K}$, the so-called crystal momentum representation, to which we now turn, works directly with $\mathbf{k}$ and *not* with $\mathbf{p}$. It will be useful to do this as $\mathbf{k}$ is a good quantum number in a perfect crystal, and operators can be written in terms of $\mathbf{k}$ rather than $\mathbf{r}$. We shall use the formalism in discussing relativistic corrections to band theory in section 1.11.9, in the $\mathbf{k}.\mathbf{p}$ method of section 1.11.7, and in the discussion of quasi-particles in insulators in Chapter 2. But it has other important applications, which we will refer to below.

1.10 Crystal momentum representation

It often happens that we wish to treat a situation in which a slowly varying perturbation is applied to the electrons of a crystal and it is then natural to expand the resultant wave function $\phi(\mathbf{r})$ in terms of the Bloch functions $\psi_{\mathbf{k}}^n$:

$$\phi(\mathbf{r}) = \sum_n \int_{\Omega_{\mathrm{B}}} d\mathbf{k}\, \phi_n(\mathbf{k})\, \psi_{\mathbf{k}}^n(\mathbf{r}). \tag{1.10.1}$$

The $\phi_n(\mathbf{k})$ may now be regarded as defining the wave function, in which case we require our operators for momentum, position, etc., in terms of $\mathbf{k}$, to operate on the $\phi_n(\mathbf{k})$. We are thus led to discuss a representation, the basis of which is formed by the $\psi_{\mathbf{k}}^n$'s. These we assume to obey the usual Schrödinger equation

$$\left\{-\frac{\hbar^2}{2m}\nabla^2+V(\mathbf{r})\right\}\psi_{\mathbf{k}}^n(\mathbf{r}) = E_n(\mathbf{k})\,\psi_{\mathbf{k}}^n(\mathbf{r}), \tag{1.10.2}$$

but we could, if we wished, add a spin–orbit coupling term to the Hamiltonian, or even choose the $\psi_{\mathbf{k}}^n$ to be solutions of the Dirac equation of the perfect crystal.

It is readily seen that the momentum operator has matrix elements of the form

$$\begin{aligned}\mathbf{p}_{n'n}(\mathbf{k}'\mathbf{k}) &= \frac{1}{\Omega_B}\int \psi_{\mathbf{k}'}^{n'*}\,\mathbf{p}\psi_{\mathbf{k}}^n\,d\mathbf{r} = \langle\psi_{\mathbf{k}'}^{n'}|\mathbf{p}|\psi_{\mathbf{k}}^n\rangle \\ &= \delta(\mathbf{k}'-\mathbf{k})\left[\hbar\mathbf{k}\,\delta_{n'n} - i\hbar\int u_{\mathbf{k}'}^{n'*}\frac{\partial u_{\mathbf{k}}^n}{\partial\mathbf{r}}\,d\mathbf{r}\right],\end{aligned} \tag{1.10.3}$$

where

$$u_{\mathbf{k}}^n(\mathbf{r}) = e^{-i\mathbf{k}.\mathbf{r}}\,\psi_{\mathbf{k}}^n(\mathbf{r}). \tag{1.10.4}$$

Matrix elements $\mathbf{r}_{n'n}(\mathbf{k}'\mathbf{k})$ of $\mathbf{r}$ need more care, for direct evaluation of $\int\psi_{\mathbf{k}'}^{n'*}\,\mathbf{r}\psi_{\mathbf{k}}^n\,d\mathbf{r}$ shows that its value is ill defined and we must therefore give a prescription for calculating it. We follow Adams (1953) in using the result

$$\frac{1}{i}\frac{\partial}{\partial\mathbf{k}}\int\psi_{\mathbf{k}'}^{n'*}\,\psi_{\mathbf{k}}^n\,d\mathbf{r} = \int\psi_{\mathbf{k}'}^{n'*}\,\mathbf{r}\psi_{\mathbf{k}}^n\,d\mathbf{r} + \int u_{\mathbf{k}'}^{n'*}(\mathbf{r})\,e^{i(\mathbf{k}-\mathbf{k}').\mathbf{r}}\frac{\partial}{\partial\mathbf{k}}u_{\mathbf{k}}^n(\mathbf{r})\,d\mathbf{r}, \tag{1.10.5}$$

which can be derived using equation (1.10.4). If we write the last term in equation (1.10.5) as a sum of integrals over unit cells, it can be seen to vanish unless $\mathbf{k} = \mathbf{k}'$, these being restricted to lie within the first BZ. Thus we have

$$\mathbf{r}_{n'n}(\mathbf{k}'\mathbf{k}) = -i\frac{\partial}{\partial\mathbf{k}}\Delta_{n'n}(\mathbf{k}'\mathbf{k}) + \delta(\mathbf{k}'-\mathbf{k})\,\mathscr{R}_{n'n}(\mathbf{k}), \tag{1.10.6}$$

where

$$\Delta_{n'n}(\mathbf{k}'\mathbf{k}) = \frac{1}{\Omega_B}\int\psi_{\mathbf{k}'}^{n'*}\,\psi_{\mathbf{k}}^n\,d\mathbf{r} \tag{1.10.7}$$

and

$$\mathscr{R}_{n'n}(\mathbf{k}) = \int_\Omega u_{\mathbf{k}}^{n'*}(\mathbf{r})\frac{\partial}{\partial\mathbf{k}}u_{\mathbf{k}}^n(\mathbf{r})\,d\mathbf{r}. \tag{1.10.8}$$

Δ, which in the limit of infinite volume reduces to the result (1.8.14), and $\mathscr{R}$ are sensitive to the phases chosen for the Bloch functions but we shall not concern ourselves with this and other formal difficulties here (for a more rigorous discussion, see Blount, 1962).

To find the result of operating on $\phi(\mathbf{k})$ with $\mathbf{r}$, we first see from equation (1.10.1) that

$$\mathbf{r}\phi(\mathbf{r}) = \sum_{nn'} \int d\mathbf{k}\, d\mathbf{k}'\, \phi_n(\mathbf{k})\, \mathbf{r}_{n'n}(\mathbf{k}'\mathbf{k})\, \psi_{\mathbf{k}'}^{n'}, \tag{1.10.9}$$

where we have used the completeness relation for the ψ's. However, from equation (1.10.5) we have

$$\begin{aligned}\frac{\partial \phi_n(\mathbf{k}')}{\partial \mathbf{k}'} &= \frac{\partial}{\partial \mathbf{k}'} \langle \psi_{\mathbf{k}'}^{n*}\, \phi(\mathbf{r}) \rangle = \sum_{n'} \int d\mathbf{k} \frac{\partial}{\partial \mathbf{k}'} \langle \psi_{\mathbf{k}'}^{n'*}\, \psi_{\mathbf{k}}^{n} \rangle\, \phi_n(\mathbf{k}) \\ &= \sum_{n} \int d\mathbf{k} \frac{\partial}{\partial \mathbf{k}'} \Delta_{n'n}(\mathbf{k}'\mathbf{k})\, \phi_n(\mathbf{k})\end{aligned} \tag{1.10.10}$$

so that the operator $\mathbf{r}$ acting on $\phi_n(\mathbf{k})$ gives us explicitly

$$\mathbf{r}\phi_n(\mathbf{k}) = i \frac{\partial}{\partial \mathbf{k}} \phi_n(\mathbf{k}) + \sum_{n} \mathscr{R}_{n'n}(\mathbf{k})\, \phi_{n'}(\mathbf{k}). \tag{1.10.11}$$

The last term would be missing if we used momentum $\mathbf{p}$ rather than crystal momentum $\mathbf{k}$.

1.10.1 Group velocity

To illustrate the use of this form of the operator $\mathbf{r}$, let us take the diagonal matrix elements of the velocity $\mathbf{v} = \mathbf{p}/m$, given by the Heisenberg equation of motion

$$\mathbf{v} = \frac{i}{\hbar}[\mathbf{r}, H], \tag{1.10.12}$$

between eigenfunctions of H. Since H is a diagonal operator, off-diagonal elements of $\mathbf{r}_{n'n}$ do not then contribute and we find

$$\begin{aligned}\mathbf{v}_{nn}(\mathbf{k}) &= \frac{1}{\hbar} \left\langle \psi_{\mathbf{k}}^{n*} \left[\frac{\partial}{\partial \mathbf{k}}, E_n(\mathbf{k})\right] \psi_{\mathbf{k}}^{n} \right\rangle \\ &= \frac{1}{\hbar} \frac{\partial E_n(\mathbf{k})}{\partial \mathbf{k}},\end{aligned} \tag{1.10.13}$$

which is just the group velocity formula found by another argument in Appendix 1.5, where the correction to (1.10.13) for the case of an energy-dependent potential is given.

1.10.2 *Effective mass tensor*

As a second example, we shall obtain a formula for the effective mass, defined in the following way. Let us assume that $E(\mathbf{k})$ has a minimum at $\mathbf{k}_0$. We can then expand

$$E(\mathbf{k}) = E(\mathbf{k}_0)+\frac{1}{2m^*}(\mathbf{k}-\mathbf{k}_0)^2+\ldots, \tag{1.10.14}$$

provided the system is isotropic. The quantity m^*, the effective mass, would be just the ordinary mass m if we had a system of free electrons (and we would also have $k_0 = 0$); m^* is a mass corrected for the influence of the lattice. More generally, we define a *reciprocal effective mass-tensor* $(1/m^*)_{ij}$ such that

$$E(\mathbf{k}) = E(\mathbf{k}_0)+\frac{1}{2}\sum_{ij}\left(\frac{1}{m^*}\right)_{ij}(\mathbf{k}-\mathbf{k}_0)_i(\mathbf{k}-\mathbf{k}_0)_j+\ldots. \tag{1.10.15}$$

Evidently

$$\left(\frac{1}{m^*}\right)_{ij} = \left.\frac{\partial^2 E(\mathbf{k})}{\partial k_i\,\partial k_j}\right|_{\mathbf{k}=\mathbf{k}_0} \tag{1.10.16}$$

For later purposes, we shall develop this expression further and in particular shall consider interband contributions to m^*. The theory will be used in the discussion of the $\mathbf{k}.\mathbf{p}$ method in section 1.11.7, and also in the treatment of crystal lattice defects in Chapter 10.

We first consider the diagonal elements of the commutator

$$[x_i, v_j] = \frac{\delta_{ij}}{m}, \quad \mathbf{r} = (x_1, x_2, x_3), \tag{1.10.17}$$

in the crystal momentum representation. To do so, we write the position operator in this representation as

$$\mathbf{r} = \mathbf{r}_d+\mathbf{X}, \tag{1.10.18}$$

where $\mathbf{r}_d$ is diagonal and $\mathbf{X}$ contains interband, or off-diagonal, terms only. We similarly define $\mathbf{v}_d$, $\mathbf{V}$ in

$$\mathbf{v} = \mathbf{v}_d+\mathbf{V}. \tag{1.10.19}$$

By using the crystal momentum representation set out above for these operators, it is readily shown that

$$\begin{aligned}\langle\psi_{\mathbf{k}}^n|\,[x_i, v_j]\,|\psi_{\mathbf{k}}^n\rangle &= [x_i, v_j]_n \\ &= [x_d^i, v_d^j]_n+[X_i, V_j]_n.\end{aligned} \tag{1.10.20}$$

In addition, since $\mathbf{r}_d$ and $\mathbf{v}_d$ are diagonal operators, we find

$$\frac{i}{\hbar}[x_d^i, v_d^j]_n = \langle\psi_{\mathbf{k}}^n|\left[\frac{\partial}{\partial k_i}, \frac{\partial E}{\partial k_j}\right]|\psi_{\mathbf{k}}^n\rangle = \frac{\partial E_n(\mathbf{k})}{\partial k_i\, \partial k_j} \tag{1.10.21}$$

and hence, at $\mathbf{k} = \mathbf{k}_0$

$$\left(\frac{1}{m^*}\right)_{ij} = \frac{\delta_{ij}}{m} - \frac{i}{\hbar}[X_i, V_j]_n. \tag{1.10.22}$$

1.10.3 *Effective mass Hamiltonian*

Finally, let us consider a slowly varying perturbation $V(\mathbf{r})$. Its matrix elements are

$$V_{n'n}(\mathbf{k}'\mathbf{k}) = \frac{1}{\Omega_B}\int d\mathbf{q}\, V(\mathbf{q}) \int \psi_{\mathbf{k}'}^{n'*}(\mathbf{r})\, e^{i\mathbf{q}.\mathbf{r}}\, \psi_{\mathbf{k}}^n(\mathbf{r})\, d\mathbf{r}, \tag{1.10.23}$$

where $V(\mathbf{q})$ is the Fourier transform of $V(\mathbf{r})$. Now

$$\int \psi_{\mathbf{k}'}^{n'*}(\mathbf{r})\, e^{i\mathbf{q}.\mathbf{r}}\, \psi_{\mathbf{k}}^n(\mathbf{r})\, d\mathbf{r} = \sum_{\mathbf{R}} e^{i(\mathbf{k}-\mathbf{k}'+\mathbf{q}).\mathbf{R}} \int_{\Omega} \psi_{\mathbf{k}'}^{n'*}(\mathbf{r})\, e^{i\mathbf{q}.\mathbf{r}}\, \psi_{\mathbf{k}}^n(\mathbf{r})\, d\mathbf{r}. \tag{1.10.24}$$

The sum over $\mathbf{R}$ restricts $\mathbf{k}$ and $\mathbf{k}'$ to values satisfying $\mathbf{q} = \mathbf{k}' - \mathbf{k}$ and hence

$$V_{n'n}(\mathbf{k}+\mathbf{q}, \mathbf{k}) = V(\mathbf{q}) \int_{\Omega} u_{\mathbf{k}+\mathbf{q}}^{n'*}\, u_{\mathbf{k}}^n\, d\mathbf{r}. \tag{1.10.25}$$

Expanding in powers of $\mathbf{q}$ we find

$$V_{n'n}(\mathbf{k}+\mathbf{q}, \mathbf{k}) = V(\mathbf{q})\, \delta_{n'n} + iV(\mathbf{q})\, \mathbf{q}\,.\,\mathscr{R}_{n'n}(\mathbf{k}) + \ldots. \tag{1.10.26}$$

Let us now suppose that $V(\mathbf{r})$ is acting on, for example, free carriers in a semiconductor, which are localized about a band minimum at $\mathbf{k}_0$. We take the expansion (1.10.15) for $E(\mathbf{k})$ to second order and, in addition, keep only the first term on the right-hand side of (1.10.26). The matrix elements of the Hamiltonian are then

$$H(\mathbf{k}+\mathbf{q}, \mathbf{k}) = \left[E(\mathbf{k}_0) + \frac{\hbar^2}{2}\sum_{ij}\left(\frac{1}{m^*}\right)_{ij}(\mathbf{k}-\mathbf{k}_0)_i(\mathbf{k}-\mathbf{k}_0)_j\right]\delta_{\mathbf{q}0} + V(\mathbf{q}), \tag{1.10.27}$$

which is an approximation valid for sufficiently small $\mathbf{q}$. It is easy to see that equation (1.10.27) gives the matrix elements of an effective Hamiltonian

$$H_{\text{eff}} = E(\mathbf{k}_0) - \frac{\hbar^2}{2}\sum_{ij}\left(\frac{1}{m^*}\right)_{ij}\frac{\partial^2}{\partial x_i\, \partial x_j} + V(\mathbf{r}) \tag{1.10.28}$$

taken between the functions $e^{-i(\mathbf{k}-\mathbf{k}_0).\mathbf{r}}$ and $e^{-i(\mathbf{k}-\mathbf{k}_0+\mathbf{q}).\mathbf{r}}$. The use of such an effective Hamiltonian in configuration space is often more convenient than use of the crystal momentum representation. We shall discuss (1.10.28) from a different standpoint in Chapter 10.

1.11 Wave function calculations

There are really only two basic methods of attack on the calculation of one-electron wave functions in crystals, their essential difference lying in the relative emphasis placed on the importance of the two requirements the wave function must meet, i.e. (a) to satisfy Schrödinger's equation; (b) to satisfy Bloch's theorem.

In a direct solution of the Schrödinger equation one can, of course, confine attention to a single unit cell of the lattice and, from this, one of the basic methods, the *cellular method*, gets its name. The usual procedure, as we shall describe in a little more detail below, is to take a particular energy, solve for the wave function expanded in combinations of spherical harmonics and choose the expansion coefficients so as to match the solution in one cell on to those in adjacent cells, in approximate accordance with Bloch's theorem. The cellular method, originated by Wigner and Seitz, was the first quantitative procedure to be used in wave function calculations for crystals.

Modern trends, however, are towards the second type of method, in which one begins with a formal expansion of the wave function such that Bloch's theorem is exactly satisfied from the outset. The most obvious expansion of this nature is the Fourier series, the coefficients in which are the momentum eigenfunctions, discussed in section 1.9. This plane-wave expansion has proved in practice to be very slowly convergent, a major reason for this being the necessary orthogonality of the wave functions of the outer electrons to the core functions. This suggested to Herring (1940) that one should orthogonalize the plane-waves $e^{i(\mathbf{k}+\mathbf{K}).\mathbf{r}}$ to the core functions, and take the initial formal expansion in terms of these *orthogonalized plane-waves* (OPW's). The effect is not only to speed convergence but also to ensure that a variational calculation yields an upper bound to the one-electron eigenvalue for the valence electrons.

We shall now discuss the two types of method in a little more detail.

1.11.1 Cellular method

We should say at once that for the cellular method to be practicable it is almost essential that the potential within a given cell is spherically symmetric.

It is then possible to factorize the Schrödinger equation in the usual way into an equation for the radial wave function and an equation satisfied by a spherical harmonic of order l. The Bloch functions are linear expansions of such wave functions and, as already remarked, the coefficients are obtained by matching across faces of the unit cell in accordance with Bloch's theorem. There are many methods, ranging from straightforward point matching

(Slater, 1937) to sophisticated least-square variational schemes (see, for example, Altmann, 1958), which have been employed for this purpose. All procedures are of necessity approximate, because of the impossibility of taking the expansion in spherical harmonics to high order. This has resulted in workers concentrating on wave functions labelled by **k**-vectors at high symmetry points in the BZ; the consequent vanishing of many of the terms of the expansion (see the explicit discussion of section (c) below) then greatly reduces the labour involved in taking it to a given order in l. Fast electronic computers are now making these calculations possible at general points in the BZ.

In considering the choice of potential, the above remarks indicate the difficulty of a proper self-consistent calculation in the Hartree sense. The simplest approach is to calculate $\psi_0(\mathbf{r}) = u_0(\mathbf{r})$, as shortly to be described, and then assume the rest of the band to be described by the *Seitz wave function*,

$$\psi_{\mathbf{k}}(\mathbf{r}) \simeq u_0(\mathbf{r})\, e^{i\mathbf{k}.\mathbf{r}}, \tag{1.11.1}$$

in which case $|u_0(\mathbf{r})|^2$ is obviously related to the total charge density. However, it is by no means certain that the Hartree potential is the best potential to take in an independent-particle framework. The potential taken by Wigner and Seitz (1930) in their work on monovalent metals was simply that due to the ion-core in the cell. This is certainly spherically symmetric, a necessity for the Wigner–Seitz method, and there are plausible arguments to justify this choice. One may expect adjacent cells to be, as a whole, electrically neutral, contributing little to the potential in our chosen cell; moreover, one expects the electron correlation effects to exclude from the cell electrons other than the one we are considering, in the case of a monovalent metal.

(a) *Calculation for* $\mathbf{k} = 0$. While the method can be used, in principle, for any wave vector **k** in the BZ, if we restrict ourselves to occupied states in the conduction bands of simple metals like the alkalis, we can obtain a fair picture on the basis of the wave function for $\mathbf{k} = 0$, the bottom of the band. In these cases, with bcc structure, the cellular polyhedron is highly symmetric (see Figure 1.2). We can therefore argue that while the wave function has the form

$$\psi_0(\mathbf{r}) = u_0(r) + \left[\frac{x^4+y^4+z^4}{r^4} - \frac{3}{5}\right] u_4(r) + \dots, \tag{1.11.2}$$

the angular term being of order $l = 4$ [see section (c) below], the dominant contribution will be $u_0(r)$. This greatly simplifies the matching problem and we need only impose the requirement that $u_0(r)$ is flat across the Wigner–Seitz sphere, of radius r_s such that its volume equals that of the cellular polyhedron; that is

$$\left(\frac{du_0(r)}{dr}\right)_{r_s}=0. \tag{1.11.3}$$

The radial Schrödinger equation must in general be integrated numerically, and when this is done for sodium with a suitable choice of potential the wave function shown in Figure 1.12 is obtained (Raimes, 1954). The method is a

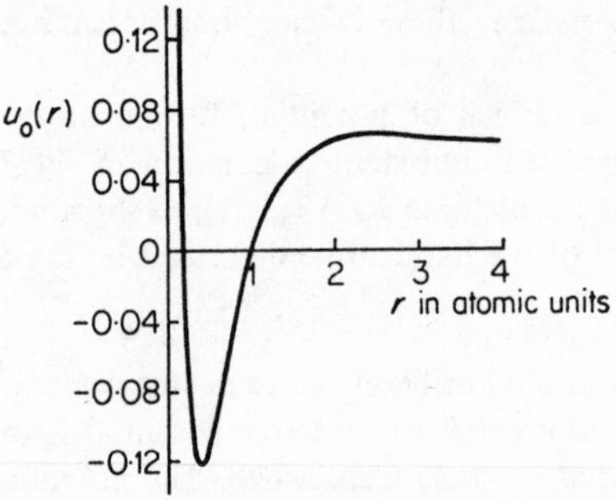

FIGURE 1.12. 3s-like Wigner–Seitz wave function for $\mathbf{k} = 0$ for metallic Na. Potential field used in this calculation was that of Prokofjew (1929). Atomic sphere radius was $3.96a_0$, and wave function is normalized to unity within this sphere. Energy of state $E(\mathbf{k} = 0)$ was -0.61 Rydberg.

good one for a wave function with a large number of radial nodes, but has not found such widespread application as the orthogonalized and the augmented plane-wave methods which we shall shortly consider.

In Problem 1.11, an application is given of the cellular method at $\mathbf{k}=0$ for metallic hydrogen. Here, the ion-core is just a proton and for sufficiently high densities we can make an expansion in the usual interelectronic spacing, r_s, and proceed analytically. In addition to the calculation of the energy and wave function of the lowest state, the mean Fermi energy can also be found by the method of Bardeen (1938) and this is set out in Problem 1.11. This procedure is a special case of the $\mathbf{k}.\mathbf{p}$ method of section 1.11.7.

For lower symmetry points, detailed consideration must be given to methods of matching the wave functions across the unit cell boundaries, and we can do no more than refer the reader to the original papers for details of this (see von der Lage and Bethe, 1947, and Altmann, 1958).

However, it is worth remarking that, for a simple band with its minimum at $\mathbf{k} = 0$, the Wigner–Seitz wave function for $\mathbf{k} = 0$ remains a pillar for any energy band calculation. In particular, we see from equation (1.9.4) that the knowledge of this wave function determines the momentum eigenfunction $v(\mathbf{k}+\mathbf{K})$ at the reciprocal lattice vectors $\mathbf{K}$; that is, at an infinite but discrete set of points.

(b) *Relation to scattering length.* Because of the importance of the $\mathbf{k} = 0$ calculation of Wigner and Seitz, we shall discuss here a simplified interpretation of the results in terms of the concept of a scattering length (see Messiah, 1969).

Let us consider a single spherical scattering potential, which has the form

$$\left.\begin{aligned} V(\mathbf{r}) &= V(r), \quad r<r_s, \\ &= 0, \quad r \geqslant r_s. \end{aligned}\right\} \tag{1.11.4}$$

Then, the spherical solution of the wave equation outside r_s is simply a linear combination of free particle solutions $j_0(\kappa r) \equiv \sin \kappa r/\kappa r$ and $n_0(\kappa r) \equiv -\cos \kappa r/\kappa r$, where $E = \kappa^2$ with energies in Rydbergs.

We can therefore write the result for $u_0(r)$ as

$$u_0(r) = j_0(\kappa r) + \kappa a(\kappa)\, n_0(\kappa r), \tag{1.11.5}$$

where we have introduced the length $a(\kappa)$ because, later, we shall see that this is related to the desired scattering length. Actually, the scattering length a is defined as

$$\lim_{\kappa \to 0} a(\kappa) = a, \tag{1.11.6}$$

but we shall defer a discussion, in terms of phase shifts, to section 1.11.5.

We must clearly choose $a(\kappa)$ to match the radial wave function inside the sphere with that outside.

The result for small κ is that

$$-\kappa a = \frac{j_0'(\kappa r_s)}{n_0'(\kappa r_s)} \tag{1.11.7}$$

which ensures continuity of the wave function derivative at r_s. For small κ, the right-hand side is readily shown to be $-\frac{1}{3}\kappa^3 r_s^3$ and hence the energy E_0 at the bottom of the band is determined by

$$E_0 = \kappa_0^2 \simeq \frac{3a}{r_s^3}, \tag{1.11.8}$$

the condition (1.11.7) ensuring the Wigner–Seitz result (1.11.3). Though the above argument is posed as though we were dealing with repulsive potentials and positive energies, the result (1.11.8) can be used for attractive potentials, when the scattering length a is negative and it turns out that the bottom of the band lies just below the zero of the atomic potential (1.11.4). In general, perturbation theory leads to a poor result (though, in the case of metallic hydrogen discussed in Problem 1.11, the perturbation theory is valid in the high density limit). The perturbation theory simply yields $E_0 = \int V(r)\,d\mathbf{r}$, which is generally much more negative for real metals than the correct Wigner–Seitz energy.

It was obvious that the result for E_0 in the Wigner–Seitz theory only depends on the atomic volume and not on the crystal structure. In general, since many simple metals will be seen later to have nearly free electron energy bands, the above calculation, which essentially fixes the position of the bottom of the band on the energy scale, is very useful.

The potential (1.11.4) used above is nowadays referred to as the "muffin tin" potential and will be used again below.

Wigner and Seitz (1955) have discussed the use of the cellular method to determine p-like bands and this type of argument has been used with some success to determine the effect of very high pressures on energy bands (see, for example, March, 1969, for a survey). But for quantitative work, the second type of method we discussed above, which starts out by satisfying Bloch's theorem, seems, on the whole, to be preferable.

Before turning to the second approach, we shall consider the question of the spherical harmonics which must be included at high symmetry points in the cellular method, by taking as a specific example the alkali halide structures.

(c) *Band structure in* NaCl *and* CsCl *lattices.* Before dealing with the specific angular dependence of the terms which enter the expansion of the wave function in the cellular method, we want to make clear the nature of the wave function about different sites in the NaCl crystal. Though the result we want to establish is general, we shall illustrate it by an argument based on an empty lattice assumption using the character table for point W given in section 1.5.7. Essentially we shall show that, if the wave function is s-like round Na for the representation W_1 (labelled alternatively W_s in Table 1.4; this notation immediately shows the atomic correspondence), it is p-like round Cl. This point has been the source of some confusion in the past: it does not, of course, arise in the case of one atom per unit cell.

Into the empty lattice description, we shall, of course, have to insert the correct symmetry of the lattice, and we shall then be able to discuss directly the form of the wave function about a nucleus of the second type (say Cl). We should emphasize that the Bravais lattice of NaCl is fcc, and hence the BZ is as shown in Figure 1.5.

Since we are dealing with the point $\mathbf{k} = (\pi/2a)(2,1,0)$, the basic plane-wave is

$$\exp[i\mathbf{k}.\mathbf{r}] = \exp[(i\pi/2a)\{2x+y\}] \tag{1.11.9}$$

and we see from Table 1.4 that, for the non-degenerate representation W_1, the linear combination of plane-waves with the correct symmetry is

$$\begin{aligned}&\exp[(i\pi/2a)\{2x+y\}]+\exp[(i\pi/2a)\{-2x+y\}]\\&\quad+\exp[(i\pi/2a)\{2z-y\}]+\exp[(i\pi/2a)\{-2z-y\}].\end{aligned} \tag{1.11.10}$$

We now consider this wave function to be centred on a Na site. To see what it looks like from a Cl site, we transfer the origin to the Cl site at $(a,0,0)$ and introducing the new coordinate $X = x-a$ the above wave function becomes

$$\begin{aligned}&-\exp[(i\pi/2a)\{2X+y\}]-\exp[(i\pi/2a)\{-2X+y\}]\\&\quad+\exp[(i\pi/2a)\{2z-y\}]+\exp[(i\pi/2a)\{-2z-y\}].\end{aligned} \tag{1.11.11}$$

We can see immediately that this wave function is zero when evaluated at the point $(X = 0, y = 0, z = 0)$ and is therefore not s-like round the Cl nucleus, whereas $(1.11.10) \neq 0$ at $x = y = z = 0$, i.e. it is s-like round the Na site. In fact the wave function (1.11.11) corresponds to the irreducible representation $W_2' \,(\equiv W_p^1)$.

We wish only to make a few further comments. One is that, in the cellular method discussed earlier, one requires the lattice harmonics up to high values of l in the general case when the empty lattice approximation is transcended.

These can be obtained by a straightforward generalization of the above plane-wave argument. We write, in place of the basic plane-wave $\exp[(i\pi/2a)\{2x+y\}]$, the function $f(x,y,z)$. It is then easily shown using the character table for W that the properly symmetrized wave function corresponding to the irreducible representation W_1 is

$$\begin{aligned}\psi = f(x,y,z)&+f(-x,y,-z)+f(z,-y,-x)+f(-z,-y,x)\\&+f(z,-y,x)+f(-z,-y,-x)+f(x,y,-z)\\&+f(-x,y,z).\end{aligned} \tag{1.11.12}$$

We now expand $f(xyz)$ in the usual form adopted in the cellular method:

$$f(xyz) = \sum c_{lm} R_l(r) P_l^m(\cos\theta)\, e^{im\phi}, \tag{1.11.13}$$

where $R_l(r)$ are the radial wave functions, and $P_l^m(\cos\theta)\, e^{im\phi}$ are the spherical harmonics. To obtain ψ in a simple form we choose y as the polar axis.

Using x, y, z for direction cosines, i.e. $y = \cos\theta$ and

$$\exp(i\phi) = (z+ix)/(1-y^2)^{\frac{1}{2}} \tag{1.11.14}$$

one can show analytically that the terms appearing in the expansion of ψ have the form

$$\left(\frac{d^m}{dy^m}\right) P_l(y)\,[(z+ix)^m+(z-ix)^m], \tag{1.11.15}$$

where, for l odd, $m = 2, 6, 10$, etc., and for l even, $m = 0, 4, 8$, etc.

It is now a straightforward matter to combine terms of a given l which lead to a function of the form (1.11.12) and to orthogonal harmonics. The results are summarized in Table 1.5, where we also show the lattice harmonics for W_p^1, which can be obtained by an obvious modification of the above argument (for a more extensive table, see Flower, March and Murray, 1960).

TABLE 1.5. Lattice harmonics for point W(NaCl lattice).
$\mathbf{k} = (\pi/2a)\,(2, 1, 0)$

	W_s	W_p^1
0	1	
1		$p = y$
2	$d = 2y^2-x^2-z^2$	$d = x^2-z^2$
3	$f = y(x^2-z^2)$	$f = y^3-\frac{3}{5}p$
4	$g_1 = x^4+y^4+z^4-\frac{3}{5}$	$g = x^4-z^4-6d/7$
	$g_2 = 2y^4-x^4-z^4-6d/7$	
5	$h = y^3(x^2-z^2)-f/3$	$h_1 = y^5-(\frac{10}{9})f-(\frac{3}{7})p$
		$h_2 = y(x^4-6x^2z^2+z^4)$
6	$i_1 = x^2y^2z^2+g_1/22-\frac{1}{105}$	$i_1 = x^6-z^6-15g/11-(\frac{5}{7})\,d$
	$i_2 = 2y^6-x^6-z^6-15g_2/11-5d/7$	$i_2 = (x^2-y^2)\,(y^2-z^2)\,(z^2-x^2)$

We briefly comment on the CsCl lattice, the Bravais lattice of which is simple cubic. This can be treated in the same way as we have discussed the NaCl case. We consider only the point Γ, corresponding to $\mathbf{k} = (\pi/a)\,(1, 1, 1)$. In this case, we can show that the representation with Γ_s symmetry round one nucleus, say Cs, has Γ_f symmetry round the other. All the harmonics for $l \leqslant 10$ are listed for Γ_s and Γ_f in Table 1.6.

Finally, for the point X, with $\mathbf{k} = (\pi/a)(1, 0, 0)$ we consider the representation X_s around one nucleus. This corresponds to X_p when viewed from a nucleus of the second kind. Independent tables are not required for this case. The results may be obtained from those for W_s in the case of X_s

by omitting the odd l values entirely, and by interchanging x and y in the even l harmonics. The results for X_p are similarly obtained from Table 1.6, by omitting the even harmonics in W_p^1 and interchanging x and y.

The empty lattice arguments used at the beginning of this section will be taken up again in section 1.11.8 in connection with the band structure of semiconducting Ge, which also has two atoms per unit cell.

TABLE 1.6. Lattice harmonics for point Γ (CsCl lattice).
$\mathbf{k} = (\pi/a)(1, 1, 1)$

	s		f
l		l	
0	1	1	...
2	...	3	$f = xyz$
4	$g = x^4+y^4+z^4-\frac{3}{5}$	5	...
6	$i = x^2y^2z^2+g/22-\frac{1}{105}$	7	$j = xyz(x^4+y^4+z^4) - 5f/11$
8	$k = x^8+y^8+z^8-28i/5-210g/143-\frac{1}{3}$	9	$x^3y^3z^3+3j/34 - 3f/143$
10	$x^{10}+y^{10}+z^{10}-45k/19-126i/17 - 210g/143-\frac{3}{11}$		

1.11.2 *Orthogonalized plane-wave (OPW) method*

The momentum eigenfunction method expands the wave function in the form (1.9.3). Atomic-like fluctuations in the wave functions which exist near nuclei in crystals (cf. Figure 1.12 for Na) evidently will require short wavelength, or equivalently high momentum, components, in this expansion, and this is the reason why the convergence is poor. To illustrate this, the Wigner–Seitz wave function for lithium at $\mathbf{k} = 0$ can be calculated using the method described above. By numerical Fourier transform, we can obtain the momentum eigenfunction v at the reciprocal lattice vectors $\mathbf{K}_n$ and the results are shown in Table 1.7. Though the component $\mathbf{K}_n = 0$ is dominant, many plane-waves are required to give a useful approximation to the wave function at the nucleus.

Construction of orthogonalized plane waves. Let us suppose that we build Bloch orbitals for the tightly bound core electrons. Then we may write these in the form (cf. section 1.8.3)

$$\psi_{\mathbf{k}}^c(\mathbf{r}) = \sum_{\mathbf{R}} e^{i\mathbf{k}.\mathbf{R}} \phi^c(\mathbf{r}-\mathbf{R}), \tag{1.11.16}$$

where the ϕ's are atomic orbitals. For core functions it is generally a good approximation to assume no overlap between the ϕ's on different sites. We can now write down the OPW's as

$$\chi_{\mathbf{k},\mathbf{K}}(\mathbf{r}) = e^{i(\mathbf{k}+\mathbf{K}).\mathbf{r}} - \sum_c \mu_{\mathbf{K}}^c(\mathbf{k})\, \psi_{\mathbf{k}}^c(\mathbf{r}), \tag{1.11.17}$$

where, for χ to be orthogonal to all the core functions, it is readily shown that

$$\mu_{\mathbf{K}}^{c*}(\mathbf{k}) = \int_\Omega e^{-i(\mathbf{k}+\mathbf{K}).\mathbf{r}}\, \psi_{\mathbf{k}}^c(\mathbf{r})\, d\mathbf{r} \tag{1.11.18}$$

for all $\mathbf{k}$.

TABLE 1.7. Momentum eigenfunction in lithium metal as calculated from Wigner–Seitz wave function for $\mathbf{k} = 0$. Fermi momentum $k_f = 0{\cdot}589$ atomic units

$\lvert \mathbf{K}_n \rvert$	Number of vectors of length $\lvert \mathbf{K}_n \rvert$	$\Omega^{\frac{1}{2}} v_{0\mathbf{K}_n}$
0	1	−0.948
1.353	12	+0.062
1.914	6	+0.041
2.344	24	+0.029
2.706	12	+0.021
3.026	24	+0.016
3.315	8	+0.013
3.580	48	+0.010
3.828	6	+0.008
4.060	36	+0.007
4.279	24	+0.006
4.488	24	+0.005
4.688	24	+0.004
4.879	72	+0.004
5.241	48	+0.003
5.413	12	+0.003
5.580	48	+0.002
5.742	30	+0.002

The procedure is now obvious. We expand the valence electron wave function $\psi_{\mathbf{k}}(\mathbf{r})$ in terms of the OPW's, to obtain

$$\psi_{\mathbf{k}}(\mathbf{r}) = \sum_{\mathbf{K}} c_{\mathbf{k}\mathbf{K}}\, \chi_{\mathbf{k},\mathbf{K}}(\mathbf{r}). \tag{1.11.19}$$

Since we expect that this expansion will converge more rapidly than the straightforward plane-wave expansion, we can assume a finite sum and regard the coefficients $c_{\mathbf{kK}}$ as variational parameters, to be determined by minimizing the expectation value of the one-electron Hamiltonian

$$H = -\frac{\hbar^2}{2m}\nabla^2 + V(\mathbf{r}). \tag{1.11.20}$$

We then find the linear equations

$$\sum_{\mathbf{K}} c_{\mathbf{kK}} \int \chi^*_{\mathbf{kK}'}[H-E]\,\chi_{\mathbf{kK}}\,d\mathbf{r} = 0, \tag{1.11.21}$$

the corresponding secular determinant being

$$\left| \int \chi^*_{\mathbf{kK}'}[H-E]\,\chi_{\mathbf{kK}}\,d\mathbf{r} \right| = 0. \tag{1.11.22}$$

The size of this secular equation is obviously determined by the number of OPW's we choose to retain in the expansion. For simple metals, even a single term gives a reasonable first approximation to the valence band. As a problem (1.7), the result of a single OPW calculation on metallic Li may be found, and is reasonably satisfactory.

For more complex solids, such a situation does not obtain, and convergence to the eigenvalue does not result until numerous waves are added. We shall not go into further detail here.

The method of extracting the atomic oscillations from a Bloch wave function, by means of linear combinations of core functions, is not unique. It will obviously be desirable to choose the linear combinations so that the remaining part of the wave function, which is now smooth, can be expanded in the most rapidly convergent set of plane-waves. This is the original motivation for the pseudopotential method, which we shall consider later.

Mention should be made that the OPW method makes no simplifications regarding the crystal potential. This is in contrast to the cellular method, where almost all work to date has had to assume $V(\mathbf{r})$ to be spherical in the unit cell, the calculation complexities being otherwise huge. Some of the other methods depend in an important way on the assumption of a muffin-tin potential.

The OPW method was used with great success by Herman for diamond-type lattices and we shall consider the results he obtained in this way for Ge in section 1.11.8(c).

1.11.3 Korringa–Kohn–Rostoker (KKR) method

This method solves the wave equation using Green functions and, in principle, is quite general. However, to allow the theory to be carried through explicitly in practice, it is necessary to assume that the potential centred on each lattice site vanishes before the boundary of the unit cell is reached. A very important simplification then occurs, for the problem can be sharply separated into two parts:

(i) The scattering properties of a single potential of the muffin-tin form (1.11.4).

(ii) The structural aspects.

This will become quite clear in the method developed originally by Korringa (1947) from general scattering theory (cf. Chapter 10) and later by Kohn and Rostoker (1954), whose approach we follow here. This method of calculating the band structure for particular metals has been found to be highly efficient even for low symmetry points in the BZ [see, for example, Faulkner, Davis and Joy (1967); some results of this work are recorded in section 1.11.8(a)].

(a) *Integral equation.* For a periodic potential $V(\mathbf{r})$, we can solve the wave equation

$$[\nabla^2+E]\psi = V(\mathbf{r})\psi \tag{1.11.23}$$

by a Green function procedure. Defining the Green function $G(\mathbf{rr}')$ through the equation

$$[\nabla^2+E]\,G(\mathbf{rr}') = \delta(\mathbf{r}-\mathbf{r}'), \tag{1.11.24}$$

we can construct a solution for the wave function ψ of the form

$$\psi(\mathbf{r}) = \int_\Omega d\mathbf{r}'\, G(\mathbf{rr}')\, V(\mathbf{r}')\,\psi(\mathbf{r}') \tag{1.11.25}$$

as is readily verified.

In order for $\psi(\mathbf{r})$ to be a Bloch function, we must choose the Green function to satisfy the relation

$$G(\mathbf{r}+\mathbf{R},\mathbf{r}') = e^{i\mathbf{k}.\mathbf{R}}\, G(\mathbf{rr}') \tag{1.11.26}$$

where $\mathbf{R}$ is any lattice vector. It is not difficult to show that the correct choice of G is

$$G(\mathbf{rr}') = \frac{-1}{\Omega}\sum_{\mathbf{K}} \frac{\exp[i(\mathbf{K}+\mathbf{k}).(\mathbf{r}-\mathbf{r}')]}{(\mathbf{K}+\mathbf{k})^2-E}. \tag{1.11.27}$$

The muffin-tin potential assumption is that the potential is taken to be spherical about each ion, within the sphere inscribed in the unit cell and constant outside. Then we can write

$$V(\mathbf{r}) = \sum_{\mathbf{R}} v(|\mathbf{r}-\mathbf{R}|), \tag{1.11.28}$$

where the v's are spherical and non-overlapping, i.e. we can choose

$$v(\mathbf{r}) = 0, \quad r>r_i, \tag{1.11.29}$$

where r_i is the radius of a sphere lying wholly within the unit cell.

It seems from detailed calculations that for simple monatomic solids, the muffin-tin potential is a reasonable approximation to the real situation in crystals. We might anticipate that deviations from such a potential could be handled by a perturbation calculation, should it ever become necessary.

The wave function within the inscribed sphere may now be written as [cf. equation (1.11.13)]

$$\psi(\mathbf{r}) = \sum_{lm} c_{lm} R_l(r) Y_{lm}(\theta, \phi), \quad r < r_i, \tag{1.11.30}$$

where the radial wave functions $R_l(r)$ satisfy the equation

$$\left[-\frac{1}{r^2}\frac{d}{dr}\left(r^2\frac{d}{dr}\right)+\frac{l(l+1)}{r^2}+V(r)-E\right] R_l(\mathbf{r}) = 0 \tag{1.11.31}$$

and the $Y_{lm}(\theta, \phi)$ are, as usual, the normalized spherical harmonics defined by

$$Y_{lm}(\theta, \phi) = \left[\frac{2l+1}{4\pi}\frac{(l-|m|)!}{(l+|m|)!}\right]^{\frac{1}{2}} P_l^m(\cos\theta)\, e^{im\phi}. \tag{1.11.32}$$

(b) *Secular equation.* The coefficients c_{lm} in the wave function (1.11.30) can be obtained from the integral equation in a manner we shall now discuss. It will be useful to define the single-centre potential by

$$\left.\begin{aligned} v(\mathbf{r}) &= v(r), \quad r \leqslant r_i - \varepsilon, \\ &= 0, \quad r > r_i - \varepsilon, \end{aligned}\right\} \tag{1.11.33}$$

where ε is a small positive quantity which we shall eventually allow to go to zero.

Equation (1.11.25) can then be written

$$\psi(\mathbf{r}) = \int_{r' < r_i - \epsilon} d\mathbf{r}' G(\mathbf{r}\mathbf{r}')\, v(\mathbf{r}')\, \psi(\mathbf{r}'). \tag{1.11.34}$$

We shall consider next a choice of the vector $\mathbf{r}$ such that $|\mathbf{r}| = r_i - 2\varepsilon$. Using the fact that $\psi(\mathbf{r})$ satisfies the exact wave equation (1.11.23), we have

$$0 = \psi(\mathbf{r}) - \int_{r' < r_i - \epsilon} d\mathbf{r}' G(\mathbf{r}\mathbf{r}')\, [\nabla^2_{\mathbf{r}'} + E]\, \psi(\mathbf{r}'). \tag{1.11.35}$$

But we have the identity

$$\int_{r' < r_i - \epsilon} d\mathbf{r}' \nabla^2_{\mathbf{r}'} G(\mathbf{r}\mathbf{r}')\, \psi - \int_{r' < r_i - \epsilon} d\mathbf{r}' G(\mathbf{r}\mathbf{r}')\, \nabla^2_{\mathbf{r}'} \psi$$

$$= \int_{r' = r_i - \epsilon} dS' \psi \frac{\partial G(\mathbf{r}\mathbf{r}')}{\partial r'} - \int_{r' = r_i - \epsilon} dS' \frac{\partial \psi}{\partial r'} G(\mathbf{r}\mathbf{r}'), \tag{1.11.36}$$

where the integrations on the right-hand side are over the surface of the sphere of radius $r_i - \varepsilon$. Hence, since G satisfies equation (1.11.24), we find from equations (1.11.35) and (1.11.36) the result

$$0 = \int_{r=r_i-\varepsilon} dS' \psi(\mathbf{r}') \frac{\partial G}{\partial r'} - \int_{r=r_i-\varepsilon} dS' G \frac{\partial \psi}{\partial r'}. \tag{1.11.37}$$

It is evident that, to obtain equations for the coefficients c_{lm} in the expansion of the wave function, we must express G in terms of spherical harmonics. To do this, we can consider the free-particle Green function

$$G_0(\mathbf{rr}') = -\frac{1}{4\pi} \frac{\cos \kappa|\mathbf{r}-\mathbf{r}'|}{|\mathbf{r}-\mathbf{r}'|}; \quad E = \kappa^2, \tag{1.11.38}$$

which satisfies the same inhomogenous wave equation as G inside the atomic polyhedron. Thus, the difference

$$D(\mathbf{rr}') = G(\mathbf{rr}') - G_0(\mathbf{rr}') \tag{1.11.39}$$

must satisfy the homogeneous wave equation. For a plane-wave, which trivially satisfies such an equation, we have Bauer's expansion

$$e^{i\mathbf{k}.\mathbf{r}} = \sum_l (2l+1)\, i^l j_l(kr)\, P_l(\cos \theta), \tag{1.11.40}$$

where θ is the angle between $\mathbf{k}$ and $\mathbf{r}$ and j_l represents as usual a spherical Bessel function of order l (see equation 1.11.43). By analogy, D has an expansion of the form

$$D(\mathbf{rr}') = \sum_{lm} \sum_{l'm'} A_{lm,l'm'} j_l(\kappa r) j_{l'}(\kappa r') Y_{lm}(\theta, \phi)\, Y_{l'm'}(\theta', \phi'), \tag{1.11.41}$$

where again $\kappa = \sqrt{E}$ and the A's are constants to be discussed later. Such an expansion is soon derived from (1.11.27) by using the addition theorem for spherical harmonics (Eyring, Walter and Kimball, 1944).

On the other hand,

$$\begin{aligned} G_0 &= -\frac{1}{4\pi} \frac{\cos(\kappa|\mathbf{r}-\mathbf{r}'|)}{|\mathbf{r}-\mathbf{r}'|} \\ &= \kappa \sum_{lm} j_l(\kappa r)\, n_l(\kappa r')\, Y_{lm}(\theta, \phi)\, Y^*_{lm}(\theta', \phi'), \quad r < r' \end{aligned} \tag{1.11.42}$$

where the spherical Bessel function j_l and the spherical Neumann function n_l are given by

$$j_l(x) = \left(\frac{\pi}{2x}\right)^{\frac{1}{2}} J_{l+\frac{1}{2}}(x) \tag{1.11.43}$$

and

$$n_l(x) = \left(\frac{\pi}{2x}\right)^{\frac{1}{2}} J_{-l-\frac{1}{2}}(x), \tag{1.11.44}$$

the J's being Bessel functions. Hence we can write the full Green function in the form

$$G(\mathbf{rr}') = \sum_{lm} \sum_{l'm'} [A_{lm,l'm'} j_l(\kappa r) j_{l'}(\kappa r) + \kappa \delta_{ll'} \delta_{mm'} j_l(\kappa r) n_l(\kappa r')] \times Y_{lm}(\theta, \phi) Y^*_{l'm'}(\theta', \phi') \quad (r < r' < r_i). \tag{1.11.45}$$

We substitute this for G, and equation (1.11.30) for ψ in equation (1.11.37), multiply through by $Y^*_{lm}(\theta, \phi)$ and integrate over the sphere $r = r_i - 2\varepsilon$. The Y's are orthonormal and we find, after letting $\varepsilon \to 0$,

$$\sum_{l'm'} j_l [A_{lm,l'm'}(j_{l'} L_{l'} - j'_{l'}) + \kappa \delta_{ll'} \delta_{mm'} (n_{l'} L_{l'} - n'_{l'})] c_{l'm'} = 0. \tag{1.11.46}$$

Here

$$j'_l(x) = \frac{dj_l(x)}{dx}, \quad n'_l = \frac{dn_l(x)}{dx} \tag{1.11.47}$$

with L as the logarithmic derivative

$$L = \frac{dR_l(r)}{dr} \bigg/ R_l(r). \tag{1.11.48}$$

All functions are evaluated at $r = r_i$.

In order that non-trivial solutions for the c's exist, the determinant of the coefficients must be zero. An equivalent condition is

$$\det \left| A_{lm,l'm'} + \kappa \delta_{ll'} \delta_{mm'} \frac{(n_l L_l - n'_l)}{j_l L_l - j'_l} \right| = 0 \tag{1.11.49}$$

since equation (1.11.46) may also be written as

$$\sum_{l'm'} \left[A_{lm,l'm'} + \kappa \delta_{ll'} \delta_{mm'} \frac{n_l L_l - n'_l}{j_l L_l - j'_l} \right] \bar{c}_{l'm'} = 0, \tag{1.11.50}$$

where

$$\bar{c}_{l'm'} = (j_{l'} L_{l'} - j'_l) c_{l'm'}. \tag{1.11.51}$$

It is evident, on comparing equations (1.11.27) and (1.11.45) for G, that the A's (called "structure constants" and calculated by methods given in Appendix 1.7) depend on the particular crystal lattice dealt with. It can be seen from equation (1.11.49) that a clear separation has been made between the effects of the geometry of the lattice and the effects of the potential, the potential entering only into the calculation of the logarithmic derivatives L and through the value r_i at which the functions j and n are evaluated.

E and $\mathbf{k}$ are, of course, contained in the A's. The method usually adopted in practice is to fix E and search for values of $\mathbf{k}$ for which (1.11.49) is satisfied. Then, automatically, constant energy surfaces in $\mathbf{k}$ space are generated.

The left-hand side of equation (1.11.49) is an infinite determinant, which, of necessity, has to be approximated by a finite one. Thus, the wave function (1.11.30) implicitly involved, while satisfying the Schrödinger equation exactly, satisfies the Bloch boundary conditions only approximately. In this respect, the KKR method is more like the cellular method of section 1.11.1 than the OPW method of the last section. The distinction we found it useful to make earlier between two types of method is not completely sharp.

Finally, it should be noted that in Appendix 1.7 we show that a variational procedure exists for solving the integral equation (1.11.25), from which it follows that the error in the energy is of second order compared with that of the wave function. We also show that the secular equation of this variational principle is precisely equation (1.11.49) when a muffin-tin potential is used. This means that, while a reasonable estimate for $E(\mathbf{k})$ may be obtained with a fairly low-order determinant of the form (1.11.49), the corresponding approximation to the wave function may be relatively poor.

1.11.4 Augmented plane-wave (APW) method

Before developing the KKR method further, it will be convenient to introduce another technique for calculating energy bands, which is somewhat similar in philosophy and, in fact, was proposed by Slater (1937, see also 1953) before the work of Korringa.

Again the method hinges on the potential having muffin-tin form, in which case it is natural to use a plane wave expansion of the wave function in the the regions ($r \geqslant r_i$) where the potential is constant. On the other hand, the natural expansion about a site is of the form of equation (1.11.30). When the local potential V is spherically symmetric, the radial wave function R_l satisfies the equation

$$\frac{d^2}{dr^2}(rR_l)+\left\{E-V(r)-\frac{l(l+1)}{r^2}\right\}rR_l = 0. \tag{1.11.52}$$

The basic idea of the APW method is to expand in functions compounded of waves of the form (1.11.30) near the nuclei and plane-waves between the nuclei. One immediate advantage is that Bloch's theorem is automatically satisfied. Thus, we begin by writing the plane-wave in the form (1.11.40). On choosing a trial value of E, (1.11.52) is solved for R_l with $r<r_i$, and the c_{lm}'s of equation (1.11.30) are chosen to match the solution on to the plane wave (1.11.40) at $r=r_i$. We then find the expectation value $\langle E\rangle$ of the crystal Hamiltonian with respect to the augmented plane wave, and also we can calculate its stationary value with respect to E, which conveniently turns out to be that value of E which makes it stationary. This is given by

$$[\Omega-\tfrac{4}{3}\pi r_i^3][E-k^2] = 4\pi\sum_l r_i^2(2l+1)\,j_l^2(kr_i)\left[\frac{d}{dr}\ln R_l(E,r)\right]_{r=r_i}. \quad (1.11.53)$$

In the general method, the expansion is in terms of APW's labelled by $\mathbf{k}_n=\mathbf{k}+\mathbf{K}_n$, $\mathbf{K}_n$ being, as usual, any reciprocal lattice vector. The secular equation formed from the matrix with elements given by

$$\Omega\langle\mathbf{k}_n|H-E|\mathbf{k}_m\rangle = (\mathbf{k}_m.\mathbf{k}_n-E)\left[\Omega\delta_{n,m}-4\pi r_i^2\frac{j_1(|\mathbf{k}_m-\mathbf{k}_n|r_i)}{|\mathbf{k}_n-\mathbf{k}_m|}\right]$$
$$+4\pi r_i^2\sum_{l=0}^{\infty}(2l+1)\,P_l(\hat{\mathbf{k}}_m.\hat{\mathbf{k}}_n)\,j_l(k_n r_i)\,j_l(k_m r_i)\frac{R_l'(E,r_i)}{R_l(E,r_i)}$$
$$(1.11.54)$$

is then solved. The notation $\hat{\mathbf{k}}=\mathbf{k}/|\mathbf{k}|$ for a unit vector is often used in this book. As with equation (1.11.53), we have assumed that E is the value of the energy obtained from the secular equation. It should be noted that only the behaviour of R_l at $r=r_i$ is required to solve for the energy.

It should be remarked that the matching procedure in the construction of an APW leaves us with a discontinuity in the derivative of this function at $r=r_i$ and this discontinuity gives a finite contribution to the right-hand side of equations (1.11.53) and (1.11.54). We have not displayed this in detail, however, because there are close parallels with the KKR method, which has already been discussed at some length.

It is noteworthy that the secular equation which can be derived from equation (1.11.54) may be written as

$$\det|\{(\mathbf{k}-\mathbf{K})^2-E\}\,\delta_{\mathbf{KK}'}+\Gamma_{\mathbf{KK}'}|=0. \quad (1.11.55)$$

The striking feature of this equation is that it is precisely similar in form to the secular equation we obtain from the treatment of section 1.9 which uses a straightforward plane-wave expansion: that is the method based on the momentum eigenfunction $v(\mathbf{k})$. In this method, the condition that equation (1.9.8) has a non-trivial solution for $v(\mathbf{k})$ is

$$\det|\langle\mathbf{k}_m|H|\mathbf{k}_n\rangle-E\delta_{mn}|=0, \quad (1.11.56)$$

where the $\langle \mathbf{k}_m | H | \mathbf{k}_n \rangle$ are matrix elements taken with respect to plane-wave states:

$$\langle \mathbf{k}_m | H | \mathbf{k}_n \rangle = \langle \mathrm{e}^{-i(\mathbf{k}+\mathbf{K}_m).\mathbf{r}}(-\nabla^2 + V(\mathbf{r}))\,\mathrm{e}^{i(\mathbf{k}+\mathbf{K}_n).\mathbf{r}} \rangle$$

$$= (\mathbf{k}+\mathbf{K}_n)^2\,\delta(\mathbf{K}_m, \mathbf{K}_n) + V_{\mathbf{K}_n - \mathbf{K}_m}, \tag{1.11.57}$$

where

$$V(\mathbf{r}) = \sum V_{\mathbf{K}}\,\mathrm{e}^{i\mathbf{K}.\mathbf{r}}. \tag{1.11.58}$$

We see that equation (1.11.56) is of precisely the form (1.11.55) and in this latter equation $\Gamma_{\mathbf{KK}'}$ is evidently playing the role of the Fourier components of an effective potential. The physical basis of the APW method suggests that replacing V by Γ will yield a much more rapidly convergent procedure.

1.11.5 Relation between KKR and APW methods

As we have stressed, equation (1.11.55) of the APW method has precisely the form of the secular equation of the method in which the Bloch wave function is expanded directly in plane waves. In order to relate the APW method to the KKR theory, we note that the secular equation (1.11.49) of this theory can also be cast into the form (1.11.55) as shown in Appendix 1.7.

In the KKR method, we emphasized that lattice structure and the scattering properties of a single muffin-tin potential separated cleanly. The most elegant way to express the single-centre scattering is in terms of the method of phase shift analysis which we shall therefore summarize briefly below, before giving the KKR result for $\Gamma_{\mathbf{KK}'}$. We stress that the use of phase shifts in this connection is convenient rather than essential.

(a) *Phase-shift analysis.* Equation (1.11.40) expresses a plane wave in terms of partial spherical waves, with a definite l value. In the presence of a scattering potential, the scattered wave can similarly be analysed into partial spherical waves at large distances from the scatterer, but the partial waves will be shifted in phase by amounts η_l which are characteristic of a given scattering potential.

Let us consider the scattering of particles coming in from the z direction. We then require a solution representing the incident plane wave e^{ikz} together with the scattered outgoing waves. Provided the condition

$$rV(r) \to 0 \quad \text{as} \quad r \to \infty \tag{1.11.59}$$

is satisfied, which is trivially the case with a muffin-tin potential, we can use the Green function method of section 1.11.3(a) to obtain solutions

$$\psi_{\pm}(\mathbf{r}) = -\frac{1}{2\pi}\int \frac{\mathrm{e}^{\pm ik|\mathbf{r}-\mathbf{r}'|}}{|\mathbf{r}-\mathbf{r}'|}\,V(\mathbf{r}')\,\psi(\mathbf{r}')\,d\mathbf{r}'. \tag{1.11.60}$$

Any solution of the complementary equation

$$(\nabla^2+k^2)\psi = 0 \tag{1.11.61}$$

may be added to the particular solution (1.11.60). Since in our case we evidently require that

$$\psi \to e^{ikz} \quad \text{as} \quad V(r)\to 0, \tag{1.11.62}$$

we find the desired solution to be

$$\psi = e^{ikz}-\frac{m}{2\pi\hbar^2}\int\frac{e^{ik|\mathbf{r}-\mathbf{r}'|}}{|\mathbf{r}-\mathbf{r}'|}V(\mathbf{r}')\psi(\mathbf{r}')\,d\mathbf{r}', \tag{1.11.63}$$

where we have taken the plus sign in equation (1.11.60) because we are only interested in outgoing scattered waves. Then taking the asymptotic form (for details see Chapter 5) we obtain, with $m = \hbar = 1$,

$$\psi = e^{ikz}-\frac{e^{ikr}}{2\pi r}\int e^{-i\mathbf{k}.\mathbf{r}'}V(\mathbf{r}')\psi(\mathbf{r}')\,d\mathbf{r}', \tag{1.11.64}$$

where $\mathbf{k}$ is in the direction of the scattering. We can write this in the form

$$\psi = e^{ikz}+f(\theta)\frac{e^{ikr}}{r}, \tag{1.11.65}$$

where θ is the scattering angle between $\mathbf{k}$ and the z-axis. Evidently, from (1.11.64) and (1.11.65) the scattering amplitude $f(\theta)$ is given by

$$f(\theta) = -\frac{1}{2\pi}\int e^{-i\mathbf{k}.\mathbf{r}'}V(\mathbf{r}')\psi(\mathbf{r}')\,d\mathbf{r}' \tag{1.11.66}$$

and the scattering cross-section $\sigma(\theta)$ is simply (see Schiff, 1949)

$$\sigma(\theta) = |f(\theta)|^2. \tag{1.11.67}$$

Let us now turn to the solution of the Schrödinger equation, assuming $V(\mathbf{r})$ to be spherically symmetric, when we can make the usual separation into spherical harmonics. For present purposes it is very convenient to write ψ in a form analogous to equation (1.11.40)

$$\psi(\mathbf{r}) = \sum_{l=0}^{\infty}(2l+1)\,i^l c_l\,R_l(r)\,P_l(\cos\theta), \tag{1.11.68}$$

where $R_l(r)$ is the real normalized solution of the radial wave equation

$$\frac{d^2}{dr^2}(rR_l)+2\left[\frac{k^2}{2}-V(r)-\frac{l(l+1)}{2r^2}\right]rR_l = 0 \tag{1.11.69}$$

and the c_l's are coefficients dependent on the particular form required for $\psi(\mathbf{r})$. In the present instance we must choose these coefficients such that we obtain asymptotically the form (1.11.65) and to do so we proceed as follows. For the free-particle spherical waves in (1.11.40), it is known that the asymptotic behaviour is (Schiff, 1949)

$$j_l(kr) \to \frac{1}{kr} \sin\left(kr - \frac{l\pi}{2}\right) \tag{1.11.70}$$

and the regular solution of (1.11.69) must have the same asymptotic form apart from a shift in phase η_l introduced above:

$$R_l(r) \to \frac{1}{kr} \sin\left(kr + \eta_l - \frac{l\pi}{2}\right). \tag{1.11.71}$$

Inserting the asymptotic forms (1.11.70) and (1.11.71) in equations (1.11.40) and (1.11.68) respectively, we obtain

$$\begin{aligned}\frac{e^{ikr}}{r} f(\theta) &= \psi(\mathbf{r}) - e^{ikz} \\ &= \sum_{l=0}^{\infty} (2l+1) \frac{i^l}{kr} \left\{ c_l \sin\left(kr + \eta_l - \frac{l\pi}{2}\right) - \sin\left(kr - \frac{l\pi}{2}\right) \right\} P_l(\cos\theta).\end{aligned} \tag{1.11.72}$$

It is obvious that c_l must be chosen to eliminate all incoming spherical waves. We have

$$\begin{aligned}&c_l \sin\left(kr + \eta_l - \frac{l\pi}{2}\right) - \sin\left(kr - \frac{l\pi}{2}\right) \\ &= \frac{c_l}{2i} \{e^{ikr} e^{i\eta_l} e^{-il\pi/2} - e^{-ikr} e^{-i\eta_l} e^{il\pi/2}\} - \frac{1}{2i} \{e^{ikr} e^{-il\pi/2} - e^{-ikr} e^{il\pi/2}\}\end{aligned} \tag{1.11.73}$$

and the absence of incoming spherical waves means that

$$c_l e^{-i\eta_l} - 1 = 0. \tag{1.11.74}$$

Hence

$$c_l = e^{i\eta} \tag{1.11.75}$$

and equation (1.11.72) becomes

$$\frac{e^{ikr}}{r} f(\theta) = \sum_{l=0}^{\infty} \frac{(2l+1)}{2i} i^l e^{-il\pi/2} (e^{2i\eta_l} - 1) \frac{e^{ikr}}{kr} P_l(\cos\theta) \tag{1.11.76}$$

so that

$$f(\theta) = \frac{1}{k} \sum_{l=0}^{\infty} (2l+1) e^{i\eta_l} \sin\eta_l \, P_l(\cos\theta) = \sum f_l P_l(\cos\theta). \tag{1.11.77}$$

The scattering length a can be defined as simply $-\lim_{k\to 0} f_0(k)$. For well behaved s-wave phase shifts, this $\sim k^{-1} \sin\eta_0 \sim k^{-1}\eta_0$ (cf. 1.11.5).

We can now readily obtain a formula for the phase shifts by inserting the forms (1.11.40) and (1.11.68) into equation (1.11.66). Remembering that

$$\int_0^\pi P_l(\cos\theta)\, P_m(\cos\theta)\sin\theta\, d\theta = \delta_{lm}\frac{2}{2l+1} \tag{1.11.78}$$

and $c_l = e^{i\eta_l}$, we obtain

$$f(\theta) = -\sum_{l=0}^{\infty}(2l+1)\, e^{i\eta_l}\int_0^\infty j_l(kr)\, V(r)\, R_l(r)\, r^2\, dr\, P_l(\cos\theta) \tag{1.11.79}$$

so that by comparison with equation (1.11.77)

$$\sin\eta_l = -k\int_0^\infty j_l(kr)\, V(r)\, R_l(r)\, r^2\, dr. \tag{1.11.80}$$

Besides their immediate use below, these results of phase shift analysis will be extensively employed when we discuss electron scattering from lattice defects in Chapter 10.

(b) *Secular determinant of KKR method in form of momentum eigenfunction method.* At this stage we return to the secular equation (1.11.49) of the KKR method. It is very convenient then that this form can be expressed in terms of the phase shifts. To see how this can be done, we note that for $r \geqslant r_i$ the radial wave function $R_l(r)$ must be a linear combination of the free particle solutions j_l and n_l. The particular linear combination can then be obtained from the results we have just derived by taking the asymptotic forms of j_l and n_l and comparing the result with equation (1.11.71). We then find

$$R_l(r) = \text{const}\,[j_l(\kappa r) - \tan\eta_l\, n_l(\kappa r)], \quad E = \kappa^2. \tag{1.11.81}$$

We can eliminate the constant by forming the logarithmic derivative needed in equation (1.11.49). We then find

$$\frac{R_l'(r)}{R_l(r)} = L_l = \frac{j_l'(\kappa r) - \tan\eta_l\, n_l'(\kappa r)}{j_l(\kappa r) - \tan\eta_l\, n_l(\kappa r)} \tag{1.11.82}$$

from which we immediately obtain

$$\tan\eta_l = \frac{L_l\, j_l(\kappa r) - j_l'(\kappa r)}{L_l\, n_l(\kappa r) - n_l'(\kappa r)}. \tag{1.11.83}$$

Thus equation (1.11.49) may be written

$$\det|A_{lm,l'm'} + \kappa\,\delta_{ll'}\,\delta_{mm'}\cot\eta_l| = 0. \tag{1.11.84}$$

(c) *Relation between effective potentials of KKR and APW methods.* We give the detailed argument by which this equation can be transformed to a form exactly like equation (1.11.55) of the APW method in Appendix A1.7. Equation (A1.7.40) gives explicitly the result for the effective potential $\Gamma_{\mathbf{KK}'}$ in terms of the phase shifts η_l (Ziman, 1965),

$$\Gamma_{\mathbf{KK}'} = \Gamma(\mathbf{k}-\mathbf{K}, \mathbf{k}-\mathbf{K}')$$
$$= \frac{-4\pi}{\Omega\kappa} \sum_l (2l+1) \tan \eta_l' \frac{j_l(|\mathbf{k}-\mathbf{K}|r_i)\, j_l(|\mathbf{k}-\mathbf{K}'|r_i)}{j_l^2(\kappa r_i)} P_l(\cos\theta_{\mathbf{K},\mathbf{K}'}) \tag{1.11.85}$$

where $\theta_{\mathbf{K},\mathbf{K}'}$ is the angle between $\mathbf{k}-\mathbf{K}$ and $\mathbf{k}-\mathbf{K}'$ and the modified phase shift η_l' is given by

$$\cot \eta_l' = \cot \eta_l - \frac{n_l(\kappa r_i)}{j_l(\kappa r_i)}. \tag{1.11.86}$$

This is the KKR result and, for comparison, we can obtain $\Gamma_{\mathbf{KK}'}$ for the APW method from equation (1.11.54). The result may be readily shown to take the form

$$\Gamma_{\mathbf{KK}'} = \frac{4\pi r_i^2}{\Omega}\Big[-(\{\mathbf{k}-\mathbf{K}\}.\{\mathbf{k}-\mathbf{K}'\}-E)\frac{j_1(|\mathbf{K}-\mathbf{K}'|r_i)}{|\mathbf{K}-\mathbf{K}'|}$$
$$+\sum_l (2l+1)\frac{R_l'(E,r_i)}{R_l(E,r_i)} j_l(|\mathbf{k}-\mathbf{K}|r_i)\, j_l(|\mathbf{k}-\mathbf{K}'|r_i)\, P_l(\cos\theta_{\mathbf{K},\mathbf{K}'})\Big]. \tag{1.11.87}$$

This can be expressed, following Morgan (1966), in terms of the modified phase shift η_l' defined in equation (1.11.86). In particular, we use for R_l'/R_l the result

$$\frac{R_l'(E,r_i)}{R_l(E,r_i)} = L_l = \frac{j_l'(\kappa r_i)}{j_l(\kappa r_i)} - \frac{\tan \eta_l'}{\kappa r_i^2 j_l^2(\kappa r_i)} \tag{1.11.88}$$

which follows by manipulation from equation (1.11.83), using the Wronskian

$$[j_l, n_l] = j_l(\kappa r_i)\, n_l'(\kappa r_i) - j_l'(\kappa r_i)\, n_l(\kappa r_i)$$
$$= \frac{1}{\kappa r_i^2}. \tag{1.11.89}$$

The term involving $\tan \eta_l'$ in (1.11.88), when substituted in (1.11.87), leads to a contribution to $\Gamma_{\mathbf{KK}'}$ which is exactly that given by the KKR result (1.11.85).

Thus we can write from (1.11.87) and (1.11.85)

$$\Gamma^{\mathrm{KKR}}_{\mathbf{KK'}} = \Gamma^{\mathrm{APW}}_{\mathbf{KK'}} - \Gamma^{0}_{\mathbf{KK'}} \tag{1.11.90}$$

where $\Gamma^{0}_{\mathbf{KK'}}$ is the value of $\Gamma^{\mathrm{APW}}_{\mathbf{KK'}}$ when $R_l(r) = j_l(\kappa r)$, that is the value of $\Gamma^{\mathrm{APW}}_{\mathbf{KK'}}$ for an empty lattice. In general, $\Gamma^{0}_{\mathbf{KK'}}$ is not zero unless $\kappa^2 = |\mathbf{k}|^2$. In contrast with this behaviour of $\Gamma^{\mathrm{APW}}_{\mathbf{KK'}}$, the effective potential $\Gamma^{\mathrm{KKR}}_{\mathbf{KK'}}$ vanishes for all values of E and $\mathbf{k}$ for an empty lattice.

1.11.6 Pseudopotentials

The fact that, when the APW technique, for example, is applied to Na metal, the energy bands are nearly free-electron like, can be understood from a number of alternative points of view. The first of these is conveniently approached via the scattering cross-section of a single ion.

If the scattering centres offer a small cross-section to free electrons with energies in the range up to the Fermi level, the metal will be free-electron like. It should be noted that a small cross-section does not imply that the potential of the ions has to be small. Thus, as we saw in the APW method, we must match the spherical Bessel function $j_l(\kappa r_i)$ with the radial wave function $R_l(r)$ calculated inside r_i, at r_i; that is, we must satisfy

$$R_l(r_i) = j_l(\kappa r_i). \tag{1.11.91}$$

A single APW can be used if

$$\left(\frac{dR_l(r)}{dr}\right)_{r_i} = \left(\frac{dj_l(\kappa r)}{dr}\right)_r \tag{1.11.92}$$

in addition to equation (1.11.91), and the two conditions may be combined into the requirement of continuity of logarithmic derivatives (Slater, 1965)

$$\frac{R_l'(r_i)}{R_l(r_i)} = \frac{j_l'(\kappa r_i)}{j_l(\kappa r_i)}. \tag{1.11.93}$$

Fues and Statz (1952) have shown that this condition is approximately satisfied when the potential is very deep and of small cross-section. Such a situation occurs near the nucleus of the inert core of an alkali metal, and explicit calculation confirms the approximate validity of equation (1.11.93) for the alkali metals.

An almost equivalent argument to this can be posed in terms of the phase shifts. Suppose with an alkali metal core, that we calculate a set of phase shifts η_l. Then, from the relations (1.11.67) and (1.11.77), we see that the same scattering cross-section can be obtained by subtracting any multiples of π from the phase shifts, thereby making them as small as possible. Then, again, there is a hope of calculating the scattering by Born approximation for the phase shifts, which is equivalent to low-order perturbation theory.

We remark in passing that the assumption that there is no interference between scattering from the different ions is not entirely equivalent to the assumption of perfectly free-electron behaviour (this is discussed in detail in Chapter 10).

But the most fundamental theoretical argument follows from the band-structure methods we have just discussed at considerable length. If the matrix elements $\Gamma_{\mathbf{KK}'}$ of the effective potential were as strong as those of the real potential, then we would be back with the old difficulties of plane-wave expansions: namely that hundreds of plane-waves are needed to reproduce the rapid atomic fluctuations of the wave functions near the nuclei. We must therefore conclude, from the success of the OPW, APW and KKR methods in calculating energy bands in real crystals, that the effective potential $\Gamma_{\mathbf{KK}'}$ is indeed, in the valence bands of such crystals, much weaker than the actual potential. We stress that this conclusion follows from first principles calculations which in no way rely on the interpretation of $\Gamma_{\mathbf{KK}'}$ as an effective potential.

Nevertheless, the way in which some weak "pseudopotential" can arise, replacing what one might have expected to be strong lattice scattering of the free-electron waves, was first suggested clearly by Phillips and Kleinman (1959). We shall discuss their form of pseudopotential in section (c) below. However, their basic idea was that, when dealing with valence electrons in crystals, the strong attractive potential of the positive ions is, at least partly, cancelled off by an effective repulsive potential, whose role is to prevent valence electrons falling down into the core levels. This weakens the effective potential felt by the valence electrons, and there is hope that the resulting potential can be treated by low-order perturbation theory. This resulting potential they termed a "pseudopotential". This goes some way towards explaining why nearly free-electron results, such as we discussed in section 1.9.2, are of much wider applicability than might have been supposed at first sight. Indeed, if we exclude noble and transition metals, the nearly free-electron approximation is quite useful for most of the remaining "simple" metals.

(a) *Coordinate space representation of matrix elements of KKR method.* It is worth while to record here an example which illustrates the discussion of pseudopotentials given above in terms of the single-centre scattering. To reiterate, any potential which has the same scattering properties as the original core potential, at the energy under discussion, may be used.

Lloyd (1965) has proposed an explicit r space form which reproduces the KKR matrix elements. Though the form is pathological, namely a constant potential with a delta function at some radius r_i, it gives back precisely the KKR result (1.11.85).

In an angular momentum representation, the matrix elements of this potential, the delta function having strength λ_l, are simply

$$\langle lm, r|\Gamma|l'm', r'\rangle = \delta_{ll'}\,\delta_{mm'}\,\delta(r-r_i)\,\lambda_l\frac{\delta(r-r')}{r'^2}. \qquad (1.11.94)$$

The l-dependent potential represented by equation (1.11.94) can now be adjusted to give the correct scattering properties, at each energy, by means of the strength parameters λ_l.

The Schrödinger equation for this potential is given by

$$\left(-\nabla^2+\frac{l(l+1)}{r^2}+\lambda_l\,\delta(r-r_i)\right)\phi_l(r) = \kappa^2\phi_l(r); \quad E=\kappa^2. \qquad (1.11.95)$$

It is not difficult to show that this pseudopotential gives the same asymptotic form of the wave function as that given by the real scattering potential, provided that the strength λ_l is chosen such that

$$\lambda_l = \frac{R_l'(r_i)}{R_l(r_i)} - \kappa\frac{j_l'(\kappa r_i)}{j_l(\kappa r_i)} \qquad (1.11.96)$$

or in terms of the modified shift η_l' of equation (1.11.86)

$$\lambda_l = -\frac{1}{\kappa}\tan\eta_l'\{r_i j_l(\kappa r_i)\}^{-2}. \qquad (1.11.97)$$

It only remains to convert this pseudopotential into a momentum representation to show that it reproduces the KKR matrix element. This leads to the result

$$\langle\mathbf{k}|\Gamma|\mathbf{k}'\rangle = e^{i\mathbf{k}-\mathbf{k}'.\mathbf{R}}\frac{(4\pi r_i)^2}{\mathscr{V}}\sum_{lm}\{\lambda_l j_l(kr_i) j_l(k'r_i)\, Y_l^m(\mathbf{k})\, Y_l^{m*}(\mathbf{k}')\}, \qquad (1.11.98)$$

where $\mathbf{R}$ is the position of the scattering potential and $\mathscr{V}$ as usual is the volume of the crystal. Summing this over all lattice sites using equation (1.3.19) then gives the total effective potential of equation (1.11.85).

It is convenient at this point to depart from first principles calculations, and use the concept of an l-dependent model potential exemplified by equation (1.11.94) in a semi-empirical manner.

(b) *Quantum defect method and model potentials.* In early work on energy bands, the quantum defect method was prominent (cf. Ham, 1955). The idea behind this method was to combine the Wigner–Seitz method, for example, with methods which allowed the radial derivative of the wave function to be obtained from atomic term values, without the necessity of calculating the self-consistent atomic potential.

This idea has been used in the so-called model potential methods. These were prompted by the philosophy of pseudopotentials and essentially take over the quantum defect ideas into this framework.

Useful prescriptions for computing such model potentials have been given by a number of authors. One of these, which often yields suitable matrix elements for band structure calculations, is due to Abarenkov and Heine (1965). They choose a potential of the rather "shocking" form shown in Figure 1.13, where for ionic charge Z

$$\left.\begin{aligned} V_l(r) &= -\frac{Ze^2}{r}, \quad r > r_c, \\ V_l(r) &= A_l(E), \quad r < r_c. \end{aligned}\right\}. \tag{1.11.99}$$

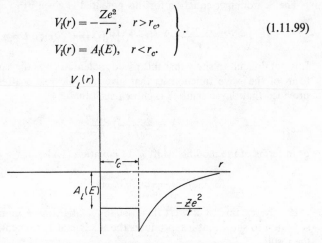

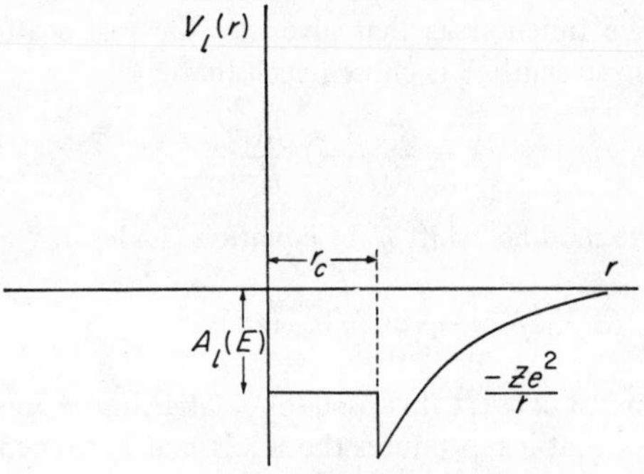

FIGURE 1.13. Heine–Abarenkov l-dependent model potential. This is a Coulomb potential with cut-off, the depth of the potential depending on l.

This incorporates the l dependence we referred to above and has the further merit that the A_l's are found to depend only weakly on energy. One can solve for the eigenstates of the free atom in this type of potential and compare with atomic term values to get $A_l(E)$ at the spectroscopic energies. One can interpolate to get values for use in the solid, at say $E = E_f$ or indeed any desired energy.

The resulting potential gives a description of the positive ion and clearly must be modified before use in a crystal by screening due to conduction or valence electrons. This screening ought really to take account of non-local effects (Harrison, 1963; Animalu, 1965), but the major effects can be dealt with via a local dielectric function discussed fully in Chapter 2.

We have seen already that, in practice, one often describes the potential in terms of its matrix elements between plane waves or OPWs. The matrix elements of interest are those of the total potential obtained by summing the products of the screened V_l's with the appropriate projection operators picking out the l component of the wave function on which the potential operates.

Though calculations using the above method have been carried through by various workers, in practice one wants a model potential which fits the band gaps. The example shown for Al in Figure 1.14 was obtained by

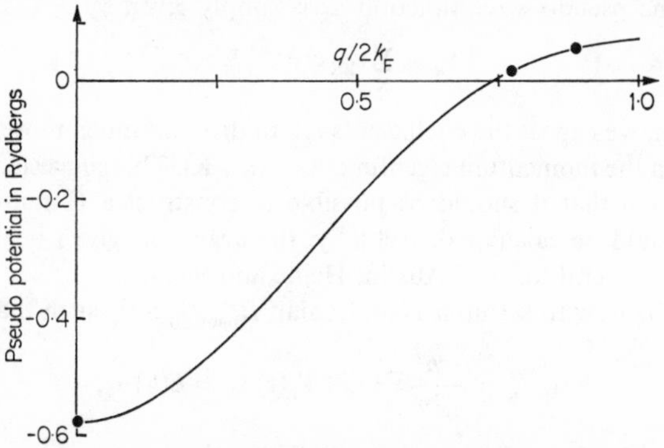

FIGURE 1.14. Pseudopotential for Al. (This is matrix element of pseudopotential between plane waves $\mathbf{k}+\mathbf{q}$ and $\mathbf{k}$, with $|\mathbf{k}+\mathbf{q}| = |\mathbf{k}| = k_f$: see Harrison (1966).)

modifying a first principles calculation to fit the observed band gaps, these being twice the corresponding Fourier component of the potential in the nearly free-electron method (see Problem 1.13 where the work of Ashcroft (1963) on the Al Fermi surface is referred to).

(c) *Other forms of pseudopotential.* We referred above to the form of pseudopotential put forward by Phillips and Kleinman (1959). It will be convenient, however, to discuss this as a special case of the formalism of Austin, Heine and Sham (1962). For a fuller discussion we must refer the reader to the book by Harrison (1966).

To motivate the introduction of these pseudopotentials, we can argue that corresponding to the weaker effective potentials with matrix elements $\Gamma_{\mathbf{KK'}}$ discussed fully above, there will be a pseudo-wave function which can be expanded in plane waves much more effectively than the original Bloch function ψ. As we saw in the OPW method of section 1.11.2, the major reason for the slow convergence of plane-wave expansions of ψ was the atomic-like oscillations of this wave function near nuclei. This behaviour is, of course, connected with the necessity for these valence electron wave functions to be orthogonal to the core functions. Without, at this stage specifying the pseudo-wave function precisely, let us write the Bloch function ψ as

$$\psi_{\mathbf{k}}(\mathbf{r}) = \alpha_{\mathbf{k}}(\mathbf{r}) - \sum_c \langle \psi_{\mathbf{k}}^c | \alpha_{\mathbf{k}} \rangle \psi_{\mathbf{k}}^c(\mathbf{r}), \tag{1.11.100}$$

a form where orthogonality of $\psi_{\mathbf{k}}(\mathbf{r})$ to the tight-binding core functions $\psi_{\mathbf{k}}^c(\mathbf{r})$ given by equation (1.11.16) is now explicit. In relation to equation (1.11.19), the pseudo-wave function $\alpha_{\mathbf{k}}$ is simply given by

$$\alpha_{\mathbf{k}} = \sum_{\mathbf{K}} c_{\mathbf{kK}} \, e^{i(\mathbf{k}+\mathbf{K}).\mathbf{r}}. \tag{1.11.101}$$

In this form, we expect the coefficients $c_{\mathbf{kK}}$ to drop off much more rapidly for large $\mathbf{K}$ than the momentum eigenfunction $v(\mathbf{k}+\mathbf{K})$. This suggested to Phillips and Kleinman that it should be possible to construct a Hamiltonian from which $\alpha_{\mathbf{k}}$ could be calculated. Actually, the argument given below is posed in the more general form of Austin, Heine and Sham.

Our task is now to set up a Hamiltonian $H_{\text{pseudo}} \equiv H_p$ such that

$$H_p \, \alpha_{\mathbf{k}} = -\frac{\hbar^2}{2m} \nabla^2 \alpha_{\mathbf{k}} + V_p(\mathbf{r}) \, \alpha_{\mathbf{k}} = E(\mathbf{k}) \, \alpha_{\mathbf{k}}, \tag{1.11.102}$$

where $V_p(\mathbf{r})$ is the pseudopotential and $E(\mathbf{k})$ is the exact energy of the Bloch electron with wave vector $\mathbf{k}$.

To find the general nature of V_p let us consider

$$\begin{aligned} E(\mathbf{k}) \langle \psi_{\mathbf{k}} | \alpha_{\mathbf{k}} \rangle &= \langle \psi_{\mathbf{k}} | H_p | \alpha_{\mathbf{k}} \rangle \\ &= \langle \psi_{\mathbf{k}} | H | \alpha_{\mathbf{k}} \rangle + \langle \psi_{\mathbf{k}} | V_p - V | \alpha_{\mathbf{k}} \rangle \\ &= E(\mathbf{k}) \langle \psi_{\mathbf{k}} | \alpha_{\mathbf{k}} \rangle + \langle \psi_{\mathbf{k}} | V_p - V | \alpha_{\mathbf{k}} \rangle. \end{aligned} \tag{1.11.103}$$

We see that

$$\langle \psi_{\mathbf{k}} | V_p - V | \alpha_{\mathbf{k}} \rangle = 0. \tag{1.11.104}$$

This equation is satisfied if

$$V_p \, \alpha_{\mathbf{k}} = V \alpha_{\mathbf{k}} + \sum_c f(\alpha_{\mathbf{k}}, c) \, \psi_{\mathbf{k}}^c(\mathbf{r}), \tag{1.11.105}$$

where the factors $f(\alpha_{\mathbf{k}}, c)$ are arbitrary. It is also satisfied if we add on to the right-hand side of equation (1.11.105) a linear combination of conduction electron wave functions, all different from $\psi_{\mathbf{k}}$; then, however, $\alpha_{\mathbf{k}}$ is not orthogonal to these functions.

Equation (1.11.105) gives the result of operating with V_p on the pseudowave function $\alpha_{\mathbf{k}}$. The pseudopotential itself is defined by generalizing equation (1.11.102) to obtain the result of operating with the pseudopotential on any function whatever. Since any function may be decomposed into plane-waves, it is particularly fruitful to define V_p by the result of operating with it on any plane-wave. By inspection of equation (1.11.105) we define such a pseudopotential by

$$V_p\, e^{i\mathbf{k}.\mathbf{r}} = Ve^{i\mathbf{k}.\mathbf{r}} + \sum_c f(\mathbf{k}, c)\, \psi_{\mathbf{k}}^c(\mathbf{r}), \tag{1.11.106}$$

$f(\mathbf{k}, c)$ being any function of $\mathbf{k}$ and c whatever. This form of pseudopotential is due to Austin, Heine and Sham.

Let us take an arbitrary function

$$\xi(\mathbf{r}) = \sum_{\mathbf{k}} e^{i\mathbf{k}.\mathbf{r}}\, \tilde{\xi}(\mathbf{k}) \tag{1.11.107}$$

and then, from equation (1.11.106), we find

$$V_p\, \xi = V(\mathbf{r})\, \xi(\mathbf{r}) + \sum_{\mathbf{k}c} f(\mathbf{k}, c)\, \tilde{\xi}(\mathbf{k})\, \phi_c(\mathbf{r}), \tag{1.11.108}$$

which may be rewritten as

$$V_p\, \xi(\mathbf{r}) = V(\mathbf{r})\, \xi(\mathbf{r}) + \sum_c \langle f(\mathbf{r}, c) | \xi(\mathbf{r}) \rangle\, \phi_c(\mathbf{r}), \tag{1.11.109}$$

where $f(\mathbf{r}, c)$ is any function whatsoever.

If, in spite of the presence of $f(\mathbf{r}, c)$ in equation (1.11.109), we assume that a particular choice of pseudopotential has been made, it is of interest to ask if the pseudo-wave function $\alpha_{\mathbf{k}}$ is then unique. To examine this point, we rewrite equation (1.11.100) as

$$\alpha_{\mathbf{k}} = \psi_{\mathbf{k}} + \sum_c \langle \psi_{\mathbf{k}}^c | \alpha_{\mathbf{k}} \rangle\, \psi_{\mathbf{k}}^c; \tag{1.11.110}$$

now with the potential (1.11.109) we have

$$H\alpha_{\mathbf{k}} + \sum_c \langle f(\mathbf{r}, c) | \alpha_{\mathbf{k}} \rangle\, \psi_{\mathbf{k}}^c = E(\mathbf{k})\, \alpha_{\mathbf{k}} \tag{1.11.111}$$

and hence

$$[E(\mathbf{k}) - E_c] \langle \psi_{\mathbf{k}}^c | \alpha_{\mathbf{k}} \rangle = \langle f(\mathbf{r}, c) | \alpha_{\mathbf{k}} \rangle \tag{1.11.112}$$

(all eigenfunctions of H are assumed to be normalized to unity). If we take $f(\mathbf{r}, c) = [E(\mathbf{k}) - E_c]\, \psi_{\mathbf{k}}^c(\mathbf{r})$, $\langle \psi_{\mathbf{k}}^c | \alpha_{\mathbf{k}} \rangle$ becomes arbitrary. If any other choice of $f(\mathbf{r}, c)$ is made, $\langle \psi_{\mathbf{k}}^c | \alpha_{\mathbf{k}} \rangle$ is uniquely given by (1.11.112) and the solution $\alpha_{\mathbf{k}}$ of equation (1.11.102) is unique.

The above choice of $f(\mathbf{r}, c)$, making $\alpha_{\mathbf{k}}$ arbitrary, gives us

$$V_p(\mathbf{r})\, \alpha_{\mathbf{k}} = V\alpha_{\mathbf{k}} + \sum_c (E(\mathbf{k}) - E_c)\, \psi_{\mathbf{k}}^c(\mathbf{r})\, \langle \psi_{\mathbf{k}}^c | \alpha_{\mathbf{k}} \rangle. \tag{1.11.113}$$

This defines the Phillips–Kleinman potential, the first pseudopotential to be set up, and since the $\alpha_{\mathbf{k}}$ it gives is arbitrary to within the addition of core functions, solutions for all pseudopotentials must be eigenfunctions of the Hamiltonian with the Phillips–Kleinman potential.

Following this formal account of pseudopotential theory, it is of interest to remark that, in spite of the apparent generality of the Austin–Heine–Sham pseudopotential, the effective potentials $\Gamma_{\mathbf{KK}'}$ of the KKR and the APW methods are not of this form. However, whereas in the discussion of the KKR and the APW methods, we focused almost entirely on the effective potentials, and not on the pseudo-wave functions, the burden of the present section has been to consider just what knowledge of these can be gained from the pseudopotential approach. While no unique results emerge, we want to stress that, outside the core, the pseudo-wave function must approximate closely to the real wave function for the theory to be useful, and also it should be possible to expand it using relatively few plane-waves. This would then validate a nearly free-electron theory.

Finally, we want to stress that the considerable successes of the pseudopotential method centre round its semi-empirical use. When attempts are made to remove the semi-empirical elements from it, obvious troubles arise. Thus, as is evident from the discussion of this section, the pseudopotential V_p is a non-local operator, which, acting on the pseudowave function can be written

$$V_p \alpha_{\mathbf{k}} = \int V_p(\mathbf{rr}') \alpha_{\mathbf{k}}(\mathbf{r}') \, d\mathbf{r}'. \tag{1.11.114}$$

Secondly, the pseudopotential is energy and momentum dependent. Thus, when we form the matrix elements of V_R between plane-wave states, say $\mathbf{k}$ and $\mathbf{k}+\mathbf{K}$, then the answer depends on both $\mathbf{k}$ and $\mathbf{K}$.

The dependence on energy is also a serious limitation: this is often expressed in terms of "core shifts" (see, for example, Lin and Phillips, 1965), but we shall not go into that here). An alternative way of coping with this is to use "l-dependent" pseudopotentials, as discussed in section 1.11.6 (b).

Thirdly, while we understand in principle the screening of "real" potentials, the screening of non-unique "pseudopotentials" is tricky, and simple free-electron gas screening (cf. Chapter 2) is almost always used.

1.11.7 k.p *method*

The methods discussed previously were designed to calculate the dispersion relation $E(\mathbf{k})$ for energy bands in crystals, and also wave functions, though these are usually given to lower accuracy. The procedure now to be described, known as the $\mathbf{k}.\mathbf{p}$ method, enables knowledge of the $E(\mathbf{k})$ relation at some given point $\mathbf{k}_0$, to be extended into the vicinity of this point, without further numerical solution of the Schrödinger equation.

We obtained in equation (1.5.26) a differential equation for the periodic part $u_{\mathbf{k}}(\mathbf{r})$ of the Bloch function $\psi_{\mathbf{k}}(\mathbf{r}) = u_{\mathbf{k}}(\mathbf{r})\,e^{i\mathbf{k}.\mathbf{r}}$. If we know the solution of (1.5.26) at some point $\mathbf{k}_0$, then we can write at point $\mathbf{k}$

$$\left\{-\frac{\hbar^2\nabla^2}{2m}+V(\mathbf{r})-\frac{i\hbar^2\mathbf{k}_0.\nabla}{m}\right\}u_{\mathbf{k}}-\frac{i\hbar^2}{m}(\mathbf{k}-\mathbf{k}_0).\nabla u_{\mathbf{k}} = \left(E-\frac{\hbar^2k^2}{2m}\right)u_{\mathbf{k}} \tag{1.11.115}$$

and we can regard the term $(i\hbar^2/m)(\mathbf{k}-\mathbf{k}_0).\nabla$ as a perturbation on the Hamiltonian for $u_{\mathbf{k}_0}$. This procedure, known as the $\mathbf{k}.\mathbf{p}$ method, enables knowledge of the $E(\mathbf{k})$ relation to be extended about the point $\mathbf{k}_0$, without further explicit numerical solution of the Schrödinger equation.

In passing, it is worth noting that the Feynman theorem

$$\frac{\partial E}{\partial\lambda} = \left\langle \psi_\lambda^* \frac{\partial H}{\partial\lambda}\psi_\lambda\right\rangle, \tag{1.11.116}$$

where H depends on the parameter λ (cf. Appendix 1.4), gives us

$$\frac{\partial E(\mathbf{k})}{\partial\mathbf{k}}-\frac{\hbar^2\mathbf{k}}{m} = \frac{\hbar^2}{im}\int_\Omega u_{\mathbf{k}}^*\frac{\partial}{\partial\mathbf{r}}u_{\mathbf{k}}\,d\mathbf{r} \tag{1.11.117}$$

if $u_{\mathbf{k}}$ is normalized in the unit cell. This equation can also be written in the form

$$\frac{1}{\hbar}\frac{\partial E}{\partial\mathbf{k}} = \frac{\hbar}{im}\int_\Omega \psi_{\mathbf{k}}^*\nabla\psi_{\mathbf{k}}\,d\mathbf{r} = \frac{1}{m}\int_\Omega \psi_{\mathbf{k}}^*\mathbf{p}\psi_{\mathbf{k}}\,d\mathbf{r}, \tag{1.11.118}$$

an equation which gives us the group velocity, as discussed in section 1.10.1.

If $\mathbf{k}_0$ is such that $(\partial E/\partial\mathbf{k})_{\mathbf{k}_0} = 0$, we must go to second order in the perturbation $\mathbf{k}.\mathbf{p}$ to obtain useful information. We may then obtain a formula for the effective mass as follows. For simplicity we shall take $\mathbf{k}_0 = 0$, and writing $\psi_{\mathbf{k}}^i \equiv |\mathbf{k}i\rangle$, where i is the band index, the eigenfunction to first order in $\mathbf{k}$ is, on putting $\hbar = 1$,

$$|\mathbf{k}i\rangle = e^{i\mathbf{k}.\mathbf{r}}\left[|0i\rangle+\frac{1}{m}\sum_j{}'\frac{|0j\rangle\langle 0j|\mathbf{k}.\mathbf{p}|0i\rangle}{E_i(0)-E_j(0)}\right]. \tag{1.11.119}$$

The energy to second order may then be written

$$E_i(\mathbf{k}) = E_i(0)+\frac{k^2}{2m}+\frac{1}{m^2}\sum_{\mu\nu}\sum_j{}'\frac{\langle 0i|p_\mu|0j\rangle\langle 0j|p_\nu|0i\rangle k_\mu k_\nu}{E_i(0)-E_j(0)}. \tag{1.11.120}$$

From the definition of the effective mass in section 1.10.2, this may also be written in the form

$$E_i(\mathbf{k}) = E_i(0) + \frac{1}{2m} \sum \left(\frac{m}{m^*}\right)_{\mu\nu} k_\mu k_\nu \qquad (1.11.121)$$

and comparison of equations (1.11.120) and (1.11.121) yields‡

$$\left(\frac{m}{m^*}\right)_{\mu\nu} = \delta_{\mu\nu} + \frac{2}{m} \sum_j{}' \frac{\langle i0|p_\mu|0j\rangle\langle j0|p_\nu|0i\rangle}{E_i(0) - E_j(0)}. \qquad (1.11.122)$$

It will be seen that if the energy gap $E_i(0) - E_j(0)$ is very small for some band E_j, the value of $(m/m^*)_{\mu\nu}$ will be dominated by the factor

$$\langle i0|p_\mu|0j\rangle\langle j0|p_\nu|0i\rangle/(E_i(0) - E_j(0)). \qquad (1.11.123)$$

Certain matrix elements $\langle i0|p_\mu|0l\rangle$ can be eliminated on symmetry grounds. As a simple example, let us suppose that $\psi_\mathbf{k}^i(-\mathbf{r}) = \psi_\mathbf{k}^i(\mathbf{r})$, but $\psi_\mathbf{k}^l(-\mathbf{r}) = -\psi_\mathbf{k}^l(\mathbf{r})$. Then we have $\langle 0i|\mathbf{p}|0l\rangle = 0$.

Also, if bands are degenerate at $\mathbf{k}_0$, we must use degenerate perturbation theory instead of the treatment given above and an example of this will be given in section 1.11.8 below when we discuss the energy bands in Ge.

1.11.8 Examples of band structure calculations

It is, of course, quite impossible in a book of this nature, to do justice to the tremendous knowledge we now possess on the energy bands of specific crystals. We shall therefore select here just a few examples to illustrate in cases of some practical interest the kind of results which can be obtained.

The first really exceptionally interesting experimental work on the Fermi surface was that of Pippard (1957) on Cu. Pippard correctly concluded that the Fermi surface in this metal contacted the zone boundary (see Figure 1.15). The details of the Fermi surface were only later determined, and we shall choose the extensive work of Davis, Faulkner and Joy (1968) as our first example. This will be a convenient point to say a little about *d*-bands in transition metals. The pioneering work of Herman (1955) using the OPW method for Ge will provide a basis for the second example, though we shall illustrate how the bands can be understood in a gross way via an empty lattice argument.

(a) *Crystal potential and band structure of copper.* It is of considerable interest that Davis, Faulkner and Joy (1968) have been able to construct a potential which reproduces the Fermi surface of Cu to really high accuracy,

‡ Equation (1.11.122) can be shown to be equivalent to equation (1.10.22).

when used in the KKR method of section 1.11.3. Just how we know this will have to await the discussion of the de Haas–van Alphen effect in Chapter 4. However, we summarize here the way in which they constructed the potential, which is, of course, the problem at the heart of any band structure work. It has to be said at the outset that a certain amount of trial and error was involved in constructing the potential. So far, no completely fundamental

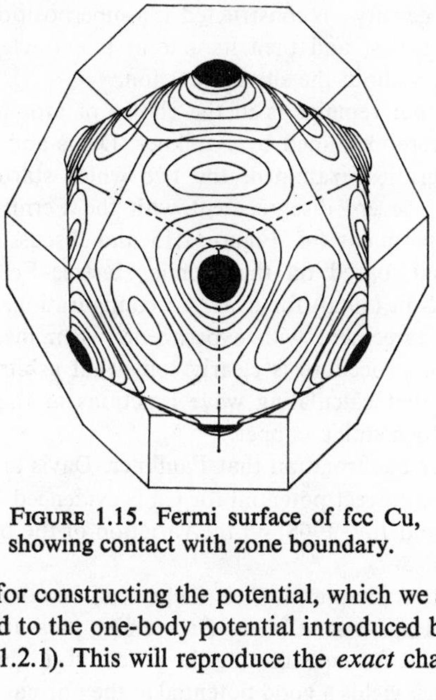

FIGURE 1.15. Fermi surface of fcc Cu, showing contact with zone boundary.

way is open to us for constructing the potential, which we anticipate to be at least closely related to the one-body potential introduced by Hohenberg and Kohn (cf. section 1.2.1). This will reproduce the *exact* charge density in the crystal.

To apply the KKR method, as we have seen, a muffin-tin potential has to be used. This means, as we discussed earlier, that the potential is taken as spherical within the unit cell, but having a finite range. Such an assumption is probably not very serious with regard to its effect on electronic band structure.

The construction of the crystal potential $V(\mathbf{r})$ was then carried out by writing

$$V(\mathbf{r}) = V_{\text{coulomb}}(\mathbf{r}) + V_{\text{exchange}}(\mathbf{r}). \tag{1.11.124}$$

Both the Coulomb and exchange terms were then calculated using wave functions of the free atom. At a chosen site in the crystal, the Coulomb part was taken from the atomic potential, plus contributions from the same potential situated on near neighbour sites.

The exchange contribution to the potential was obtained using Slater's prescription (see Chapter 2, section 2.11.5, for details), based on the ideas of Dirac for introducing exchange into the Thomas–Fermi theory. Then we write

$$V_{\text{exchange}}(\mathbf{r}) = -6\left[\frac{3\rho(\mathbf{r})}{8\pi}\right]^{\frac{1}{3}}, \tag{1.11.125}$$

where the electron density ρ is constructed as a superposition of Hartree–Fock atomic charge densities, and then its s term is extracted in the spherical harmonic expansion about the site in question.

The ambiguity that remains is in the choice of atomic charge densities. Various choices were examined by Faulkner, Davis and Joy (1967). Of the three different densities examined, the two which started from a $3d^{10}4s^1$ configuration gave severe disagreement with the Fermi surface of Cu as obtained from experiment (cf. Figure 1.15 and discussion of Chapter 4). The third potential, based on the atomic Hartree–Fock wave functions calculated by Watson (1960) for a $3d^9 4s^2$ configuration, was found to give quite satisfactory agreement with experiment, when inserted in a $3d^{10}4s^1$ configuration! This procedure is clearly somewhat arbitrary as there is no reason to believe that calculating wave functions in this manner has any special relevance to metallic copper.

However, further confirmation that Faulkner, Davis and Joy (1967) have "hit on" almost the correct potential for Cu is evidenced by their later work (Davis, Faulkner and Joy, 1968) on the variation of the band structure with lattice constant.

As we shall see, in Chapter 4, their theory can be compared with experiments via the de Haas–van Alphen effect under pressure. Using the same prescription as above for constructing the potential, i.e. using the "atomic" charge density which yields a good potential at the normal lattice spacing, the effects of small changes in the lattice spacings can be predicted using the same charge density in the same equations. Similar results for Cu cannot be obtained by a simple scaling of the potential or a free-electron treatment.

We shall not discuss the predictions of their theory in detail here. Suffice it to say that, whereas we have described in this chapter how a whole variety of methods now exist which will all yield the same (and essentially the exact) energies for a given periodic potential, some trial and error is still involved in getting this potential.

If X-ray intensities, as we discuss in Chapter 5, could be obtained to really high accuracy, it would be possible to work back to a one-body potential, which should be closely related to that needed to generate the measured Fermi surface of a metal. To calculate this potential from theory, however, is a many-body problem and we shall therefore discuss the interacting electron system in a crystal in Chapter 2. Needless to say, to make progress, we shall have to try to isolate the essential features introduced by the electron–electron interactions. It is clearly not going to be possible to obtain an exact solution with a lattice potential plus electron–electron interactions.

(b) *Treatment of d-bands in transition metals*. The dispersion relations $E(\mathbf{k})$ of metals like Cu and Ni consist in essence of five narrow d-bands. These bands, however, cross and mix with a nearly free electron band, formed largely from s- and p-atomic states.

The secular equation derived earlier, namely

$$\det|\{|\mathbf{k}-\mathbf{K}|^2-E\}\,\delta_{\mathbf{KK'}}+\Gamma_{\mathbf{KK'}}|=0 \tag{1.11.126}$$

still holds, but it is no longer reasonable to suppose that Γ can be treated as a weak pseudopotential.

But the basic theory given above for $\Gamma_{\mathbf{KK'}}$ in either the augmented plane wave framework, or in the KKR method, can deal, of course, with quite general problems in band structure.

The problem can be viewed in a simplified manner by noting that the scattering becomes strong, and very dependent on energy, when passing through a resonance around the energy of the d-band.

In the KKR method, Ziman has shown (see Appendix 1.7) that $\Gamma_{\mathbf{KK'}}$ is given by equation (1.11.85). As we remarked earlier, the $\Gamma_{\mathbf{KK'}}$ given above is not identical with Γ (APW) (see section 1.11.5(c) where we show that the difference between them is an "empty lattice" term, which is not zero).

The d-wave phase shift in the expression (1.11.85) obviously is the focal point of our discussion. If we consider a transition metal atom placed in a free electron gas, with the d-band level E_d of the atom a little above the bottom of the plane wave band, as it is in the solid, then the d-state is a resonant state, that is, it has a finite lifetime, because an electron can "escape" into a plane-wave state of the same energy.

This is the case of an incoming plane-wave showing a resonance in the $l=2$ phase shift, which is given by (cf. Messiah, p. 399)

$$\tan\eta_2=\frac{\frac{1}{2}W}{E_d-E}. \tag{1.11.127}$$

The width W is related to the lifetime of the d-resonance by the uncertainty principle.

It seems therefore that one should write

$$\Gamma=\Gamma_{nfe}+\Gamma_{\text{resonance}}, \tag{1.11.128}$$

where all the features associated with the d-band are lumped into the second term. Clearly $\Gamma_{\text{resonance}}$ is related to tan η_2 in equation (1.11.127) and we note that the width and shape of the d-band, as well as the hybridization of the d-bands with the free-electron-like s–p-band, must all be related to the single constant W in (1.11.127), since E_d simply specifies the position of the d-bands relative to the s–p-band.

Figure 1.16 shows an example of the band structure of Cu, taken from the work of Segall (1962). The $E(\mathbf{k})$ curves are shown along ΓL etc. in the Brillouin

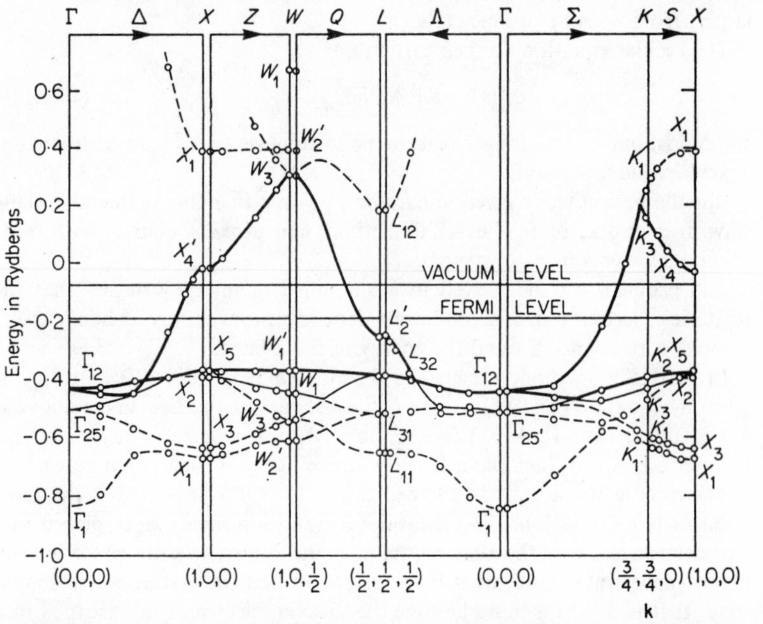

FIGURE 1.16. $E(\mathbf{k})$ curves for Cu, as calculated by Segall (1962). Symmetry points Γ, L, etc. are as shown in Figure 1.5. Symmetry directions are also indicated at top of figure. Notation of Bouchaert, Smoluchowski and Wigner (1936) is used in labelling the levels. Fermi level is shown just above -0.2 Rydberg with the potential used by Segall. Vacuum level shown, and the heavily drawn bands, are relevant to a later discussion of optical properties.

zone of Figure 1.5. The two bands along Λ joining Γ'_{25} and L_{31} and Γ_{12} and L_{32} are each doubly degenerate and are evidently very flat.

If there were no mixing of d-states with the s–p-states, then the fifth d-band of Λ symmetry would follow a curve which would join Γ'_{25} to L_{11} and the nfe-band of the same symmetry from Γ_1 to L'_2 and then from the next highest

L_1 point. The effect of mixing is to split bands apart where they would otherwise cross, the resultant band splitting being as in Figure 1.16.

It turns out that in this hybridization gap seen in Figure 1.16, the density of *sp*-states is very low, these states being effectively excluded over this energy region in the lower part of the *d*-band.

Then it is possible, making the split of Γ into two parts as in (1.11.128), to transform the secular equation (1.11.126) into a form which separates the *d*-band from the nearly-free-electron ones

$$\det|H_{m,ij} - E\delta_{ij}| = 0, \tag{1.11.129}$$

where H_m has the structure

$$H_m = \left|\begin{array}{c|c} \text{Pseudo } sp\text{-Pseudo } sp & sp\text{-}d \\ \hline d\text{-}sp & d\text{-}d \end{array}\right|. \tag{1.11.130}$$

The lower right 5×5 piece describes the five *d*-bands, the upper left part the *nfe* bands and the off-diagonal part labelled *d-sp* the mixing between them. Needless to say, this has approximated the infinite determinant (1.11.126) by the finite determinant (1.1.129). H_m is referred to as the model Hamiltonian, due originally to Mueller (1967).

It is worth noting that such a model Hamiltonian can be fitted to the band structures of the noble metals, Cu, Ag, Au, and the magnetic materials Fe and Ni, whose band structures have been calculated previously by the APW or KKR methods, using about a dozen parameters.

We wish to note here a few further points of some qualitative or semi-quantitative significance which follows from the general ideas sketched above.

First of all, the width of the *d*-bands can be roughly estimated as a function of lattice spacing (see, for example, Ziman (1969)).

Secondly, Hubbard and Dalton (1968) have had some success in parameterizing *d*-bands and calculating densities of states.

Thirdly, connected with the density of states, the structure is crucial, and it is not too bad to think of a common density of states for, say, the bcc transition metals. A curve appropriate for non-magnetic Fe is shown in Figure 1.17.

Furthermore, Watson, Ehrenreich and Hodges (1970) have observed that by renormalizing atomic charge distributions the positions of the *d*-bands relative to the *s–p*-bands can be rather well determined.

Having introduced Mueller's model Hamiltonian, and noted how it leads to a classification of band structures in transition metals in terms of

(i) the width W of the *d*-band, which is related to the lifetime $\hbar/W$ of the *d*-resonance at a given site;

(ii) the energy E_d of the centre of gravity of the d-bands, which specifies where the d-bands lie relative to the bottom of the conduction band $E(\Gamma_1)$ say,

we shall briefly record a microscopic interpretation of E_d here, as proposed by Watson, Ehrenreich and Hodges (1970).

It is clear from what we have already said that W can be rather directly related to atomic properties. But until the work of Watson and his colleagues, a satisfactory description of E_d on the basis of purely atomic considerations had not emerged. Their method can also be used to estimate $E(\Gamma_1)$.

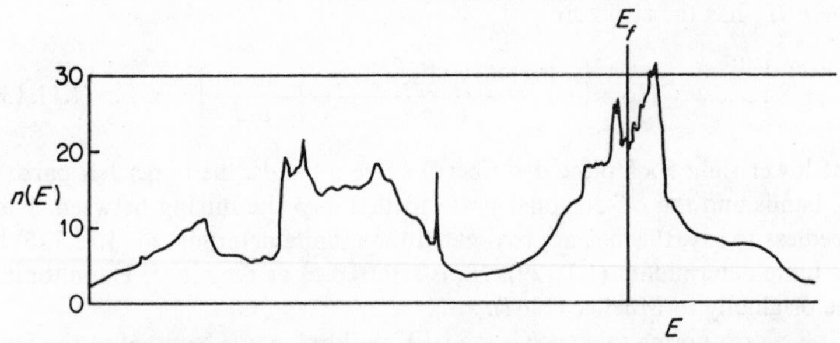

FIGURE 1.17. Density of states for bcc non-magnetic Fe (after Dalton, 1970), using the potential of Wood (1962). The general features of this curve are also evident for bcc non-magnetic Cr (cf. Dalton, 1970), though, of course, the Fermi level in Cr is placed well below the highest energy peak, in contrast to Fe.

We saw, for Cu, that a successful method of constructing a potential could be based on overlapping atomic charge densities. Watson, Ehrenreich and Hodges use a "renormalized atom", for which l-dependent Hartree–Fock crystal potentials can be obtained. The prescription is to cut off the free atom s- and d-wave functions at the Wigner–Seitz radius r_w and renormalize in this sphere: for the fcc transition metals in $d^n s'$ configurations.

The effect of such renormalization is to increase the d-electron charge inside the Wigner–Seitz sphere by less than 5%, but to increase the s-charge by a factor of 2 to 3. As we discuss in Chapter 2, the large compression of the s-charge is primarily responsible for the cohesion of the alkali metals.

We show in Table 1.8 the way in which E_d is affected by this renormalization. It is seen that for Co, Ni, Cu and Ag, an almost constant increase of about $\frac{1}{2}$ Rydberg occurs from the free atom values.

Because of the large distortion of the s-charge, it is less relevant to examine the renormalized s electron's one-electron energy than that of an OPW, ϕ say, at $\mathbf{k} = 0$, in the renormalized potential. Table 1.8 shows the free atom energies with

$$E(\Gamma_1) \equiv \int_{WS\,\text{sphere}} \phi^*(\mathbf{r})\,[-\tfrac{1}{2}\nabla^2 + V^{l=0}_{\text{renorm}}]\,\phi(\mathbf{r})\,d\mathbf{r} \qquad (1.11.131)$$

and with the value of $E(\Gamma_1)$ determined by a first principles APW calculation based on the l-dependent renormalized atom potential, with the outer region replaced by a muffin-tin form. $E(\Gamma_1)$ can be seen to lie considerably below the atomic value E_s, but the "renormalized atom" model gives excellent agreement with the energy band calculations.

TABLE 1.8. Free atom, renormalized atom and band theory energies, in Rydbergs

		Co ($d^8\,s$)	Ni ($d^9\,s$)	Cu ($d^{10}\,s$)	Ag ($d^{10}\,s$)
Free atom	E_d	−0.718	−0.797	−0.914	−1.037
	E_s	−0.43	−0.44	−0.46	−0.43
Renormalized atom	E_d	−0.185	−0.278	−0.399	−0.625
	$E(\Gamma_1)$	−0.75	−0.79	−0.83	−0.74
Centre of gravity					
Γ		−0.183	−0.265	−0.400	−0.606
X		−0.199	−0.276	−0.410	−0.598
Band calculation	$E(\Gamma_1)$	−0.71	−0.76	−0.81	−0.72

The same band calculations can be used to infer the relationship between E_d, obtained for a renormalized d-function in the renormalized atom potential and those of the d-bands. Table 1.8 shows that the centre of gravity of the d-band levels at Γ and X, defined as (cf. Figure 1.16)

$$\tfrac{3}{5}E(\Gamma_{25'}) + \tfrac{2}{5}E(\Gamma_{12}) \quad \text{and as} \quad \tfrac{1}{5}E(X_1) + \tfrac{1}{5}E(X_3) + \tfrac{1}{5}E(X_2) + \tfrac{2}{5}E(X_5)$$

respectively, differ by less than 0.03 Rydberg from the renormalized atom values of E_d.

Thus, some of the gross features of the band structures of the transition metals can be understood in this way, though, to some extent, this is hindsight, the first principles calculations having already been available.

(c) *Band structure of* Ge *and empty lattice treatment*. We referred to the band structure of NaCl and CsCl crystals in connection with the cellular method in section 1.11.1. We want here to examine the band structure of Ge, which again has two atoms per unit cell. The Brillouin zone for this crystal is the same as for Al and Cu and is shown in Figure 1.5.

We show in Figure 1.18 the schematic form of the valence and low-lying conduction bands, along symmetry directions in the BZ. In this diagram (cf. Herman, 1958), spin–orbit splittings (see following section) are omitted, to keep the discussion at a simpler level.

The band structure of Figure 1.18 is established from a variety of experimental results, plus the pioneering calculations of Herman (1955), using the OPW method.

It is known from experiment that the conduction band edge lies at the points L and is defined by a non-degenerate state at each of these points. Although

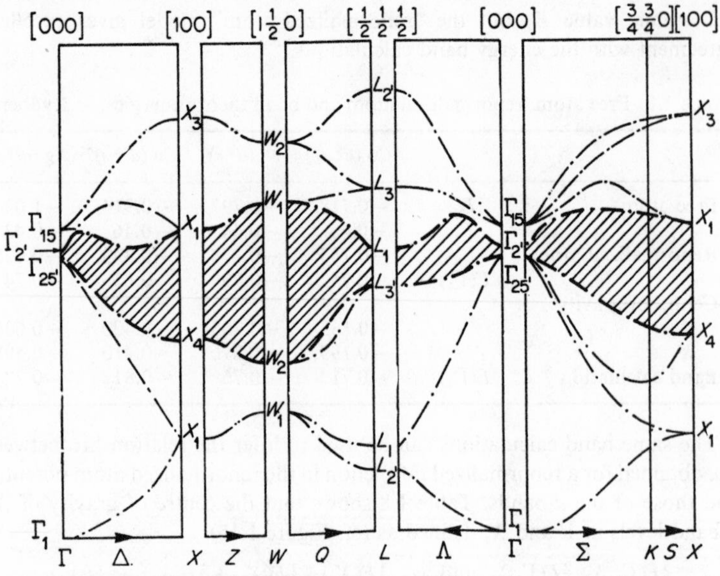

FIGURE 1.18. Schematic form of valence and low-lying conduction bands of Ge. Shaded region indicates separation between valence and conduction bands (after Herman, 1958).

there are eight points L, the pairs lying at the centres of opposite hexagonal faces are physically equivalent, being separated by a reciprocal lattice vector. Hence the conduction band edge of Ge is defined by four non-degenerate minima.

It is readily shown by perturbation theory that the surfaces of constant energy in the vicinity of each of the four minima are ellipsoids of revolution. This would apply whether the symmetry classification at the band edge were L_1, as seems to be the case, or $L_{2'}$. On the other hand, if the conduction band edge were defined by the doubly degenerate states L_3, the surfaces of constant energy near L would be warped or fluted surfaces.

Herman's calculations predicted that there should be three sets of minima in the lowest conduction band edge of Ge. In addition to the four [111] minima discussed above, which define the band edge, there are six along the [100] axes, i.e. the line Δ, and one at [000] labelled $\Gamma_{2'}$ in Figure 1.18. This has been confirmed experimentally.

Cyclotron resonance experiments (cf. Chapter 8) indicate that the valence band edge in Ge lies at [000]. If the spin–orbit interaction is neglected, the band edge is defined by the triply degenerate state $\Gamma_{25'}$. Each of the three valence bands which touch at [000] have fluted or warped constant energy surfaces near [000]. Second-order degenerate perturbation theory (Shockley, 1950) shows that $E(\mathbf{k})$ is given by solving a secular equation of the form

$$\begin{vmatrix} Lk_x^2+M(k_y^2+k_z^2)-E(\mathbf{k}) & Nk_xk_y & Nk_zk_x \\ Nk_xk_y & Lk_y^2+M(k_z^2+k_x^2)-E(\mathbf{k}) & Nk_yk_z \\ Nk_zk_x & & Lk_z^2+M(k_x^2+k_y^2)-E(\mathbf{k}) \end{vmatrix} = 0, \tag{1.11.132}$$

where the band edge has been chosen as the zero of energy.

$E(\mathbf{k})$ is always an analytic function of $\mathbf{k}$ at a non-degenerate band edge (see section 1.4.1). Hence $E(\mathbf{k})$ can always be expanded in a Taylor series in k_x, k_y and k_z in the vicinity of such an edge. But $E(\mathbf{k})$ is not analytic at a degenerate band edge and must be described by a secular equation like (1.11.132).

When the twofold spin degeneracy is taken into account, it is clear that $\Gamma_{25'}$ represents a sixfold degenerate state. But if we now introduce spin–orbit coupling, along lines laid down in 1.11.9(b), then this state splits into an upper fourfold and a lower doubly degenerate state. In all crystals having a centre of inversion, as in Ge, the double degeneracy due to spin is not removed by the spin–orbit interaction. Hence the valence band edge in Ge is defined by two doubly degenerate bands which touch at [000].

The energy surface for these two bands near [000] are described by the following non-analytic function

$$E(\mathbf{k}) = -Ak^2 \pm [Bk^4 + C(k_x^2k_y^2 + k_y^2k_z^2 + k_z^2k_x^2]^{\frac{1}{2}}, \tag{1.11.133}$$

where the zero of energy has again been taken at the band edge. The split-off band is described by

$$E(\mathbf{k}) = -\Delta - Ak^2, \tag{1.11.134}$$

where Δ is the spin–orbit splitting (0.29 eV in Ge; see Kane, 1956).

(d) *Empty lattice band structure for* Ge. The band structure of Figure 1.18, which neglects spin–orbit splitting, was based on Herman's OPW calculations. However, as Herman (1958) points out, it is very instructive to compare this band structure with that of a free-electron gas, whose electron density is equal to the average valence electron density in Ge.

When we are in the empty lattice, then, of course, the valence and conduction bands are described by

$$E(\mathbf{k}) = |\mathbf{k}+\mathbf{K}|^2, \tag{1.11.135}$$

where $\mathbf{K}$ is as usual a reciprocal lattice vector. When this is depicted in the BZ for Ge, the empty lattice band structure shown in Figure 1.19 is obtained. The Fermi level corresponding to four electrons per atom (eight per unit cell) is about 2.4. The spatial degeneracy of each band is also indicated.

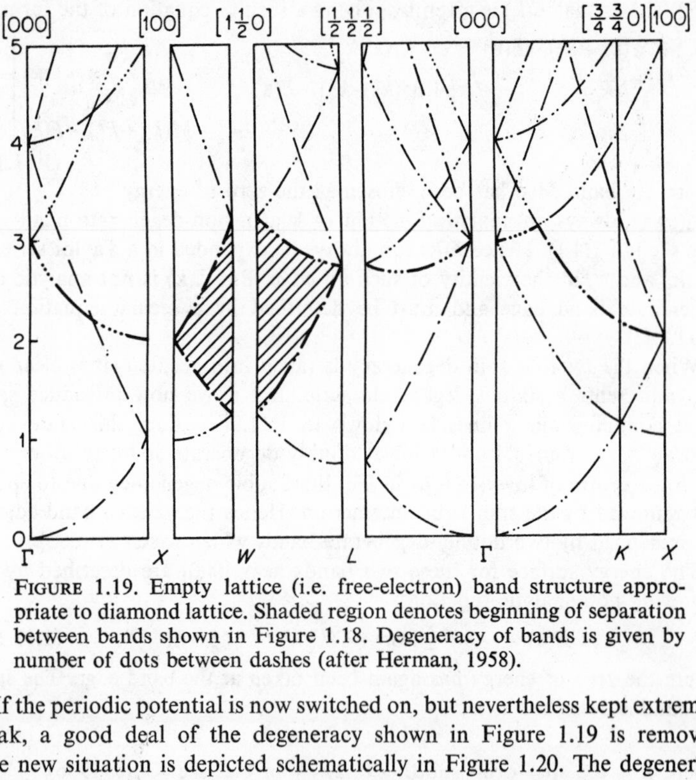

FIGURE 1.19. Empty lattice (i.e. free-electron) band structure appropriate to diamond lattice. Shaded region denotes beginning of separation between bands shown in Figure 1.18. Degeneracy of bands is given by number of dots between dashes (after Herman, 1958).

If the periodic potential is now switched on, but nevertheless kept extremely weak, a good deal of the degeneracy shown in Figure 1.19 is removed. The new situation is depicted schematically in Figure 1.20. The degeneracy has been removed, in drawing Figure 1.20, in such a way that the ordering of the realistic band structure of Figure 1.18 is regained: this is, of course, hindsight and in no sense a unique prediction from the "nearly empty" lattice theory.

When the realistic bands in Figure 1.18 are now compared with Figures 1.19 and 1.20, many of the features are readily understood. Although there is clearly significant distortion in passing from the empty lattice of Figure 1.20 to the band structure in Figure 1.18, a lot of the basic features persist.

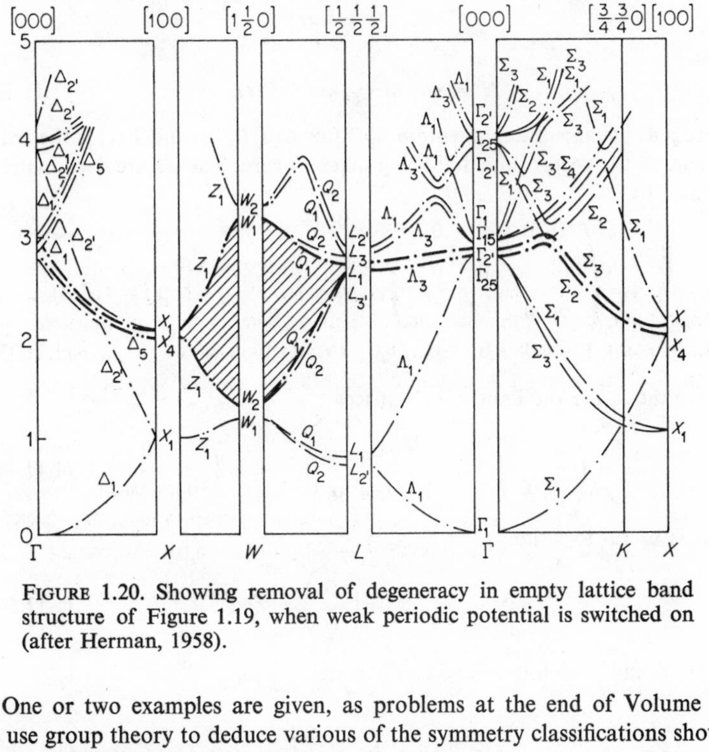

FIGURE 1.20. Showing removal of degeneracy in empty lattice band structure of Figure 1.19, when weak periodic potential is switched on (after Herman, 1958).

One or two examples are given, as problems at the end of Volume II, to use group theory to deduce various of the symmetry classifications shown in the nearly free electron scheme of Figure 1.20.

1.11.9 Relativistic band structure calculations

The Dirac equation must be used in crystals built from heavy atoms, instead of the non-relativistic Schrödinger equation, and then some interesting new features arise, which we shall summarize below.

In Hamiltonian form, the Dirac equation may be written as

$$i\hbar\frac{\partial\Psi}{\partial t} = H\Psi, \tag{1.11.136}$$

where Ψ denotes the four-component wave function

$$\Psi(\mathbf{r}) = \begin{pmatrix} \chi_1(\mathbf{r}) \\ \chi_2(\mathbf{r}) \\ \chi_3(\mathbf{r}) \\ \chi_4(\mathbf{r}) \end{pmatrix} \tag{1.11.137}$$

and

$$H = -c\boldsymbol{\gamma}.\mathbf{p} - \gamma_4 mc^2 + V(\mathbf{r}). \tag{1.11.138}$$

Here $\mathbf{p}$ as usual is the momentum operator $(\hbar/i)(\partial/\partial\mathbf{r})$ and $V(\mathbf{r})$ is the scalar potential, the vector potential being taken as zero. The γ's are 4×4 matrices defined by

$$\gamma_4 = \begin{pmatrix} 1 & 0 & 0 & 0 \\ 0 & 1 & 0 & 0 \\ 0 & 0 & -1 & 0 \\ 0 & 0 & 0 & -1 \end{pmatrix}, \quad \gamma_j = \begin{pmatrix} 0 & \sigma_j \\ \sigma_j & 0 \end{pmatrix} \quad (j = 1, 2, 3), \tag{1.11.139}$$

where the σ_j are the Pauli spin matrices

$$\sigma_1 = \begin{pmatrix} 0 & 1 \\ 1 & 0 \end{pmatrix}, \quad \sigma_2 = \begin{pmatrix} 0 & -i \\ i & 0 \end{pmatrix}, \quad \sigma_3 = \begin{pmatrix} 1 & 0 \\ 0 & -1 \end{pmatrix}. \tag{1.11.140}$$

It is useful to write

$$\Psi = \begin{pmatrix} \phi \\ \psi \end{pmatrix}, \tag{1.11.141}$$

where ϕ and ψ each have two components:

$$\phi = \begin{pmatrix} \chi_1 \\ \chi_2 \end{pmatrix}, \quad \psi = \begin{pmatrix} \chi_3 \\ \chi_4 \end{pmatrix}. \tag{1.11.142}$$

Then if we write (see Schiff (1949), p. 332)

$$i\hbar\frac{\partial\Psi}{\partial t} = [E + mc^2]\Psi, \tag{1.11.143}$$

we find

$$[E + 2mc^2 - V]\phi + c\boldsymbol{\sigma}.\mathbf{p}\psi = 0 \tag{1.11.144}$$

and

$$[E - V]\psi + c\boldsymbol{\sigma}.\mathbf{p}\phi = 0. \tag{1.11.145}$$

(a) *Low-energy limit*. From equation (1.11.144) we see that, at low energies, $\phi \ll \psi$. In fact $\phi \sim (v/c)\psi$ where v is the velocity and the electron wave function is almost entirely described by ψ. We therefore eliminate ϕ in favour of ψ to obtain

$$E\psi = \frac{1}{2m}(\boldsymbol{\sigma}.\mathbf{p})\left[1+\frac{E-V}{2mc^2}\right]^{-1}(\boldsymbol{\sigma}.\mathbf{p})\psi + V\psi. \tag{1.11.146}$$

We next use the relations

$$\left(1+\frac{E-V}{2mc^2}\right)^{-1} \simeq 1-\frac{E-V}{2mc^2}, \tag{1.11.147}$$

$$\mathbf{p}V = V\mathbf{p} - i\hbar\nabla V \tag{1.11.148}$$

and

$$(\boldsymbol{\sigma}.\nabla V)(\boldsymbol{\sigma}.\mathbf{p}) = \nabla V.\mathbf{p} + i\boldsymbol{\sigma}.(\nabla V\times\mathbf{p}), \tag{1.11.149}$$

in order to rewrite equation (1.11.146) as

$$E\psi = \left(1-\frac{E-V}{2mc^2}\right)\frac{p^2\psi}{2m} + V(\mathbf{r})\psi - \frac{\hbar^2}{4m^2c^2}\nabla V.\nabla\psi + \frac{\hbar}{4m^2c^2}\boldsymbol{\sigma}.(\nabla V\times\mathbf{p}\psi). \tag{1.11.150}$$

To eliminate E from the right-hand side, we note that the last two terms are of order c^{-2}. The fine structure constant α equals c^{-1} in a.u.:

$$\alpha = \frac{e^2}{\hbar c} = \frac{1}{137}. \tag{1.11.151}$$

Hence, to order α^2, we may write $E-V = p^2/2m$ to finally obtain

$$E\psi = \frac{p^2}{2m}\psi - \frac{p^4\psi}{8m^3c^2} + V(\mathbf{r})\psi - \frac{\hbar^2}{4m^2c^2}\nabla V.\nabla\psi + \frac{\hbar}{4m^2c^2}\boldsymbol{\sigma}.(\nabla V\times\mathbf{p}\psi). \tag{1.11.152}$$

(b) *Time-reversal symmetry and spin–orbit coupling*. The final term on the right-hand side is the spin–orbit coupling term. If we assume for a moment that $V(\mathbf{r})$ is spherically symmetric, we have

$$\nabla V(\mathbf{r}) = \frac{1}{r}\frac{dV(r)}{dr}\mathbf{r} \tag{1.11.153}$$

and thus, denoting the orbital angular momentum by $\mathbf{l}$,

$$\frac{1}{r}\frac{dV}{dr}(\mathbf{r}\times\mathbf{p}).\boldsymbol{\sigma} = \frac{1}{r}\frac{dV}{dr}\mathbf{l}.\boldsymbol{\sigma}. \tag{1.11.154}$$

The spin–orbit coupling is then seen to have no effect on s-states.

All the relativistic corrections are most important in the heavier atoms and near the nucleus. The corrections from the second and fourth terms of equation (1.11.152) are often omitted, leaving the simpler Hamiltonian

$$H = \frac{p^2}{2m} + V(\mathbf{r}) + \frac{\hbar}{4m^2c^2}\boldsymbol{\sigma}.(\nabla V \times \mathbf{p}). \tag{1.11.155}$$

Then, only the spin–orbit coupling term modifies the original non-relativistic Hamiltonian. In particular, the classification of energy levels by symmetry is altered, since the spin-orbit term involves the 2×2 Pauli matrices. (An example of this is given in Figure 1.21 below.) These must be taken into account in forming the irreducible matrix representations of the symmetry

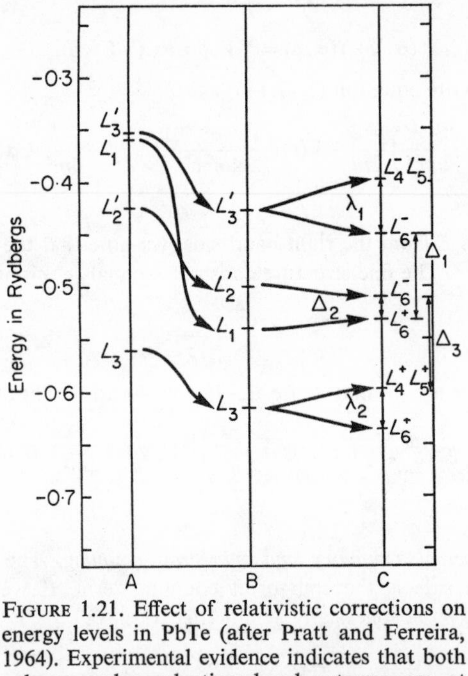

FIGURE 1.21. Effect of relativistic corrections on energy levels in PbTe (after Pratt and Ferreira, 1964). Experimental evidence indicates that both valence and conduction band extrema are at point L on BZ face. Line A. Levels in absence of relativistic effects. Line B. Modifications of A introduced by mass-velocity and $\boldsymbol{\epsilon}.\mathbf{p}$ terms. Line C. Modifications of B due to spin–orbit correction. Final band gap is between upper level and lower level. Value found theoretically is -0.31 Rydberg, in good agreement with experiment.

group of the crystal. The way this is done is set out fully in Appendix GAII. One conclusion from the discussion given there which is worth noting here is that whereas in the absence of spin–orbit coupling there is double spin-degeneracy at any point **k** in the BZ, the eigenfunctions being

$$\begin{pmatrix} u_{\mathbf{k}}(\mathbf{r}) \\ 0 \end{pmatrix} e^{i\mathbf{k}.\mathbf{r}}, \quad \begin{pmatrix} 0 \\ u_{\mathbf{k}}(\mathbf{r}) \end{pmatrix} e^{i\mathbf{k}.\mathbf{r}}, \tag{1.11.156}$$

this degeneracy persists in the presence of spin–orbit coupling only if the crystal has an inversion centre.

Nonetheless, the time-reversal degeneracy $E(\mathbf{k}) = E(-\mathbf{k})$, discussed in section 1.5.6, persists for any crystal. We first remark that if

$$H\Psi = E\Psi, \tag{1.11.157}$$

$$H^*\Psi^* = E\Psi^*, \tag{1.11.158}$$

and therefore, provided the Hamiltonian is real, $\Psi^* = e^{-i\mathbf{k}.\mathbf{r}}\, U_k^*$ is an eigenfunction of H provided $\Psi = e^{i\mathbf{k}.\mathbf{r}} U_{\mathbf{k}}$ is. Now the Hamiltonian of equation (1.11.155) is *not* real. However, it is easy to prove that

$$H\sigma_2 = \sigma_2 H^* \tag{1.11.159}$$

and hence

$$H\sigma_2\Psi^* = \sigma_2 H^*\Psi^* = E\sigma_2\Psi^*. \tag{1.11.160}$$

Further, if

$$\Psi_{\mathbf{k}} = \begin{pmatrix} u_{\mathbf{k}}^1(\mathbf{r}) \\ u_{\mathbf{k}}^2(\mathbf{r}) \end{pmatrix} e^{i\mathbf{k}.\mathbf{r}} \tag{1.11.161}$$

we have

$$\sigma_2\Psi_{\mathbf{k}}^* = i\begin{pmatrix} -u_{\mathbf{k}}^{2*}(\mathbf{r}) \\ u_{\mathbf{k}}^{1*}(\mathbf{r}) \end{pmatrix} e^{-i\mathbf{k}.\mathbf{r}}. \tag{1.11.162}$$

Hence the presence of spin–orbit coupling does not alter the degeneracy

$$E(\mathbf{k}) = E(-\mathbf{k}). \tag{1.11.163}$$

The correction to the energy due to spin–orbit coupling is often found by first-order perturbation theory.

We may note that $\mathbf{k}.\mathbf{p}$ perturbation theory given in section 1.11.7 is formally unchanged by the spin–orbit term. Instead of $H_1 = \mathbf{k}.\mathbf{p}$, we now have

$$H_1 = \mathbf{k}.\boldsymbol{\pi}, \tag{1.11.164}$$

where, in atomic units,

$$\boldsymbol{\pi} = \mathbf{p}+\frac{\alpha^2}{4}(\boldsymbol{\sigma}\times\nabla V). \qquad (1.11.165)$$

The procedure of section 1.11.7 may now be followed, with $\boldsymbol{\pi}$ replacing $\mathbf{p}$, and the effective masses thereby deduced.

(c) *Mass–velocity and* $\mathscr{E}.\mathbf{p}$ *terms.* From the derivation of equation (1.11.152), it can be seen that the second term on the right-hand side represents the relativistic mass–velocity correction. The fourth term has no classical analogue; we shall refer to it as the $\mathscr{E}.\mathbf{p}$ term, $\mathscr{E}$ being the effective field seen by an electron. While these two terms introduce no further changes in the scheme by which states are classified according to symmetry, they can cause important corrections to the band structure, as was shown by Johnson, Conklin and Pratt (1963) for PbTe (see Figure 1.21). These workers used the APW method of section 1.11.6(c) to obtain the energy levels in the absence of relativistic and spin–orbit effects. These effects were then incorporated by solving the secular equation obtained by taking matrix elements of the relevant terms between matrix elements of the APW functions of the various bands.

It is also possible to include relativistic effects in the $\mathbf{k}.\mathbf{p}$ method, to obtain effective masses. For this purpose, it is convenient to use a Hamiltonian of the form

$$H = -\nabla^2-\frac{\alpha^2}{4}\nabla^4+V(\mathbf{r})+\frac{\alpha^2}{8}\nabla^2 V-\frac{i\alpha^2}{4}\boldsymbol{\sigma}.(\nabla V\times\nabla) \qquad (1.11.166)$$

if the energy is measured in Rydbergs. This Hamiltonian differs from that of equation (1.11.152) only in the Darwin correction term $\frac{1}{8}\alpha^2\nabla^2 V$. That this and the $\mathscr{E}.\mathbf{p}$ term are equivalent can be seen by taking matrix elements between eigenfunctions of H, for we have that, with boundary surface s,

$$\int\nabla\{\psi_1^*\psi_2\nabla V\}\,d\mathbf{r} = \int_s \psi_1^*\psi_2\nabla V.d\mathbf{s}. \qquad (1.11.167)$$

Taking boundary conditions normally invoked to prove the Hermiticity of $\hbar\nabla/i$, the surface integral vanishes and the equivalence follows.

Adopting the procedure of section 1.11.7, we find from equation (1.11.167) that the term involving $\mathbf{k}$ in the Hamiltonian (cf. equation 1.11.115) for $u_{\mathbf{k}}(\mathbf{r}) = \psi_{\mathbf{k}}e^{-i\mathbf{k}.\mathbf{r}}$ is

$$H_1 = -2i\nabla.\mathbf{k}+\frac{\alpha^2}{4}(\boldsymbol{\sigma}\times\nabla V).\mathbf{k}-i\alpha^2(\mathbf{k}.\nabla^2)\nabla+\alpha^2(\mathbf{k}.\nabla)^2+\frac{\alpha^2}{2}k^2\nabla^2+O(k^3). \qquad (1.11.168)$$

We can write this in the form

$$H_1 = \mathbf{k}.\boldsymbol{\pi} \tag{1.11.169}$$

and then we can again modify the $\mathbf{k}.\mathbf{p}$ method in the way we have already described when only spin–orbit coupling is included.

(d) *Relativistic APW method.* Although the Dirac equation for electrons in crystals is usually solved in the approximate form of equation (1.11.152) or equivalently (1.11.168) the influence of the smaller component ϕ of Ψ can be appreciable near the nucleus. The full formulation of a relativistic OPW theory has been given by Soven (1965) and for the APW method by Loucks (1965). The basic procedure in each case remains the same as set out in sections 1.11.2 and 1.11.4, but to illustrate the modifications involved in a relativistic treatment, we shall briefly outline the relativistic APW method as formulated by Loucks.

Outside the sphere r_i, the potential V is zero and from equation (1.11.144)

$$\phi^0 = -\frac{c\boldsymbol{\sigma}.\mathbf{p}\psi^0}{E+2mc^2}, \tag{1.11.170}$$

where, apart from a normalizing factor,

$$\psi^0 = \chi(m)\, e^{i\mathbf{k}_n.\mathbf{r}} \quad (\mathbf{k}_n = \mathbf{k}+\mathbf{K}_n), \tag{1.11.171}$$

the superscripts indicating the outer field-free region. Here $\chi(m)$ is the Pauli spinor

$$\chi(+\tfrac{1}{2}) = \begin{pmatrix} 1 \\ 0 \end{pmatrix}, \quad \chi(-\tfrac{1}{2}) = \begin{pmatrix} 0 \\ 1 \end{pmatrix}. \tag{1.11.172}$$

Since the *total* energy E_{T} is such that

$$(E+mc^2)^2 = E_{\mathrm{T}}^2 = c^2p^2+m^2c^4, \tag{1.11.173}$$

we have, in units such that $m = c = \hbar = 1$,

$$(E+mc^2)^2 = k_n^2+1 \tag{1.11.174}$$

and hence

$$\phi^0 = -\frac{\boldsymbol{\sigma}.\mathbf{p}}{k_n^*+1}\psi^0, \tag{1.11.175}$$

where

$$k_n^* = (k_n^2+1)^{\frac{1}{2}}. \tag{1.11.176}$$

We therefore write the total wave function‡ as

$$\Psi'^{0}_{nm} = \left(\frac{k^*_n+1}{2k^*_n}\right)^{\frac{1}{2}} \begin{pmatrix} \chi(m) \\ -\dfrac{\boldsymbol{\sigma}.\mathbf{k}_n}{k^*_n+1}\chi(m) \end{pmatrix} e^{i\mathbf{k}_n.\mathbf{r}} \quad (m=\pm\tfrac{1}{2}). \quad (1.11.177)$$

Here n labels k_n as shown in equation (1.11.171), $m = \pm\frac{1}{2}$ denotes the spin state and the factor $[(k^*_n+1)/2k_n]^{\frac{1}{2}}$ normalizes the function such that there is one particle per unit volume.

Inside the sphere, we have a linear combination of central field orbitals

$$\Psi'^{i}_{nm} = \sum_{\kappa\mu} A^{nm}_{\kappa\mu}\Psi'^{\mu}_{\kappa}, \quad (1.11.178)$$

where the $A^{nm}_{\kappa\mu}$ are to be chosen to match Ψ'^{i}_{nm} on to Ψ'^{0}_{nm} at the surface of the sphere, and the Ψ'^{μ}_{κ} are solutions of the Dirac equation, in the form

$$\Psi'^{\mu}_{\kappa} = \begin{pmatrix} g_\kappa(\mathbf{r})\,\chi^{\mu}_{\kappa} \\ if_\kappa(\mathbf{r})\,\chi^{\mu}_{-\kappa} \end{pmatrix}. \quad (1.11.179)$$

It may be verified that, for this to be so, f and g must satisfy the coupled differential equations

$$\left.\begin{aligned} \frac{df}{dr} &= \left(\frac{\kappa-1}{r}\right)f-(E-V)g, \\ \frac{dg}{dr} &= (E-V+2)f-\left(\frac{\kappa+1}{r}\right)g, \end{aligned}\right\} \quad (1.11.180)$$

while the spin functions χ^{μ}_{κ} have the properties

$$\left.\begin{aligned} (\boldsymbol{\sigma}.\boldsymbol{l}+1)\,\chi^{\mu}_{\kappa} &= -\kappa\chi^{\mu}_{\kappa}, \\ \boldsymbol{\sigma}.\hat{\mathbf{r}}\chi^{\mu}_{\kappa} &= -\chi^{\mu}_{-\kappa}. \end{aligned}\right\} \quad (1.11.181)$$

Explicitly, the spin functions are

$$\begin{aligned} \chi^{m}_{\kappa} = {} & c(l\tfrac{1}{2}j,\, \mu-\tfrac{1}{2},\, \tfrac{1}{2})\, Y^{\mu-\frac{1}{2}}_{l}(\hat{\mathbf{r}})\,\chi(\tfrac{1}{2}) \\ & + c(l\tfrac{1}{2}j,\, \mu+\tfrac{1}{2},\, -\tfrac{1}{2})\, Y^{\mu+\frac{1}{2}}_{l}(\hat{\mathbf{r}})\,\chi(-\tfrac{1}{2}), \end{aligned} \quad (1.11.182)$$

where the coefficients c are defined generally by, for example, Rose (1961). In our case they take the explicit form

$$\begin{aligned} c(l\tfrac{1}{2}j,\, m-m_2,\, m_2) &= \pm\left(\frac{l\pm m+\frac{1}{2}}{2l+1}\right)^{\frac{1}{2}} \quad \text{if } j=l\pm\tfrac{1}{2}, m_2=\tfrac{1}{2}, \\ &= \left(\frac{l\mp m+\frac{1}{2}}{2l+1}\right)^{\frac{1}{2}} \quad \text{if } j=l\pm\tfrac{1}{2},\ m_2=-\tfrac{1}{2}. \end{aligned} \quad (1.11.183)$$

‡ Since the work of Loucks (1965) is widely used in practical calculations, we interchange the components of the spinor in what follows, in accord with his discussion.

As defined by equation (1.11.182), the χ_κ^m are orthonormal in the sense

$$\int \chi_\kappa^{\mu\dagger} \chi_{\kappa'}^{\mu'} \sin\theta \, d\theta \, d\phi = \delta_{\mu\mu'} \delta_{\kappa\kappa'}. \tag{1.11.184}$$

In order to effect the matching of ψ^i and ψ^0, we also expand the latter quantity in the spin functions, with the result [cf. Rose (1961)]

$$\psi_{nm}^0 = \sum_{\kappa\mu} a_{\kappa\mu}^{nm} \begin{pmatrix} j_l(k_n r)\chi_\kappa^\mu \\ \dfrac{ik_n S_\kappa}{k_n^*+1} j_{l'}(k_n r)\chi_{-\kappa}^\mu \end{pmatrix}, \tag{1.11.185}$$

where

$$a_{\kappa\mu}^{nm} = 4\pi i^l \left(\frac{k_n^*+1}{2k_n^*}\right)^{\frac{1}{2}} c(l\tfrac{1}{2}j, \mu-m, m)\, Y_l^{(\mu-m)\dagger}(\hat{\mathbf{k}}_n). \tag{1.11.186}$$

In equation (1.11.185)

$$S_\kappa = \text{sign of } \kappa \tag{1.11.187}$$

and κ runs over all positive and negative integers but not zero. j, l and l' are found according to the rules

$$l = \kappa, \quad j = l - \tfrac{1}{2}, \quad l' = \kappa - 1 \quad (\kappa > 0), \tag{1.11.188}$$

$$l = -\kappa - 1, \quad j = l + \tfrac{1}{2}, \quad l' = -\kappa \quad (\kappa < 0),$$

and, for fixed κ, μ runs from $-j$ to j.

It is now evident that the choice

$$A_{\kappa\mu}^{nm} = a_{\kappa\mu}^{nm} j_l(k_n r_i)/g_n(r_i) \tag{1.11.189}$$

ensures the continuity of ψ, the upper component of Ψ as shown in equation (1.11.141) at r_i. We cannot simultaneously make ϕ continuous, but we have already seen that ϕ is of importance only near the nucleus, and that in the outer regions of the cell $\psi \gg \phi$.

Thus, we have now set up a single relativistic APW. The procedure is to expand a general wave function Ψ in terms of these, and insert it in the variational expression for the energy

$$E + mc^2 = \int_\Omega \Psi^\dagger H \Psi \, d\mathbf{r} - \frac{i}{2}\int_s (\Psi^o + \Psi^i)^\dagger \boldsymbol{\gamma}.\hat{\mathbf{r}}(\Psi^o - \Psi^i)\, ds, \tag{1.11.190}$$

where the surface integration is over the sphere of radius r_i. This equation is derived in Appendix 1.8, where we also show that the matrix elements for the secular equation for E, in Rydberg units, are

$$M\begin{pmatrix} N & M \\ n & m \end{pmatrix} = \left(\frac{k_N^2+k_n^2}{2}-E\right)\delta_{mN}\Omega_{nN}+4\pi r_i^2\sum_{\kappa} D_{\kappa}\begin{pmatrix} N & M \\ n & m \end{pmatrix}$$
$$\times\left\{j_l(k_n r_i)j_l(k_N r_i)\left(\frac{cf_{\kappa}(r_i E)}{g_{\kappa}(r_i E)}\right)\right.$$
$$\left.-\tfrac{1}{2}S_{\kappa}[k_N j_l(k_n r_i)j_{l'}(k_N r_i)+k_n j_l(k_N r_i)j_{l'}(k_n r_i)]\right\}, \quad (1.11.191)$$

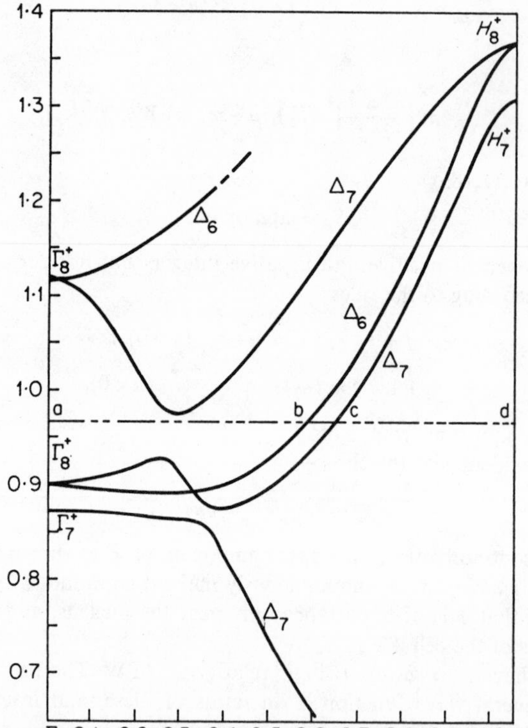

FIGURE 1.22(a)

FIGURE 1.22(a) shows results of a full relativistic APW calculation using the Dirac equation for the energy bands of body-centred-cubic tungsten. Bands are along symmetry direction Δ, which joins points Γ and H shown in BZ for body-centred cubic lattice, Figure 1.22(b) below (after Loucks, 1965).

where D_κ is defined in equation (A1.9.64). Here

$$\Omega_{nN} = \Omega\delta_{nN} - 4\pi r_i^2 \frac{j_1(|\mathbf{k}_N - \mathbf{k}_n| r_i)}{|\mathbf{k}_N - \mathbf{k}_n|}. \tag{1.11.192}$$

Equation (1.11.191) should be compared with equation (1.11.54) of the non-relativistic APW theory. It will be noted then that the sum over l is replaced by the sum over κ, which includes both spin orientations.

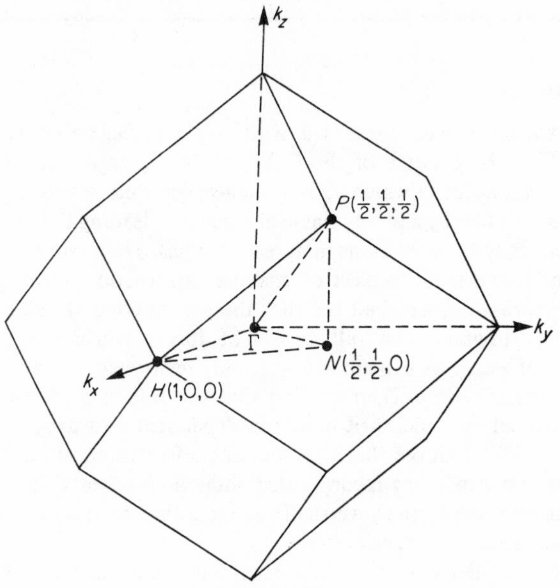

FIGURE 1.22(b)

An example of the results of an application of the above procedure is shown in Figure 1.22, which gives the relativistic energy bands for tungsten, along with the BZ for the body-centred cubic lattice. Only the symmetry axis ΓH is shown. A check of the separation bc marked on Figure 1.22 is provided by comparison with the size-effect measurements of Walsh and Grimes (1964). There is good agreement between theory and experiment but we must refer the reader to the original papers for further details.

Finally, we recommend the reader to consult the article by Ziman (1971) on the calculation of Bloch functions, in *Solid State Physics*, Volume 26, and also Volume 24 of this same series on pseudopotentials, which are discussed by Heine and his colleagues.

CHAPTER 2

Collective Effects in Solids

2.1 Introduction

So far we have been concerned with single particle theory. However, because of the long range of the Coulomb force, organized or collective oscillations can occur in which all the valence electrons in a solid participate. We shall see this first by a purely classical argument, leading to the well-known plasma frequency of oscillations in an electron gas. We then turn to examine in detail the response of the electron gas to a perturbation. Dealing first with a static perturbation, we shall see that the gas responds to screen out the disturbance. However, the full power of this perturbation theory, the significance of which we explain below, is only seen when we consider a time-dependent perturbation. Then we find that the response of the Fermi gas can be described by a dielectric constant, dependent on wave-number q and frequency ω. We shall see that this contains information about the plasma oscillations and hence has incorporated some account of collective effects. To demonstrate this further, we shall calculate the correlation energy of an electron gas in the high density limit.

An alternative derivation of these results is presented using a simplified Hamiltonian due to Sawada, for which some exact results can be obtained.

The next part of the chapter is concerned with techniques we need in the discussion of the many-body problem.

We then turn to a discussion of the phenomenological theory of Fermi liquids due to Landau. Here we see how an approximate solution of the many-body problem can be found in terms of a finite (and small) number of parameters. This also affords an opportunity to discuss the quasi-particle picture followed by a microscopic foundation to the Landau theory. Finally we combine the perturbation theory with the ideas of Landau theory to make a calculation of the specific heat of an interacting electron gas.

The theory of the inhomogeneous electron gas, which is, of course, the basic problem we meet with in any crystal, is then considered. This ties in with the discussion of one-body potential theory and indeed has as one of its prime objectives the calculation of such a potential from first principles.

The chapter concludes with a discussion of the dielectric function of a semi-conductor and quasi-particles in insulators.

2.2 Classical plasma frequency

We shall begin the discussion of collective motions with a purely classical and elementary treatment of plasma oscillations. Let the uniform background density of positive charge be $e\rho_0$, and $\rho(\mathbf{r}t)$ be the density of electrons at position $\mathbf{r}$ at time t. The excess positive charge is given by $e(\rho_0-\rho)$ and hence from Maxwell's equations

$$\operatorname{div}\mathscr{E} = 4\pi e(\rho_0-\rho) \tag{2.2.1}$$

where $\mathscr{E}$ is the electric field.

Now we displace the electron gas by $\mathbf{x}$ to give a current density $\rho\dot{\mathbf{x}}$, and we have the equation of continuity

$$\operatorname{div}(\rho\dot{\mathbf{x}}) = -\frac{\partial\rho}{\partial t}. \tag{2.2.2}$$

If we assume the displacement $\mathbf{x}$ to be small, then the plasma oscillations are small in amplitude and we may write (2.2.2) as

$$\rho_0\operatorname{div}\dot{\mathbf{x}} = -\frac{\partial\rho}{\partial t} \tag{2.2.3}$$

which can be integrated to give

$$\rho_0-\rho = \rho_0\operatorname{div}\mathbf{x} \tag{2.2.4}$$

since $\rho = \rho_0$, when $\mathbf{x} = 0$. Thus, we have the result

$$\operatorname{div}\mathscr{E} = 4\pi e\rho\operatorname{div}\mathbf{x} \tag{2.2.5}$$

when we substitute (2.2.4) into (2.2.1), and hence

$$\mathscr{E} = 4\pi e\rho_0\mathbf{x} \tag{2.2.6}$$

which satisfies the correct boundary condition that $\mathscr{E} = 0$ when $\mathbf{x} = 0$. Combining this with the Newtonian equation of motion for an electron in an electric field $\mathscr{E}$, namely

$$m\ddot{\mathbf{x}} = -e\mathscr{E} \tag{2.2.7}$$

we have

$$m\ddot{\mathbf{x}}+4\pi e^2\rho_0\mathbf{x} = 0. \tag{2.2.8}$$

This immediately shows that oscillations in the electron gas can occur, with angular frequency ω_p given by

$$\omega_p = (4\pi\rho_0 e^2/m)^{\frac{1}{2}}. \tag{2.2.9}$$

The plasma frequency ω_p for electrons in a metal is readily shown from (2.2.9) to be of the order of 10^{16} sec^{-1}.

While equation (2.2.9) is a basic result in treating collective motions of an electron gas, we shall need to go further and discuss the form of the dispersion relation of the plasma oscillations. We shall return to this problem, which is conveniently approached via the frequency and wave-number-dependent dielectric function [see equation (2.3.60) below], in section 2.6.3, but it will be helpful, prior to that, to ignore the frequency dependence to get a simple picture of the way the electron gas responds to a test charge.

2.3 k and ω-dependent dielectric constant of Fermi gas

2.3.1 Static dielectric constant $\varepsilon(k)$

We consider a static charge Z embedded in a Fermi gas, and we shall give a direct and elementary derivation of the screened field around it from the Schrödinger equation, even though the result can, in fact, be obtained from results in Chapter 1 (see Appendix A1.2). Thus if the self-consistent potential energy in which all electrons move is $V(\mathbf{r})$ as above, then the original plane waves $e^{i\mathbf{k}.\mathbf{r}}$ of the conduction electrons are distorted into wave functions $\psi_\mathbf{k}(\mathbf{r})$, $\mathbf{k}$ labelling the unperturbed state from which $\psi_\mathbf{k}$ derives when the ion is introduced. We must then solve

$$\nabla^2\psi_\mathbf{k}+\frac{2m}{\hbar^2}[E_\mathbf{k}-V(\mathbf{r})]\psi_\mathbf{k}=0. \tag{2.3.1}$$

It is convenient to rewrite this, with $E_\mathbf{k}=\hbar^2k^2/2m$, as

$$\nabla^2\psi_\mathbf{k}+k^2\psi_\mathbf{k}=\frac{2m}{\hbar^2}V(\mathbf{r})\psi_\mathbf{k}. \tag{2.3.2}$$

By analogy with equations (1.11.23) and (1.11.24), we now seek a formal solution using the Green function obeying the following equation

$$\nabla^2G(\mathbf{r}\mathbf{r}')+k^2G=-4\pi\delta(\mathbf{r}-\mathbf{r}'). \tag{2.3.3}$$

For $\mathbf{k}=0$, an obvious solution is $1/|\mathbf{r}-\mathbf{r}'|$. To effect the desired generalization of this we notice that asymptotically we require the wave function to take the form

$$\psi_\mathbf{k}(\mathbf{r})\sim e^{i\mathbf{k}.\mathbf{r}}+f(\theta)\frac{e^{ikr}}{r}, \tag{2.3.4}$$

the first term being the incident plane wave, θ the angle of scattering, and e^{ikr}/r representing the outgoing spherical wave. It then becomes clear that the desired Green function is

$$G(\mathbf{r}\mathbf{r}') = \frac{e^{ik|\mathbf{r}-\mathbf{r}'|}}{|\mathbf{r}-\mathbf{r}'|}, \tag{2.3.5}$$

which obviously reduces to $1/|\mathbf{r}-\mathbf{r}'|$ when $\mathbf{k} = 0$. Then we have, for the full solution,

$$\psi_{\mathbf{k}}(\mathbf{r}) = \mathscr{V}^{-\frac{1}{2}} e^{i\mathbf{k}.\mathbf{r}} - \frac{m}{2\pi\hbar^2}\int d\mathbf{r}' G(\mathbf{r}\mathbf{r}') V(\mathbf{r}') \psi_{\mathbf{k}}(\mathbf{r}'), \tag{2.3.6}$$

where $\mathscr{V}$ is the volume of the crystal.

Now we approximate in (2.3.6) by noting that in the "correction" term proportional to V, we can write $\psi_{\mathbf{k}}(\mathbf{r}) = \mathscr{V}^{-\frac{1}{2}} e^{i\mathbf{k}.\mathbf{r}}$ and work simply to $O(V)$. Then we find

$$\psi_{\mathbf{k}}(\mathbf{r}) = \mathscr{V}^{-\frac{1}{2}}\left[e^{i\mathbf{k}.\mathbf{r}} - \frac{m}{2\pi\hbar^2}\int d\mathbf{r}' G(\mathbf{r}\mathbf{r}') V(\mathbf{r}') e^{i\mathbf{k}.\mathbf{r}'}\right]. \tag{2.3.7}$$

To form the electron density $\rho(\mathbf{r})$, we must sum $\psi_{\mathbf{k}}^*(\mathbf{r})\psi_{\mathbf{k}}(\mathbf{r})$ over all $\mathbf{k}$ out to the Fermi surface. Remembering that to be consistent, we must only retain terms of $O(V)$, we find

$$\sum_{|\mathbf{k}|<k_f} \psi_{\mathbf{k}}^*(\mathbf{r})\psi_{\mathbf{k}}(\mathbf{r}) = \sum_{|\mathbf{k}|<k_f} \mathscr{V}^{-1} - \sum_{|\mathbf{k}|<k_f} \mathscr{V}^{-1} \frac{m}{2\pi\hbar^2}\int d\mathbf{r}' V(\mathbf{r}') \times [G(\mathbf{r}\mathbf{r}') e^{i\mathbf{k}.\mathbf{r}'-\mathbf{r}} + G^*(\mathbf{r}\mathbf{r}') e^{-i\mathbf{k}.\mathbf{r}'-\mathbf{r}}]. \tag{2.3.8}$$

But the summation over $\mathbf{k}$ can be replaced by an integration, and we find, remembering that there are $\mathscr{V}/(2\pi)^3$ states per unit volume of $\mathbf{k}$ space and two spin directions

$$\rho = \rho_0 - [2m/(2\pi)^4\hbar^2]\int d\mathbf{r}' V(\mathbf{r}') \times \int_{|\mathbf{k}|<k_f} d\mathbf{k}[G(\mathbf{r}\mathbf{r}') e^{i\mathbf{k}.(\mathbf{r}'-\mathbf{r})} + G^*(\mathbf{r}\mathbf{r}') e^{-i\mathbf{k}(\mathbf{r}'.\mathbf{r})}]. \tag{2.3.9}$$

Integrating over the angles of $\mathbf{k}$ simply replaces $e^{i\mathbf{k}.(\mathbf{r}'-\mathbf{r})}$ by

$$\frac{\sin k|\mathbf{r}-\mathbf{r}'|}{k|\mathbf{r}-\mathbf{r}'|},$$

the s term in the expansion of a plane wave into spherical waves (cf. equation 1.11.40). We then obtain, combining G^* and G from equation (2.3.5)

$$\rho(\mathbf{r}) = \rho_0 - \frac{2m}{(2\pi)^4\hbar^2}\int d\mathbf{r}' V(\mathbf{r}')\int_0^{k_f} dk\, 4\pi k^2 \left[\frac{\sin k|\mathbf{r}-\mathbf{r}'|}{k|\mathbf{r}-\mathbf{r}'|}.\frac{2\cos k|\mathbf{r}-\mathbf{r}'|}{|\mathbf{r}-\mathbf{r}'|}\right]. \tag{2.3.10}$$

Performing the integration over k, the final result for the displaced charge $\rho(\mathbf{r})-\rho_0$ may be written as

$$\rho(\mathbf{r})-\rho_0 = -\frac{mk_f^2}{2\pi^3\hbar^2}\int d\mathbf{r}'\, V(\mathbf{r}')\frac{j_1(2k_f|\mathbf{r}-\mathbf{r}'|)}{|\mathbf{r}-\mathbf{r}'|^2}, \tag{2.3.11}$$

where $j_1(x) = x^{-2}[\sin x - x\cos x]$, the first-order spherical Bessel function. If we can assume that $V(\mathbf{r}')$ varies sufficiently slowly in space so that we can replace it approximately by $V(\mathbf{r})$ then we find

$$\rho(\mathbf{r})-\rho_0 = -\frac{mk_f^2}{2\pi^3\hbar^2}V(\mathbf{r})\int d\mathbf{r}'\frac{j_1(2k_f|\mathbf{r}-\mathbf{r}'|)}{|\mathbf{r}-\mathbf{r}'|^2} \tag{2.3.12}$$

and performing the integration over $\mathbf{r}'$ we obtain

$$\rho(\mathbf{r})-\rho_0 = -\frac{q^2}{4\pi}V(\mathbf{r}): \quad q^2 = \frac{4k_f}{\pi a_0}, \tag{2.3.13}$$

where $a_0 = \hbar^2/me^2$ is the first Bohr radius, which is just the Thomas–Fermi approximation in a linear framework. Using equation (2.3.11) we avoid the semi-classical approximation, and obtain from the Poisson equation

$$\nabla^2 V = \frac{me^2}{\hbar^2}\cdot\frac{2k_f^2}{\pi^2}\int d\mathbf{r}'\, V(\mathbf{r}')\frac{j_1(2k_f|\mathbf{r}-\mathbf{r}'|)}{|\mathbf{r}-\mathbf{r}'|^2}. \tag{2.3.14}$$

Without solving this equation, we can see that the displaced charge can have a very different character at large r, depending on whether we use the wave theory result (2.3.11) or the semi-classical result (2.3.13). Thus, let us take a simple case (not self-consistent) when V is very short range, i.e. write $V(\mathbf{r}) = \lambda\delta(\mathbf{r})$. Then the displaced charge according to (2.3.13) is of similar short range while (2.3.11) gives, on the contrary,

$$\rho(\mathbf{r})-\rho_0 \sim \frac{j_1(2k_f r)}{r^2}$$

and thus, at large $\mathbf{r}$

$$\rho(\mathbf{r})-\rho_0 \sim \frac{\cos 2k_f r}{r^3}, \tag{2.3.15}$$

using the form of j_1 given after equation (2.3.11). Whereas equation (2.3.13) leads to a screened Coulomb potential $V(r) = -(Ze^2/r)\exp(-qr)$ when combined with Poisson's equation, we find that if we solve (2.3.14) the self-consistent potential $V(r)$ also goes as $\cos 2k_f r/r^3$, exactly of the form (2.3.15), at sufficiently large r. Numerical solutions of (2.3.14) for $V(r)$ are tabulated by March and Murray (1961). Alternatively, (2.3.14) can be solved analytically in **k** space, and yields, if we define the Fourier components of the potential $V(\mathbf{r})$ by

$$\tilde{V}(\mathbf{k}) = \int d\mathbf{r}\, e^{i\mathbf{k}.\mathbf{r}}\, V(\mathbf{r}), \tag{2.3.16}$$

$$\tilde{V}(\mathbf{k}) = -4\pi Ze^2 \Big/ \left[k^2 + \frac{k_f}{\pi a_0} g\left(\frac{k}{2k_f}\right)\right]: \quad g(x) = 2 + \frac{x^2-1}{x} \ln\left|\frac{1-x}{1+x}\right|. \tag{2.3.17}$$

This reproduces the Thomas–Fermi screened Coulomb result if we note that as $k \to 0$, $g(x) \to 4$ and hence

$$\tilde{V}(k) \doteqdot -\frac{4\pi Ze^2}{k^2+q^2} \tag{2.3.18}$$

which, on transforming into **r** space, gives back $V(r) = -(Ze^2/r)e^{-qr}$ which was quoted above.

Defining the static dielectric constant $\varepsilon(k)$ by

$$\tilde{V}(k) = -\frac{4\pi Ze^2}{k^2\,\varepsilon(k)}, \tag{2.3.19}$$

$\varepsilon(k)$ representing the "shielding" of the bare-ion potential $-4\pi Ze^2/k^2$,

$$\varepsilon(k) = \frac{k^2 + (k_f/\pi a_0)\, g(k/2k_f)}{k^2} \tag{2.3.20}$$

while the semi-classical result is

$$\underset{\text{semiclassical}}{\varepsilon(k)} = \frac{k^2+q^2}{k^2}. \tag{2.3.21}$$

Equation (2.3.20) appears first to have been given by Lindhard (1954), though it is at least implicit in a paper by Bardeen (1937) on electron–phonon interaction.

2.3.2 *Frequency dependent dielectric constant* $\varepsilon(\mathbf{k}\omega)$

Having given a very simple physical introduction to static dielectric screening theory, we now turn to discuss the frequency dependence of the dielectric constant. Starting from Maxwell's equations we can define the dielectric constant $\varepsilon(\mathbf{k}\omega)$ generally through

$$\mathbf{D} = \varepsilon\mathscr{E}, \tag{2.3.22}$$

where

$$\operatorname{div}\mathbf{D} = 4\pi\rho_t \tag{2.3.23}$$

and

$$\operatorname{div}\mathscr{E} = 4\pi(\rho_t + \rho_m). \tag{2.3.24}$$

Here ρ_t is the density of a test charge and ρ_m is the charge density of the medium itself. Since there is assumed to be a uniform positive background charge, of number density ρ_0 just as in the previous section, this becomes

$$\rho_m = \rho - \rho_0, \tag{2.3.25}$$

where ρ is the electron density.

We now "switch on" a density fluctuation given by

$$\rho_t = e\rho_t(\mathbf{k})\,e^{i(\mathbf{k}.\mathbf{r}-\omega t)} + \text{complex conjugate} \tag{2.3.26}$$

and then we find from (2.3.22)–(2.3.24)

$$4\pi e\rho_t(\mathbf{k})\,e^{-i\omega t} = i\mathbf{k}.\mathbf{D}(\mathbf{k}) = i\varepsilon(k\omega)\,\mathbf{k}.\mathscr{E}(\mathbf{k}) \tag{2.3.27}$$

and similarly, Fourier analysing ρ_m by analogy with equation (2.3.26) we obtain

$$4\pi e\,e^{-i\omega t}[\rho_t(\mathbf{k}) + \rho_m(\mathbf{k})] = i\mathbf{k}.\mathscr{E}(\mathbf{k}), \tag{2.3.28}$$

where $\rho_m(\mathbf{k})$ is the expectation value of the density of the medium.

If the second of these equations is divided by the other, we have

$$\frac{1}{\varepsilon(\mathbf{k}\omega)} = \frac{e^{-i\omega t}[\rho_t(\mathbf{k}) + \rho_m(\mathbf{k})]}{\rho_t(\mathbf{k})\,e^{-i\omega t}} = \frac{\rho_t + \rho_m}{\rho_t}. \tag{2.3.29}$$

We can see from (2.3.29) that the dielectric constant is just the result of division of the relevant Fourier component of the test charge by that of the total charge.

Let us now find an expression for ε in terms of the eigenstates of the system. For this purpose, it is convenient to write the Hamiltonian in terms of the Fourier components $\rho_\mathbf{k}$ of the density operator $\rho(\mathbf{r})$. If we assume periodic boundary conditions so that $\mathbf{k}$ takes on a discrete set of allowed values then we have‡

$$\rho(\mathbf{r}) = \sum_i \delta(\mathbf{r} - \mathbf{r}_i) = \sum_\mathbf{k} \rho_\mathbf{k}\,e^{i\mathbf{k}.\mathbf{r}} \tag{2.3.30}$$

and hence the Hamiltonian H_0 for the interacting homogeneous electron gas is

$$H_0 = \sum_i \frac{p_i^2}{2m} + \sum_\mathbf{k} \frac{2\pi e^2}{k^2}(\rho_\mathbf{k}^\dagger \rho_\mathbf{k} - \rho_0) \tag{2.3.31}$$

as is easily shown from the self energy $(e^2/2)\int \rho(\mathbf{r}_1)\,\rho(\mathbf{r}_2)/|\mathbf{r}_1 - \mathbf{r}_2|\,d\mathbf{r}_1\,d\mathbf{r}_2$.

The interaction between the system and the test charge is

$$e^2 \int \rho(\mathbf{r}_1)\,\rho_t(\mathbf{r}_2)/|\mathbf{r}_1 - \mathbf{r}_2|\,d\mathbf{r}_1\,d\mathbf{r}_2$$

or

$$H_1 = \frac{4\pi e^2}{k^2}[\rho_\mathbf{k}^\dagger \rho_t(\mathbf{k})\,e^{-i(\omega+i\delta)t} + \rho_\mathbf{k}\,\rho_t^*(\mathbf{k})\,e^{i(\omega-i\delta)t}]. \tag{2.3.32}$$

‡ In this section, and also in some subsequent sections, it will be very convenient notationally to take a crystal of unit volume.

Here δ is an infinitesimally small positive definite quantity enabling us to immediately apply time-dependent perturbation theory to obtain the first-order wave function

$$\psi = \psi_0 - \sum_n{}' \frac{4\pi e^2}{k^2}\rho_t(\mathbf{k})\left[\frac{\langle n|\rho_{\mathbf{k}}^\dagger|0\rangle\, e^{-i\omega t+\delta t}}{\omega_n-\omega-i\delta}+\frac{\langle n|\rho_{\mathbf{k}}|0\rangle\, e^{i\omega t-\delta t}}{\omega_n+\omega+i\delta}\right]\psi_n. \tag{2.3.33}$$

In equation (2.3.33) $\psi_n \equiv |n\rangle$ is the eigenfunction of H_0 with eigenvalue E_n and

$$\omega_n = E_n - E_0. \tag{2.3.34}$$

To derive the frequency-dependent dielectric constant, we obtain from (2.3.33), to first order in ρ_t,

$$\langle\psi|\rho_{\mathbf{k}}|\psi\rangle \propto \rho_m(\mathbf{k}) = -\frac{4\pi e^2}{k^2}\rho_t(\mathbf{k})\sum_n\left|\langle n|\rho_{\mathbf{k}}|0\rangle\right|^2\left[\frac{1}{\omega+\omega_n+i\delta}+\frac{1}{\omega_n-\omega-i\delta}\right] \tag{2.3.35}$$

so that from equation (2.3.29) we have

$$\frac{1}{\varepsilon(\mathbf{k}\omega)} = 1-\frac{4\pi e^2}{k^2}\sum_n\left|\langle n|\rho_{\mathbf{k}}|0\rangle\right|^2\left[\frac{1}{\omega+\omega_n+i\delta}+\frac{1}{\omega_n-\omega-i\delta}\right]. \tag{2.3.36}$$

We see from equation (2.3.36) that the resonances of the system occur at the roots of $\varepsilon(\mathbf{k}\omega) = 0$.

We shall discuss below an approximate method of explicitly calculating $\varepsilon(\mathbf{k}\omega)$. However, before doing so, we want to show that $\varepsilon(\mathbf{k}\omega)$ can be used to calculate the total energy of the electron gas.

From equation (2.3.31), we can write immediately for the quantum mechanical average of the potential energy V,

$$\langle V\rangle = 2\pi e^2\sum_{\mathbf{k}}\left[\frac{\langle\rho_{\mathbf{k}}^\dagger\rho_{\mathbf{k}}\rangle}{k^2}-\frac{\rho_0}{k^2}\right]. \tag{2.3.37}$$

To calculate the averages of the $\rho_{\mathbf{k}}$'s involved in this equation, let us take the imaginary part of the reciprocal of $\varepsilon(k\omega)$, using the formal identity

$$\lim_{\delta\to 0}\frac{1}{x\pm i\delta} = P\left(\frac{1}{x}\right)\mp i\pi\delta(x), \tag{2.3.38}$$

where P stands for the principal part; that is, in any contour integration, we avoid the pole at the origin. We then obtain

$$\operatorname{Im}\frac{1}{\varepsilon(\mathbf{k}\omega)} = \frac{4\pi^2 e^2}{k^2}\sum_n\langle n|\rho_{\mathbf{k}}|0\rangle^2[\delta(\omega+\omega_n)-\delta(\omega-\omega_n)]. \tag{2.3.39}$$

Integrating this over ω from 0 to ∞ we find

$$\langle V\rangle = -\sum_{\mathbf{k}}\left[\frac{2\pi\rho_0 e^2}{k^2}+\frac{1}{2\pi}\int_0^\infty \operatorname{Im}\frac{1}{\varepsilon(\mathbf{k}\omega)}\,d\omega\right]. \tag{2.3.40}$$

Let us now generalize the Hamiltonian $T+V$, T being the kinetic energy, by introducing a coupling constant λ so that the new Hamiltonian, $H(\lambda)$ say, is defined by

$$H(\lambda) = T+\lambda V. \tag{2.3.41}$$

The introduction of λ in the electron gas allows us to formally reduce all possible densities to a common density (say unit density). We evidently have a new ground-state wave function ψ_λ, and from Feynman's theorem discussed in Appendix 1.4 we can write

$$\frac{\partial H(\lambda)}{\partial\lambda} = \langle\psi_\lambda|V|\psi_\lambda\rangle. \tag{2.3.42}$$

It is therefore clear that

$$\langle H_0\rangle = \langle H(1)\rangle = \langle H(0)\rangle + \int_0^1 d\lambda\,\langle\psi_\lambda|V|\psi_\lambda\rangle \tag{2.3.43}$$

with (the energies being taken per unit volume)

$$\langle H(0)\rangle = \tfrac{3}{5}\rho_0 E_{\mathrm{f}}, \tag{2.3.44}$$

the energy of a system of free electrons without Coulombic interaction, E_{f} being simply the free-electron Fermi energy. This gives the total energy of the electron gas in terms of $\varepsilon(\mathbf{k}\omega)$, when this is known as a function of λ, as we shall see quite explicitly in equation (2.4.1) below.

Self-consistent-field dielectric constant. Let us now extend the Hartree method of section 2.3.1 and consider the one-particle Hamiltonian

$$H = T+V(\mathbf{r}t) \tag{2.3.45}$$

where $V(\mathbf{r}t)$ is a time-dependent potential calculated from the self-consistent field of the electrons and a possible external perturbation. We allow for correlation effects in the sense that these are essentially due to the dynamical aspects of the motion and even in the absence of an external potential we shall consider the variation of the self-consistent field with time.

We shall consider non-zero temperatures, since this introduces little complication of formalism. We work with a temperature-dependent generalization of the electron density $\rho(\mathbf{r})$, namely $\rho(\mathbf{r}'\mathbf{r})$ defined by

$$\rho(\mathbf{r}'\mathbf{r}) = \sum_i \psi_i^*(\mathbf{r}')\,\psi_i(\mathbf{r})\,f(E_i) \tag{2.3.46}$$

where $H\psi_i = E_i\psi_i$ and $f(E)$ is the Fermi–Dirac function‡. But when the Hamiltonian H depends explicitly on time, as in equation (2.3.45), the proper tool is a time-dependent density matrix operator ρ defined by its matrix elements

$$\langle\phi_i|\rho|\phi_f\rangle = \langle\phi_i|f(H)|\phi_f\rangle, \tag{2.3.47}$$

where ϕ_i and ϕ_f are any two members of a complete orthonormal set. If $\{\psi_i\}$ represents the complete set of solutions of the time-dependent Schrödinger equation

$$H\psi_i = i\hbar\frac{\partial\psi_i}{\partial t}, \tag{2.3.48}$$

then

$$\rho(\mathbf{r}'\mathbf{r}) = \sum_i \psi_i^* f(H)\psi_i = \sum_{ijl} \langle\psi_i|\phi_j\rangle\langle\phi_i|\psi_i\rangle\phi^*(\mathbf{r}')f(H)\phi_i(\mathbf{r})$$

$$= \sum_j \phi_j^*(\mathbf{r}')f(H)\phi_j(\mathbf{r}) \tag{2.3.49}$$

and if we expand the density $\rho(\mathbf{r}) = \rho(\mathbf{r},\mathbf{r})$ in the ϕ's we find

$$\rho(\mathbf{r}) = \sum_i \langle\phi_i^*\rho(\mathbf{r})\rangle\phi_i(\mathbf{r}) = \sum_i \phi_i(\mathbf{r})\sum_j \langle\phi_i^*\phi_j^* f(H)\phi_j(\mathbf{r})\rangle. \tag{2.3.50}$$

It is convenient to take the set $\{\phi_i\}$ as plane-waves $\mathscr{V}^{-\frac{1}{2}}e^{i\mathbf{k}\cdot\mathbf{r}} \equiv |\mathbf{k}\rangle$ and we then obtain the Fourier series

$$\rho(\mathbf{r}) = \sum_{\mathbf{k}}\sum_{\mathbf{k}'} \langle\mathbf{k}+\mathbf{k}'|\rho|\mathbf{k}'\rangle|\mathbf{k}\rangle. \tag{2.3.51}$$

Let us now linearize the equation of motion for the density matrix operator (see, for example, Appendix 2.2, equation A2.2.1), with $\hbar = 1$,

$$i\frac{\partial\rho}{\partial t} = [H, \rho] \tag{2.3.52}$$

by writing $\rho = \rho_0 + \Delta\rho$, where ρ_0 is the matrix operator when $V = 0$, and then neglecting the product $V\Delta\rho$. We obtain

$$i\frac{\partial}{\partial t}\Delta\rho = [T, \Delta\rho] + [V, \rho_0]. \tag{2.3.53}$$

We now take matrix elements of this equation between $\langle\boldsymbol{\kappa}|$ and $|\boldsymbol{\kappa}+\mathbf{k}\rangle$. Using the result

$$\rho_0|\mathbf{k}\rangle = f(E_{\mathbf{k}})|\mathbf{k}\rangle, \tag{2.3.54}$$

‡ When the temperature T tends to zero, $f(E)$ becomes unity for $E < E_f$ and otherwise zero. $\rho(\mathbf{r}'\mathbf{r})$ in equation (2.3.46) then reduces to the Dirac density matrix of section 2.7.5.

where

$$E_{\mathbf{k}} = \frac{k^2}{2m}, \tag{2.3.55}$$

we have

$$i\frac{\partial}{\partial t}\langle \boldsymbol{\kappa}|\Delta\rho|\mathbf{k}+\boldsymbol{\kappa}\rangle = (E_{\kappa}-E_{\kappa+\mathbf{k}})\langle \boldsymbol{\kappa}|\Delta\rho|\boldsymbol{\kappa}+\mathbf{k}\rangle$$
$$+\{f(E_{\kappa+\mathbf{k}})-f(E_{\kappa})\}\langle \boldsymbol{\kappa}|V|\boldsymbol{\kappa}+\mathbf{k}\rangle. \tag{2.3.56}$$

At this point, we regard $V(\mathbf{r}t)$ as composed of two parts: V_{p} due to an external potential and V_{s} due to the screening by the electron gas. Clearly V_{s} will be related to the change in electron density through Poisson's equation, just as before. Expanding V_{s} in a Fourier series and using equation (2.3.51) we find

$$-\nabla^2 V_{\mathrm{s}} = \sum_{\mathbf{k}} k^2\langle \boldsymbol{\kappa}|V_{\mathrm{s}}|\boldsymbol{\kappa}+\mathbf{k}\rangle|\mathbf{k}\rangle$$
$$= 4\pi e^2 \sum_{\mathbf{k}}\langle \boldsymbol{\kappa}+\mathbf{k}|\Delta\rho|\boldsymbol{\kappa}\rangle|\mathbf{k}\rangle \tag{2.3.57}$$

or

$$\langle \boldsymbol{\kappa}|V_{\mathrm{s}}|\boldsymbol{\kappa}+\mathbf{k}\rangle = \frac{4\pi e^2}{k^2}\sum_{\boldsymbol{\kappa}}\langle \boldsymbol{\kappa}+\mathbf{k}|\Delta\rho|\boldsymbol{\kappa}\rangle. \tag{2.3.58}$$

Thus equation (2.3.56) is non-trivial, even if the external potential is zero. Since $\langle \boldsymbol{\kappa}+\mathbf{k}|\Delta\rho|\boldsymbol{\kappa}\rangle$ is independent of $\boldsymbol{\kappa}$ in the plane-wave basis, this equation is simply

$$\tilde{V}_{\mathrm{s}}(\mathbf{k}\omega) = \frac{4\pi e^2}{k^2}\tilde{\Delta}\rho(\mathbf{k}\omega). \tag{2.3.59}$$

Now we note from equation (2.3.29) that the definition of $\varepsilon(\mathbf{k}\omega)$ may be written, in terms of the Fourier components $V(\mathbf{k})$ and $V_{\mathrm{p}}(\mathbf{k})$ of the effective and external potentials respectively, as

$$V(\mathbf{k}) = \frac{V_{\mathrm{p}}(\mathbf{k})}{\varepsilon(\mathbf{k},\omega)}, \tag{2.3.60}$$

where the external potential V_{p} is taken to have the time dependence $V_{\mathrm{p}}(t) = V_{\mathrm{p}}\exp(i\omega t)$. We may alternatively write

$$\varepsilon(\mathbf{k}\omega) = \frac{V_{\mathrm{p}}}{V} = \frac{V-V_{\mathrm{s}}}{V} = 1-\frac{V_{\mathrm{s}}(\mathbf{k})}{V(\mathbf{k})}, \tag{2.3.61}$$

and using equation (2.3.58) this becomes

$$\varepsilon(\mathbf{k},\omega) = 1-\frac{4\pi e^2}{k^2}\sum_{\boldsymbol{\kappa}}\frac{\langle \boldsymbol{\kappa}+\mathbf{k}|\Delta\rho|\boldsymbol{\kappa}\rangle}{V(\mathbf{k})}. \tag{2.3.62}$$

Since for a uniform gas $V(\mathbf{k}) = \langle\boldsymbol{\kappa}|V|\boldsymbol{\kappa}+\mathbf{k}\rangle$, if we now take $\Delta\rho$ to have time dependence $e^{i\omega t}$ equation (2.3.56) finally yields

$$\varepsilon(\mathbf{k}, \omega) = 1-\frac{4\pi e^2}{k^2}\sum_{\boldsymbol{\kappa}}\frac{f(E_{\mathbf{k}+\boldsymbol{\kappa}})-f(E_{\boldsymbol{\kappa}})}{E_{\mathbf{k}+\boldsymbol{\kappa}}-E_{\mathbf{k}}+\omega-i\delta} \tag{2.3.63}$$

for the self-consistent field (SCF) dielectric function of the homogeneous gas.

2.4 Correlation energy in electron gas

2.4.1 *Use of SCF dielectric constant*

We can calculate the correlation energy per unit volume using the formula obtained by combination of equations (2.3.40), (2.3.43) and (2.3.44)

$$\frac{E_{\text{total}}}{\mathscr{V}} = \tfrac{3}{5}\rho_0 E_{\mathrm{f}}-\frac{2\pi^2 e^2}{\mathscr{V}}\sum_{k}\int_0^1\frac{d\lambda}{\lambda}\left[\frac{\lambda}{k^2}+\frac{1}{4\pi^3 e^2}\int_0^\infty d\omega\,\mathrm{Im}\left\{\frac{1}{\varepsilon_\lambda(k\omega)}\right\}\right]. \tag{2.4.1}$$

[Here λe^2 is the coupling parameter, cf. equation (2.3.41).] This is exact. If we use for $\varepsilon(\mathbf{k}\omega)$ the SCF approximate form already calculated with e^2 replaced by λe^2 we obtain, with automatic inclusion of an exchange term, which we shall display in equation (2.4.3) below,

$$\frac{E_{\text{total}}}{N} = \frac{2.21}{r_s^2}-\frac{0.916}{r_s}+0.062\ln r_s-0.096+0(r_s\ln r_s), \tag{2.4.2}$$

where r_s is the mean interelectronic spacing defined by $\rho_0 = 3/4\pi r_s^3 a_0^3$ and the energy units are Rydbergs.

This result is exact to order $\ln r_s$ and $(r_s)^0$. The first two terms constitute the Hartree–Fock energy, the first being the mean Fermi energy and the second the exchange energy [see equation (2.7.49) below] and so the rest of the expression is the correlation energy, which, by definition, is the difference between the exact and Hartree–Fock (HF) energies.

The expression (2.4.2) was first given by Gell-Mann and Brueckner (1957) though the earlier work of Macke (1950) was central in the development of the theory, and we shall briefly discuss the Gell-Mann–Brueckner method as a preliminary to an explanation in the following two sections why the SCF method should be exact in the high density limit.

2.4.2 *Gell-Mann and Brueckner method*

Let us suppose we carry out a standard perturbation expansion on free electrons, the perturbation being the electron–electron interaction. The first two terms are finite, being just the matrix elements of the Hamiltonian between plane-wave HF determinants, and constitute the HF energy. If we go to second order in perturbation theory [order $(r_s)^0$] we find there are another two terms, the second-order exchange energy

$$\frac{E_{ex}^{(2)}}{N} = \frac{3}{16\pi^5}\int\frac{d\mathbf{q}}{q^2}\int_{\substack{p_1<1\\|\mathbf{p}_1+\mathbf{q}|>1}} d\mathbf{p}_1\int_{\substack{p_2<1\\|\mathbf{p}_2+\mathbf{q}|>1}} d\mathbf{p}_2\frac{1}{(\mathbf{p}_1+\mathbf{p}_2+\mathbf{q})^2}\cdot\frac{1}{(\mathbf{p}_1+\mathbf{p}_2).\mathbf{q}+q^2}$$

$$= 0.046 \text{ Rydberg} \tag{2.4.3}$$

and a second-order direct term having the form

$$\frac{E_d^{(2)}}{N} = \frac{-3}{8\pi^5}\int\frac{d\mathbf{q}}{q^4}\int_{\substack{p_1<1\\|\mathbf{p}_1+\mathbf{q}|>1}} d\mathbf{p}_1\int_{\substack{p_2<1\\|\mathbf{p}_2+\mathbf{q}|>1}}\frac{d\mathbf{p}_2}{(\mathbf{p}_1+\mathbf{p}_2).\mathbf{q}+q^2} \tag{2.4.4}$$

where we measure momenta in units of the Fermi momentum. As we have indicated by giving its numerical value, the first term is finite; the integration of the second term, however, is logarithmically divergent. This, of course, immediately poses a problem, that of extracting a meaningful finite result. The work of Bohm and Pines, already referred to, indicates the physics of the matter: the collective motions act to screen out the Coulomb field at distances greater than the order of $r_{\max}\sim(k_f/a_0)^{-\frac{1}{2}}$ (cf. equation 2.3.13) so that one cuts off the $\mathbf{q}$ integration below a limit $q_{\min}\sim r_s^{-\frac{1}{2}}$ since $r_s k_f$ is a constant. One then obtains the result for the correlation energy

$$\frac{E_{corr}}{N} = A\ln r_s + C + \text{terms vanishing as } r_s\to 0. \tag{2.4.5}$$

Introducing the low q cut-off is sufficient to obtain A from equation (2.4.4) and yields the value 0.062, but C is naturally unobtainable by this means. Gell-Mann and Brueckner sought to obtain the full expression for the energy to order $(r_s)^0$ and $\ln r_s$ by achieving a formal summation of divergent terms to all orders in perturbation theory. To this end they isolated in each order the most divergent terms and were indeed able to sum the resulting subseries under the integral sign to obtain a finite answer.

We shall content ourselves with the above brief summary of Gell-Mann and Brueckner's work (the relevant diagrammatic techniques, however, are discussed later in section 2.10.5) and shall develop an effective Hamiltonian which can be shown by these techniques to give the same results for the correlation energy as Gell-Mann and Brueckner's. This effective Hamiltonian method has the advantage of being intimately connected with the physically appealing self-consistent field method already discussed.

2.5 Hamiltonian in second-quantized form

In much of the rest of this chapter we shall use the occupation number formalism, and so there follows a brief résumé of this formalism specialized to fermions.

Consider a complete set of normalized Slater determinants of one particle functions $|\mathbf{k}\rangle \equiv e^{i\mathbf{k}.\mathbf{r}}$. One may consider spin to be included, but the quantum number absorbed into the label $\mathbf{k}$. With each determinant we associate a state vector

$$|n_{\mathbf{k}_1}, n_{\mathbf{k}_2}, \ldots, n_{\mathbf{k}_i}, \ldots\rangle$$

where the $\mathbf{k}_i$ run over all allowed values of $\mathbf{k}$. If the associated determinant contains the function $e^{i\mathbf{k}.\mathbf{r}}$, $n_{\mathbf{k}} = 1$; otherwise it is zero. This occupation-number representation is not limited to any particular number of particles. In fact, let us now define creation and annihilation operators $a_{\mathbf{k}}^{\dagger}$ and $a_{\mathbf{k}}$ such that

$$a_{\mathbf{k}_i}^{\dagger}|n_{\mathbf{k}_1}, n_{\mathbf{k}_2}, \ldots, n_{\mathbf{k}_i}, \ldots\rangle = f_c|n_{\mathbf{k}_1}, \ldots, n_{\mathbf{k}_i}+1, \ldots\rangle, \tag{2.5.1}$$

$$a_{\mathbf{k}_i}|n_{\mathbf{k}_1}, n_{\mathbf{k}_2}, \ldots, n_{\mathbf{k}_i}, \ldots\rangle = f_d|n_{\mathbf{k}_1}, \ldots, n_{\mathbf{k}_i}-1, \ldots\rangle, \tag{2.5.2}$$

where f_c and f_d are, at present, unspecified constants which will depend on the state vector.

Since the $n_{\mathbf{k}_i}$ are necessarily 0 or 1, we must have

$$\left.\begin{aligned} a_{\mathbf{k}_i}^{\dagger}|n_{\mathbf{k}_1}, n_{\mathbf{k}_2}, \ldots, n_{\mathbf{k}_i}, \ldots\rangle &= 0 \quad \text{if} \quad n_{\mathbf{k}_i} = 1, \\ a_{\mathbf{k}_i}|n_{\mathbf{k}_1}, n_{\mathbf{k}_2}, \ldots, n_{\mathbf{k}_i}, \ldots\rangle &= 0 \quad \text{if} \quad n_{\mathbf{k}_i} = 0, \end{aligned}\right\}. \tag{2.5.3}$$

f_c and f_d will be partly specified by the requirement that unless

$$a_{\mathbf{k}_i}|n_{\mathbf{k}_1}, n_{\mathbf{k}_2}, \ldots\rangle = 0$$

the new vector has the same normalization as the old, which implies that

$$\langle n_{\mathbf{k}_1}, n_{\mathbf{k}_2}, \ldots|a_{\mathbf{k}_i}^{\dagger} a_{\mathbf{k}_i}|n_{\mathbf{k}_1}, n_{\mathbf{k}_2}\ldots\rangle = n_{\mathbf{k}_i}. \tag{2.5.4}$$

We therefore say $a_{\mathbf{k}_i}^{\dagger} a_{\mathbf{k}_i}$ is the *number operator* for the state $|\mathbf{k}_i\rangle$.

For similar reasons, we also write

$$\langle n_{\mathbf{k}_1}, n_{\mathbf{k}_2}, \ldots|a_{\mathbf{k}_i} a_{\mathbf{k}_i}^{\dagger}|n_{\mathbf{k}_1}, n_{\mathbf{k}_2}\ldots\rangle = 1 - n_{\mathbf{k}_i}. \tag{2.5.5}$$

Let us now define the *vacuum state vector* $|0\rangle$ for which $n_{\mathbf{k}_i} = 0$ for all $\mathbf{k}_i$ and which we specify to be normalized to unity:

$$\langle 0|0\rangle = 1. \tag{2.5.6}$$

From equation (2.5.4), we can see that any normalized state vector for which N $n_{\mathbf{k}}$'s namely $n_{\mathbf{k}_1'}, n_{\mathbf{k}_2'}, \ldots, n_{\mathbf{k}_N'}$ are non-zero may be written

$$a_{\mathbf{k}_1'}^{\dagger} a_{\mathbf{k}_2'}^{\dagger} \ldots a_{\mathbf{k}_N'}^{\dagger}|0\rangle \leftrightarrow \Phi_{\mathbf{k}_1'\ldots\mathbf{k}_N'}, \tag{2.5.7}$$

where in this expression we have indicated the correspondence with the $N \times N$ normalized determinant

$$\Phi_{\mathbf{k}_1'\ldots\mathbf{k}_N'} = \frac{1}{\sqrt{N!}}\begin{vmatrix} e^{i\mathbf{k}_1'.\mathbf{r}_1} & e^{i\mathbf{k}_2'.\mathbf{r}_1} & \ldots & e^{i\mathbf{k}_N'.\mathbf{r}_1} \\ e^{i\mathbf{k}_1'.\mathbf{r}_2} & e^{i\mathbf{k}_2'.\mathbf{r}_2} & \ldots & e^{i\mathbf{k}_N'.\mathbf{r}_2} \\ \cdot & \cdot & \cdot & \cdot \end{vmatrix}. \tag{2.5.8}$$

Since a determinant is antisymmetric with respect to interchange of columns, $\Phi_{\mathbf{k}_1'\ldots\mathbf{k}_N'}$ is antisymmetric in the labelling with respect to the $\mathbf{k}_1', \ldots, \mathbf{k}_N'$. We carry this over into the occupation number representation by specifying that $a_{\mathbf{k}_1'}^\dagger a_{\mathbf{k}_2'}^\dagger \ldots a_{\mathbf{k}_N'}^\dagger |0\rangle$ is antisymmetric with respect to interchange of the $a_{\mathbf{k}_i}^\dagger$'s.

The creation and annihilation operators are of the utmost importance in the occupation number representation, because not only can all vectors be generated according to equation (2.5.7), but all operators may be expressed in terms of them.

Our next task must be to obtain the Hamiltonian in terms of the creation and annihilation operators. We do this using the principle that the form of any operator in a representation is completely specified by its matrix elements with respect to all states. As a first application we can obtain anticommutation rules for the $a_{\mathbf{k}}^\dagger$'s and $a_{\mathbf{k}}$'s. One may verify that as a consequence of the antisymmetry requirement on $a_{\mathbf{k}_1'} \ldots a_{\mathbf{k}_N'}|0\rangle$ and the normalization relations (2.5.4) and (2.5.5) we obtain

$$\left.\begin{aligned} [a_{\mathbf{k}'}^\dagger, a_{\mathbf{k}}^\dagger]_+ &= [a_{\mathbf{k}'}, a_{\mathbf{k}}]_+ = 0, \\ [a_{\mathbf{k}'}^\dagger, a_{\mathbf{k}}]_+ &= \delta_{\mathbf{k}'\mathbf{k}}. \end{aligned}\right\} \tag{2.5.9}$$

Turning now to the Hamiltonian, we know already that acting on any determinant of plane-waves the kinetic energy operator gives

$$T\Phi_{\mathbf{k}_1'\ldots\mathbf{k}_N'} = \sum_{i=1}^{N} \frac{\hbar^2 k_i'^2}{2m} (\Phi_{\mathbf{k}_1'\ldots\mathbf{k}_N'}). \tag{2.5.10}$$

It is also readily verified that

$$\sum_{\mathbf{k}} \frac{\hbar^2 k^2}{2m} a_{\mathbf{k}}^\dagger a_{\mathbf{k}} (a_{\mathbf{k}_1'}^\dagger \ldots a_{\mathbf{k}_N'})|0\rangle = \sum_{i=1}^{N} \frac{\hbar^2 k_i'^2}{2m} (a_{\mathbf{k}_1'}^\dagger \ldots a_{\mathbf{k}_N'}^\dagger)|0\rangle \tag{2.5.11}$$

so that in view of the correspondence (2.5.7) we can write

$$T = \sum_{\mathbf{k}} E_{\mathbf{k}} a_{\mathbf{k}}^\dagger a_{\mathbf{k}} \quad \left(E_{\mathbf{k}} = \frac{\hbar^2 k^2}{2m}\right). \tag{2.5.12}$$

It will be useful later to have the kinetic energy in a more general form than the diagonal result (2.5.12). Thus, if we choose a new basis set $|\mu\rangle$, rather than $\mathbf{k}$, we can write

$$a_{\mathbf{k}} = \sum_{\mu} \langle \mu | \mathbf{k} \rangle a_{\mu}, \tag{2.5.13}$$

where a_μ annihilates a particle in state μ. Then if h_1 is the one-electron Hamiltonian, being simply the kinetic energy operator with the form (2.5.12), we can write

$$E_{\mathbf{k}} = \langle \mathbf{k} | h_1 | \mathbf{k} \rangle \tag{2.5.14}$$

and hence, from (2.5.12), we have

$$T = \sum_{\mu\nu\mathbf{k}\mathbf{k}'} \langle \mathbf{k}|h_1|\mathbf{k}'\rangle \langle \mu|\mathbf{k}\rangle \langle \mathbf{k}'|\nu\rangle a_\mu^\dagger a_\nu \tag{2.5.15}$$

since h_1 is taken as diagonal in the representation $|\mathbf{k}\rangle$. Using the completeness theorem for the eigenfunctions $\mathbf{k}$, we obtain

$$T = \sum_{\mu\nu} \langle \mu|h_1|\nu\rangle a_\mu^\dagger a_\nu. \tag{2.5.16}$$

We next examine the form of the potential energy starting from the result, already given in (2.3.31), that

$$V = 2\pi e^2 \sum_{\mathbf{q}} \frac{1}{q^2}(\rho_{\mathbf{q}}^\dagger \rho_{\mathbf{q}} - \rho_0), \tag{2.5.17}$$

where $\rho_{\mathbf{q}}$ is defined in terms of the density operator

$$\rho_{\text{op}}(\mathbf{r}) = \sum_{\mathbf{q}} \rho_{\mathbf{q}}^\dagger e^{i\mathbf{q}\cdot\mathbf{r}}. \tag{2.5.18}$$

Now between the Slater determinants of normalized plane-waves we have

$$\left.\begin{aligned} \langle \Phi_{\mathbf{k}_1'\ldots\mathbf{k}_N'}|\rho_{\text{op}}(\mathbf{r})|\Phi_{\mathbf{k}_1'\ldots\mathbf{k}_N'}\rangle &= \rho_0 = \frac{N}{\Omega}, \\ \langle \Phi_{\mathbf{k}''\mathbf{k}_2'\ldots\mathbf{k}_N'}|\rho_{\text{op}}(\mathbf{r})|\Phi_{\mathbf{k}'\mathbf{k}_2'\ldots\mathbf{k}_N'}\rangle &= e^{i(\mathbf{k}''-\mathbf{k}')\cdot\mathbf{r}}, \quad \mathbf{k}'' \neq \mathbf{k}'. \end{aligned}\right\} \tag{2.5.19}$$

Other matrix elements between essentially different determinants from these (i.e. those which do not differ merely by rearrangement of the $\mathbf{k}$'s) are zero.

Now it is easy to verify that

$$\left.\begin{aligned} \frac{1}{\Omega}\left\langle a_{\mathbf{k}''} \ldots a_{\mathbf{k}_N'} \middle| \sum_{\mathbf{k}_1\mathbf{k}_2} a_{\mathbf{k}_1}^\dagger a_{\mathbf{k}_2} e^{i(\mathbf{k}_1-\mathbf{k}_2)\cdot\mathbf{r}} \middle| a_{\mathbf{k}_1'} \ldots a_{\mathbf{k}_N'}\right\rangle &= \rho_0 \quad \text{if} \quad \mathbf{k}'' = \mathbf{k}', \\ &= \frac{e^{i(\mathbf{k}''-\mathbf{k}')\cdot\mathbf{r}}}{\Omega} \quad \text{if} \quad \mathbf{k}'' \neq \mathbf{k}'. \end{aligned}\right\} \tag{2.5.20}$$

Again, other essentially different matrix elements are zero. From the correspondence between equations (2.5.19) and (2.5.20) we can now say that

$$\rho_{\text{op}}(\mathbf{r}) = \sum_{\mathbf{k}_1\mathbf{k}_2} a_{\mathbf{k}_1}^\dagger a_{\mathbf{k}_2} e^{i(\mathbf{k}_1-\mathbf{k}_2)\cdot\mathbf{r}} \tag{2.5.21}$$

and from equation (2.5.18)

$$\rho_{\mathbf{q}} = \sum_{\mathbf{k}} a_{\mathbf{k}}^\dagger a_{\mathbf{k}+\mathbf{q}} = \rho_{-\mathbf{q}}^\dagger. \tag{2.5.22}$$

We insert this into equation (2.5.17) to obtain the potential energy, and finally, using the anticommutation relations of equation (2.5.9), we can write

$$H = \sum_{\mathbf{k}} E_{\mathbf{k}} a_{\mathbf{k}}^{\dagger} a_{\mathbf{k}} + \sum_{\mathbf{q}}{}' \frac{2\pi e^2}{q^2} \sum_{\mathbf{k}_1\mathbf{k}_2} a_{\mathbf{k}_1+\mathbf{q}}^{\dagger} a_{\mathbf{k}_2-\mathbf{q}}^{\dagger} a_{\mathbf{k}_2} a_{\mathbf{k}_1}. \tag{2.5.23}$$

In a similar fashion to that in which we rewrote the first term in (2.5.23) in the form (2.5.16), we can express the interaction term in (2.5.23) more generally as

$$\frac{1}{2} \sum_{\mu\nu\tau\sigma} \langle \mu\nu | v | \sigma\tau \rangle a_{\mu}^{\dagger} a_{\nu}^{\dagger} a_{\tau} a_{\sigma}, \tag{2.5.24}$$

where v is the interaction, replacing $4\pi e^2/q^2$ in (2.5.23). We shall need this general form of the interaction energy in Chapter 4, but for the moment we shall work with the result (2.5.23).

2.6 Sawada method and plasma dispersion relations

As we saw in section 2.4.2, Gell-Mann and Brueckner formally summed divergent series to obtain their result for the correlation energy. In this section we shall discuss the use of an effective Hamiltonian which, while again exact as $r_s \to 0$, enables us to calculate the correlation energy without the necessity of such a summation. By these means we shall also show why the SCF dielectric constant gives the correlation energy of Gell-Mann and Brueckner and also that the zeros of this dielectric constant give the correct excitation energies in the high density régime.

2.6.1 Effective Hamiltonian

In Chapter 1, we discussed the Fermi sphere for a non-interacting uniform electron gas, with particles filling states for $\mathbf{k} < \mathbf{k}_f$, the Fermi momentum, and, at $T = 0$, no particles outside. As we switch on the interactions, the effect is to excite some particles from inside the Fermi sphere into higher momentum states, leaving vacant states or holes inside. This formation of particle–hole pairs is focussed on, following Sawada (1957), by retaining in the general form equation (2.5.23) only those terms in the potential energy which correspond to particle–hole creation or annihilation.

We can obtain the resulting effective Hamiltonian in an illuminating form if we specify that in the relevant terms of (2.5.23) all momenta written as a sum (e.g. $\mathbf{k}+\mathbf{q}$) shall refer to particles above the Fermi surface, and all single momenta indices refer to particles below (e.g. $\mathbf{k}$), so that $a_{\mathbf{k}}$ is always a creation operator for a hole. To obtain such a form we note that, for example,

$$\sum_{\mathbf{k}_1\mathbf{k}_2} a^\dagger_{\mathbf{k}_1+\mathbf{q}}\, a^\dagger_{\mathbf{k}_2-\mathbf{q}}\, a_{\mathbf{k}_2}\, a_{\mathbf{k}_1} = \sum_{\mathbf{k}_1\mathbf{k}_2} a^\dagger_{\mathbf{k}_1}\, a^\dagger_{\mathbf{k}_2}\, a_{\mathbf{k}_2+\mathbf{q}}\, a_{\mathbf{k}_1-\mathbf{q}}. \tag{2.6.1}$$

We then obtain, the summation now being over particle–hole pairs,

$$V_{\text{Sawada}} \equiv V_{\text{S}} = \sum_q \frac{2\pi e^2}{q^2} \sum^{\text{(pairs)}} (a^\dagger_{\mathbf{k}_1}\, a^\dagger_{\mathbf{k}_2}\, a_{\mathbf{k}_2+\mathbf{q}}\, a_{\mathbf{k}_1-\mathbf{q}} + a^\dagger_{\mathbf{k}_1}\, a^\dagger_{\mathbf{k}_2-\mathbf{q}}\, a_{\mathbf{k}_2}\, a_{\mathbf{k}_1-\mathbf{q}}$$
$$+ a^\dagger_{\mathbf{k}_1+\mathbf{q}}\, a^\dagger_{\mathbf{k}_2}\, a_{\mathbf{k}_2+\mathbf{q}}\, a_{\mathbf{k}_1} + a^\dagger_{\mathbf{k}_1+\mathbf{q}}\, a^\dagger_{\mathbf{k}_2-\mathbf{q}}\, a_{\mathbf{k}_2}\, a_{\mathbf{k}_1}). \tag{2.6.2}$$

As we remarked in the last section, spin variables have been absorbed into the **k**'s; this should be remembered, for the exclusion principle must be obeyed in the sense that, for example, if in the first term $\sigma_1 = \sigma_2$, we cannot have $\mathbf{k}_1 = \mathbf{k}_2$.

It will be convenient for us to define creation and annihilation operators $d^\dagger$ and d for electron–hole pairs by

$$d_{\mathbf{q}}(\mathbf{k}) = a^\dagger_{\mathbf{k}}\, a_{\mathbf{k}+\mathbf{q}}, \tag{2.6.3}$$

and then in terms of these operators we can rewrite equation (2.6.2) as

$$V_{\text{S}} = 2\pi e^2 \sum_{\mathbf{q}} \frac{1}{q^2} \sum_{\mathbf{k}_1\mathbf{k}_2} [d_{\mathbf{q}}(\mathbf{k}_1) + d^\dagger_{-\mathbf{q}}(-\mathbf{k}_1)]\,[d_{-\mathbf{q}}(-\mathbf{k}_2) + d^\dagger_{\mathbf{q}}(\mathbf{k}_2)]. \tag{2.6.4}$$

This is basically the form of the potential we shall use, but does not actually specify our effective Hamiltonian, which will be defined by commutation relations. In the high density limit we can take

$$[V_{\text{S}}, d^\dagger_{\mathbf{q}}(\mathbf{k})] = \frac{4\pi e^2}{q^2} \sum_{\mathbf{k}'} [d^\dagger_{\mathbf{q}}(\mathbf{k}') + d_{-\mathbf{q}}(-\mathbf{k}')] \tag{2.6.5}$$

by noting that the ground-state wave function tends to a determinant of plane-waves in the limit as $r_s \to 0$. This is discussed in more detail in Appendix 2.1. We ought to note here, however, that in such an argument we can miss terms which contribute a constant (independent of r_s) to the total energy. This is in fact what happens here, for it may readily be verified that in reaching, for example, equation (2.6.5) from equation (2.6.4) we neglect the terms giving the second-order exchange energy (equation 2.4.3).

Hence, we define our effective Hamiltonian as

$$H_{\text{S}} = T + V_{\text{S}}, \tag{2.6.6}$$

where V_{S} is determined by equation (2.6.5). This effective Hamiltonian includes all terms in the series summed by Gell-Mann and Brueckner, except for the second-order exchange energy.

2.6.2 *Excited states*

We turn now to the problem of calculating the eigenvalues of the effective Hamiltonian. The basic idea of the method is to assume knowledge of the ground-state wavefunction Ψ and energy E of the Schrödinger equation

$$H\Psi = E\Psi \tag{2.6.7}$$

and from these to construct the excited states of the system by means of operators $\Omega_{\mathbf{k}}$ such that

$$H\Omega_{\mathbf{k}}^{\dagger}\Psi = [E+\omega_{\mathbf{k}}]\Omega_{\mathbf{k}}^{\dagger}\Psi. \tag{2.6.8}$$

Equations (2.6.7) and (2.6.8) lead to the equation of motion

$$[H, \Omega_{\mathbf{k}}^{\dagger}] = \omega_{\mathbf{k}}\Omega_{\mathbf{k}}^{\dagger}. \tag{2.6.9}$$

The Hermitian conjugate of equation (2.6.9) gives us

$$H\Omega_{\mathbf{k}}\Psi = (E-\omega_{\mathbf{k}})\Omega_{\mathbf{k}}\Psi \tag{2.6.10}$$

and so, in terms of $\Omega_{\mathbf{k}}$, the ground-state wavefunction must satisfy the condition

$$\Omega_{\mathbf{k}}\Psi = 0. \tag{2.6.11}$$

(a) *Example of independent particles.* The kinetic energy operator in second-quantized form has been displayed in equation (2.5.12) and it follows almost immediately that

$$[T, d_{\mathbf{q}}^{\dagger}(\mathbf{k})] = [E_{\mathbf{k}+\mathbf{q}} - E_{\mathbf{k}}]\, d_{\mathbf{q}}^{\dagger}(\mathbf{k}) = \omega_{\mathbf{q}}(\mathbf{k})\, d_{\mathbf{q}}^{\dagger}(\mathbf{k}) \tag{2.6.12}$$

and, as we know, $E_{\mathbf{k}+\mathbf{q}} - E_{\mathbf{k}}$ is the excitation energy associated with that state defined by operating with $d_{\mathbf{q}}^{\dagger}(\mathbf{k})$ on the unperturbed Fermi gas.

(b) *Application to Sawada Hamiltonian.* We assume at this stage that we can write the operator $\Omega_{\mathbf{q}}$ in the form

$$\Omega_{\mathbf{q}}(\omega) = \sum_{\substack{k<k_f \\ |\mathbf{k}+\mathbf{q}|>k_f}} \{C_{\mathbf{q}}(\mathbf{k}\omega)\, d_{\mathbf{q}}(\mathbf{k}) + D_{\mathbf{q}}(\mathbf{k}\omega)\, d_{-\mathbf{q}}^{\dagger}(-\mathbf{k})\}. \tag{2.6.13}$$

By requiring that (cf. equation (2.6.9))

$$[H_{\mathrm{S}}, \Omega_{\mathbf{q}}^{\dagger}(\omega)] = \omega\Omega_{\mathbf{q}}^{\dagger}(\omega) \tag{2.6.14}$$

we find, using equations (2.6.5) and (2.6.12) and collecting terms,

$$\sum_{\substack{k<k_f \\ \mathbf{k}+\mathbf{q}|>k_f}} \{[\omega - \omega_{\mathbf{q}}(\mathbf{k})]\, C_{\mathbf{q}}^{*}(\mathbf{k}\omega)\, d_{\mathbf{q}}^{\dagger}(\mathbf{k}) + [\omega + \omega_{\mathbf{q}}(\mathbf{k})]\, D_{\mathbf{q}}^{*}(\mathbf{k}\omega)\, d_{-\mathbf{q}}(-\mathbf{k})$$

$$-\frac{4\pi e^2}{q^2}[C_{\mathbf{q}}^{*}(\mathbf{k}\omega) - D_{\mathbf{q}}^{*}(\mathbf{k}\omega)] \sum_{\substack{k'<k_f \\ |\mathbf{k}'+\mathbf{q}|>k_f}} [d_{\mathbf{q}}(\mathbf{k}') + d_{-\mathbf{q}}^{\dagger}(-\mathbf{k}')]\} = 0. \tag{2.6.15}$$

This is an eigenvalue equation for ω, and it is easily seen that it is satisfied by choosing

$$C_{\mathbf{q}}^*(\mathbf{k}\omega) = \frac{1}{\omega - \omega_{\mathbf{q}}(\mathbf{k})}; \quad D_{\mathbf{q}}^*(\mathbf{k}\omega) = \frac{1}{\omega + \omega_{\mathbf{q}}(\mathbf{k})} \tag{2.6.16}$$

provided

$$1 - \frac{4\pi e^2}{q^2} \sum_{\substack{k<k_f \\ |\mathbf{k}+\mathbf{q}|>k_f}} \left[\frac{1}{\omega - \omega_{\mathbf{q}}(\mathbf{k})} - \frac{1}{\omega + \omega_{\mathbf{q}}(\mathbf{k})}\right] = 0. \tag{2.6.17}$$

If we look back to equation (2.3.63) we find exactly the same condition, from the requirement that $\mathrm{Re}\,[\varepsilon_{\mathrm{SCF}}(\mathbf{q}\omega)] = 0$. Therefore the excitation energies of the Sawada effective Hamiltonian and those of the SCF method are exactly the same. We shall see precisely why this is so in Appendix 2.2.

2.6.3 *Plasma dispersion relation*

We can rewrite equation (2.6.17) in the form (using equation (2.6.12))

$$\mathrm{Re}\,\varepsilon_{\mathrm{SCF}}(\mathbf{q}\omega) - 1 = \alpha_q(\omega) \equiv \frac{4\pi e^2}{q^2} \sum_{\substack{k<k_f \\ |\mathbf{k}+\mathbf{q}|>k_f}} \left\{\frac{1}{\omega - [(q^2/2) + \mathbf{q}.\mathbf{k}]} - \frac{1}{\omega + [(q^2/2) + \mathbf{q}.\mathbf{k}]}\right\} = 1 \tag{2.6.18}$$

and we can now plot $\alpha_q(\omega)$ as a function of ω. The result is shown in Figure 2.1, the circles showing the intersections with the value unity. The main

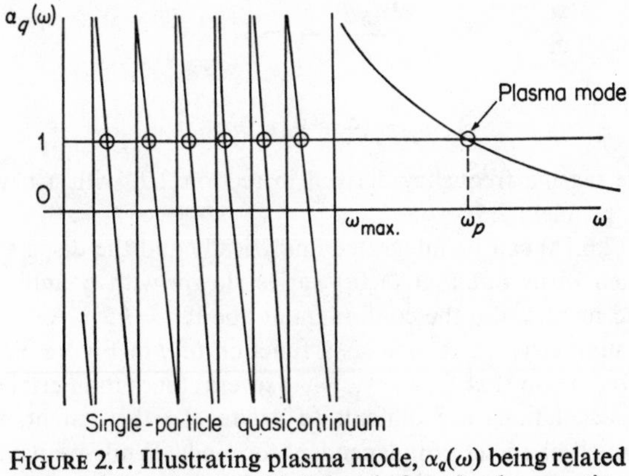

FIGURE 2.1. Illustrating plasma mode, $\alpha_q(\omega)$ being related to the self-consistent field dielectric function by equation (2.6.18).

point to be made is that the solutions are only slightly displaced from the independent-particle values $\omega_{\mathbf{q}}(\mathbf{k})$ except for one. This exceptional solution is marked at $\omega = \omega_p$.

Thus, we get a whole set of excited states in a quasi-continuum which can be thought of as single-particle excitations. However, the exceptional solution outside the continuum has a quite different character and all the particles contribute more or less equally to it. It corresponds to a coherent movement of the electrons due to a change in all wave-vectors by almost the same amount **q**.

The long wavelength limit of the plasma mode can be obtained almost immediately. We expand (2.6.17) in powers of $\omega_{\mathbf{q}}(\mathbf{k})/\omega_p$ when we obtain

$$\frac{4\pi e^2}{q^2} \sum_{\substack{k<k_f \\ |\mathbf{k}+\mathbf{q}|>k_f}} \left[\frac{2\omega_{\mathbf{q}}(\mathbf{k})}{\omega_p^2} + \frac{2\omega_{\mathbf{q}}^3(\mathbf{k})}{\omega_p^4} + \dots\right] = 1. \tag{2.6.19}$$

Then, with $\omega_{\mathbf{q}}(\mathbf{k}) = (q^2/2) + \mathbf{k}.\mathbf{q}$, the left-hand side of this equation can be evaluated to $O(\omega_{\mathbf{q}}^2)$ neglecting $|\mathbf{k}+\mathbf{q}| > k_f$, to obtain

$$\frac{4\pi e^2}{q^2}\left[\sum_{k<k_f} \left(\frac{\mathbf{q}+\mathbf{k}.\mathbf{q}}{\omega_p}\right)^2 + \sum_{\substack{k<k_f \\ |\mathbf{k}+\mathbf{q}|>k_f}} \left(\frac{\mathbf{q}.\mathbf{k}}{\omega_p^4}\right)^3\right] = 1 \tag{2.6.20}$$

or

$$4\pi e^2 \sum_{k<k_f} \frac{1}{\omega_p^2} + (\text{higher terms in } \mathbf{q}) = 1. \tag{2.6.21}$$

As $q \to 0$ we have from (2.6.21)

$$\frac{4\pi e^2}{m} \cdot \frac{N}{\omega_p^2} = 1 \tag{2.6.22}$$

or

$$\omega_p = (4\pi e^2)^{\frac{1}{2}} (\rho_0/m)^{\frac{1}{2}}, \tag{2.6.23}$$

which is the plasma frequency derived in section 2.2, with a trivial change of units (cf. problem 2.5).

Actually (2.6.18) can be integrated analytically and the dispersion investigated. It then turns out that $\omega_p(\mathbf{q})$ varies slowly with **q** until it becomes unstable and merges into the continuum at about $q = k_f$.

Thus, in summary, we show ω as a function of q in Figure 2.2. It should be emphasized again that for $q \ll k_f$, ω_p is several times the Fermi energy and the plasma oscillations are difficult to excite. To this extent, we have a behaviour much like a set of independent fermions. While we have discussed here a weak coupling (or high density) approximation, in practice it is found to contain many of the features which appear in experiments on metals.

From this very specific discussion of the Sawada Hamiltonian we turn next to deal with some further tools of considerable importance in many-body theory. This is a necessary preliminary to our discussion of quasi-particles and Fermi liquid theory.

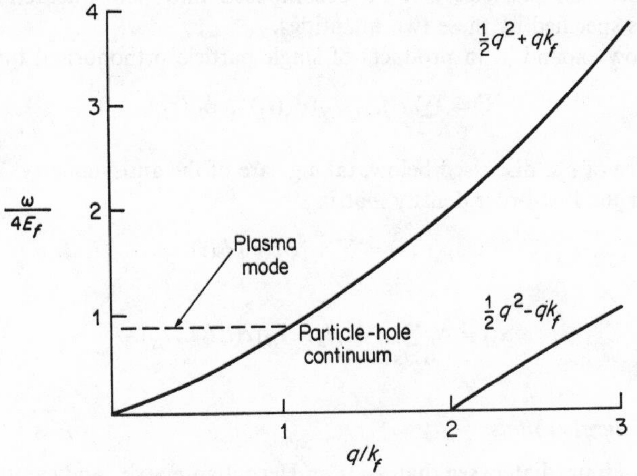

FIGURE 2.2. Excitation spectrum, indicating schematically plasma mode and continuum. Pair excitations occur between the lines labelled $\frac{1}{2}q^2+qk_f$ and $\frac{1}{2}q^2-qk_f$.

2.7 Density matrices and Green functions

2.7.1 Definitions and properties of density matrices

We start from the many-body ground state wave function $\Psi(\mathbf{r}_1 \ldots \mathbf{r}_N)$. We shall specialize to fermions, though much of what we shall say has wider relevance. Furthermore, we shall ignore spin, which may be included at the cost of slightly more complicated formalism. The nth-order density matrix is defined by

$$\Gamma^{(n)}(\mathbf{r}' \ldots \mathbf{r}'_n; \mathbf{r}_1 \ldots \mathbf{r}_n) = \binom{N}{n} \int \Psi^*(\mathbf{r}'_1 \ldots \mathbf{r}'_n \mathbf{r}_{n+1} \ldots \mathbf{r}_N) \times \Psi(\mathbf{r}_1 \ldots \mathbf{r}_N)\, d\mathbf{r}_{n+1} \ldots d\mathbf{r}_N. \quad (2.7.1)$$

We employ a special notation for two particular quantities. They are the pair function

$$P(\mathbf{r}_1 \mathbf{r}_2) = \Gamma^{(2)}(\mathbf{r}_1 \mathbf{r}_2; \mathbf{r}_1 \mathbf{r}_2), \quad (2.7.2)$$

the probability of simultaneously finding particles at $\mathbf{r}_1$ and $\mathbf{r}_2$, and the first-order density matrix

$$\rho(\mathbf{r}'\mathbf{r}) = \Gamma^{(1)}(\mathbf{r}'\mathbf{r}), \tag{2.7.3}$$

the diagonal element of which is just the total particle density. If the interaction between particles can be decomposed into pair interactions, the energy is specified by these two quantities.

We now expand Ψ in products of single-particle orthonormal functions:

$$\Psi = \sum_{l_1 \ldots l_N} c(l_1 \ldots l_N)\, \phi_{l_1}(\mathbf{r}_1) \ldots \phi_{l_N}(\mathbf{r}_N), \tag{2.7.4}$$

the choice of c's, discussed below, taking care of the antisymmetry. We may write for the first-order density matrix

$$\rho(\mathbf{r}', \mathbf{r}) = \sum_{kl} \alpha_{kl}\, \phi_k^*(\mathbf{r}')\, \phi_l(\mathbf{r}), \tag{2.7.5}$$

where

$$\alpha_{kl} = N \sum_{l_2 \ldots l_N} c^*(k, l_2 \ldots l_N)\, c(l, l_2 \ldots l_N). \tag{2.7.6}$$

2.7.2 *Natural orbitals*

We can immediately see that α_{kl} is an Hermitian matrix, and can therefore be diagonalized. If we suppose that b_{mn} is the diagonalizing matrix, and defining the orthonormal set

$$\psi_l = \sum_m b_{lm}\, \phi_m, \tag{2.7.7}$$

we see that equation (2.7.5) becomes

$$\rho(\mathbf{r}', \mathbf{r}) = \sum_l a_l\, \psi_l^*(\mathbf{r}')\, \psi_l(\mathbf{r}). \tag{2.7.8}$$

The orbitals ψ which bring the first-order density matrix $\rho(\mathbf{r}'\mathbf{r})$ into this diagonal form are termed the "natural orbitals" and the a_l are termed "occupation numbers". The Dirac density matrix referred to earlier is characterized by a_l's that are 1 or 0. We can expand Ψ in the ψ's in a manner exactly analogous to equation (2.7.4). Let us therefore replace $\phi_l(\mathbf{r})$ by $\psi_l(\mathbf{r})$ in the following.

It should be noted that, if Ψ is normalized to unity,

$$\sum_{l_1 \ldots l_N} |c(l_1 \ldots l_N)|^2 = 1 \tag{2.7.9}$$

and

$$\sum_l a_l = N. \tag{2.7.10}$$

We further note that the coefficients $c(l_1 \ldots l_N)$ are antisymmetric in $l_1 \ldots l_N$, since Ψ is antisymmetric in $\mathbf{r}_1, \ldots, \mathbf{r}_N$; this implies that we can write

$$\Psi = \sum_K c_K \Psi_K, \tag{2.7.11}$$

where K denotes a "configuration" $l_1 < l_2 < \ldots < l_N$, and Ψ_K is a normalized determinant:

$$\Psi_K = \left(\frac{1}{N!}\right)^{\frac{1}{2}} \det \psi_{l_i}(\mathbf{r}_j). \tag{2.7.12}$$

We can see that $c_K = (N!)^{\frac{1}{2}} c(l \ldots l_N)$, and that

$$\sum_K |c_K|^2 = 1. \tag{2.7.13}$$

As we have already seen, in the HF approximation, Ψ is a single determinant of the form (2.7.12), and the a_l of (2.7.8) are either zero or 1. Generalizing this, it is easy to see that

$$a_l = \sum_{K(l)} |c_K|^2,$$

the summation being over all configurations including ψ_l. But since equation (2.7.13) holds, it follows that

$$0 \leqslant a_l \leqslant 1 \quad (\text{all } l). \tag{2.7.14}$$

This is known as the Pauli condition on the first-order density matrix (see Problem 2.6 for Pauli conditions on the second-order density matrix).

We can show that for a perfect crystal the natural orbitals must be the Bloch functions introduced in Chapter 1, that is, they must be such that

$$\psi_{\mathbf{k}}(\mathbf{r}+\mathbf{R}_\mu) = \psi_{\mathbf{k}}(\mathbf{r})\, e^{i\mathbf{k}.\mathbf{R}_\mu}, \tag{2.7.15}$$

$$\int_\Omega |\psi_{\mathbf{k}}(\mathbf{r})|^2\, d\mathbf{r} = 1. \tag{2.7.16}$$

To normalize to unity over the unit cell is of course not obligatory, but the normalization *must* be independent of **k**.

To prove the Bloch character of the natural orbitals, we first notice that, taking periodic boundary conditions,

$$|\Psi(\mathbf{r}_1+\mathbf{R}_\mu, \ldots, \mathbf{r}_N+\mathbf{R}_\mu)|^2 = |\Psi(\mathbf{r}_1, \ldots, \mathbf{r}_N)|^2. \tag{2.7.17}$$

This expresses the physical equivalence of every cell of the lattice. From equation (2.7.17), it follows that

$$\Psi(\mathbf{r}_1+\mathbf{R}_\mu, \ldots, \mathbf{r}_N+\mathbf{R}_\mu) = \Psi(\mathbf{r}_1, \ldots, \mathbf{r}_N)\, e^{i\mathbf{K}.\mathbf{R}_\mu}. \tag{2.7.18}$$

Let us now expand in orthogonal Bloch functions $\phi_{\mathbf{k}}^{l_i}$ (l_i being an index corresponding to the band index of the independent-particle model):

$$\Psi = \sum_{l,\mathbf{k}} c(l_1 \ldots l_N;\, \mathbf{k}_1 \ldots \mathbf{k}_N)\, \phi_{\mathbf{k}_1}^{l_1}(\mathbf{r}_1) \ldots \phi_{\mathbf{k}_N}^{l_N}(\mathbf{r}_N). \tag{2.7.19}$$

We can see from equations (2.7.15) and (2.7.18) that the sum over $\mathbf{k}$'s is restricted by the condition

$$\sum_{i=1}^{N} \mathbf{k}_i = \mathbf{K}, \tag{2.7.20}$$

so that we may write

$$\rho(\mathbf{r}', \mathbf{r}) = \sum_{\substack{l,m \\ \mathbf{k}}} \alpha_{lm}(\mathbf{k})\, \phi_{\mathbf{k}}^{l*}(\mathbf{r}')\, \phi_{\mathbf{k}}^{m}(\mathbf{r}). \tag{2.7.21}$$

Thus, for any set of Bloch functions, the occupation numbers are always diagonal in $\mathbf{k}$. We now diagonalize in the "band-indices" l and m; the diagonalizing matrix $b_{lm}(12)$ is dependent on $\mathbf{k}$, but, because it is unitary, the natural orbitals

$$\psi_{\mathbf{k}}^{l}(\mathbf{r}) = \sum_{m} b_{lm}(\mathbf{k})\, \phi_{\mathbf{k}}^{m}(\mathbf{r}) \tag{2.7.22}$$

obey condition (2.7.16) since the $\phi_{\mathbf{k}}^{m}$ do. It can also be seen that the $\psi_{\mathbf{k}}^{l}(\mathbf{r})$ obey condition (2.7.15), again since the $\phi_{\mathbf{k}}^{m}$ do so. We may note, as a special case, that the natural orbitals of a uniform interacting electron gas must be plane-waves.

2.7.3 *Van Hove correlation function*

The pair function $P(\mathbf{r}'\mathbf{r})$ defined in equation (2.7.2) can be usefully generalized to describe time-dependent correlations. To see how to effect this generalization, we start from the identity

$$\sum_{ij}^{(N)} \langle \delta(\mathbf{r}' - \mathbf{r}_i)\, \delta(\mathbf{r} - \mathbf{r}_j) \rangle = \rho(\mathbf{r}', \mathbf{r})\, \delta(\mathbf{r}' - \mathbf{r}) + P(\mathbf{r}', \mathbf{r}). \tag{2.7.23}$$

The left-hand side is the expectation value‡ of the product of the density operators $\sum_i \delta(\mathbf{r}' - \mathbf{r}_i)$ and $\sum_j \delta(\mathbf{r} - \mathbf{r}_i)$ introduced in equation (2.3.30). Generalizing this, we can say that the probability of finding a particle at $\mathbf{r}'$ and a particle at $\mathbf{r}$ at an interval of time t later is

$$\sum_{ij} \langle \delta(\mathbf{r}' - \mathbf{r}_i(0))\, \delta[\mathbf{r} - \mathbf{r}_j(t)] \rangle = g(\mathbf{r}', \mathbf{r};\, t) = g_1(\mathbf{r}', \mathbf{r};\, t) + g_2(\mathbf{r}', \mathbf{r};\, t), \tag{2.7.24}$$

where $\sum_j \delta[\mathbf{r} - \mathbf{r}_j(t)]$ is the Heisenberg operator $e^{iHt} \sum_j \delta(\mathbf{r} - \mathbf{r}_j)\, e^{-iHt}$. $g_1(\mathbf{r}', \mathbf{r};\, t)$ is the probability of finding a particle at $\mathbf{r}'$, and then *the same*

‡ This is taken specifically with respect to the ground-state wave function here. However, the elevated temperature case is similarly dealt with by averaging over the appropriate ensemble.

particle at $\mathbf{r}$ an interval of time t later. $g_2(\mathbf{r}', \mathbf{r}; t)$ is similarly defined, but the particles at $\mathbf{r}$ and $\mathbf{r}'$ are different ones. (This division is not experimentally possible when the indistinguishability principle is operative.) g is known as the (time-dependent) correlation function; g_1 and g_2 are known as the self- and distinct-particle correlation functions. These were first introduced by van Hove (1954).

In the case of the uniform gas, which we discuss in considerable detail in this chapter, the van Hove function $g(\mathbf{r}', \mathbf{r}; t)$ depends only on $\mathbf{r}' - \mathbf{r}$. If we take a Fourier transform with respect to this variable and also with respect to t, the transformed variables being $\mathbf{k}$ and ω respectively, we obtain the so-called structure factor $S(k, \omega)$, which will be important in a number of respects later, so that we turn now to discuss it in some detail.

2.7.4 Structure factor, dielectric function and energy in uniform electron gas

Writing the delta functions in equation (2.7.24) in terms of their constant Fourier components and using equation (2.3.30), it follows readily that

$$S(k\omega) = \sum_n |\langle n| \rho_{\mathbf{k}}^\dagger |0\rangle|^2 \, \delta(\omega - \omega_n) \tag{2.7.25}$$

and we shall now derive interesting and useful relations between the structure factor, energy and dielectric constant. It follows immediately from (2.7.25) that

$$\int_0^\infty d\omega S(k\omega) = \sum_n |\langle n| \rho_{\mathbf{k}} |0\rangle|^2 = \langle \Psi_0 | \rho_{\mathbf{k}}^\dagger \rho_{\mathbf{k}} | \Psi_0 \rangle \tag{2.7.26}$$

and so if we split H_0 into kinetic and potential energy operators:

$$H_0 = T + V \tag{2.7.27}$$

it is evident from (2.3.31) that

$$\langle V \rangle = \sum_{\mathbf{k}} \frac{2\pi e^2}{k^2} \left\{ \int_0^\infty [S(k\omega) + S(k, -\omega)] \, d\omega - \rho_0 \right\}, \tag{2.7.28}$$

showing us that knowledge of the structure factor determines the potential energy of the uniform electron gas. Finally, we see from equation (2.3.39) that the imaginary part of $1/\varepsilon(k\omega)$ is given by

$$\mathrm{Im}\left(\frac{1}{\varepsilon(k\omega)}\right) = \frac{4\pi^2 e^2}{k^2} [S(k, -\omega) - S(k\omega)]. \tag{2.7.29}$$

Clearly, knowledge of an approximate form of $\varepsilon(k\omega)$, such as that given by the self-consistent field method of section 2.4.1, is giving information on the structure factor also. $S(k, -\omega)$ is evidently zero in the ground state.

2.7.5 *Density matrices and correlation functions in Hartree–Fock theory*

In the Hartree and HF approximations, the first-order density matrix is of the form we have used earlier,

$$\rho(\mathbf{r}',\mathbf{r}) = \sum_{l=1}^{N} \psi_l^*(\mathbf{r}')\,\psi_l(\mathbf{r}) \tag{2.7.30}$$

and is known as the Dirac density matrix (see Appendix 1.2 for a perturbative expansion of this).

To obtain the total energy we also require the pair function. To find this we note that we may write the HF ground-state wave function as

$$\Psi_0 = \sum_{l_1\ldots l_N}^{(N)} c(l_1\ldots l_N)\,\psi_{l_1}(\mathbf{r}_1)\ldots\psi_{l_N}(\mathbf{r}_N) = \frac{1}{2}\sum_{l_1\ldots l_N}^{(N)} c(l_1\ldots l_N)$$
$$\times\,[\psi_{l_1}(\mathbf{r}_1)\,\psi_{l_2}(\mathbf{r}_2)-\psi_{l_1}(\mathbf{r}_2)\,\psi_{l_2}(\mathbf{r}_1)]\,\psi_{l_3}(\mathbf{r}_3)\ldots\psi_{l_N}(\mathbf{r}_N), \tag{2.7.31}$$

where $c(l_1\ldots l_N) = \pm 1/N!$ when non-zero (it is antisymmetric in the l's).

From the definition (2.7.2) we readily see that

$$P(\mathbf{r}_1,\mathbf{r}_2) = \sum_{l,m}^{(N)} \psi_l^*(\mathbf{r}_1)\,\psi_m^*(\mathbf{r}_2)\,\psi_l(\mathbf{r}_1)\,\psi_m(\mathbf{r}_2) - \sum_{l,m}^{(N)} \psi_l^*(\mathbf{r}_2)\,\psi_m^*(\mathbf{r}_1)\,\psi_l(\mathbf{r}_1)\,\psi_m(\mathbf{r}_2), \tag{2.7.32}$$

which may also be written

$$P(\mathbf{r}_1,\mathbf{r}_2) = \rho(\mathbf{r}_1,\mathbf{r}_1)\,\rho(\mathbf{r}_2,\mathbf{r}_2) - \rho(\mathbf{r}_2,\mathbf{r}_1)\,\rho(\mathbf{r}_1,\mathbf{r}_2). \tag{2.7.33}$$

In the Hartree method, in contrast, only the first term of the right-hand side of equation (2.7.33) appears. (The last term $\times\frac{1}{2}$ is correct for paired spins.)

Let us also obtain the time-dependent generalization $g(\mathbf{r}',\mathbf{r};t)$. In the Hartree method applied to a large system, the single-particle functions all obey the same Schrödinger equation, the Hamiltonian of which may be used to define time-dependent operators. This cannot be done in the HF framework, however, and we need to recast the general form (2.7.24) before making approximations. Using the theorem that the many-body wave functions Ψ_m satisfy

$$\sum_m |\Psi_m\rangle\langle\Psi_m| = 1, \tag{2.7.34}$$

which expresses the completeness relation, we have

$$g(\mathbf{r}',\mathbf{r};t) = \sum_n \langle\Psi_0|\sum_i \delta(\mathbf{r}'-\mathbf{r}_i)|\Psi_n\rangle\langle\Psi_n|\mathrm{e}^{+iHt}\sum_i \delta(\mathbf{r}-\mathbf{r}_i)\,\mathrm{e}^{-iHt}|\Psi_0\rangle \tag{2.7.35}$$

or

$$g(\mathbf{r}',\mathbf{r};t) = \sum_n \langle\Psi_0|\sum_i \delta(\mathbf{r}'-\mathbf{r}_i)|\Psi_n\rangle\langle\Psi_n|\sum_i \delta(\mathbf{r}-\mathbf{r}_i)|\Psi_0\rangle\,\mathrm{e}^{i(E_n-E_0)t}. \tag{2.7.36}$$

In the HF approximation, Ψ_0 and Ψ_n become determinants, which do not differ by more than one single-particle function if they contribute non-zero terms to equation (2.7.36). Suppose the different functions are ψ_p and ψ_q. Now we remember Koopman's theorem of Appendix 1.1: the energy required to remove an electron in a HF state ψ_s from the solid is just E_s, the eigenvalue of the HF equation for ψ_s. It then follows that $(E_0 - E_n) = E_q - E_p$. Separating off the terms in which $\Psi_n = \Psi_0$, equation (2.7.36) then becomes

$$g(\mathbf{r}', \mathbf{r};\, t) = \rho(\mathbf{r}', \mathbf{r}')\, \rho(\mathbf{r}, \mathbf{r}) + \sum_{p=1}^{(N)} \sum_{q=N+1}^{\infty} \psi_p^*(\mathbf{r}')\, \psi_q(\mathbf{r}')\, \psi_p(\mathbf{r})\, \psi_q^*(\mathbf{r})\, e^{i(E_p - E_q)t}. \tag{2.7.37}$$

The second term of this may be rewritten in the form

$$\sum_{p=1}^{N} \sum_{q=1}^{\infty} \psi_p^*(\mathbf{r}')\, \psi_p(\mathbf{r})\, \psi_q^*(\mathbf{r})\, \psi_q(\mathbf{r}')\, e^{i(E_p - E_q)t}$$
$$- \sum_{p=1}^{N} \sum_{q=1}^{N} \psi_p^*(\mathbf{r}')\, \psi_p(\mathbf{r})\, \psi_q^*(\mathbf{r})\, \psi_q(\mathbf{r}')\, e^{i(E_p - E_q)t}$$

and so equation (2.7.37) may be written as

$$g(\mathbf{r}', \mathbf{r};\, t) = \rho(\mathbf{r}', \mathbf{r}')\, \rho(\mathbf{r}, \mathbf{r}) + \rho(\mathbf{r}', \mathbf{r};\, t)\, G(\mathbf{r}', \mathbf{r};\, t) - \rho(\mathbf{r}', \mathbf{r};\, t)\, \rho^*(\mathbf{r}', \mathbf{r}, t), \tag{2.7.38}$$

where

$$\rho(\mathbf{r}', \mathbf{r};\, t) = \sum_{i=1}^{(N)} \psi_i^*(\mathbf{r}')\, \psi_i(\mathbf{r})\, e^{iE_i t}, \tag{2.7.39}$$

$$G(\mathbf{r}', \mathbf{r};\, t) = \sum_{i=1}^{\infty} \psi_i^*(\mathbf{r}')\, \psi_i(\mathbf{r})\, e^{-iE_i t}. \tag{2.7.40}$$

$\rho(\mathbf{r}', \mathbf{r};\, t)$ is the time-dependent generalization of the Dirac matrix; $G(\mathbf{r}', \mathbf{r};\, t)$ is known as the time-dependent *Green function* of the system (cf. the next section).

The result (2.7.38) has been derived for use in X-ray scattering theory in Chapter 5. For the moment we turn our attention to ρ and G. In the Hartree and HF approximations, $\rho(\mathbf{r}', \mathbf{r};\, t)$ specifies the system completely. For example, the Fourier transform

$$\sigma(\mathbf{r}', \mathbf{r};\, \omega) = \frac{1}{2\pi} \int_{-\infty}^{\infty} \rho(\mathbf{r}', \mathbf{r};\, t)\, e^{-i\omega t}\, dt \tag{2.7.41}$$

can be used on the diagonal, to obtain the charge distribution of all electrons with energy $\hbar\omega$. In particular the total charge distribution is

$$\rho(\mathbf{r}) = \int_{-\infty}^{+\infty} \sigma(\mathbf{r}, \mathbf{r};\, \omega)\, d\omega. \tag{2.7.42}$$

In fact, the restriction on the sum over states in equation (2.7.39) makes $\rho(\mathbf{r}',\mathbf{r};\,t)$ very difficult to calculate directly. We therefore often calculate the Green function or an intimately related function, the Bloch matrix of Appendix 1.2; we can derive $\rho(\mathbf{r}',\mathbf{r};\,t)$ from either of these.

(a) *Fermi hole and pair function of uniform electron gas.* As a very simple, though important, example of the calculation of a HF pair function, let us consider the uniform gas again. Then, the Dirac density matrix is built from plane-waves $\mathscr{V}^{-\frac{1}{2}}e^{i\mathbf{k}.\mathbf{r}}$ with occupation numbers 1 or 0, and we may write

$$\rho(\mathbf{r}_1\,\mathbf{r}_2) = \sum_{|\mathbf{k}|<k_f} \frac{1}{\mathscr{V}} e^{-i\mathbf{k}.\mathbf{r}_1-\mathbf{r}_2}. \tag{2.7.43}$$

Replacing the summation over $\mathbf{k}$ by an integration, and recalling from Chapter 1 that there are $\mathscr{V}/8\pi^3$ states per unit volume of $\mathbf{k}$ space, we find almost immediately (cf. Appendix 1.2)

$$\rho(\mathbf{r}_1\,\mathbf{r}_2) = \frac{k_f^2}{\pi^2}\frac{j_1(k_f|\mathbf{r}_1-\mathbf{r}_2|)}{|\mathbf{r}_1-\mathbf{r}_2|}, \tag{2.7.44}$$

where we have introduced a factor of two to take account of the spin degeneracy. Hence we obtain from equation (2.7.33) the desired result, after again accounting for the spin,

$$P(\mathbf{r}_1\,\mathbf{r}_2) = \left(\frac{k_f^3}{3\pi^2}\right)^2 - \frac{k_f^4}{2\pi^4}\frac{\{j_1(k_f|\mathbf{r}_1-\mathbf{r}_2|)\}^2}{|\mathbf{r}_1-\mathbf{r}_2|^2}. \tag{2.7.45}$$

As expected, this is a function only of the distance $|\mathbf{r}_1-\mathbf{r}_2 \equiv r$ between particles and we have plotted $P(r)/\rho_0^2$ in Figure 2.3, this quantity being $g(r)$,

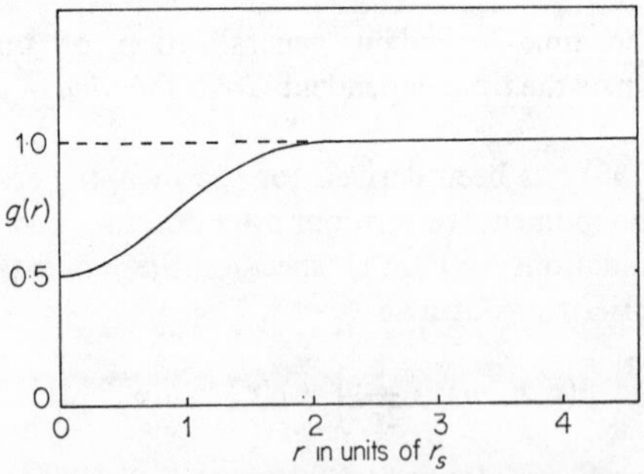

FIGURE 2.3. Fermi hole round electron in uniform gas given by equation (2.7.45). Oscillations at large r exist, but are not seen to graphical accuracy. r_s is mean interelectronic spacing.

the radial distribution function. We can view the quantity $\rho_0 g(r)$ as the electron density which would be observed if one chose to "sit" on the electron at the origin. The fact that this electron "digs a hole" around itself, actually only for parallel spin particles in this approximation, is a consequence of Fermi statistics. When we include the Coulombic interaction between electrons, the Fermi hole shown in Figure 2.3, characterized by $g(0) = \frac{1}{2}$, will clearly become a deeper "exchange plus correlation hole" and then, at the origin, we must have

$$0 < g(0) < \tfrac{1}{2}. \tag{2.7.46}$$

This is a useful condition, and gives us some criterion on the range of validity of high-density theories such as we dealt with in section 2.4. The radial distribution function $g(r)$ is, of course, a positive definite quantity. We shall return to the effect of correlations on the pair function later in this chapter.

(b) *Free-electron exchange energy.* In equation (2.4.2), we have already quoted the potential energy, or exchange energy, which results from calculating the expectation value of the Hamiltonian H_0 of equation (2.3.31) with respect to a determinant of plane-waves. However, the Fermi hole just discussed affords a more physical way of deriving this result, and we shall briefly summarize the argument here therefore. Since we can regard $\rho_0 g(r)$ as the charge distribution we see, at distance r, when we "sit" on an electron at the origin, we can write down the energy of interaction of the electron at the origin with the "displaced charge" $\rho_0[g(r)-1]$ as

$$\frac{\langle V \rangle}{N} = \frac{1}{2}\int \frac{\rho_0[g(r)-1]}{r}\,d\mathbf{r}, \tag{2.7.47}$$

the factor $\frac{1}{2}$ avoiding counting the electron–electron interactions twice over. Inserting the result

$$g(r) = 1 - \frac{9}{2}\left[\frac{j_1(k_f r)}{k_f r}\right]^2, \tag{2.7.48}$$

which follows immediately from equation (2.7.45), into equation (2.7.47) we find for the energy per particle in Rydbergs

$$\frac{\langle V \rangle}{N} = -\frac{0.916}{r_s}, \quad r_s a_0 = \{3/4\pi\rho_0\}^{\frac{1}{3}} \tag{2.7.49}$$

which shows us that the exchange energy per unit volume in a uniform electron gas is proportional to the four-thirds power of the electron density. This is to be compared with the kinetic energy density discussed in Chapter 1, which is proportional to $\rho_0^{\frac{5}{3}}$. The result quoted in equation (2.4.2) also follows immediately from equation (2.7.49).

Having illustrated the use of the pair correlation function in this way, we shall now go on to an extensive discussion of Green functions, which are very powerful tools in many-body theory and, indeed, forms of which we have already encountered in section 2.3.1 and in equation (2.7.40) above.

2.7.6 *Green functions*

(a) *Non-interacting systems.* Green functions were originally introduced to solve differential equations and we shall initially consider them from this point of view. This will prove particularly appropriate for scattering problems, and indeed we have already made use of them in this way in the discussion of the Korringa–Kohn–Rostoker method in Chapter 1, section 1.11.3, and in the derivation of the static dielectric constant $\varepsilon(\mathbf{k})$ in section 2.3.1 above.

Let us consider the solution of the Schrödinger equation with Hamiltonian $H+V(\mathbf{r})$,

$$H\psi(\mathbf{r})-E\psi(\mathbf{r}) = -V(\mathbf{r})\,\psi(\mathbf{r}). \tag{2.7.50}$$

If we introduce the Green function G obeying the equation

$$H(\mathbf{r})\,G(\mathbf{r}'\mathbf{r})-EG(\mathbf{r}'\mathbf{r}) = -\delta(\mathbf{r}'-\mathbf{r}) \tag{2.7.51}$$

then it is readily shown that equation (2.7.50) has the formal solution

$$\psi(\mathbf{r}) = \phi(\mathbf{r})+\int G(\mathbf{r}\mathbf{r}')\,V(\mathbf{r}')\,\psi(\mathbf{r}')\,d\mathbf{r}', \tag{2.7.52}$$

where we have included, in this expression for ψ, the incident wave ϕ which obeys

$$H\phi(\mathbf{r})-E\phi(\mathbf{r}) = 0. \tag{2.7.53}$$

We use the above approach to scattering problems elsewhere in the book, but, at the moment, we are interested in the Green function in its own right. It obviously contains information on the unperturbed system and, in fact, one may readily verify that G can be written as

$$G(\mathbf{r}'\mathbf{r}E) = \sum_{i=0}^{\infty}\frac{\phi_i^*(\mathbf{r}')\,\phi_i(\mathbf{r})}{E-E_i}, \tag{2.7.54}$$

where

$$H\phi_i = E_i\,\phi_i. \tag{2.7.55}$$

It is often useful to introduce a positive infinitesimal δ into the denominator of equation (2.7.54). We therefore define

$$G^{\pm}(\mathbf{r}'\mathbf{r}E) = \sum_i \frac{\phi_i^*(\mathbf{r}')\,\phi_i(\mathbf{r})}{E-E_i \pm i\delta}, \quad \delta>0. \tag{2.7.56}$$

Now we have the identity (2.3.38) and hence it follows that

$$\frac{1}{\pi} \operatorname*{Im}_{\delta\to 0} G^{-}(\mathbf{r}'\mathbf{r}E) = \sum_i \phi_i^*(\mathbf{r}')\,\phi_i(\mathbf{r})\,\delta(E-E_i). \tag{2.7.57}$$

It is now clear that we can obtain quantities of very direct interest from G. For example, the Dirac density matrix we introduced earlier is simply given by

$$\pi\rho(\mathbf{r}'\mathbf{r}\zeta) = \operatorname*{Im}_{\delta\to 0} \int_{-\infty}^{\zeta} G^{-}(\mathbf{r}'\mathbf{r}E)\,dE. \tag{2.7.58}$$

Also, the important density of electronic states, which we saw in Chapter 1 could be obtained from the dispersion relation $E(\mathbf{k})$, can be directly calculated from the Green function as

$$n(E) = \operatorname*{Im}_{\delta\to 0} \int \frac{G^{-}(\mathbf{r}\mathbf{r}E)}{\pi}\,d\mathbf{r}. \tag{2.7.59}$$

(b) *Time-dependent perturbations.* When we generalize to a time-dependent perturbation and we wish to solve

$$\left(\frac{\hbar}{i}\frac{\partial}{\partial t} + H\right)\psi(\mathbf{r}t) = -V(\mathbf{r}t)\,\psi(\mathbf{r}t) \tag{2.7.60}$$

it is natural to introduce time-dependent Green functions satisfying the equation

$$\left[\frac{\hbar}{i}\frac{\partial}{\partial t} + H(\mathbf{r})\right] G^{\pm}(\mathbf{r}'\mathbf{r}t) = -\hbar\delta(\mathbf{r}'-\mathbf{r})\,\delta(t). \tag{2.7.61}$$

It is readily shown that these quantities are simply the Fourier transforms of $G^{\pm}(\mathbf{r}'\mathbf{r}E)$, given by

$$2\pi G^{\pm}(\mathbf{r}'\mathbf{r}t) = \int_{-\infty}^{\infty} e^{-iEt/\hbar}\, G^{\pm}(\mathbf{r}'\mathbf{r}E)\,dE. \tag{2.7.62}$$

For the case $t<0$, we close the contour by means of a semi-circle at infinity in the upper half-plane, whereas if $t>0$ we close it in the lower half-plane. It is then not difficult to see that

$$G^{+} = 0, \quad t<0; \quad G^{-} = 0, \quad t>0. \tag{2.7.63}$$

These conditions have physical significance for the solution of equation (2.7.60) in terms of G. We have the solution

$$\psi(\mathbf{r}, t) = \phi(\mathbf{r}t) + \int G^{+}(\mathbf{r}, \mathbf{r}', t-t')\, V(\mathbf{r}', t')\, \psi(\mathbf{r}' t')\, d\mathbf{r}'\, dt' \qquad (2.7.64)$$

and the conditions (2.7.63) on G^{+} imply that values of $V(\mathbf{r}', t')\,\psi(\mathbf{r}', t')$ for $t' > t$ do not enter in the determination of $\psi(\mathbf{r}, t)$. In this sense, G^{+} embodies the principle of causality and is termed a "causal" Green function. Alternatively, G^{+} is termed the "retarded" Green function, while G^{-} is called the "advanced" Green function.

An important application of the Green functions is to describe the time evolution of a system. If the system is described by the wave function $\phi(\mathbf{r}, t_0)$ at time t_0, it will develop in time according to

$$\phi(\mathbf{r}, t) = \mathrm{e}^{-iH(t-t_0)/\hbar}\, \phi(\mathbf{r}, t_0) \qquad (2.7.65)$$

which may also be written

$$\phi(\mathbf{r}, t) = \sum_i \phi_i(\mathbf{r}) \langle \phi_i^* \,\mathrm{e}^{-iH(t-t_0)/\hbar}\, \phi(\mathbf{r}, t_0)\rangle$$

$$= \sum_i \phi_i(\mathbf{r}) \langle \phi_i^* \,\phi(\mathbf{r}, t_0)\rangle\, \mathrm{e}^{-iE_i(t-t_0)/\hbar}, \qquad (2.7.66)$$

that is

$$\phi(\mathbf{r}, t) = \int G(\mathbf{r}, \mathbf{r}', t-t_0)\, \phi(\mathbf{r}', t_0)\, d\mathbf{r}'. \qquad (2.7.67)$$

Combining this with equation (2.7.64), the following integral equation for the Green functions can be obtained

$$G^{\pm}(\mathbf{r}, \mathbf{r}', t) = G_0^{\pm}(\mathbf{r}, \mathbf{r}', t) + \int G_0^{\pm}(\mathbf{r}, \mathbf{r}'', t'-t)\, V(\mathbf{r}'' t')\, G^{\pm}(\mathbf{r}'', \mathbf{r}', t')\, d\mathbf{r}''\, dt', \qquad (2.7.68)$$

where G_0 is now written for the unperturbed Green function corresponding to $V = 0$. This may be Fourier transformed to obtain equations for $G^{\pm}(\mathbf{r}, \mathbf{r}', E)$ and when V is independent of time we find, in particular,

$$G^{\pm}(\mathbf{r}, \mathbf{r}', E) = G_0^{\pm}(\mathbf{r}, \mathbf{r}', E) + \int d\mathbf{r}''\, G_0^{\pm}(\mathbf{r}, \mathbf{r}'', E)\, V(\mathbf{r}'')\, G^{\pm}(\mathbf{r}'', \mathbf{r}', E). \qquad (2.7.69)$$

(c) *Interacting systems in ground state.*

(i) *Single-particle Green function*: Regarding the Green function as a description of the development in time of a system, it may be generalized to apply to interacting systems in a very natural way. We use the occupation number formalism and suppose that the ground state is described by $|\psi_0\rangle$. If at time $t = 0$ we add a particle with momentum $\mathbf{k}$, the system is described, immediately after this addition, by $a_{\mathbf{k}}^\dagger|\psi_0\rangle$, where $a_{\mathbf{k}}^\dagger$ is the creation operator for the plane-wave state $\exp(i\mathbf{k}.\mathbf{r})$. The development of the system in time will now proceed according to $e^{-iHt/\hbar}a_{\mathbf{k}}^\dagger|\psi_0\rangle$ and if we require the probability amplitude for the persistence of the added particle in the precise momentum state $\mathbf{k}$, we must take the scalar product of $e^{-iHt/\hbar}|\psi_0\rangle$ with the function describing the ground state plus a particle added in a state $\mathbf{k}$ at time t. At this time, the ground state is described by $e^{-iHt/\hbar}|\psi_0\rangle$, and immediately after the addition of the particle is described by $a_{\mathbf{k}}^\dagger e^{-iHt/\hbar}|\psi_0\rangle$. The probability amplitude we require is thus $\langle\psi_0|e^{iHt/\hbar}a_{\mathbf{k}}e^{-iHt/\hbar}a_{\mathbf{k}}^\dagger|\psi_0\rangle$. We can in an analogous way describe the persistence of an empty state, and these descriptions are combined in the following definition of the single-particle Green function

$$G(\mathbf{k},\mathbf{k}_0,t-t_0) = -i\langle\psi_0|T[a_{\mathbf{k}}(t)\,a_{\mathbf{k}_0}^\dagger(t_0)]|\psi_0\rangle, \tag{2.7.70}$$

where we have included off-diagonal elements in the definition. T is an operator which orders earlier times to the right and simultaneously multiplies that on which it operates by $(-1)^p$, where p is the number of exchanges of Fermion operators needed to accomplish the desired ordering.

Equation (2.7.70) gives the Green function in momentum space, whereas we originally introduced Green functions in configuration space. We may define a configuration-space single-particle Green function in the occupation-number representation as

$$G(\mathbf{r},\mathbf{r}_0,t-t_0) = -i\langle\psi_0|T[\psi(\mathbf{r},t)\,\psi^\dagger(\mathbf{r}_0,t_0)]|\psi_0\rangle, \tag{2.7.71}$$

where $\psi^\dagger(\mathbf{r})$ is a field operator destroying a particle at $\mathbf{r}$. In terms of the annihilation operators for plane-wave states it may be written as

$$\psi(\mathbf{r}) = \sum_{\mathbf{k}} a_{\mathbf{k}}e^{i\mathbf{k}.\mathbf{r}} \tag{2.7.72}$$

and it can be seen that $G(\mathbf{r},\mathbf{r}_0,t-t_0)$ is just the Fourier transform of $G(\mathbf{k},\mathbf{k}_0,t-t_0)$. Actually, our present definitions do not reduce precisely to our original ones when the interactions are zero; equation (2.7.70) is chosen as the definition most suitable when the many-body problem is formulated in the language of second quantization. The correspondence is close, however. In particular, we shall find later on that the single-particle Green function obeys an equation analogous to equations (2.7.68) or (2.7.69) and we shall prove immediately that the energies of elementary excitations of the system appear in denominators in a similar way to the appearance of the E_i in equation (2.7.56).

To do so, we introduce as intermediate states all the exact levels of the $(N+1)$-particle system. Then the diagonal element $\mathbf{k}_0 = \mathbf{k}$ of equation (2.7.70) becomes

$$\begin{aligned} G(\mathbf{k}, t) &= -i \sum_n \langle \psi_0 | a_{\mathbf{k}} | \psi_n \rangle \langle \psi_n | e^{-iHt} a_{\mathbf{k}}^\dagger | \psi_0 \rangle e^{+iE_0 t} \\ &= -i \sum_n |(a_{\mathbf{k}}^\dagger)_{n0}|^2 e^{-i(E_n - E_0)t}, \quad t > 0, \end{aligned} \tag{2.7.73}$$

where the latter equation defines $(a_{\mathbf{k}}^\dagger)_{n0}$.

We now set

$$E_n(N+1) - E_0(N) = E_n(N+1) - E_0(N+1) + E_0(N+1) - E_0(N) = \omega_{n0} + \mu, \tag{2.7.74}$$

where ω_{n0} is an excitation energy in the $(N+1)$-particle system and μ is the chemical potential of the N-particle system. We also have

$$G(\mathbf{k}, t) = i \sum_n |(a_{\mathbf{k}})_{n0}|^2 e^{i(E_n - E_0)t}, \quad t < 0, \tag{2.7.75}$$

with intermediate states of $N-1$ particles and we can again write

$$E_n(N-1) - E_0(N) = \omega'_{n0} - \mu'. \tag{2.7.76}$$

For a very large system, we can put

$$\omega'_{n0} = \omega_{n0}, \quad \mu' = \mu, \tag{2.7.77}$$

with an error of order $1/N$. Hence

$$G(\mathbf{k}, t) = \begin{cases} -i \sum_n |(a_{\mathbf{k}}^\dagger)_{n0}|^2 e^{-i(\omega_{n0}+\mu)t}, & t > 0, \\ i \sum_n |(a_{\mathbf{k}})_{n0}|^2 e^{i(\omega_{n0}-\mu)t}, & t < 0. \end{cases} \tag{2.7.78}$$

(ii) *Spectral density functions*: As $N \to \infty$, we convert the sums to integrals by introducing the spectral density functions

$$A^+(\mathbf{k}, \omega) = \sum_n |(a_{\mathbf{k}}^\dagger)_{n0}|^2 \, \delta(\omega - \omega_{n0}), \tag{2.7.79}$$

$$A^-(\mathbf{k}, \omega) = \sum_n |(a_{\mathbf{k}})_{n0}|^2 \, \delta(\omega - \omega_{n0}). \tag{2.7.80}$$

Then we may write

$$G(\mathbf{k}, t) = \begin{cases} -i \int_0^\infty d\omega \, A^+(\mathbf{k}, \omega) \, e^{-i(\omega+\mu)t}, & t > 0, \\ i \int_0^\infty d\omega \, A^-(\mathbf{k}, \omega) \, e^{i(\omega-\mu)t}, & t < 0, \end{cases} \tag{2.7.81}$$

and the Fourier transform with respect to time is

$$G(\mathbf{k},\omega) = \int_0^\infty d\omega' \left[\frac{A^+(\mathbf{k},\omega')}{\omega-(\omega'+\mu)+i\delta} + \frac{A^-(\mathbf{k},\omega')}{\omega'+\omega-\mu-i\delta}\right]. \tag{2.7.82}$$

In quantum field theory, this is known as the Lehmann representation. It should be noted that the A's are real and that

$$A^{\pm}(\mathbf{k},\omega) \geqslant 0. \tag{2.7.83}$$

Furthermore, we have the relations

$$\left.\begin{aligned} A^+(\mathbf{k},\omega-\mu) &= -\frac{1}{\pi}\operatorname{Im} G(\mathbf{k},\omega), \quad \omega>\mu, \\ A^-(\mathbf{k},\mu-\omega) &= \frac{1}{\pi}\operatorname{Im} G(\mathbf{k},\omega), \quad \omega<\mu. \end{aligned}\right\} \tag{2.7.84}$$

The spectral density function $A^+(\mathbf{k}\omega)$ gives the probability that the original system plus the particle added into state $\mathbf{k}$ will be found in an exact eigenstate of the $(N+1)$-body system, as can be seen from equation (2.7.79).

(iii) *Green functions in terms of natural orbitals*: Although we originally included off-diagonal elements in equation (2.7.70), these must actually be zero if the system described is the homogeneous electron gas. To see this, we first remark that the wave functions of the system can all be chosen to obey a condition of the form

$$\psi_n(\mathbf{r}_1+\boldsymbol{\rho}, \mathbf{r}_2+\boldsymbol{\rho}, \ldots, \mathbf{r}_N+\boldsymbol{\rho}) = e^{i\boldsymbol{\kappa}_n\cdot\boldsymbol{\rho}}\psi_n(\mathbf{r}_1, \mathbf{r}_2, \ldots, \mathbf{r}_N), \tag{2.7.85}$$

where $\boldsymbol{\rho}$ is any translation. It must therefore follow that in the expansion

$$\psi_n = \sum_{\mathbf{k}_1\ldots\mathbf{k}_N} c_n(\mathbf{k}_1, \ldots, \mathbf{k}_N)\,\psi_S(\mathbf{k}_1, \ldots, \mathbf{k}_N), \tag{2.7.86}$$

where ψ_S is a Slater determinant of plane-waves, the c's are non-zero only if

$$\sum_{i=1}^{N} \mathbf{k}_i = \boldsymbol{\kappa}_n. \tag{2.7.87}$$

The corresponding expression in the occupation-number representation is

$$|\psi_n\rangle = \sum_{\mathbf{k}_1\ldots\mathbf{k}_N} c_n(\mathbf{k}_1, \ldots, \mathbf{k}_N) \prod_{i=1}^{N} a_{\mathbf{k}_i}^\dagger |0\rangle \tag{2.7.88}$$

with the $\mathbf{k}$'s again obeying equation (2.7.87). $\boldsymbol{\kappa}_n$ is, of course, simply the total momentum. With this in mind, we now examine the expression

$$\langle\psi_0|\,a_{\mathbf{k}_0}^\dagger e^{iHt} a_{\mathbf{k}} e^{-iHt}|\psi_0\rangle = \sum_n \langle\psi_0|\,a_{\mathbf{k}_0}^\dagger|\psi_n\rangle\langle\psi_n|\,a_{\mathbf{k}}|\psi_0\rangle\, e^{i(E_n-E_0)t}, \tag{2.7.89}$$

where the ψ_n are eigenstates of the $(N-1)$-particle system. These eigenstates obey equation (2.7.85), so that, in order that $\langle\psi_n|a_{\mathbf{k}}|\psi_0\rangle$ is non-zero, we must have

$$\boldsymbol{\kappa}_n = \boldsymbol{\kappa}_0 - \mathbf{k}. \tag{2.7.90}$$

Similarly, $\langle\psi_0|a^\dagger_{\mathbf{k}_0}|\psi_n\rangle$ is non-zero only if

$$\boldsymbol{\kappa}_n = \boldsymbol{\kappa}_0 - \mathbf{k}_0 \tag{2.7.91}$$

and so the left-hand side of equation (2.7.89) is non-zero only if $\mathbf{k} = \mathbf{k}_0$. It is thus clear that $G_0(\mathbf{k}, \mathbf{k}_0, t)$ is a diagonal matrix.

It must be remarked that we can define Green functions with respect to creation and annihilation operators for any set of one-particle functions obeying the correct boundary conditions. For example, in discussing Bloch electrons, we may regard the $a_{\mathbf{k}}$ of equation (2.7.70) as the annihilation operator for a Bloch state $\mathbf{k}$. The Green function so defined is diagonal, the proof being exactly the same as that used for the homogeneous electron gas, with $\boldsymbol{\rho}$ now representing any lattice vector. We may also note that, if a Green function is diagonal with respect to some set of single-particle functions, these functions are the natural orbitals of the problem (cf. section 2.7.2).

(iv) *Energy of ground state in terms of single-particle Green function*: We shall now prove the very interesting result that provided the system has only two-particle interactions the ground-state energy may be written solely in terms of the single-particle Green function introduced above.

We consider a Hamiltonian of the form [cf. equation (2.5.23)]

$$H = H_0 + H_1, \tag{2.7.92}$$

where

$$H_0 = \sum_{\mathbf{k}} E_{\mathbf{k}}\, a^\dagger_{\mathbf{k}} a_{\mathbf{k}} \tag{2.7.93}$$

and

$$H_1 = \sum_{\mathbf{k}_1\mathbf{k}_2\mathbf{k}_3\mathbf{k}_4} V_{\mathbf{k}_1\mathbf{k}_2\mathbf{k}_3\mathbf{k}_4}\, a^\dagger_{\mathbf{k}_1} a^\dagger_{\mathbf{k}_2} a_{\mathbf{k}_3} a_{\mathbf{k}_4}. \tag{2.7.94}$$

Now with these forms of H_0 and H_1 we find

$$\sum_{\mathbf{k}} a^\dagger_{\mathbf{k}}[H_0, a_{\mathbf{k}}] = -H_0 \tag{2.7.95}$$

and

$$\sum_{\mathbf{k}} a^\dagger_{\mathbf{k}}[H_1, a_{\mathbf{k}}] = -2H_1. \tag{2.7.96}$$

Hence we may write

$$\sum_{\mathbf{k}} \langle\psi_0|a^\dagger_{\mathbf{k}}[H, a_{\mathbf{k}}]|\psi_0\rangle = -\langle\psi_0|H_0 + 2H_1|\psi_0\rangle \tag{2.7.97}$$

so that the ground-state energy may be written as

$$E_0 = \langle\psi_0|\,H_0+H_1\,|\psi_0\rangle = \frac{1}{2}\Big(\langle H_0\rangle - \sum_{\mathbf{k}}\langle\psi_0|\,a_{\mathbf{k}}^{\dagger}[H, a_{\mathbf{k}}]\,|\psi_0\rangle\Big). \qquad (2.7.98)$$

But from the definition (2.7.80) of the spectral-density function,

$$\langle\psi_0|\,a_{\mathbf{k}}^{\dagger}[H, a_{\mathbf{k}}]\,|\psi_0\rangle = \int_0^{\infty} d\omega\, A^{-}(\mathbf{k}, \omega)\,(\omega-\mu) \qquad (2.7.99)$$

and the right-hand side of this equation is readily rewritten as

$$\int_0^{\infty} d\omega\, A^{-}(\mathbf{k}, \omega)\,(\omega-\mu) = -\left[\frac{d}{dt}G(\mathbf{k}t)\right]_{t\to -0} = \frac{i}{2\pi}\int_c d\omega\,\omega G(\mathbf{k}\omega), \qquad (2.7.100)$$

whence we obtain

$$E_0 = \frac{1}{4\pi i}\sum_{\mathbf{k}}\int_c d\omega\, G(\mathbf{k}\omega)\,[E_{\mathbf{k}}+\omega], \qquad (2.7.101)$$

where c is a contour consisting of the real axis and a semi-circle at infinity in the upper-half plane. Here we have made use of the expressions

$$\langle H_0\rangle = \int P(\mathbf{k})\, E_{\mathbf{k}}\, d\mathbf{k} \qquad (2.7.102)$$

and

$$P(\mathbf{k}) = \langle\psi_0|\,a_{\mathbf{k}}^{\dagger} a_{\mathbf{k}}\,|\psi_0\rangle = \frac{1}{2\pi i}\int_c d\omega\, G(\mathbf{k}\omega). \qquad (2.7.103)$$

For a uniform electron gas, $\langle H_0\rangle$ is the kinetic energy and $P(\mathbf{k})$ the momentum distribution. The above result for the energy in terms of G is useful in discussing the cohesive energy of simple metals.

(v) *Higher-order Green functions*: In a similar manner to the definition of the one-particle Green function, we can define higher-order functions. For example, the two-particle Green function is defined as

$$G_2(1234) = -i\langle\psi_0|\,T\,[\psi(\mathbf{r}_1 t_1)\,\psi(\mathbf{r}_2 t_2)\,\psi^{\dagger}(\mathbf{r}_3 t_3)\,\psi^{\dagger}(\mathbf{r}_4 t_4)]\,|\psi_0\rangle. \qquad (2.7.104)$$

Though we shall discuss this in considerable detail later, we shall merely indicate here its relation to the van Hove correlation function of section 2.7.3. To do so, we put $1 \equiv 3$ and $2 \equiv 4$ in (2.7.104) and if we impose the requirement that T orders a creation operator to the left of an annihilation operator when the two operators are labelled by the same times, we have

$$\begin{aligned} G_2(1212) &= -i\langle\psi_0|\,T\,[\psi^{\dagger}(\mathbf{r}_1 t_1)\,\psi(\mathbf{r}_1 t_1)\,\psi^{\dagger}(\mathbf{r}_2 t_2)\,\psi(\mathbf{r}_2 t_2)]\,|\psi_0\rangle \\ &= -i\langle\psi_0|\,T\,[\rho(\mathbf{r}_1 t_1)\,\rho(\mathbf{r}_2 t_2)]\,|\psi_0\rangle, \end{aligned} \qquad (2.7.105)$$

where

$$\rho(\mathbf{r}t) = \psi^\dagger(\mathbf{r}t)\,\psi(\mathbf{r}t) = e^{iHt}\,\psi^\dagger(\mathbf{r})\,\psi(\mathbf{r})\,e^{-iHt}. \tag{2.7.106}$$

But this is simply the density operator, as will be seen by referring to equations (2.5.21) and (2.7.72). Hence $G_2(1212)$ is seen to be identical to the van Hove correlation function defined in equation (2.7.24).

At this point, we want to briefly consider the equations of motion for the Green functions of a many-body system. For example, it can be shown that for a uniform gas of electrons interacting through a potential v, the equation of motion for the one-particle Green function is (to be proved in Chapter 10)

$$\left(i\frac{\partial}{\partial t_1}+\frac{\hbar^2}{2m}\nabla_1^2\right)G(12) = -\,\delta(12)+\int d1'\,v(11')\,G_2(1'1,1'2), \tag{2.7.107}$$

which evidently involves the two-particle Green function. In general, the $(n+1)$th-order Green function enters into the equation of motion for the nth-order Green function, and to make progress in solving this hierarchy of equations some "decoupling" procedure must be employed. This means, for example, that we try to get approximate expressions for the $(n+1)$th-order Green function in terms of lower-order functions. This can usefully be done for certain problems in magnetism, as we shall see in Chapter 4, and also in the work of Singwi and co-workers (1968) which we discuss in section 2.11.3 below. Mainly we shall adopt here the alternative of perturbation expansions for the Green functions. In these expansions, the higher-order Green functions do not appear explicitly.

2.8 Quasi-particles

2.8.1 Intuitive discussion

We have seen that, in the self-consistent field or random-phase approximation‡ schemes, the response of the electron gas to an external potential is that of independent particles responding to a *screened* external field. This leads us to the picture of the long-range Coulombic forces being screened out round an individual electron by a surrounding cloud of other electrons (actually a hole in the electron distribution). We therefore can think of the electron gas as composed of quasi-particles, each of these being the entity of electron plus screening cloud. This helps us to envisage the description of the low-lying excitations of the system in terms of an independent particle model.

Now let us suppose that we add n extra particles to the ground state of an $(N-n)$-body system. We shall add them one by one, each time leaving the

‡ See section 2.3 and Appendix 2.1.

system in its ground state. Let us denote the increase in energy on adding the first particle by $\tilde{\varepsilon}_1$, the increase due to the addition of the second by $\tilde{\varepsilon}_2$ and so on. The final energy is then, of course,

$$E_0(N) = E_0(N-n) + \sum_{i=1}^{n} \tilde{\varepsilon}_i. \tag{2.8.1}$$

We can argue that each added particle will itself provide "a nucleus" for a quasi-particle and if $n/N \ll 1$ we can expect little interference between the additional quasi-particles. Hence the $\tilde{\varepsilon}_i$ will, to a first approximation, refer to independent quasi-particle states. Let us suppose, therefore, that we omit the jth particle. Then we will have an excited state of the $(N-1)$-body system, with energy

$$E_{\text{ex}}(N-1) = E_0(N-n) + \sum_{i \neq j}' \tilde{\varepsilon}_i. \tag{2.8.2}$$

Furthermore, we can now obtain an excited state of the N-body system by adding a quasi-particle of energy $\tilde{\varepsilon}_l$ $(l > n)$ to this $(N-1)$-body state; namely

$$E_{\text{ex}}(N) = E_0(N-n) + \sum_{i \neq j}' \tilde{\varepsilon}_i + \tilde{\varepsilon}_l. \tag{2.8.3}$$

We can usefully refer to the vacant state of energy $\tilde{\varepsilon}_j$ as a "quasi-hole" and thus equation (2.8.3) expresses the energy of the system on creation of a particle–hole pair.

It must be remarked that, for a translationally invariant system, we can choose the eigenstates such that they each correspond to a definite momentum. The argument we have given for the energies of the low-lying excitations applies equally well to their momenta. We therefore conclude that each quasi-particle is labelled with a definite momentum. To find the values of these momenta, we turn to examine the Green function of the system.

2.8.2 *Description in terms of Green functions*

Let us suppose that an $(N+1)$-body eigenstate can be obtained by the addition of a quasi-particle of momentum $\mathbf{k}$. If, instead, we add a bare particle of momentum $\mathbf{k}$, it will gather a screening cloud round it, and become a quasi-particle. Of course, the state $a_{\mathbf{k}}^{\dagger}|\psi_0\rangle$, where $|\psi_0\rangle$ is the ground-state wave function, will not, in general, be an *exact* eigenstate, but we expect it to have finite overlap with the corresponding quasi-particle state. This leads us to consider the single-particle Green function

$$G(\mathbf{k}t) = \begin{cases} -i\langle\psi_0|\, a_{\mathbf{k}}(t)\, a_{\mathbf{k}}^{\dagger}(0)\,|\psi_0\rangle, & t > 0, \\ i\langle\psi_0|\, a_{\mathbf{k}}^{\dagger}(0)\, a_{\mathbf{k}}(t)\,|\psi_0\rangle, & t < 0, \end{cases} \tag{2.8.4}$$

which tells us about the way in which single-particle excitations of the system can propagate and interact.

(a) *Non-interacting particles.* For a system of non-interacting particles in its ground state, we readily find that, with $\varepsilon(\mathbf{k})$ as the one-electron eigenvalues

$$G_0(\mathbf{k}t) = \begin{cases} -i(1-P(\mathbf{k}))\,e^{-i\varepsilon(\mathbf{k})t}, & t>0, \\ iP(\mathbf{k})\,e^{-i\varepsilon(\mathbf{k})t}, & t<0, \end{cases} \tag{2.8.5}$$

where $P(\mathbf{k})$ is the momentum distribution given by

$$P(\mathbf{k}) = \begin{cases} 1, & k<k_f, \\ 0, & k>k_f. \end{cases} \tag{2.8.6}$$

The Fourier transform of equation (2.8.5) is simply

$$G_0(\mathbf{k}\omega) = \frac{1}{\omega-\varepsilon(\mathbf{k})+i\delta_{\mathbf{k}}}, \tag{2.8.7}$$

where $\delta_{\mathbf{k}}$ is an infinitesimal, such that

$$\delta_{\mathbf{k}}>0, \quad k>k_f; \quad \delta_{\mathbf{k}}<0, \quad k<k_f. \tag{2.8.8}$$

(b) *Interacting particles.* We have seen that the spectral density function $A^{+}(\mathbf{k}\omega)$ gives the probability that $a_{\mathbf{k}}^{\dagger}|\psi_0\rangle$ is found in an exact eigenstate

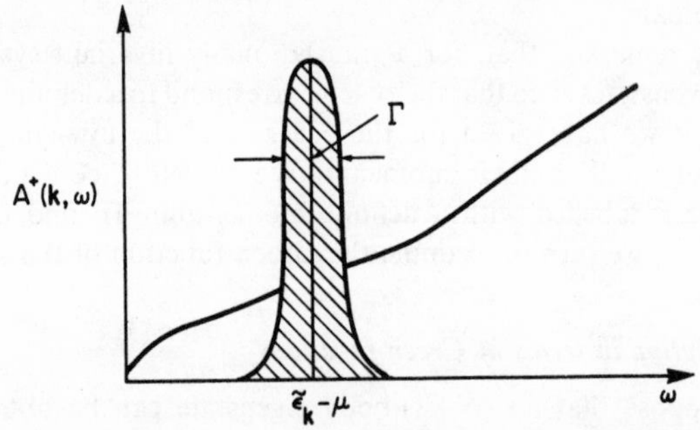

FIGURE 2.4. Spectral density function $A^{+}(\mathbf{k}\omega)$ against ω. Peak defines quasi-particle state. If this state had infinite lifetime, this would be a delta function peak.

with excitation energy ω. Hence, if $\mathbf{k}$ labels a quasi-particle state, $A^{+}(\mathbf{k}\omega)$ will have a strong maximum at $\tilde{\varepsilon}_{\mathbf{k}} = \omega+\mu$, the energy of the quasi-particle state, as shown in Figure 2.4. If $a_{\mathbf{k}}^{\dagger}|\psi_0\rangle$ were an exact eigenstate, this peak would be a δ-function, as we can see from equation (2.8.7) when we recall that A^{+} is, essentially, the imaginary part of G.

We therefore expect that instead of equation (2.8.7), with a pole infinitesimally displaced from the real axis, we must write

$$G(\mathbf{k}\omega) = \frac{z(\mathbf{k})}{\omega - \tilde{\varepsilon}(k) + i\Gamma(\mathbf{k})} + \text{well-behaved correction terms}, \qquad (2.8.9)$$

where $\Gamma(\mathbf{k})$ is in general finite and $z(\mathbf{k}) \leqslant 1$. Taking the Fourier transform of equation (2.8.9) we find

$$iG(\mathbf{k}t) = z(\mathbf{k}) \exp\left[-i\tilde{\varepsilon}(\mathbf{k})\, t - \Gamma(\mathbf{k})\, t\right] + G_{\text{inc}}, \quad t > 0, \qquad (2.8.10)$$

if $\Gamma(\mathbf{k}) > 0$. This shows us immediately that $\Gamma(\mathbf{k})$ is the inverse lifetime of the particle. We recall from section 2.7.6(c) that $G(\mathbf{k}t)$ is the probability amplitude for the persistence of an added particle in state $\mathbf{k}$.

We can expect $G(\mathbf{k}\omega)$ to have other poles than the one shown explicitly in (2.8.9), but the pole nearest to the real axis will dominate in defining the quasi-particle. This means, however, that we must wait until the contribution of the correction term, or incoherent part G_{inc} of G in (2.8.10), is sufficiently damped. We might argue that the quasi-particle must be given time to form: the time it takes the screening cloud to gather round the particle. Calling this time τ_c, which, in a metal, we can expect to be of the order of the period of the plasma oscillations discussed in section 2.2, the quasi-particle is well defined in the region of time t such that

$$\tau_c \ll t \ll \frac{1}{\Gamma}. \qquad (2.8.11)$$

We have been speaking so far about quasi-particles. If, however, Γ is negative, the simple pole of (2.8.9) will contribute to $G(\mathbf{k}t)$ only if $t < 0$. Equation (2.8.4) then shows that $G(\mathbf{k}t)$ must describe a quasi-hole rather than a quasi-particle.

2.8.3 Fermi surface

We shall now suppose that $\Gamma(\mathbf{k})$ changes sign as we cross a surface S_F, which we define as the Fermi surface. According to the above discussion, on one side of this surface, $\mathbf{k}$ describes a quasi-particle, and on the other side it describes a quasi-hole. It is not possible to add a quasi-particle to the system, with a $\mathbf{k}$ vector lying within the quasi-hole region.

Let us now add δN particles to the N-body ground state to form the $(N+\delta N)$-body ground state; the particles being added into quasi-particle states $\mathbf{k}_i$ $(i = 1, \ldots, \delta N)$. We assume that the quasi-particles are Fermions: we shall prove this in section 2.10.7(b). Then, since the states described by the $\mathbf{k}_i$ are now occupied, the $\mathbf{k}_i$ must describe *quasi-hole* states of the $(N+\delta N)$-body system and the Fermi surface will shift accordingly. Since there are $1/4\pi^3$ states per unit volume in momentum space, the volume enclosed between the old and the new Fermi surfaces is

$$\delta V_F = \frac{4\pi^3}{\mathscr{V}}\,\delta N, \tag{2.8.12}$$

where $\mathscr{V}$ is the volume of the system.

If the density ρ_0 were very small, then for suitable short-range interactions the particles would be virtually non-interacting, so that as $\rho_0 \to 0$ the Fermi surface would shrink to a point. Hence if S_F is a continuous function of particle number, equation (2.8.12) can be integrated to show that S_F encloses just N states in momentum space and so is just the Fermi surface of the non-interacting system. We shall assume this here, though we shall encounter an exception to this rule in Chapter 4.

Figure 2.5(a) shows the momentum distribution of non-interacting particles, with its discontinuity at $k = k_f$. We can argue that the Fermi

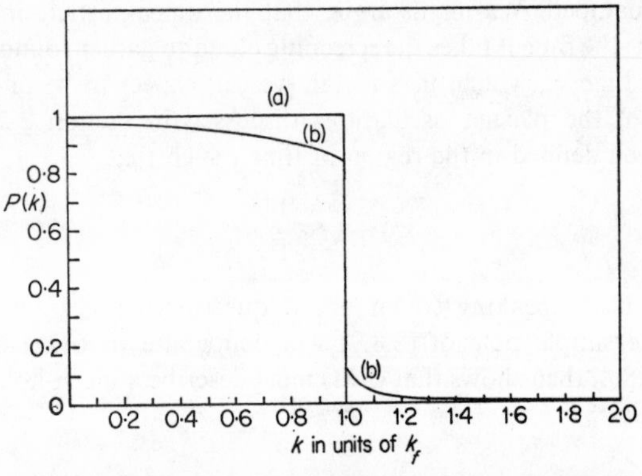

FIGURE 2.5. Momentum distribution $P(\mathbf{k})$ of uniform electron gas. Curve (a) Non-interacting particles. Curve (b) Including interactions.

surface will persist, as a discontinuity in the momentum distribution at k_f,‡ even when the interactions are switched on. Formally, this is a consequence of the change of sign of $\Gamma(\mathbf{k})$ when $k = k_f$. Since

$$P(\mathbf{k}) = \langle \psi_0 | a_\mathbf{k}^\dagger a_\mathbf{k} | \psi_0 \rangle = -iG(\mathbf{k}, -0) \tag{2.8.13}$$

‡ In section 2.8.4, it is argued that this is true, as we switch on interactions in the high-density limit. However, when a transition occurs to the Wigner lattice state, the discontinuity is expected to vanish.

we may write

$$P(\mathbf{k}) = \frac{1}{2\pi i}\int_c d\omega\, G(\mathbf{k}\omega), \tag{2.8.14}$$

where the contour c consists of the real axis closed by a semi-circle at infinity in the upper half-plane. Hence, when $k > k_f$, so that the pole of equation (2.8.9) is in the lower half-plane, the contribution to $P(\mathbf{k})$ will come solely from the correction terms. On the other hand, when $k < k_f$, the pole contributes $z(\mathbf{k})$ to $P(\mathbf{k})$. The momentum distribution is therefore of the form of Figure 2.5(b). Calculations by Daniel and Vosko (1960) indicate that $z(k_f) \simeq 0.5$ for sodium, but this is certainly extrapolating their theory well outside its proper range of validity.

The argument given above relies on the system being in a *normal* state. By this, we mean that the functions we deal with are well behaved, in the sense that the states of the non-interacting system transform smoothly into those of the interacting system, as the interactions are switched on. This is at the basis of the Landau quasi-particle theory (to be discussed fully in section 2.9. below), in which Figure 2.5(a) gives the distribution in $\mathbf{k}$ of the quasi-particles. The correspondence with the non-interacting system can be seen in the present discussion by comparing equations (2.8.7) and (2.8.9), and noting the persistence of a peak in $A^{\dagger}(\mathbf{k}\omega)$ at $\tilde{\varepsilon}_{\mathbf{k}} - \mu$ when interactions take place.

It must be emphasized that we are *not*, in this way, excluding the possibility of new phenomena, such as collective excitations, which disappear as the interaction is reduced to zero. All that we are requiring is that certain characteristics of the non-interacting system remain when the interaction is increased from zero to its full value. Thus, we exclude the possibility of a phase change, such as a change to the superconducting state.

While Figure 2.5(a) gives the quasi-particle distribution over the entire range of $\mathbf{k}$, the quasi-particles will decay, and probably will have no more than formal significance except near the Fermi surface, where, because of the change of sign across S_F

$$\Gamma(\mathbf{k}) \to 0 \quad \text{as} \quad k \to k_f. \tag{2.8.15}$$

Thus, the quasi-particle lifetime approaches infinity as $\mathbf{k}$ approaches the Fermi surface. Fortunately, many physical properties depend only on the quasi-particle behaviour near S_F, as we shall see when we discuss the Landau Fermi-liquid theory.

2.8.4 Quasi-particle lifetime

To investigate the way in which Γ tends to zero, let us consider a system with a single quasi-particle above the Fermi surface, in state $\mathbf{k}$ ($k > k_f$). Let us now focus attention on the scattering of this quasi-particle with a quasi-particle inside the Fermi surface. The transition probability for such two-particle scattering of the added quasi-particle will depend on the product of the number of particles beneath the Fermi surface available for scattering with the number of states into which the particle can scatter. Thus we have

$$W_{\mathbf{k}} \propto \int d\mathbf{k}' \int d\mathbf{k}_1, \tag{2.8.16}$$

where $\mathbf{k}, \mathbf{k}'$ are the initial states and $\mathbf{k}_1, \mathbf{k}_1'$ are the final states. To obtain the limits of integration in (2.8.16), we must take account first of momentum conservation through

$$\mathbf{k}+\mathbf{k}' = \mathbf{k}_1+\mathbf{k}_1'. \tag{2.8.17}$$

The exclusion principle, which must, of course, operate because the quasi-particles are Fermions, insists that $\mathbf{k}_1, \mathbf{k}_1'$ are such that $\mathbf{k}_1, \mathbf{k}_1' > k_f$. Let us assume then that $k = k_f + \delta$. We have then

$$k_f \geqslant k' \geqslant k_f - \delta, \quad k_f < k_1 \leqslant k_f + \delta, \tag{2.8.18}$$

which gives the required regions of integration to be shells of thickness δ about the Fermi surface. Hence we find

$$W_{\mathbf{k}} \propto \delta^2 = (k-k_f)^2, \tag{2.8.19}$$

a result which Luttinger (1961) has shown to follow to all orders in perturbation theory.

Taking the quasi-particle energies proportional to k^2, the scattering probability, and hence the inverse lifetime, goes as

$$\Gamma_{\mathbf{k}} \propto (\varepsilon_{\mathbf{k}} - \mu)^2, \tag{2.8.20}$$

where μ is the chemical potential. This quantity is the energy ε_f of a quasi-particle at the Fermi surface, because, since ε_f is the minimum energy an added quasi-particle can have,

$$\mu = E(N+1) - E(N) = \varepsilon_f. \tag{2.8.21}$$

Multiple scattering of the added particle may be treated in a similar manner but it is evident that equation (2.8.20) gives the principal term in $(\varepsilon_{\mathbf{k}} - \mu)$ as $\varepsilon_{\mathbf{k}} \to \mu$.

The above argument clearly demonstrates that it is the Pauli exclusion principle that enables us to transform the problem of a strongly interacting gas of electrons into one of weakly interacting quasi-particles.

2.8.5 Operational definition of quasi-particle

Let us suppose that $q_{\mathbf{k}}$ is an operator such that, acting on the ground-state $|\psi_0\rangle$,

$$|\psi_{\mathbf{k}}\rangle = q_{\mathbf{k}}^{\dagger}|\psi_0\rangle \tag{2.8.22}$$

represents the quasi-particle state. If such a state can be defined, we will have

$$\begin{aligned} q_{\mathbf{k}}^{\dagger}(t)|\psi_0\rangle &= \exp(iHt)q_{\mathbf{k}}^{\dagger}\exp(-iHt)|\psi_0\rangle \\ &= \exp(i\bar{\varepsilon}_{\mathbf{k}}t)q_{\mathbf{k}}^{\dagger}|\psi_0\rangle \end{aligned} \tag{2.8.23}$$

and

$$\begin{aligned} \langle\psi_0|q_{\mathbf{k}}(t)q_{\mathbf{k}}^{\dagger}(0)|\psi_0\rangle &= \langle\psi_0|q_{\mathbf{k}}q_{\mathbf{k}}^{\dagger}|\psi_0\rangle\exp(-i\bar{\varepsilon}_{\mathbf{k}}t) \\ &= \exp(-i\bar{\varepsilon}_{\mathbf{k}}t) \end{aligned} \tag{2.8.24}$$

for times $t \ll \Gamma^{-1}$.

In order to establish the quasi-particle state, we wish to filter out of the superposition of states

$$a_{\mathbf{k}}^{\dagger}(t)|\psi_0\rangle = \sum_n \langle\psi_n|a_{\mathbf{k}}^{\dagger}|\psi_0\rangle\exp(+i\omega_n t)|\psi_n\rangle \tag{2.8.25}$$

those states $|\psi_n\rangle$ not contributing to the quasi-particle state. To do so, we define

$$q_{\mathbf{k}}^{\dagger} = C\int_{-\infty}^{\infty} dt'\exp(-i\bar{\varepsilon}_{\mathbf{k}}t')f(t')\,a_{\mathbf{k}}^{\dagger}(t'), \tag{2.8.26}$$

where C is a constant to be determined and $f(t')$ is a function which is negligible outside a region we write as $\tau \leqslant t' \leqslant \tau+\alpha^{-1}$. Then, inserting (2.8.26) into (2.8.25) we find that the contributions of $|\psi_n\rangle$ such that $[\omega_n - \bar{\varepsilon}(\mathbf{k})] \gg \alpha$ are negligible. The function f is often chosen as

$$f(t) = \begin{cases} 0, & t>0, \\ e^{\alpha t}, & t<0, \end{cases} \tag{2.8.27}$$

but its exact form is of no consequence, as can readily be verified. We shall therefore make another choice, for the following reason. If we calculate $\langle\psi_0|q_{\mathbf{k}}(t)q_{\mathbf{k}}^{\dagger}(0)|\psi_0\rangle$ from equation (2.8.26), we must evidently obtain a factor

$$\langle\psi_0|a_{\mathbf{k}}(t+t')a_{\mathbf{k}}^{\dagger}(t'')|\psi_0\rangle = \langle\psi_0|a_{\mathbf{k}}(t+t'-t'')a_{\mathbf{k}}^{\dagger}(0)|\psi_0\rangle, \tag{2.8.28}$$

which is proportional to $G(\mathbf{k}, t+t'-t'')$ for $t'+t-t''>0$, but not otherwise. Thus, values of t', t'' for which $t'+t-t''<0$ introduce a complication we wish to avoid. We shall therefore define (inserting $\alpha/\sqrt{z}$ for C, the correctness of which we shall verify shortly)

$$q_{\mathbf{k}}^{\dagger} = \frac{\alpha}{\sqrt{[z(k)]}}\int_{-\infty}^{0} dt''\exp(-i\bar{\varepsilon}_{\mathbf{k}}t'')\exp(\alpha t'')\,a_{\mathbf{k}}^{\dagger}(t'') \tag{2.8.29}$$

which is just the combination of equations (2.8.26) and (2.8.27), but, on the other hand, we shall choose

$$q_{\mathbf{k}} = \frac{\alpha}{\sqrt{[z(\mathbf{k})]}} \int_0^\infty dt' \exp(i\tilde{\varepsilon}_{\mathbf{k}} t') \exp(-\alpha t') a_{\mathbf{k}}(t'). \tag{2.8.30}$$

We now find

$$\langle \psi_0 | q_{\mathbf{k}}(t) q_{\mathbf{k}}^\dagger(0) | \psi_0 \rangle = \left(\frac{\alpha}{z}\right)^2 \int_{-\infty}^0 dt'' \int_0^\infty dt' \exp[i\tilde{\varepsilon}(\mathbf{k})(t'-t'')] \times \exp[\alpha(t''-t')] G(\mathbf{k}, t+t'-t''). \tag{2.8.31}$$

$G(\mathbf{k}t)$ may be expressed in terms of the spectral density as

$$G(\mathbf{k}t) = -i \int_0^\infty d\omega\, A^+(\mathbf{k}\omega) \exp[-i(\omega+\mu)t], \quad t>0, \tag{2.8.32}$$

in which case equation (2.8.31) becomes

$$\langle \psi_0 | q_{\mathbf{k}}(t) q_{\mathbf{k}}^\dagger(0) | \psi_0 \rangle = -i\left(\frac{1}{z}\right)^2 \int_0^\infty d\omega \exp[-i(\omega+\mu)t] \frac{A^+(\mathbf{k}\omega)\,\alpha^2}{\alpha^2 + [\tilde{\varepsilon}(\mathbf{k}) - \omega - \mu]^2}. \tag{2.8.33}$$

Now the quasi-particle is a superposition of exact eigenstates, defined by the Lorentzian peak of A^+ shaded in Figure 2.4:

$$A_L^+(\mathbf{k}\omega) = \frac{z(\mathbf{k})\,\Gamma(\mathbf{k})}{\Gamma^2 + (\omega - \tilde{\varepsilon} + \mu)^2}. \tag{2.8.34}$$

We must therefore have $\alpha \gg \Gamma$, on the one hand but, on the other hand, we must require α to be sufficiently small to cut out the part of A^+ not required in defining the quasi-particle. We need $\Gamma \to 0$ for these conditions to be satisfied; if they are, we have, after replacing A^+ by the right-hand side of (2.8.34) in (2.8.33),

$$\langle \psi_0 | q_{\mathbf{k}}(t) q_{\mathbf{k}}^\dagger(0) | \psi_0 \rangle = \exp(-i\tilde{\varepsilon}_{\mathbf{k}} t) \exp[-\Gamma(\mathbf{k}) t] \tag{2.8.35}$$

in conformity with equation (2.8.24).

These, then are the tools we shall use in section 2.10, to give a microscopic foundation to the famous Landau Fermi-liquid theory.

2.9 Landau Fermi-liquid theory

We stressed in Chapter 1 that, while the calculation of the ground-state energy of the crystal is a matter of considerable importance, many of the physical properties of the system depend on the low-lying excitations. The density functional theory is appropriate for the ground state and the discussion below, in a sense, has a similar philosophy for the excited states.

Here, we shall describe the Landau theory of Fermi liquids, introduced initially by Landau (1956) to deal with the neutral Fermi liquid He^3. For electrons in metals, proper account must be taken of the effects of the long-range Coulomb interactions and the Landau theory was suitably extended by Silin (1957) for this case. All the evidence points to the fact that this theory is rather directly applicable to simple metals like the alkalis, where the Fermi surfaces deviate but little from the Fermi sphere (see Chapter 4, section 4.2).

This theory, as hinted above, says nothing about the ground state of the system and it fails when we are dealing with excitation energies which are becoming comparable with the Fermi energy. However, a description of the low-lying excitations of such a strongly interacting electron system as we find in the alkali metals ($r_s \sim 4a_0$) is adequate for the calculation of the $T = 0$ response of the system to slowly varying external fields of wave number $q \ll k_f$ and $\hbar\omega \ll E_f$. It can also be used to calculate the low temperature ($k_B T \ll E_f$) thermodynamic properties of the system.

Though the theory as set out below is built up phenomenologically, Landau saw that it must be intimately connected with the full microscopic theory. Subsequently, Luttinger and Nozières have shown that Fermi liquid theory is equivalent to a perturbation treatment of the Coulomb interactions to all orders in the gas parameter r_s. The equivalence is true to all orders in r_s and at first sight it is not at all clear that any simple truncation of the perturbation series will lead to good results for real metals. However, calculations based on a selective summation of the terms in such a perturbation series appear to give a fairly consistent description of the electron–electron interactions, as we shall discuss briefly later.

Landau takes then, as starting-point, the concept of quasi-particles which we introduced earlier and he develops his theory without, at least in the first instance, inquiry as to the *exact* nature of the quasi-particles. The results are then obtained in terms of a small number of quantities which are, in general, unknown.‡ We shall, at this stage, illustrate the power of the theory by taking two examples, namely specific heat and electric current. However, a more extensive discussion of Landau theory, as applied to a description of waves in metals, will be given in Chapter 7.

2.9.1 Distribution function and specific heat

We start out from the assumption that we are dealing with a state perturbed only slightly from the ground state, and so the quasi-particle distribution function n_k is perturbed only slightly from that shown in Figure 2.5(a).

‡ As indicated above, attempts have, of course, been made to calculate these unknown quantities from a basic microscopic theory. An example of this is the effective mass, which appears in various formulae, for example, that for the specific heat. We shall calculate the self-consistent-field value of the effective mass elsewhere.

We therefore put

$$n_{\mathbf{k}} = n_{\mathbf{k}}^0 + \delta n_{\mathbf{k}}. \tag{2.9.1}$$

The energy of the perturbed state is then written in the form, to second order in $\delta n_{\mathbf{k}}$,

$$E = E_0 + \sum_{\mathbf{k}} \varepsilon_{\mathbf{k}}^0 \, \delta n_{\mathbf{k}} + \frac{1}{2} \sum_{\mathbf{k} \neq \mathbf{k}'}' f(\mathbf{k}\mathbf{k}') \, \delta n_{\mathbf{k}} \, \delta n_{\mathbf{k}'}. \tag{2.9.2}$$

Here E_0 is the exact energy of the ground state, $\varepsilon_{\mathbf{k}}^0$ is the energy of a single-particle excitation when only *one* quasi-particle is excited, and $f(\mathbf{k}\mathbf{k}')$ represents the interaction energy between two excitations $\mathbf{k}'$ and $\mathbf{k}$. Thus, in general, the single-particle excitation spectrum is given by

$$\varepsilon_{\mathbf{k}} = \varepsilon_{\mathbf{k}}^0 + \sum_{\mathbf{k}'}^{(\mathbf{k}' \neq \mathbf{k})} f(\mathbf{k}\mathbf{k}') \, \delta n_{\mathbf{k}'}. \tag{2.9.3}$$

Just as in section 1.10.1 of Chapter 1, we can calculate a group velocity for the quasi-particle by

$$\mathbf{v}_{\mathbf{k}} = \frac{1}{\hbar} \frac{\partial \varepsilon_{\mathbf{k}}}{\partial \mathbf{k}}. \tag{2.9.4}$$

For an isotropic system we write this as

$$\mathbf{v}_{\mathbf{k}} = \frac{\hbar \mathbf{k}}{m^*}, \tag{2.9.5}$$

defining the effective mass m^* of the quasi-particle. We shall only be interested in its value at k_{f}. The density of states for the quasi-particles is easily calculated, and we find in the neighbourhood of the Fermi energy ε_{f},

$$n(\varepsilon_{\mathrm{f}}) = \frac{\mathscr{V} k_{\mathrm{f}} m^*}{\pi^2 \hbar^2}, \tag{2.9.6}$$

which is exactly the result of the non-interacting gas, with m^* replacing m.

Turning now to the calculation of the specific heat c_v, we explicitly take, since the quasi-particles are Fermions,

$$n_{\mathbf{k}} = [1 + \mathrm{e}^{(\varepsilon_{\mathbf{k}} - \varepsilon_{\mathrm{f}})/k_{\mathrm{B}} T}]^{-1}. \tag{2.9.7}$$

Now since

$$\delta E = c_v \, \delta T = \sum_{\mathbf{k}} \varepsilon_{\mathbf{k}} \, \delta n_{\mathbf{k}} \tag{2.9.8}$$

we find

$$c_v = \frac{\pi^2 k_{\mathrm{B}}^2 T}{3\mathscr{V}} n(\varepsilon_{\mathrm{f}}) = \frac{\frac{1}{3} m^* k_{\mathrm{f}} k_{\mathrm{B}}^2 T}{\hbar^2}. \tag{2.9.9}$$

Again, this is the free-electron gas result, with the use of an effective mass. This calculation does not require the explicit retention of the second-order term in equation (2.9.2), but in certain other cases it contributes a term of the same order as the first.

We also note that $n^0_{\mathbf{k}}$ has only formal significance except for k near k_f, as was shown in the last section. We could in fact rewrite all our expressions to avoid these other regions, but this is hardly necessary. We saw from the content of the last section that we must limit ourselves to regions of k near k_f for all purposes, in order to avoid the complication of quasi-particles of finite lifetime.

2.9.2 Particle current

Let $\mathbf{J}_{\mathbf{k}}$ be the current produced if a single quasi-particle is excited. This, as we shall see, *cannot* be equated with the velocity $\mathbf{v}_{\mathbf{k}}$, because the movement of any particle through a medium with which it interacts produces "back-flow" in this medium. A simple example which comes to mind straightaway is the motion of a ship: as the ship moves forwards there is a net flow of water backwards. The total current in the electron gas can, however, be written as

$$\mathbf{J} = \sum_{\mathbf{k}} n_{\mathbf{k}} \mathbf{v}_{\mathbf{k}}. \tag{2.9.10}$$

In the ground state of a translationally invariant system, $n^0_{\mathbf{k}}$ is equal to

$$n^0_{\mathbf{k}} = \begin{cases} 1, & k<k_f, \\ 0, & k>k_f, \end{cases} \tag{2.9.11}$$

and $\mathbf{J} = 0$, so that we can see from (2.9.10) to first order in $\delta n_{\mathbf{k}}$, that

$$\mathbf{J} = \sum_{\mathbf{k}} \delta n_{\mathbf{k}} \mathbf{v}_{\mathbf{k}} + \sum_{\mathbf{k}} n^0_{\mathbf{k}} \delta \mathbf{v}_{\mathbf{k}} \tag{2.9.12}$$

with

$$\delta \mathbf{v}_{\mathbf{k}} = \frac{1}{\hbar} \frac{\partial}{\partial \mathbf{k}} \sum_{\mathbf{k}'} f(\mathbf{k}\mathbf{k}')\, \delta n_{\mathbf{k}'} \tag{2.9.13}$$

from equations (2.9.3) and (2.9.4). Integrating the second term in equation (2.9.12) by parts (the summation being converted to an integration as usual) we find

$$\mathbf{J} = \sum_{\mathbf{k}} \delta n_{\mathbf{k}} \mathbf{v}_{\mathbf{k}} - \sum_{\mathbf{k}\mathbf{k}'} \frac{\partial n^0}{\partial \mathbf{k}} \frac{1}{\hbar} f(\mathbf{k}\mathbf{k}')\, \delta n_{\mathbf{k}'}. \tag{2.9.14}$$

From equation (2.9.11) we see that

$$\frac{\partial n^0}{\partial \mathbf{k}} = -\hbar \mathbf{v}_{\mathbf{k}}\, \delta(\varepsilon_{\mathbf{k}} - \varepsilon_f). \tag{2.9.15}$$

If we wish to consider the current due to a single quasi-particle, we put

$$\delta n_{\mathbf{k}'} = \sum_{\mathbf{k}} \delta_{\mathbf{k}'\mathbf{k}} \tag{2.9.16}$$

whereupon we find

$$\mathbf{J}_{\mathbf{k}} = \mathbf{v}_{\mathbf{k}} + \sum_{\mathbf{k}'} f(\mathbf{k}\mathbf{k}')\, \mathbf{v}_{\mathbf{k}'}\, \delta(\varepsilon_{\mathbf{k}'} - \varepsilon_{\mathfrak{f}}) \tag{2.9.17}$$

or, in integral form, and displaying the spin σ,

$$\mathbf{J}_{\mathbf{k}} = \mathbf{v}_{\mathbf{k}} + \frac{1}{8\pi^3} \sum_{\sigma} \int d\mathbf{k}'\, \mathbf{v}_{\mathbf{k}'}\, \delta(\varepsilon_{\mathbf{k}'} - \varepsilon_{\mathfrak{f}}) f(\mathbf{k}\mathbf{k}'). \tag{2.9.18}$$

We see that there is indeed "back-flow", represented by the second term of equation (2.9.18).

We now note that in our translationally invariant system the potential energy depends only on the *relative positions* of the particles, and so the total current is a constant of the motion, the relevant operator commuting with the Hamiltonian. Since the current operator does not involve explicitly the strength λ of the interaction, it can be shown that its expectation value is independent of λ. Now, in the non-interacting system, the quasi-particles are just the electrons and so

$$\mathbf{J}_{\mathbf{k}} = \frac{\hbar \mathbf{k}}{m}. \tag{2.9.19}$$

The correspondence between levels in non-interacting and interacting systems (see section 2.8.1) means that if a single electron is excited in state $\mathbf{k}$ in the non-interacting system, the system becomes one in which a single quasi-particle is excited in state $\mathbf{k}$ when the interaction is switched on. Thus equation (2.9.19) gives the current of a quasi-particle for any interaction strength, and equations (2.9.5), (2.9.18) and (2.9.19) give us an equation relating m^* and f when the integration over the δ-function is performed at the Fermi surface. The explicit result is given in equation (2.10.41) below.

2.9.3 Boltzmann equation

We need to obtain now the transport equation, from which, in principle, the quasi-particle distribution function $n(\mathbf{p}, \mathbf{r}, t)$ can be obtained. This is, essentially, the Boltzmann equation, and since this plays such a crucial role in many problems in solid-state physics, we shall digress at this point to give a derivation of it which is rather generally applicable. We shall then, of course, use it immediately in the framework of Landau theory.

The transport properties of major interest to us are concerned with charge, mass, energy and momentum, and the corresponding macroscopic coefficients, discussed fully in Chapter 6, are, in principle, calculable from the microscopic motion. We now tackle the transport problem in a general fashion, the approach we initially adopt being characteristic of kinetic theory, in that we start with a distribution n such that $n(\mathbf{p}, \mathbf{r}, t)$ represents the number of particles, electrons to be specific, in the volume $d\mathbf{r}$ about $\mathbf{r}$ which have momenta in the region $d\mathbf{p}$ about $\mathbf{p}$. The equation governing the rate of change of this distribution function is known as the Boltzmann equation, and may be written in the form

$$\frac{\partial n}{\partial t}+\mathbf{v}.\frac{\partial n}{\partial \mathbf{r}}+\mathbf{F}.\frac{\partial n}{\partial \mathbf{p}}=\left[\frac{\partial n}{\partial t}\right]_{\text{collision}}, \tag{2.9.20}$$

where $\mathbf{v} = \mathbf{p}/m$ and $\mathbf{F}$ the force on a particle having momentum $\mathbf{p}$ and position $\mathbf{r}$. The origins of the various terms may be understood as follows. Suppose in time δt, a representative particle at $\mathbf{r}$ goes to $\mathbf{r}+\delta\mathbf{r}$ and its momentum $\mathbf{p}$ becomes $\mathbf{p}+\delta\mathbf{p}$, and the particles previously occupying the volume $d\mathbf{r}\,d\mathbf{p}$ of phase space now occupy the volume $d\mathbf{r}'\,d\mathbf{p}'$. Particle numbers are conserved, and so, neglecting at first collisions,

$$n(\mathbf{p}+\delta\mathbf{p}, \mathbf{r}+\delta\mathbf{r}, t+\delta t)\,d\mathbf{r}'d\mathbf{p}' = n(\mathbf{p}, \mathbf{r}, t)\,d\mathbf{r}\,d\mathbf{p}. \tag{2.9.21}$$

The volumes of phase space $d\mathbf{r}'d\mathbf{p}'$ and $d\mathbf{r}\,d\mathbf{p}$ will be very nearly equal for small δt, any divergence from the (moving) volume being small since the positions and momenta of all particles occupying them differ negligibly, and we shall take them as equal in the following.

When collisions are taken into account, equation (2.9.21) will no longer be true and on the right-hand side we therefore add $(\partial n/\partial t)_{\text{coll}}\,d\mathbf{r}\,d\mathbf{p}\,\delta t$ to represent the number of particles forced into $d\mathbf{r}\,d\mathbf{p}$ by collisions. Hence

$$n(\mathbf{p}+\delta\mathbf{p}, \mathbf{r}+\delta\mathbf{r}, t+\delta t) = n(\mathbf{p}, \mathbf{r}, t)+\left[\frac{\partial n}{\partial t}\right]_{\text{coll}}\delta t. \tag{2.9.22}$$

Writing the left-hand side as

$$n(\mathbf{p}+\delta\mathbf{p}, \mathbf{r}+\delta\mathbf{r}, t+\delta t) = n(\mathbf{r}, \mathbf{p}, t)+\frac{\partial n}{\partial \mathbf{p}}.\,\delta\mathbf{p}+\frac{\partial n}{\partial \mathbf{r}}.\,\delta\mathbf{r}+\frac{\partial n}{\partial t}\,\delta t \tag{2.9.23}$$

and

$$\mathbf{F}=\frac{\delta\mathbf{p}}{\delta t}, \quad \mathbf{v}=\frac{\delta\mathbf{r}}{\delta t}, \tag{2.9.24}$$

we obtain equation (2.9.20). An obvious restriction is that $\mathbf{F}$ must vary slowly over distances of the order of the mean free path between collisions.

2.9.4 *Use of transport equation in Fermi-liquid theory*

We now specialize to the Fermi-liquid theory and we can evidently write

$$\frac{\partial n}{\partial t}+\frac{1}{\hbar}\sum_{\alpha}\left\{\frac{\partial n}{\partial x_\alpha}\frac{\partial E}{\partial k_\alpha}-\frac{\partial n}{\partial k_\alpha}\frac{\partial E}{\partial x_\alpha}\right\}=\left.\frac{\partial n}{\partial n}\right|_{\text{coll}} \tag{2.9.25}$$

Now we suppose that external electromagnetic fields are present, described as usual by a scalar potential ϕ and a vector potential $\mathscr{A}$. Then, since $m\dot{\mathbf{r}} = \mathbf{p}-(e/c)\mathscr{A}$, we expect that for fields which vary sufficiently slowly in space and time we can write that the distribution function n is given by

$$n = n_0\left(\mathbf{p}-\frac{e}{c}\mathscr{A}\right)+\delta n\left(\mathbf{p}-\frac{e}{c}\mathscr{A}(rt)\right)$$

where $\mathbf{p}\to\mathbf{p}-(e/c)\mathscr{A}$ (see section 4.15.2 below) and

$$E = E\left[\mathbf{p}-\frac{e}{c}\mathscr{A}(\mathbf{r}t)\right]+e\phi. \tag{2.9.26}$$

At this stage, we expand E and retain only terms linear in δn. We then find

$$\frac{-c}{e}\frac{\partial n_0}{\partial t}=\frac{\partial n_0}{\partial E}\frac{\partial E}{\partial \mathbf{p}}.\frac{\partial \mathscr{A}}{\partial t}=\frac{\partial n_0}{\partial E}\left(\mathbf{v}.\frac{\partial \mathscr{A}}{\partial t}\right) \tag{2.9.27}$$

and, with $E = E^0+\delta E$,

$$\{\delta n, E^0\} = \frac{e}{c}(\mathbf{v}\times\mathscr{H}).\frac{\partial \delta n}{\partial \mathbf{p}}, \tag{2.9.28}$$

$$\{n_0, \delta E\} = \frac{e}{c}(\mathbf{v}\times\mathscr{H}).\frac{\partial}{\partial \mathbf{p}}(\delta E)\frac{\partial n_0}{\partial E}, \tag{2.9.29}$$

where { } denotes the Poisson brackets as in equation (2.9.25).

Defining the quantity g via

$$n = n_0+\frac{\partial n_0}{\partial E}g, \tag{2.9.30}$$

we obtain as the linear transport equation for g the result, with $\mathscr{H}$ and $\mathscr{E}$ the magnetic and electric fields respectively (we use the $\mathscr{H}\neq 0$ case in Chapter 7),

$$\frac{\partial g}{\partial t}+\left[\mathbf{v}.\nabla+\frac{e}{c}(\mathbf{v}\times\mathscr{H}).\nabla_p\right][g+\delta\varepsilon_1] = e\mathscr{E}.\mathbf{v}+\left.\frac{\partial g}{\partial t}\right|_{\text{coll}}, \tag{2.9.31}$$

where, referring back to equations (2.9.2) and (2.9.3),

$$\delta\varepsilon_1 = \frac{2}{(2\pi)^3}\int d\mathbf{p}' f(\mathbf{p}\mathbf{p}')\,g(\mathbf{p}'\mathbf{r}t)\,\delta(\mu-\varepsilon^0_{p'}). \tag{2.9.32}$$

Assuming that the number of particles is conserved, we can integrate both sides of equation (2.9.31) to obtain the continuity equation

$$\frac{\partial \rho}{\partial t}+\nabla . \mathbf{j}=0, \tag{2.9.33}$$

where the components of the current density vector $\mathbf{j}$ are given by

$$\begin{aligned} j_\alpha &= \frac{e}{(2\pi)^3}\int d\mathbf{p}\,\frac{p_\alpha}{m^*}(g+\delta\varepsilon_1)\,\delta(E_0-\mu) \\ &= \frac{e}{(2\pi)^3}\int d\mathbf{p}\,J_\alpha(\mathbf{p})\,g(E_0-\mu) \end{aligned} \tag{2.9.34}$$

and the density ρ is simply

$$\rho=\frac{e}{(2\pi)^3}\int d\mathbf{p}\,g\,\delta(E_0-\mu). \tag{2.9.35}$$

The component $J_\alpha(\mathbf{p})$ is the current carried by a quasi-particle, namely

$$\begin{aligned} J_\alpha(\mathbf{p}) &= \frac{\partial E}{\partial p_\alpha} \\ &= \frac{p_\alpha}{m^*}+\frac{2}{(2\pi)^3}\int d\mathbf{p}' f(\mathbf{p}\mathbf{p}')\frac{p'_\alpha}{m^*}\,\delta(\mu-E^0_{p'}). \end{aligned} \tag{2.9.36}$$

This, then, is a rather different route to equation (2.9.18) and has the advantage of showing how the expressions for charge and current follow from the linearized transport equation, as indeed does the equation of continuity.

2.9.5 Landau scattering function in Hartree–Fock theory

It is already perfectly clear that the Landau two-particle function is quite central in the theory. But the theory, of itself, does not contain a prescription whereby it can be calculated. There are two routes open to us:

(i) We can try to devise a microscopic theory for f.

(ii) We can attempt to get information on f semi-empirically.

Both routes will be considered in detail below.

To make a start along route (i), let us consider what Hartree–Fock theory has to say about the scattering function f. In Hartree–Fock theory, the ground-state energy per particle is given by

$$E_{\rm HF} = N^{-1}\langle 0|\sum_{\mathbf{p}\sigma}\varepsilon^0_{\mathbf{p}} a^\dagger_{\mathbf{p}\sigma} a_{\mathbf{p}\sigma}|0\rangle$$

$$+N^{-1}\langle 0|\sum_{\substack{\mathbf{pqk}\\ \sigma\sigma'}}\left(\frac{V_{\mathbf{k}}}{2}\right) a^\dagger_{\mathbf{p+k},\sigma} a^\dagger_{\mathbf{q-k},\sigma'} a_{\mathbf{q}\sigma'} a_{\mathbf{p}\sigma}|0\rangle$$

$$= N^{-1}\sum_{\mathbf{p}\sigma}\varepsilon^0_{\mathbf{p}} n_{\mathbf{p}} - N^{-1}\sum_{\substack{\mathbf{pk}\\ \sigma}}\left(\frac{V_{\mathbf{k}}}{2}\right) n_{\mathbf{p+k},\sigma} n_{\mathbf{p}\sigma}, \tag{2.9.37}$$

where, as usual, $a^\dagger_{\mathbf{p}\sigma}$ is the electron creation operator. $V_{\mathbf{k}}$ represents the Coulomb repulsion,

$$V_{\mathbf{k}} = \frac{4\pi e^2}{k^2}, \quad \mathbf{k} \neq 0, \tag{2.9.38}$$

and $V(0) = 0$ as a consequence of charge neutrality.

The definitions (2.9.3) and (2.9.2) of ε and f respectively then lead to the results

$$\varepsilon_{\rm HF}(k) = \frac{k^2}{2m} - \sum_{\mathbf{q}} V_{\mathbf{q}} n_{\mathbf{q+k},\sigma}, \tag{2.9.39}$$

which is the customary one-particle energy in Hartree–Fock theory, while differentiating again we find

$$\left.\begin{aligned} f_{\mathbf{pp}'} &= -\frac{4\pi e^2}{|\mathbf{p}-\mathbf{p}'|^2} \quad \text{(parallel spins)} \\ \text{and} \quad f_{\mathbf{pp}'} &= 0 \quad \text{(antiparallel spins).} \end{aligned}\right\} \tag{2.9.40}$$

2.9.6 Some general properties of scattering function

This latter consequence of Hartree–Fock theory is of course not appropriate to a fully correlated theory in which $f \neq 0$ for antiparallel spins.

However, some restrictions still exist on the possible forms for f in the case when we can neglect spin–orbit coupling and any other spin-dependent forces.

Then it follows from inversion symmetry, time-reversal invariance and the symmetry that f must have when $\mathbf{p}$ and $\mathbf{p}'$ are interchanged that the most general form of f is

$$f(\mathbf{p}\sigma, \mathbf{p}'\sigma') = f(\mathbf{pp}') + \zeta(\mathbf{pp}')\,\boldsymbol{\sigma}.\boldsymbol{\sigma}'. \tag{2.9.41}$$

With the inclusion of spin, we see that there are now two symmetric functions f and ζ whereas only a single function is needed to describe spinless particles.

Rotational invariance assumption. To keep the theory in a simple form, we shall now discuss the forms of f and ζ when we have full rotational invariance. This assumption is clearly inexact for real metals, but for systems like Na and K, where the Fermi surface is accurately spherical, the lattice has rather small effects on the single-electron band structure of these metals.

Then, the functions f and ζ are simply functions of $|\mathbf{p}|, |\mathbf{p}'|$ and the angle θ between $\mathbf{p}$ and $\mathbf{p}'$ and since we are interested in the Fermi surface we need only consider $|\mathbf{p}| = |\mathbf{p}'| = p_{\text{f}}$.

Then we can expand f and ζ in Legendre polynomials in the form

$$f(\mathbf{pp}') = \sum_n f_n P_n(\cos\theta) \tag{2.9.42}$$

and

$$\zeta(\mathbf{pp}') = \sum_n \zeta_n P_n(\cos\theta). \tag{2.9.43}$$

The conventional form of the Landau parameters is then obtained by writing

$$A_n = m^* p_{\text{f}} f_n \Big/ \pi^2(2n+1) \tag{2.9.44}$$

and

$$B_n = m^* p_{\text{f}} \zeta_n \Big/ \pi^2(2n+1), \tag{2.9.45}$$

which, along with m^*/m, are the set of parameters characterizing the interacting electron gas. In fact, these parameters provide us with the starting-point to route (ii) above. We shall see, especially in Chapter 7, how experiments on spin-wave resonance can give us rather direct information about a few of the Landau parameters A_n and B_n for small n.

To proceed further along route (i), we must now go on to develop many-body methods, and, in particular, the definition and use of Feynman graphs. We shall then return to deal with the Landau scattering function in full detail.

2.10 Diagrammatic techniques and microscopic form of Landau theory

2.10.1 Feynman graphs

Suppose we observe the motion of a single free particle of definite momentum $\mathbf{k}$ from time t_0, say, to a later time t. We can represent the propagation of the particle by a single directed line labelled by its momentum [Figure 2.6(a)] and we also note that this motion will be described by the free-particle Green function $G_0(\mathbf{k}, t-t_0)$ given in equation (2.8.5), which is termed in this connection the free-particle propagator.

Let us next assume that two particles having momenta $\mathbf{k}_1$ and $\mathbf{k}_2$ are present, and that, because of the Coulombic interaction between them, they scatter off one another at time t_1 into other definite momentum states. This is represented in Figure 2.6(b), where their interaction is shown by a dotted line, termed an "interaction line". It will be observed that we have labelled the diagram to show the conservation of momentum.

We must also take account of the fact that, if a particle is moving through an electron gas, it may scatter off a hole. We shall indicate the motion of a hole by a line distinguished from a particle line by a reversed arrow. This

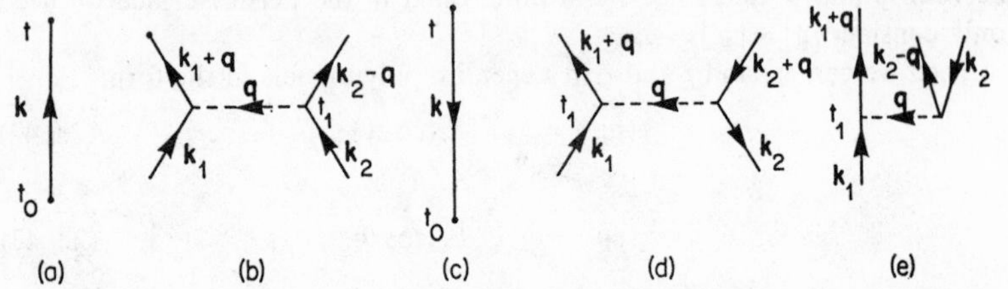

FIGURE 2.6. Feynman graphs. As examples, graph (a) shows electron propagation without interaction and graph (b) shows mutual scattering of two particles with momenta $\mathbf{k}_1$ and $\mathbf{k}_2$ while graph (c) shows hole propagation.

means that the line is directed from a later to an earlier time [Figure 2.6(c)]. If we always associate a line labelled by $\mathbf{k}$ and directed from t_1 to t_2 with $G_0(\mathbf{k}, t_2-t_1)$, then this is consistent with the property

$$G_0(\mathbf{k}, t) = 0 \quad \text{if} \quad t>0, \quad k<k_f \quad \text{or} \quad t<0, \quad k>k_f. \tag{2.10.1}$$

Figure 2.6(e) shows the creation of a particle–hole pair, resulting from the Coulombic interaction. It should be noted that all diagrams showing a single interaction must have the property in evidence in Figure 2.6(b), (d) and (e): there must be one arrow leading to, and one arrow leading from, each vertex (end of an interaction line), because a particle can never scatter to become a hole (or vice versa) and the creation or annihilation of a particle must always be accompanied by the creation or annihilation of a hole.

We now go back to a single particle and allow it to move through the free-electron gas in its ground state. If we then switch on the Coulombic interaction $V_{\mathbf{q}}$, various things can happen. For example, the particle may interact with the gas at time t_1 to produce a particle–hole pair, scattering into momentum state $\mathbf{k}+\mathbf{q}$ before interacting with the particle–hole pair again to return to its original state. This is shown in Figure 2.7(a).

A slightly more complicated case is shown in Figure 2.7(b), where the Coulombic interaction produces an excitation (two particle–hole pairs) within the gas, before the particle we have singled out suffers scattering. Of the other examples shown in Figure 2.7, example (c) calls for some comment. It represents the simultaneous creation and destruction of a particle, and, since the interaction is accompanied by no transfer of momentum, i.e. $\mathbf{q} = 0$, it describes forward scattering. While, classically, with $\mathbf{q} = 0$ nothing would have actually changed, this is not the case in a quantum-mechanical system, because the principle of indistinguishability does not allow us to assume the final particle to be the original one.

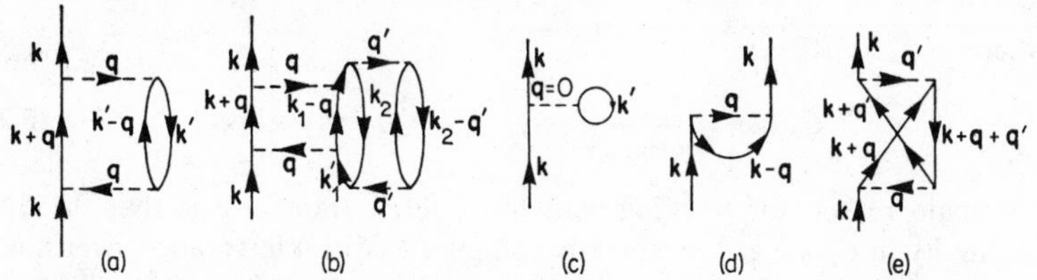

FIGURE 2.7. Examples of Feynman graphs for interaction of electron with rest of electron gas.

Rules for relating graphs to terms in perturbation series

When we allow the particle we have singled out to suffer all possible interactions with the rest of the gas before finally returning to its original state $\mathbf{k}$, it becomes a *quasi-particle* (at least when near the Fermi surface), with its motion described by the single-particle Green function $G(\mathbf{k}, t-t_0)$ of the interacting system. Furthermore, each diagram we can draw corresponds to a term in a perturbation series of G based on the application of the time-development operator (see Appendix 2.2). This is very fortunate, as the purely algebraic manipulation of other than quite low orders of this series is extremely complicated. As it is, we draw every possible topologically distinct diagram which shows just one particle, of momentum $\mathbf{k}$, at the initial and final times t_0 and t, and the scatterings of this particle at intermediate times. To obtain the corresponding term of the perturbation series, we then associate a factor $G_0(\mathbf{k}', t_1 - t_2)$ with each Fermion (full) line labelled by $\mathbf{k}'$ and directed from t_2 to t_1, and associate a factor $V_\mathbf{q}$ (the Fourier transform of the Coulombic interaction) with each interaction line. Finally, we integrate over all intermediate times and momenta, including the $\mathbf{q}$'s, subject to the restrictions imposed by the momentum conservation at each vertex. There are factors other than the ones we have mentioned here in the terms, and the full set of rules is derived in Appendix 2.2. But these factors need not concern us in detail here, for the main point we wish to emphasize is that the very existence of these rules, and the fact that there exists a diagram for each term, implies that for many purposes we need not work with the perturbation series directly, but confine our attention very largely to the Feynman diagrams. To illustrate this, we turn now to derive Dyson's equation for the one-particle Green function.

2.10.2 *Dyson's equation*

To facilitate the discussion, we define the Fourier transform of the free-particle propagator $G_0(\mathbf{k}t)$ through

$$G_0(\mathbf{k}t) = \frac{1}{2\pi}\int_{-\infty}^{\infty} d\omega\, e^{-i\omega t} G_0(\mathbf{k}\omega), \tag{2.10.2}$$

where

$$G_0(\mathbf{k}\omega) = \frac{1}{\omega - \varepsilon_{\mathbf{k}} + i\delta_{\mathbf{k}}}, \quad \delta_{\mathbf{k}} \to \pm 0 \quad \text{as} \quad k \gtrless k_f. \tag{2.10.3}$$

The main reason for working with the Fourier transform is that the discontinuity in $G_0(\mathbf{k}t)$ at $t = 0$ greatly complicates direct integration over time.

Now we shall apply the diagrammatic technique set out above to illustrate the formal summation of the perturbation series for $G(\mathbf{k}, t-t_0)$ in a very simple way.

Let us consider as an example the factor corresponding to the interaction shown in Figure 2.8. We have then an integral of the form

$$\int_{-\infty}^{\infty} dt'\, G_0(\mathbf{k}_1, t'-t_1)\, G_0(\mathbf{k}_2, t'-t_2)\, G_0(\mathbf{k}_1+\mathbf{q}, t_3-t')\, G_0(\mathbf{k}_2-\mathbf{q}, t_4-t'). \tag{2.10.4}$$

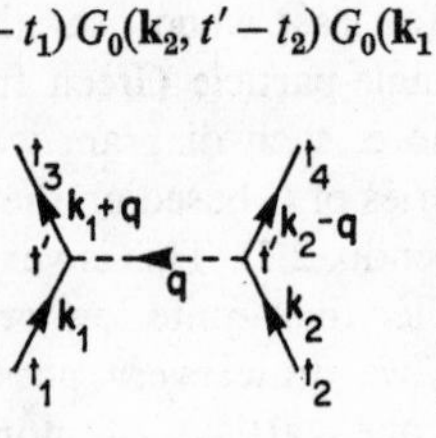

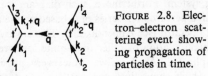

FIGURE 2.8. Electron–electron scattering event showing propagation of particles in time.

Inserting expressions of the form (2.10.2) into this integral we obtain

$$\begin{aligned}\frac{1}{(2\pi)^3}\int d\omega_1\, d\omega_2\, d\omega_3\, d\omega_4\, G_0(\mathbf{k}_1\,\omega_1)\, G_0(\mathbf{k}_2\,\omega_2)\, G_0(\mathbf{k}_1+\mathbf{q}, \omega_3)\\ \times G_0(\mathbf{k}_2-\mathbf{q}, \omega_4)\,\delta(\omega_1+\omega_2-\omega_3-\omega_4)\\ \times e^{i\omega_1 t_1} e^{i\omega_2 t_2} e^{i\omega_3 t_3} e^{i\omega_4 t_4}.\end{aligned} \tag{2.10.5}$$

The diagram can now be relabelled as in Figure 2.9 where, because of the δ-function appearing above, we have *conservation of energy parameter at each vertex*. We now associate each Fermion line labelled by $(\mathbf{k}', \omega')$ with the

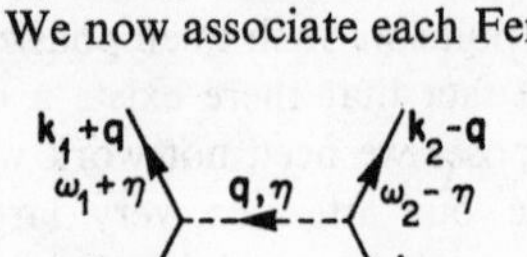

FIGURE 2.9. Diagram as in Figure 2.8, but showing energies and energy transfer.

factor $G_0(\mathbf{k}'\omega')$ and our relabelled diagrams correspond to a perturbation series for $G(\mathbf{k}\omega)$.

We have, of course, to consider the sum of the entire perturbation expansion. To do this, we divide all the diagrams into two classes, irreducible and reducible. Reducible diagrams can be divided into two by cutting one internal Fermion line, whereas irreducible ones cannot. Figure 2.10 shows two reducible diagrams, whereas those of Figure 2.7 are all irreducible. We shall denote the sum of the contributions of all irreducible diagrams by Σ. This is *not* the sum of the entire series, for we still have to include the contribution of the reducible diagrams.

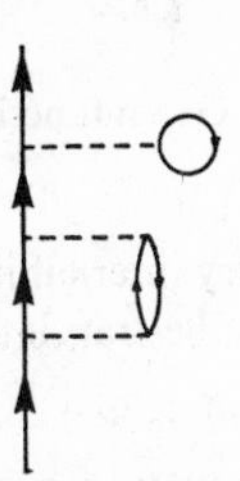

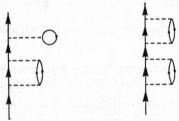

FIGURE 2.10. Examples of reducible diagrams. A single cut at the appropriate point will separate diagrams into two distinct parts.

To help us to do this, we introduce a diagram to represent the factor Σ. This is simply the sum of all irreducible diagrams, as indicated in Figure 2.11. For convenience in drawing this figure, we have distorted some of the original diagrams. For example, (b) and (c) of Figure 2.11 are respectively equivalent to (d) and (e) of Figure 2.7. There is no longer any point in drawing interaction lines horizontally, time having been removed from the problem.

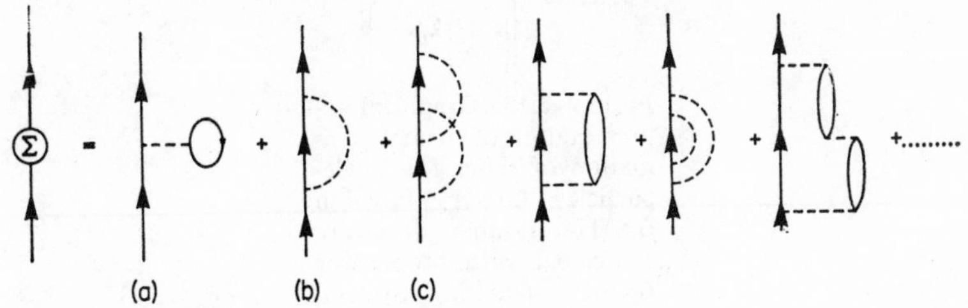

FIGURE 2.11. Diagrams for self-energy Σ. (a), (b) and (c) are explicitly discussed in text.

Reducible diagrams can now be taken into account if we redraw the series for $G(\mathbf{k}\omega)$ as in Figure 2.12. Here, an internal line carries the same indices $\mathbf{k}, \omega$ as those which label the external free lines, because of the conservation

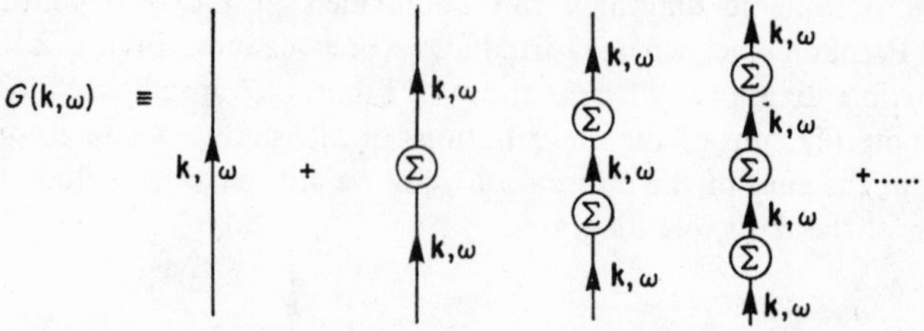

FIGURE 2.12. Series for single-particle Green function G in terms of self-energy Σ.

of these quantities at each vertex. Hence every internal line can be associated with the factor $G_0(\mathbf{k}\omega)$ and Figure 2.12 may be translated into the series

$$G(\mathbf{k}\omega) = G_0(\mathbf{k}\omega) + G_0(\mathbf{k}\omega)\,\Sigma(\mathbf{k}\omega)\,G_0(\mathbf{k}\omega) + G_0(\mathbf{k}\omega)\,\Sigma(\mathbf{k}\omega)\,G_0(\mathbf{k}\omega)\,\Sigma(\mathbf{k}\omega)\,G_0(\mathbf{k}\omega) + \ldots. \tag{2.10.6}$$

But this is just the iterative solution to the equation

$$G(\mathbf{k}\omega) = G_0(\mathbf{k}\omega) + G_0(\mathbf{k}\omega)\,\Sigma(\mathbf{k}\omega)\,G(\mathbf{k}\omega), \tag{2.10.7}$$

which is Dyson's equation for the special case of no external potential. This formal summation of the diagrammatic series may itself be represented graphically, as shown in Figure 2.13. The double line represents the propagation of a quasi-particle and is associated with the factor $G(\mathbf{k}\omega)$.

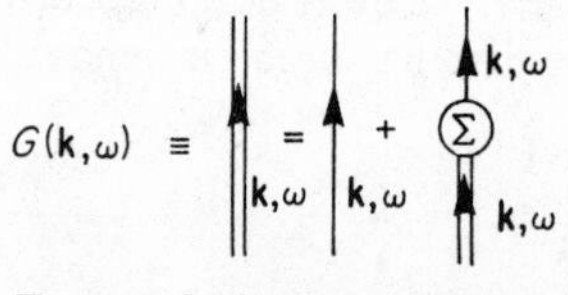

FIGURE 2.13. Graphical representation of formal summation of series for single-particle Green function G. The double lines are associated with propagator G of interacting system, while, as usual, single line denotes unperturbed propagator G_0.

Transformation of equation (2.10.7) to $\mathbf{r}t$ space is now straightforward. The result is

$$G(12) = G_0(12) + \int d3\, d4\, G_0(13)\, \Sigma(34)\, G(42) \qquad (2.10.8)$$

with $1 \equiv \mathbf{r}_1 t_1$, etc. For the uniform electron gas we have

$$\left.\begin{aligned} G_0(12) = G_0(\mathbf{r}_1-\mathbf{r}_2, t_1-t_2), \quad G(12) &= G(\mathbf{r}_1-\mathbf{r}_2, t_1-t_2), \\ \Sigma(12) &= \Sigma(\mathbf{r}_1-\mathbf{r}_2, t_1-t_2). \end{aligned}\right\} \qquad (2.10.9)$$

All the diagrams may be considered equally well in $\mathbf{r}t$ space, when we shall label them in accordance with Figure 2.14, drawing the interaction lines horizontally. An interaction line from $\mathbf{r}_1'$ to $\mathbf{r}_2'$ will be associated with the

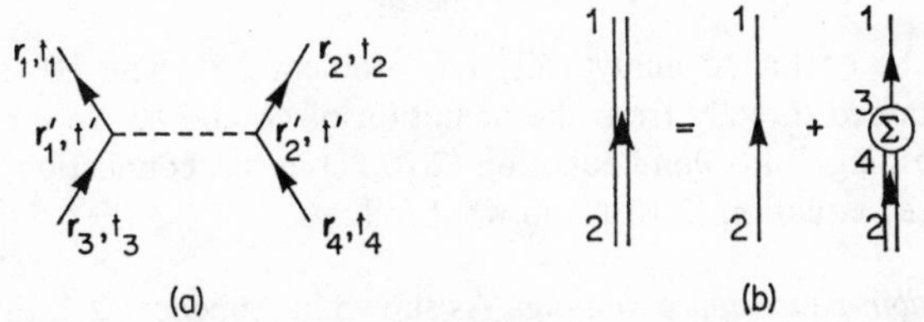

FIGURE 2.14. Feynman graphs in $(\mathbf{r}, t)$ space. (a) Single scattering event. (b) Representation of Dyson equation (2.10.8).

Coulomb potential $v(|\mathbf{r}_1'-\mathbf{r}_2'|)$, and a Fermion line directed from $\mathbf{r}_1' t_1'$ to $\mathbf{r}_1 t$ will be associated with the factor $G(\mathbf{r}_1-\mathbf{r}_1', t_1-t_1')$. In Figure 2.14(b) we show the Dyson equation in diagrammatic form in $\mathbf{r}t$ space, labelled to correspond with equation (2.10.8).

2.10.3 Self energy

Equation (2.10.7) may be rewritten as

$$G(\mathbf{k}\omega) = \frac{G_0(\mathbf{k}\omega)}{1-\Sigma(\mathbf{k}\omega)\, G_0(\mathbf{k}\omega)} = \frac{1}{\omega-\varepsilon_\mathbf{k}-\Sigma(\mathbf{k}\omega)+i\delta_\mathbf{k}}, \qquad (2.10.10)$$

where Σ is known as the self-energy. Dyson's equation is now seen to change the problem of calculating the single-particle Green function G into one of calculating the self-energy. We therefore focus attention next on the quantity Σ, which is represented by the diagrammatic series in Figure 2.11.

(a) *Hartree–Fock approximation.* The Hartree–Fock (HF) correction to the free-electron result is, in fact, the lowest-order correction in perturbation theory and so corresponds to diagrams (a) and (b) of Figure 2.11. The forward scattering part, given by (a), corresponds to the Hartree term and is exactly cancelled by the contribution of the uniform positive background.

This leaves us with diagram (b) of Figure 2.11 and, to evaluate this, appeal can be made to the detailed rules set out in Appendix 2.2, when we obtain

$$\Sigma_{\mathrm{HF}} = \int \frac{d\mathbf{q}}{4\pi^3} V_{\mathbf{q}} \int_{-\infty}^{\infty} G_0^-(\mathbf{k}+\mathbf{q}, \omega+\eta)\, d\eta. \tag{2.10.11}$$

The superscript $-$ on G_0 indicates that we must take $\delta \to -0$ in equation (2.8.7). Setting $V_{\mathbf{q}} = 4\pi e^2/q^2$, we find

$$\Sigma_{\mathrm{HF}} = \frac{e^2}{\pi^2} \int_{|\mathbf{k}+\mathbf{q}|<k_f} \frac{d\mathbf{q}}{q^2} \tag{2.10.12}$$

which can be evaluated analytically (cf. problem 2.7). The HF self-energy can be obtained directly from the definition of G, and so we do not really have to use diagram (b) and equation (2.10.11) in this connection. However, we shall need equation (2.10.11) in what follows.

(b) *Random-phase approximation.* As shown in Appendix 2.2, the random-phase approximation (RPA), or, as will turn out to be an alternative description, the time-dependent SCF approximation, is given by the infinite sub-series of Figure 2.15. This can also be represented schematically by Figure 2.16 which is just a HF diagram with a double-dotted line representing a screened

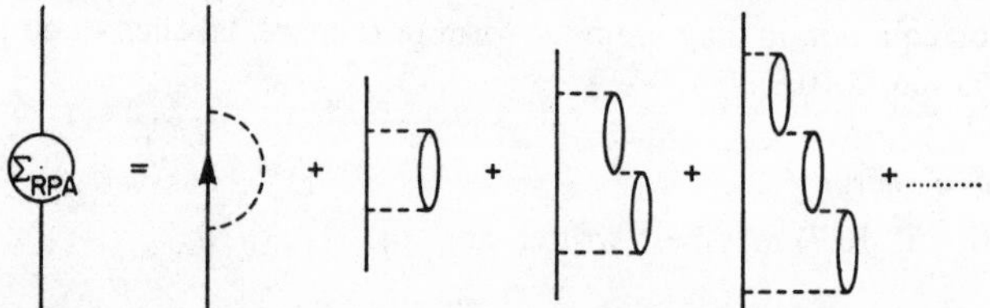

FIGURE 2.15. Random phase approximation to self-energy Σ.

interaction $V_{\mathrm{scf}}(\mathbf{q}, \eta)$, the result of the summation being indicated in Figure 2.16(b). From equation (2.10.11) we have

$$\Sigma_{\mathrm{RPA}} = \int d\mathbf{q} \int_{-\infty}^{\infty} V_{\mathrm{RPA}}(\mathbf{q}, \eta)\, G_0^-(\mathbf{k}+\mathbf{q}, \omega+\eta)\, d\eta. \tag{2.10.13}$$

V_{RPA} is represented by the diagrams of Figure 2.16(b) and the form of this diagrammatic series suggests it can be formally summed in just the way we obtained equation (2.10.7) from Figure 2.12. This is indeed so and we have the result

$$V_{\text{RPA}}(\mathbf{q}, \eta) = \frac{V_{\mathbf{q}}}{1+V_{\mathbf{q}}\Pi_{\text{RPA}}(\mathbf{q}\eta)}, \tag{2.10.14}$$

where Π_{RPA} is the factor associated with the "polarization loop" shown in Figure 2.16(c). Equation (2.10.14) obviously gives the result

$$\varepsilon_{\text{RPA}}(\mathbf{q}\eta) = 1+V_{\mathbf{q}}\Pi_{\text{RPA}}(\mathbf{q}\eta). \tag{2.10.15}$$

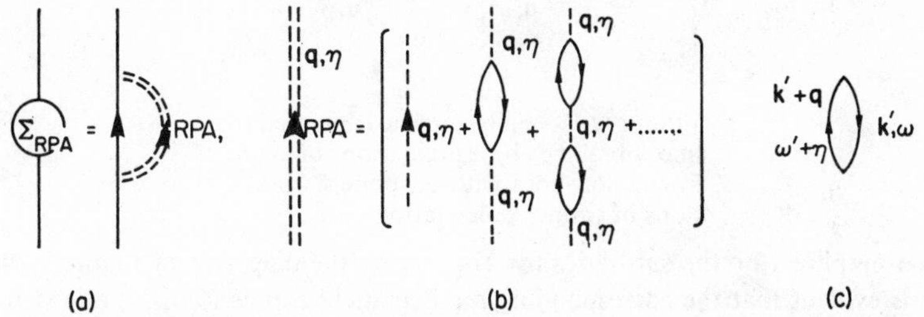

FIGURE 2.16. Illustrating dressing of interaction lines in random phase approximation. (a) Self energy. (b) Screened or dressed interaction line. (c) Polarization loop.

(c) *Formal summation of entire series.* We can now effect a generalization of the above argument in a fairly obvious way. Thus, we define a "proper" (or "irreducible") polarization part of any diagram which can be inserted into an interaction line and which cannot be split into two by cutting an internal interaction line. We denote the sum of all such terms by Π (see Figure 2.17), as the generalization of Π_{RPA} above.

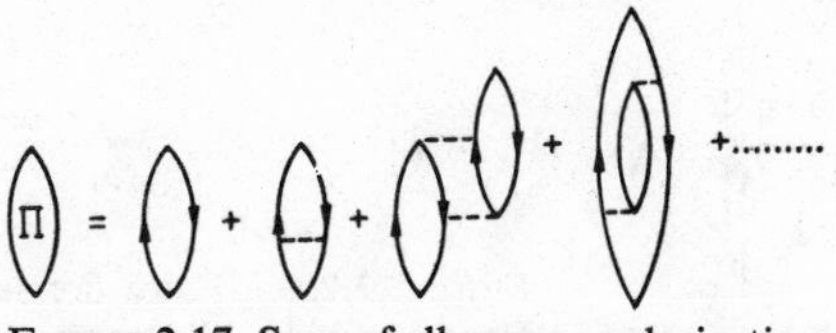

FIGURE 2.17. Sum of all proper polarization parts. Π is related to dielectric function by equation (2.10.16).

The fully screened interaction is given by Figure 2.18, which can be summed just as before to give the full frequency-dependent dielectric constant as

$$\varepsilon(\mathbf{q}\eta) = 1+\frac{4\pi e^2}{q^2}\Pi(\mathbf{q}, \eta) \tag{2.10.16}$$

with Π given by Figure 2.17. This procedure is termed the "dressing" or "renormalization" of the interaction. Also, we can dress Fermion lines, as

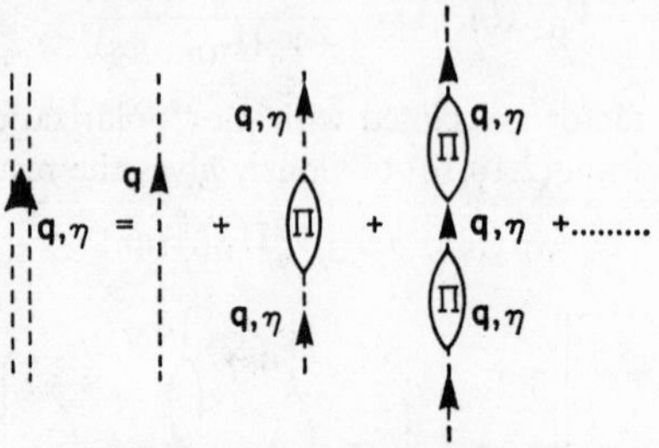

FIGURE 2.18. Fully screened interaction line obtained by summation of bare interaction lines with all possible insertions of proper polarization parts.

is exemplified by the sub-series for Σ given by the diagrams of Figure 2.19. It is evident that the corresponding mathematical expression is obtained by replacing G_0 by G in the HF result (2.10.11).

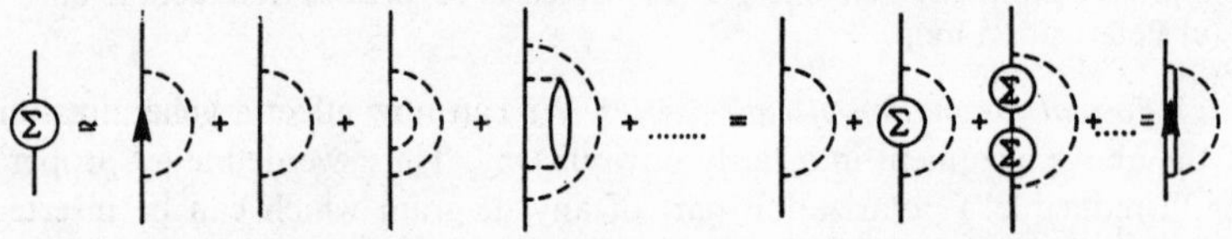

FIGURE 2.19. Dressing of Fermion line in sub-series for self-energy Σ.

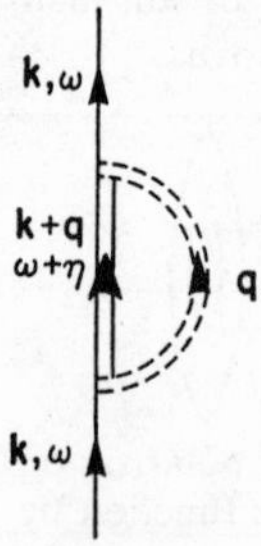

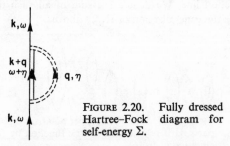

FIGURE 2.20. Fully dressed Hartree–Fock diagram for self-energy Σ.

The dressing of interaction lines can now be combined with the dressing of Fermion lines as we now shall describe. For example, in Figure 2.20 we fully dress the HF diagram to obtain the approximation‡

$$\Sigma \approx \int \frac{d\mathbf{q}}{4\pi^3} \int \frac{V_{\mathbf{q}}}{\varepsilon(\mathbf{q}\eta)} G(\mathbf{k}+\mathbf{q}, \omega+\eta)\, d\eta. \tag{2.10.17}$$

‡ This integral must be carried out as described in Appendix 2.2, p. 576.

Such a procedure can be made the basis of a self-consistent perturbation theory (see Appendix 2.2).

We shall finally write down diagrams representing the complete formal summation of the infinite series for Σ. We begin by noting that every undressed diagram, other than a forward scattering part, must consist of a Fermion line, going up the page, from which an interaction splits off to rejoin the line further up via some complex C (Figure 2.21(a)). Examples of C,

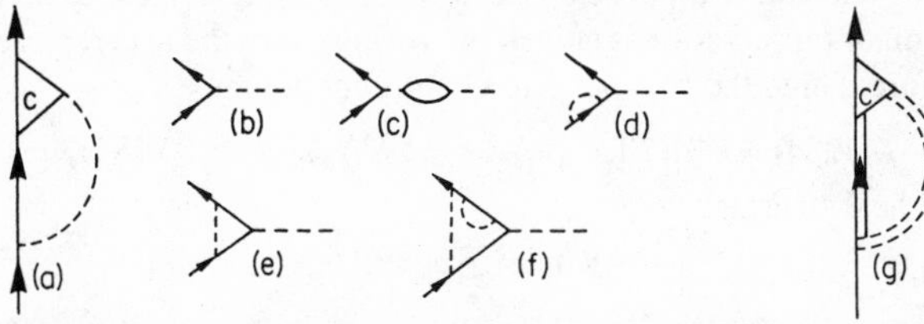

FIGURE 2.21. Vertex parts and their appearance in diagrams for self-energy. Vertex part C shown in Figure (a) could be, for example, any one of forms, (a) to (f). (g) is obtained by dressing Fermion line and interaction line of (a). Complex C' is not necessarily the same as C, as explained in text.

a *vertex part*, are shown in Figure 2.21, diagrams (b) to (f). Now let us suppose that we dress the interaction line and inner Fermion line of (a) to obtain (g). The complex C' will not always be identical to C. For example, if C is the diagram (c), its polarization part will be absorbed into the dressed interaction line and C' will be the diagram (b). Again, if C is the diagram (d), the self-energy part will be combined in the dressed Fermion line and C' will again be the diagram (b). Furthermore, in this last example the self-energy part could not have been on the free line which emerges from C towards the top of the page, by the definition of Σ, the diagrams for which are irreducible. It is therefore clear that C' must be a "proper" or irreducible vertex part; that is, a vertex part that cannot be divided into two by cutting either an inner Fermion line or an inner interaction line. Finally, by summing over all proper vertex parts to obtain Λ, as defined in Figure 2.22(a), we obtain the

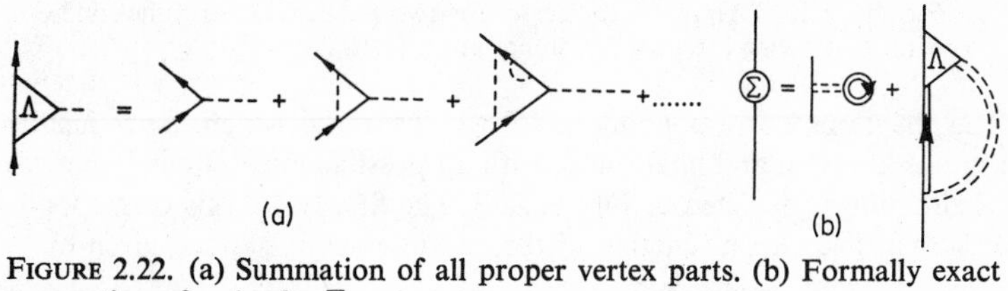

FIGURE 2.22. (a) Summation of all proper vertex parts. (b) Formally exact summation of series for Σ.

formally exact result for Σ, which is shown in Figure 2.22(b). For completeness, we have shown the fully dressed forward scattering part, although its contribution is cancelled by the uniform background when there is no external potential.

2.10.4 Two-particle Green function and Bethe–Salpeter equation

We shall treat the two-particle Green function in $\mathbf{r}, t$ space, as no essential simplication of presentation is gained by working with the $\mathbf{k}, \omega$ representation. We therefore define the two-particle Green function as

$$G(1234) = \langle \Psi | T[\psi(\mathbf{r}_1 t_1)\psi(\mathbf{r}_2 t_2)\psi^\dagger(\mathbf{r}_3 t_3)\psi^\dagger(\mathbf{r}_4 t_4)] | \Psi \rangle, \tag{2.10.18}$$

where the field operator

$$\psi(\mathbf{r}) = \sum_{\mathbf{k}} a_{\mathbf{k}} e^{i\mathbf{k}.\mathbf{r}} \tag{2.10.19}$$

destroys a particle at point $\mathbf{r}$. $|\Psi\rangle$ is the exact ground-state wave function. The two-particle Green function in momentum space would be obtained by substituting $a_{\mathbf{k}_i}(t_i)$ for $\psi(\mathbf{r}_i t_i)$ in equation (2.10.18).

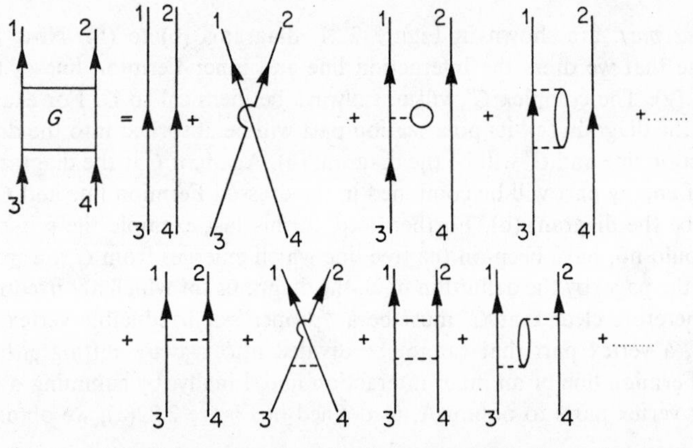

FIGURE 2.23. Diagrammatic series for two-particle Green function. First two terms are as given in equation (2.10.20).

The diagrams corresponding to G will be those which show just two Fermions at initial and final times, with all possible interactions in between. This situation is depicted in Figure 2.23. The first two terms correspond to the two-particle Green function of the non-interacting system, given by

$$G_0(1234) = G_0(13)\,G_0(24) - G_0(14)\,G_0(23). \tag{2.10.20}$$

We may effect a partial summation of the series for G by dressing Fermion lines to obtain the result of Figure 2.24. From this, it is evident that $G(1234)$ describes all possible interactions which two renormalized (i.e. fully dressed) particles can have with each other. We should also note that by reversal of time ordering we describe particle–hole scattering or hole–hole scattering.

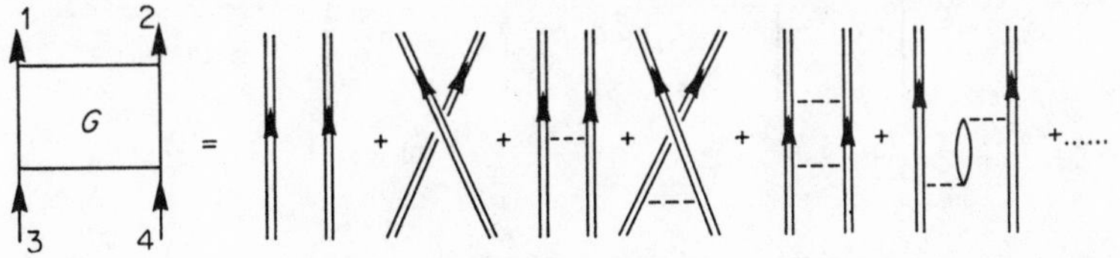

FIGURE 2.24. Two-particle Green function, showing interactions of two dressed or renormalized Fermi particles. This is an alternative, and generally more useful, way of representing series in Figure 2.23.

Equivalently, we can interchange labels in equation (2.10.18). For example, Figure 2.25 shows the result of interchange of 2 and 4 (particle–hole scattering).

To obtain a complete formal summation, let J represent the sum of all irreducible interaction diagrams between two particles, an irreducible diagram being one which cannot be divided into two parts, each having interaction lines, by cutting two internal Fermion lines. The sum is shown in

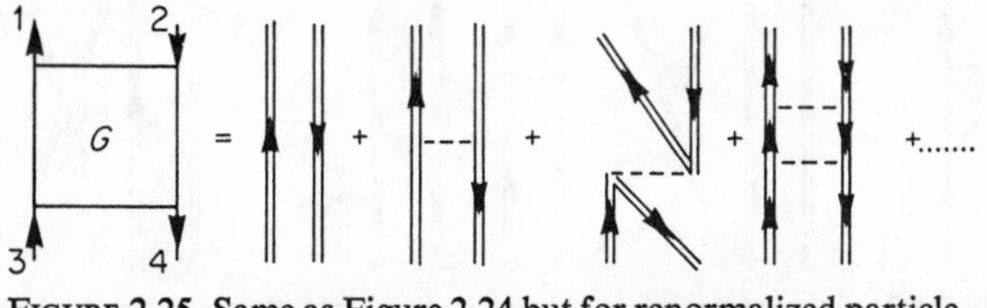

FIGURE 2.25. Same as Figure 2.24 but for renormalized particle–hole scattering. Note disappearance of second and fourth diagrams on right-hand side of Figure 2.24 and new diagram (third on right-hand side) in this figure.

Figure 2.26(c); (a) and (b) illustrate what is meant by reducible interaction diagrams. We want to note here that the outer Fermion lines of any diagram contributing to J are there merely to show how J is inserted into a pair of Fermion lines.

$G(1234)$ may now be obtained by allowing two quasi-particles to repeatedly scatter through the interaction described by J, as shown in Figure 2.27.

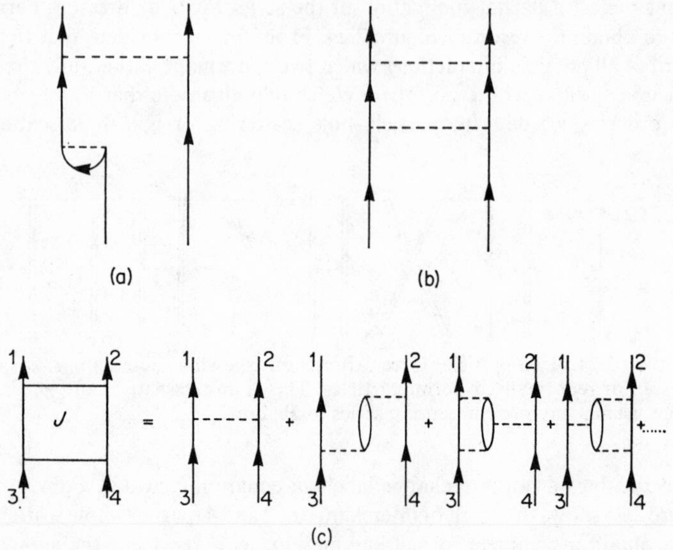

FIGURE 2.26. Irreducible interactions between two particles are shown in (c), while (a) and (b) are simple examples of diagrams reducible by just two cuts of Fermion lines. Antisymmetrizing (c) gives full J.

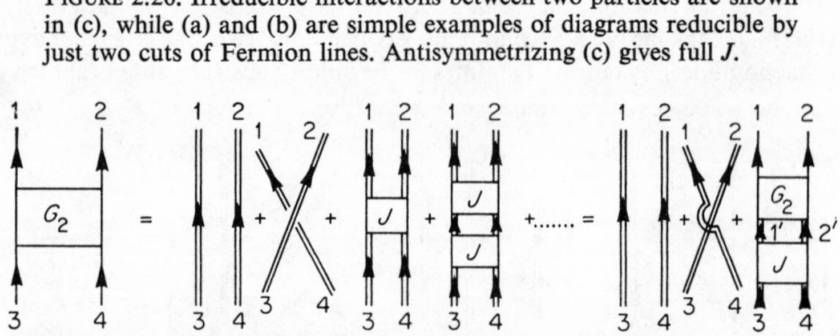

FIGURE 2.27. Diagrammatic representation of Bethe–Salpeter equation. First representation of series on right-hand side shows renormalized particles interacting through J. This series can be summed as shown.

The corresponding integral equation is (with time ordering 33′1′1)

$$G(1234) = G(13)\,G(24) - G(14)\,G(23)$$

$$+\frac{1}{2}\int G(3'3)\,G(4'4)\,J(1'2'3'4')\,G(121'2')\,d1'\,d2'\,d3'\,d4'. \qquad (2.10.21)$$

This is known as the Bethe–Salpeter equation. Omission of the factor $\frac{1}{2}$ on the right-hand side would mean we had counted every interaction diagram twice. Both $G_2 = G(1234)$ and J are antisymmetrical.

The lowest-order approximation for J, giving the HF result, is obtained by taking the first diagram on the right-hand side of Figure 2.26(c). We have then

$$J(1234) \simeq [\delta(13)\,\delta(24) - \delta(14)\,\delta(23)]\,v(|\mathbf{r}_1 - \mathbf{r}_2|), \qquad (2.10.22)$$

where v is the bare Coulombic interaction. This approximation may be improved by dressing the interaction line, so that v becomes a screened Coulomb potential.

In connection with Figure 2.27 there is an alternative way of writing the Bethe–Salpeter equation. Figure 2.27 can be translated into the algebraic form

$$G(1234) = G(13)\,G(24) - G(14)\,G(23) + \int G(3'3)\,G(4'4)\,\Gamma(1'2'3'4')\,G(11')\,G(22')\,d1'\,d2'\,d3'\,d4' \qquad (2.10.23)$$

where Γ is represented in Figure 2.28 and evidently obeys the integral equation shown there. As remarked for J, the external Fermion lines are shown only to indicate how Γ takes its place in a complete diagram. In fact,

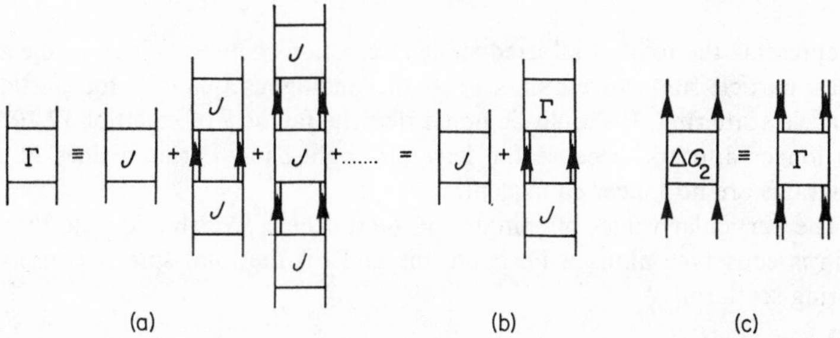

FIGURE 2.28. Γ in Bethe–Salpeter equation (2.10.23). (a) Representation of Γ as all repeated scatterings of quasi-particles through J, the total effective interaction in a single scattering. (b) Summation of this series. (c) Bound part of two-particle Green function in terms of Γ.

Figure 2.28(c) represents the part of the right-hand side of equation (2.10.23) involving Γ. This part, ΔG_2, is often called the "bound" part of G_2, in contradistinction to the term $G(13)\,G(24) - G(14)\,G(23)$, which describes the propagation of two quasi-particles without interaction. Figure 2.28(a) shows that Γ represents all repeated scatterings of the two quasi-particles through J; thus J represents the total effective interaction which operates in a single scattering between quasi-particles.

So far, we have treated only the scattering between two quasi-particles, but an equally important problem concerns the scattering of a quasi-particle with a quasi-hole. This obtains on assuming $t_1 > t_3$ but $t_2 < t_4$, as shown in Figure 2.25. Figure 2.29 gives the diagrammatic form of the Bethe–Salpeter equation for the corresponding Γ, which has the algebraic form in $(\mathbf{k}\omega)$ space

$$\Gamma(1234) = I(1234) + \sum_{56} I(1256)\, G(5)\, G(6)\, \Gamma(5634). \qquad (2.10.24)$$

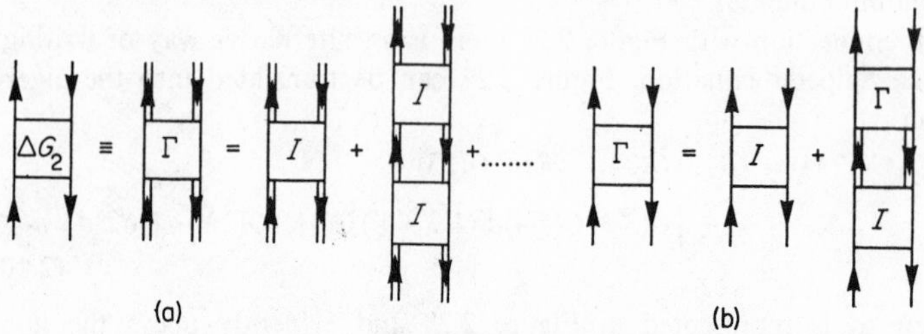

FIGURE 2.29. (a) Bound part of G_2 and series for Γ in terms of I for particle–hole scattering. (b) Integral equation for Γ (Bethe–Salpeter equation).

I represents the total of all irreducible interactions between a quasi-hole and quasi-particle and corresponds to J, the analogous quantity for particle–particle scattering. It should be noted that the factor $\frac{1}{2}$ of equation (2.10.21) no longer appears, because the left- and right-hand Fermion lines of the diagrams are no longer equivalent.

The particular values of spin are of interest here. We should note that (i) spin is conserved along a Fermion line and (ii) the total spin is conserved during scattering.

(a) *Total spin unity*. Here $\sigma_1 = -\sigma_2 = \sigma$, say, because the spin associated with a hole line labelled by σ is $-\sigma$. We therefore obtain only diagrams of the type shown in Figure 2.30. The dotted lines in the circle A of Figure 2.30

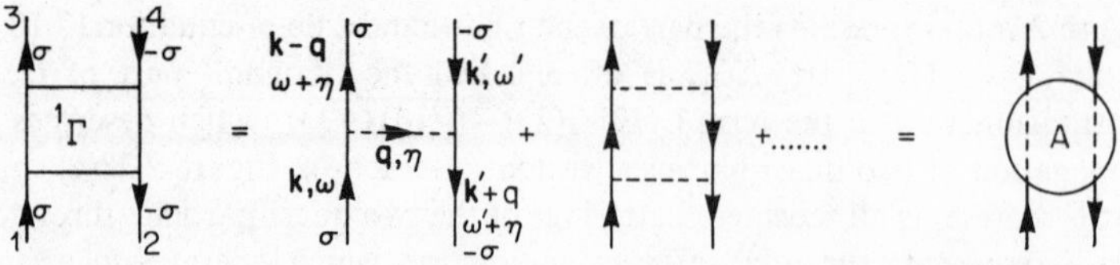

FIGURE 2.30. Γ for total spin unity denoted by $^1\Gamma$. Dotted lines in last part A of diagram indicate way lines running in and out of diagram are to be internally connected. Diagrams of form B in Figure 2.31 are excluded because of requirement that spin is conserved along a Fermion line.

indicate that the Fermion line starting from point 1 must always run to point 3.

(b) *Total spin zero*. Here $\sigma_1 = \sigma_2 = \sigma$ and since the total spin is conserved in the scattering, we must then have $\sigma_3 = \sigma_4 = \pm\sigma$. Taking $\sigma_3 = -\sigma$ first, conservation of spin along a Fermion line implies that we now have only diagrams of the type shown in Figure 2.31. On the other hand, if $\sigma_3 = \sigma_4 = \sigma$, we have diagrams of the type shown in Figure 2.32, i.e. both diagrams of type *A* and *B* appear. Hence we have, the superscripts indicating the total spin,

$$^0\Gamma(\mathbf{k}\omega\sigma;\ \mathbf{k}'\omega'\sigma;\ \mathbf{q},\eta) = {}^1\Gamma(\mathbf{k}\omega;\ \mathbf{k}'\omega';\ \mathbf{q},\eta)+{}^0\Gamma(\mathbf{k}\omega\sigma;\ \mathbf{k}'\omega',-\sigma;\ \mathbf{q}\eta), \tag{2.10.25}$$

where $\mathbf{q}$ and η are respectively momentum and energy transfer as indicated in Figure 2.30. In this equation, we have displayed the spin explicitly. Elsewhere, it will be absorbed in $\mathbf{k}$, the momentum parameter; hence if we put $\mathbf{k}_1 = \mathbf{k}_3$, for example, it is to be understood that $\sigma_1 = \sigma_3$.

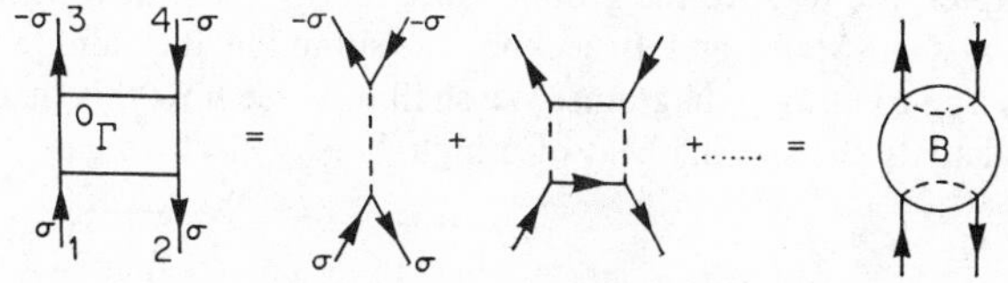

FIGURE 2.31. $^0\Gamma$ for total spin zero ($\sigma_1 = -\sigma_3$). Conservation of spin along a Fermion line now excludes diagrams of type A of Figure 2.30.

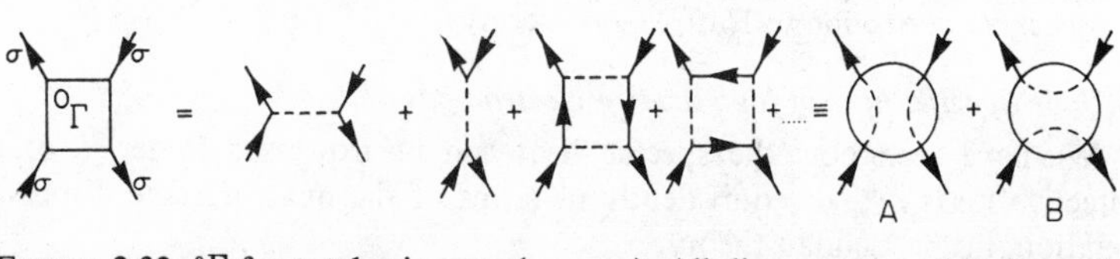

FIGURE 2.32. $^0\Gamma$ for total spin zero ($\sigma_1 = \sigma_3$). All diagrams A and B shown in Figures 2.30 and 2.31 now appear.

2.10.5 Gell-Mann and Brueckner calculation of ground-state energy

Before going on to use the two-particle Green function in discussing the Landau theory, we note that the ground-state energy can also be obtained from a graphical series, as we have shown in detail in Appendix 2.2. The corresponding diagrams have no external lines, as shown in Figure 2.33.

We may regard these diagrams as showing the excitations of the non-interacting gas which result from "switching on" the Coulombic interaction.

The Hartree–Fock result is the lowest-order correction to the non-interacting energy and so the corresponding diagrams are evidently Figure 2.33(a) and (b). For reasons given in Appendix 2.2, Gell-Mann and Brueckner chose the "ring-diagrams" exemplified in Figure 2.34 as giving the dominant

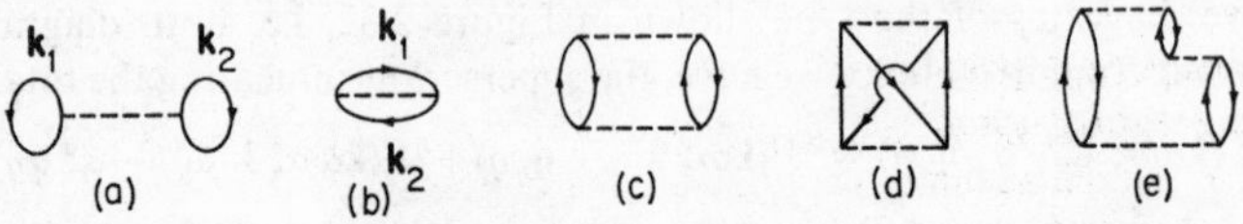

FIGURE 2.33. Examples of diagrams for total energy. Sum of (a) and (b) is Hartree–Fock energy. (c), (d) and (e) represent corrections to Hartree–Fock energy, (d) being exchange graph corresponding to (c).

correction to the HF energy. The correspondence with the diagrams of the RPA, shown in Figure 2.16, is evidently close, and, in fact, as noted elsewhere in this chapter, the RPA to the ground-state energy is identical with the result obtained by Gell-Mann and Brueckner on summing the infinite sub-series corresponding to all ring diagrams. We shall now see how this theory can be used to calculate the specific heat of a high-density electron gas.

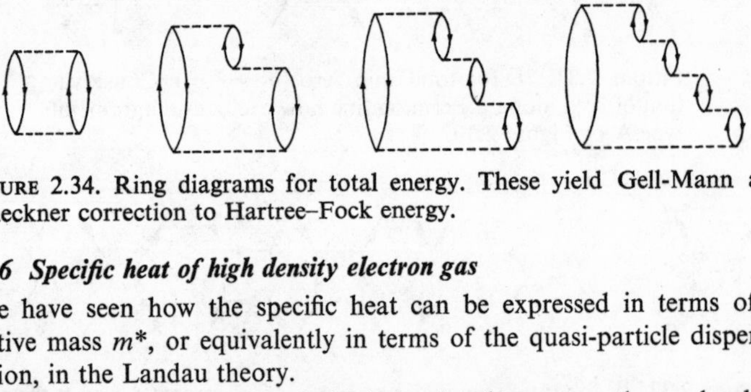

FIGURE 2.34. Ring diagrams for total energy. These yield Gell-Mann and Brueckner correction to Hartree–Fock energy.

2.10.6 Specific heat of high density electron gas

We have seen how the specific heat can be expressed in terms of the effective mass m^*, or equivalently in terms of the quasi-particle dispersion relation, in the Landau theory.

It is of interest to show how the high-density electron gas theory developed in section 2.4.2 enables us to evaluate directly the quasi-particle excitation energy (Gell-Mann, 1957; Silverstein, 1962; Rice, 1965).

We consider those states of the electron gas in which only a small number of particles ν are in excited states. Then, the energy of this state may be written, with k referring to particles and p to holes, as

$$E = E_0 + \sum_{j=1}^{\nu} [\varepsilon(k_j) - \varepsilon(p_j)] + O(\nu/N). \tag{2.10.26}$$

For free electrons, we have from the elementary theory of section 1.2,

$$\varepsilon(p_{\mathrm{f}}) = \frac{p_{\mathrm{f}}^2}{\alpha^2 r_{\mathrm{s}}^2}, \quad \alpha = \left(\frac{4}{9\pi}\right)^{\frac{1}{3}} \tag{2.10.27}$$

and

$$\left(\frac{d\varepsilon}{dp}\right)_{p=p_{\mathrm{f}}} = \frac{2}{\alpha^2 r_{\mathrm{s}}^2}. \tag{2.10.28}$$

The objective now must be to calculate $(d\varepsilon/dp)_{p_{\mathrm{f}}}$ in the high-density limit, for an interacting electron gas. $\varepsilon(p)$ is, in fact, the energy lost when a particle, with the system in its ground state, is annihilated. Clearly $p \leqslant 1$ here, where p is measured in units of p_{f} above. Similarly, starting again from the ground state and creating an electron of momentum $\mathbf{k}$ ($k \geqslant 1$), $\varepsilon(k)$ is the energy gain.

To see how these quantities can be calculated, consider a second-order term in the ground-state energy, namely

$$-\frac{1}{\pi^4}\int \frac{d\mathbf{q}}{q^4} \sum_{\substack{p_1<1 \\ |\mathbf{p}_1+\mathbf{q}|>1, |\mathbf{p}_2+\mathbf{q}|>1}} \int_{p_2<1} d\mathbf{p}_2 \frac{1}{q^2+\mathbf{q}\cdot(\mathbf{p}_1+\mathbf{p}_2)}. \tag{2.10.29}$$

Let us now remove a particle with upward spin, and having momentum $\mathbf{p}$. There are four contributions to the energy which is lost:

(i) The single-particle state with momentum $\mathbf{p}$ and upward spin is no longer occupied and must be left out of the summation.

(ii) For one of the spin states, a term $\mathbf{p}_1+\mathbf{q} = \mathbf{p}$ is added to the sum.

(iii) and (iv) Corresponding contributions from the sum or integral over $\mathbf{p}_2$ rather than $\mathbf{p}_1$. These are equal to (i) and (ii).

Thus, the contribution of this term to the energy lost by annihilating a particle is

$$-\frac{1}{\pi^4}\left[\int_{|\mathbf{p}+\mathbf{q}|>1} \frac{d\mathbf{q}}{q^4} \int_{\substack{p_2<1 \\ |\mathbf{p}_2+\mathbf{q}|>1}} d\mathbf{p}_2 \frac{1}{q^2+\mathbf{q}\cdot(\mathbf{p}_2+\mathbf{p}_2)} - \int_{|\mathbf{p}-\mathbf{q}|<1} \frac{d\mathbf{q}}{q^4} \times \int_{\substack{p_2<1 \\ |\mathbf{p}_2+\mathbf{q}|>1}} d\mathbf{p}_2 \frac{1}{\mathbf{q}\cdot(\mathbf{p}+\mathbf{p}_2)}\right]. \tag{2.10.30}$$

The contribution of each term in the ground-state energy can be similarly considered.

If we deal specifically with $(d\varepsilon/dp)_{p_{\mathrm{f}}}$ then we can write for the exchange energy

$$\varepsilon(p) = -\frac{1}{\pi^2 \alpha r_{\mathrm{s}}} \int_{p_2<1} d\mathbf{p}_2 \frac{1}{(\mathbf{p}+\mathbf{p}_2)^2}. \tag{2.10.31}$$

We then obtain

$$\left(\frac{d\varepsilon}{dp}\right)_{p_f} = \frac{2}{\pi\alpha r_s}\int_{-1}^{1}\frac{x\,dx}{2(1-x)} \equiv \frac{1}{\pi\alpha r_s}\int_0^2 dq\frac{(1-q)}{q} \tag{2.10.32}$$

and this diverges logarithmically (Bardeen, 1936).

The contribution in (2.10.30), after differentiation, gives a contribution to $(d\varepsilon/dp)_{p_f}$ of

$$-\frac{8}{\pi^2}\int_{-1}^{1}\frac{x\,dx}{[2(1-x)]^2}. \tag{2.10.33}$$

Collecting these contributions together, we find

$$\left(\frac{d\varepsilon}{dp}\right)_{p_f} = \frac{2}{\alpha^2 r_s^2}+\frac{2}{\pi\alpha r_s}\int_{-1}^{1}\frac{x\,dx}{2(1-x)}\left[1-\frac{4\alpha r_s}{\pi}\frac{1}{2(1-x)}+\ldots\right]. \tag{2.10.34}$$

Following the arguments used in calculating the ground state energy, it can be shown that the quantity in the brackets [ ] in (2.10.34) is simply the first two terms in the expansion of

$$\left[1+\frac{4\alpha r_s}{\pi}\frac{1}{2(1-x)}\right]^{-1}. \tag{2.10.35}$$

Thus

$$\left(\frac{d\varepsilon}{dp}\right)_{p_f} = \frac{2}{\alpha^2 r_s^2}+\frac{1}{\pi\alpha r_s}\left\{-2+\left(1+\frac{2\alpha r_s}{\pi}\right)\ln\left[\frac{(2\alpha r_s/\pi)+2}{2\alpha r_s/\pi}\right]\right\}, \tag{2.10.36}$$

which is the result we wished to obtain. At low temperatures, this leads to the ratio of the specific heat of the interacting electron gas to that of the non-interacting system c_v^0 as

$$\frac{c_v}{c_v^0} = \left\{1+\frac{\alpha r_s}{2\pi}\left[-\ln r_s+\ln\left(\frac{\pi}{\alpha}\right)-2\right]+\ldots\right\}. \tag{2.10.37}$$

Thus, the Landau parameter A_1, which clearly is a function of r_s in a homogeneous gas, is given for small $r_s (r_s \ll a_0)$ by

$$A_1 = \frac{\alpha r_s}{2\pi}\left\{\left[-\ln r_s+\ln\left(\frac{\pi}{\alpha}\right)-2\right]+\ldots\right\} \tag{2.10.38}$$

This latter result follows because we have that

$$\frac{m^*}{m} = 1+A_1 \tag{2.10.39}$$

which Landau obtained originally by the following argument. If the interacting system is translationally invariant, as we are assuming, then the total momentum, that is current induced by a uniform field, is conserved in collisions between quasi-particles. The current associated with a single excited quasi-particle must be equal to the current when the interactions are turned off, so that j must be given by

$$j_\alpha = e\int \frac{d\mathbf{p}}{(2\pi)^3}\frac{p_\alpha}{m} g\,\delta(E^0-\mu), \tag{2.10.40}$$

g being defined in equation (2.9.30). It follows then, straightforwardly, that equation (2.10.39) must be obeyed. We do not want to give the impression that the high-density theory of A_1 given above is useful in real alkali metals where the condition $r_s \ll a_0$ is completely violated (for example, in caesium metal, the lowest density alkali, $r_s \simeq 5.5a_0$).

To conclude this section, we note that the result for m^*/m is simply a special case of the general relation

$$\frac{1}{m} = \frac{1}{m^*} + \frac{2\pi k_f}{16\pi^3}\int_0^\pi \sum_{\sigma\sigma'} f(\theta,\sigma\sigma')\cos\theta\sin\theta\,d\theta, \tag{2.10.41}$$

where we have written

$$f(\mathbf{k}\mathbf{k}')\big|_{k=k'=k_f} = f(\theta,\sigma\sigma') \quad \left(\cos\theta = \frac{\mathbf{k}.\mathbf{k}'}{kk'}\right). \tag{2.10.42}$$

We shall now go on to deal with the microscopic theory of the Landau scattering function in full detail.

2.10.7 *Microscopic basis of Landau theory*

We have previously discussed the existence of quasi-particles in terms of Green functions. In fact, for normal systems, the entire phenomenological Landau theory can be justified using the formalism introduced in sections 2.7 and 2.8 (Nozières, 1964). Further, it is possible to obtain expressions from the *microscopic* theory for the basic quantities characterizing the Landau theory.

Here we shall illustrate the microscopic theory by:

(i) Giving a proof that the quasi-particles are Fermions (this has been an assumption we made previously, of course).

(ii) Calculating the interaction energy $f(\mathbf{k}'\mathbf{k})$.

(iii) Deriving a sum rule on the Landau parameters A_l.

(a) *Excited-state Green function.* We shall suppose that the quasi-particle state $\mathbf{k}'$ is long-lived, so that, to a good approximation, we may write the excited-state eigenfunction as

$$|\psi_{\mathbf{k}'}\rangle = q_{\mathbf{k}'}^{\dagger}|\psi_0\rangle, \tag{2.10.43}$$

where $|\psi_0\rangle$ represents the ground state, and $q_{\mathbf{k}'}$ is the quasi-particle operator introduced in section 2.8.5. Then the excited-state Green function

$$G_{\mathbf{k}'}(\mathbf{k}t) = -i\langle\psi_{\mathbf{k}'}|T[a_{\mathbf{k}}(t)\,a_{\mathbf{k}}^{\dagger}(0)]|\psi_{\mathbf{k}'}\rangle \tag{2.10.44}$$

may evidently be expressed in terms of quantities of the form

$$\langle\psi_0|q_{\mathbf{k}'}\,a_{\mathbf{k}}(t)\,a_{\mathbf{k}}^{\dagger}q_{\mathbf{k}'}^{\dagger}|\psi_0\rangle$$

and, referring to the definition of $q_{\mathbf{k}'}$, we see that the two-particle Green function G_2 enters, provided the operators $a_{\mathbf{k}}^{\dagger}(t')$ entering into $q_{\mathbf{k}'}^{\dagger}$ all correspond to $t'<(0,t)$ and the operators $a_{\mathbf{k}'}(t'')$ entering into $q_{\mathbf{k}}^{\dagger}$, all correspond to $t''>(0,t)$. This leads us to choose the filtering functions $f(t)$ of equation (2.8.26) such that

$$|\psi_{\mathbf{k}'}\rangle = \frac{\alpha}{\sqrt{[z(\mathbf{k}')]}}\int_{-\infty}^{0} dt'\exp[-i\tilde{\varepsilon}_{\mathbf{k}'}(t'+\tau')]\exp(\alpha t')\,a_{\mathbf{k}'}^{\dagger}(t'+\tau')|\psi_0\rangle \tag{2.10.45}$$

and

$$\langle\psi_{\mathbf{k}'}| = \frac{\alpha}{\sqrt{[z(\mathbf{k}')]}}\int_{0}^{\infty} dt''\exp[i\tilde{\varepsilon}_{\mathbf{k}'}(t''+\tau'')]\exp(-\alpha t'')\langle\psi_0|\,a_{\mathbf{k}'}(t''+\tau''), \tag{2.10.46}$$

where

$$\tau' \ll (0,t) \ll \tau'', \tag{2.10.47}$$

so that in the regions where the required time ordering does not obtain the integrands are negligible anyway.

Then

$$G_{\mathbf{k}'}(\mathbf{k}t) = -\frac{i\alpha^2}{z(\mathbf{k}')}\int_{-\infty}^{0} dt'\int_{0}^{\infty} dt''\exp[i\tilde{\varepsilon}_{\mathbf{k}'}(t''+\tau''-t'-\tau')]\exp[\alpha(t'-t'')]$$
$$\times\langle\psi_0|T[a_{\mathbf{k}'}(t''+\tau'')\,a_{\mathbf{k}}(t)\,a_{\mathbf{k}}^{\dagger}(0)\,a_{\mathbf{k}'}^{\dagger}(t'+\tau')]|\psi_0\rangle. \tag{2.10.48}$$

The last factor of the integrand is simply G_2.

(b) *Proof that quasi-particles are Fermions.* We have already seen that the Fermi surface divides **k**-space into two regions, the quasi-particle region and the quasi-hole region, and that quasi-particles can be added only into the former region. In order to prove that the quasi-particles are Fermions, therefore, it is necessary to show that, on addition of a quasi-particle into state **k**, the Fermi surface moves so that the point **k** is now included in the quasi-hole region.

Thus, we set $\mathbf{k} = \mathbf{k}'$ to obtain

$$G_{\mathbf{k}}(\mathbf{k}, t) = -\frac{i\alpha^2}{z(\mathbf{k})}\int_{-\infty}^{0} dt' \int_{0}^{\infty} dt'' \exp[i\tilde{\varepsilon}_{\mathbf{k}}(t''+\tau''-t'-\tau')]\exp[\alpha(t'-t'')] \times G_2(\mathbf{k}, t''+\tau''; \mathbf{k}, t; \mathbf{k}, 0; \mathbf{k}, t'+\tau'). \quad (2.10.49)$$

Now we recall from section 2.10.4 that we may write

$$G_2(1234) = G(13)\,G(24) - G(14)\,G(23) + \Delta G_2(1234), \quad (2.10.50)$$

where the first two terms, called the "free part" of G_2, represent propagation of the two particles without interaction and ΔG_2 is the bound part, taking account of the interactions between the quasi-particles. These interactions are of finite range, whereas the two particles are free to run throughout the entire system; hence ΔG_2 is of order $1/\mathscr{V}$, $\mathscr{V}$ being as usual the volume of the system. We therefore keep only the free part of G_2 and insert into equation (2.10.49)

$$G_{\text{free}} = G(\mathbf{k}, t''+\tau'')\,G(\mathbf{k}, t-\tau'-t') - G(\mathbf{k}, t''+\tau''-t'-\tau')\,G(\mathbf{k}, t). \quad (2.10.51)$$

The second term contributes simply

$$\langle\psi_{\mathbf{k}}|\psi_{\mathbf{k}}\rangle\, G(\mathbf{k}, t) = G(\mathbf{k}, t). \quad (2.10.52)$$

As for the first term, because of the filtering action of the integration, only the coherent part

$$G_{\text{coh}} = \begin{cases} iz(\mathbf{k})\exp(-i\tilde{\varepsilon}_{\mathbf{k}}t), & t>0, \\ 0, & t<0, \end{cases} \quad (2.10.53)$$

contributes; in fact we readily verify that the result is just $-iz(\mathbf{k})\exp(-i\tilde{\varepsilon}_{\mathbf{k}}t)$ for both $t>0$ and $t<0$. The full result is then

$$G_{\mathbf{k}}(\mathbf{k}, t) = G(\mathbf{k}, t) - iz(\mathbf{k})\exp(-i\tilde{\varepsilon}_{\mathbf{k}}t) \quad (2.10.54)$$

and the coherent part of this is evidently

$$G_{\mathbf{k}}(\mathbf{k}, t) - G_{\text{inc}}(\mathbf{k}, t) = \begin{cases} 0, & t>0, \\ -iz(\mathbf{k})\exp(-i\tilde{\varepsilon}_{\mathbf{k}}t), & t<0, \end{cases} \quad (2.10.55)$$

which, by definition, describes a quasi-hole. Hence we cannot add a further quasi-particle into state $\mathbf{k}$ and the quasi-particles are Fermions.

(c) *Interaction function* $f(\mathbf{k}, \mathbf{k}')$. Let us now take $\mathbf{k} \neq \mathbf{k}'$, when the free part of G_2 is

$$G(\mathbf{k}', \mathbf{k};\, t''+\tau'')\,G(\mathbf{k}, \mathbf{k}';\, t-\tau'-t') - G(\mathbf{k}', \mathbf{k}', t''+\tau''-t'-\tau')\,G(\mathbf{k}, \mathbf{k}, t) = -G(\mathbf{k}', t''+\tau''-t'-\tau')\,G(\mathbf{k}, t), \quad (2.10.56)$$

the single-particle Green function being diagonal in **k**. On insertion of this term into equation (2.10.49), we obtain

$$\langle\psi_{\mathbf{k}'}|\psi_{\mathbf{k}'}\rangle\, G(\mathbf{k}t) = G_{\mathbf{k}'}(\mathbf{k}, t), \tag{2.10.57}$$

the Green function being unchanged. To this order of approximation, the excitations are independent. Thus, if we write

$$G_{\mathbf{k}'}(\mathbf{k}, t) = G(\mathbf{k}, t) + \delta_{\mathbf{k}'}\, G(\mathbf{k}, t) \tag{2.10.58}$$

the correction to $G(\mathbf{k}t)$ comes entirely from the bound part of G_2:

$$\delta_{\mathbf{k}'}\, G(\mathbf{k}, t) = -\frac{i\alpha^2}{z(\mathbf{k}')}\int_{-\infty}^{0} dt' \int_{0}^{\infty} dt'' \exp\left[i\tilde{\varepsilon}_{\mathbf{k}'}(t''+\tau''-t'-\tau')\right]\exp\left[\alpha(t'-t'')\right]$$
$$\times\, \Delta G_2(\mathbf{k}', t''+\tau''; \mathbf{k}, t; \mathbf{k}, 0; \mathbf{k}', t'+\tau'). \tag{2.10.59}$$

This is of order $(1/\mathscr{V})$, but so is $f(\mathbf{k}', \mathbf{k})$, which we aim to calculate.

Since $f(\mathbf{k}', \mathbf{k})$ is, by definition, the change in energy of quasi-particle **k** on injecting a quasi-particle **k**′ into the system, we may write

$$G_{\mathbf{k}'}(\mathbf{k}, t) = i[z(\mathbf{k}) + \delta_{\mathbf{k}'}\, z(\mathbf{k})]\exp\{-i[\tilde{\varepsilon}(\mathbf{k}) + f(\mathbf{k}', \mathbf{k})]\, t\} \tag{2.10.60}$$

at times sufficiently large that only the coherent part of G matters. Since f is of order $1/\mathscr{V}$, we rewrite equation (2.10.60) as

$$\delta_{\mathbf{k}'}\, G(\mathbf{k}, t) = [i\delta_{\mathbf{k}'}\, z(\mathbf{k}) + z(\mathbf{k}) f(\mathbf{k}', \mathbf{k})\, t]\exp(-i\tilde{\varepsilon}_{\mathbf{k}}\, t) \tag{2.10.61}$$

or, for long times,

$$\delta_{\mathbf{k}'}\, G(\mathbf{k}, t) = z(\mathbf{k}) f(\mathbf{k}', \mathbf{k})\, t \exp(-i\tilde{\varepsilon}_{\mathbf{k}}\, t). \tag{2.10.62}$$

Hence we have

$$f(\mathbf{k}', \mathbf{k}) = \frac{i\alpha^2}{z(\mathbf{k})\, z(\mathbf{k}')\, t}\int_{-\infty}^{0} dt' \int_{0}^{\infty} dt'' \exp\left[i\tilde{\varepsilon}_{\mathbf{k}'}(t''+\tau''-t'-\tau') + i\tilde{\varepsilon}_{\mathbf{k}}\, t\right]$$
$$\times \exp\left[\alpha(t'-t'')\right]\Delta G_2(\mathbf{k}', t''+\tau''; \mathbf{k}, t; \mathbf{k}, 0; \mathbf{k}', t'+\tau'). \tag{2.10.63}$$

It should be noted that the values of the times in ΔG_2 imply that it describes a particle–hole pair. Using equation (2.10.23) we find

$$f(\mathbf{k}', \mathbf{k}) = 2\pi i z(\mathbf{k})\, z(\mathbf{k}')\, \Gamma(\mathbf{k}', \tilde{\varepsilon}_{\mathbf{k}'}; \mathbf{k}, \tilde{\varepsilon}_{\mathbf{k}}; \mathbf{k}, \tilde{\varepsilon}_{\mathbf{k}}; \mathbf{k}', \tilde{\varepsilon}_{\mathbf{k}'}). \tag{2.10.64}$$

Thus the interaction energy is directly related to Γ, which represents scattering between a particle–hole pair. (The relationship between dispersion relations and forward scattering amplitudes is a general one; we shall give a particularly simple example of this in Chapter 10.)

Equation (2.10.64) is not our final expression, for the value of Γ contained therein must be calculated by a limiting process. For liquids with short-range forces, such as He[3]

$$\Gamma(\mathbf{k}\omega'; \mathbf{k}\omega; \mathbf{k}, \omega; \mathbf{k}'\omega') = \lim_{\substack{\mathbf{q}\to 0\\ \eta\to 0}} \Gamma(\mathbf{k}, \omega, \mathbf{k}'\omega'; \mathbf{q}, \eta) \tag{2.10.65}$$

where the Γ on the right-hand side is labelled in accordance with Figure 2.35, from which we see that $\mathbf{q}, \eta$ are the total momentum and energy (conserved

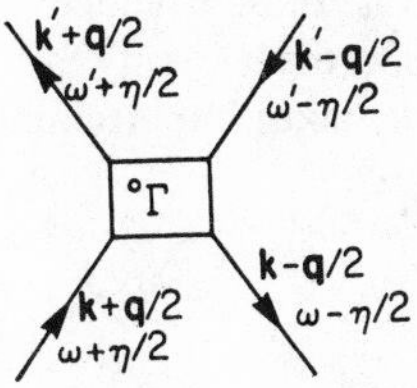

FIGURE 2.35. Particle–hole scattering through $^\circ\Gamma$. This figure is labelled according to notation used in text. Spins are absorbed through notation $(\mathbf{k}, \sigma) \equiv \mathbf{k}$. Hence total spin is zero (cf. Figures 2.31 and 2.32).

in the scattering). We shall see shortly that the limit is not unambiguously defined, but depends on the ratio

$$\mathbf{r} = \mathbf{q}/\eta \tag{2.10.66}$$

as $\mathbf{q} \to 0$. However, the limit we want is clear. $\mathbf{k}'$ and $\mathbf{k}$ are precisely defined in equation (2.10.64) and so $\mathbf{q} = 0$, whereas the values of $\tilde{\varepsilon}_{\mathbf{k}}$ and $\tilde{\varepsilon}_{\mathbf{k}'}$ are uncertain to the extent of the level widths of the quasi-particle states. Hence we put

$$f(\mathbf{k}', \mathbf{k}) = 2\pi i z(\mathbf{k})\, z(\mathbf{k}')\, \Gamma^0(\mathbf{k}', \mathbf{k}), \tag{2.10.67}$$

where the superscript on Γ denotes the limit $\mathbf{r} = 0$. We have omitted $\tilde{\varepsilon}_{\mathbf{k}}$ and $\tilde{\varepsilon}_{\mathbf{k}'}$, assuming them to be equal to μ, the chemical potential, for in the microscopic theory quasi-particles can be adequately defined only very near the Fermi surface.

Equation (2.10.67) must be modified for the systems with which we are primarily concerned, because of the long-range character of the Coulomb interaction. Γ contains diagrams such as Figure 2.36(a) in which we have an internal interaction line labelled by $\mathbf{q}$. For the Γ on the right-hand side of equation (2.10.64) we have, quite precisely, $\mathbf{q} = 0$ and the contributions of such diagrams are cancelled by that of the uniform background charge. In fact, because of the presence of the background charge, the Coulombic interaction is strictly given by (for the discrete allowed values of $\mathbf{q}$)

$$V_{\mathbf{q}} = \begin{cases} 0, & \mathbf{q} = 0, \\ \dfrac{4\pi e^2}{q^2}, & \mathbf{q} \neq 0. \end{cases} \tag{2.10.68}$$

The difficulty we face is that if we smoothly take the limit defined on the right-hand side of equation (2.10.65), this discontinuous nature of $V_{\mathbf{q}}$ at the origin will not be taken into account. We must therefore remove the offending diagrams before we take the limit and accordingly rewrite equation (2.10.67) as

$$f(\mathbf{k}', \mathbf{k}) = 2\pi i z(\mathbf{k})\, z(\mathbf{k}')\, \tilde{\Gamma}^0(\mathbf{k}', \mathbf{k}), \tag{2.10.69}$$

where $\tilde{\Gamma}$ is the sum of proper diagrams, that is, those irreducible with respect to internal interaction lines. We can look upon the necessity for the modification as a consequence of the fact that $f(\mathbf{k}', \mathbf{k})$ takes into account short-range

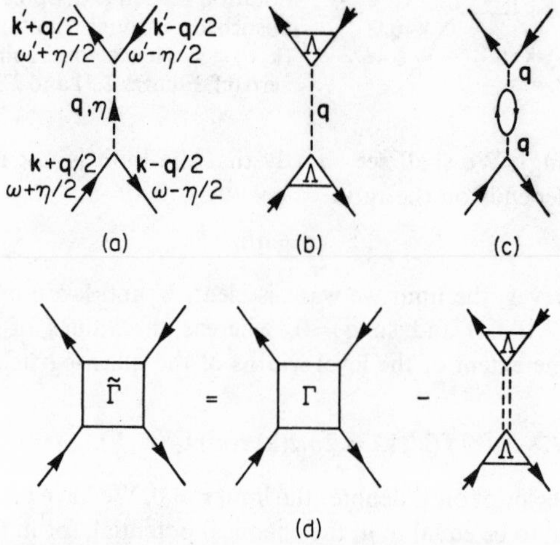

FIGURE 2.36. (a), (b), (c) Diagrams contributing to Γ, showing, in particular, contributions of vertex parts. (d) Showing relation between Γ and its irreducible part $\tilde{\Gamma}$.

interactions only; in the application of Landau's theory to charged fluids, the long-range part of the interaction is incorporated into a Hartree term, modifying the local Hamiltonian and thus terms with $\mathbf{q} \to 0$ cannot enter into $f(\mathbf{k}', \mathbf{k})$.

To relate $\tilde{\Gamma}$ to Γ we note that the most general diagram of Γ with one internal interaction line with respect to which it is reducible is of the form of Figure 2.36(b), with proper vertex parts at each end of the interaction line. We also note that we have improper diagrams such as Figure 2.36(c), in which proper polarization parts are introduced into the interaction line; in other words we can replace the single interaction line by a screened interaction.

The net result of all this is the diagrammatic relation of Figure 2.36(d), from which we obtain

$$\tilde{\Gamma}(\mathbf{k}\omega;\,\mathbf{k}'\omega';\,\mathbf{q},\eta) = \Gamma(\mathbf{k},\omega;\,\mathbf{k}',\omega';\,\mathbf{q},\eta) + \frac{iV_{\mathbf{q}}}{2\pi}\,\frac{\Lambda(\mathbf{k},\omega;\,\mathbf{q},\eta)\,\Lambda(\mathbf{k}',\omega';\,\mathbf{q},\eta)}{\varepsilon(\mathbf{q},\eta)}. \tag{2.10.70}$$

To obtain f entirely in terms of quantities fundamental in the microscopic theory, we must find an expression for the $z(\mathbf{k})$ appearing in equation (2.10.69). To do so, it is only necessary to compare equation (2.10.53) with Dyson's equation expressed as

$$G(\mathbf{k},\omega) = \frac{1}{(\mathbf{k}^2/2m) - \omega - \Sigma(\mathbf{k}\omega)}. \tag{2.10.71}$$

When the self-energy is expanded to first-order in $\omega - \mu$ about $\omega = \mu$, we find

$$z(\mathbf{k}_{\mathrm{f}}) = \left.\frac{1}{1 + (\partial\Sigma(\mathbf{k},\omega)/\partial\omega)}\right|_{\omega=\varepsilon_{\mathrm{f}}=\mu} \tag{2.10.72}$$

(d) *Sum rule for Landau parameters.* As a final illustration of the formalism of the Landau theory, we shall derive a sum rule (Brinkman and co-workers, 1968‡), which will be used when we come to derive the Landau parameters from experiment in Chapter 7. To do so, we investigate the limit $\mathbf{k}' \to \mathbf{k}$ in the previous section. We first consider this limit applied to Γ, but before doing so it will be convenient to discuss the reasons why this limit $\mathbf{q} \to 0$, $\eta \to 0$ on this function is ambiguous.

(i) *Limits* Γ^0 and Γ^∞: Let us first examine Figure 2.37, a typical diagram for Γ. The internal dressed lines running between the complexes A and B give us the product $G(\mathbf{k}''+\mathbf{q}/2, \omega''+\eta/2)\,G(\mathbf{k}''-\mathbf{q}/2, \omega''-\eta/2)$ and it is this factor which causes the trouble. We next note that, very near the Fermi surface,

$$G(\mathbf{k},\omega) = \frac{-z(\mathbf{k})}{\tilde{\varepsilon}(\mathbf{k}) - \omega - i\delta_{\mathbf{k}}} + G_{\mathrm{inc}} \begin{cases} \delta_{\mathbf{k}} > 0, & k > k_{\mathrm{f}} \\ \delta_{\mathbf{k}} < 0, & k < k_{\mathrm{f}} \end{cases} \tag{2.10.73}$$

where G_{inc}, arising from the continuous background in A^+ (see equation (2.7.81)), is regular in ω. Hence the singular term in the product of Green functions arises from

$$\frac{z(\mathbf{k}''+(\mathbf{q}/2))}{\tilde{\varepsilon}_{\mathbf{k}''+(\mathbf{q}/2)} - \omega'' - (\eta/2) - i\delta_{\mathbf{k}''+(\mathbf{q}/2)}} \cdot \frac{z(\mathbf{k}''-(\mathbf{q}/2))}{\tilde{\varepsilon}_{\mathbf{k}''-(\mathbf{q}/2)} - \omega'' + (\eta/2) + i\delta_{\mathbf{k}''-(\mathbf{q}/2)}}, \tag{2.10.74}$$

‡ In addition to this reference, unpublished notes made available by T. M. Rice have been valuable to us in this chapter.

when $\mathbf{k}''$ is at the Fermi surface so that $\delta_{\mathbf{k}''+(\mathbf{q}/2)}$ and $\delta_{\mathbf{k}''-(\mathbf{q}/2)}$ are of different signs. It can readily be shown that

$$\lim_{\substack{\mathbf{q}\to 0\\ \eta\to 0}} G\left(\mathbf{k}''+\frac{\mathbf{q}}{2},\omega''+\frac{\eta}{2}\right) G\left(\mathbf{k}''-\frac{\mathbf{q}}{2},\omega''-\frac{\eta}{2}\right) = G^2(\mathbf{k}'',\omega'')+R^{\mathbf{r}}(\mathbf{k}'',\omega''), \quad \mathbf{r}=\frac{\mathbf{q}}{\eta}, \tag{2.10.75a}$$

where

$$R^{\mathbf{r}}(\mathbf{k},\omega) = \frac{2\pi i z^2(\mathbf{k})\,\mathbf{r}.\mathbf{v_k}\,\delta(\mu-\omega)\,\delta(\mu-\tilde{\varepsilon}_{\mathbf{k}})}{1-\mathbf{r}.\mathbf{v_k}+i\alpha} \tag{2.10.75b}$$

with $v_{\mathbf{k}}$ the quasi-particle velocity $\partial\tilde{\varepsilon}_{\mathbf{k}}/\partial\mathbf{k}$ and α an infinitesimal with the sign of $\mathbf{q}.\mathbf{n_k}$, where $\mathbf{n_k}$ is the normal to the Fermi surface at $\mathbf{k}$. We can note that

$$R^0 = 0. \tag{2.10.76}$$

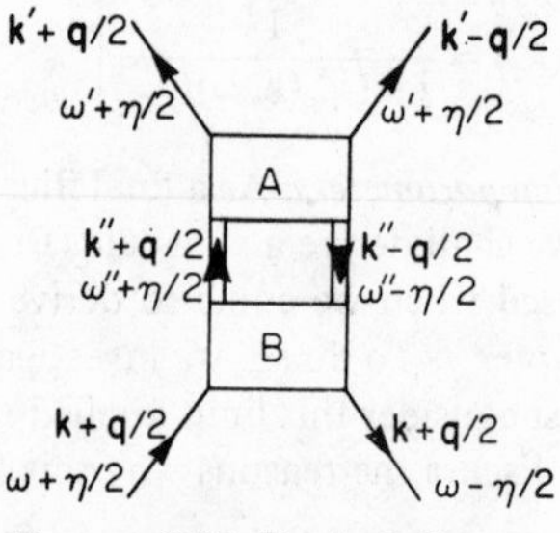

FIGURE 2.37. Typical diagram for Γ showing decomposition into two complexes A and B with internal dressed lines between.

We see from the above that as $\mathbf{q},\eta\to 0$ we associate every pair of internal Fermion lines, running between irreducible interactions I, with the factor $G^2+R^{\mathbf{r}}$. The nth-order term therefore can be decomposed into new terms containing products $\prod_{i=1}^{m} G^2(i) \prod_{j=1}^{n-m} R^{\mathbf{r}}(j)$. If $m=n$ we have a term of Γ^0; let us consider the other terms. To do this, we draw a diagram to represent one of the new terms. The diagram will be as for the original term, but each pair of internal Fermion lines will now be associated with either G^2 or $R^{\mathbf{r}}$. Suppose that, starting from the bottom of the diagram, every pair of internal Fermion lines in a certain complex B (see Figure 2.37) is associated with a factor G^2, the pair of internal lines shown explicitly in Figure 2.37 is associated with the first factor $R^{\mathbf{r}}$ encountered and the complex B can contain any combination of G^2's and $R^{\mathbf{r}}$'s. On summing over all diagrams, we then have a factor Γ^r from B, whereas we obtain Γ^0 from A, because of equation (2.10.76). We finally reach the conclusion that

$$\Gamma^r(\mathbf{k},\omega',\mathbf{k}',\omega') = \Gamma^0(\mathbf{k},\omega;\,\mathbf{k}',\omega') + \sum_{\mathbf{k}'',\omega''} \Gamma^0(\mathbf{k},\omega;\,\mathbf{k}''\omega'')\,R^r(\mathbf{k}'',\omega'') \times \Gamma^r(\mathbf{k}'',\omega'';\,\mathbf{k}'\omega'). \qquad (2.10.77)$$

Γ^∞, like Γ^0, does not depend on the direction of $\mathbf{r}$. From equation (2.10.77)

$$\Gamma^\infty(\mathbf{k},\omega;\,\mathbf{k}'\omega') = \Gamma^0(\mathbf{k}\omega;\,\mathbf{k}'\omega') + \sum_{\mathbf{k}'',\omega''} \Gamma^0(\mathbf{k},\omega',\mathbf{k}'',\omega'') \times [-2\pi i z^2(\mathbf{k}'')\,\delta(\omega''-\mu)\,\delta(\tilde{\varepsilon}_{\mathbf{k}''}-\mu)\,\Gamma^\infty(\mathbf{k}'',\omega'';\,\mathbf{k}'\omega')]. \qquad (2.10.78)$$

It should be emphasized that our arguments hold equally well for $\tilde{\Gamma}^r$ and so equation (2.10.78) may be rewritten for $\tilde{\Gamma}^\infty$ and $\tilde{\Gamma}^0$.

(ii) *Limit* $\mathbf{k}'\to\mathbf{k}$: Because G_2 is antisymmetric, Γ is also antisymmetric: thus diagrams (a) and (b) of Figure 2.38 represent the same term, apart from a change of sign. The diagram makes it clear that $\Gamma(1234)$, representing

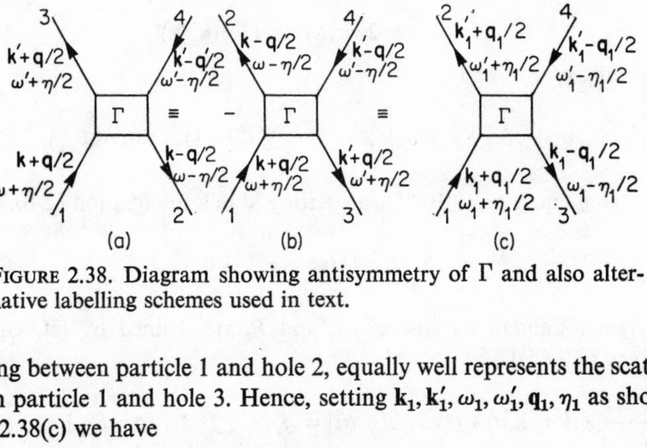

FIGURE 2.38. Diagram showing antisymmetry of Γ and also alternative labelling schemes used in text.

scattering between particle 1 and hole 2, equally well represents the scattering between particle 1 and hole 3. Hence, setting $\mathbf{k}_1, \mathbf{k}_1', \omega_1, \omega_1', \mathbf{q}_1, \eta_1$ as shown in Figure 2.38(c) we have

$$\Gamma(\mathbf{k},\omega;\,\mathbf{k}',\omega';\,\mathbf{q},\eta) = -\Gamma(\mathbf{k}_1,\omega_1;\mathbf{k}_1',\omega_1';\mathbf{q}_1,\eta_1), \qquad (2.10.79)$$

that is,

$$\Gamma(\mathbf{k},\omega;\,\mathbf{k}',\omega';\mathbf{q},\eta) = -\Gamma\left(\frac{\mathbf{k}+\mathbf{k}'+\mathbf{q}}{2},\ \frac{\omega+\omega'+\eta}{2};\ \frac{\mathbf{k}+\mathbf{k}'-\mathbf{q}}{2};\ \frac{\omega+\omega'-\eta}{2};\ \frac{\mathbf{k}-\mathbf{k}'}{2};\ \frac{\omega-\omega'}{2}\right). \qquad (2.10.80)$$

We shall choose the arguments of Γ so that they are the same on both sides of this equation in order to prove that for the particular values chosen Γ is zero. However, in view of our discussion about limiting values above, we must recognize the possibility that, if we choose the arguments such that they approach one another from opposite sides of the Fermi surface, we find not the value zero but a discontinuity at k_{f}. To avoid this complication, we shall set $\omega = \omega' = \mu$ and $\mathbf{k}, \mathbf{k}'$ on the Fermi surface, before taking the limit $\mathbf{k} \to \mathbf{k}'$. This being settled, let us put $\mathbf{q} = \mathbf{k} - \mathbf{k}', \eta = \omega - \omega'$ and then from equation (2.10.80) we find

$$\Gamma(\mathbf{k}, \omega, \mathbf{k}', \omega', \mathbf{k}-\mathbf{k}', \omega-\omega') = 0. \tag{2.10.81}$$

Then setting $\omega = \omega'$ and taking the limit $\mathbf{q} = \mathbf{k} - \mathbf{k}' = 0$ we have

$$\Gamma^{\infty}(\mathbf{k}, \mathbf{k}) = 0. \tag{2.10.82}$$

The limit evidently corresponds to $\mathbf{r} = \infty$ because we set $\eta = 0$ before taking $\mathbf{q} \to 0$.

We may obtain the desired sum rule for short-range forces without further ado. Corresponding to equation (2.10.67), we define a function dual to $f(\mathbf{k}, \mathbf{k}')$

$$g(\mathbf{k}\mathbf{k}') = 2\pi i z(\mathbf{k})\, z(\mathbf{k}')\, \Gamma^{\infty}(\mathbf{k}', \mathbf{k}) \tag{2.10.83}$$

and defining $g_l^{\pm}$ through

$$\frac{m^* k_{\mathrm{f}}}{2\pi^2}\{g(\mathbf{k}, \sigma; \mathbf{k}', \sigma') \pm g(\mathbf{k}, \sigma, \mathbf{k}', -\sigma)\} = \sum_l (2l+1)\, g_l^{\pm} P_l(\cos\theta_{\mathbf{k}'\mathbf{k}}), \tag{2.10.84}$$

we find from equation (2.10.82), on setting $\mathbf{k}' = \mathbf{k}$ in equation (2.10.83),

$$\sum_{l=0}^{\infty} (2l+1)\{g_l^+ + g_l^-\} = 0. \tag{2.10.85}$$

The original Landau parameters A_l and B_l are defined by [cf. equations (2.9.42) to (2.9.45)]

$$\frac{m^* k_{\mathrm{f}}}{2\pi^2}[f(\mathbf{k}\sigma; \mathbf{k}'\sigma) + f(\mathbf{k}\sigma; \mathbf{k}' - \sigma)] = \sum_l (2l+1)\, A_l P_l(\cos\theta_{\mathbf{k}'\mathbf{k}}), \tag{2.10.86}$$

$$\frac{m^* k_{\mathrm{f}}}{2\pi^2}[f(\mathbf{k}\sigma; \mathbf{k}'\sigma) - f(\mathbf{k}\sigma, \mathbf{k} - \sigma)] = \sum_l (2l+1)\, B_l P_l(\cos\theta_{\mathbf{k}'\mathbf{k}}). \tag{2.10.87}$$

By substituting f and g as defined by equations (2.10.67) and (2.10.83) into equation (2.10.84) we find

$$g_l^+ = \frac{A_l}{1+A_l}, \quad g_l^- = \frac{B_l}{1+B_l} \tag{2.10.88}$$

and so the sum rule (2.10.85) may also be written in the form

$$\sum_{l=0}^{\infty}(2l+1)\left\{\frac{A_l}{1+A_l}+\frac{B_l}{1+B_l}\right\}=0. \tag{2.10.89}$$

This holds for short-range forces only. For Coulombic forces, we define

$$g(\mathbf{k},\mathbf{k}')=2\pi i z(\mathbf{k})\,z(\mathbf{k}')\,\tilde{\Gamma}^{\infty}(\mathbf{k}\mathbf{k}') \tag{2.10.90}$$

and then

$$\frac{\pi^2}{m^* k_{\mathrm{f}}}\sum_{l=0}^{\infty}(2l+1)\left\{\frac{A_l}{1+A_l}+\frac{B_l}{1+B_l}\right\}=\tilde{\Gamma}^{\infty}(\mathbf{k},\mathbf{k})\,2\pi i z^2(\mathbf{k}). \tag{2.10.91}$$

To relate $\tilde{\Gamma}^{\infty}$ to Γ^{∞} we use equation (2.10.70). Because of equation (2.10.82),

$$\tilde{\Gamma}^{\infty}(\mathbf{k},\mathbf{k})=\lim_{q\to 0}\frac{i}{2\pi}\frac{\Lambda^{\infty}(\mathbf{k},\mu)\,\Lambda^{\infty}(\mathbf{k},\mu)}{q^2\,\varepsilon^{\infty}}. \tag{2.10.92}$$

Since $\mathbf{k}$ is on the Fermi surface, we can use the Ward identity

$$\Lambda^{\infty}(\mathbf{k},\mu)=\frac{v_{\mathrm{f}}}{z(\mathbf{k})}\frac{dk_{\mathrm{f}}}{d\mu}, \tag{2.10.93}$$

which is proved in Appendix A2.3. Combining this with the result for the compressibility (see the result of Problem 2.13)

$$\lim_{\mathbf{q}\to 0}\frac{\varepsilon^{\infty}}{V_{\mathbf{q}}}=\frac{d\rho_0}{d\mu}, \tag{2.10.94}$$

ρ_0 being as usual the density, we find from equation (2.10.92) that

$$2\pi i z^2(\mathbf{k})\,\tilde{\Gamma}^{\infty}(\mathbf{k},\mathbf{k})=-\frac{\pi^2}{m^*}\frac{dk_{\mathrm{f}}}{d\mu}. \tag{2.10.95}$$

Then since

$$\frac{dk_{\mathrm{f}}}{d\mu}=\frac{m^*}{k_{\mathrm{f}}(1+A_0)}, \tag{2.10.96}$$

we have the following sum rule for the electron gas

$$\sum_{l=0}^{\infty}(2l+1)\left[\frac{A_l}{1+A_l}+\frac{B_l}{1+B_l}\right]=-\frac{1}{1+A_0} \tag{2.10.97}$$

or, on taking the right-hand side over to the left,

$$\frac{B_0}{1+B_0}+\sum_{l=1}^{\infty}(2l+1)\left[\frac{A_l}{1+A_l}+\frac{B_l}{1+B_l}\right]=-1 \tag{2.10.98}$$

showing that A_0 does not enter the sum rule. Comparing equation (2.10.98) with equation (2.10.89), it can be seen that the effect of the long-range force is to replace $A_0/1+A_0$ by unity.

2.11 Inhomogeneous electron gas

The theory of correlations so far developed will work accurately only in a uniform electron gas. In practice, of course, we are dealing with periodic lattices, where the valence electrons are piled up round the nuclei, and it is the influence of correlations in such an inhomogeneous gas that we would like to include.

There is some hope that the recasting of the many-body problem for the ground state into one-body language can lead to substantial progress here.

Essentially then, the problem boils down to finding the one-body potential which includes exchange and correlation effects (see section 1.2.1 in Chapter 1) and the corresponding exchange and correlation energies. To do this exactly would necessitate an exact solution of the many-body problem, which, at present at any rate, is out of the question.

It will be convenient to introduce the discussion by taking a more careful look at the homogeneous gas, in terms of the density functional approach discussed in Chapter 1. The results of this are sufficiently encouraging for us to attempt their generalization to the inhomogeneous gas. We take first then the problem of the dielectric function of the uniform gas, including exchange and correlation, which we have already considered in sections 2.3 and 2.4.

2.11.1 Dielectric function of homogeneous gas including exchange and correlation

We have seen that the first-order perturbation theory for a potential V in a Fermi gas leads to a density change given by

$$\Delta\rho(\mathbf{r}) = -\frac{k_f^2}{2\pi^3}\int \frac{j_1(2k_f|\mathbf{r}-\mathbf{r}'|)}{|\mathbf{r}-\mathbf{r}'|^2}\, V(\mathbf{r}')\,d\mathbf{r}' \tag{2.11.1}$$

which we shall write in a formal way as

$$\Delta\rho(\mathbf{r}) = \int F(\mathbf{r}-\mathbf{r}')\, V(\mathbf{r}')\,d\mathbf{r}'. \tag{2.11.2}$$

Now, from one-body potential theory, including exchange and correlation, we know that the change in the one-body potential is a unique functional of the electron density and hence we have for the change in the exchange and correlation part V_{XC} of the potential

$$\Delta V_{XC}(\mathbf{r}) = \int U(\mathbf{r}-\mathbf{r}')\,\delta\rho(\mathbf{r}')\,d\mathbf{r}'. \tag{2.11.3}$$

Furthermore, if we assume a "test charge" represented by a potential $V_{\text{bare}} \equiv V_b$ is introduced into the gas, then

$$V(\mathbf{r}) = \int U(\mathbf{r}-\mathbf{r}')\,\delta\rho(\mathbf{r}')\,d\mathbf{r}' + V_b(\mathbf{r}) + \int \frac{\rho(\mathbf{r}')}{|\mathbf{r}-\mathbf{r}'|}\,d\mathbf{r}'. \tag{2.11.4}$$

Taking the Fourier transform and using the general definition of $\varepsilon^{-1}(\mathbf{k})$ as total charge divided by test charge, we find

$$\varepsilon^{-1}(k) = \frac{1-F(\mathbf{k})\,U(\mathbf{k})}{1-F(k)\,U'(k)} \tag{2.11.5}$$

where $U'(k)$ is simply $U+(4\pi/k^2)$. We regain the Lindhard result (2.3.20) by putting $U = 0$. Improved approximations to $\varepsilon(\mathbf{k})$ obviously involve approximating to the exchange and correlation function $U(\mathbf{k})$.

We shall generalize the free-electron response function $F(\mathbf{r}-\mathbf{r}')$ and the correlation function $U(\mathbf{r}-\mathbf{r}')$ to the case of a Bloch lattice in section 2.11.4 below. Their Fourier transforms must then, of course, involve reciprocal lattice vectors $\mathbf{K}$.

2.11.2 *Functional derivative approach to compressibility*

We now recall from Chapter 1 that the one-body potential $V(\mathbf{r})$ is the functional derivative with respect to the charge density of $V[\rho]$ defined through

$$E[\rho] = T[\rho] + V[\rho], \tag{2.11.6}$$

where $E[\rho]$ is the ground-state energy and $T[\rho]$ is the ground-state kinetic energy of a system of non-interacting particles of the same density. Thus we have

$$\delta V[\rho] = \int V(\mathbf{r})\,\delta\rho(\mathbf{r})\,d\mathbf{r} \tag{2.11.7}$$

giving the change in $V[\rho]$ to first-order in the density change. We may write the second-order change as

$$\delta^{(2)} V[\rho] = \int U'(\mathbf{r}\mathbf{r}')\,\delta\rho(\mathbf{r})\,\delta\rho(\mathbf{r}')\,d\mathbf{r}\,d\mathbf{r}'. \tag{2.11.8}$$

It is easy to prove that the function U' defined in this way is identical with that introduced following equation (2.11.5). We can thus calculate an approximate value of U', given an approximation to $V[\rho]$.

As a simple example of this procedure, let us take the Hartree–Fock form for $V[\rho]$, given, in terms of the Dirac density matrix $\rho(\mathbf{r}\mathbf{r}')$ defined in equation (2.7.30), by

$$V[\rho] = \frac{e^2}{2}\int \frac{\rho(\mathbf{r})\,\rho(\mathbf{r}')}{|\mathbf{r}'-\mathbf{r}|}\,d\mathbf{r}\,d\mathbf{r}' - \frac{e^2}{2}\int \frac{\rho(\mathbf{r}'\mathbf{r})\,\rho(\mathbf{r}\mathbf{r}')}{|\mathbf{r}'-\mathbf{r}|}\,d\mathbf{r}\,d\mathbf{r}'. \tag{2.11.9}$$

If we now insert for $\rho(\mathbf{rr}')$ a Dirac density matrix formed from the single-particle eigenfunction of the Hamiltonian $(-\hbar^2/2m)\nabla^2 + V(\mathbf{r})$, $V(\mathbf{r})$ being the one-body potential, the static dielectric function can in principle be found (see Problem 2.12). But here we wish to illustrate the approach by calculating the compressibility from equations given in Problems 5.13 and 5.14. The present method gives us

$$\varepsilon(q0) = 1 + \frac{1}{q^2}\left(\frac{4k_f}{\pi}\right)\frac{\pi k_f}{\pi k_f - 1} \quad \text{as} \quad q \to 0 \tag{2.11.10}$$

and thus

$$\frac{K_{\text{free}}}{K} = \frac{\pi k_f - 1}{\pi k_f}, \tag{2.11.11}$$

where K_{free} is the compressibility of the non-interacting gas. In this equation, k_f is to be obtained from

$$k_f = \frac{\alpha}{r_s}, \quad \alpha = \frac{3(3\pi^2)^{\frac{1}{2}}}{(4\pi)^{\frac{1}{3}}}. \tag{2.11.12}$$

This is, in fact, the result of the Dirac–Slater exchange theory. Table 2.1 shows a comparison of the results of equation (2.11.11) compared with an accurate result from Hubbard's theory of the ground-state energy as a function of r_s (see section 2.11.3 below). More precisely

$$\frac{K_{\text{free}}}{K} = \left(\frac{4}{9\pi}\right)^{\frac{2}{3}}\frac{r_s}{6}\left[\frac{d^2E}{dr_s^2} - \frac{2}{r_s}\frac{dE}{dr_s}\right]. \tag{2.11.13}$$

TABLE 2.1

r_s	K_{free}/K from Hubbard theory	$\left(\frac{\pi k_f - 1}{\pi k_f}\right)$
1	0.83	0.83
2	0.65	0.67
3	0.46	0.50
4	0.26	0.34
5	0.06	0.17
6	−0.15	0.034

Agreement is seen to be quite good, especially when one notes that Hubbard's theory, for example, used to obtain the $E(r_s)$ from which Table 2.1 results, gives a dielectric constant from which one obtains a compressibility which goes negative at $r_s \geqslant 3$. This is a characteristic of a number of theories (for example, the theory of Singwi and co-workers, 1968, discussed below); the $q \to 0$ limit of the dielectric constant represents a lower order of approximation than that obtained by differentiating the energy with respect to r_s. The latter is often given quite satisfactorily by approximate theories, whereas $\varepsilon(q)$ in the limit $q = 0$ is quite a stringent test of a theory.

The reasonable result obtained over quite a wide range of r_s from equation (2.11.11) is probably due to the fact that we have used a functional derivative method here. Another related method applied to this problem is that of Singwi and co-workers, which we shall now develop in some detail.

2.11.3 Effective field approach to electron gas

It is quite clear that the central problem of an approach based on the density functional theory is to calculate the effective one-body potential yielding the exact density. Or, in linear response theory, we need to calculate the appropriate change in this one-body potential. We shall accordingly discuss here an approach due to Singwi and colleagues (1968, 1970). The potential they obtain is, in general ω-dependent and we should note that equations (2.11.2) and (2.11.3) for linear response theory in the one-body potential framework are very easily formally generalized to cover this case.

(a) *Equation of motion for one-particle density matrix.* We begin with the equation of motion for ρ introduced in section 2.3.2. Defining the density fluctuation $\rho_{\mathbf{k}}$ by

$$\rho_{\mathbf{k}} = \sum_i e^{i\mathbf{k}.\mathbf{r}_i}, \tag{2.11.14}$$

we have

$$\dot{\rho}_{\mathbf{k}} = [\rho_{\mathbf{k}}, H]/i\hbar \tag{2.11.15}$$

and when we take the Hamiltonian (cf. equation 2.3.31) as

$$H = \sum_i \frac{p_i^2}{2m} + \sum_{\mathbf{k}\neq 0} v(\mathbf{k})\,(\rho_{\mathbf{k}}^{\dagger}\,\rho_{\mathbf{k}} - N) \tag{2.11.16}$$

then we find from equation (2.11.15) that

$$\dot{\rho}_{\mathbf{k}} = -i \sum_i \left(\frac{\mathbf{k}.\mathbf{p}_i}{m} + \frac{\hbar k^2}{2m}\right) e^{-i\mathbf{k}.\mathbf{r}_i} \tag{2.11.17}$$

and using the Heisenberg equation of motion for $\mathbf{p}$, we obtain

$$\ddot{\rho}_{\mathbf{k}} = -\sum_i \left(\frac{\mathbf{k}.\mathbf{p}_i}{m} + \frac{\hbar k^2}{2m}\right) e^{-i\mathbf{k}.\mathbf{r}_i} - \sum_{\mathbf{q}} v(\mathbf{q}) \frac{\mathbf{k}.\mathbf{q}}{m}\,\rho_{\mathbf{k}-\mathbf{q}}\,\rho_{\mathbf{q}}. \tag{2.11.18}$$

This last equation can be rewritten as

$$\ddot{\rho}_{\mathbf{k}} = -\sum_i \left(\frac{\mathbf{k}.\mathbf{p}_i}{m}+\frac{\hbar k^2}{2m}\right) e^{-i\mathbf{k}.\mathbf{r}_i} - v(\mathbf{k})\frac{k^2}{m}\rho_0\,\rho_{\mathbf{k}} - \sum_{\mathbf{q}\neq\mathbf{k}} v(\mathbf{q})\frac{\mathbf{k}.\mathbf{q}}{m}\rho_{\mathbf{k}-\mathbf{q}}\,\rho_{\mathbf{q}}. \tag{2.11.19}$$

(b) *Random phase approximation.* The last term on the right-hand side of equation (2.11.19) involves the product of two density fluctuations. Since $\rho_{\mathbf{q}}$ given by equation (2.11.14) is a sum of complex exponential terms with differing phases of $\mathbf{q} \neq 0$, and since $\langle\rho_{\mathbf{q}}\rangle$ $(q \neq 0)$ vanishes if the system is homogeneous, we expect destructive interference to occur in this term and we drop it from the equation, this being the random phase assumption.

The first term on the right-hand side may be estimated to be of the order of $k^2 v_{\mathrm{f}}^2 \rho_{\mathbf{k}}$ and so, introducing the plasma frequency ω_{p} defined by equation (2.2.9) we have, if

$$\frac{k^2 v_{\mathrm{f}}^2}{\omega_{\mathrm{p}}^2} \ll 1, \tag{2.11.20}$$

$$\ddot{\rho}_{\mathbf{k}} + \omega_{\mathrm{p}}^2\,\rho_{\mathbf{k}} = 0. \tag{2.11.21}$$

The connection with the self-consistent field approximation of section 2.3.2 is readily seen, because the neglect of the last term in equation (2.11.19) leaves only the one term $v(\mathbf{k})\,(k^2/m)\,\rho_0\,\rho_{\mathbf{k}}$ involving the potential, and the factor $v(\mathbf{k})\,\rho_{\mathbf{k}}$ of this term simply represents the time-dependent Hartree field. Calculation of the dielectric constant naturally then reduces to that in section 2.3.2.

(c) *Effective field of Singwi and co-workers.* Instead of neglecting the last term of equation (2.11.19) entirely, we now rewrite it in the form

$$m^{-1}\sum_{\mathbf{q}\neq\mathbf{k}} \mathbf{k}.\mathbf{q}v(\mathbf{q}) \sum_i e^{i\mathbf{k}.\mathbf{r}_i} \sum_j e^{i(\mathbf{q}-\mathbf{k}).(\mathbf{r}_i-\mathbf{r}_j)} \tag{2.11.22}$$

and replace the sum over j by its static average value, that is,

$$\left\langle \sum_j e^{i(\mathbf{q}-\mathbf{k}).(\mathbf{r}_i-\mathbf{r}_j)} \right\rangle = N^{-1}\left\langle \sum_{ij} e^{i(\mathbf{q}-\mathbf{k}).(\mathbf{r}_i-\mathbf{r}_j)} \right\rangle$$

$$= S(\mathbf{q}-\mathbf{k}), \tag{2.11.23}$$

where $S(\mathbf{q})$ is the structure factor defined in section 2.7.4. Equation (2.11.19) can now be rewritten as

$$\ddot{\rho}_{\mathbf{k}} = -\sum_i \left(\frac{\mathbf{k}.\mathbf{p}_i}{m}+\frac{\hbar k^2}{2m}\right) e^{-i\mathbf{k}.\mathbf{r}_i} - U'(\mathbf{k})\frac{k^2}{m}\rho_0\,\rho_{\mathbf{k}}, \tag{2.11.24}$$

where

$$U'(\mathbf{k}) = v(\mathbf{k})\left[1+\frac{\rho_0^{-1}}{8\pi^3}\int \frac{\mathbf{k}.\mathbf{q}}{q^2}\,[S(\mathbf{k}-\mathbf{q})-1]\,d\mathbf{q}\right]. \tag{2.11.25}$$

Equation (2.11.24) is of the same form as the equation of motion in the random-phase approximation, but the Coulomb potential $v(\mathbf{k})$ is replaced by an effective interaction $U(\mathbf{k})$ and we may now derive the dielectric constant in the same manner. We obtain, following the notation of Singwi and colleagues,

$$\varepsilon(\mathbf{k},\omega) = 1+\frac{Q_0(\mathbf{k},\omega)}{1-G(\mathbf{k})\,Q_0(\mathbf{k},\omega)}, \tag{2.11.26}$$

where

$$Q_0(\mathbf{k},\omega) = \varepsilon_{\mathrm{SCF}}-1 \tag{2.11.27}$$

and

$$G(\mathbf{k}) = -\frac{\rho_0^{-1}}{8\pi^3}\int \frac{\mathbf{k}.\mathbf{q}}{q^2}\,[S(\mathbf{k}-\mathbf{q})-1]\,d\mathbf{q}. \tag{2.11.28}$$

Since, as we can see from section 2.7.4,

$$S(\mathbf{k}) = -\frac{\hbar k^3}{4\pi e^2\,\rho_0}\int_0^\infty \mathrm{Im}\,[\varepsilon(\mathbf{k},\omega)]^{-1}\,d\omega, \tag{2.11.29}$$

equations (2.11.28) and (2.11.29) can be solved self-consistently to yield values of $S(\mathbf{k})$ and hence the pair function.

In terms of our formulation of section 2.11.1 the present approximation to the one-body potential generating the exact density is given by

$$V(\mathbf{k}\omega) = V_{\mathrm{external}}(\mathbf{k}\omega)+V_{\mathrm{H}}(\mathbf{k},\omega)+U(k)\,\rho(\mathbf{k},\omega) \tag{2.11.30}$$

where V_{H} is the Hartree potential and

$$U(\mathbf{k}) = \frac{v(\mathbf{k})}{8\pi^3\,\rho_0}\int d\mathbf{q}\frac{\mathbf{k}.\mathbf{q}}{q^2}\,[S(\mathbf{q}-\mathbf{k})-1]. \tag{2.11.31}$$

By using the value of $\varepsilon(\mathbf{k},\omega)$ thus obtained, it is straightforward to calculate the correlation energy from equation (2.4.1) and good results are obtained in this way. On the other hand, the compressibility limit of $\varepsilon(\mathbf{k},\omega)$ [see equation (2.10.94)] is not very good. In fact, while the static approximation of equation (2.11.25) may be expected to give reasonable values at large wave-vectors $\mathbf{k}$, since one can expect the pair-correlation function $g(r)$ to be unaffected by a rapid spatial variation of external field, at smaller wave-vectors the adjustment of the pair-correlation function to the external disturbance will lead to a screening of the Coulomb potential. Instead of equation (2.11.31), we then write

$$U(\mathbf{k}) = \frac{v(\mathbf{k})}{8\pi^3 \rho_0} \int d\mathbf{q} \frac{\mathbf{k}.\mathbf{q}}{q^2} \frac{[S(\mathbf{q}-\mathbf{k})-1]}{\varepsilon(\mathbf{q})}. \tag{2.11.32}$$

Some formal justification of this modification has been given by Singwi and co-workers (1970), by examination of the properties of the density–density correlation function $\langle \rho(\mathbf{r}, t)\, \rho(\mathbf{r}', t') \rangle$.

Using equation (2.11.32) the compressibility results are much improved. On the other hand, the form of $g(r)$ near the origin is poor in that $g(r)$ ought not, of course, to go negative, and Singwi and co-workers (1970) give an interpolation scheme for $G(k)$ which preserves the behaviour of $g(r)$ when U' is given by equation (2.11.31) and the better compressibility results when U' is given by equation (2.11.32). They find that the self-consistent values of $G(k)$ can be quite accurately fitted by the form

$$G(k) = A\left[1 - \exp\left(-B\frac{k^2}{k_f^2}\right)\right], \tag{2.11.33}$$

where the parameters are listed in the range of metallic densities in Table 2.2.

TABLE 2.2

(r_s/a_0)	1	2	3	4	5	6
A	0.7756	0.8894	0.9629	0.9959	1.0138	1.0218
B	0.4307	0.3401	0.2924	0.2612	0.2377	0.2189

For small k, equation (2.11.33) reproduces the values given by equation (2.11.32) and, for large k, the values given by equation (2.11.31). The interpolation takes place in the range between $k \approx k_f/2$ and $k \simeq k_f$. Though some difficulties remain at the foundations of this approach, it is clear that the parametrization afforded above is very useful indeed.

(d) *Hubbard dielectric function.* We want here to record two other forms of the dielectric function which have been of value in the development of the theory. Let us suppose that the Hartree–Fock result for $S(\mathbf{k}-\mathbf{q})$ is used to calculate $G(\mathbf{k})$, which is equivalent to taking the pair correlation function $g(r)$ to be the Fermi hole shown in Figure 2.3. This form of $S(\mathbf{k})$ may be expressed as

$$S(\mathbf{k}) = -\frac{2}{8\pi^3 \rho_0} \int_{q \leqslant k_f} d\mathbf{q} \int_{q' \leqslant k_f} d\mathbf{q}'\, \delta(\mathbf{q}-\mathbf{q}'+\mathbf{k}) \tag{2.11.34}$$

and inserting this into equation (2.11.28) we find

$$G(\mathbf{k}) = \frac{2}{(8\pi^3)^2 \rho_0^2} \int_{q \leqslant k_f} d\mathbf{q} \int_{q' \leqslant k_f} d\mathbf{q}' \frac{\mathbf{k}.(\mathbf{k}+\mathbf{q}-\mathbf{q}')}{|\mathbf{k}+\mathbf{q}-\mathbf{q}'|^2}. \tag{2.11.35}$$

The dielectric constant proposed by Hubbard (1957) now follows if we approximate $G(\mathbf{k})$ in equation (2.11.35) by replacing the denominator $|\mathbf{k}+\mathbf{q}-\mathbf{q}'|^2$ by $(k^2+k_f^2)$, when one finds

$$G(\mathbf{k}) = \tfrac{1}{2}[k^2/(\mathbf{k}^2+\mathbf{k}_f^2)]. \tag{2.11.36}$$

This form, inserted into equation (2.11.26) gives Hubbard's result, though the above does not do full justice to the merits of his theory. Actually, equation (2.11.35) for $G(\mathbf{k})$ can be evaluated exactly, but very little change from Hubbard's results occurs.

A similar approximation can be made in equation (2.11.32) which includes screening. If we insert, for example, the semi-classical dielectric constant given by equation (2.3.21) then a form of $G(\mathbf{k})$ given by

$$G(\mathbf{k}) = \tfrac{1}{2}[k^2/(k^2+k_f^2+q^2)] \tag{2.11.37}$$

results, where in this equation q^2 is the Thomas–Fermi result $4k_f/\pi a_0$. This form has been employed by Sham (1965).

(e) *Low-density electron gas.* Even for the lowest-density alkali metal Cs, with $r_s = 5.5a_0$, there is evidence that most of the general features predicted by high-density electron gas theory still persist.

Still, it is of considerable theoretical interest to ask just what happens when the electron density is progressively reduced. Wigner (1934, 1938) pointed out that in the case when the electron–electron interactions become strong compared with the kinetic energy (and since the kinetic energy scales as r_s^{-2} and the potential energy as r_s^{-1} (cf. equation (2.4.2)), this is clearly the case at sufficiently low densities) the problem will become one of minimizing the potential energy. If we therefore single out an electron and look at the distribution around it, it is clear that an effective way for electrons to avoid one another is to go on to the sites of a lattice. From purely electrostatic considerations, the lowest energy found for the structures so far investigated is for a body-centred-cubic lattice (see, for example, Fuchs, 1935). For such a "Wigner lattice", the energy per particle can be calculated as

$$E/N = -\frac{1.792}{r_s} \text{ Rydberg}, \tag{2.11.38}$$

which is substantially lower than the exchange energy of equation (2.4.2), which yields $-0.916r_s^{-1}$ Rydberg. If we now increase the density, and relax therefore the coupling between electrons somewhat, we expect the electrons to vibrate about these lattice sites. This problem has been carefully discussed by Carr (1961), but, for our present purposes, we shall merely refer to the localized orbital description which follows from such a picture. Essentially, we use below an Einstein independent oscillator model of the lattice vibrations.

(i) *Localized orbitals*: We can construct the field in which the electrons vibrate, by examining a Wigner–Seitz cell (see Figure 1.2 for a body-centred-cubic lattice). Since it has quite high symmetry, it can be replaced by a sphere to a fair approximation and then, within one such sphere, the only important contribution to the potential felt by the electron within that sphere is from the positive background charge within this sphere. Other cells, due to their high symmetry, make only multipole contributions to the potential within the cell under consideration. Then it is an elementary calculation in electrostatics to show that the potential energy in which the electron moves is given by

$$-\frac{e^2 r^2}{2r_s^3}+\text{constant}. \tag{2.11.39}$$

It follows that the ground-state wave function is that of an isotropic harmonic oscillator and is simply

$$\psi=\left(\frac{\alpha}{\pi}\right)^{\frac{3}{4}}\exp\left(-\tfrac{1}{2}\alpha r^2\right),\quad \alpha=r_s^{-\frac{3}{2}}. \tag{2.11.40}$$

From this wave function, we immediately obtain the kinetic energy per particle as $T/N=3/(2r_s^{\frac{3}{2}})$ Rydberg and hence, since the total energy for a harmonic oscillator is equally divided between kinetic and potential energy, we can write

$$\frac{E}{N}=-\frac{1.792}{r_s}+\frac{3}{r_s^{\frac{3}{2}}}. \tag{2.11.41}$$

The pair function for a low-density gas derived from $\psi(r)$ in equation (2.11.40) is given by March and co-workers (1967). A correct lattice dynamical calculation (Carr, 1961; Coldwell-Horsfall and Maradudin, 1963) gives 2.66 for the coefficient of the term in $r_s^{-\frac{3}{2}}$, in equation (2.11.41).

We want only to make two further comments here. The first is that, since we shall use it later in Chapter 10, we shall record an interpolation formula due to Wigner for the correlation energy, using the above low-density argument. Though we now know from the Gell-Mann/Brueckner theory that his formula is too crude as r_s tends to zero, it still appears useful at intermediate densities. It is simply

$$E_{\text{correlation}} = -\frac{0.88}{r_s + 7.8} \text{ Rydberg.} \tag{2.11.42}$$

Secondly, some alternative to the Wigner lattice low-density argument has been proposed by Singwi and co-workers (1970), but we must refer to their original paper for that matter.

2.11.4 Response function for inhomogeneous gas

For the general case of an inhomogeneous gas, the linear response function $\mathscr{F}(\mathbf{rr}'\omega)$ is defined by

$$\delta\rho(\mathbf{r}\omega) = \int \mathscr{F}(\mathbf{rr}'\omega)\, \delta V_{\text{ext}}(\mathbf{r}'\omega)\, d\mathbf{r}', \tag{2.11.43}$$

where $\delta\rho(\mathbf{r}\omega)$ is the density change, to first order, when the small external perturbation $\delta V_{\text{ext}}(\mathbf{r}\omega) = \delta V(\mathbf{r})\, e^{i\omega t}$ is applied to the system under consideration. We should note that the periodicity of the lattice implies that the response to a potential $\delta V(\mathbf{r}-\mathbf{R})$ is just the response to $\delta V(\mathbf{r})$ shifted through the lattice vector $\mathbf{R}$. That is,

$$\delta\rho(\mathbf{r}-\mathbf{R}) = \int \mathscr{F}(\mathbf{r}, \mathbf{r}')\, \delta V_{\text{ext}}(\mathbf{r}'-\mathbf{R})\, d\mathbf{r}' \tag{2.11.44}$$

or

$$\delta\rho(\mathbf{r}) = \int \mathscr{F}(\mathbf{r}+\mathbf{R}, \mathbf{r}')\, \delta V_{\text{ext}}(\mathbf{r}'-\mathbf{R})\, d\mathbf{r}'. \tag{2.11.45}$$

But from equation (2.11.43), we also obtain, on changing the variable of integration,

$$\delta\rho(\mathbf{r}) = \int \mathscr{F}(\mathbf{r}, \mathbf{r}'-\mathbf{R})\, \delta V_{\text{ext}}(\mathbf{r}'-\mathbf{R})\, d\mathbf{r}' \tag{2.11.46}$$

and, since $\delta V(\mathbf{r})$ is arbitrary, comparison of this equation with equation (2.11.45) yields

$$\mathscr{F}(\mathbf{r}+\mathbf{R}, \mathbf{r}') = \mathscr{F}(\mathbf{r}, \mathbf{r}'-\mathbf{R}). \tag{2.11.47}$$

The Fourier transform of the density change is evidently given by

$$\delta\rho(\mathbf{k}) = \frac{1}{8\pi^3} \int \mathscr{F}(\mathbf{k}, \mathbf{k}')\, \delta V_{\text{ext}}(\mathbf{k}')\, d\mathbf{k}', \tag{2.11.48}$$

where $\delta V_{\text{ext}}(\mathbf{k})$ is the Fourier transform of the potential and

$$\mathscr{F}(\mathbf{k}, \mathbf{k}') = \int \mathscr{F}(\mathbf{r}, \mathbf{r}')\, e^{i\mathbf{k}.\mathbf{r}}\, e^{-i\mathbf{k}'.\mathbf{r}'}\, d\mathbf{r}\, d\mathbf{r}'. \tag{2.11.49}$$

From equation (2.11.47), it can readily be shown that

$$\mathscr{F}(\mathbf{k},\mathbf{k}+\boldsymbol{\kappa}) = 0 \quad \text{unless} \quad \boldsymbol{\kappa} = \mathbf{K}, \tag{2.11.50}$$

K being as usual a reciprocal lattice vector.

We can, if we wish, see the relationship of $\mathscr{F}$ to the dielectric constant ε if the system is a uniform electron gas by referring back to section 2.11.1. Let us, however, calculate the change of the electrostatic potential in terms of $\mathscr{F}$ in the presence of a lattice.

We obtain

$$\delta V_{\text{electrostatic}}(\mathbf{r}) = \delta V_{\text{ext}}(\mathbf{r}) + e^2 \int \frac{\delta\rho(\mathbf{r}')}{|\mathbf{r}-\mathbf{r}'|}\, d\mathbf{r}' \tag{2.11.51}$$

or, in Fourier transform,

$$\delta V_{\text{el}}(\mathbf{k}) = \delta V_{\text{ext}}(\mathbf{k}) + \frac{4\pi e^2}{k^2}\,\delta\rho(\mathbf{k}). \tag{2.11.52}$$

If we now use equation (2.11.48), this can be written in the form

$$\delta V_{\text{el}}(\mathbf{k}) = \delta V_{\text{ext}}(\mathbf{k}) + \frac{4\pi e^2}{k^2}\sum_{\mathbf{K}} \mathscr{F}(\mathbf{k},\mathbf{k}+\mathbf{K})\,\delta V_{\text{ext}}(\mathbf{k}+\mathbf{K}). \tag{2.11.53}$$

Introducing the reciprocal dielectric tensor $\varepsilon^{-1}(\mathbf{k},\mathbf{k}+\mathbf{K})$ we may obviously put

$$\delta V_{\text{el}}(\mathbf{k}) = \sum_{\mathbf{K}} \frac{\delta V_{\text{ext}}(\mathbf{k}+\mathbf{K})}{\varepsilon(\mathbf{k},\mathbf{k}+\mathbf{K})}, \tag{2.11.54}$$

where

$$\varepsilon^{-1}(\mathbf{k},\mathbf{k}+\mathbf{K}) = \delta_{\mathbf{K}0} + \frac{4\pi e^2}{k^2}\mathscr{F}(\mathbf{k},\mathbf{k}+\mathbf{K}). \tag{2.11.55}$$

Expressions for response function. As we discuss in more detail in Appendix 2.2, external potentials can be included in the discussion of the behaviour of a system of electrons, in terms of Feynman graphs by representing the (first-order) interaction of an external field $V(\mathbf{q},\eta) = V(\mathbf{q})\,e^{i\eta t}$ with an electron by Figure 2.39(a). We can think of the electron being scattered with momentum transfer $\mathbf{q}$ and energy transfer η. As before, each Fermion line is associated with a free-particle propagator G_0. To illustrate this, the one-particle Green function for the non-interacting gas in the presence of the external field is obtained by drawing lines showing no scattering; one scattering event, two events and so on, as shown in Figure 2.39(b).

One associates a factor $iV_{\mathbf{q}}$ with each external interaction line, integrates over intermediate momenta and frequencies as in section 2.10 and sums the series shown diagrammatically in Figure 2.39(b), to obtain equation (2.7.69), the integral equation for the Green function of non-interacting particles, just as we obtained Dyson's equation in section 2.10.2.

If we are assuming the presence of a periodic potential, we can either include this in the external potential or associate each line running between vertices with the propagator $G(\mathbf{k}', \mathbf{k}'_0, \omega)$ for Bloch electrons. Since $\mathscr{F}$ describes linear response, the latter alternative is the more appropriate here,

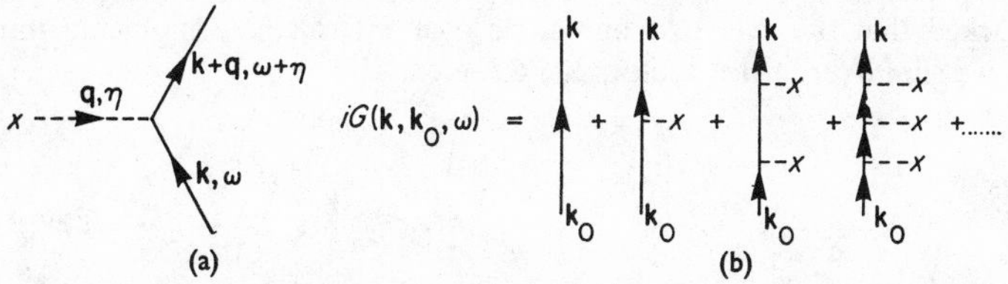

FIGURE 2.39. Graphs for interaction of Fermion with external field. (a) Representation of single scattering event. (b) Green function for system of non-interacting Fermions.

for we need a description of the system in which the periodic potential due to the lattice appears to all orders while the external perturbation appears only to first order.

To obtain $\mathscr{F}$ in diagrammatic terms, we first consider diagrams for the density. From equation (2.5.22) for the density-fluctuation operator, and the definition (2.7.70) of the Green function, we see that

$$\rho_{\mathbf{q}} = \sum_{\mathbf{k}} G(\mathbf{k}, \mathbf{k}+\mathbf{q}, t)\big|_{t=+0} = \int \sum_{\mathbf{k}} G(\mathbf{k}, \mathbf{k}+\mathbf{q})\, d\omega. \tag{2.11.56}$$

We can express the integration over $\mathbf{k}$ and ω by bending the line representing the Green function round, so that its two ends meet (see Figure 2.40(a)).

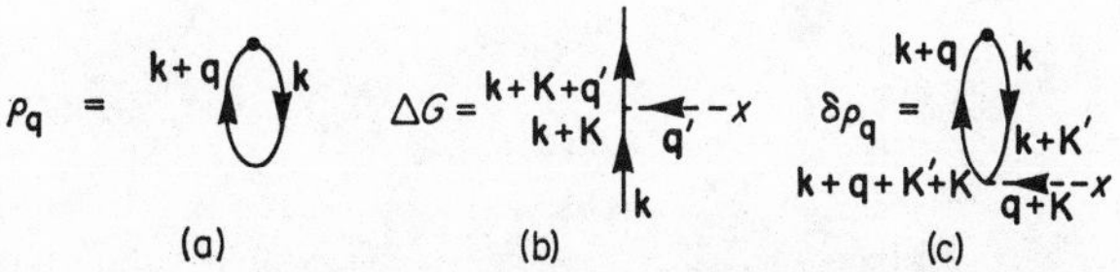

FIGURE 2.40. Graphs to calculate first-order density change in non-interacting system. (a) Unperturbed density. (b) First-order correction to single-particle Green function. (c) First-order density change.

We next enquire what is the first-order density change on applying $V(\mathbf{q}+\mathbf{K})$. In view of our earlier remarks, the first-order change in the Green function is as shown in Figure 2.40(b) and hence $\delta\rho(\mathbf{q})$ is given in Figure 2.40(c).

This diagram gives us for non-interacting systems, $\delta\rho = \Pi_0\,\delta V$ where Π_0 is the RPA approximation to the polarization part [cf. Figure 2.16(c)]. It should be noted that $\Pi_0 \equiv \Pi_0(\mathbf{k}, \mathbf{k}+\mathbf{K})$ and that it now has off-diagonal elements. If, on the other hand, we explicitly include interactions in our diagrams, we obtain Figure 2.41(a), where Π represents the sum of all proper polarization diagrams, and the interaction line has been dressed. The full series in terms of Π is shown in Figure 2.41(b). It should also be remarked that the rule that one sums over intermediate momenta implies that we sum over lattice vectors $\mathbf{K}'$, $\mathbf{K}''$, etc.

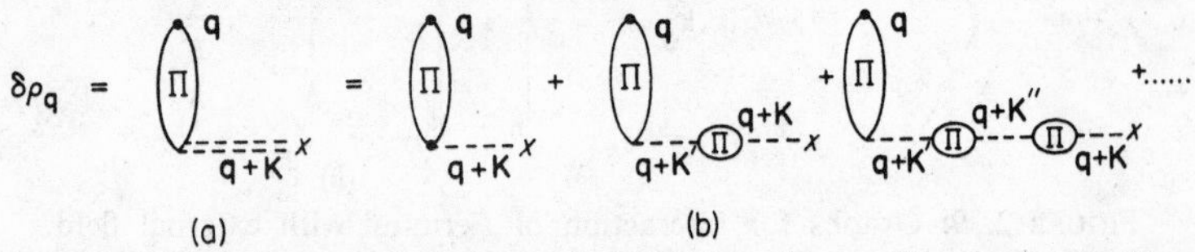

FIGURE 2.41. Diagrams for first-order density change $\delta\rho_q$ in interacting system. (a) Showing dressed external interaction line. (b) Full series in terms of sum of proper polarization parts.

The simplest approximation to Π that we can take in the resulting series for $\mathscr{F}(\mathbf{q}, \mathbf{q}+\mathbf{K})$ is $\Pi_0(\mathbf{q}, \mathbf{q}+\mathbf{K})$, when we obtain $\mathscr{F}$ in the RPA. Claesson (1969) has calculated the diagonal element $\mathscr{F}(\mathbf{qq})$ in this way for the conduction electrons in aluminium, with the result shown in Figure 2.42. The

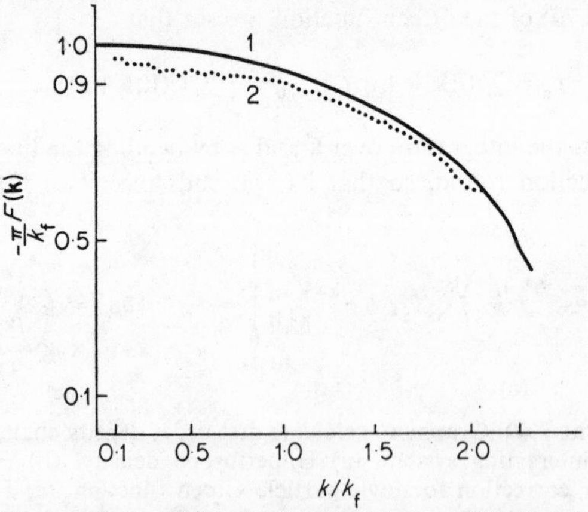

FIGURE 2.42. Dielectric function in metallic aluminium. Curve 1. Homogeneous gas. Curve 2. Result for fcc lattice, along direction ΓL of BZ shown in Figure 1.5. The response function F for non-interacting electrons is the same as Π_0 (after Claesson, 1969).

lattice has the effect of decreasing the free-electron result by quite a small amount, some 4%, in this "nearly free electron" metal.

From Figure 2.43(a) for $\mathscr{F}$, and equation (2.11.55), we immediately deduce that ε^{-1} may be expressed diagrammatically as shown in Figure 2.43(b). Comparison of this with, for example, Figure 2.18, shows that the

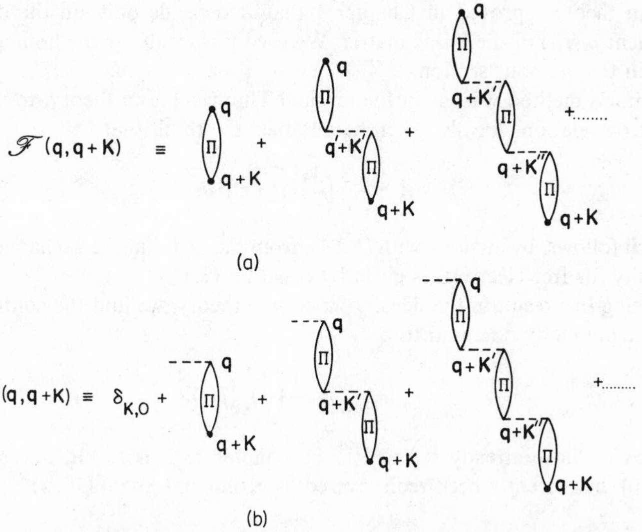

FIGURE 2.43. Diagrams for response function $\mathscr{F}$ and dielectric function ε. **K**'s are reciprocal lattice vectors as usual. Diagrams involving **K**′; **K**′ and **K**″ etc. imply summation over these primed reciprocal lattice vectors.

dressing of interaction lines is indeed equivalent to multiplying $V(\mathbf{q})$ by ε^{-1}, as we assumed in section 2.10, when discussing graphical techniques earlier.

We shall return to the formal properties of $\mathscr{F}$ in discussing insulators, a little later on.

2.11.5 Exchange potential

A lot of effort has gone into finding a local potential which would simulate exchange effects. Following the approach of Thomas and Fermi, it was Dirac (1930) who showed how exchange could be introduced into the theory and his method has really been the forerunner of most of the successful calculations which include exchange. While, to obtain the exchange energy A in Hartree–Fock theory exactly we require (see equations (2.7.32) and (2.7.47))

$$A = -\frac{1}{4}\int \frac{[\rho(\mathbf{r}\mathbf{r}_0\mu)]^2}{|\mathbf{r}-\mathbf{r}_0|}\,d\mathbf{r}\,d\mathbf{r}_0 \tag{2.11.57}$$

which involves the Dirac density matrix ρ, we now know from the Hohenberg–Kohn theorem proved in Chapter 1 that A depends only on the diagonal element $\rho(\mathbf{r}\mathbf{r}\mu)$ of the Dirac matrix. We used this result for the homogeneous gas in the previous section.

Dirac's method was, as in the original Thomas–Fermi theory, to use free-electron relations locally. Then he obtained the result that

$$A \approx -\frac{3}{4}\left(\frac{3}{\pi}\right)^{\frac{1}{3}}\int [\rho(\mathbf{r})]^{\frac{4}{3}}\,d\mathbf{r} \tag{2.11.58}$$

which follows, by analogy with (1.2.1), from the fact that the exchange energy density for free electrons is given by equation (2.4.2).

Using this result in the density-functional theory, we find the contribution to the one-body potential to be

$$V_{\text{exchange}} = -\left(\frac{3}{\pi}\right)^{\frac{1}{3}}[\rho(\mathbf{r})]^{\frac{1}{3}} \tag{2.11.59}$$

which we have already referred to in Chapter 1. This result, due to Dirac (1930), has recently been rediscovered by Kohn and Sham (1965).

2.11.6 Gradient corrections to slowly varying density formulae

In the density-functional theory, we really would like the one-body potential as an explicit *function* of the density $\rho(\mathbf{r})$. To take the simplest example, the original Thomas–Fermi theory (see section 1.2.1) yields

$$V(\mathbf{r}) = -\tfrac{1}{2}(3\pi^2)^{\frac{2}{3}}\,\rho^{\frac{2}{3}}. \tag{2.11.60}$$

Adding the exchange potential (2.11.59) yields

$$V(\mathbf{r}) = -\tfrac{1}{2}(3\pi^2)^{\frac{2}{3}}\,\rho^{\frac{2}{3}} - \left(\frac{3}{\pi}\right)^{\frac{1}{3}}\rho^{\frac{1}{3}}. \tag{2.11.61}$$

The problem with these formulae is that they are valid for slowly varying densities only, whereas, in real systems, there are regions near nuclei when the gradients of the density become substantial. However, in Chapter 6, we shall see they are already valuable near the surface of a metal.

An attempt was made by Von Weizsäcker (1935) to correct the Thomas–Fermi form of kinetic energy for density gradients. Von Weizsäcker's method is now known to be only qualitatively valid and is, in fact, equivalent to arguing that we add a term to the "Fermi gas" kinetic energy which would be valid, alone, for Bose particles. Then, if all N particles drop into the lowest energy level, we have

$$\rho = N\psi^2, \tag{2.11.62}$$

where ψ is the ground-state wave function. The corresponding energy is

$$\begin{aligned} &+\frac{\hbar^2}{2m}N\int(\nabla\psi)^2\,d\mathbf{r} \\ &=\frac{\hbar^2}{2m}N\int\left[\nabla\left(\frac{\rho}{N}\right)^{\frac{1}{2}}\right]^2 d\mathbf{r} \\ &=\frac{\hbar^2}{2m}\int(\tfrac{1}{2}\rho^{-\frac{1}{2}}\nabla\rho)^2\,d\mathbf{r} \\ &=\frac{\hbar^2}{8m}\int\frac{(\nabla\rho)^2}{\rho}\,d\mathbf{r}. \end{aligned} \tag{2.11.63}$$

Actually, as Kirschnitz (1957) was the first to show, the coefficient $\frac{1}{8}$ should be replaced by $\frac{1}{72}$ in an inhomogeneous electron gas. We give an argument which leads to this result in Appendix 2.4.

It has to be said that this procedure for correcting homogeneous gas theory has not been notably successful when applied to atoms, molecules and solids, except for the interesting problems that arise in the surface of a metal which we referred to above.

However, it is perfectly possible, by solving the Schrödinger equation, to avoid making any approximations in the kinetic energy term of the one-body density theory. This we shall assume to be done, in what follows. Then, the question arises as to what form the gradient corrections to the exchange and correlation contributions to the one-body potential take. These have been discussed by Herman and co-workers (1969; see also Stoddart and co-workers, 1971) for exchange and by Ma and Brueckner (1968) for the correlation energy.

(a) *Exchange energy*. In their calculations on atoms, Herman and co-workers have argued on dimensional grounds, by analogy with the Von Weizsäcker term (2.11.63) in the kinetic energy, that the gradient correction to the Dirac–Slater exchange energy (2.11.58) must have the form

$$\text{const}\int\frac{(\nabla\rho)^2}{\rho^{\frac{4}{3}}}\,d\mathbf{r}. \tag{2.11.64}$$

Their work determines the constant essentially empirically, by comparison with Hartree–Fock theory. Certainly, however, the question arises as to whether the constant would be the same in a local one-body potential theory and in a Hartree–Fock theory with an energy-dependent potential.

Stoddart and colleagues (1971) have examined the theory using the exchange energy (2.11.57), with the perturbation expansion of Appendix 1.1 inserted for the Dirac density matrix. Their conclusion is that, in addition to the term of Herman and co-workers (1969), there are terms of the form

$$\int \frac{(\nabla \rho)^2}{\rho^{\frac{4}{3}}} \ln \rho \, d\mathbf{r}, \quad \int \frac{(\nabla \rho)^2}{\rho^{\frac{4}{3}}} (\ln \rho)^2 \, d\mathbf{r}, \quad \text{etc.,} \tag{2.11.65}$$

which could not be detected, of course, by dimensional analysis. Preliminary estimates of these constants have been obtained, but the arguments are far from rigorous.

(b) *Correlation energy.* Quite a number of workers proposed, before the density-functional theory was placed on a firm basis as an approach to the many-body problem by Hohenberg and Kohn, that one should take over from the uniform interacting gas theory the correlation energy as a function of gas density. A much-used form was that of Wigner (1938), which we gave in equation (2.11.42). This form will be used when we apply inhomogeneous electron gas theory to the problem of a metal surface.

However, we shall assume, for the moment, that the uniform gas correlation energy is known. Then, we shall follow Ma and Brueckner (1968) in attempting to calculate gradient corrections to this term.

The merit of the work of Ma and Brueckner is that it combines the density functional philosophy which we set out fully in Chapter 1 with the diagrammatic techniques to obtain the first term in a gradient expansion of the correlation energy of an electron gas with a slowly varying density. As usual, the correlation energy is defined as the difference between the exact energy $E[\rho]$ and the Hartree–Fock energy $E_{\mathrm{HF}}[\rho]$. Thus we write

$$E[\rho] = E_{\mathrm{HF}}[\rho] + E_{\mathrm{c}}[\rho], \tag{2.11.66}$$

where $E_{\mathrm{c}}[\rho]$ is the correlation energy as a functional of the density.

Within the spirit of the Von Weizsäcker-like approach referred to above and in Appendix 2.4, we write

$$E_{\mathrm{c}} = \int d\mathbf{r} \{\varepsilon_{\mathrm{c}}[\rho(\mathbf{r})] + B(\rho(\mathbf{r}) | \nabla \rho(\mathbf{r}) |^2\} + \ldots, \tag{2.11.67}$$

where we have anticipated that the first term will take over the correlation energy density $\varepsilon_{\mathrm{c}}(\rho_0)$ of the uniform gas as a function of the density ρ_0 into a slowly varying density theory, simply by replacing ρ_0 by $\rho(\mathbf{r})$, the inhomogeneous gas density, just as for kinetic [equation (1.2.1)] and exchange [equation (2.11.58)] energies. The object of the Ma–Brueckner work was then to calculate the quantity $B(\rho)$ as the functional of the density ρ. We shall outline the way this is done here, and some details of the procedure can be found in Appendix 2.5.

We consider an electron gas, in unit volume for convenience, perturbed by an external static potential

$$\phi(\mathbf{r}) = \sum \tilde{\phi}_{\mathbf{k}} e^{i\mathbf{k}.\mathbf{r}}. \tag{2.11.68}$$

The perturbing Hamiltonian is then

$$H' = \sum_{\mathbf{k}\neq 0} \tilde{\phi}_{\mathbf{k}} \rho_{-\mathbf{k}} \tag{2.11.69}$$

where $\rho_{\mathbf{k}}$ is the density operator. To second order in the perturbing potential, the energy is then given by

$$E = E^0(\rho_0) + \sum_{\mathbf{k}\neq 0} \tilde{\phi}_{\mathbf{k}} \tilde{\phi}_{-\mathbf{k}} \sum_m \frac{\langle 0|\rho_{\mathbf{k}}|m\rangle\langle m|\rho_{-\mathbf{k}}|0\rangle}{E_0 - E_m}$$

$$= E^0(\rho_0) + \frac{1}{2} \sum_{\mathbf{k}\neq 0} \tilde{\phi}_{\mathbf{k}} \tilde{\phi}_{-\mathbf{k}} \mathscr{F}(\mathbf{k}, 0). \tag{2.11.70}$$

In equation (2.11.70), $|0\rangle$, E_0, $|m\rangle$ and E_m are the exact ground and excited state wave functions and energies of the uniform electron gas. $\mathscr{F}$ as usual is the density response function of a uniform electron gas, given by‡

$$\mathscr{F}(\mathbf{k}, \omega) = -i \int dt\, d\mathbf{r}\, e^{-i\mathbf{k}.\mathbf{r}+i\omega t} \langle T\{\rho(\mathbf{r}t)\,\rho(0)\}\rangle$$

$$= \mathscr{F}(-\mathbf{k}, \omega) = \mathscr{F}(\mathbf{k}, -\omega), \tag{2.11.71}$$

the average to be taken over the ground state of a uniform electron gas. Then evidently the density, to first order in ϕ, is given by

$$\rho_{\mathbf{k}} = \tilde{\phi}_{\mathbf{k}} \mathscr{F}(\mathbf{k}, 0) \quad \mathbf{k} \neq 0. \tag{2.11.72}$$

The Hartree–Fock energy can now be separated from the total energy, following Ma and Brueckner, in the manner outlined below. First, the results (2.11.70) and (2.11.72) can be obtained by minimizing the functional

$$E[\rho] = E^0[\rho_0] + \sum_{\mathbf{k}\neq 0} \tilde{\phi}_{-\mathbf{k}} \rho_{\mathbf{k}} - \frac{1}{2} \sum_{\mathbf{k}\neq 0} \rho_{\mathbf{k}} \rho_{-\mathbf{k}} \mathscr{F}^{-1}(\mathbf{k}, 0). \tag{2.11.73}$$

Similarly, minimizing the functional

$$E_{\mathrm{HF}}[\rho] = E_{\mathrm{HF}}[\rho_0] + \sum_{\mathbf{k}\neq 0} \tilde{\phi}_{-\mathbf{k}} \rho_{\mathbf{k}} - \frac{1}{2} \sum_{\mathbf{k}\neq 0} \rho_{\mathbf{k}} \rho_{-\mathbf{k}} \mathscr{F}_{\mathrm{HF}}^{-1}(\mathbf{k}, 0) \tag{2.11.74}$$

‡ This equation can be obtained by relating $\mathscr{F}(\mathbf{k}, \omega)$ to the dielectric function $\varepsilon(\mathbf{k}, \omega)$, for which a general expression has already been obtained in equation (2.3.36).

yields the density and energy in the Hartree–Fock approximation. Here, $\mathscr{F}_{\mathrm{HF}}$ is the density response function in the Hartree–Fock approximation. Subtracting equation (2.11.74) from (2.11.73) we find

$$E[\rho]-E_{\mathrm{HF}}[\rho] = E_c[\rho] = E_c^0[\rho_0]-\frac{1}{2}\sum_{\mathbf{k}\neq 0}\rho_{\mathbf{k}}\,\rho_{-\mathbf{k}}[\mathscr{F}^{-1}(\mathbf{k},0)-\mathscr{F}_{\mathrm{HF}}^{-1}(\mathbf{k},0)], \tag{2.11.75}$$

where E_c^0 is the correlation energy of the uniform gas of density ρ_0.

From equations (2.10.6) relating Π, V and ε, and equation (2.11.5) applied to a homogeneous gas, we find immediately

$$\mathscr{F}^{-1} = \Pi^{-1}-V, \tag{2.11.76}$$

where $V = 4\pi e^2/k^2$ and Π contains no isolated interaction line.

Equation (2.11.75) now takes the form

$$E_c[\rho] = E_c^0[\rho_0]-\frac{1}{2}\sum_{\mathbf{k}\neq 0}\rho_{\mathbf{k}}\,\rho_{-\mathbf{k}}[\Pi^{-1}(\mathbf{k},0)-\Pi_{\mathrm{HF}}^{-1}(\mathbf{k},0)]. \tag{2.11.77}$$

The contribution of the term $-V$ of $\mathscr{F}^{-1}$, namely

$$\frac{1}{2}\sum_{\mathbf{k}\neq 0}\rho_{\mathbf{k}}\,\rho_{-\mathbf{k}}(4\pi e^2/k^2), \tag{2.11.78}$$

cancels that in $\mathscr{F}_{\mathrm{HF}}^{-1}$ and means that the electrostatic energy of the charge distribution plays no part in E_c, as was to be expected.

To compare the forms (2.11.77) and (2.11.67), we expand the integrand of (2.11.67) in powers of the "displaced charge" $\Delta(\mathbf{r}) = \rho(\mathbf{r})-\rho_0$ just as in Appendix A1.2, and the quantity $\Pi^{-1}-\Pi_{\mathrm{HF}}^{-1}$ in equation (2.11.77) in powers of k^2. If we write

$$\Pi(\mathbf{k},0) = a^{-1}+bk^2+\ \ldots, \tag{2.11.79}$$

then

$$\Pi^{-1}(\mathbf{k},0) = a-a^2bk^2+\ldots. \tag{2.11.80}$$

a^{-1} is given by the relation to the compressibility (see equation 2.10.94)

$$a^{-1} = \lim_{\mathbf{k}\to 0}\Pi(\mathbf{k},0) = -\frac{d\rho_0}{d\mu} \tag{2.11.81}$$

and therefore

$$a = -\frac{d^2}{d\rho_0^2}E^0(\rho_0). \tag{2.11.82}$$

Expanding equation (2.11.67) gives the result

$$E_c[\rho] = \varepsilon_c(\rho_0)+\frac{d^2}{d\rho_0^2}\varepsilon_c(\rho_0)\frac{1}{2}\sum_{\mathbf{k}\neq 0}\rho_{\mathbf{k}}\,\rho_{-\mathbf{k}}+B(\rho_0)\sum_{\mathbf{k}\neq 0}k^2\,\rho_{\mathbf{k}}\,\rho_{-\mathbf{k}}. \tag{2.11.83}$$

Substituting equation (2.11.80) in equation (2.11.77) and comparing with equation (2.11.83) we find

$$B(\rho) = \tfrac{1}{2}(a^2 b - a_{\mathrm{HF}}^2 b_{\mathrm{HF}}). \tag{2.11.84}$$

The function $\Pi(\mathbf{k}, 0)$ and hence b can be calculated from the perturbation expansion in Figure 2.44. The diagrams for the function Π_{HF}, shown in Figure 2.44(a), can be obtained as follows. We go back to the self-energy Σ in Figure 2.11. The first two diagrams (a) and (b) of this figure give the HF

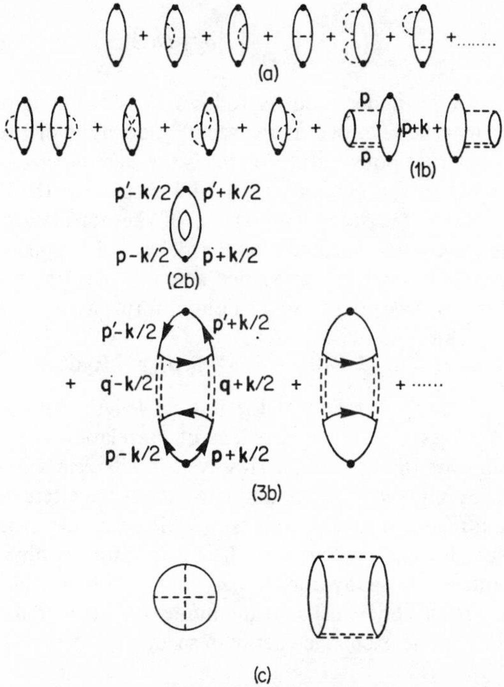

FIGURE 2.44. (a) Diagrams for $\Pi_{\mathrm{HF}}(\mathbf{k}, 0)$. (b) Terms to add to Π_{HF} to give Π. Contributions (1b) to (3b) are discussed in Appendix 2.4. The specific diagrams shown are related to Gell–Mann and Brueckner diagrams for the energy shown in (c) below. (c) These Gell–Mann and Brueckner diagrams for the correlation energy of a uniform gas, in which the interaction lines are screened in RPA, when modified to include second-order perturbation by the external field, become the diagrams shown in (b), provided the screened interaction lines are again calculated in RPA.

approximation to the self-energy. This result Σ_{HF} can be inserted into the diagrammatic series for the Green function G shown in Figure 2.12. We can now obtain $\mathscr{F}_{\mathrm{HF}}$ and hence Π_{HF} from G by following the procedure set out in section 2.11.4(a). Similarly, Figure 2.44(b) is obtained from the diagrams in other than the two HF diagrams, the complete form of Π thus being the sum of all the diagrams shown. Care is needed, because some of the diagrams in Figure 2.44(b) diverge unless the interaction lines are properly modified by the dielectric constant. Then it turns out that, with $b = b_{\mathrm{HF}} + b_{\mathrm{c}}$

$$B(\rho_0) = \left(\frac{\pi^2}{k_{\mathrm{f}} m}\right)^2 \tfrac{1}{2} b_{\mathrm{c}} + O(e^4). \tag{2.11.85}$$

This formula can be understood as follows. b is the long-range part of $\Pi(\mathbf{k}, 0)$, which may be regarded as the second derivative of the energy with respect to the external potential. $(\pi^2/k_{\mathrm{f}} m)$ is the derivative of the chemical potential (roughly of the external field) with respect to the density. Thus $B(\rho_0)$ is qualitatively the second derivative of the correlation energy with respect to the long-range variation of the density. In Appendix A2.4, we summarize the way Ma and Brueckner actually perform the calculation of b_{c}. The result, when we go over to the non-uniform gas, may be summarized as

$$B(\rho) = [8.47 \times 10^{-3} + O(\rho^{-\frac{1}{3}} \ln \rho)]\, \rho^{-\frac{4}{3}} \text{ Rydberg}, \tag{2.11.86}$$

where the Bohr radius is taken as the unit of length. Ma and Brueckner discuss this with respect to the problem of correlations in atoms, where their conclusions are disappointing. However, it will clearly have a range of validity, as for example in a calculation like that of the electron density near the surface of a metal where the density is falling from its mean value ρ_0 to zero as we pass through the surface. Their calculation is obviously of very considerable interest in many-body theory, even though it is likely that to get quantitative results one must sum an infinite series in a gradient expansion (cf. the remarks on the exchange energy in section 2.10).

2.12 Quasi-particles in insulators

We have seen earlier that the many-body wave functions of a system of interacting particles moving in a periodic potential may be chosen to obey Bloch's theorem. We therefore choose the electronic wave function of an insulator with one additional electron such that

$$\Psi_{\mathbf{k}}(\mathbf{r}_1 + \mathbf{R}, \mathbf{r}_2 + \mathbf{R}, \ldots, \mathbf{r}_{N+1} + \mathbf{R}) = e^{i\mathbf{k}.\mathbf{R}} \Psi_{\mathbf{k}}(\mathbf{r}_1, \mathbf{r}_2, \ldots, \mathbf{r}_{N+1}) \tag{2.12.1}$$

and write the total energy as $E(\mathbf{k})$. To examine the structure of the energy spectrum for a given $\mathbf{k}$, we make the independent particle approximation;

because **k** remains a constant of the motion when interactions are switched on, this structure will, in normal circumstances, persist during and after the switching on process.

In the absence of the extra particle, the N electrons of the insulator just suffice to completely fill the lowest bands, and the $(N+1)$th particle must be added into a single-particle state $\mathbf{k}'$ in an unoccupied band. We can then form a continuum of states, all labelled by **k**, by promoting a particle, from the single-particle state $\mathbf{k}_1$ of a filled band to single-particle state $\mathbf{k}_1'$ of the unoccupied band, such that $\mathbf{k} = \mathbf{k}' + \mathbf{k}_1' - \mathbf{k}_1$. However, this promotion will cost at least the energy E_g, this being the gap between the uppermost occupied band and the lowest unoccupied one. We can therefore see that under normal circumstances there will be at least one isolated energy state below this continuum, this being formed by choosing $\mathbf{k}' = \mathbf{k}$. We therefore obtain an energy spectrum such as shown in Figure 2.45, E_0 being the energy of the N-particle ground state.

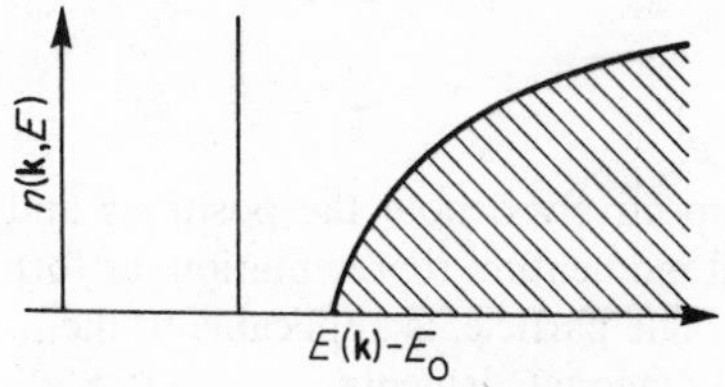

FIGURE 2.45. Energy spectrum of insulator. E_0 is energy of N particle ground state. $n(\mathbf{k}, E)$ is density of states in energy for a fixed total wave vector **k** for $N+1$ particle system.

In the switching-on process, the added particle in single-particle state $\phi_\mathbf{k}$ (contributing the isolated δ-function spike in Figure 2.45) becomes a quasi-particle, interacting with and polarizing the electrons of the filled bands as it moves through the crystal. Unlike the quasi-particles of the electron gas, its lifetime is infinite, there being no other quasi-particles against which it can scatter. Thus the spectral density of Figure 2.4 is replaced by that of Figure 2.46 shown on p. 212.

The quasi-particle operator $q_\mathbf{k}$ can be constructed as in section 2.8 by filtering out the continuum. Great care is no longer necessary in constructing the filtering function $f(t)$, the quasi-particle now being represented by a δ-function peak separated from the continuum, as shown in Figure 2.45.

A number of such quasi-particles, sufficiently few in number that their interactions are negligible, will form the free carriers of, for example, a doped semiconductor. These will move under the influence of an external perturbing potential $V(\mathbf{r})$ and it is evident on physical grounds that the effective potential acting upon the quasi-particles will be $V(\mathbf{r})/\varepsilon$ where ε is the dielectric constant of the insulator. ε is different from unity because of the presence of the electrons in the filled bands.

Following Blount (1962), we can construct an effective Hamiltonian for the quasi-particle using the crystal momentum representation of section 1.10. We can write

$$\Psi_{\mathbf{k}}(\mathbf{r}_1, \mathbf{r}_2, \ldots, \mathbf{r}_{N+1}) = e^{i\mathbf{k}.\Sigma_i \mathbf{r}_i}\, u_{\mathbf{k}}(\mathbf{r}_1, \ldots, \mathbf{r}_{N+1}) \tag{2.12.2}$$

where $u_{\mathbf{k}}$ is periodic in the $\mathbf{r}_i$ since we then obtain

$$\Psi_{\mathbf{k}}(\mathbf{r}_1+\mathbf{R}, \ldots, \mathbf{r}_{N+1}+\mathbf{R}) = e^{i(N+1)\mathbf{k}.\mathbf{R}}\Psi_{\mathbf{k}}(\mathbf{r}_1, \ldots, \mathbf{r}_{N+1}). \tag{2.12.3}$$

Since $\mathbf{k}$ is chosen such that $e^{iN\mathbf{k}.\mathbf{R}} = 1$, equations (2.12.1) and (2.12.3) are identical.

If we now define

$$\left.\begin{aligned} \mathbf{x} &= \sum_i \mathbf{r}_i \\ \mathbf{v} &= \sum_i \mathbf{v}_i \end{aligned}\right\} \tag{2.12.4}$$

so that $\mathbf{r}$ and $\mathbf{v}$ respectively denote the positions and velocities of all the particles, the crystal momentum representation, as formulated in Chapter 1, section 1.10, for a single particle, is applicable to the present case also. Thus we again obtain the diagonal elements

$$\mathbf{v}_{nn}(\mathbf{k}) = \frac{1}{\hbar}\frac{\partial E_n(\mathbf{k})}{\partial \mathbf{k}} \tag{2.12.5}$$

and

$$[x_d^i, v_d^j]_n = \frac{\partial^2 E_n}{\partial k_i\, \partial k_j}, \tag{2.12.6}$$

which should be compared with equations (1.10.13) and (1.10.21). On the other hand, instead of equation (1.10.22) we find

$$-\frac{i}{\hbar}[x_d^i, v^i] = \frac{N+1}{m} \tag{2.12.7}$$

and hence equation (1.10.21) is replaced by

$$\left(\frac{1}{m^*}\right)_{ij} = \frac{\partial^2 E}{\partial k_i\, \partial k_j} = \frac{N+1}{m} - \frac{i}{\hbar}[X_i, V_j]_n. \tag{2.12.8}$$

What would be most helpful would be if we could replace equation (2.12.8) by an equation showing the correction one must apply to the electron mass m to obtain the effective mass. To do so, we note that, for the ground state of the N-particle system

$$0 = \frac{N}{m} + \frac{i}{\hbar}[X_i, V_j]_0 \qquad (2.12.9)$$

because of the property

$$0 = \mathbf{v} = \frac{\partial E_0}{\partial \mathbf{k}} \qquad (2.12.10)$$

of the insulator. Hence equation (2.12.8) may be rewritten as

$$\left(\frac{1}{m^*}\right)_{ij} = \frac{1}{m} + \frac{i}{\hbar}[V_i, X_j]_n - \frac{i}{\hbar}[V_i, X_j]_0, \qquad (2.12.11)$$

Thus, we see that we must subtract off the filled-band contribution from the commutator.

Finally, we shall set up the effective Hamiltonian for the quasi-particle. We write the total potential applied to the electrons as

$$\sum_i V(\mathbf{r}_i) = \int V(\mathbf{q}) \sum_i e^{i\mathbf{q}\cdot\mathbf{r}_i}\, d\mathbf{q} \qquad (2.12.12)$$

so that, if we write as usual

$$\rho(\mathbf{q}) = \sum_{i=0}^{N} e^{i\mathbf{q}\cdot\mathbf{r}_i}, \qquad (2.12.13)$$

the matrix elements of the potential take the form

$$V_{nn'}(\mathbf{k}, \mathbf{k}') = \int d\mathbf{q}\, V(\mathbf{q})\, \delta(\mathbf{k} - \mathbf{k}' + \mathbf{q}) \langle n\mathbf{k} | \rho(\mathbf{q}) | n'\mathbf{k} + \mathbf{q} \rangle. \qquad (2.12.14)$$

We earlier remarked that, on physical grounds, it is to be expected that the quasi-particle will move in a field represented by $V(\mathbf{r})/\varepsilon$ where ε is the dielectric constant. In fact, Kohn (1958) has shown that, at small $\mathbf{q}$, the matrix elements of $\rho(\mathbf{q})$ approach

$$\lim \rho_{nn'}(\mathbf{q}) = \frac{\delta_{nn'}}{\varepsilon} \qquad (2.12.15)$$

where ε is the static dielectric constant of the insulator.

Thus, for small $\mathbf{q}$, we may write

$$V_{nn'}(\mathbf{k}, \mathbf{k} + \mathbf{q}) = \frac{V(\mathbf{q})}{\varepsilon} \delta_{n'n} \qquad (2.12.16)$$

and, following the method of section 1.10, we find the effective Hamiltonian

$$H_{\text{eff}} = -\frac{\hbar^2}{2} \sum_{ij} \left(\frac{1}{m^*}\right)_{ij} \frac{\partial^2}{\partial x_i\, \partial x_j} + \frac{V(\mathbf{r})}{\varepsilon}. \qquad (2.12.17)$$

However, this simple screening of the potential obtains only to lowest order. The term in the integrand of equation (2.12.14) which is linear in $\mathbf{q}$ is

$$V(\mathbf{q})\,\delta(\mathbf{k}-\mathbf{k}'+\mathbf{q})\left[\frac{\partial}{\partial\mathbf{q}}\langle n\mathbf{k}|\,\rho(\mathbf{q})\,|n'\mathbf{k}+\mathbf{q}\rangle\right]_{\mathbf{q}=0}. \tag{2.12.18}$$

Comparison with equation (1.10.26) shows that for an exact parallel to hold between the independent-particle and quasi-particle cases we must have

$$\frac{\partial}{\partial\mathbf{q}}\langle n\mathbf{k}|\,\rho(\mathbf{q})\,|n'\mathbf{k}+\mathbf{q}\rangle|_{\mathbf{q}=0} = \frac{X_{n'n}}{\varepsilon} \tag{2.12.19}$$

which is not in general true, since the polarization of the core charge varies over the unit cell. We shall now conclude this brief discussion of insulators by discussing the behaviour of the response function, especially in the long wavelength limit.

2.13 Response function for insulators

Diagrammatic techniques such as we have discussed extensively in Appendix A2.2 are readily applicable to the electrons in an insulator (Ambegaokar and Kohn, 1960). In Appendix A2.3, we show that, as a consequence of the Ward identities for electrons moving in a perfect lattice (these identities relating derivatives of the self-energy to small q and ω limits of the vertex functions),

$$\frac{\partial\rho_{\mathbf{K}}}{\partial\mu} = \mathscr{F}(\mathbf{K},0) \tag{2.13.1}$$

where $\rho_{\mathbf{K}}$ is a Fourier component of the electron density, $\mathbf{K}$ being a reciprocal lattice vector and μ the chemical potential. The fact that a material is an insulator implies filled bands and an energy gap and hence

$$\frac{\partial\rho_{\mathbf{K}}}{\partial\mu} = 0. \tag{2.13.2}$$

Equation (2.13.1), together with (2.13.2), enables us to place restrictions on the small q expansions of the response functions of an insulator. However, when $\mathbf{q} = 0$ precisely, we have a special case, because diagrams contributing to $\mathscr{F}$, which have internal interaction lines labelled by $\mathbf{q}$, will then give vanishing contributions, since we define the Fourier transform of the potential such that

$$V(\mathbf{q}=0) = 0 \tag{2.13.3}$$

(see section 2.7.4). Some care is therefore needed, and we shall, in fact, first examine the small $\mathbf{q}$ expansion of $\tilde{\mathscr{F}}$, the sum of all proper diagrams (irreducible with respect to an internal interaction line) for which the limit $\mathbf{q} \to 0$ can certainly be taken smoothly.

Now

$$\mathscr{F}(\mathbf{k}+\mathbf{K},\mathbf{k}) = \int \mathscr{F}(\mathbf{r},\mathbf{r}')\, e^{i\mathbf{k}.(\mathbf{r}-\mathbf{r}')}\, e^{i\mathbf{K}.\mathbf{r}'}\, d\mathbf{r}\, d\mathbf{r}' \tag{2.13.4}$$

and it therefore follows that

$$\frac{\partial \mathscr{F}(\mathbf{K},0)}{\partial \mathbf{k}} = \int \mathscr{F}(\mathbf{r},\mathbf{r}')\, e^{i\mathbf{K}.\mathbf{r}'}(\mathbf{r}-\mathbf{r}')\, d\mathbf{r}\, d\mathbf{r}'. \tag{2.13.5}$$

If $\mathbf{K} = 0$

$$\frac{\partial \mathscr{F}(0,0)}{\partial \mathbf{k}} = \int \mathscr{F}(\mathbf{r}\mathbf{r}')\, \mathbf{r}\, d\mathbf{r}\, d\mathbf{r}' - \int \mathscr{F}(\mathbf{r}\mathbf{r}')\, \mathbf{r}'\, d\mathbf{r}\, d\mathbf{r}' \tag{2.13.6}$$

and since we have

$$\mathscr{F}(\mathbf{r}\mathbf{r}') = \mathscr{F}(\mathbf{r}'\mathbf{r}) \tag{2.13.7}$$

we obtain the result

$$\frac{\partial \mathscr{F}(00)}{\partial \mathbf{k}} = 0. \tag{2.13.8}$$

However, as is evident from equation (2.13.5), this conclusion cannot be drawn for $\partial \mathscr{F}(\mathbf{K},0)/\partial \mathbf{k}$ if $\mathbf{K} \neq 0$. From equations (2.13.1) and (2.13.2) we also have that

$$\mathscr{F}(\mathbf{K},0) = 0 \tag{2.13.9}$$

for an insulator and we can now draw the following conclusions

$$\mathscr{F}(\mathbf{K}+\mathbf{q},\mathbf{q}) = \mathbf{q}.\mathscr{F}'(\mathbf{K},0) + O(q^2), \tag{2.13.10}$$

$$\mathscr{F}(\mathbf{q},\mathbf{q}+\mathbf{K}) = \mathbf{q}.\mathscr{F}'(0,\mathbf{K}) + O(q^2), \tag{2.13.11}$$

$$\mathscr{F}(\mathbf{q},\mathbf{q}) = \sum_{\alpha\beta} \mathscr{F}^{(2)}_{\alpha\beta}(0,0)\, q_\alpha q_\beta + O(q^4), \tag{2.13.12}$$

$$\mathscr{F}(\mathbf{q}+\mathbf{K},\mathbf{q}+\mathbf{K}') = \mathscr{F}(\mathbf{K},\mathbf{K}') + O(q). \tag{2.13.13}$$

We also note that the function $P(\mathbf{q},\mathbf{q}+\mathbf{K})$, defined, following Ambegaokar and Kohn (1960), as the sum of all those diagrams contributing to $\mathscr{F}$ which contain no internal interaction line labelled by $\mathbf{q}$, has a similar small $\mathbf{q}$ expansion to $\mathscr{F}$, as is evident from the diagrams for P shown in Figure 2.47. The total response function $\mathscr{F}$ is very easily expressed in terms of P, as one sees by comparing Figures 2.47 and 2.43 (it should be noted that $\mathscr{F} = \Pi$). For example,

$$\mathscr{F}(\mathbf{q},\mathbf{q}) = P(\mathbf{q},\mathbf{q}) + P(\mathbf{q},\mathbf{q})\, V(\mathbf{q})\mathscr{F}(\mathbf{q},\mathbf{q}). \tag{2.13.14}$$

The small **q** expansions of $\mathscr{F}$ may thus be derived from those of P and we shall adopt this procedure in section 3.14, where we discuss the lattice dynamics of insulators.

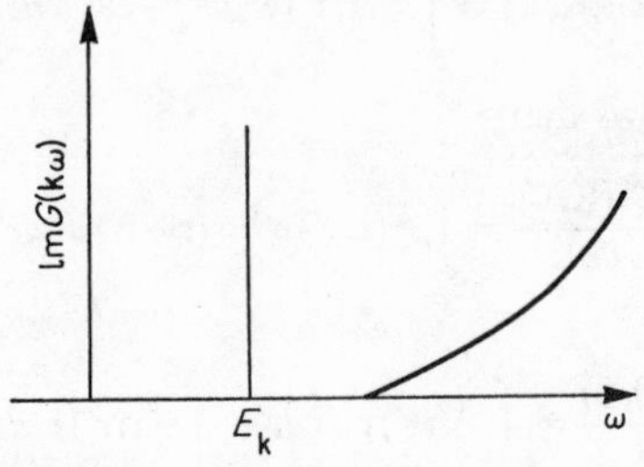

FIGURE 2.46. Spectral density in insulator.

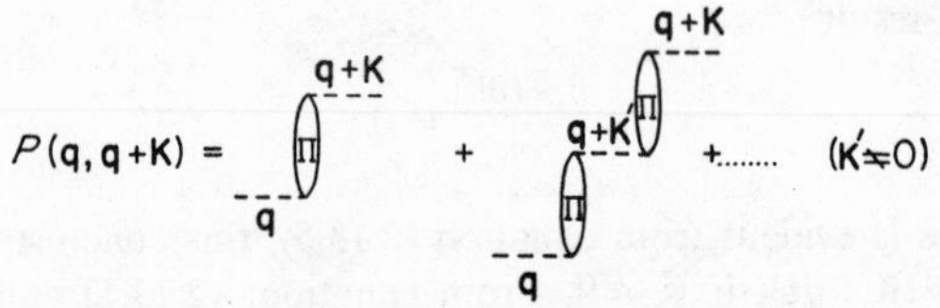

FIGURE 2.47. Diagrammatic series for P in terms of irreducible part of response function $\mathscr{F}$.

2.14 Dielectric response function in semi-conductors

While, in principle, a knowledge of the one-electron energy levels $E(\mathbf{k})$ and the corresponding wave functions $\psi_{\mathbf{k}}(\mathbf{r})$ will allow the calculation of the dielectric response function in a semi-conductor, in practice simplified models are often very instructive in displaying the gross features.

Models due to Callaway (1959) and Penn (1962) which are useful in this sense in a semi-conductor, start from an electron gas, with an energy gap E_g. The model, in essence, depends on the magnitude of E_g/E_f and the mean electron density.

The detailed calculations for this model are summarized in Appendix 2.5. However, a few general results of some importance can be summarized as follows:

(i) The main effect of the gap is to reduce the screening. Essentially, the presence of the gap suppresses electron–hole excitations of energies less than the gap width and therefore reduces the screening of the test charge.

(ii) For small energy gaps, the model gives Friedel oscillations.

(iii) These oscillations decrease in amplitude as the gap increases and tend to disappear when the gap is about twice the Fermi energy.

Useful results can be obtained from such a one-parameter model, results for the wave-number dependent dielectric function $\varepsilon(\mathbf{q})$ being given in Appendix 2.6 (see especially Figure A2.6.3). A reasonable choice of E_g correctly reproduces the measured static dielectric constant as $\mathbf{q} \to 0$.

The fundamental status of the response functions for the different classes of solids, metals, insulators and small gap semi-conductors should be clear from the above discussion. Though, of course, very different physical situations are being dealt with, there is a good deal of unity brought into the theory in this way. Furthermore, by separating, as we did, for example, quite clearly in the case of metals, one-body and many-body effects, we can build in the energy-band structure from the outset, from a suitable one-body potential, which, however, includes a contribution from exchange and correlation effects. This part of the work, we stress again though, only involves solving a one-body problem, given this one-body potential. Obviously, then, the electron interactions can be included at a second stage, and in simple metals, the homogeneous electron gas, to which we have devoted a great deal of attention in this chapter, affords a most useful starting-point. An admirable review of the effects of electron–electron interactions on one-electron states in solids has been given by Hedin and Lundquist (1969) and the reader who wishes to have further details should refer to this article.

CHAPTER 3

Lattice Waves

3.1 Introduction

So far, the problem of electrons moving in a static lattice of ions has been the focal point of our discussion: independent electron motion in Chapter 1 and correlated behaviour in the last chapter.

In reality, the ions vibrate about their equilibrium positions, and the description of the crystal given so far is oversimplified. Indeed to make contact with numerous important properties of crystal lattices, for example, specific heats, thermal expansion, elastic constants and so on, it is essential to consider these atomic vibrations. Einstein, in a very early attempt to do this, assumed each atom in the crystal lattice to vibrate independently. While this model is occasionally still useful, we now know from the work of Debye, Born and von Karman and many other authors that this again is a problem in collective motion, rather than individual particle behaviour.

Thus, at low temperatures, as Debye showed, it is a much better approximation to treat the crystal lattice as an elastic continuum than as an assembly of Einstein oscillators, by discussing the normal modes of an elastic continuum. Needless to say, such a model must break down when the wavelengths associated with the lattice waves become comparable with the lattice spacing, and then we can anticipate that the Brillouin zone structure, which we saw was of such crucial importance for the propagation of electron waves in a periodic potential, will again dominate the problem.

Indeed, from the classic work of Brillouin, and many others, we know that waves propagating in periodic structures have well-defined characteristics. Thus, the appearance in Chapter 1 of allowed and forbidden energy regions for electrons in a periodic potential is not so much due to the use of Schrödinger's equation as to the periodicity of the structure in which the electron waves travel. We shall, therefore, in this chapter, see a very close connection between the problem of lattice waves and that of electron wave propagation. As in Chapter 1 the basic problem was to determine the one-electron energy E, as a function of wave vector $\mathbf{k}$, so in the present chapter a central theme will be the determination of the frequency of a lattice vibration as a function of wave vector **k**. But to do this, we have, as emphasized above, to recognize the fundamental collective nature of the problem of lattice waves in crystals.

3.1.1 Plasma description of velocity of sound in metals

To illustrate this collective behaviour, without the full framework needed to discuss the discrete crystal lattice, let us notice that in the long wavelength limit, i.e. wave vector $\mathbf{k} \to 0$, we have modes such that

$$\omega = v_s k \tag{3.1.1}$$

where v_s is the velocity of sound, in, say, a simple metal like Na. We shall see later that there are three modes of this kind, but, at this stage, we consider only longitudinal waves.

In Chapter 2, we dealt with the problem of the plasma oscillations of the electrons in a uniform positive background of ionic charge.

Let us now apply the formula (2.2.9) derived there for the electron plasma frequency to the ions in a simple metal, with a well-defined valency Z (say Na, with $Z = 1$). Then if the ionic mass is M, each ion carries a charge Ze and the *ionic plasma frequency* may be written

$$\omega_p^{\text{ion}} = \left[\frac{4\pi n_i (Ze)^2}{M}\right]^{\frac{1}{2}}, \tag{3.1.2}$$

where n_i is the number of ions per unit volume.

But clearly n_i is related to ρ_0, the number of electrons per unit volume, by

$$\rho_0 = Z n_i \tag{3.1.3}$$

and hence we find from (3.1.3) and (3.1.2) that

$$\omega_p^{\text{ion}} = \left(\frac{4\pi \rho_0 Z e^2}{M}\right)^{\frac{1}{2}}. \tag{3.1.4}$$

This formula (3.1.4) does not agree with the observed form (3.1.1) for whereas this form shows us that $\omega \to 0$ as $k \to 0$, the result (3.1.4) gives a frequency independent of wave vector **k**.

It should be clear from the discussion of screening in an electron gas where the argument needs modification. In using (3.1.2), it is assumed that the background electronic charge is uniform, whereas the discussion of Chapter 2 shows us that the electrons will pile up round the ions, and screen out the Coulomb field in a distance $\sim q^{-1}$ where q is given by equation (2.3.13). Thus, we note that, because of screening, we can write in semiclassical theory that the Coulomb potential Ze/r of an ion goes over into a screened Coulomb potential $(Ze/r)\exp(-qr)$. In **k**-space, we can write that

$$\frac{Ze}{k^2} \to \frac{Ze}{k^2+q^2}, \tag{3.1.5}$$

where the quantities written in (3.1.5) are simply the Fourier transforms of the Coulomb and screened Coulomb potentials respectively. Hence, from (3.1.5), we can see that, in the long wavelength limit $k \to 0$, the effect of screening is to replace the charge Ze in plasma theory by the "effective charge" defined by

$$Ze \to \frac{Zek^2}{q^2}. \tag{3.1.6}$$

If we now insert this result into (3.1.4) we find

$$\omega = \frac{(4\pi\rho_0 Ze^2)^{\frac{1}{2}}}{M^{\frac{1}{2}} q} k. \tag{3.1.7}$$

Expressing ρ_0 and q in terms of the Fermi energy E_f, we obtain immediately for the velocity of sound v_s, by comparing (3.1.1) and (3.1.7), the result

$$Mv_s^2 = \tfrac{2}{3} ZE_f, \tag{3.1.8}$$

which is the so-called Bohm–Staver (1952) formula [see also Bardeen and Pines (1955)].

This already gives us a useful order of magnitude estimate of the velocity of sound in metals like Na, with small cores and low valence. However, the main purpose of our discussion is to illustrate that a result like (3.1.1), which is found experimentally from the everyday knowledge that we can propagate sound through a crystal, involves, at very least, in a conducting crystal:

(i) A collective treatment of ionic motions.

(ii) A proper account of electron screening.

Though the existence of modes like (3.1.1), which are referred to as acoustic modes, can now be understood in a metal, they also exist in insulating crystals, but a rather more sophisticated theory is needed in this case (see section 3.12).

We turn, then, from this elementary introduction, to discuss the customary treatment of lattice waves for an arbitrary wave vector **k**. Because, as we stressed above, the main features arise from the propagation of waves in periodic structures, we shall begin the discussion from a classical standpoint. Later, we will give the quantum theory of lattice vibrations, though for many purposes the classical treatment remains adequate.

3.2 Harmonic approximation and dynamical matrix

We shall begin our discussion with the case of solids having one atom per unit cell and later generalize this. Taking as basis vectors, $\mathbf{R}_1, \mathbf{R}_2, \mathbf{R}_3$, we denote the equilibrium positions by

$$\mathbf{l} = l_1 \mathbf{R}_1 + l_2 \mathbf{R}_2 + l_3 \mathbf{R}_3, \tag{3.2.1}$$

where l_1, l_2, l_3 are integers.

The displacement of the lth atom from equilibrium will be denoted by $\mathbf{u}_l$, so that the position vector of the lth atom becomes

$$\mathbf{R}_l = \mathbf{l} + \mathbf{u}_l. \tag{3.2.2}$$

We proceed classically and then the total kinetic energy of the lattice may evidently be written

$$T_{\text{ions}} = \tfrac{1}{2} \sum_l \sum_\alpha M \dot{u}_l^\alpha \dot{u}_l^\alpha, \tag{3.2.3}$$

where M is the ionic mass and u_l^α is the α-Cartesian component of $\mathbf{u}_l$, α being x, y or z.

We now expand the many-body potential Φ, governing the motion of the ions, in powers of the atomic displacements $\mathbf{u}_l$:

$$\Phi = \Phi_0 + \sum_l \sum_\alpha \Phi_\alpha(\mathbf{l})\, u_l^\alpha + \tfrac{1}{2} \sum_{l,l'} \sum_{\alpha,\beta} \Phi_{\alpha\beta}(\mathbf{l}, \mathbf{l}')\, u_l^\alpha u_{l'}^\beta + \ldots, \tag{3.2.4}$$

where Φ_0 is the equilibrium value,

$$\Phi_\alpha(\mathbf{l}) = \left(\frac{\partial \Phi}{\partial u_l^\alpha}\right)_0 \tag{3.2.5}$$

and

$$\Phi_{\alpha\beta}(\mathbf{l}, \mathbf{l}') = \left(\frac{\partial^2 \Phi}{\partial u_l^\alpha \partial u_{l'}^\beta}\right)_0. \tag{3.2.6}$$

The subscript zero implies that the derivatives are evaluated at the equilibrium configuration. In the equilibrium position the second term on the right-hand side of (3.2.4) vanishes. The condition of translational invariance applied to $\Phi_{\alpha\beta}(\mathbf{l}, \mathbf{l}')$ yields the following results:

$$\Phi_{\beta\alpha}(\mathbf{l}', \mathbf{l}) = \Phi_{\alpha\beta}(\mathbf{l}, \mathbf{l}') = \Phi_{\alpha\beta}(\mathbf{l} - \mathbf{l}') \tag{3.2.7}$$

and

$$\sum_{l'} \Phi_{\alpha\beta}(\mathbf{l}, \mathbf{l}') = 0. \tag{3.2.8}$$

Equation (3.2.8) simply expresses the fact that the forces on any atom are zero if each atom is displaced from equilibrium by the same amount.

In the harmonic approximation, we neglect terms not explicitly displayed on the right-hand side of (3.2.4), and the equations of motion may then be written as

$$M\ddot{u}_l^\alpha = -\sum_{l'}\sum_{\beta}\Phi_{\alpha\beta}(\mathbf{l}-\mathbf{l}')\,u_{l'}^\beta. \tag{3.2.9}$$

The solution now proceeds in the usual way. To avoid considering surface effects, we take the familiar periodic boundary conditions: the crystal, with its N lattice vectors $\mathbf{l}$, is part of an infinite crystal composed of crystals vibrating in an identical way to the one we consider.

We now define normal coordinates $Q_\mathbf{k}$ by

$$u_l^\alpha = \frac{1}{(NM)^{\frac{1}{2}}}\sum_{\mathbf{k}}\varepsilon_\mathbf{k}^\alpha Q_\mathbf{k}\,e^{i\mathbf{k}.\mathbf{l}}, \tag{3.2.10}$$

where the vectors $\mathbf{k}$ are all those obeying equation (1.4.11) and lying within the BZ defined in Chapter 1 and $\boldsymbol{\varepsilon}_\mathbf{k}$ are polarization vectors, discussed more fully below. For this choice of vectors, there exist the orthogonality and closure relations

$$\sum_l e^{i\mathbf{k}.\mathbf{l}} = N\delta(\mathbf{k},\mathbf{K}) \tag{3.2.11}$$

and

$$\sum_\mathbf{k} e^{i\mathbf{k}.(\mathbf{l}-\mathbf{l}')} = N\delta(\mathbf{l},\mathbf{l}'), \tag{3.2.12}$$

$\mathbf{K}$ being any reciprocal lattice vector. Using these relations, substitution of the transformation (3.2.10) into (3.2.9) yields the result

$$-\ddot{Q}_\mathbf{k}\,\varepsilon_\mathbf{k}^\alpha = Q_\mathbf{k}\sum_\beta D_{\alpha\beta}(\mathbf{k})\,\varepsilon_\mathbf{k}^\beta, \tag{3.2.13}$$

where $\mathbf{D}(\mathbf{k})$ is the *dynamical matrix* with elements

$$D_{\alpha\beta}(\mathbf{k}) = M^{-1}\sum_\mathbf{l}\Phi_{\alpha\beta}(\mathbf{l})\,e^{-i\mathbf{k}.\mathbf{l}}. \tag{3.2.14}$$

This is easily shown to be Hermitian if one remembers (3.2.7).

The frequencies of the running waves described by the $Q_\mathbf{k}$ are now obtained from the eigenvalue equation for the polarization vectors $\boldsymbol{\varepsilon}_\mathbf{k}$:

$$\sum_\beta D_{\alpha\beta}(\mathbf{k})\,\varepsilon_\mathbf{k}^\beta = \omega^2\,\varepsilon_\mathbf{k}^\alpha. \tag{3.2.15}$$

The three eigenvectors of $\mathbf{D}(\mathbf{k})$ and the corresponding normal coordinates will be denoted by $\boldsymbol{\varepsilon}_{\mathbf{k}\sigma}$ and $Q_{\mathbf{k}\sigma}$ ($\sigma = 1, 2, 3$).

The eigenvectors satisfy the orthogonality and closure relations

$$\boldsymbol{\varepsilon}_{\mathbf{k}\sigma} \cdot \boldsymbol{\varepsilon}_{\mathbf{k}\sigma'} = \delta_{\sigma\sigma'}, \tag{3.2.16}$$

$$\sum_{\sigma} \varepsilon^{\alpha}_{\mathbf{k}\sigma} \varepsilon^{\beta}_{\mathbf{k}\sigma} = \delta_{\alpha\beta}. \tag{3.2.17}$$

If we now adopt the phase convention

$$\boldsymbol{\varepsilon}_{\mathbf{k}\sigma} = \boldsymbol{\varepsilon}_{-\mathbf{k}\sigma} \tag{3.2.18}$$

then it follows that

$$Q_{\mathbf{k}\sigma} = Q^{*}_{-\mathbf{k}\sigma}. \tag{3.2.19}$$

Equation (3.2.10) must now be modified to take explicit account of the polarization. This change is essentially trivial and the new equation is readily seen to be

$$\mathbf{u}_{\mathbf{l}} = (MN)^{-\frac{1}{2}} \sum_{\mathbf{k}\sigma} Q_{\mathbf{k}\sigma} \boldsymbol{\varepsilon}_{\mathbf{k}\sigma} e^{i\mathbf{k}.\mathbf{l}}. \tag{3.2.20}$$

Using equations (3.2.11), (3.2.12), (3.2.16) and (3.2.17), the effective Hamiltonian corresponding to the equations of motion (3.2.9), or equivalently (3.2.13), can be written in the diagonal form

$$H_{\text{lattice}} = \tfrac{1}{2} \sum_{\mathbf{k}\sigma} [\dot{Q}^{*}_{\mathbf{k}\sigma} \dot{Q}_{\mathbf{k}\sigma} + Q^{*}_{\mathbf{k}\sigma} Q_{\mathbf{k}\sigma} \omega^{2}(\mathbf{k}\sigma)], \tag{3.2.21}$$

where

$$\omega^{2}(\mathbf{k}\sigma) = \langle \boldsymbol{\varepsilon}_{\mathbf{k}\sigma} | D(\mathbf{k}) | \boldsymbol{\varepsilon}_{\mathbf{k}\sigma} \rangle = \sum_{\alpha\beta} \varepsilon^{\alpha}_{\mathbf{k}\sigma} D_{\alpha\beta}(\mathbf{k}) \varepsilon^{\beta}_{\mathbf{k}\sigma}. \tag{3.2.22}$$

One notes from the definition of $D(\mathbf{k})$ that

$$D(\mathbf{k}+\mathbf{K}) = D(\mathbf{k}), \tag{3.2.23}$$

where $\mathbf{K}$ is a reciprocal lattice vector. Thus $\omega^{2}(\mathbf{k})$ is periodic in the reciprocal lattice.

In the elastic continuum model which is discussed fully in Appendix 3.1, the three polarization vectors $\boldsymbol{\varepsilon}_{\mathbf{k}\sigma}$ correspond to two transverse modes and one longitudinal mode.

Whereas in the isotropic continuum, the eigenvectors $\boldsymbol{\varepsilon}_{\mathbf{k}\sigma}$ are either exactly perpendicular or exactly parallel to $\mathbf{k}$, this is not true in general in the discrete lattice, but only when $\mathbf{k}$ is along symmetry directions of the BZ.

3.3 Acoustic and optical modes

The three modes we found above are degenerate at $\mathbf{k} = 0$, independently of the symmetry of the crystal (the consequences of which will be discussed later), because $\mathbf{k} = 0$ corresponds to identical displacements of all atoms of the crystal, which cannot cause any energy change, and so

$$\omega^2(\mathbf{k}\sigma) = 0 \quad \text{at } \mathbf{k} = 0, \tag{3.3.1}$$

which, of course, agrees with the Bohm–Staver result (3.1.7) for longitudinal waves at $k = 0$.

However, if we have more than one kind of atom (we include in this description cases where atoms occupy physically inequivalent sites) the different kinds of atom can vibrate out of phase with one another, so that equation (3.3.1) need not hold for all modes of vibration. It *must* hold for three modes, however, for just the same reason as it holds when there is one atom per unit cell only. These three modes are called "acoustic modes", for at low $\mathbf{k}$ they have frequencies of the order of those of sound (cf. equations (3.1.7) and (3.1.8)). The remaining modes, for which equation (3.3.1) is not true, have frequencies in the infra-red region of the spectrum and are referred to as optical modes.

The extension of our formalism to crystals with optical modes is readily carried out. If s is the site of an ion in the unit cell about the origin, we write (M_s being the mass of the ion on site $\mathbf{s}$)

$$\mathbf{u}_s(\mathbf{l}) = \mathbf{u}(\mathbf{l}+\mathbf{s}) = \frac{1}{(M_s N)^{\frac{1}{2}}} \sum_{\mathbf{k}} Q_{\mathbf{k}\sigma}\, \boldsymbol{\varepsilon}^s_{\mathbf{k}\sigma}\, e^{i\mathbf{k}.(\mathbf{l}+\mathbf{s})} \tag{3.3.2}$$

and can readily show that

$$\sum_{\beta s'} D^{ss'}_{\alpha\beta}(\mathbf{k})\, \varepsilon^{\beta s'}_{\mathbf{k}\sigma} = \omega^2_{\mathbf{k}\sigma}\, \varepsilon^{\alpha s}_{\mathbf{k}\sigma}. \tag{3.3.3}$$

Here s and s' ($= 1, \ldots, r$) label the different kinds of ionic sites, and

$$D^{ss'}_{\alpha\beta}(\mathbf{k}) = \frac{1}{(M_s M_{s'})^{\frac{1}{2}}} \sum_{\mathbf{l}} \Phi_{\alpha\beta}(\mathbf{l}+\mathbf{s}, \mathbf{l}'+\mathbf{s}')\, e^{i\mathbf{k}.\mathbf{l}-\mathbf{l}'+\mathbf{s}-\mathbf{s}'} \tag{3.3.4}$$

where

$$\Phi_{\alpha\beta}(\mathbf{l}+\mathbf{s}, \mathbf{l}'+\mathbf{s}') = \frac{\partial^2 \Phi}{\partial u^\alpha_{\mathbf{l}s}\, u^\beta_{\mathbf{l}'s'}}. \tag{3.3.5}$$

Evidently $D(\mathbf{k})$ is now a $3r \times 3r$ matrix with $3r$ independent eigenvalues and there are $3r-3$ optical modes. The time reversal symmetry‡ $\omega^2(\mathbf{k}) = \omega^2(-\mathbf{k})$ is valid [see equations (3.2.18) and (3.2.19)].

‡ See section 1.5.6 of Chapter 1 where this topic is discussed.

We now write the $3r$ polarization vectors as

$$\begin{pmatrix} \boldsymbol{\varepsilon}^1_{\mathbf{k}\sigma} \\ \vdots \\ \boldsymbol{\varepsilon}^r_{\mathbf{k}\sigma} \end{pmatrix}$$

since they have $3r$ components and may be regarded as compounded of r polarization vectors $\boldsymbol{\varepsilon}^s_{\mathbf{k}\sigma}$, one for each kind of ion. In Appendix 3.3, we show how the group theory of section 3.5 can be generalized to this case.

3.4 Nature and range of forces

Early work on lattice dynamics made the starting assumption of short-range forces. While such an approach was of great value in the development of lattice dynamics, it is now quite clear that, in a variety of crystals, and in particular in metals and valence semi-conductors, the interionic forces are frequently long range. Since to demonstrate this involves detailed consideration of the effects of the electrons on the lattice, we shall first approach the problem by asking what information on the character and range of the forces can be derived from experiment.

3.4.1 Dispersion relation and force range

Combination of (3.2.14) and (3.2.22) yields the series

$$\omega^2(\mathbf{k}\sigma) = \sum_{\mathbf{l}} \Omega(\mathbf{l}\sigma)\, e^{i\mathbf{k}.\mathbf{l}}, \tag{3.4.1}$$

where

$$M\Omega(\mathbf{l}\sigma) = \sum_{\alpha\beta} \varepsilon^{\alpha}_{\mathbf{k}\sigma}\, \Phi_{\alpha\beta}(\mathbf{l})\, \varepsilon^{\beta}_{\mathbf{k}\sigma}. \tag{3.4.2}$$

Now we can see that if the spectrum of $\omega^2(\mathbf{k},\sigma)$ is approximately "sinusoidal" and periodic in the BZ along a symmetry direction, the forces must be of short range, the first harmonic $\Omega(\mathbf{l},\sigma)|_{\mathbf{l}=\mathbf{l}_1}$ being dominant, $\mathbf{l}_1$ being the shortest direct lattice vector. But, for example, the transverse acoustic (TA) branch‡ in germanium is flat over much of the BZ (see Figure 3.1); we must therefore expect forces of quite long range to be present.

It must be remarked that one must be careful in the interpretation of (3.4.1) as a Fourier series. $\omega^2(\mathbf{k},\sigma)$ *may* be written as a Fourier series with the form of (3.4.1), since it is periodic in the reciprocal lattice. It should be

‡ See the previous section for the distinction between acoustic and optical modes.

noted, however, that, as is evident from (3.4.2), $\Omega(\mathbf{l}, \sigma)$ *depends on* $\mathbf{k}$. One may, therefore, use (3.4.1) as a simple Fourier series *only along directions where the polarization vectors do not change.*

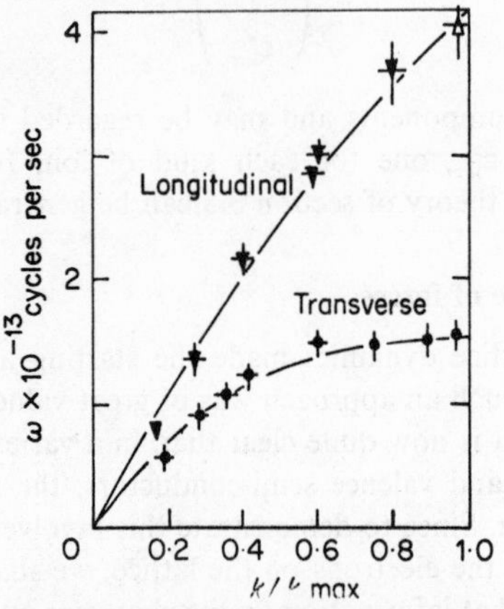

FIGURE 3.1. Acoustic branches of phonon dispersion relations in Ge, with wave vector along [100]. Solid curves are obtained by theory to be discussed later. The important point here is the flatness of the transverse acoustic branch, which reveals long range forces.

3.4.2 Central forces

The theory of electrons in simple metals which we discussed in Chapter 1 provides some justification for assuming the potential energy to be a sum of pair potentials $\phi(\mathbf{R}_1 - \mathbf{R}_2)$, depending only on the distance $|\mathbf{R}_1 - \mathbf{R}_2|$ between the ions. The atomic force constants $\Phi_{\alpha\beta}(\mathbf{l})$ are then

$$\Phi_{\alpha\beta}(\mathbf{l}) = -\left(\frac{\partial^2 \phi}{\partial x_\alpha \, \partial x_\beta}\right)_{\mathbf{r}=\mathbf{l}}. \tag{3.4.3}$$

Using this result in the definition (3.2.14) of the dynamical matrix, and explicitly satisfying (3.2.8), we find

$$D_{\alpha\beta}(\mathbf{k}) = M^{-1} \sum_{\mathbf{l}}{}' (1 - \mathrm{e}^{-i\mathbf{k}.\mathbf{l}}) \left(\frac{\partial^2 \phi}{\partial x_\alpha \, \partial x_\beta}\right)_{\mathbf{r}=\mathbf{l}}, \tag{3.4.4}$$

where the prime on the sum of (3.3.4) indicates that the term $\mathbf{l} = 0$ is to be omitted. This is the basic result for the dynamical matrix in terms of the force law.

We may note that

$$\left.\begin{aligned}\frac{\partial^2\phi(r)}{\partial x^2}&=\frac{x^2}{r^2}\left(\frac{\partial^2\phi}{\partial r^2}-\frac{1}{r}\frac{\partial\phi}{\partial r}\right)+\frac{1}{r}\frac{\partial\phi}{\partial r},\\ \frac{\partial^2\phi(r)}{\partial x\,\partial y}&=\frac{xy}{r^2}\left(\frac{\partial^2\phi}{\partial r^2}-\frac{1}{r}\frac{\partial\phi}{\partial r}\right).\end{aligned}\right\} \tag{3.4.5}$$

Thus, assuming central forces to hold, we can in principle derive their first and second derivatives at the lattice points from experiment, using the relations (3.4.2) and (3.4.3). In equation (3.4.2) it is, of course, necessary to know the direction of the polarization vector. This can be found from symmetry arguments for certain values of **k**, as we now show.

3.5 Consequences of symmetry

We may expect $\omega^2(\mathbf{k})$ to have the symmetry of the BZ. However, the branches of the phonon spectrum which are in general non-degenerate may touch on points or on lines. In order to predict when this happens, we must investigate the consequences of symmetry further, bearing in mind the group theory outlined in Chapter 1. In the following, the total potential energy Φ is not assumed to be built from pair potentials.

Let us first look at the derivative $\Phi_{\alpha\beta}(\mathbf{l})$, which is a component of the matrix $[\Phi_{\alpha\beta}(\mathbf{l})]$. It is evident that if we consider an equivalent vector $\mathbf{m} = a\mathbf{l}$, the energy change on moving the ion at $\mathbf{m}$ by au_l, will be the same as moving the ion at $\mathbf{l}$ by u_l. This physical requirement is easily shown to imply the relation

$$[\Phi_{\alpha\beta}(a\mathbf{l})] = (a_{ij})\,[\Phi_{\alpha\beta}(\mathbf{l})]\,(a_{ij})^{\dagger}, \tag{3.5.1}$$

where (a_{ij}) is the 3×3 matrix representing the unitary transformation a.

Let us now look at equation (3.2.15), and investigate the consequence of transforming **k** to $a\mathbf{k}$, a again being a unitary transformation of the rotation group of the lattice (see Chapter 1, section 1.6).

From equation (3.2.14),

$$\begin{aligned}D_{\alpha\beta}(a\mathbf{k}) &= \frac{1}{M}\sum_{\mathbf{l}}\Phi_{\alpha\beta}(\mathbf{l})\,e^{-ia\mathbf{k}.\mathbf{l}}\\ &= \frac{1}{M}\sum_{\mathbf{l}}\Phi_{\alpha\beta}(\mathbf{l})\,e^{-i\mathbf{k}.a^{\dagger}\mathbf{l}}\\ &= \frac{1}{M}\sum_{\mathbf{l}}\Phi_{\alpha\beta}(a\mathbf{l})\,e^{-i\mathbf{k}.\mathbf{l}},\end{aligned} \tag{3.5.2}$$

the last line following since $a\mathbf{l}$ is a lattice vector.

It may be seen from equations (3.5.1) and (3.5.2) that

$$[D(a\mathbf{k})] = (a)\,[D(\mathbf{k})]\,(a)^{\dagger}. \tag{3.5.3}$$

Thus $D(a\mathbf{k})$ and $D(\mathbf{k})$ are connected by a unitary transformation, and so have the same eigenvalues. This tells us that $\omega^2(\mathbf{k})$ has the symmetry of the BZ,‡ which we had already guessed; but it also tells us that at symmetry points for which there are operators a such that $a\mathbf{k} = \mathbf{k}+\mathbf{K}$, we may expect degeneracy, exactly as was found in the band theory given in Chapter 1. For these operators,

$$(a)\,[D(\mathbf{k})] = [D(\mathbf{k})]\,(a)^{\dagger}. \tag{3.5.4}$$

The polarization vectors appear in equation (3.2.15) as eigenvectors of $[D(\mathbf{k})]$, and so, in accordance with the group theory argument of Chapter 1, span invariant manifolds of the subgroup to which a belongs. Hence, the polarization vectors may be obtained by finding the irreducible representations of a, this giving us, at the same time, the essential degeneracy in $\omega^2(\mathbf{k})$.

3.5.1 Modes of high symmetry in face-centred cubic lattice

We will take three examples for the fcc lattice (see Figure 1.5):

(i) The line $\Lambda = (1,1,1)$ in the BZ, showing that for propagation in this direction there is an exactly longitudinal mode and two degenerate transverse modes.

(ii) The line $\Sigma = (1,1,0)$, where there is again an exactly longitudinal mode, and the two transverse modes are, in general, non-degenerate.

(iii) The point $W = \pi/a\,(1,\tfrac{1}{2},0)$, where two modes are degenerate, but no mode is exactly longitudinal or transverse.

Example (i) *The line* Λ is characterized by the vector (k_x, k_x, k_x). As a first step in calculating the polarization vectors, we may easily verify that the transformations, forming a subgroup of the full point group, which leave this vector unchanged, are represented by the matrices

$$E = \begin{pmatrix} 1 & 0 & 0 \\ 0 & 1 & 0 \\ 0 & 0 & 1 \end{pmatrix}, \quad a = \begin{pmatrix} 1 & 0 & 0 \\ 0 & 0 & 1 \\ 0 & 1 & 0 \end{pmatrix}, \quad b = \begin{pmatrix} 0 & 1 & 0 \\ 1 & 0 & 0 \\ 0 & 0 & 1 \end{pmatrix},$$

$$c = \begin{pmatrix} 0 & 1 & 0 \\ 0 & 0 & 1 \\ 1 & 0 & 0 \end{pmatrix}, \quad d = \begin{pmatrix} 0 & 0 & 1 \\ 1 & 0 & 0 \\ 0 & 1 & 0 \end{pmatrix}, \quad e = \begin{pmatrix} 0 & 0 & 1 \\ 0 & 1 & 0 \\ 1 & 0 & 0 \end{pmatrix}. \tag{3.5.5}$$

‡ This is always true when the Bravais lattice is equivalent to the crystal lattice. When optical modes are present, then it does not always hold.

These form a group as can be seen by constructing the multiplication table. We should note at this point that there are other matrix representations of this group, but, as will be seen from the argument leading to (3.5.4), *we must only consider the matrix representations of the unitary transformations of a vector.*

Let us seek a one-dimensional invariant manifold of the group, i.e. a vector which does no more than change sign under the operations of the group. If we write

$$\mathbf{v} = \begin{pmatrix} \alpha \\ \beta \\ \gamma \end{pmatrix},$$

then, from equation (3.5.5),

$$a\mathbf{v} = \begin{pmatrix} \alpha \\ \gamma \\ \beta \end{pmatrix}$$

and thus $\beta = \gamma$. Furthermore,

$$b\mathbf{v} = \begin{pmatrix} \beta \\ \alpha \\ \gamma \end{pmatrix}$$

and thus $\alpha = \beta$. Hence, one of the polarization vectors has components (k_x, k_x, k_x) and therefore lies along Λ. Obviously this corresponds to a pure longitudinal mode. From its construction, we see that it is the only one-dimensional invariant vector. There is thus a two-dimensional irreducible representation, which must be spanned by the polarization vectors of the degenerate transverse modes (cf. Chapter 1, section 1.5).

Example (ii) *The line* Σ has general vector $(k_x, k_x, 0)$. We have matrices

$$E = \begin{pmatrix} 1 & 0 & 0 \\ 0 & 1 & 0 \\ 0 & 0 & 1 \end{pmatrix}, \quad a = \begin{pmatrix} 0 & 1 & 0 \\ 1 & 0 & 0 \\ 0 & 0 & 1 \end{pmatrix}, \quad b = \begin{pmatrix} 0 & 1 & 0 \\ 1 & 0 & 0 \\ 0 & 0 & -1 \end{pmatrix}. \tag{3.5.6}$$

Repeating our procedure by putting

$$\mathbf{v} = \begin{pmatrix} \alpha \\ \beta \\ \gamma \end{pmatrix},$$

we have from equation (3.5.6) that

$$a\begin{pmatrix}\alpha\\ \beta\\ \gamma\end{pmatrix}=\begin{pmatrix}\beta\\ \alpha\\ \gamma\end{pmatrix} \quad\text{and}\quad b\begin{pmatrix}\alpha\\ \beta\\ \gamma\end{pmatrix}=\begin{pmatrix}\beta\\ \alpha\\ -\gamma\end{pmatrix}.$$

It is readily seen that we have three orthogonal vectors satisfying these relations: $(1,1,0)$, $(1,-1,0)$ and $(0,0,1)$. The first is the polarization vector for the longitudinal mode. The other two, exactly transverse, belong to modes which will be, in general, non-degenerate, because each of them forms itself an irreducible representation.

Example (iii) *The point W*. There are four equivalent points given by

$$\frac{\pi}{a}(\tfrac{1}{2},1,0),\quad \frac{\pi}{a}(\tfrac{1}{2},-1,0),\quad \frac{\pi}{a}(-\tfrac{1}{2},0,1)\quad\text{and}\quad \frac{\pi}{a}(-\tfrac{1}{2},0,-1)$$

where $2a$ is the lattice parameter. One may easily verify that the required sub-group consists of the matrices

$$E=\begin{pmatrix}1&0&0\\0&1&0\\0&0&1\end{pmatrix},\quad a=\begin{pmatrix}1&0&0\\0&1&0\\0&0&-1\end{pmatrix},\quad b=\begin{pmatrix}1&0&0\\0&-1&0\\0&0&1\end{pmatrix},$$

$$c=\begin{pmatrix}1&0&0\\0&-1&0\\0&0&-1\end{pmatrix},\quad d=\begin{pmatrix}-1&0&0\\0&0&1\\0&1&0\end{pmatrix},\quad e=\begin{pmatrix}-1&0&0\\0&0&-1\\0&1&0\end{pmatrix},$$

$$f=\begin{pmatrix}-1&0&0\\0&0&1\\0&-1&0\end{pmatrix},\quad g=\begin{pmatrix}-1&0&0\\0&0&-1\\0&-1&0\end{pmatrix}. \tag{3.5.7}$$

Again we assume a one-dimensional irreducible manifold

$$\mathbf{v}=\begin{pmatrix}\alpha\\ \beta\\ \gamma\end{pmatrix},$$

and then from equation (3.5.7) we find

$$a\mathbf{v}=\begin{pmatrix}\alpha\\ \beta\\ -\gamma\end{pmatrix}.$$

Thus, either $\gamma = 0$ or $\alpha = \beta = 0$. If γ were non-zero, then we would have

$$d\mathbf{v} = \begin{pmatrix} 0 \\ \gamma \\ 0 \end{pmatrix}.$$

But this violates our invariance requirement unless γ is zero. Thus,

$$\mathbf{v} = \begin{pmatrix} \alpha \\ \beta \\ 0 \end{pmatrix},$$

and we then find

$$b\mathbf{v} = \begin{pmatrix} \alpha \\ -\beta \\ 0 \end{pmatrix}.$$

Thus either $\beta = 0$ or $\alpha = 0$. If β were non-zero, we would obtain

$$d\mathbf{v} = \begin{pmatrix} 0 \\ 0 \\ \beta \end{pmatrix}.$$

Hence β must in fact be zero. This leaves us with the vector

$$\mathbf{v} = \begin{pmatrix} \alpha \\ 0 \\ 0 \end{pmatrix}.$$

We thus have one non-degenerate mode, since the above procedure led to a unique polarization vector and two degenerate ones. But it should be noted that no polarization vector is exactly along the direction $(-1, -2, 0)$ in which W lies. The modes are not exactly longitudinal or transverse.

3.5.2 Face-centred cubic lattice with short range forces

Let us consider at this stage a particular model with central forces of such short range that only nearest-neighbour interactions need be considered. This is not of purely academic interest as it is relevant to the lattice spectrum of crystalline argon where the attractive forces are van der Waals in character. If we had a simple-cubic lattice rather than an fcc structure, we should have found that no transverse waves are propagated when only nearest-neighbour central interactions are included in the model, for the lattice then offers no resistance to shear. This therefore focuses attention on possible limitations of such a short-range force model.

Let us write the derivatives at the nearest neighbours as

$$\xi = \left(\frac{\partial^2\phi}{\partial r^2} - \frac{1}{r}\frac{\partial\phi}{\partial r}\right), \quad \eta = \frac{1}{r}\frac{\partial\phi}{\partial r}.\ddagger \tag{3.5.8}$$

The dynamical matrix is now readily obtained from (3.4.4). The reader is left to verify that D can be put into the form

$$D(\mathbf{k}) = \begin{pmatrix} d_{11} & d_{12} & d_{13} \\ d_{21} & d_{22} & d_{23} \\ d_{31} & d_{32} & d_{33} \end{pmatrix}, \tag{3.5.9}$$

where

$$d_{11}(\mathbf{k}) = (2\xi+4\eta)(2-\cos k_x a\cos k_y a - \cos k_x a\cos k_z a), \tag{3.5.10}$$

$$d_{22}(\mathbf{k}) = (2\xi+4\eta)(2-\cos k_x a\cos k_y a - \cos k_y a\cos k_z a), \tag{3.5.11}$$

$$d_{33}(\mathbf{k}) = (2\xi+4\eta)(2-\cos k_x a\cos k_z a - \cos k_y a\cos k_z a), \tag{3.5.12}$$

and the off-diagonal elements are given by

$$d_{21}(\mathbf{k}) = d_{12}(\mathbf{k}) = -2\xi\sin k_x a\sin k_y a, \tag{3.5.13}$$

$$d_{31}(\mathbf{k}) = d_{13}(\mathbf{k}) = -2\xi\sin k_x a\sin k_z a, \tag{3.5.14}$$

$$d_{32}(\mathbf{k}) = d_{23}(\mathbf{k}) = -2\xi\sin k_y a\sin k_z a. \tag{3.5.15}$$

We shall find the eigensolutions in the symmetry directions, Δ, Λ and Σ.

(i) $\Delta \equiv (1,0,0)$. The matrix becomes

$$\begin{pmatrix} (2\xi+4\eta)(2-2\cos k_x a) & 0 & 0 \\ 0 & (2\xi+4\eta)(1-\cos k_x a) & 0 \\ 0 & 0 & (2\xi+4\eta)(1-\cos k_x a) \end{pmatrix}. \tag{3.5.16}$$

The eigenvalues are evident.

‡ The first derivative $\partial\phi/\partial r$ is zero for equilibrium under ordinary pair forces. But in metals, this is not so, an additional pressure term being needed for equilibrium.

(ii) $\Lambda \equiv (1,1,1)$. The matrix becomes

$$\begin{pmatrix} (2\xi+4\eta)(2-2\cos^2 k_x a) & -2\xi\sin^2 k_x a & -2\xi\sin^2 k_x a \\ -2\xi\sin^2 k_x a & (2\xi+4\eta)(2-2\cos^2 k_x a) & -2\xi\sin^2 k_x a \\ -2\xi\sin^2 k_x a & -2\xi\sin^2 k_x a & (2\xi+4\eta)(2-2\cos^2 k_x a) \end{pmatrix}, \tag{3.5.17}$$

which has the form

$$\begin{pmatrix} a & b & b \\ b & a & b \\ b & b & a \end{pmatrix}.$$

The longitudinal mode has polarization vector $\begin{pmatrix} 1 \\ 1 \\ 1 \end{pmatrix}$;

$$\begin{pmatrix} a & b & b \\ b & a & b \\ b & b & a \end{pmatrix}\begin{pmatrix} 1 \\ 1 \\ 1 \end{pmatrix} = (a+2b)\begin{pmatrix} 1 \\ 1 \\ 1 \end{pmatrix}. \tag{3.5.18}$$

To find the degenerate transverse eigenvalue, we have

$$\begin{pmatrix} a & b & b \\ b & a & b \\ b & b & a \end{pmatrix}\begin{pmatrix} 1 \\ 1 \\ -2 \end{pmatrix} = (a-b)\begin{pmatrix} 1 \\ 1 \\ -2 \end{pmatrix}. \tag{3.5.19}$$

Thus

$$\omega_l^2(k_x, k_x, k_x) = (2\xi+4\eta)(2-2\cos^2 k_x a) - 4\xi\sin^2 k_x a, \tag{3.5.20}$$

$$\omega_t^2(k_x, k_x, k_x) = (2\xi+4\eta)(2-2\cos^2 k_x a) + 2\xi\sin^2 k_x a. \tag{3.5.21}$$

(iii) $\Sigma = (1,1,0)$. The matrix becomes

$$\begin{pmatrix} a & c & 0 \\ c & a & 0 \\ 0 & 0 & b \end{pmatrix},$$

where

$$a = (2\xi+4\eta)(2-\cos^2 k_x a - \cos k_x a), \tag{3.5.22}$$

$$b = (2\xi+4\eta)(2-2\cos k_x a), \tag{3.5.23}$$

$$c = -2\xi\sin^2 k_x a. \tag{3.5.24}$$

One eigenvalue is immediately evident, and the other two are easily found, remembering the polarization vectors already given for Σ. We have

$$\omega_l^2(k_x, k_x, 0) = a+c = (2\xi+4\eta)(2-\cos^2 k_x a-\cos k_x a)-2\xi \sin^2 k_x a, \quad (3.5.25)$$

$$\omega_{t1}^2(k_x, k_x, 0) = a-c = (2\xi+4\eta)(2-\cos^2 k_x a-\cos k_x a)+2\xi \sin^2 k_x a, \quad (3.5.26)$$

$$\omega_{t2}^2(k_x, k_x, 0) = b = (2\xi+4\eta)(2-2\cos k_x a). \quad (3.5.27)$$

Thus, in each case, we have obtained quite explicit expressions for the dispersion relations $\omega(\mathbf{k})$ in terms of the lattice constant and the derivatives of the pair potential representing the central force.

To avoid the reader supposing that the above theory will work quantitatively for solid argon, we want to stress that many-body forces play a role there. A useful way of seeing whether a central force model is good comes from the Cauchy relations between elastic constants to which we shall now turn.

3.6 Equilibrium and Cauchy relations

We earlier put $\Phi_\alpha = 0$, obtaining the harmonic approximation. This does not mean that the lattice has to be in its equilibrium position (under no external constraint). One easily sees that in the fcc lattice we discussed, regardless of the value of the lattice parameter a, one can change the coordinate of any atom *in the interior* of the crystal without any change in energy to first order, the surrounding atoms exerting forces cancelling one another. This may be seen to be a general result; we avoid the problem of the surface by, effectively, considering an infinite crystal. Then Φ_α vanishes whatever the pressure, and we can apply the harmonic approximation at any value of the lattice parameter we wish.

The condition for equilibrium in the harmonic approximation may be obtained by a simple argument using the virial theorem. This theorem tells us that the kinetic and potential energies of harmonic oscillators are equal, and that they make no contribution to the virial expression for the pressure: equilibrium is determined *statically* so that the virial theorem gives

$$\sum_{\mathbf{l}} \mathbf{l}.\frac{\partial \Phi}{\partial \mathbf{l}} = 0. \quad (3.6.1)$$

We may alternatively derive the equilibrium condition by the thermodynamic relation

$$P = -\left(\frac{\partial F}{\partial V}\right)_T, \quad (3.6.2)$$

F being the Helmholtz free energy and V the volume. Now F will be of the form

$$F = \Phi_0 + F_v, \quad (3.6.3)$$

where F_v will be the vibrational part of the free energy. If we regard the l's about which we expand (3.2.4) as parameters in the Hamiltonian, equilibrium is achieved by $\partial F/\partial \mathbf{l} = 0$, which becomes just the static condition

$$\frac{\partial \Phi_0}{\partial a} = 0 \tag{3.6.4}$$

(if we keep the form of the lattice the same, only varying the lattice parameter) provided the ω's are independent of a. Let us denote the lattice points obtained statically by $\tilde{\mathbf{l}}$, $\tilde{\mathbf{m}}$, etc., with the corresponding quantities of (3.2.4) similarly denoted by tildes. Now if $\Delta\mathbf{l} = \tilde{\mathbf{l}} - \mathbf{l}$,

$$\begin{aligned}\tilde{\Phi}_{\alpha\beta}\tilde{u}_l^\alpha \tilde{u}_m^\beta &= \tilde{\Phi}_{\alpha\beta}(\mathbf{R}_l - \tilde{\mathbf{l}})_\alpha(\mathbf{R}_m - \tilde{\mathbf{m}})_\beta = \tilde{\Phi}_{\alpha\beta}(\mathbf{R}_l - \mathbf{l} + \Delta\mathbf{l})_\alpha(\mathbf{R}_m - \mathbf{m} + \Delta\mathbf{m})_\beta \\ &= \tilde{\Phi}_{\alpha\beta}(\Delta\mathbf{l})_\alpha(\Delta\mathbf{m})_\beta - \tilde{\Phi}_{\alpha\beta}[(\Delta\mathbf{m})_\beta u_l^\alpha + (\Delta\mathbf{l})_\alpha u_m^\beta] + \tilde{\Phi}_{\alpha\beta}u_l^\alpha u_m^\beta.\end{aligned} \tag{3.6.5}$$

Thus

$$\Phi_0 = \tilde{\Phi}_0 + \sum_{\alpha\beta} \tilde{\Phi}_{\alpha\beta}(\Delta\mathbf{l})_\alpha(\Delta\mathbf{m})_\beta \tag{3.6.6}$$

and

$$\Phi_{\alpha\beta}(\mathbf{l}, \mathbf{m}) = \tilde{\Phi}_{\alpha\beta}(\tilde{\mathbf{l}}, \tilde{\mathbf{m}}). \tag{3.6.7}$$

The dynamical matrix is

$$D(\mathbf{k}) = \sum_{\mathbf{l}} \tilde{\Phi}_{\alpha\beta}(\mathbf{l})\, e^{-i\mathbf{k}.\mathbf{l}}. \tag{3.6.8}$$

Thus, although the coupling parameters $\Phi_{\alpha\beta}(\mathbf{l})$ do not change with lattice spacing, the dynamical matrix *does* alter.

However, it is to be remembered that if the lattice expands, the reciprocal lattice contracts in the same proportion. Hence, if $\mathbf{l} = \alpha\tilde{\mathbf{l}}$, for any allowed value of $\mathbf{k}$ in the "static" lattice, there is an allowed value $\mathbf{k}/\alpha$ in the expanded lattice. Further, from (3.6.8) we see that

$$D(\mathbf{k}/\alpha) = \tilde{D}(\mathbf{k}). \tag{3.6.9}$$

The eigenvectors and eigenvalues of both matrices are the same: the allowed values of ω do not vary, only the $\mathbf{k}$-labelling is different.

Our simple argument involving the use of the virial theorem having been confirmed, let us return to the condition (3.6.1). If the crystal is in equilibrium under central pair forces obtainable from a pair potential $\phi(\mathbf{r})$, this becomes

$$\sum_{\mathbf{l}} \mathbf{l}.\frac{\partial\phi(l)}{\partial\mathbf{l}} = \sum_{\mathbf{l}} l^2\left[\frac{1}{l}\frac{\partial\phi(l)}{\partial l}\right] = 0. \tag{3.6.10}$$

We shall now show that this condition for equilibrium implies a relation between macroscopic elastic constants. We must therefore briefly digress to define these. This is done in its full generality in Appendix 3.1, but we can argue about the nature of the wave equation describing the elastic waves on a largely qualitative basis.

Consider an element of the elastic solid at position $\mathbf{r}$ given by

$$\mathbf{r} = u\hat{\mathbf{x}}+v\hat{\mathbf{y}}+w\hat{\mathbf{z}}, \tag{3.6.11}$$

$\hat{\mathbf{x}}$, $\hat{\mathbf{y}}$ and $\hat{\mathbf{z}}$ being unit vectors along the x, y and z directions. When a stress is applied, suppose this element goes to $\mathbf{r}'$, where (see Appendix 3.1)

$$\mathbf{r}' = (u+\delta u)\,\hat{\mathbf{x}}+(v+\delta v)\,\hat{\mathbf{y}}+(w+\delta w)\,\hat{\mathbf{z}}. \tag{3.6.12}$$

We expect that in discussing waves the quantities u, v and w will satisfy classical equations of wave motion. For example, if u, v and w were independent and the medium were isotropic, then we could write

$$\frac{1}{v_s^2}\frac{\partial^2 u}{\partial t^2} = \nabla^2 u. \tag{3.6.13}$$

In general, in considering real crystals, the direction x will appear on a different footing from y and z. Also, from coupling terms between u, v and w, cross-derivatives appear and we can write

$$\frac{\partial^2 u}{\partial t^2} = k_1\frac{\partial^2 u}{\partial x^2}+k_2\left(\frac{\partial^2 u}{\partial y^2}+\frac{\partial^2 u}{\partial z^2}\right)+k_3\left(\frac{\partial^2 v}{\partial x\,\partial y}+\frac{\partial^2 w}{\partial x\,\partial z}\right). \tag{3.6.14}$$

This equation is derived from elasticity theory in Appendix 3.1 for a cubic crystal. The constants k_1, k_2 and k_3 are there identified with the three independent elastic constants c_{11}, c_{12} and c_{44} of a cubic crystal through the equation

$$k_1 = \frac{c_{11}}{\rho},\quad k_2 = \frac{c_{44}}{\rho},\quad k_3 = \frac{c_{12}+c_{44}}{\rho}. \tag{3.6.15}$$

Equilibrium under central forces also means that certain relations, the Cauchy relations, between these elastic constants are satisfied for crystals having an inversion centre; the general proof of this is somewhat involved (such a proof may be found in Born and Huang, 1954) but for cubic crystals the Cauchy relations reduce to

$$c_{12} = c_{44} \tag{3.6.16}$$

and a relatively simple proof is then possible, which we will now demonstrate.

3.6.1 Elastic waves in (110) direction

The wave equations for the Cartesian displacements u, v, w in a cubic crystal are given by equation (3.6.14) with two other corresponding equations for $\partial^2 v/\partial t^2$ and $\partial^2 w/\partial t^2$. Equation (3.6.14) is valid, of course, only at very long wavelengths, when the solid behaves as an anisotropic continuum. We seek a solution of the form

$$\mathbf{u} = \boldsymbol{\varepsilon}_{\mathbf{k}}^{\sigma} e^{i(\mathbf{k}.\mathbf{r}-\omega t)}. \tag{3.6.17}$$

We already know the polarization directions for propagation in the (110) direction from section 3.5. Substituting the three possibilities for $\mathbf{u} = (u, v, w)$ in equation (3.6.14) and the other two corresponding equations we find that for the longitudinal wave,

$$\omega^2(110) = \frac{1}{2\rho}(c_{11}+c_{12}+2c_{44})\,k^2 \tag{3.6.18}$$

and for the two transverse waves

$$\omega^2(1-10) = \frac{1}{2\rho}(c_{11}-c_{12})\,k^2, \tag{3.6.19}$$

$$\omega^2(001) = \frac{c_{44}\,k^2}{\rho}. \tag{3.6.20}$$

The Cauchy relation (3.6.16) is satisfied if

$$\omega^2(110)-\omega^2(1-10) = 2\omega^2(001) \tag{3.6.21}$$

which we shall prove to be true below.

Let us first find the three frequencies from the dynamical matrix. From equations (3.4.4) and (3.4.5)

$$\omega^2 = \sum_{\mathbf{l}}{}'(1-e^{-i\mathbf{k}.\mathbf{l}})\langle \varepsilon_{\mathbf{k}\sigma}|\,\xi(\mathbf{l})\,|\,\varepsilon_{\mathbf{k}\sigma}\rangle, \tag{3.6.22}$$

where

$$\xi_{xx}(\mathbf{r}) = \frac{x^2}{r^2}\alpha(r)+\beta(r), \quad \xi_{xy}(\mathbf{r}) = \frac{xy}{r^2}\alpha, \quad \text{etc.,} \tag{3.6.23}$$

with

$$\alpha(r) = \frac{\partial^2\phi(r)}{\partial r^2}-\frac{1}{r}\frac{\partial\phi(r)}{\partial r}, \tag{3.6.24}$$

$$\beta(r) = \frac{1}{r}\frac{\partial\phi(r)}{\partial r}. \tag{3.6.25}$$

For very small k, equation (3.6.22) gives

$$\omega^2 = \frac{1}{2}\sum_{\mathbf{l}}{}'(\mathbf{k}.\mathbf{l})^2\langle \varepsilon_{\mathbf{k}\sigma}|\,\xi(\mathbf{l})\,|\,\varepsilon_{\mathbf{k}\sigma}\rangle. \tag{3.6.26}$$

Remembering we are dealing with a cubic crystal with an inversion centre and putting $\mathbf{l} = (x, y, z)$, it is easy to show from (3.6.23) that

$$\omega^2(110) = \frac{k^2}{2} \sum_{x,y,z} (x+y)^2 \xi_{xx} + k^2 \sum_{x,y,z} xy\xi_{xy}. \tag{3.6.27}$$

$$\omega^2(1-10) = \frac{k^2}{2} \sum (x+y)^2 \xi_{xx} - k^2 \sum xy\, \xi_{xy}, \tag{3.6.28}$$

$$\omega^2(001) = \frac{k^2}{2} \sum (x+y)^2 \xi_{zz} = k^2 \sum x^2 \xi_{zz}. \tag{3.6.29}$$

3.6.2 *Relations between elastic constants*

From these equations we find

$$\omega^2(110) - \omega^2(1-10) = 2k^2 \sum_{\mathbf{l}} \frac{x^2 y^2}{r^2} \alpha(\mathbf{l}), \tag{3.6.30}$$

$$\omega^2(001) = k^2 \sum_{\mathbf{l}} \frac{x^2 y^2}{r^2} \alpha(l) + k^2 \sum_{\mathbf{l}} x^2 \beta. \tag{3.6.31}$$

From (3.6.16) we see that the Cauchy relation will be satisfied if

$$\sum_{\mathbf{l}} x_i^2 \beta(\mathbf{l}) = \sum_{\mathbf{l}} x_i^2 \left[\frac{1}{l}\frac{\partial \phi(l)}{\partial l}\right] = \frac{1}{3} \sum_{\mathbf{l}} l^2 \left[\frac{1}{l}\frac{\partial \phi(l)}{\partial l}\right] = 0. \tag{3.6.32}$$

This is just the condition (3.6.21) for equilibrium under central forces, proving that Cauchy relations hold in these circumstances.

The Cauchy relation is *not* satisfied in metals. For copper, for instance, $c_{12} = 11.8 \times 10^{11}$, $c_{44} = 6.1 \times 10^{11}$ dyne/cm^2. We easily verify from what has gone before that

$$c_{12} - c_{44} = -\tfrac{2}{3}\rho \sum_{\mathbf{l}} l \frac{\partial \phi(l)}{\partial l}. \tag{3.6.33}$$

This need *not*, however, stop us analysing dispersion relations in terms of pair forces. According to the theory we shall outline later in this chapter, the total crystal potential energy is composed of pair potential terms plus a volume-dependent term (essentially the energy of a free electron gas in which we immerse our ions). We may say that the total potential energy consists of a sum of pair potentials plus many-body terms, with the many-body terms only depending on the ionic coordinates through the volume (to a good approximation) for simple methods.

This discussion shows how phonon dispersion relations can be calculated from a force law, and how they relate to macroscopic elastic constants in the long wavelength limit.

The next step is to calculate thermodynamic properties of the crystal in terms of these frequencies. In order to do this, it will be helpful at this point to pose the problem of lattice waves in quantum mechanical terms.

3.7 Second quantization and energy spectrum

We start from the lattice Hamiltonian (3.2.21) which displays H as a sum of Hamiltonians of harmonic oscillator form.

We quantize by the usual condition: $Q_{\mathbf{k}\sigma}$ commutes with other variables in H except the one canonically conjugate to it (see Dirac, 1958)

$$\left[\frac{\partial H}{\partial \dot{Q}_{\mathbf{k}\sigma}}, Q_{\mathbf{k}\sigma}\right] = \frac{\hbar}{i}. \tag{3.7.1}$$

Evidently

$$[\dot{Q}^*_{\mathbf{k}'\sigma'}, Q_{\mathbf{k}\sigma}] = \frac{\hbar}{i}\,\delta_{\sigma'\sigma}\,\delta(\mathbf{k}, \mathbf{k}'). \tag{3.7.2}$$

We now second-quantize for harmonic oscillators (cf. Chapter 2, section 2.5).

$$a_{\mathbf{k}\sigma} = (2\omega\hbar)^{-\frac{1}{2}}(\omega Q_{\mathbf{k}\sigma} + i\dot{Q}_{\mathbf{k}\sigma}) \tag{3.7.3}$$

and

$$a^\dagger_{\mathbf{k}\sigma} = (2\omega\hbar)^{-\frac{1}{2}}(\omega Q^*_{\mathbf{k}\sigma} - i\dot{Q}^*_{\mathbf{k}\sigma}), \tag{3.7.4}$$

these equations defining the annihilation operator a_k and the creation operator $a^\dagger_k$ of the phonons.

It is readily verified from equation (3.7.2) that

$$[a_{\mathbf{k}'\sigma'}, a^\dagger_{\mathbf{k}\sigma}] = \delta(\mathbf{k}, \mathbf{k}')\,\delta_{\sigma',\sigma}. \tag{3.7.5}$$

From equations (3.7.3) and (3.7.4) we find

$$\begin{aligned}\sum_{\mathbf{k}\sigma} \hbar\omega(\mathbf{k}\sigma)\, a^\dagger_{\mathbf{k}\sigma} a_{\mathbf{k}\sigma} &= \frac{1}{2}\sum_{\mathbf{k}\sigma}(\omega Q^*_{\mathbf{k}\sigma} - i\dot{Q}^*_{\mathbf{k}\sigma})(\omega Q_{\mathbf{k}\sigma} + i\dot{Q}_{\mathbf{k}\sigma}) \\ &= \frac{1}{2}\sum_{\mathbf{k}\sigma}[\omega^2(\mathbf{k}\sigma)\, Q^*_{\mathbf{k}\sigma} Q_{\mathbf{k}\sigma} + \dot{Q}^*_{\mathbf{k}}\dot{Q}_{\mathbf{k}}] - \sum_{\mathbf{k}\sigma}\frac{i\omega}{2}[\dot{Q}_{\mathbf{k}\sigma}, Q^*_{\mathbf{k}\sigma}] \\ &= H - \sum_{\mathbf{k}\sigma}\tfrac{1}{2}\hbar\omega(\mathbf{k}\sigma),\end{aligned} \tag{3.7.6}$$

and thus we can write the Hamiltonian in the form

$$H = \sum_{\mathbf{k}\sigma}\hbar\omega(\mathbf{k}\sigma)\,(a^\dagger_{\mathbf{k}\sigma} a_{\mathbf{k}\sigma} + \tfrac{1}{2}). \tag{3.7.7}$$

If we denote the ground state, with no phonons, as $|0\rangle$, then an excited state with just one phonon is $a^\dagger_{\mathbf{k}\sigma}|0\rangle$, as, of course, follows mathematically from the Hamiltonian (3.7.7). We can similarly show that a general eigenket representing n_1 phonons in the state $k_1\sigma_1$ and n_2 in the state $k_2\sigma_2$, etc. is

$$(a^\dagger_{\mathbf{k}_1\sigma_1})^{n_1}(a^\dagger_{\mathbf{k}_2\sigma_2})^{n_2}\ldots|0\rangle. \tag{3.7.8}$$

The eigenvalue corresponding to this ket is

$$(n_1+\tfrac{1}{2})\,\hbar\omega(\mathbf{k}_1\sigma_1)+(n_2+\tfrac{1}{2})\,\hbar\omega(\mathbf{k}_2\sigma_2)+\ldots. \tag{3.7.9}$$

It follows in general that $n(k\sigma)$, representing the number of phonons in state $k\sigma$, corresponds to the operator

$$\hat{n}(\mathbf{k}\sigma) = a^\dagger_{\mathbf{k}\sigma}\,a_{\mathbf{k}\sigma}, \tag{3.7.10}$$

referred to as the number operator.

In summary, the above formula reveals the phonons to be quasi-particles corresponding to waves in the crystal similar to the relation between photons and electro-magnetic waves. Like the photons, the phonons are bosons, as the commutator (3.7.5) shows. In the absence of any phonons, the energy is that of the "zero point vibrations", $\sum_{\mathbf{k}\sigma}\frac{1}{2}\hbar\omega(\mathbf{k}\sigma)$. As the temperature is increased from zero, phonons are excited into their energy states $(n+\frac{1}{2})\,\hbar\omega(\mathbf{k}\sigma)$.

3.7.1 Partition function and thermodynamic properties

From equation (3.7.9), we readily obtain the partition function for the dynamical part of the system as

$$Z = \sum_{n(\mathbf{k}\sigma)} \exp\{-\beta E[n(\mathbf{k}\sigma)]\}$$

$$= \prod_{\mathbf{k}\sigma} \frac{\exp[-\frac{1}{2}\beta\hbar\omega(\mathbf{k}\sigma)]}{1-\exp[-\beta\hbar\omega(\mathbf{k}\sigma)]}, \tag{3.7.11}$$

where $\beta = 1/k_B T$, k_B is Boltzmann's constant and T the absolute temperature. Equation (3.7.11) is an immediate consequence of analysing the system into independent harmonic oscillators.

The Helmholtz free energy of the oscillators is

$$F_v = -k_B T \ln Z = k_B T \sum_{\mathbf{k}\sigma} \ln\left\{2\sinh\left[\frac{\hbar\omega(\mathbf{k}\sigma)}{2k_B T}\right]\right\}. \tag{3.7.12}$$

As we saw in section 3.6, the ω's are independent of lattice parameter a, and there is no thermal expansion in the harmonic approximation. The ω's are thus independent of temperature, and the expressions for vibrational internal energy E_v, specific heat C (which is independent of pressure and volume) and entropy S of the crystal become

$$E_v = F_v - T\left(\frac{\partial F_v}{\partial T}\right)_\Omega$$

$$= \sum_{\mathbf{k}\sigma}\left\{\frac{\hbar\omega(\mathbf{k}\sigma)}{2} + \frac{\hbar\omega(\mathbf{k}\sigma)}{\exp[\hbar\omega(\mathbf{k}\sigma)\beta]-1}\right\}, \tag{3.7.13}$$

$$C = \left(\frac{\partial E}{\partial T}\right)_\Omega$$

$$= k_B \sum_{\mathbf{k}\sigma}\left[\frac{\hbar\omega(\mathbf{k}\sigma)}{2k_B T}\right]^2 \Big/ \sinh^2\left[\frac{\hbar\omega(\mathbf{k}\sigma)}{2k_B T}\right], \tag{3.7.14}$$

$$S = -\left(\frac{\partial F}{\partial T}\right)_\Omega$$

$$= k_B \sum_{\mathbf{k}\sigma}\left[\frac{\hbar\omega(\mathbf{k}\sigma)}{2k_B T}\coth\left[\frac{\hbar\omega(\mathbf{k}\sigma)}{2k_B T}\right] - \ln\left\{2\sinh\left[\frac{\hbar\omega(\mathbf{k}\sigma)}{2k_B T}\right]\right\}\right]. \tag{3.7.15}$$

We see that in the harmonic approximation the thermodynamic functions are additive functions of the normal mode frequencies. This has the consequence that all these functions are expressible as averages over the frequency distribution function $g(\omega)$, defined such that $g(\omega)\,d\omega$ is the number of allowable frequencies in the interval $(\omega, \omega+d\omega)$. We obtain

$$F_v = 3rNk_B T \int \ln\left[2\sinh\left(\frac{\hbar\omega}{2k_B T}\right)\right] g(\omega)\,d\omega, \tag{3.7.16}$$

$$E_v = 3rN\frac{\hbar}{2}\int \coth\left(\frac{\hbar\omega}{2k_B T}\right)\omega g(\omega)\,d\omega, \tag{3.7.17}$$

$$C = 3rNk_B \int \left(\frac{\hbar\omega}{2k_B T}\right)^2 \operatorname{csch}^2\left(\frac{\hbar\omega}{2k_B T}\right) g(\omega)\,d\omega \tag{3.7.18}$$

and

$$S = 3rNk_B \int \left[\frac{\hbar\omega}{2k_B T}\coth\left(\frac{\hbar\omega}{2k_B T}\right) - \ln\left(2\sinh\frac{\hbar\omega}{2k_B T}\right)\right] g(\omega)\,d\omega, \tag{3.7.19}$$

giving the thermodynamic properties in terms of the frequency distribution. Clearly, therefore, to evaluate the thermodynamic properties we must have precise information on the frequency spectrum $g(\omega)$ and to this problem we now turn.

3.8 Frequency spectra

The number of allowed frequencies in the range ω to $\omega+d\omega$, namely $3Ng(\omega)\,d\omega$, may be analysed into contributions $g_\sigma(\omega)$ from each branch, through

$$g(\omega) = \sum_\sigma g_\sigma(\omega). \tag{3.8.1}$$

The normalization for each branch is given by

$$\int_0^{\omega_L(\sigma)} g_\sigma(\omega)\,d\omega = \tfrac{1}{3}, \tag{3.8.2}$$

where $\omega_L(\sigma)$ is the maximum frequency of the branch.

Since ω^2 appears as the eigenvalue of the dynamical matrix, it is also useful to define a second function $G(\omega^2)$ such that $G(\omega^2)\,d\omega^2$ is the fraction of allowed squared frequencies in the region between ω^2 and $\omega^2+d\omega^2$. The relationship between the two functions is

$$g(\omega) = 2\omega G(\omega^2). \tag{3.8.3}$$

The obvious way of calculating $g(\omega)$ is the so-called "root sampling method", in which one generates a large number of frequencies by solving the secular equation (3.2.15) at a large number of uniformly distributed points in the BZ, and then approximates the spectrum by a normalized histogram‡ (this cannot reproduce singularities in the spectrum; cf. section 3.8.3 below). This method obviously involves great labour, and since results have been well reviewed elsewhere (see Maradudin, Montroll and Weiss, 1963) we will content ourselves here with discussing first the Debye model spectrum, which we referred to in section 3.1, the results of which we rely on elsewhere in the book, and following that by an account of certain general properties of real frequency spectra.

3.8.1 Debye spectrum

The Debye spectrum of a solid is obtained assuming continuum behaviour throughout the frequency range; because the frequency increases without limit for a continuum [as was seen in (3.1.1)] we cut the spectrum off at a maximum frequency, below which there are $3N$ normal modes (there being N allowed values of $\mathbf{k}$ in the BZ, if N is the number of atoms). The Debye spectrum is not very like any real spectrum, except in its coarse features. However, thermodynamical properties obtained through equations (3.7.16) to (3.7.19) are often quite insensitive to fine detail in $g(\omega)$; further, the specific heat at low temperatures is dominated by the behaviour of $g(\omega)$ for small ω, and we have already remarked that for long wavelengths the discrete lattice

‡ An effective interpolation method of Gilat and Raubenheimer (1966) is now widely used.

behaves as a continuum. In addition the so-called Debye temperature θ_D defined by

$$k_B \theta_D = \hbar\omega_D, \tag{3.8.4}$$

where ω_D is the cut-off frequency referred to above, turns out to play a fundamental role in the physics of lattice vibrations. We shall see later that in fact the Debye temperature can be given a quite precise meaning with regard to the low temperature lattice specific heat.

By studying the standing waves in an elastic continuum, the number of longitudinal modes with frequencies less than ω is just the number of positive integer lattice points (l_1, l_2, l_3) which obey

$$4\pi^2 v_l^2\left(\frac{l_1^2}{L_1^2}+\frac{l_2^2}{L_2^2}+\frac{l_3^2}{L_3^2}\right) \leqslant \omega^2, \tag{3.8.5}$$

where v_l is the velocity of the longitudinal waves, i.e. we want the number of lattice points in an octant of a sphere of radius ω. If ω is very large compared with $2\pi v_l/L_1$, we can say this is just the number of parallelopipeds of volume $2\pi v_l/\mathscr{V}$ ($\mathscr{V} = L_1 L_2 L_3$, the volume of the specimen) within the octant: we have the number

$$\mathscr{N}_l(\omega) = \frac{\pi\mathscr{V}}{6}\left(\frac{\omega}{\pi v_l}\right)^3. \tag{3.8.6}$$

Similarly, there being two transverse modes, with velocity v_t we have

$$\mathscr{N}_t(\omega) = \frac{\pi\mathscr{V}}{3}\left(\frac{\omega}{\pi v_t}\right)^3. \tag{3.8.7}$$

The Debye cut-off frequency ω_D we require is now evidently given by

$$3N = \mathscr{N}_l(\omega_D)+\mathscr{N}_t(\omega_D) \tag{3.8.8}$$

and using equations (3.8.6) and (3.8.7) this becomes

$$\frac{4\pi}{3}\mathscr{V}\left(\frac{2}{v_t^3}+\frac{1}{v_l^3}\right)\left(\frac{\omega_D}{2\pi}\right)^3 = 3N. \tag{3.8.9}$$

It is easy to verify that in terms of this frequency we have

$$\mathscr{N}(\omega) = \mathscr{N}_l(\omega)+\mathscr{N}_t(\omega),$$

$$\mathscr{N}(\omega) = \begin{cases} 3\left(\dfrac{\omega}{\omega_D}\right)^3 N, & 0\leqslant\omega\leqslant\omega_D, \\ 3N, & \omega>\omega_D. \end{cases} \tag{3.8.10}$$

From its definition, the frequency spectrum is given by

$$g(\omega)\,d\omega = \frac{1}{3N}[\mathcal{N}(\omega+d\omega)-\mathcal{N}(\omega)] = \frac{1}{3N}\mathcal{N}'(\omega)\,d\omega. \tag{3.8.11}$$

Thus the Debye approximation for the frequency spectrum of a solid is

$$g(\omega) = \begin{cases} \dfrac{3\omega^2}{\omega_D^3}, & 0 \leqslant \omega \leqslant \omega_D, \\ 0, & \omega > \omega_D. \end{cases} \tag{3.8.12}$$

For the specific heat, we have, using equation (3.7.18),

$$C_v = 3rNk_B\left(\frac{T}{\theta_D}\right)^3 \int_0^{\theta_D/T} \frac{x^4 e^x\,dx}{(e^x-1)^2}, \tag{3.8.13}$$

θ_D being defined, we recall, by equation (3.8.4). For low temperatures, that is $T \ll \theta_D$, the upper limit of the integral may be taken as infinity, and we find

$$\frac{C_v}{3Nk_B} = \frac{4\pi^4 r}{5}\left(\frac{T}{\theta_D}\right)^3. \tag{3.8.14}$$

This relation works well at low temperatures for only modes of low frequency are involved there, and for these the Debye model is correct. However, the inadequacy of this model becomes evident when, at higher temperatures, one attempts to fit the specific heats of all solids to the universal curve which equation (3.8.13) implies. Also equation (3.8.2) is not exactly obeyed.

Nevertheless, the simplicity of the Debye model makes it very attractive. We shall use it, for example, in estimating the effect of lattice vibrations on X-ray scattering by electrons (see Chapter 5).

3.8.2 Moments of frequency spectra

The nth moment of a general frequency spectrum is defined as

$$\mu_n = \int_0^{\omega_L} \omega^n g(\omega)\,d\omega, \tag{3.8.15}$$

where ω_L is the maximum allowed frequency. Given the dynamical matrix, the even moments may be obtained without solution for the ω's. In fact,

$$\mu_{2n} = \frac{1}{3N}\sum_{\mathbf{k}} \chi[D^n(\mathbf{k})], \tag{3.8.16}$$

where χ is the *trace* of the matrix (the sum of its elements down the diagonal).

To prove equation (3.8.16) we first recall that

$$D(\mathbf{k})\,\varepsilon_{\mathbf{k}}^{\sigma} = \omega^2(\mathbf{k},\sigma)\,\varepsilon_{\mathbf{k}}^{\sigma}, \tag{3.8.17}$$

and hence

$$D^n(\mathbf{k})\,\varepsilon_{\mathbf{k}\sigma} = \omega^{2n}(\mathbf{k},\sigma)\,\varepsilon_{\mathbf{k}\sigma}. \tag{3.8.18}$$

Thus $[D^n(k)]$ may be diagonalized by unitary transformation into the form

$$\begin{pmatrix} \omega^{2n}(\mathbf{k}\sigma_1) & 0 & 0 \\ 0 & \omega^{2n}(\mathbf{k}\sigma_2) & 0 \\ 0 & 0 & \omega^{2n}(\mathbf{k}\sigma_3) \end{pmatrix}. \tag{3.8.19}$$

However, as is easily proved, the trace of a matrix is invariant under unitary transformation; thus

$$\chi[D^n(\mathbf{k})] = \sum_{\sigma} \omega^{2n}(\mathbf{k}\sigma) \tag{3.8.20}$$

and

$$\frac{1}{3N}\sum_{\mathbf{k}} \chi[D^n(\mathbf{k})] = \frac{1}{3N}\sum_{\mathbf{k}\sigma} \omega^{2n}(\mathbf{k}\sigma). \tag{3.8.21}$$

From equation (3.8.15) it is evident that the right-hand side of equation (3.8.21) is just the moment μ_{2n}.

Examples of the use of moments to obtain thermodynamic properties are given later. Montroll (1960) has also evolved a method of calculating frequency spectra using equation (3.8.13), the so-called "moment trace method". In this method one expands $g(\omega)$ as a series of Legendre polynomials:

$$g(\omega) = \sum_{n=0}^{\infty} a_{2n}\,P_{2n}\left(\frac{\omega}{\omega_{\mathrm{L}}}\right) \tag{3.8.22}$$

with

$$a_{2n} = \frac{1}{(4n+1)} \int_{-1}^{1} g(\omega_{\mathrm{L}}\,x)\,P_{2n}(x)\,dx. \tag{3.8.23}$$

Only even polynomials are used, since $g(\omega)$ is an even function of ω. Now the even moments of the spectrum are

$$\mu_{2n} = \omega_L^{2n+1} \int_0^1 g(\omega_{\mathrm{L}}\,x)\,x^{2n}\,dx, \tag{3.8.24}$$

so that, in terms of "dimensionless moments" u_{2n},

$$\int_0^1 x^{2n}\,g(\omega_{\mathrm{L}}\,x)\,dx = \frac{u_{2n}}{\omega_{\mathrm{L}}}. \tag{3.8.25}$$

$P_{2n}(x)$ contains even powers of x only, and, by combining equations (3.8.22) and (3.8.24), the symbolic expression for the coefficient a_{2n}:

$$\omega_{\mathrm{L}}\, a_{2n} = \frac{2}{4n+1} P_{2n}(x)\bigg|_{x^{2n}=u_{2n}}, \tag{3.8.26}$$

is obtained. The moment trace method has been studied for one-dimensional and square lattices and for simple-cubic and body-centred lattices by Montroll and others. In these studies the method fails near singularities which we shall see below are present in the frequency spectrum, as is only to be expected. It may, however, be used in conjunction with the methods evolved by Phillips (1956) for dealing with these singularities, which we now discuss.

3.8.3 Singularities in frequency spectra

$G(\omega^2)$ is a function analogous to the density of states $n(E)$ for electrons in band theory (see Chapter 1), and so, among other forms, we can express it as

$$8\pi^3 G(\omega^2) = \frac{\Omega}{3r} \sum_{\sigma} \int_{s} \frac{ds}{|\nabla_{\mathbf{k}}\, \omega^2(\mathbf{k}, \sigma)|} \tag{3.8.27}$$

where the integration is over a constant frequency surface in the BZ, and as usual Ω is the volume of a unit cell of the crystal.

$$8\pi^3 g(\omega) = \frac{2\omega\Omega}{3r} \sum_{\sigma} \int_{s} \frac{ds}{|\nabla_{\mathbf{k}}\, \omega^2(\mathbf{k}, \sigma)|}. \tag{3.8.28}$$

Now one can predict singularities in $g(\omega)$ by methods exactly analogous to the ones we used to predict singularities in $n(E)$, with ω^2 replacing $n(E)$. We may mention an exhaustive analysis made by Phillips (1956) for the fcc lattice.

If, from such an analysis, the structure of the singularities is known, the moment trace method just discussed can be extended to include them, as was done by Lax and Lebowitz (1954). They wrote $G(\omega^2) = G_s(\omega^2) + R(\omega^2)$, where $G_s(\omega^2)$ is the singular part of G; $R(\omega^2)$ is analytic and may be expanded in Legendre polynomials.

A simple variant of the method has been used by Phillips for aluminium. The shape (but not its magnitude) of each singularity in each branch was obtained from topological and symmetry considerations. Phillips left the height of the main peak in each branch undetermined and added enough linear and quadratic terms in the expansion of the spectrum about each singularity to interpolate smoothly between singularities. The parameter in each branch was then determined by normalization. The moments were obtained from the secular matrix derived by Walker (1956). The success of

Phillips' method may be gauged from Figure 3.2, the histogram being obtained by Walker. It would appear that the method is very powerful; but it must be emphasized that the Figure 3.2 does *not* represent the true spectrum of aluminium: the dynamical matrix was constructed using forces of shorter range than are now known to be present. We shall give further information about the range of the forces in aluminium in Chapter 5.

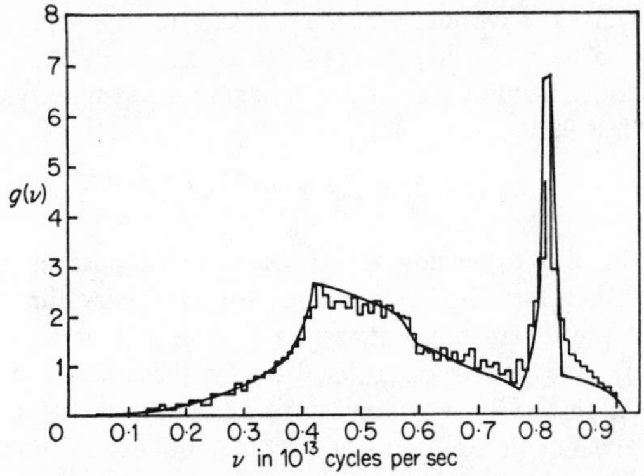

FIGURE 3.2. Short-range force results for a model of aluminium frequency spectrum. The histogram, due to Walker, is directly calculated from dynamical matrix. Full curve, due to Phillips, is obtained by interpolating between singularities predicted on topological grounds. Actual frequency spectrum of Al is different, for the interionic forces in this metal are long-range.

3.8.4 Spectral distribution per solid angle

Suppose we take a particular direction $\hat{\mathbf{k}} = (\theta, \phi)$ in the BZ. Then we can define a distribution function $\gamma_\sigma(\omega)$ for this direction, just as if we were treating a one-dimensional problem. We can write, with r the number of atoms per unit cell,

$$\gamma_\sigma(\omega)\,d\omega = \frac{1}{3r\Omega_{\mathrm{B}}}\left(\frac{dk}{d\omega}\right)^{-1} dk, \tag{3.8.29}$$

the density of states of ω being constant along this line. We can easily see that

$$g_\sigma(\omega) = \int_0^\pi \int_0^{2\pi} k_\sigma^2(\theta, \phi, \omega)\,\gamma_\sigma(\omega) \sin\theta\, d\theta\, d\phi, \tag{3.8.30}$$

where $k_\sigma^2(\omega, \theta, \phi)$ is defined through

$$\omega[\mathbf{k}(\omega, \theta, \phi), \sigma] = \omega. \tag{3.8.31}$$

We define

$$g_\sigma(\omega, \theta, \phi) = \frac{1}{3r\Omega_B} k_\sigma^2(\omega, \theta, \phi) \left[\frac{dk}{d\omega(\theta, \phi)}\right] \tag{3.8.32}$$

to be the spectrum distribution per solid angle for the direction (θ, ϕ). There may be more than one **k** satisfying equation (3.8.31), in which case we sum the right-hand side of equation (3.8.32) over the solutions.

Since $\omega(\mathbf{k}, \sigma)$ has the symmetry of the lattice, $g_\sigma(\omega, \theta, \phi)$ also has. We may therefore expand in lattice harmonics $L_m(\theta, \phi), L_m$ being the linear combination of spherical harmonics of order m having the symmetry of the lattice (cf. Table 1.6 in Chapter 1)

$$g_\sigma(\omega, \theta, \phi) = \sum_m a_m(\omega, \sigma) L_m(\theta, \phi). \tag{3.8.33}$$

The existence of this expansion led Houston to propose an interpolation scheme. One takes the dispersion curves for symmetry directions, either obtained from the dynamical matrix (see section 3.2) or from experiment (cf. Chapter 5), and constructs $g_\sigma(\omega, \theta, \phi)$ for these directions. One now retains in equation (3.8.33) as many terms as the symmetry directions; the a_m's are then given as the solutions of a set of simultaneous linear equations (see Maradudin *et al*, 1963).

The integral of $g_\sigma(\omega, \theta, \phi)$ over θ and ϕ equation (3.8.33) is just

$$g_\sigma(\omega) = a_0(\omega, \sigma), \tag{3.8.34}$$

because of the orthogonality of lattice harmonics. Houston's method may obviously be used to evaluate any integration over angles if the integrand has the symmetry of the lattice. It must be emphasized, however, that Houston's method is not suitable for determining the entire frequency spectrum. Symmetry directions lead to points giving rise to singularities, and in one dimension the singularities in frequency spectra are infinities [see van Hove (1954) and Problem 3.3]. $g_\sigma(\omega, \theta, \phi)$ has such infinities in the symmetry directions, and if one obtains $g_\sigma(\omega)$ from equation (3.8.34), with the a_m's obtained by Houston's method, these infinities remain. One may mention, however, attempts by Nakamura and Hwang (1955) to remove this defect in the method.

3.9 Calculation of specific heat

The presence of the exponential terms in the integrands of the expressions (3.7.12) to (3.7.14) for thermodynamic functions means that, roughly speaking, quantities at low temperatures depend on $g(\omega)$ at low ω, and quantities at high temperatures depend on the high ω end of the spectrum. This means that if we work with $g(\omega)$ directly, low temperatures offer the least difficulty, the troublesome singularities in $g(\omega)$ which we have discussed earlier not being present at small ω. However, because of these singularities, we are led to seek methods of calculating quantities in terms of the moments of the spectrum if we wish to obtain results for higher temperatures.

3.9.1 Low temperatures and Debye θ

Since the Debye theory always applies in the long wavelength limit, $g(\omega)$ is quadratic in ω for small ω, and more generally we may write

$$g(\omega) = b_2\omega^2 + b_4\omega^4 + \ldots. \tag{3.9.1}$$

Allowing the limit of the integrand in equation (3.7.18) to tend to infinity, which is permissible at sufficiently low temperatures, we obtain, with r atoms per unit cell, the asymptotic expansion

$$C(T) = 3Nk_B r\left[\frac{4\pi^4}{15}\left(\frac{k_B T}{\hbar}\right)^3 b_2 + \frac{16\pi^6}{21}\left(\frac{k_B T}{\hbar}\right)^5 b_4 + \ldots\right]. \tag{3.9.2}$$

It is customary to write the first term of this in a form exactly like that of equation (3.8.14), namely

$$C = \frac{12\pi^4 rNk_B}{5}\left(\frac{T}{\theta_D}\right)^3, \tag{3.9.3}$$

θ_D being termed the *Debye temperature* [cf. equation (3.8.4)]. We can easily see that

$$\frac{1}{\theta_D^3} = \frac{k_B^3}{3\hbar^3} b_2. \tag{3.9.4}$$

(a) *Expression for Debye temperature in terms of moments.* From (3.8.22) and (3.9.4) it is easy to write down an expression for θ_D^3 in terms of moments. We merely pick out the coefficients of ω^2 on the right-hand side of (3.8.22). Writing the Legendre polynomial as

$$P_n(x) = \sum_{\nu=0}^{n} p_n^\nu x^\nu, \tag{3.9.5}$$

we have

$$b_2 = \frac{1}{\omega_L^3} 2 \sum_{n=0}^{\infty} \sum_{\nu=0}^{n} u_{2\nu} p_{2n}^{2\nu} p_{2n}^2 (4n+1)^{-1}$$

and so

$$\frac{1}{\theta_D^3} = 2\left(\frac{k_B}{\hbar\omega_L}\right)^3 \sum_{n=0}^{\infty} \sum_{\nu=0}^{n} u_{2\nu} p_{2n}^{2\nu} p_{2n}^2 (4n+1)^{-1}. \tag{3.9.6}$$

This series may not be very suitable for determining the Debye θ; we can only remark that convergence is unlikely to be rapid, the moments $u_{2\nu}$ being insensitive to the low end of $g(\omega)$, which determines θ_D.

(b) *Debye temperature evaluated from frequency spectrum.* We shall now see that, in contrast to what might appear at first sight from (3.9.6), b_2 is determined completely by knowledge of the elastic constants.

The Debye temperature may be evaluated by the same method as we used for the Debye spectrum, but allowing for the anisotropy of the medium. For low values of ω, the correct expansion in the direction $\mathbf{k} \equiv (k, \theta, \phi)$ is

$$\omega^2(\mathbf{k}, \sigma) = c_\sigma^2(\theta, \phi)k^2 + d_\sigma^2(\theta, \phi)k^4 + \ldots, \tag{3.9.7}$$

where $c_\sigma(\theta, \phi)$ is just the velocity of sound for the σ-mode. Inversion of (3.8.8) yields for the distribution function per unit solid angle

$$g_\sigma(\omega, \theta, \phi) = \frac{1}{3r\Omega_B}\left[\frac{\omega^2}{c_\sigma^7(\theta, \phi)} - \frac{5}{2}\frac{d_\sigma^2(\theta, \phi)}{c_\sigma(\theta, \phi)}\omega^4 + \ldots\right]. \tag{3.9.8}$$

Thus, from equations (3.8.30) and (3.8.32), the coefficients in the expansion (3.9.1) are

$$b_2 = \frac{1}{3r\Omega_B}\sum_{\sigma=1}^{3}\int_0^{2\pi} d\phi \int_0^{\pi} \sin\theta \, d\theta \, c_\sigma^{-3}(\theta, \phi) \tag{3.9.9}$$

and

$$b_4 = \frac{-5}{6r\Omega_B}\sum_{\sigma}\int_0^{2\pi} d\phi \int_0^{\pi} \sin\theta \, d\theta \, c_\sigma^{-7}(\theta, \phi) d_\sigma^2(\theta, \phi). \tag{3.9.10}$$

From equation (3.9.4) we obtain

$$\frac{1}{\theta_D^3} = \frac{1}{9r\Omega_B}\left(\frac{k_B}{\hbar}\right)^3 \sum_{\sigma}\int_0^{2\pi} d\phi \int_0^{\pi} \sin\theta \, d\theta \, c_\sigma^{-3}(\theta, \phi) \tag{3.9.11}$$

expressing the Debye temperature in terms of the velocities of sound. The propagation velocities $c_\sigma(\theta, \phi)$ are expressible in terms of the elastic constants. This fact has been used by Betts, Bhatia and Wyman (1956) to obtain θ_D for nine cubic crystals; in this work Houston's method [see section (3.8.4)] was used. $c_\sigma(\theta, \phi)$ was expressed in lattice harmonics and their coefficients approximately found from the values of c_σ in terms of the elastic constants along symmetry directions of the crystal. More accurate work is that by de Launay (1956). The dynamical matrix is simplest at small $\mathbf{k}$, being expressible in terms of the elastic constants, as we have seen, and de Launay solved for $c_\sigma(\theta, \phi)$ for a large number of directions in the BZ and for different elastic constants. He then integrated (3.9.9) by direct numerical attack, from his results preparing a table of an auxiliary function $f(s,t)$ $[s = (c_{11}-c_{44})/(c_{12}+c_{44}), t = (c_{12}-c_{44})/c_{44}]$ in terms of which we may write

$$\theta_{\mathrm{D}}^3 = \frac{9}{4\pi\Omega}\left(\frac{h}{k_{\mathrm{B}}}\right)^3 \left(\frac{c_{44}}{\rho}\right)^{\frac{3}{2}} \frac{9}{18+\sqrt{3}} f(s,t). \tag{3.9.12}$$

De Launay has developed an interpolation scheme for use with his table, and hence θ_{D} is readily found from experimentally known elastic constants, giving a check on the low temperature specific heat.

The coefficient b_4 in equation (3.9.1) (given by 3.9.10) has not been widely studied since, unlike b_2, or, equivalently, θ_{D}, its evaluation requires an assumption as to the nature of the forces in the crystal as well as knowledge of the elastic constants.

Some calculations are available, however. Marcus has investigated the temperature dependence of the Debye θ_{D} [regarding equation (3.8.14) as an exact form]. He writes

$$\theta_{\mathrm{D}}(T) = \theta_{\mathrm{D}}(0)\left[1 - c(r_1, r_2)\left(\frac{T}{\theta_{\mathrm{D}}(0)}\right)^2 + \ldots\right], \tag{3.9.13}$$

where $r_1 = (c_{11}-c_{12})/2c_{11}$ and $r_2 = c_{44}/c_{11}$. $c(r_1, r_2)$ is given by

$$c(r_1, r_2) = \left(\frac{20\pi^2}{21}\right)\left[\frac{k_{\mathrm{B}}\,\theta_{\mathrm{D}}(0)}{\hbar}\right]^2 \frac{b_4}{b_2} \tag{3.9.14}$$

and Marcus (see Maradudin, Montroll and Weiss, 1963) has calculated this for general nearest-neighbour interactions in an fcc lattice.

3.9.2 *Calculation at high temperatures*

From Problem 3.3, the specific heat C_{v} is given in terms of the moments of the frequency spectrum $g(\omega)$.

We mention here an alternative procedure which does not work directly from the frequency spectrum. We return to equation (3.7.14), which we can rewrite in the form

$$C_{\mathrm{v}} = k_{\mathrm{B}} \sum_{\mathbf{k}\sigma} E[\beta\hbar\omega(\mathbf{k}\sigma)], \tag{3.9.15}$$

where

$$E(x) = \frac{x^2 \mathrm{e}^x}{(\mathrm{e}^x - 1)^2}. \tag{3.9.16}$$

Replacing the summation by integration in the usual way, (3.9.15) becomes

$$C_{\mathrm{v}} = \frac{N k_{\mathrm{B}}}{\Omega_{\mathrm{B}}} \sum_{\sigma} \int_{\Omega_{\mathrm{B}}} E[\beta\hbar\omega(\mathbf{k}, \sigma)]\, d\mathbf{k}. \tag{3.9.17}$$

This expression has been employed by Horton and Schiff (1956) in conjunction with measured dispersion curves for $\omega(\mathbf{k}\sigma)$. They replaced the BZ by the equal volume sphere, and used Houston's method (section 3.8.4) to evaluate the integral.

They corrected their result in two ways: (i) by an estimate of the error incurred by replacing the BZ by a sphere—the correction to C_v was about 1%; (ii) they simply multiplied their resulting expression by a factor which ensures their expression gives the correct result $C_v = 3R$ at very high temperatures.

We ought to remark that the Debye T^3 law is fundamentally related to the three-dimensional nature of real crystals. In a layer structure, like graphite, de Sorbo and Tyler (1953) and others have explained the observed T^2 law for the specific heat in terms of a two-dimensional model, the binding between layers being quite weak.

3.10 Electron–ion interaction and phonon frequencies in metals

Apart from the discussion of the Bohm–Staver formula for the velocity of sound in metals in section 3.1.1, much of the theory so far described in this chapter is quite general, applying to any type of crystal.

However, to turn the theory into specific calculations for particular crystals, it is very important to classify the nature of the force law of interaction between the vibrating atoms. The Bohm–Staver formula represents an attempt to do this, but is valid only in the long wavelength limit and only for simple metals.

In this section, we shall consider how the concepts of electron theory developed in Chapter 1 will enable us to make very specific calculations of phonon dispersion relations in simple metals. The approach adopted here was pioneered by Toya (1958), and developed especially by Cochran (1963) and Sham (1965). Very many subsequent papers have added important detailed knowledge of phonon spectra. We shall follow here the discussion of Vosko, Taylor and Keech (1965) quite closely. Essentially, we shall generalize the Bohm–Staver theory in a fully quantitative way, removing at the same time its restriction to the long wavelength limit.

We shall deal with the electronic contribution to the lattice dynamics of semi-conductors and insulators after developing the theory for metals.

3.10.1 Adiabatic approximation

In section 3.2 we treated the vibrations of ions composing a lattice as a general problem in dynamics; we assumed there exists a many-body potential Φ governing the ionic motions, but the physical origins of this potential were not discussed except indirectly in the Bohm–Staver theory for a metal, which recognizes that a metal consists of ions plus valence electrons. Our assumption then is tantamount to the supposition that the dynamical aspects of ion and electron motions may be decoupled. This is the *Born–Oppenheimer* or adiabatic approximation. For its mathematical discussion we refer the reader to Appendix 3.3. Strictly, such an approach should treat insulators and metals separately. We confine ourselves here in the main text to physical arguments.

It is easy to visualize the electronic charge distribution of the ion core moving with the nucleus as it vibrates. We have already tacitly assumed this picture to be correct, as our previous language shows, we have talked about the vibrations of the *ions*. The behaviour of the valence electrons cannot be so simple, of course, but, nevertheless, we might expect something analogous to happen. For example, when a longitudinal wave progresses through the lattice, causing local rarefactions and compressions in the positive-ion density, we may suppose the electrons move so as to screen out these fluctuations; that is, we expect the response of the valence electrons to be such that the system is kept locally electrically neutral, and this is equivalent to the Bohm–Staver argument. The maximum lattice frequency is of the order of 10^{13} $\sec^{-1}$, quite small compared with the plasma frequency, of the order of 10^{16} $\sec^{-1}$. Hence if the valence electrons are able to respond to perturbations in times of the order of such plasma times they will effectively be following the motion of the lattice instantaneously at all frequencies of vibration. This is the essence of the adiabatic approximation. To calculate Φ as defined in section 3.2 we freeze the ions in each instantaneous configuration and calculate the energy of the electrons. Φ is this energy plus the electrostatic energy of the nuclear–nuclear interaction.

One of the oldest theories is that of "rigid ions" in which Φ is supposed to be a sum of central pair forces. In this theory one imagines each ion as it moves about in a solid (or liquid) to rigidly take with it a charge distribution of valence electrons such as we saw envisaged in the pseudopotential approach discussed in Chapter 1. In principle these inter-ionic forces are calculated in a purely electro-static manner, and one then calculates phonon modes in the way discussed in the previous part of this chapter. Though the rigid-ion model is very useful it is, of course, an approximation, and we shall see later that it is necessary to transcend it in some cases and introduce many-body forces.

At the present time, however, the charge distributions can hardly be said to be calculable even from the "rigid-ion" model, unless the pseudopotential is workable, and in fact the "rigid-ion" or "neutral pseudoatom" picture is often used as no more than a guide in extracting pair forces from experiment.

It seems probable that to suppose Φ to be a sum of pair potentials is inadequate for a really accurate description of lattice vibrations but at least it gives us many of the prominent features found from experiment (cf. Chapter 5).

3.10.2 *Effect of electron gas on phonon frequencies*

Using the electron theory developed in Chapter 1, we shall now investigate the first-principles calculation of the dynamical matrix of section 3.2, with a view to calculating the phonon dispersion relations of simple metals.

Neglecting fine details, the essential result of the presence of the electron gas in the metal solid is a renormalization of the longitudinal frequencies of vibration of a set of bare ions as we have already seen in section 3.1.1. Essentially the electron gas in a metal is unresistant to shear, and so, to a good first approximation, the frequencies of the transverse vibrations are just those of a set of bare ions with the electron gas absent. The presence of the gas is all important for the longitudinal vibrations, however.

We retain only the essential features of the problem, leaving a more detailed description for later sections. In particular we shall assume a homogeneous electron gas.

From equation (3.2.10) we obtain the expression for the normal mode

$$Q_{\mathbf{k}}^{\alpha} = \sum u_{\mathbf{l}}^{\alpha} e^{i\mathbf{k}.\mathbf{l}} \tag{3.10.1}$$

where $u_{\mathbf{l}}$ is the departure of the vector $\mathbf{l}$ from equilibrium. We here suppose all motions to be longitudinal in some particular direction, and have absorbed the polarization vector in $\mathbf{Q}$.

From section 3.2, to obtain energy changes we must go to second order in the u's. Let $V(\mathbf{r}-\mathbf{l}-\mathbf{u})$ be the potential due to the lth ion. From the theorem

$$\frac{\partial E}{\partial \lambda} = \langle \psi | \frac{\partial H}{\partial \lambda} | \psi \rangle \tag{3.10.2}$$

(Appendix A1.4) we obtain

$$\delta^{(1)} E = \sum_{\mathbf{l}} \int \rho(\mathbf{r}+\mathbf{l}) \nabla V(\mathbf{r}) . \mathbf{u}_{\mathbf{l}} \, d\mathbf{r} \tag{3.10.3}$$

and

$$\delta^{(2)} E = \sum_{\mathbf{l}} \int \delta\rho(\mathbf{r}+\mathbf{l}) \nabla V(\mathbf{r}) . \mathbf{u}_{\mathbf{l}} \, d\mathbf{r} + \sum_{l\alpha} \int \rho(\mathbf{r}+\mathbf{l}) \frac{\partial^2 V(\mathbf{r})}{\partial x_{\alpha}^2} (u_{\mathbf{l}}^{\alpha})^2 \, d\mathbf{r}. \tag{3.10.4}$$

The term involving second derivatives of V may be ignored when the condition (3.2.8) is implied; $\delta\rho$ in equation (3.10.4) is, of course, the density change caused by displacing the ions. This expression (3.10.4), in passing, is essentially the electron–phonon part of the Hamiltonian (cf. Chapter 6). Thus the electronic energy change on creation of the mode $Q_{\mathbf{q}}$ is

$$\delta^{(2)} E = M(\mathbf{q})\, \rho_{-\mathbf{q}}\, Q_{\mathbf{q}}, \tag{3.10.5}$$

where $\rho_{\mathbf{q}}$ is the Fourier transform of the density change $\delta\rho$ and $M(\mathbf{q})$ is defined by

$$M(\mathbf{q}) = \left(\frac{\rho_0}{M}\right)^{\frac{1}{2}} \int \nabla V(\mathbf{r})\, e^{i\mathbf{q}\cdot\mathbf{r}}\, d\mathbf{r}, \tag{3.10.6}$$

which is, in fact, the electron–ion matrix element for plane-waves.

The equation for $Q_{\mathbf{q}}$ in the absence of the electron gas is

$$\ddot{Q}_{\mathbf{q}} + \omega_{\mathrm{p}}^2\, Q_{\mathbf{q}} = 0, \tag{3.10.7}$$

where ω_{p} is the plasma frequency of the ions given in equation (3.1.4) and the additional term required when the electron gas is present is obtained by differentiating equation (3.10.5) with respect to Q.‡ We obtain then

$$\ddot{Q}_{\mathbf{q}} + \omega_q^2\, Q_{\mathbf{q}} = -M(-\mathbf{q})\, \rho_{\mathbf{q}}. \tag{3.10.8}$$

To calculate $M(q)$ we assume point ions; the integral (3.10.6) is then easily performed, and we obtain

$$\lim_{R\to\infty} \frac{4\pi}{q}\left(1 - \frac{\sin qR}{qR}\right) = \frac{4\pi}{q}. \tag{3.10.9}$$

We use linear response theory to obtain $\rho_{\mathbf{q}}$ and we find, for valency $Z = 1$,

$$\rho_{\mathbf{q}} = \frac{1-\varepsilon(\mathbf{q})}{\varepsilon(\mathbf{q})}\, \rho_{\mathbf{q}}^{i}. \tag{3.10.10}$$

Now the Fourier transform of the ionic density change $\rho_{\mathbf{q}}^{i}$ is, from equation (3.10.1),

$$\rho_{\mathbf{q}}^{i} = -i(\rho_0/M)^{\frac{1}{2}}\, q Q_{\mathbf{q}} \tag{3.10.11}$$

and so equation (3.10.8) becomes

$$\ddot{Q}_{\mathbf{q}} + \left\{\omega_q^2 + i\left(\frac{\rho_0}{M}\right)^{\frac{1}{2}} q \left[\frac{1-\varepsilon(\mathbf{q})}{\varepsilon(\mathbf{q})}\right] V_{-q}\right\} Q_{\mathbf{q}} = 0. \tag{3.10.12}$$

We now take the $q \to 0$ limit, and use the Thomas–Fermi approximation (2.3.21) for $\varepsilon(q)$. We then find

$$\ddot{Q}_{\mathbf{q}} + \left(\omega_q^2 - \frac{4\pi e^2 \rho_0}{M} \frac{q_{\mathrm{TF}}^2}{q^2 + q_{\mathrm{TF}}^2}\right) Q_{\mathbf{q}} = 0; \quad q_{\mathrm{TF}}^2 = \frac{4k_{\mathrm{f}}}{\pi a_0} \tag{3.10.13}$$

‡ It is not difficult to show that we must assume the force to be given by MQ_q, where M is the ionic mass.

i.e. as $q \to 0$

$$\ddot{Q}_{\mathbf{q}} + v_s^2 q^2 Q_{\mathbf{q}} = 0 \tag{3.10.14}$$

where

$$v_s^2 = \frac{m}{3M} v_f^2 \tag{3.10.15}$$

which is the Bohm–Staver formula for $Z = 1$ already obtained in section 3.1.1 by a less careful argument.

3.10.3 Second order energy and dynamical matrix

Improvements on the above immediately suggest themselves: more precise treatment of the ion–ion interaction—more realistic choice of electron–ion potential—in a Hartree framework an improvement in the treatment of screening—extension to include electron-correlation effects. It is the latter two aspects of the problem we shall concentrate on here. Of the other matters, the core–core overlap can be taken as of the Born–Mayer type, and the choice of electron–ion potential has already been discussed at length. Below we seek expressions for $D^{e}_{\alpha\beta}(\mathbf{k})$, the electronic contribution to the dynamical matrix by which the phonon frequencies are determined (equation 3.2.15). We note from equations (3.2.6), (3.2.14) and (3.10.4) that a formal expression can immediately be given:

$$D^{e}_{\alpha\beta}(\mathbf{q}) = -\sum_{l} \left\langle \frac{\partial \rho(r)}{\partial l_\alpha} \frac{\partial V(r)}{\partial x_\beta} \right\rangle e^{i\mathbf{q}.\mathbf{l}}, \tag{3.10.16}$$

where we have used the relation

$$\left\langle \frac{\partial \rho}{\partial l_\alpha} \frac{\partial V(r-l')}{\partial l'_\beta} \right\rangle = -\left\langle \frac{\partial \rho(\mathbf{r})}{\partial (l_\alpha - l'_\alpha)} \frac{\partial V(r)}{\partial x_\beta} \right\rangle. \tag{3.10.17}$$

(a) *One-body treatment*. To calculate $\partial\rho/\partial l_\alpha$ we now use the results that if

$$H_0 \psi_m = E_m \psi_m,$$

$$(H_0 + H_1)\psi' = E\psi'; \psi' = \psi_{\text{pert}} \tag{3.10.18}$$

then the change in the first-order density matrix $\rho(\mathbf{r}'\mathbf{r})$ is

$$\delta\rho(\mathbf{r}'\mathbf{r}) = \sum_{\mathbf{kq}} \frac{f(\mathbf{k}+\mathbf{q}) - f(\mathbf{k})}{E(\mathbf{k}+\mathbf{q}) - E(\mathbf{k})} \langle \psi^*_{\mathbf{k}} | H_1 | \psi_{\mathbf{k}+\mathbf{q}} \rangle \psi^*_{\mathbf{k}+\mathbf{q}}(\mathbf{r}') \psi_{\mathbf{k}}(\mathbf{r}), \tag{3.10.19}$$

where $\psi_{\mathbf{k}}$ are the Bloch orbitals. From equation (3.10.4) the second-order energy change is then easily shown to be

$$2\delta E^{(2)} = \langle \delta\rho \delta V \rangle = \sum_{\mathbf{kq}} \frac{f(\mathbf{k}+\mathbf{q}) - f(\mathbf{k})}{E(\mathbf{k}+\mathbf{q}) - E(\mathbf{k})} \langle \psi_{\mathbf{k}} | H_1 | \psi_{\mathbf{k}+\mathbf{q}} \rangle \langle \psi^*_{\mathbf{k}+\mathbf{q}} | V_1 | \psi_{\mathbf{q}} \rangle \tag{3.10.20}$$

where now H_1 is taken as the change in the *one-body Hamiltonian H* when the ionic potential changes by V_1. We should note that equation (3.10.20) is exact if H, with eigenfunctions $\psi_{\mathbf{k}}$, is the Hamiltonian including the one-body potential reproducing the exact density, and H_1 is the change in this one-body potential (see section 1.2). Writing

$$H_1 = u_\alpha \frac{\partial H}{\partial l_\alpha} \quad \text{and} \quad V_1 = \frac{\partial V}{\partial x_\beta} u_\beta \tag{3.10.21}$$

and using the periodicity of H, we find from equations (3.10.16) and (3.10.20)

$$D^{\mathrm{e}}_{\alpha\beta}(\mathbf{q}) = -N \sum_{\mathbf{k}} \frac{[f(\mathbf{k})-f(\mathbf{k}+\mathbf{q})]}{E(\mathbf{k})-E(\mathbf{k}+\mathbf{q})} \langle \psi_{\mathbf{k}+\mathbf{q}} | \frac{\partial H}{\partial l_\alpha} | \psi_{\mathbf{k}} \rangle \langle \psi_{\mathbf{k}} | \frac{\partial V}{\partial x_\beta} | \psi_{\mathbf{k}+\mathbf{q}} \rangle \tag{3.10.22a}$$

(b) *Kohn anomaly*. Inspection of the above equation leads us to expect that large contributions to $D^{\mathrm{e}}_{\alpha\beta}(\mathbf{q})$ will come when $E(\mathbf{k}) \simeq E(\mathbf{k}+\mathbf{q})$. However, the factor $f(\mathbf{k})-f(\mathbf{k}+\mathbf{q})$ in the numerator means then that we are restricted to a region very near the Fermi surface: that is, both $\mathbf{k}$ and $\mathbf{k}+\mathbf{q}$ are almost on the Fermi surface. When q exceeds the maximum width of the Fermi surface, parallel to $\mathbf{q}$, this condition can no longer be fulfilled and we expect $D^{\mathrm{e}}_{\alpha\beta}(\mathbf{q})$ to suffer a discontinuity in its derivative. This will in turn produce a cusp‡ in the phonon dispersion relation at q_m, where q_m is the maximum width of the Fermi surface in a direction parallel to $\mathbf{q}$. This is known as the *Kohn anomaly*. To make the argument quite specific, let us take $\psi_{\mathbf{k}}$ as a single plane-wave, or (if we wish) an OPW or APW. Then the matrix elements in equation (3.10.21) are independent of $\mathbf{k}$, and we find

$$D^{\mathrm{e}}_{\alpha\beta}(\mathbf{q}) = N \langle \psi_{\mathbf{k}+\mathbf{q}} | \frac{\partial H}{\partial l_\alpha} | \psi_{\mathbf{k}} \rangle \langle \psi_{\mathbf{k}} | \frac{\partial V}{\partial x_\beta} | \psi_{\mathbf{k}+\mathbf{q}} \rangle \Pi_0(\mathbf{q}) \tag{3.10.22b}$$

where, using the Lindhard expression (2.3.20) for the dielectric constant,

$$\Pi_0(\mathbf{q}) = \frac{q^2}{4\pi e^2} [\varepsilon(\mathbf{q})-1]. \tag{3.10.23}$$

The cusp in $\omega(\mathbf{q})$ predicted above is evidently related to the "kink" in $\varepsilon(\mathbf{q})$ at $|\mathbf{q}| = 2k_{\mathrm{f}}$.

(c) *Many-body treatment*. To discuss the effect of correlations we seek some way in which to utilize the perturbation technique discussed in Chapter 2, suitably modified so that the external perturbing potential is due to departures of the ions from equilibrium, rather than the total ionic potential itself. We therefore follow Vosko, Taylor and Keech (1965) and write the

‡ More precisely, the form of the singularity depends on the dimensionality of the opposite pieces of the Fermi surface (spherical, cylindrical or parallel planes).

Hamiltonian, in second quantized form, in terms of Bloch functions which are solutions of the Schrödinger equation

$$\frac{-\hbar^2}{2m}\nabla^2\psi_{\mathbf{k}}+V_{\mathrm{H}}(\mathbf{r})\psi_{\mathbf{k}} = E(\mathbf{k})\psi_{\mathbf{k}}. \tag{3.10.24}$$

Here $V_{\mathrm{H}}(\mathbf{r})$ is the Hartree potential

$$V_{\mathrm{H}}(\mathbf{r}) = V_b(\mathbf{r})+V_e(\mathbf{r}), \tag{3.10.25}$$

where

$$V_e(\mathbf{r}) = e^2\int\frac{\rho(\mathbf{r}')}{|\mathbf{r}-\mathbf{r}'|}\,d\mathbf{r}' \tag{3.10.26}$$

and $V_b(\mathbf{r})$ is the potential of the bare ions. Our reasons for choosing the Hartree equations to define the one-particle Bloch states will become apparent a little later. The Hamiltonian including perturbation due to ionic displacements is

$$H = -\frac{\hbar^2}{2m}\nabla^2+\sum_i V_b(\mathbf{r}_i)+\sum_{i<j}\frac{e^2}{r_{ij}}+\sum_i \mathbf{u}_i.\frac{\partial V_b}{\partial \mathbf{l}_i}, \tag{3.10.27}$$

now becomes, in second quantized form,

$$H = \sum_{\mathbf{k}} E(\mathbf{k})\,a_{\mathbf{k}}^\dagger a_{\mathbf{k}}+\sum_{\mathbf{k}_1\mathbf{k}_2} V_b'(\mathbf{k}_1,\mathbf{k}_2)\,a_{\mathbf{k}_1}^\dagger a_{\mathbf{k}_2}+\frac{1}{2}\sum_{\substack{\mathbf{k}_1\mathbf{k}_2\\ \mathbf{k}_3\mathbf{k}_4}} V_e(\mathbf{k}_1\,\mathbf{k}_2\,\mathbf{k}_3\,\mathbf{k}_4)\,a_{\mathbf{k}_1}^\dagger a_{\mathbf{k}_2}^\dagger a_{\mathbf{k}_3} a_{\mathbf{k}_4}, \tag{3.10.28}$$

where

$$V_b'(\mathbf{k}_1\,\mathbf{k}_2) = -\sum_i\int d\mathbf{r}\,\mathbf{u}.\frac{\partial V_b(\mathbf{r}-\mathbf{l})}{\partial \mathbf{r}}\,\psi_{\mathbf{k}_1}^*(\mathbf{r})\,\psi_{\mathbf{k}_2}(\mathbf{r}) \tag{3.10.29}$$

and

$$V_e(\mathbf{k}_1\,\mathbf{k}_2\,\mathbf{k}_3\,\mathbf{k}_4) = e^2\int d\mathbf{r}_1,d\mathbf{r}_2\,\frac{\psi_{\mathbf{k}_1}^*(\mathbf{r}_1)\,\psi_{\mathbf{k}_2}^*(\mathbf{r}_2)\,\psi_{\mathbf{k}_3}(\mathbf{r}_1)\,\psi_{\mathbf{k}_4}(\mathbf{r}_2)}{r_{12}}. \tag{3.10.30}$$

$a_{\mathbf{k}}^\dagger$ and $a_{\mathbf{k}}$ are creation and annihilation operators for electrons in the Bloch states $\psi_{\mathbf{k}}$ (*not*, in general, plane-wave states). There is an obvious omission in equation (3.10.28), that of the term corresponding to $V_e(\mathbf{r})$ defined in equation (3.10.26). We must here bear this in mind: its effect will be considered a little later.

Now if we use diagrammatic techniques, and represent the interaction with the external potential $V_b'(\mathbf{k}_1\,\mathbf{k}_2)$ by a dashed line as in the last chapter, this line must enter twice in any diagram we need to consider, since it is the energy change to second order in the $\mathbf{u}_i$'s we wish to calculate. Ignoring the electron terms in equation (3.10.27), therefore, there is only one diagram: that shown in Figure 3.3. Its contribution is calculated by the rules given in Chapter 2, where the Fermion lines are associated with the propagator

$$G(\mathbf{k}, \omega) = \frac{1}{\omega - E(\mathbf{k}) + i\delta_{\mathbf{k}}}. \tag{3.10.31}$$

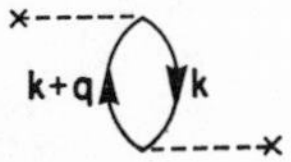

FIGURE 3.3. Second-order contribution to energy in absence of interactions.

Here $\delta_k \to \pm 0$ according as $\mathbf{k}$ is outside or inside the *true Fermi surface* (not, in general, a sphere) and

$$\Delta E_1^{(2)} = \frac{2}{2!} \sum_{\mathbf{kq}} V'(\mathbf{k}, \mathbf{k}+\mathbf{q})\, V_b'(\mathbf{k}+\mathbf{q}, \mathbf{k}) \int \frac{d\omega}{2\pi i}\, G(\mathbf{k}, \omega)\, G(\mathbf{k}+\mathbf{q}, \omega). \tag{3.10.32}$$

Here the factor 2 in the numerator takes account of spin. We shall suppose equation (3.10.31) describes the valence electrons only, for we are not interested in the lower-lying energy bands. We can separate off the part of the propagator corresponding to these latter states and forget about it. This, of course, corresponds to the assumption that the valence electrons have little influence on the ion-core distributions, and the entire influence of these cores is contained in the definition of the one-particle states, e.g. equation (3.10.24).

We now also note that Bloch's theorem implies that

$$V(\mathbf{k}_1 \mathbf{k}_2 \mathbf{k}_3 \mathbf{k}_4) = 0 \quad \text{unless} \quad \begin{cases} \mathbf{k}_1 = \mathbf{k}_4 + \mathbf{q} + \mathbf{K}, \\ \mathbf{k}_2 = \mathbf{k}_3 - \mathbf{q} + \mathbf{K}'. \end{cases} \tag{3.10.33}$$

$\mathbf{K}$ or $\mathbf{K}' \neq 0$ corresponds to umklapp processes which we will neglect. Further, we shall suppose in order to calculate the matrix elements for the electron–electron interaction it is adequate to approximate the Bloch functions of the valence electrons by plane-waves. Then the last term of equation (3.10.28) is

$$V_{ee} = \frac{1}{2} \sum_q \frac{4\pi e^2}{q^2} \sum_{\mathbf{k}_1 \mathbf{k}_2} a^\dagger_{\mathbf{k}_1+\mathbf{q}}\, a^\dagger_{\mathbf{k}_2-\mathbf{q}}\, a_{\mathbf{k}_2}\, a_{\mathbf{k}_1} \tag{3.10.34}$$

which is exactly the form taken earlier in Chapter 2 for the electron–electron interaction in a free-electron gas. With this in mind, we again look at Figure 3.3. We see that between the two dotted lines we have a polarization loop, and the effect of equation (3.10.34) on the diagrams is to replace this by the sum of all proper polarization parts, for the free-electron gas, the sum being shown in Figure 3.4.

From the diagrams of Figure 3.4 are missing all forward scattering parts and the diagrams due to the Hartree potential, which should be subtracted off equation (3.10.28) because it was used in defining the one-particle Bloch states. The simplest diagrams containing these are shown in Figure 3.5.

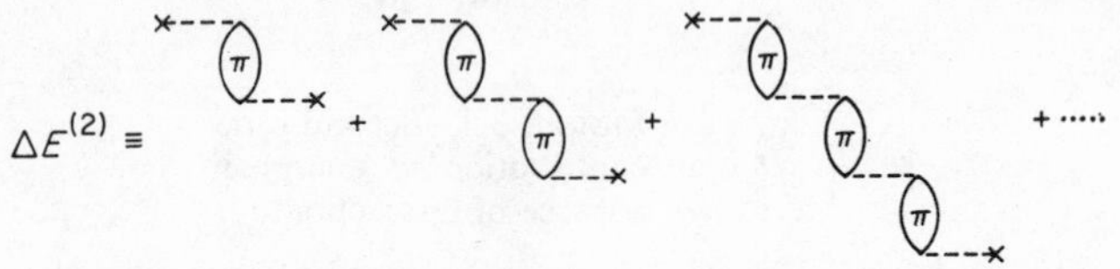

FIGURE 3.4. Second-order contribution to energy when electron–electron interactions are present.

We said in Chapter 2 that in a uniform electron gas the effect of the forward scattering parts was cancelled by the uniform background charge, and it is not hard to show that this result carries over to the inhomogeneous system considered here.

Returning now to Figure 3.4, we note that the diagrammatic series looks very like that for the screened Coulombic interaction of the electron gas. We cannot simply take over corresponding expressions because the dashed

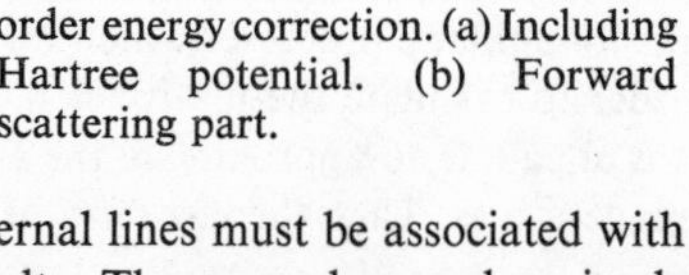

FIGURE 3.5. Examples of diagrams omitted from calculation of second-order energy correction. (a) Including Hartree potential. (b) Forward scattering part.

lines representing external lines must be associated with **k**-dependent factors if we want exact results. The general procedure is clear, however, and in illustration we shall approximate equation (3.10.29) by taking the $\psi_{\mathbf{k}}$'s as single OPW's, when the **k**-dependence vanishes. This approximation is certainly good for small q. Thus, we simply write

$$V_1'(\mathbf{q}, \mathbf{q}+\mathbf{k}) \simeq V_b'(\mathbf{q}). \tag{3.10.35}$$

We can now take the loop marked Π to represent

$$-\Pi(\mathbf{q}, 0) = -\frac{q^2}{4\pi^2}[\varepsilon(\mathbf{q}, 0) - 1], \tag{3.10.36}$$

where $\varepsilon(\mathbf{q}, 0)$ is the *static* dielectric constant, because the energy or frequency parameter ε must be conserved at each vertex and V_b' carries zero frequency parameter according to the adiabatic assumption. Translating Figure 3.4 accordingly, we obtain, before summing over $\mathbf{q}$,

$$\Delta E^{(2)} = -|V_b'(\mathbf{q})|^2\left(\Pi(\mathbf{q},0) - \frac{4\pi e^2}{q^2}\Pi^2(\mathbf{q},0) + \left(\frac{4\pi e^2}{q^2}\right)^2\Pi^3(\mathbf{q},0) + \ldots\right)$$

$$= -|V_b'(\mathbf{q})|^2 \frac{\Pi(\mathbf{q},0)}{1+(4\pi e^2/q^2)\Pi(\mathbf{q},0)} \tag{3.10.37}$$

or, completing the sum,

$$\Delta E^{(2)} = -\mathscr{V}\sum_{\mathbf{q}}|V_b'(\mathbf{q})|^2\frac{\Pi(\mathbf{q},0)}{\varepsilon(\mathbf{q},0)}, \tag{3.10.38}$$

where as usual $\mathscr{V}$ is the volume of the system.

If one turns now to equations (3.10.22) and (3.10.23) of the one-body treatment with the same approximations concerning the Bloch waves, one sees that the results are formally precisely the same if we regard $\partial H/\partial l$ as the change in a screened ionic potential, the screening being calculated using a scalar dielectric constant $\varepsilon(\mathbf{q}, 0)$. Note then in both one-body and many-body treatments, $\Pi(\mathbf{q}, 0)$ is defined relative to the real Fermi surface of the system, and, further, the choice of energies $E(\mathbf{k})$ in equation (3.10.31) is not automatically $\hbar^2 k^2/2m$; thus the dielectric constant $\varepsilon(\mathbf{q}, 0)$ is not that of the ground-state of the uniform electron gas, except as an approximation. It is, however, an important point that it is possible that $E(\mathbf{k}) \simeq \hbar^2 k^2/2\mathrm{m}$. Intuitively we might expect that the $\varepsilon(\mathbf{q})$ appearing in the linear response theory can be approximated by the $\varepsilon(\mathbf{q})$ of the ground state of the uniform electron gas. The valence electrons are obviously *not* uniform in density and their energy lies way above the core levels which are formally included in a direct perturbation treatment such as discussed in Chapter 2. We have seen in this section, however, how to reformulate the perturbation treatment using Bloch functions rather than plane-waves as the one-body states for which to define creation and annihilation operators. This constitutes an important difference of principle, for the plane-waves which are perturbed into the valence Bloch states by the external potential lie way outside the first Brillouin zone in a repeated zone scheme, with very obvious difficulties resulting. The net result for "free-electron-like" metals of the present treatment is that the valence electrons *do* behave rather as if they are the constituents of the ground state of a uniform electron gas.

Finally, we return briefly to the Kohn anomaly which is discussed above in a one-body context. This anomaly appears in the many-body treatment above through the dielectric constant in equation (3.10.38). We have already mentioned, in Chapter 2, the singularity in the dielectric constant at $2k_f$. Now, however, $\varepsilon(\mathbf{q}, 0)$ is directionally dependent, since the *true* Fermi surface (that surface in **k** space on which the quasi-particle lifetimes become infinite) was built into the treatment leading to equation (3.10.38). Thus the "Kohn cusp" in the vibration spectrum reflects the true Fermi surface (for an experiment on the Kohn anomaly of Pb, see Brockhouse *et al*, 1962).

3.10.4 Calculations of phonon spectra

It was first pointed out by Toya (1958) that Bardeen's work on electrical resistance (for which he derived the electron–phonon matrix element) formed a basis for a calculation of phonon frequencies. Toya's work (on sodium) came before an appreciation of the effect of singularities in $\varepsilon(\mathbf{q}, 0)$, and so we will discuss more recent work than his below.

We note, however, that after including correlation effects by the method shown in 3.10.3(c), so that screening is inserted using linear response theory, the calculation is reduced to that of finding the Bardeen–Hartree matrix element (V_1 being the bare electron–ion potential)

$$\mathbf{M}(\mathbf{k}', \mathbf{k}) = \int d\mathbf{r}\, \psi^*_{\mathbf{k}'} \nabla V_1(\mathbf{r})\, \psi_{\mathbf{k}}(\mathbf{r}) \tag{3.10.39}$$

since the definition (3.10.29) may be written

$$V'_b(\mathbf{k}', \mathbf{k}) = -\sum_l \mathbf{u}_l . \mathbf{M}(\mathbf{k}', \mathbf{k})\, e^{i(\mathbf{k}-\mathbf{k}').\mathbf{l}}. \tag{3.10.40}$$

In equation (3.10.29) the $\psi_{\mathbf{k}}$'s are exact solutions of the Hartree equation (3.10.24), and so it is natural to follow Bardeen in making the separation of M into $M_H + M_e$, where M_H is the matrix element of the gradient of the Hartree potential and M_e the corresponding quantity for the potential due to the electrons alone. In Problem 3.6, the reader is asked to calculate the Bardeen matrix element in this way.

We may remark immediately that while $M_e(\mathbf{k}', \mathbf{k})$ is the more important of the two terms, no accurate correspondence with experiment can be obtained unless $M_H(\mathbf{k}', \mathbf{k})$, which contains the whole of the direct dependence on the ionic potential, is retained.

By adjustment of the pseudo-potential V_l defined in equation (P3.6) one can obtain fair agreement with experimental results for the phonon dispersion relationships, the agreement being quite good for all high symmetry lines in aluminium, but less so in sodium and lead (Figure 3.6). The principal failure is in the predicted magnitude of the Kohn effect, the magnitude of the relevant matrix element being far too small, for Pb.

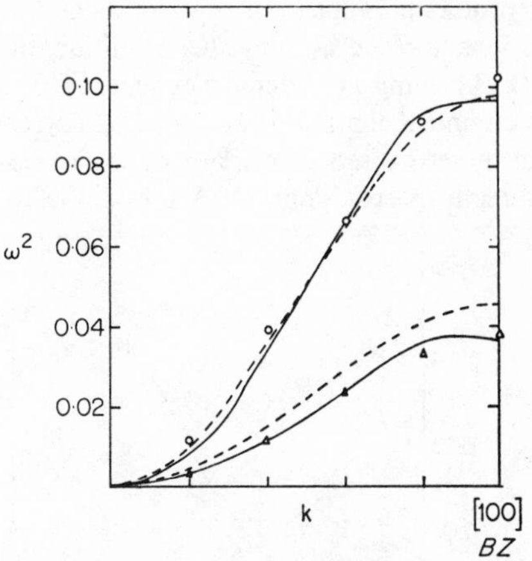

FIGURE 3.6(a) Phonon dispersion curves for simple metals. Al, along [100]. Longitudinal and transverse branches shown. Experimental points are from Yarnell and co-workers (1963). Solid lines. Method of Vosko, Taylor and Keech. Dashed lines. Method of Bardeen.

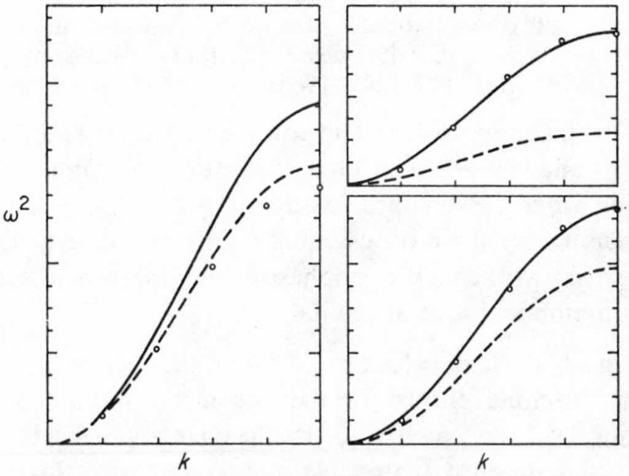

FIGURE 3.6(b) Na along [110]. Curves as in (a). Large figure shows longitudinal branches. Top-right figure shows transverse branch, with polarization $(1/\sqrt{2})(1, -1, 0)$. Bottom-right figure shows transverse branch with polarization $(0, 0, 1)$.

The natural step in an attempt to justify and refine Bardeen's theory is to use OPW expansions for the $\psi_{\mathbf{k}}$'s in the definition of M_{H}, the Bardeen approach to $M_{\mathrm{e}}(\mathbf{k}', \mathbf{k})$ being considered adequate. This has been done by, among others, Woll and Kohn (1962) and Vosko, Taylor and Keech (1965). The treatments of these two sets of workers differ, however, in that whereas Woll and Kohn directly insert a single OPW for $\psi_{\mathbf{k}}$, Vosko, Taylor and Keech

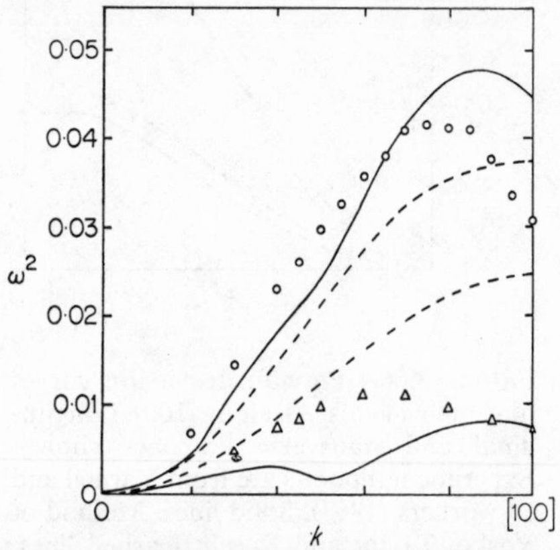

FIGURE 3.6(c) Pb, along [100]. Curves as in (a). Kohn anomalies are more pronounced in [110] dispersion curve (Brockhouse *et al*, 1962). Figure 3.6 reproduced by permission of the National Research Council of Canada from the *Canadian Journal of Physics*, **43**, pp. 1187–1247 (1965).

insert formally exact expansions and use the fact that the $\psi_{\mathbf{k}}$'s are formally exact solutions of a Hartree–Schrödinger equation to simplify the expressions before making a one OPW approximation.

We now summarize the most essential points that emerge from the work of Vosko and co-workers. We emphasize that the remarks relate only to sodium, aluminium and lead in general.

(a) *Structure.* Detailed calculation shows the better part of the large difference in screening effects for the bcc metal sodium and fcc metals aluminium and lead is accounted for by the change of structure, rather than valence, and the Bardeen matrix element shows why this should be so. Figure 3.7 illustrates the shape factor $G(qr_s)$, defined in Problem (P3.6), plotted against q for bcc and fcc metals. In the latter it is longer ranged, implying a greater screening dependence.

It should be noted that we have eliminated the direct effect of lattice parameters in this comparison by a choice of "natural" units for q. More rigorously, we should find $G(q)$ by taking

$$\int_{\Omega} G(\mathbf{r})\, e^{i\mathbf{q}\cdot\mathbf{r}}\, d\mathbf{r} = \rho_0 \int_{\Omega} e^{i\mathbf{q}\cdot\mathbf{r}}\, d\mathbf{r}$$
$$= 0 \quad \text{if} \quad \mathbf{q} = \mathbf{K}. \tag{3.10.41}$$

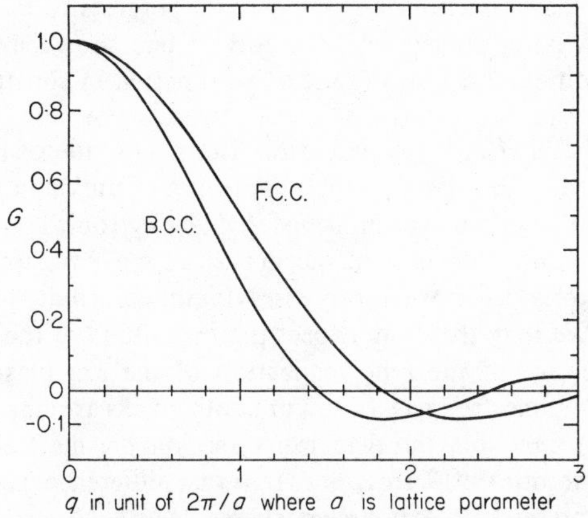

FIGURE 3.7. Shape factor for Bardeen electron–phonon matrix element. Different curves refer to bcc and fcc lattices. Reproduced by permission of the National Research Council of Canada from the *Canadian Journal of Physics*, **43**, pp. 1187–1247 (1965).

(b) *Choice of potential and core functions.* For sodium, the Prokofiev potential is available, and this may be defined as the best one-body potential available for the valence electrons, including as it does correlation effects between core and valence electrons. An alternative choice may be obtained from HF self-consistent field calculations. A full HF calculation has been made for aluminium by Heine (1957), and Vosko and colleagues also took HF–Slater potentials for the three metals calculated. The agreement between HFS and Heine potentials is good for aluminium; on the other hand, the HFS and Prokofiev Na potentials are very different in the small and intermediate q regions. This is explained by the correlations between core and valence electrons being very important in the outer regions of the unit cell, where the kinetic energy is low and the nuclear charge is almost entirely screened out. This led Vosko and associates to adjust their potentials for aluminium and lead, to obtain agreement with experiment while maintaining equivalence with HFS potentials at high q.

Another significant element in calculations is the choice of core functions to which the valence states are taken orthogonal. In principle these core functions should be precisely the functions obtained from the same Hartree equations as used for the valence electrons, as we saw in Chapter 1. However, expediency very often dictates other choices: for example Sham (1965), in work discussed further below, took the Fock–Petrashen (1934) wave functions for Na^+ since these functions enable analytic progress to be made. A comparison of this latter choice with the correct one shows surprisingly large effects due to different choice of core wave function in constructing OPW's.

(c) *Kohn effect and umklapp processes.* The magnitude of the Kohn effect may be expected to be related to the importance of umklapp processes, since both depend on large momentum transfer. The electronic contribution to the transverse vibrations depends on the magnitude of umklapp processes—in a one OPW framework, normal processes do not contribute to the transverse modes at all. We may therefore expect the magnitude of the Kohn effect to be tied to the size of the renormalization of the ion–plasma‡ transverse frequencies by the electron gas. In sodium both effects are small. In particular, for the (1, 1, 0) direction, the two transverse ion–plasma frequencies are in the approximate ratio of 7 to 1, and this vast difference persists when the electrons are present, as experiment shows. Although the Kohn effect in sodium is so small as to evade direct experimental test, it was found by Woods and co-workers from a Fourier analysis that the force constants oscillate in sign like the long-range oscillations in the screening charge (Chapter 2). These effects are intimately related, for both the Kohn effect and the long-range oscillations may be regarded as reflecting the singularities in $\varepsilon(\mathbf{q}, 0)$. Conditions for the direct observation of the Kohn anomaly would seem more favourable in the polyvalent metals aluminium and lead. In fact it is much larger in lead than in aluminium, although Vosko and co-workers calculate the relevant electron–ion matrix element to be much the same in the two metals. The explanation of this would appear to be that the effect, superposed on a smooth background, shows up more in lead because the magnitude of the background is small. We should add some comments on the choice of $\varepsilon(\mathbf{q}, 0)$ made by Vosko and co-workers, which they took as given by equation (2.11.24), with

$$\frac{G(q)}{q^2} = \frac{1}{2(q^2+\xi k_f^2)}, \tag{3.10.42}$$

‡ A hypothetical system in which the ions oscillate against a uniform neutralizing background of negative charge.

ξ being chosen so that the relation of equation (P2.14.4) between $\lim_{q\to 0} \varepsilon(\mathbf{q}, 0)$ and the compressibility of the electron gas is satisfied at the order of approximation to which they worked. This choice of $G(q)$ produced better agreement with experimental results for sodium than the original Hubbard choice

$$\frac{G(q)}{q^2} = \frac{1}{2(q^2+k_f^2)} \tag{3.10.43}$$

or the later modification

$$\frac{G(q)}{q^2} = \frac{1}{2(q^2+k_f^2+q_{TF}^2)} \tag{3.10.44}$$

where q_{TF}^2 is the Thomas–Fermi screening parameter, $4k_f/\pi a_0$. Sodium was chosen for comparison by Vosko and associates, since they were most confident of the weak $\mathbf{k}$ dependence of $M(\mathbf{k}', \mathbf{k})$ for that metal.

To go beyond linear-screening theory, using diagrammatic techniques leads to considerable complications. We should note that to validate linear screening we must not only assume weak $\mathbf{k}$-dependence of $V_b(\mathbf{k}', \mathbf{k})$, or equivalently, $M(\mathbf{k}', \mathbf{k})$, but also neglect the orthogonalization to the core in obtaining the energy change in equation (3.10.38). We cannot expect, however, to obtain the redistribution of valence electrons in the core region by linear-response theory. This difficulty suggests we examine the use of independent particle equations giving the exact density (see section 3.10.3 above). We can examine work such as that by Sham (1965) on sodium, in this spirit. In Chapter 2 we suggested that the density might be given rather well by the HF equations with screened exchange.

These are the basic equations of Sham's approach, who treats the screening problem as follows.

The one-body wave function of a valence electron is written in the form

$$\psi_{\mathbf{k}} = Z_{\mathbf{k}}^{\frac{1}{2}}\{\phi_{\mathbf{k}} - \phi_c\langle\phi_{\mathbf{k}}\phi_c\rangle\}, \tag{3.10.45}$$

where ϕ_c represents the core wave functions and $\phi_{\mathbf{k}}$ the "smooth" part of $\psi_{\mathbf{k}}$. $\phi_{\mathbf{k}}$ is given by a Schrödinger equation with a pseudopotential and is approximated rather well by a plane-wave everywhere. We require the factor $Z_{\mathbf{k}}$ in order to normalize $\phi_{\mathbf{k}}$ and $\psi_{\mathbf{k}}$ to the same value.

Now as the ions move, the "core part" $\psi_{\mathbf{k}} - Z_{\mathbf{k}}^{\frac{1}{2}}\phi_{\mathbf{k}}$ moves rigidly with the core, and if the cores are small, should not contribute to the energy change. We therefore consider only energy changes caused by changes in the $\phi_{\mathbf{k}}$'s, and this leads to a treatment in which linear screening is adequate but the screening charge is the smooth part

$$\rho_s(\mathbf{r}) = \sum_{\mathbf{k}} Z_{\mathbf{k}} \phi_{\mathbf{k}}^*(\mathbf{r}) \phi_{\mathbf{k}}(\mathbf{r}). \tag{3.10.46}$$

This procedure has not been justified rigorously, but the fact that Sham obtains poorer agreement with experiment than Vosko and co-workers may well be due to other approximations. For example, Sham took the Fock–Petrashen wave functions for the core, which we have already seen to be a rather inaccurate choice; neither did he avail himself of a separation of $M(\mathbf{k}',\mathbf{k})$ analogous to that in Problem 3.6. Such a separation is not certain to improve the accuracy of the treatment.

3.10.5 Cauchy relations and pair forces

If we choose a local pseudopotential, then results are obtained in which the dynamics of the ions of the lattice are governed by pair forces, and we shall see in Chapter 5 that experimentally determined phonon-dispersion relations in metals can often be fitted by this assumption. This situation is rather perplexing since we saw in section 3.6 that the Cauchy relations must be obeyed by a crystal in equilibrium under central forces, with no external pressures, whereas they certainly are not in metals. One can imagine a "fictitious" pressure on the metal to maintain an interpretation in terms of central forces, a possible source of the fictitious pressure in the electron gas. However, this does not solve the problem, for it leads us to write the total energy as (the ϕ's being the pair potentials)

$$E = \sum_{\mu<\nu} \phi(R_{\mu\nu})+f(\Omega), \tag{3.10.47}$$

where f is a function only of volume. On the other hand, by direct application of the virial theorem in equation (3.10.47) we obtain a virial (at absolute zero and neglecting zero-point energy vibrations)

$$\sum_{\mathbf{R}} \frac{\partial E}{\partial \mathbf{R}}.\mathbf{R} = \sum_{\mu<\nu} \frac{\partial \phi(\mathbf{R}_{\mu\nu})}{\partial \mathbf{R}_{\mu\nu}}.\mathbf{R}_{\mu}. \tag{3.10.48}$$

This gives rise to a fictitious pressure (cf. Cochran, 1963) whereas the total contribution to the virial is normally zero.

The answer, of course, must lie in the inexactitude of the model, which we may regard as obtained in the following way. $E(\mathbf{R}_1, \ldots, \mathbf{R}_N)$ may be split into two-body terms and many-body (three, four, ..., -body) terms. The two-body terms, it seems, can often be obtained by second-order perturbation theory with a suitable pseudopotential, to good accuracy; we have then to estimate the many-body terms. These, we say, are calculable to a good degree of approximation by smearing out the ions into a uniform background charge. Then $f(\Omega)$ is effectively the energy of a uniform electron gas.

We can very well appreciate, however, that if we calculate $\sum_{\mu}(\partial E/\partial \mathbf{R}_{\mu}).\mathbf{R}_{\mu}$ *before* approximating the many-body terms by smearing out the ions we might get a rather different answer from equation (3.10.48). We can effectively do this by using a scale-transformation.

Let

$$\mathbf{R}_\mu \to \mathbf{R}_\mu\left(1+\frac{\delta l}{l}\right). \tag{3.10.49}$$

Then

$$\delta E = \sum_\mu \frac{\partial E}{\partial \mathbf{R}_\mu}.\delta\mathbf{R}_\mu = \frac{\delta l}{l}\sum \frac{\partial E}{\partial \mathbf{R}_\mu}.\mathbf{R}_\mu, \tag{3.10.50}$$

i.e.

$$3\mathscr{V}p = \sum_{\mathbf{R}} \frac{\partial E}{\partial \mathbf{R}_\mu}.\mathbf{R}_\mu = l\frac{\partial E}{\partial l} = r_s\frac{\partial E}{\partial r_s} \tag{3.10.51}$$

where r_s is the mean inter-particle spacing. Thus the expression (3.10.48) for the pressure should be replaced by

$$3\mathscr{V}p = \sum_{\mu<\nu} \frac{\partial \phi}{\partial \mathbf{R}_\mu}.\mathbf{R}_\mu + r_s\frac{df}{dr_s} + \sum_{\mu\nu}\frac{\partial\phi(\mathbf{R}_{\mu\nu})}{\partial r_s}. \tag{3.10.52}$$

Thus the fictitious pressure actually comes not just from the electron gas, but also from the r_s dependence of the pair potentials ϕ.

Later calculations on simple metals have developed further along these lines, and we might mention here particularly the work of Shaw and Pynn (1969).

(a) *General inclusion of many-body forces and connection with rigid-ion model.* Methods based on rigid ions or neutral pseudoatoms clearly lead to simple pair-wise interactions. It is possible, of course, using pseudopotential theory, as we remarked above, to go to higher order, and in this way three-body, etc. forces can be included.

However, for general crystals, it would seem important to try to treat many-body forces from the outset, rather than classifying the forces into three-body, four-body, etc. We want to summarize briefly here the way in which this can be achieved (Jones and March, 1970).

In the adiabatic and harmonic approximation, the basic quantity we need in a lattice dynamical theory is the change in density from the perfect ionic lattice when a phonon is "frozen" in the system. It is perfectly clear that since we are only doing a small displacements calculation, we can get this from linear response theory. Furthermore, we know from the density-functional theory of Chapter 1 that, both in the perfect lattice and in the lattice with a phonon "frozen" in, we can calculate the ground-state electron density in the many-body system exactly from one-body potentials.

Let us suppose that the first-order density change on displacing the ions is given by

$$\rho_1(\mathbf{r}) = \sum \mathbf{u}_l . \frac{\partial \rho(\mathbf{r})}{\partial \mathbf{u}_l}. \tag{3.10.53}$$

Then, if the one-body potential changes by an amount $\Delta V^{(1)}$ as we go from the perfect to the phonon perturbed lattice, we can write

$$\rho_1(\mathbf{r}E) = \int \Delta V^{(1)}(\mathbf{r}') F(\mathbf{r}\mathbf{r}'E) \, d\mathbf{r}'. \tag{3.10.54}$$

The one-body response function $F(\mathbf{r}\mathbf{r}'E)$ can in fact (see Stoddart, March and Stott, 1969) be expressed in the form

$$\frac{\partial F}{\partial E} = 2\mathrm{Re}\left[G_0(\mathbf{r}\mathbf{r}_1 E_+) \frac{\partial \rho_0(\mathbf{r}\mathbf{r}'E)}{\partial E} \right], \tag{3.10.55}$$

where G_0 is the perfect lattice Green function and $\rho_0(\mathbf{r}\mathbf{r}'E)$ is a Dirac density matrix whose diagonal element $\rho_0(\mathbf{r})$ is the exact crystal density. However, from our point of view, we note that we saw how to write down F in terms of diagrams in Chapter 2, and we can, alternatively, express the theory in terms of the linear response function $\mathscr{F}$ introduced in section 2.11.4.

Let us next notice that $\Delta V^{(1)}(\mathbf{r})$ is the change occasioned by ionic displacements in the one-body potential incorporating electron–electron interactions. Since we can always write for a given ionic configuration

$$V'(\mathbf{r}) = V_{\text{electro-static}} + V_{\text{exchange}} + V_{\text{correlation}}, \tag{3.10.56}$$

where these are solely functionals of the electron density, from the Hohenberg–Kohn theorem discussed in section 1.2.1 of Chapter 1, we can express $\Delta V^{(1)}$ to first order in ρ_1 as

$$\Delta V^{(1)} = \Delta V^{(1)}_{\text{electro-static}} + \int U(\mathbf{r}\mathbf{r}') \rho_1(\mathbf{r}') \, d\mathbf{r}', \tag{3.10.57}$$

where

$$\Delta V^{(1)}_{\text{electro-static}} = \int \frac{\rho_1(\mathbf{r}') \, d\mathbf{r}'}{|\mathbf{r}' - \mathbf{r}|} + \sum_l \left[\frac{Ze}{|\mathbf{r} - \mathbf{l} - \mathbf{u}_l|} - \frac{Ze}{|\mathbf{r} - \mathbf{l}|} \right] \tag{3.10.58}$$

and Ze represents the charge on each ion. As in the discussion above, we could take a potential V_b describing the nucleus plus core electrons instead of Ze/r, with trivial changes in the theory below. The quantity U, introduced in section 2.11.4, could be approximated, if we wished, by taking V_{exchange} and $V_{\text{correlation}}$ to have the Dirac–Slater and Wigner forms (2.11.59) and (2.11.42) respectively, corrected when necessary by gradient terms like equation (2.11.64) of Herman, van Dyke and Ortenburger (1969) and equation (2.11.67) of Ma and Brueckner (1968).

Equations (3.10.54) and (3.10.57) are now to be solved simultaneously and to do so we first write equation (3.10.53) in the form (cf. Johnson, 1969)

$$\rho_1(\mathbf{r}) = \sum \mathbf{u}_l . \mathbf{R}_l(\mathbf{r}). \tag{3.10.59}$$

Now if $\mathbf{u}_0 . \mathbf{R}(\mathbf{r})$ represents the density change resulting from a displacement $\mathbf{u}_0$ of the ion at the origin, say, the physical equivalence of every site implies that the displacement of the ion at $\mathbf{l}$ must result in the same density change referred to $\mathbf{l}$ as a new origin, that is $\mathbf{R}_l = \mathbf{R}(\mathbf{r}-\mathbf{l})$. Using this result in equation (3.10.59), we can write from equations (3.10.57) and (3.10.58)

$$\Delta V^{(1)}(\mathbf{r}) = \sum_l \int \frac{\mathbf{u}_l . \mathbf{R}(\mathbf{r}'-\mathbf{l})}{|\mathbf{r}'-\mathbf{r}|} d\mathbf{r}' + \sum_l \left[\frac{Ze}{|\mathbf{r}-\mathbf{l}-\mathbf{u}_l|} - \frac{Ze}{|\mathbf{r}-\mathbf{l}|}\right] + \sum_l \mathbf{u}_l . \int U(\mathbf{r}\mathbf{r}')\, \mathbf{R}(\mathbf{r}'-\mathbf{l})\, d\mathbf{r}'. \tag{3.10.60}$$

We wish now to show that, when ionic terms are expanded to order $\mathbf{u}_l$, this may be expressed in a form analogous to equation (3.10.59), namely

$$\Delta V^{(1)}(\mathbf{r}) = \sum_l \mathbf{u}_l . \mathbf{P}(\mathbf{r}-\mathbf{l}). \tag{3.10.61}$$

The first two terms on the right-hand side present no difficulty. As for the exchange and correlation terms, it is easy to prove that

$$U(\mathbf{r}+\mathbf{l}, \mathbf{r}'+\mathbf{l}) = U(\mathbf{r}\mathbf{r}') \tag{3.10.62}$$

[cf. equation (2.11.47) of section 2.11.4]. Hence we can write

$$\int U(\mathbf{r}\mathbf{r}')\, \mathbf{R}(\mathbf{r}'-\mathbf{l})\, d\mathbf{r}' = \int U(\mathbf{r}, \mathbf{r}'+\mathbf{l})\, \mathbf{R}(\mathbf{r}')\, d\mathbf{r}' = \int U(\mathbf{r}-\mathbf{l}, \mathbf{r}')\, \mathbf{R}(\mathbf{r}')\, d\mathbf{r}'. \tag{3.10.63}$$

Thus equation (3.10.61) follows, with $\mathbf{P}(\mathbf{r})$ given by

$$\mathbf{P}(\mathbf{r}) = \int \frac{\mathbf{R}(\mathbf{r}')\, d\mathbf{r}'}{|\mathbf{r}-\mathbf{r}'|} - \frac{Ze\mathbf{r}}{r^2} + \int U(\mathbf{r}\mathbf{r}')\, \mathbf{R}(\mathbf{r}')\, d\mathbf{r}'. \tag{3.10.64}$$

As remarked above, we can generalize equation (3.10.64) by replacing $Ze\mathbf{r}/r^2$ by the gradient of an appropriate ionic potential $V_b(r)$. But using the property of the one-body response function F analogous to that of equation (2.11.47) for the linear response function $\mathscr{F}$, we find when we insert equation (3.10.61) into equation (3.10.54) that

$$\rho_1(\mathbf{r}) = \sum \mathbf{u}_l \cdot \int \mathbf{P}(\mathbf{r}'-\mathbf{l})\, F(\mathbf{r}\mathbf{r}')\, d\mathbf{r}'$$

$$= \sum \mathbf{u}_l \cdot \int \mathbf{P}(\mathbf{r}')\, F(\mathbf{r}\mathbf{r}'+\mathbf{l})\, d\mathbf{r}'$$

$$= \sum \mathbf{u}_l \cdot \int \mathbf{P}(\mathbf{r}')\, F(\mathbf{r}-\mathbf{l},\mathbf{r}')\, d\mathbf{r}'. \tag{3.10.65}$$

Comparing this equation with equation (3.10.59) we obtain the basic integral equation

$$\mathbf{R}(\mathbf{r}) = \int \mathbf{P}(\mathbf{r}')\, F(\mathbf{r}\mathbf{r}')\, d\mathbf{r}' \tag{3.10.66}$$

as a consequence of equations (3.10.62) and (2.11.47). If the exchange and correlation function $U(\mathbf{r}\mathbf{r}')$ were known, known of course, for the inhomogeneous gas in the perfect lattice, then equations (3.10.64) and (3.10.66) could be solved iteratively to yield $\mathbf{R}(\mathbf{r})$. Alternatively, we could solve an equation relating the response function $\mathscr{F}$ to F, namely [cf. equation (2.11.76) of Chapter 2]

$$\mathscr{F}(\mathbf{r}\mathbf{r}') = F(\mathbf{r}\mathbf{r}') + \int F(\mathbf{r}\mathbf{r}''')\, U'(\mathbf{r}'''\mathbf{r}'')\, \mathscr{F}(\mathbf{r}''\mathbf{r}')\, d\mathbf{r}''\, d\mathbf{r}''', \tag{3.10.67}$$

where U' now includes the electrostatic interaction between electrons. If we rewrite equation (3.10.66) in $\mathbf{k}$ space, then we find

$$\mathbf{R}(\mathbf{k}) = \sum_{\mathbf{K}'}\sum_{\mathbf{K}} F(\mathbf{k},\mathbf{k}+\mathbf{K}')\, U'(\mathbf{k}+\mathbf{K},\mathbf{k}+\mathbf{K}'+\mathbf{K})\, \mathbf{R}(\mathbf{k}+\mathbf{K}'')$$
$$+ i \sum_{\mathbf{K}} F(\mathbf{k},\mathbf{k}+\mathbf{K})\, V_b(\mathbf{k}+\mathbf{K})(\mathbf{k}+\mathbf{K}). \tag{3.10.68}$$

Clearly, given information from a band structure calculation on the one-body response function F, and knowledge of exchange and correlation in the inhomogeneous gas, which we summarized in section 2.11 of Chapter 2, we can, in principle, solve equation (3.10.68) to obtain $\mathbf{R}(\mathbf{k})$.

(a) *Connection with rigid-ion model.* At this point, it is helpful to relate the above treatment to the ideas underlying the rigid-ion model. This model, essentially, thinks of the perfect crystal density ρ_0 being decomposed into a sum of localized densities, $\sigma(\mathbf{r})$, centred on each lattice site. Thus we may write

$$\rho_0(\mathbf{r}) = \sum_{\mathbf{l}} \sigma(\mathbf{r}-\mathbf{l}). \tag{3.10.69}$$

As we discuss fully in Chapter 5, such a decomposition is far from unique, an infinity of σ's reproducing the same crystal density. However, the idea of the rigid-ion model is that a choice of σ *can* be made such that, when an ion is moved by an infinitesimal amount from a lattice site, this localized distribution is carried rigidly with it.

The above argument shows that there is a quantity which can be associated with each ion, the quantity $\mathbf{R}(\mathbf{r})$ above, but it is a vector quantity, and not a scalar. Just as in the rigid-ion model, this is indeed closely connected with the decomposition of the perfect lattice density into localized objects. We can see this by putting all the $\mathbf{u}_l$ in equation (3.10.59) to be the same, and then we must obtain the change in electron density on uniform translation of the lattice through $\mathbf{u}_l$, and hence

$$\nabla \rho_0(\mathbf{r}) = \sum_{l} \mathbf{R}(\mathbf{r}-\mathbf{l}). \tag{3.10.70}$$

Thus, we have a vectorial generalization of the rigid-ion model which is *exact*. We can decompose the gradient of the perfect lattice density into localized vector functions, one on each site, and these then move rigidly with the nuclei when these undergo small displacements.

It might be asked what is the difference between equations (3.10.69) and (3.10.70) and in particular why we cannot make them identical by writing $\mathbf{R}(\mathbf{r}) = \operatorname{grad} \sigma(\mathbf{r})$. The answer to this is that, from equation (3.10.70), we have immediately

$$\operatorname{curl} \nabla \rho_0(\mathbf{r}) = 0 = \operatorname{curl} \sum_{l} \mathbf{R}(\mathbf{r}-\mathbf{l}) \tag{3.10.71}$$

whereas the assumption $\mathbf{R}(\mathbf{r}) = \operatorname{grad} \sigma(\mathbf{r})$ would satisfy this by making every term in the sum vanish. This is not, in general, a consequence of equation (3.10.66) for $\mathbf{R}(\mathbf{r})$.

However, if we go into $\mathbf{k}$ space, the connection with the rigid-ion model becomes even more clear. If $\rho_{\mathbf{K}}$ are the Fourier components of the periodic charge density introduced in equation (1.3.10) of Chapter 1 then at reciprocal lattice vectors, $\mathbf{R}_{\mathbf{K}} = i\mathbf{K}\rho_{\mathbf{K}}$. This is an exact result, and since in principle the Fourier components of the charge density can be got from the measured Bragg reflexion intensities (cf. Chapter 5) we can get direct information on $\mathbf{R}$ from experiment.

(b) *Representation in* **k** *space of many-body forces.* We can now see, very directly, the effect of many-body forces. In a rigid-ion or pair-force model, the result that $\mathbf{R}_{\mathbf{K}}$ is parallel to $\mathbf{K}$ when $\mathbf{K}$ is a reciprocal lattice vector is the result for all $\mathbf{k}$. But in a correct theory, based on equation (3.10.68) for $\mathbf{R}$, we find that $\mathbf{R}_{\mathbf{k}}$ is not parallel to $\mathbf{k}$, and the deviations of $\mathbf{R}$ from the direction of $\mathbf{k}$ is a direct measure of the strength of the many-body forces. It is of obvious interest for the future to try to map out quantitatively $\mathbf{R}$ as a function of $\mathbf{k}$ between lattice vectors. Clearly, knowledge of $\mathbf{R}$ will give us the density change due to a phonon, and hence dispersion relations can be calculated along similar lines to those already discussed. Direct check of the dielectric behaviour of an inhomogeneous electron gas ought therefore, in principle, to be possible from the above argument, though it is not yet known how sensitive the phonon dispersion relations are to changes in the exchange and correlation function $U(\mathbf{r}, \mathbf{r}')$.

However, we expect the many-body force contributions to be significant in some metals in the long-wavelength limit, and pseudopotential calculations of Skolt (1969), carried to higher order, have already indicated, from a different approach to the one above, that this is indeed the case in grey tin.

3.11 Lattice dynamics of ionic and covalent crystals

Having discussed at some length the problem of calculating from first principles the phonon dispersion relations in simple metals, we want now to consider the rather different problems posed by ionic and covalent crystals.

Though we shall concentrate mainly on covalently bonded structures like Ge here, it will be helpful to comment briefly on the properties of alkali halide crystals first of all. Important progress in the theory of the dielectric properties of these crystals stemmed from the work of Dick and Overhauser (1958), in which they gave a plausible wave-mechanical basis for a simple model of an atom with a closed electron configuration.

The model of Dick and Overhauser is that the atom may be represented by a core consisting of the nucleus and inner electrons and a shell representing the outer electrons. The shell and core are taken as coupled to one another by an isotropic force constant, and each is assumed to retain spherical symmetry although a dipole moment may be generated by displacing them relative to one another. In an alkali halide crystal the short-range forces are assumed to act through the shells.

That account had to be taken of the deformation of the electron distributions as the ions moved was clear not only to Dick and Overhauser, but also it is the basic idea behind the work of Tolpygo and co-workers (1957) and the "deformation dipole" model of Hardy (1961).

With regard to lattice dynamics, Cochran (1959a) used the shell model in the theory of the lattice dynamics of alkali halides. The dispersion curves for sodium iodide, in particular, were found to fit well with such a shell model (see, however, section (3.11.1) below), which contains no parameters which could be adjusted to fit the neutron measurements (Woods, Cochran and Brockhouse, 1960). Though we shall not refer to this work in more detail we want here to deal with the application of the shell model to covalent Ge (Cochran, 1959b).

The use of a shell model gives the atom the property that it is polarized in an electric field [see Figure 3.8(b)], and also that it possesses "distortion polarizability" under the influence of short-range forces acting through both cores and shells. The lattice vibrations therefore induce electric fields in the crystal. We shall refer in Chapter 7 to a microscopic theory of infra-red absorption in Ge as well as in ionic crystals.

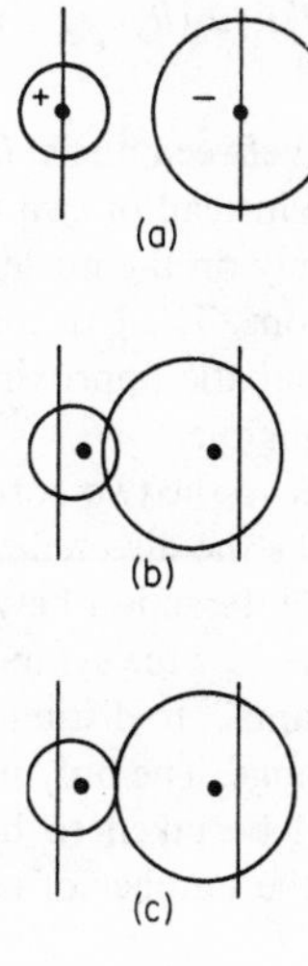

FIGURE 3.8. Illustrating shell model in an ionic crystal. (a) Showing adjacent undisturbed ions. (b) Showing independent displacement polarization of ions, when electric field is applied. (c) Showing final positions, allowing for repulsion due to the overlap of charge clouds shown in (b).

3.11.1 Shell model for Ge

Early attempts to explain the dispersion relations for Ge, determined experimentally by Brockhouse and Iyengar (1958), required the assumption of force constants between atoms which were rather far apart. The theory then was posed in terms of a large number of undetermined parameters, which had no obvious physical interpretation. Furthermore, the fact that the elastic constants of Ge obey closely a relation due to Born (1914) namely

$$4c_{11}(c_{11}-c_{44}) = (c_{11}+c_{12})^2 \tag{3.11.1}$$

was not explained by such a theory. This condition is a necessary, though not a sufficient, condition that only first neighbours are important. One of the things we shall see below is that it is possible to understand equation (3.11.1) on a shell model, without invoking short-range forces.

(a) *Lattice dynamics of shell model.* The theory outlined earlier, based on

$$\Phi_2 = -\frac{1}{2}\sum_{ii'}\sum_{ll'}\sum_{\alpha\beta}\Phi_{\alpha\beta}(li, l'i')\,u_\alpha(li)\,u_\beta(l'i'), \tag{3.11.2}$$

where $\Phi_{\alpha\beta}$ is force constant for interaction between nuclei li and $l'i'$, has now to have a summation over four values of i instead of two to take account of shells, so that the potential depends not only on the nuclear coordinates but also on the coordinates of the outer electrons. The harmonic approximation is still used, and the equivalent of the adiabatic approximation is achieved by taking the mass of each shell to be negligible.

We shall use indices 1 and 2 for i, to refer to the two cores per unit cell and 3 and 4 to the corresponding shells. We take the directed bonds into account by having the force constants defining the interaction between neighbouring units (cores or shells) as general as is allowed by the symmetry of the crystal. These will be termed "bonding force constants", to distinguish them from the electro-static forces, which are not short range. The only interaction between a shell and a core of the same atom will be taken to be described by an isotropic force constant k. Z will denote the number of electrons in a shell, $-e$ the electronic charge.

The usual dynamical matrix for Ge contains 6×6 coefficients. Since we now have $i = 1:4$, the dynamical matrix for the shell model of Ge contains 12×12 coefficients.

But for $\mathbf{q}$ along [100] or [111], the usual sixth-order equation factors into three equations each of order 2, of which one gives the dispersion relation for the longitudinal modes and the others give two identical solutions for the two transverse modes. We shall consider only 4×4 matrices which refer to a particular mode, with $\mathbf{q}$ in a symmetry direction. With this restriction we can write

$$M_i\,\omega^2\,U(i) = \sum_{i'=1}^{4} D(i,i')\,U(i'), \tag{3.11.3}$$

where $D(ii')$ is a linear combination of D_{xx}, D_{xy}, etc. It is shown in Appendix 3.4 that the shell model leads then to

$$\left.\begin{aligned}
M_1\omega^2 U(1) &= \{aC_1+B(11)\}U(1)+\{aC_2+D\}U(2)\\
&\quad+\{-aC_1-k\}U(3)+\{-aC_2+F\}U(4),\\
M_2\omega^2 U(2) &= \{aC_2^*+D^*\}U(1)+\{aC_1+B(22)\}U(2)+\{-aC_2^*+F^*\}\\
&\quad\times U(3)+\{-aC_1-k\}U(4),\\
M_3\omega^2 U(3) &= \{-aC_1-k\}U(1)+\{-aC_2+F\}U(2)\\
&\quad+\{aC_1+B(33)\}U(3)+\{aC_2+S\}U(4),\\
M_4\omega^2 U(4) &= \{-aC_2^*+F^*\}U(1)+\{-aC_1-k\}U_2+\{aC_2^*+S^*\}U(3)\\
&\quad+\{aC_1+B(44)\}U(4),\\
u &= U e^{i(\mathbf{q}.\mathbf{r}-\omega t)}
\end{aligned}\right\} \tag{3.11.4}$$

where $a = Z^2e^2/v$, $v = 2r_0^3$ with the atoms per unit cell at $(0,0,0)$ and $(-r_0/2, -r_0/2, -r_0/2)$.

The bonding coefficients F, S and D are shown in Figure 3.9; compare Appendix 3.4.

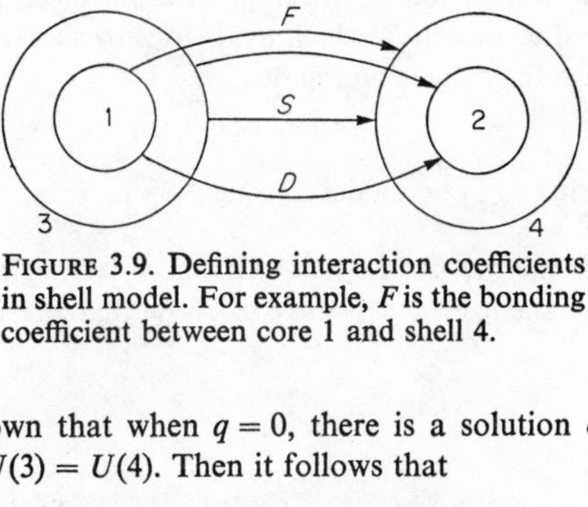

FIGURE 3.9. Defining interaction coefficients in shell model. For example, F is the bonding coefficient between core 1 and shell 4.

Now it is known that when $q = 0$, there is a solution $\omega = 0$ for which $U(1) = U(2) = U(3) = U(4)$. Then it follows that

$$B(11) = B(22) = k+D_0+F_0 \tag{3.11.5}$$

and

$$B(33) = B(44) = k+S_0+F_0, \tag{3.11.6}$$

where $-D_0 = (D)_{q=0}$ and similarly for S_0 and F_0. We now put $M_1 = M_2 = M$ and $M_3 = M_4 = 0$. Then we find after some algebra

$$M\omega^2 U(1) = A_0 U(1)+AU(2), \tag{3.11.7}$$

$$M\omega^2 U(2) = A^* U(1)+A_0 U(2), \tag{3.11.8}$$

where

$$A_0 = R_0 + \frac{-(aC_1+k+T_0)(T_0^2+|T|^2)+T_0[T(aC_2^*+S^*)+T^*(aC_2+S)]}{(aC_1+k+T_0)^2-|aC_2+S|^2} \tag{3.11.9}$$

and

$$A = R + \frac{-2TT_0(aC_1+k+T_0)+T_0^2(aC_2+S)+T^2(aC_2^*+S^*)}{(aC_1+k+T_0)^2-|aC_2+S|^2}. \tag{3.11.10}$$

We have here used the shorthand notation

$$R = D+S+2F = B(12)+B(34)+B(14)+B(32) \tag{3.11.11}$$

and

$$T = S+F = B(34)+B(14). \tag{3.11.12}$$

However, T and R have some physical significance, in that if we put $k = \infty$ in equations (3.11.9) and (3.11.10) we see that T is the coefficient of the bonding interaction between a shell and the surrounding atoms, while R is the coefficient of interaction between "rigid" atoms. From Figure 3.9 we can see that $R = D+S+2F$ is simply the sum of the coefficients of bonding interactions between units composing the atoms. A is to be interpreted as the coefficient of interaction between polarizable atoms, consisting of a "rigid" part R and a correction which tends to zero as $k \to \infty$.

Finally, we have the dispersion relation

$$M\omega^2 = A_0 \pm |A|, \tag{3.11.13}$$

the + sign referring to optic modes and the − sign to acoustic modes.

(b) *Dielectric properties and elastic constants.* If we neglect the interaction between core and shell, then it is easy to show that the polarizability α is given by

$$\alpha = \frac{Z^2 e^2}{k}. \tag{3.11.14}$$

Considering the interaction between a uniform electric field and the cores and shells, the correct polarizability in the present model is, in fact,

$$\alpha = Z^2 e^2/(k+F_0). \tag{3.11.15}$$

The Clausius–Mossotti formula now relates the atomic polarizability to the dielectric constant, the relation being

$$\frac{4\pi}{3}\left(\frac{2\alpha}{v}\right) = \frac{\varepsilon-1}{\varepsilon+2}, \tag{3.11.16}$$

the factor of 2 accounting for the fact that in Ge there are two atoms/unit cell. Hence, the model dielectric constant is known in terms of the parameters Z, k and F_0.

We referred earlier to the Born relation between elastic constants. To see how this can still be valid in the shell model we write down the velocity of sound v_s from equation (3.11.13) as

$$Mv_s^2 = \lim_{q\to 0} (A_0 - |A|)/q^2 \tag{3.11.17}$$

and the quantity on the right-hand side is proportional to a linear combination of c_{11}, c_{12} and c_{14}. For example, for a longitudinal mode with q along [100], the appropriate elastic constant c is simply c_{11}, that is

$$c = c_{11} = (2/v) \lim_{q\to 0} [A_0 - |A|]/q^2, \tag{3.11.18}$$

where we have used equation (3.6.15).

We have seen already that R represents the coefficient of interaction between rigid atoms, and so, in the limit $k \to \infty$, we have the result

$$c = (2/v) \lim_{q\to 0} [R_0 - |R|]/q^2. \tag{3.11.19}$$

However, this last result leads us to the Born relation between elastic constants, which is closely satisfied for Ge. It is therefore of interest that equation (3.11.19) reduces to equation (3.11.1) for any finite k provided only that if we write

$$R = |R| \exp(i\alpha) \quad \text{and} \quad T = |T| \exp(i\beta), \tag{3.11.20}$$

$\alpha = \beta$. This gives immediate information then on the shell model parameters for Ge (Cochran, 1959b).

Brockhouse and Iyengar (1958) have used neutron spectroscopy to get the phonon dispersion relations in Ge (see Chapter 5). A set of parameters can be found which gives a good fit to the elastic constants and to this neutron data, and we conclude by referring to Figure 3.10 which shows the nature of the agreement.

(c) *Difficulties of shell model applied to ionic crystals.* The predictions of the shell model are far superior to the "rigid-ion" class of models. In particular the rigid-ion model gives the high-frequency dielectric constant ε_∞ as unity. Using the Lyddane–Sachs–Teller relation given below in equation (3.11.31), this last result implies that the longitudinal optic frequencies $\omega_{LO}(\mathbf{q})$ will be considerably in error. This is not only true at $\mathbf{q} = 0$, but throughout a substantial portion of the BZ as well.

The shell model, as we have seen, allows a fit to ε_∞ to be made. However, in cubic lattices, the rigid-ion and the shell models may predict the same frequencies at specified wave vectors **q** (see, for example, Lidiard, 1969). If we consider the case of NaCl, to be definite then, viewed along a $\langle 111 \rangle$ direction, this is composed of layers of anions alternating with layers of

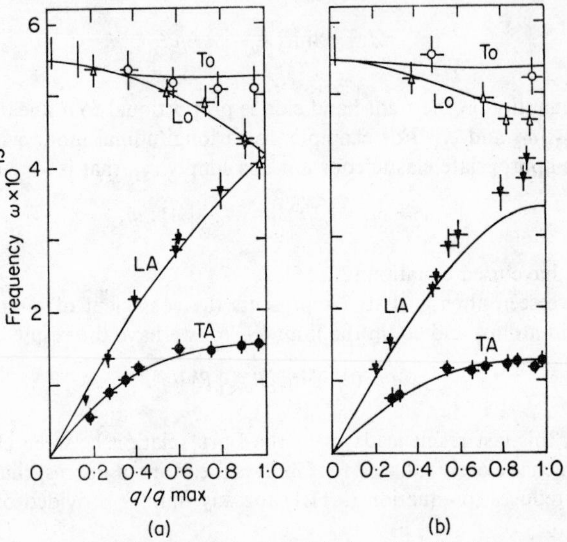

FIGURE 3.10. Phonon dispersion relations for Ge along symmetry directions. (a) [100], (b) [111]. Neutron experimental data is shown, together with solid curves predicted by shell model. Reproduced by permission of the National Research Council of Canada from the *Canadian Journal of Physics*, **43**, pp. 1187–1247 (1965).

cations, and for modes propagating along this direction, the lattice dynamics is just like that of a diatomic linear chain. In particular, at the zone boundary

$$\mathbf{q} = \frac{\pi}{r_0}(\tfrac{1}{2}, \tfrac{1}{2}, \tfrac{1}{2}),$$

r_0 being the smallest anion–cation separation, the longitudinal modes are such that the ions of one kind are stationary, while the two neighbouring layers of the other kind move in antiphase, the wavelength being twice the spacing between layers of the same kind.

Suppose now that the cations are lighter than the anions. Then the higher-frequency mode (LO) is associated with the anions at rest and the cations moving in antiphase, and vice versa for the lower (LA) frequency. The situation is reversed if the cations are heavier. Therefore, in these modes, the stationary ions are subject to compressional forces due to the antiphase vibrations of neighbouring planes, but *not* to forces which cause ionic deformation. Only the moving ions can be polarized and deformed in this mode, and if they are not deformable the rigid-ion and the shell models must agree in their predictions. For example, the cations can safely be assumed rigid in the Li and Na halides, and the rigid-ion and the shell models therefore give the same LO frequency (or LA for NaF).

The experimentally observed frequencies in this region often differ considerably from those calculated by these models, and we shall refer to some results on NaI in Chapter 5 to illustrate this.

On this basis, it has been proposed that this trouble is due to the omission from the model of any compressibility of the individual ions. Another degree of freedom, radial deformability, has therefore been added (Schröder, 1966; Melvin, Pirie and Smith, 1968); this has been termed the "breathing shell model". The predictions are much better, the LO modes being substantially improved. Such a model, in which one internal degree of freedom is such that the shell displaces relative to the core and the other degree of freedom is one in which the shell expands or contracts radially, appears to give quite a good representation of the lattice vibrations of ionic lattices. We shall see below just what a "shell model" implies about the microscopic theory.

(d) *Relation to microscopic theory*. The microscopic basis for the shell model, as Sinha (1969) has emphasized, can be expressed in terms of an approximation to the microscopic polarizability tensor which we shall now define.

It follows from the discussion of section 2.3.2 that

$$\operatorname{div}\mathbf{P} = -\rho_{\text{medium}} \equiv -\rho_m, \tag{3.11.21}$$

where $\mathbf{P}$ is the polarization. The microscopic polarizability tensor Π is now defined by

$$P_\alpha(\mathbf{q}+\mathbf{K}) = \sum_{\beta\mathbf{K}'} \Pi_{\alpha\beta}(\mathbf{q}+\mathbf{K}, \mathbf{q}+\mathbf{K}')\mathscr{E}_\beta(\mathbf{q}+\mathbf{K}'). \tag{3.11.22}$$

This is the generalization of the elementary relation $\mathbf{P} = \alpha\mathscr{E}$ used earlier in this section. The sum over $\mathbf{K}'$ is due to the fact that the application of an electric field, of wave-vector $\mathbf{q}$, to a periodic lattice leads to polarization contributions differing from the wave-vector of the applied field by reciprocal lattice-vectors $\mathbf{K}$ (cf. section 2.11.4).

Though Sinha's approximations leading to the shell model can now be summarized in terms of Π, we shall take this opportunity to relate it to the linear response function $\mathscr{F}$ which we have used frequently.

Applying a longitudinal field of a particular wave-vector $\mathbf{q}+\mathbf{K}$, related to the external potential V_{ext} by

$$\mathscr{E}_\lambda = -i(\mathbf{q}+\mathbf{K})\,V_{\text{ext}}(\mathbf{q}+\mathbf{K}) \tag{3.11.23}$$

we find from equation (3.11.22) the result

$$i\sum_\alpha (\mathbf{q}+\mathbf{K}')_\alpha P_\alpha(\mathbf{q}+\mathbf{K}') = -\rho_m(\mathbf{q}+\mathbf{K}'). \tag{3.11.24}$$

But ρ_m is related to the external potential via the response function $\mathscr{F}$ by

$$\rho_m(\mathbf{q}+\mathbf{K}') = \mathscr{F}(\mathbf{q}+\mathbf{K}',\mathbf{q}+\mathbf{K})\,V_{\text{ext}}(\mathbf{q}+\mathbf{K}) \tag{3.11.25}$$

and hence we have

$$i\sum_\alpha (\mathbf{q}+\mathbf{K}')_\alpha P_\alpha(\mathbf{q}+\mathbf{K}') = -\mathscr{F}(\mathbf{q}+\mathbf{K}',\mathbf{q}+\mathbf{K})\,V_{\text{ext}}(\mathbf{q}+\mathbf{K}). \tag{3.11.26}$$

Now we substitute for P_α from equation (3.11.22), when we find

$$i\sum_{\alpha\beta} (\mathbf{q}+\mathbf{K}')_\alpha \Pi_{\alpha\beta}(\mathbf{q}+\mathbf{K}',\mathbf{q}+\mathbf{K})\,\mathscr{E}_\beta(\mathbf{q}+\mathbf{K}) = -\mathscr{F}(\mathbf{q}+\mathbf{K}',\mathbf{q}+\mathbf{K})\,V_{\text{ext}}(\mathbf{q}+\mathbf{K}), \tag{3.11.27}$$

where the sum over reciprocal lattice vectors in equation (3.11.22) has been dropped because we have only a single Fourier component of the field.

Finally, we insert equation (3.11.23) for $\mathscr{E}_\beta$ and we find

$$\mathscr{F}(\mathbf{q}+\mathbf{K},\mathbf{q}+\mathbf{K}') = -\sum_{\alpha\beta} (\mathbf{q}+\mathbf{K})_\alpha \Pi_{\alpha\beta}(\mathbf{q}+\mathbf{K},\mathbf{q}+\mathbf{K}')(\mathbf{q}+\mathbf{K}')_\beta(\mathbf{q}+\mathbf{K}')_\beta. \tag{3.11.28}$$

Since we can write $\mathscr{F}$ in the form (in RPA)

$$\mathscr{F}(\mathbf{q}+\mathbf{K},\mathbf{q}+\mathbf{K}') = \sum_{\mathbf{k}\mathbf{k}'}{}' \frac{n(\mathbf{k})-n(\mathbf{k}')}{E(\mathbf{k})-E(\mathbf{k}')} \langle \psi_{\mathbf{k}}|\,\mathrm{e}^{-i(\mathbf{q}+\mathbf{K}).\mathbf{r}}\,|\psi_{\mathbf{k}'}\rangle \langle \psi_{\mathbf{k}'}|\,\mathrm{e}^{i(\mathbf{q}+\mathbf{K}').\mathbf{r}}\,|\psi_{\mathbf{k}}\rangle, \tag{3.11.29}$$

the prime indicating that terms $\mathbf{k} = \mathbf{k}'$ are to be excluded from the sum, it is easily seen that these relations lead to‡

$$\begin{aligned}\Pi_{\alpha\beta}(\mathbf{q}+\mathbf{K},\mathbf{q}+\mathbf{K}') &= \frac{\hbar^2}{m}\sum_{\mathbf{k}\mathbf{k}'}{}' \frac{n(\mathbf{k})-n(\mathbf{k}')}{E(\mathbf{k})-E(\mathbf{k}')} \\ &\times \left\{\left[E(\mathbf{k})-E(\mathbf{k}')+\frac{\hbar^2(\mathbf{q}+\mathbf{K})^2}{2m}\right]\right. \\ &\times \left.\left[E(\mathbf{k}')-E(\mathbf{k})+\frac{\hbar^2(\mathbf{q}+\mathbf{K}')^2}{2m}\right]\right\}^{-1} \\ &\times \langle \psi_{\mathbf{k}}|\,P_\alpha \mathrm{e}^{-i(\mathbf{q}+\mathbf{K}).\mathbf{r}}\,|\psi_{\mathbf{k}'}\rangle \langle \psi_{\mathbf{k}'}|\,P_\beta \mathrm{e}^{i(\mathbf{q}+\mathbf{K}).\mathbf{r}}\,|\psi_{\mathbf{k}}\rangle.\end{aligned} \tag{3.11.30}$$

‡ We have used the result of Pines (1963) that

$$\langle \psi_{\mathbf{k}}\,|\,\mathrm{e}^{i\mathbf{K}.\mathbf{r}}\,|\,\psi_{\mathbf{k}'}\rangle = \frac{\hbar}{m}\sum_\alpha \frac{k_\alpha \langle \psi_{\mathbf{k}}\,|\,P_\alpha\,\mathrm{e}^{i\mathbf{K}.\mathbf{r}}\,|\,\psi_{\mathbf{k}'}\rangle}{E(\mathbf{k})-E(\mathbf{k}')+\hbar^2 k^2/2m},$$

where P_α is the momentum operator. This is valid for any eigenfunctions of a crystal Hamiltonian.

In the case $\mathbf{K} = \mathbf{K}' = 0$ and $\mathbf{q} \to 0$, this reduces to the static polarizability tensor $\Pi_{\alpha\beta}(0)$ for the crystal. This equation is quite general, applying to the case of crystals with more than one atom per unit cell.

The central approximation of the shell model, which allows a solution to be obtained without explicitly inverting a dielectric matrix, which is a very difficult task, is to write $\Pi_{\alpha\beta}(\mathbf{q}+\mathbf{K}, \mathbf{q}+\mathbf{K}')$ as a product of three functions, which depend only on variables $\mathbf{q}+\mathbf{K}$, $\mathbf{q}$ and $\mathbf{q}+\mathbf{K}'$ respectively.

The validity of such an approximation depends, clearly, on the band structure of the crystal under discussion.

3.11.2 Lyddane–Sachs–Teller relation

An important result relating the transverse optic and longitudinal optic frequencies to the static and high-frequency dielectric constants in ionic crystals was first given by Lyddane, Sachs and Teller (1941). The result is

$$\frac{\omega_{l0}^2}{\omega_{t0}^2} = \frac{\varepsilon_s}{\varepsilon_\infty} \tag{3.11.31}$$

and the proof is given in Appendix 3.5.

Since it is easier experimentally to obtain ω_{t0}, this is a valuable way to determine ω_{l0}. In ionic crystals, as we have seen, the optical modes result in the positive ions moving in opposition to the negative ions.

In the case of covalent crystals, $\varepsilon_{\text{static}} = \varepsilon_\infty$ and thus the relation (3.11.31) predicts that the longitudinal and transverse optical modes are degenerate at $\mathbf{q} = 0$. This can be shown to be true on very general grounds. Cochran has shown that the Lyddane–Sachs–Teller relation (3.11.31) can be obtained from the shell model described above.

3.12 Electronic contribution to lattice dynamics in insulators

We have already taken up the detailed calculation from electron theory of dispersion relations in simple metals in section 3.10. We wish now to discuss the electronic contribution to the phonon spectrum of insulators, which can be formulated in terms of the microscopic response function (Sham, 1969).

From the previous discussion of the shell model in section 3.11.1, we have seen that allowance has to be made for the electronic contribution. But one of the troubles of approaches like the shell model, in spite of very obvious qualitative successes, is that the values of the parameters obtained by fitting the model to experiment do not always agree with the physical interpretation of the parameters. Thus, a more fundamental approach is called for and this has been provided by Sham.

What we shall be concerned with here is to obtain the long-wavelength behaviour of the lattice dynamical properties, from the microscopic formulation of the lattice motion in terms of the ions and the response of the electrons.

Such a microscopic formulation leads to the Lyddane–Sachs–Teller relation for the optical modes which we have just discussed. Again the existence of an acoustic mode, with $\omega \propto q$ in the small q limit, which we proved above in the case of metals, needs separate proof in an insulator.

We shall use this theory in Chapter 7 to obtain formulae for the macroscopic optical constants (near infra-red) in terms of microscopic quantities.

The discussion will be restricted to normal insulators. Thus the excitonic insulator, which we shall refer to briefly in Chapter 9, will not be included in the discussion.

3.12.1 *Dynamical matrix*

To extend our previous discussion of the dynamical matrix, we again separate it into two contributions:

(i) The direct Coulomb interaction between nuclei, Φ^N

(ii) The interaction via the electrons Φ^E.

The direct interaction Φ^N can be handled by well-established methods due to Ewald and many others (cf. Born and Huang, p. 254, 1954).

If the li nucleus has charge $Z_i e$ and mass M_i, then we can write

$$\phi^N_{\lambda\lambda'}(\mathbf{q}ii') = e^{i\mathbf{q}\cdot(\mathbf{R}_i-\mathbf{R}_{i'})}(Z_i Z_{i'} e^2)/(M_i M_{i'})^{\frac{1}{2}}$$

$$\times\left[\frac{4\pi q_\lambda q_{\lambda'}}{\Omega q^2} - Q_{\lambda\lambda'}(\mathbf{q}ii') + \delta_{ii'}\sum_{i''}(Z_{i''}/Z_i)\,Q_{\lambda\lambda'}(0, ii'')\right], \tag{3.12.1}$$

where Ω as usual is the volume of a unit cell. When a phonon of wave vector $\mathbf{q}$ is present, the most singular part of the dynamical matrix Φ^N, due to the long-range Coulomb singularity as $q \to 0$, is isolated in the first term in the square brackets. $Q_{\lambda\lambda'}(\mathbf{q}ii')$ contains the rest of the response from the bare nucleus. This therefore includes Bragg diffraction terms (see Chapter 5 for a full discussion) but is continuous at $q = 0$. The dynamical matrix for $\mathbf{q} = 0$ is taken *not* to include the first divergent term. The last term in the square brackets is a consequence of the infinitesimal invariance relation for the force constants due to the bare nuclei and only, in turn, this guarantees zero contribution to ω^2 for the acoustic modes at $\mathbf{q} = 0$.

3.12.2 *Effects due to electrons*

In the adiabatic and harmonic approximations we have so far adopted, the force constants between nuclei due to the electrons are to be found from the effective nuclear potential energy due to the presence of electrons, of second order in the ion displacements $\mathbf{u}_{li}$. These are due to the deviation of the electron–nucleus potential from the perfect crystal and consist of two types of terms:

(i) The second-order contribution from the deviation of the electron–nucleus potential of the first order in $\mathbf{u}$.

(ii) The first-order contribution from the deviation of the second order in $\mathbf{u}$.

The electronic contribution to the force constants has been given previously in equation (3.10.16). Inserting into this equation

$$\frac{\partial\rho(\mathbf{r})}{\partial x_\lambda} = \int d\mathbf{r}'\mathscr{F}(\mathbf{r}\mathbf{r}')\sum_{li}\frac{\partial V(\mathbf{r}'-\mathbf{R}_{li};i)}{\partial x'_\lambda} \tag{3.12.2}$$

obtained from equation (2.11.43) of Chapter 2, we find for the electronic contribution to the force constants

$$\begin{aligned}\Phi^{e}_{\lambda\lambda'}(li,l'i') &\\ = \iint d\mathbf{r}\,d\mathbf{r}'\Bigg[&\frac{\partial V(\mathbf{r}-\mathbf{R}_{li};i)}{\partial x_\lambda}\mathscr{F}(\mathbf{r}\mathbf{r}')\frac{\partial V(\mathbf{r}'-\mathbf{R}_{l'i'})}{\partial x'_{\lambda'}}\\ &-\delta_{ll'}\,\delta_{ii'}\sum_{l''i''}\frac{\partial V(\mathbf{r}-\mathbf{R}_{li};i)}{\partial x_\lambda}\mathscr{F}(\mathbf{r}\mathbf{r}')\frac{\partial V(\mathbf{r}'-\mathbf{R}_{l''i''};i'')}{\partial x'_{\lambda'}}\Bigg],\end{aligned} \tag{3.12.3}$$

where the interaction between an electron at $\mathbf{r}$ and the nucleus at $\mathbf{R}_{li}+\mathbf{u}_{li}$ is given by

$$V(\mathbf{r}-\mathbf{R}_{li}-\mathbf{u}_{li};i) = -Z_i e^2\Big/|\mathbf{r}-\mathbf{R}_{li}-\mathbf{u}_{li}|. \tag{3.12.4}$$

By Fourier transform, we find that the electronic contribution to the force constants is given by

$$\phi^{E}_{\lambda\lambda'}(\mathbf{q}kk') = (M_k M_{k'})^{-\frac{1}{2}}\sum_{\mathbf{K}\mathbf{K}'}[X_{\lambda\lambda'}(\mathbf{q}+\mathbf{K}k,\mathbf{q}+\mathbf{K}'k')-\delta_{kk'}\sum_{k''}X_{\lambda\lambda'}(\mathbf{K}k,\mathbf{K}'k''), \tag{3.12.5}$$

where $\mathbf{K}$ and $\mathbf{K}'$, as usual, are reciprocal lattice vectors and

$$\begin{aligned}X_{\lambda\lambda'}(\mathbf{q}+\mathbf{K}k,\mathbf{q}+\mathbf{K}'k') &= \Omega^{-1}e^{i(\mathbf{q}+\mathbf{K}).\mathbf{R}_k}V(-\mathbf{q}-\mathbf{K},k)(\mathbf{q}+\mathbf{K})_\lambda\\ &\times\mathscr{F}(\mathbf{q}+\mathbf{K},\mathbf{q}+\mathbf{K}')\,V(\mathbf{q}+\mathbf{K}',k')(\mathbf{q}+\mathbf{K}')_{\lambda'}e^{i(\mathbf{q}+\mathbf{K}').\mathbf{R}_{k'}},\end{aligned} \tag{3.12.6}$$

$V(\mathbf{q}+\mathbf{K}, i)$ being the Fourier transform of the electron–nucleus potential $V(\mathbf{r}i)$ of equation (3.12.4), and $\mathbf{R}_k$ being $\mathbf{R}_{lk}-\mathbf{l}$.

Thus, we get the dynamical matrix for phonons in terms of the basic interactions.

3.12.3 *Acoustic phonons as* $q \to 0$

The discussion given above is quite general. But we now wish to examine the limit of the dynamical matrix as $q \to 0$, $q \neq 0$ for insulators, including intrinsic semi-conductors at low temperatures.

In order to isolate singular terms, we must consider the behaviour of the response function as $q \to 0$ in an insulator (see Sham, 1966). The density response function $\mathscr{F}(\mathbf{q}+\mathbf{K}, \mathbf{q}+\mathbf{K}')$ used above is the sum of all polarization diagrams. Suppose the proper polarization part is denoted by $\tilde{\mathscr{F}}$. In an insulator at $T = 0$ and $q \to 0$, we can write (see Chapter 2, section 2.13)

$$\left.\begin{aligned}
\tilde{\mathscr{F}}(\mathbf{q}, \mathbf{q}) &= \sum_{\lambda\lambda'} q_\lambda \tilde{\mathscr{F}}^{(2)}_{\lambda\lambda'} q_{\lambda'} + O(q^4), \\
\tilde{\mathscr{F}}(\mathbf{q}, \mathbf{q}+\mathbf{K}) &= \sum_{\lambda} q_\lambda \tilde{\mathscr{F}}^{(1)}_{\lambda}(0, \mathbf{K}) + O(q^2), \\
\tilde{\mathscr{F}}(\mathbf{q}+\mathbf{K}, \mathbf{q}) &= \sum_{\lambda} \tilde{\mathscr{F}}^{(1)}_{\lambda}(\mathbf{K}, 0)\, q_\lambda + O(q^2), \\
\tilde{\mathscr{F}}(\mathbf{q}+\mathbf{K}, \mathbf{q}+\mathbf{K}') &= \tilde{\mathscr{F}}(\mathbf{K}, \mathbf{K}') + O(q),
\end{aligned}\right\} \tag{3.12.7}$$

where $\mathbf{K}$ and $\mathbf{K}'$ are not equal to zero.

To find the leading term in $\tilde{\mathscr{F}}$ for small q, we use the work of Ambegaokar and Kohn (1960) in defining $P(\mathbf{q}+\mathbf{K}, \mathbf{q}+\mathbf{K}')$ as in section 2.13 to be the sum of all polarization diagrams not containing the Coulomb line $v(q)$.

Since P has the same small q behaviour as $\tilde{\mathscr{F}}$, by expressing $\mathscr{F}$ in terms of P we can find the long-wavelength behaviour of $\mathscr{F}$. For example, we have for the rigid lattice on which the theory has been based

$$\mathscr{F}(\mathbf{q}, \mathbf{q}) = P(\mathbf{q}, \mathbf{q})/[1 - v(\mathbf{q})\, P(\mathbf{q}, \mathbf{q})]. \tag{3.12.8}$$

It is also readily shown that the relation to the true dielectric function ε of the entire system, in which, of course, the ions are free to move, is

$$\varepsilon_\infty(\mathbf{q}) = 1 - v(\mathbf{q})\, P(\mathbf{q}, \mathbf{q}), \tag{3.12.9}$$

where the subscript infinity indicates that ε is evaluated at a frequency sufficiently high that the lattice vibrations do not contribute. At small nonzero $\mathbf{q}$ equal to $q\hat{\mathbf{q}}$, say, we have, with $P^{(n)}$ the nth derivative of P,

$$\varepsilon_\infty(q\hat{\mathbf{q}}) = 1 - 4\pi e^2 \sum_{\lambda\lambda'} \hat{q}_\lambda\, P^{(2)}_{\lambda\lambda'}(0)\, \hat{q}_{\lambda'}, \tag{3.12.10}$$

where this is seen to be independent of q, and we shall therefore denote it by $\varepsilon_\infty(\hat{q})$. Thus we may write, at long wavelengths,

$$\mathscr{F}(\mathbf{q},\mathbf{q}) = \frac{q^2}{4\pi e^2}\left[\frac{1}{\varepsilon_\infty(\hat{q})}-1\right], \tag{3.12.11}$$

an equation one could obtain directly from the general relation (2.11.55) between $\mathscr{F}$ and ε. However, we can also show, at small q, that

$$\mathscr{F}(\mathbf{q},\mathbf{q}+\mathbf{K}) \simeq \sum_\lambda q_\lambda P_\lambda^{(1)}(0,\mathbf{K})/\varepsilon_\infty(\hat{q}),$$

$$\mathscr{F}(\mathbf{q}+\mathbf{K},q) = \sum_\lambda P_\lambda^{(1)}(\mathbf{K},0)\, q_\lambda/\varepsilon_\infty(\hat{q}),$$

$$\mathscr{F}(\mathbf{q}+\mathbf{K},\mathbf{q}+\mathbf{K}') \simeq \mathscr{F}(\mathbf{K},\mathbf{K}')+\sum_{\lambda\lambda'} P_\lambda^{(1)}(\mathbf{K},0)\,\hat{q}_\lambda \frac{4\pi e^2}{\varepsilon(\hat{q})}\,\hat{q}_{\lambda'} P_{\lambda'}^{(1)}(0,\mathbf{K}). \tag{3.12.12}$$

For small q, we can therefore get, by summing equation (3.12.6) over $\mathbf{K}$ and $\mathbf{K}'$, the result

$$\sum_{\mathbf{K}\mathbf{K}'} X_{\lambda\lambda'}(\mathbf{q}+\mathbf{K},i;\mathbf{q}+\mathbf{K}',i')$$

$$= \sum_{\mu\mu'} e^{i\mathbf{q}\cdot(\mathbf{R}_i-\mathbf{R}_{i'})}\left\{-\frac{4\pi e^2 Z_i Z_{i'}}{\Omega}\frac{q_\lambda q_{\lambda'}}{q^2}+\frac{4\pi e^2}{\Omega\varepsilon_\infty(\hat{\mathbf{q}})}[Z_i\,\delta_{\lambda\mu}-Z^\dagger_{\lambda\mu}(i)]\right.$$

$$\left.\times\frac{q_\mu q_{\mu'}}{q^2}[Z_{i'}\,\delta_{\mu'\lambda'}-Z_{\mu'\lambda'}(\mathbf{k}')]+S_{\lambda\lambda'}(\mathbf{q},ii')\right\}. \tag{3.12.13}$$

The first two terms on the right are irregular, since they depend on the direction of $\mathbf{q}$ as $q\to 0$. However, since we use the convention $v(q)=0$, $q=0$ (see Chapter 2) they vanish in this limit.

The matrix $Z_{\mu\lambda}(i)$ is the effective charge acquired by the nuclei by dragging the electrons with them as they move. Specifically

$$Z_{\mu\lambda}(i) = \sum_{\mathbf{K}} P_\mu^{(1)}(0,K)\, V(\mathbf{K}i)\, K_\lambda\, e^{-i\mathbf{K}\cdot\mathbf{R}_i}. \tag{3.12.14}$$

In the tight-binding limit (cf. Problem 3.5) $Z_i\,\delta_{\mu\lambda}-Z_{\mu\lambda}(i)$ is simply the ionic charge and $\varepsilon_\infty(\hat{\mathbf{q}})$ is unity. The third term on the right-hand side of equation (3.12.13) is the regular part. As $q\to 0$,

$$S_{\lambda\lambda'}(0,ii')\to \sum_{\mathbf{K}\mathbf{K}'} X_{\lambda\lambda'}(\mathbf{K}i,\mathbf{K}'i'). \tag{3.12.15}$$

We may finally combine all our previous results to rewrite the total dynamical matrix as

$$\Phi_{\lambda\lambda'}(\mathbf{q}, ii') = (M_i M_{i'})^{-\frac{1}{2}} e^{i\mathbf{q}.(\mathbf{R}_i-\mathbf{R}_{i'})}$$

$$\times \sum_{\mu\mu'} \left\{ \frac{4\pi e^2}{\Omega q^2 \varepsilon_\infty(\hat{\mathbf{q}})} [Z_i \delta_{\lambda\mu} - Z^\dagger_{\lambda\mu}(i)] q_{\mu\mu'} [Z_{i'} \delta_{\mu'\lambda'} - Z_{\mu'\lambda'}(i')] \right.$$

$$\left. + T_{\lambda\lambda'}(\mathbf{q}, ii') - \delta_{ii'} \sum_j T_{\lambda\lambda'}(0, ij) \right\}. \tag{3.12.16}$$

Here

$$T_{\lambda\lambda'}(\mathbf{q}, ii') = -Z_i Z_{i'} Q_{\lambda\lambda'}(\mathbf{q}, ii') + S_{\lambda\lambda'}(\mathbf{q}, ii'), \tag{3.12.17}$$

this being continuous as $q \to 0$. Thus the long-range and short-range parts of the ion–ion interaction have been separated from one another, the last two terms on the right-hand side of equation (3.12.16) contributing terms of order q^2 to the square of the frequency of the acoustic modes as $q \to 0$.

In order that the phonon frequencies be renormalized such that $\omega \propto q$ at small q, the following sum rule must be valid

$$\sum_i Z_{\mu\lambda}(i) = \sum_i Z_i \delta_{\mu\lambda} \tag{3.12.18}$$

in which case the long-range part contributes to ω^2 no term of order unity.

To prove the sum rule, we take the derivative of the expression

$$\tilde{\mathscr{F}}(\mathbf{k}, \mathbf{k}+\mathbf{K}) = \mathscr{V}^{-1} \iint d\mathbf{r}\, \mathbf{r}'\, e^{-i\mathbf{k}.\mathbf{r}} \mathscr{F}(\mathbf{r}\mathbf{r}') e^{i(\mathbf{k}+\mathbf{K}).\mathbf{r}'}, \tag{3.12.19}$$

$\mathscr{V}$ being as usual the volume of the crystal. Then we have

$$\tilde{\mathscr{F}}^{(1)}_\lambda(0, K) = -i\mathscr{V}^{-1} \iint d\mathbf{r}\, d\mathbf{r}'\, x_\lambda \tilde{\mathscr{F}}(\mathbf{r}\mathbf{r}') e^{i\mathbf{K}.\mathbf{r}}. \tag{3.12.20}$$

Since

$$P^{(1)}_\mu(0, \mathbf{K}) = \sum_{\mathbf{K}'} \tilde{\mathscr{F}}_\mu(0, \mathbf{K}')/\varepsilon(\mathbf{K}', \mathbf{K}) \tag{3.12.21}$$

we find from equation (3.12.14) that

$$Z_{\mu\lambda}(i) = \sum_{\mathbf{K}\mathbf{K}'} \int \tilde{\mathscr{F}}^{(1)}_\mu(0, \mathbf{K})\, \varepsilon^{-1}(\mathbf{K}, \mathbf{K}')\, v(\mathbf{K}', i)\, K'_\lambda e^{-i\mathbf{K}'.\mathbf{R}_i}.$$

$$= -N^{-1} \iint d\mathbf{r}\, d\mathbf{r}'\, X_\mu \mathscr{F}(\mathbf{r}\mathbf{r}') \sum_i \frac{\partial V}{\partial x_{\lambda'}} (\mathbf{r}' - \mathbf{R}_{li}; i) \tag{3.12.22}$$

which we write as

$$Z_{\mu\lambda}(i) = N^{-1} \int d\mathbf{r}\, x_\mu\, \delta\rho_\lambda(\mathbf{r}, i). \tag{3.12.23}$$

On the other hand, in terms of the quantities $\delta\rho_\lambda$, we see from equation (3.12.2) that

$$\frac{\partial\rho(\mathbf{r})}{\partial x_\lambda} = \sum_i \delta\rho_i(\mathbf{r}, i). \tag{3.12.24}$$

It then follows that we can also write

$$\sum_i Z_{\mu\lambda}(i) = \delta_{\mu\lambda}\int \rho(\mathbf{r})\, d\mathbf{r} = \delta_{\mu\lambda}\sum_i Z_i \tag{3.12.25}$$

which completes the proof.

In more physical terms, $-e\sum_\lambda Z_{\mu\lambda}(i)\,U_\lambda$ is the dipole moment of the displaced electronic charge when the sub-lattice of type i nuclei is translated rigidly through a vector $\mathbf{U}$, while $e[Z_i\,\delta_{\mu\lambda} - Z_{\mu\lambda}(i)]$ is the effective charge associated with the i-type nuclei when their sub-lattice is moved bodily relative to the rest of the crystal.

This formalism evidently provides a route by which the effective charge can be calculated from a knowledge of the electronic band structure.

In Chapter 7, we shall return to this formalism in connection with infra-red optical properties, which evidently requires the above argument to be used for the optical modes. We shall see there, quite explicitly, that the reason why we get no contribution of order 1 to ω^2 for the acoustic modes is because the components of the polarization vectors are independent of i, the atoms vibrating in phase. Then the sum rule (3.12.18) suffices to prove that ω^2 is $O(q^2)$ as $q \to 0$ for the acoustic modes.

3.12.4 Finite phonon lifetimes due to electron–phonon interactions

For most of the remainder of this chapter, we shall discuss the effect of anharmonicity on the vibrational spectrum of crystals. But we ought to note here that one of the effects of anharmonicity, viz. the finite lifetime of phonons, also is present within a harmonic theory if the adiabatic approximation is transcended. This is seen most simply using the formalism developed in the present section. To go beyond the adiabatic approximation, the response function $\mathscr{F}$ of equation (3.12.2) is taken to depend on frequency ω. In general, it will then have an imaginary part, so that the dynamical matrix also becomes complex. The inverse lifetime τ^{-1} can be identified with the imaginary part of the phonon frequency, just as for the quasi-particles of an electronic system, discussed in Chapter 2.

We therefore obtain

$$\frac{1}{\tau_\sigma(\mathbf{q})} = -\frac{1}{2\omega_\mathbf{q}}\langle \boldsymbol{\varepsilon}_\sigma | \operatorname{Im} D(\mathbf{q}, \omega_\mathbf{q}) | \boldsymbol{\varepsilon}_\sigma\rangle \tag{3.12.26}$$

on the understanding that the real part of D is much greater than the imaginary part and is taken to define the real phonon frequencies $\omega_{\mathbf{q}}$.

This result, like that of equation (3.12.26), applies to any crystal. A calculation of the phonon lifetimes in a metal, namely Al, has been made, within the harmonic approximation, by Björkman, Lundqvist and Sjölander (1967).

3.13 Interaction between phonons

The beautiful theory of independent phonons, which we have discussed rather fully in the earlier sections of this chapter, has had many successes, and its quantitative agreement with experiment in connection with neutron-scattering measurements will be discussed further in Chapter 5. The neutron experiments reveal, however, that phonons really interact with one another and that this becomes increasingly important at high temperatures. This should occasion no great surprise: we had omitted such effects by expanding the crystal potential energy in powers of the displacements and retaining only the harmonic terms.

Two routes lie open to us for transcending this approximation. The obvious one is to include higher terms in the displacements in the crystal potential energy. This we might expect to work for properties that are already represented by the harmonic theory. On the other hand, there is no very basic reason for expecting those effects which depend entirely on "anharmonic" terms to be well given by including a few more terms.

Thus, a second route which has recently been considered is to tackle the lattice dynamics problem on a computer, by a method which has had great success in dealing with the behaviour of atoms in liquids: the method of molecular dynamics. The trouble here is, of course, that, without having available analytic approximations, one must rely on the numerical results being able to show up the main features of the problem. Though we shall report in section 3.15 below some results of such machine calculations, we shall begin with a discussion of "corrections" to the harmonic approximation, bearing in mind the possible limitations of such an approach.

3.13.1 Inclusion of anharmonicity in Hamiltonian

The crystal potential energy is, when expanded in powers of the displacements from equilibrium positions,

$$\Phi = \Phi_0+\Phi_2+\Phi_3+\Phi_4+\ldots \tag{3.13.1}$$

Φ_0 and Φ_2 have already been given (equation 3.2.4). Analogously,

$$\Phi_3 = \sum_{\substack{\mathbf{lmn}\\ \alpha\beta\gamma}} \Phi_{\alpha\beta\gamma}(\mathbf{lmn})\, u_l^\alpha u_m^\beta u_n^\gamma, \tag{3.13.2}$$

$$\Phi_4 = \sum_{\substack{\mathbf{lmnp} \\ \alpha\beta\gamma\delta}} \Phi_{\alpha\beta\gamma\delta}(\mathbf{lmnp})\, u_l^{\alpha} n_m^{\beta} u_n^{\gamma} u_p^{\delta}, \tag{3.13.3}$$

where the "coupling parameters" are given by

$$\Phi_{\alpha\beta\gamma}(\mathbf{lmn}) = \frac{\partial^3 \Phi}{\partial u_l^{\alpha}\, \partial u_m^{\beta}\, \partial u_n^{\gamma}}, \quad \text{etc.} \tag{3.13.4}$$

We have found that if it is assumed that Φ_3, Φ_4, etc. are all zero, equilibrium is determined *statically*. We shall, where necessary, denote expansion about such equilibrium points by tildes:

$$\Phi = \tilde{\Phi}_0 + \tilde{\Phi}_2 + \tilde{\Phi}_3 + \tilde{\Phi}_4 + \dots. \tag{3.13.5}$$

Now, although they may be small, it can be shown that the deviations from harmonicity $\tilde{\Phi}_3$, $\tilde{\Phi}_4$ are never *exactly* zero; consideration of rotational invariance producing relations between coupling parameters of nth and $(n+1)$th orders. Such anharmonicity is referred to as *induced*. An example is easily given for central forces. Equation (3.4.5) shows that the harmonic terms entail $\partial^2 \phi/\partial r^2$ and $(1/r)\, \partial\phi/\partial r$, and a further differentiation shows that Φ_3 depends on $\partial^2 \phi/\partial r^2$ as well as $\partial^3 \phi/\partial r^3$. Thus, there is an anharmonic term even if $\partial^3 \phi/\partial r^3 = 0$. We shall content ourselves with this single example of induced anharmonicity, referring the reader to Leibfried and Ludwig (1961) for full details, since it turns out that such anharmonicity is usually only a small part of the total.

The effect of the perturbations Φ_3 and Φ_4 (we shall consider no others) is most elegantly treated by the many-body techniques originally developed in quantum field theory. The treatment reveals the picture of an interacting phonon field: of phonons colliding, splitting or combining because of the presence of anharmonic terms. Such a picture is very valuable and will be presented fully below. First, however, we shall set up the Hamiltonian in second-quantized form. It will be necessary to briefly discuss thermal expansion; because the lattice displacements must be considered about the true equilibrium positions, not the static ones, the effect of thermal expansion will thereby be included in our Hamiltonian before applying many-body perturbation theory.

Before going further, we mention the *quasi-harmonic* approximation. In this approximation we expand Φ about the actual equilibrium positions, cutting the series off at Φ_2, so obtaining a dynamical Hamiltonian harmonic in character. Such an approximation is hardly consistent because application of the virial theorem will still give us the static equilibrium condition.‡

‡ It is optimistic to expect the virial theorem to give us precisely the same equilibrium condition as minimization of the free energy in *any* approximate (perturbative) theory. The two methods ought usually to give similar results, however, and indeed do so to $O(\Phi_4)$ and Φ_3^2.

However, it is a physically very reasonable approximation—for example, we would expect the frequencies measured by neutron diffraction techniques (see Chapter 5) to be closer to the quasi-harmonic frequencies than those of the purely harmonic approximation. We also see that it allows thermal expansion to be included in the theory.

3.13.2 Normal coordinates and creation operators

In general, the crystal is "strained" away from the static equilibrium condition. Because we wish to perturb the harmonic solution it is still highly desirable to define normal coordinates by the relation

$$\mathbf{u}_l = \sum_{\mathbf{k}\sigma} Q_{\mathbf{k}\sigma}\, \varepsilon_{\mathbf{k}\sigma}\, e^{i\mathbf{k}.\mathbf{l}}, \tag{3.13.6}$$

but it must now be understood that the $\mathbf{u}_l$'s refer to displacements from the true equilibrium positions, not the static ones. We can see that unless we are to be content with the quasi-harmonic approximation, we must re-examine the significance of the normal coordinates when anharmonicity is present.

Let us refer back to the basic Hamiltonian

$$H = \frac{1}{2M}\sum_{\mathbf{l}\alpha} \dot{u}_l^{\alpha^2} + \Phi_0 + \Phi_2 + \Phi_3 + \Phi_4 + \ldots. \tag{3.13.7}$$

If we do so, we see that the analysis beginning with equation (3.2.10) and leading to equation (3.2.21) is independent of the presence of Φ_3, Φ_4, etc., despite the fact that (3.10.9) no longer holds. Equation (3.13.7) may thus be rewritten

$$H = \frac{1}{2}\sum_{\mathbf{k}\sigma} [\dot{Q}^*_{\mathbf{k}\sigma}\dot{Q}_{\mathbf{k}\sigma} + Q^*_{\mathbf{k}\sigma}Q_{\mathbf{k}\sigma}\,\omega^2(\mathbf{k},\sigma)] + \Phi_3 + \Phi_4 + \ldots. \tag{3.13.8}$$

The annihilation operators $a_{\mathbf{k}\sigma}$ are again defined by (3.7.3):

$$a_{\mathbf{k}\sigma} = (2\omega\hbar)^{-\frac{1}{2}}[\omega(\mathbf{k}\sigma)\,Q_{\mathbf{k}\sigma} + i\dot{Q}_{\mathbf{k}\sigma}]. \tag{3.13.9}$$

Hamilton's equations may no longer be used to obtain the commutation rule (3.7.5) because of the correction terms in (3.13.8). But we note that the simpler argument of section 3.7 to obtain (3.7.5) may be replaced by an argument using

$$\left[\hbar\frac{\partial}{\partial \dot{u}_l^{\alpha}}, \dot{u}_l^{\alpha}\right] = \hbar, \tag{3.13.10}$$

which leads to the same result and is applicable in the present case also. Thus the commutator

$$[a_{\mathbf{k}'\sigma'}, a^{\dagger}_{\mathbf{k}\sigma}] = \delta(\mathbf{k}'\mathbf{k})\,\delta_{\sigma'\sigma} \tag{3.13.11}$$

stands, and in second-quantized form,

$$H = \sum_{\mathbf{k}\sigma} \hbar\omega(\mathbf{k}\sigma)\,[a^{\dagger}_{\mathbf{k}\sigma}a_{\mathbf{k}\sigma}+\tfrac{1}{2}]+\Phi_3+\Phi_4+\dots \tag{3.13.12}$$

We shall see what form, explicit in $a_{\mathbf{k}\sigma}$ and $a^{\dagger}_{\mathbf{k}\sigma}$, Φ_3 and Φ_4 take, in the next section.

3.13.3 *Anharmonic terms in second quantization*

By using equation (3.13.6) it is possible for us to express (3.13.2) in terms of the normal coordinates $Q_{\mathbf{k}\sigma}$. Let us therefore again use the operators

$$q_{\mathbf{k}\sigma} = a_{\mathbf{k}\sigma}+a^{\dagger}_{-\mathbf{k}\sigma}. \tag{3.13.13}$$

It will be seen from equations (3.7.3) and (3.7.4) that we then have

$$Q_{\mathbf{k}\sigma} = \left(\frac{\hbar}{2\omega}\right)^{\frac{1}{2}} q_{\mathbf{k}\sigma}, \tag{3.13.14}$$

and the operators for the displacements are

$$u^{\alpha}_{\mathbf{l}} = \frac{1}{\sqrt{(NM)}}\sum_{\mathbf{k}\sigma}\left(\frac{\hbar}{2\omega}\right)^{\frac{1}{2}} \varepsilon^{\alpha}_{\mathbf{k}\sigma}q_{\mathbf{k}\sigma}e^{i\mathbf{k}.\mathbf{l}}. \tag{3.13.15}$$

Thus

$$u^{\alpha}_{\mathbf{l}}u^{\beta}_{\mathbf{m}}u^{\gamma}_{\mathbf{n}} = \frac{\hbar^{\frac{3}{2}}}{(NM)^{\frac{3}{2}}}\sum_{\sigma_1\sigma_2\sigma_3}\sum_{\mathbf{k}_1\mathbf{k}_2\mathbf{k}_3} \varepsilon^{\alpha}_{\mathbf{k}_1\sigma_1}\varepsilon^{\beta}_{\mathbf{k}_2\sigma_2}\varepsilon^{\gamma}_{\mathbf{k}_3\sigma_3}\exp[i(\mathbf{k}_1.\mathbf{l}+\mathbf{k}_2.\mathbf{m}+\mathbf{k}_3.\mathbf{n})]$$
$$\times\left(\frac{\hbar}{2\omega}\right)^{\frac{1}{2}} q_{\mathbf{k}_1\sigma_1}q_{\mathbf{k}_2\sigma_2}q_{\mathbf{k}_3\sigma_3}$$

and we can write

$$\Phi_3 = \sum_{\mathbf{k}_1\mathbf{k}_2\mathbf{k}_3}\sum_{\sigma_1\sigma_2\sigma_3} V\begin{pmatrix}\sigma_1\,\sigma_2\,\sigma_3\\ \mathbf{k}_1\,\mathbf{k}_2\,\mathbf{k}_3\end{pmatrix} q_{\mathbf{k}_1\sigma}q_{\mathbf{k}_2\sigma}q_{\mathbf{k}_3\sigma} \tag{3.13.16}$$

with

$$V\begin{pmatrix}\sigma_1\,\sigma_2\,\sigma_3\\ \mathbf{k}_1\,\mathbf{k}_2\,\mathbf{k}_3\end{pmatrix} = \sum_{\substack{\mathbf{lmn}\\ \alpha\beta\gamma}}\Phi_{\alpha\beta\gamma}(\mathbf{lmn})\,\varepsilon^{\alpha}_{\mathbf{k}_1\sigma_1}\varepsilon^{\beta}_{\mathbf{k}_2\sigma_2}\varepsilon^{\gamma}_{\mathbf{k}_3\sigma_3}\exp[i(\mathbf{k}_1.\mathbf{l}+\mathbf{k}_2.\mathbf{m}+\mathbf{k}_3.\mathbf{n})]. \tag{3.13.17}$$

Now it must be noticed that since we sum over $\mathbf{l}$, $\mathbf{m}$, and $\mathbf{n}$, we may add any lattice vector, say $\mathbf{p}$, to $\mathbf{l}$, $\mathbf{m}$ and $\mathbf{n}$ in equation (3.13.17) without altering the result. But the equivalence of every lattice cell means that

$$\Phi_{\alpha\beta\gamma}(\mathbf{l}+\mathbf{p},\mathbf{m}+\mathbf{p},\mathbf{n}+\mathbf{p}) = \Phi_{\alpha\beta\gamma}(\mathbf{l},\mathbf{m},\mathbf{n}). \tag{3.13.18}$$

Thus, summing over $\mathbf{p}$,

$$V\begin{pmatrix}\sigma_1 \sigma_2 \sigma_3 \\ \mathbf{k}_1 \mathbf{k}_2 \mathbf{k}_3\end{pmatrix} \sum_{\mathbf{p}} 1 = \sum_{\substack{\mathbf{lmn} \\ \alpha\beta\gamma}} \Phi_{\alpha\beta\gamma}(\mathbf{lmn})\, \varepsilon^{\alpha}_{\mathbf{k}_1\sigma_1} \varepsilon^{\beta}_{\mathbf{k}_2\sigma_2} \varepsilon^{\gamma}_{\mathbf{k}_3\sigma_3} \exp[i(\mathbf{k}_1.\mathbf{l}+\mathbf{k}_2.\mathbf{m}+\mathbf{k}_3.\mathbf{n})] \times \sum_{\mathbf{p}} e^{i(\mathbf{k}_1+\mathbf{k}_2+\mathbf{k}_3).\mathbf{p}}. \tag{3.13.19}$$

Now

$$\sum_{\mathbf{p}} e^{i(\mathbf{k}_1+\mathbf{k}_2+\mathbf{k}_3).\mathbf{p}} = \Delta(\mathbf{k}_1+\mathbf{k}_2+\mathbf{k}_3), \tag{3.13.20}$$

where

$$\Delta(\mathbf{k}) = \frac{1}{\Omega_{\mathrm{B}}} \sum_{\mathbf{K}} \delta(\mathbf{k}+\mathbf{K}). \tag{3.13.21}$$

Thus $V\begin{pmatrix}\sigma_1 \sigma_2 \sigma_3 \\ \mathbf{k}_1 \mathbf{k}_2 \mathbf{k}_3\end{pmatrix}$ is zero unless $\mathbf{k}_1+\mathbf{k}_2+\mathbf{k}_3 = \mathbf{K}$, a reciprocal lattice vector.

A similar analysis may be carried out for Φ_4, etc., combining the results of this section and the last, we see that the Hamiltonian may be written

$$H = \sum_{\mathbf{k}\sigma} \hbar\omega(\mathbf{k}\sigma)(a^{\dagger}_{\mathbf{k}} a_{\mathbf{k}} + \tfrac{1}{2}) + \sum_{\substack{\mathbf{k}_1\mathbf{k}_2\mathbf{k}_3 \\ \sigma_1\sigma_2\sigma_3}} V\begin{pmatrix}\sigma_1 \sigma_2 \sigma_3 \\ \mathbf{k}_1 \mathbf{k}_2 \mathbf{k}_3\end{pmatrix} q_{\mathbf{k}_1\sigma_1} q_{\mathbf{k}_2\sigma_2} q_{\mathbf{k}_3\sigma_3} + \sum_{\substack{\mathbf{k}_1\mathbf{k}_2\mathbf{k}_3\mathbf{k}_4 \\ \sigma_1\sigma_2\sigma_3\sigma_4}} V\begin{pmatrix}\sigma_1 \sigma_2 \sigma_3 \sigma_4 \\ \mathbf{k}_1 \mathbf{k}_2 \mathbf{k}_3 \mathbf{k}_4\end{pmatrix} q_{\mathbf{k}_1\sigma_1} q_{\mathbf{k}_2\sigma_2} q_{\mathbf{k}_3\sigma_3} q_{\mathbf{k}_4\sigma_4}, \tag{3.13.22}$$

where

$$V\begin{pmatrix}\sigma_1 \sigma_2 \ldots \sigma_N \\ \mathbf{k}_1 \mathbf{k}_2 \ldots \mathbf{k}_N\end{pmatrix} = 0, \quad \text{unless} \quad \mathbf{k}_1+\mathbf{k}_2+\ldots+\mathbf{k}_N = \mathbf{K}. \tag{3.13.23}$$

This corresponds to the *conservation of crystal momentum*. With the interpretation of the a's as creation and annihilation operators, the interactions caused by $V\begin{pmatrix}\sigma_1 \sigma_2 \sigma_3 \\ \mathbf{k}_1 \mathbf{k}_2 \mathbf{k}_3\end{pmatrix}$ may be represented graphically (foreshadowing the analysis in section 3.14) as in Figure 3.11. Four processes are shown. The other four have similar graphs.

If $\mathbf{K} = 0$ we say we have a *normal process*; if $\mathbf{K} \neq 0$ we have an *umklapp* process (a name introduced by Peierls—meaning that the phonon "flips over"; one may think of it as normal scattering plus Bragg reflexion).

The condition (3.13.22) and the way it is derived reveal the significance of the name "crystal momentum" for $\mathbf{k}$, the phonon wave number. The law of conservation of momentum follows from invariance of the Hamiltonian under *infinitesimal* translations (Dirac, 1957) whereas the law for conservation of momentum expressed in (3.13.23) has been derived from invariance of the Hamiltonian under translation of a lattice vector; equation (3.13.23) shows that crystal momentum *can* change—but only by a reciprocal lattice vector **K**, and points in reciprocal space differing by **K** are equivalent.

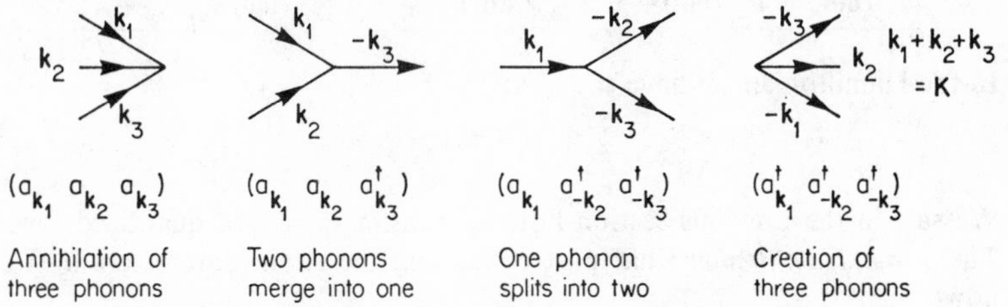

FIGURE 3.11. Illustrating splitting and combining of phonons.

3.13.4 Strain Hamiltonian

There are two procedures open to us in treating anharmonicity. We can expand Φ about the known equilibrium positions, taking the frequencies $\omega(\mathbf{k}\sigma)$ in (3.2.21) to be those of the quasi-harmonic approximation. Alternatively, in an *a priori* calculation, it is most convenient to perturb the purely harmonic result and expand about the static equilibrium positions. Now in either case we may wish to consider the effect of strains away from the original configurations. The formalism is similar for both, and we shall consider the perturbation of the purely harmonic solutions.

Since (3.13.6) is now taken to represent displacements from the true equilibrium positions, in addition to the potential given by (3.13.5) we must consider the effect of "straining" away the crystal from the static equilibrium position. That is, we must find the true coupling parameters in terms of the static condition ones. We do this by Taylor expansion and, as example, consider a Hamiltonian containing derivatives of Φ no higher than the third. Thus $\tilde{\Phi}_4$ is neglected and $\tilde{\Phi}_3$ unaffected. We only Taylor-expand $\Phi_{\alpha\beta}$ and Φ_0.

We represent the homogeneous deformation of the Bravais lattice by

$$l_\alpha = \bar{l}_\alpha + \sum_\beta u_{\alpha\beta}\,\bar{l}_\beta. \tag{3.13.24}$$

The Taylor expansion of the second-order coupling parameters is then

$$\Phi_{\alpha\beta}(\mathbf{mn}) = \tilde{\Phi}_{\alpha\beta}(\mathbf{mn}) + \sum_{\mu\mathbf{l}} \tilde{\Phi}_{\mu\alpha\beta}(\mathbf{lmn}) \left(\sum_\gamma u_{\mu\gamma}\,\bar{l}_\gamma\right) + \sum_{\substack{\mu\nu\\ \mathbf{hl}}} \tilde{\Phi}_{\mu\nu\alpha\beta}(\mathbf{hlmn}) \left(\sum_\gamma u_{\mu\gamma}\,\bar{l}_\gamma\right)\left(\sum_\lambda u_{\nu\lambda}\,\bar{l}_\lambda\right) + \ldots \tag{3.13.25}$$

It is convenient to transfer to second-quantization form in two stages. Let us first rid the expression of the explicit appearance of $\bar{l}_\gamma$, etc., introducing coefficients Ψ' which are the sums of derivatives of $\tilde{\Phi}_{\alpha\beta}$ multiplied by lattice components $\bar{l}_\gamma$. The relation of the Ψ''s to the Φ's is perhaps seen most easily if we merely rewrite equation (3.13.25) in terms of the new quantities:

$$\Phi_{\alpha\beta}(\mathbf{mn}) = \tilde{\Phi}_{\alpha\beta}(\mathbf{mn}) + \sum_{\mu\gamma} \Psi'^{\gamma}_{\mu\alpha\beta}(\mathbf{mn})\, u_{\mu\gamma} + \sum_{\substack{\mu\nu\\ \gamma\lambda}} \Psi'^{\gamma\lambda}_{\mu\nu\alpha\beta}(\mathbf{mn})\, u_{\mu\gamma}\, u_{\nu\lambda} + \ldots. \tag{3.13.26}$$

In the Hamiltonian we have the term

$$\sum_{\substack{mn\\ \sigma\beta}} \Phi_{\alpha\beta}(\mathbf{m},\mathbf{n})\, u^{\alpha}_{m}\, u^{\beta}_{n}.$$

We saw in the previous section how to transfer to second-quantized form. The *strain Hamiltonian* (that part containing the strain parameters $u_{\mu\gamma}$) is now‡

$$H_s = \sum_{\mu\gamma} u_{\mu\gamma}\, V_{\mu\gamma}\begin{pmatrix}\sigma_1\,\sigma_2\\ \mathbf{k}_1\,\mathbf{k}_2\end{pmatrix} q_{\mathbf{k}_1\sigma_1} q_{\mathbf{k}_2\sigma_2} + \sum_{\substack{\mu\nu\\ \gamma\lambda}} V_{\mu\nu\gamma\lambda}\begin{pmatrix}\sigma_1\,\sigma_2\\ \mathbf{k}_1\,\mathbf{k}_2\end{pmatrix} u_{\mu\gamma}\, u_{\gamma\lambda}\, q_{\mathbf{k}_1\sigma_1} q_{\mathbf{k}_2\sigma_2}, \tag{3.13.27}$$

where

$$V_{\mu\gamma}\begin{pmatrix}\sigma_1\,\sigma_2\\ \mathbf{k}_1\,\mathbf{k}_2\end{pmatrix} = \sum_{\substack{\alpha\beta\\ mn}} \Psi'^{\gamma}_{\mu\alpha\beta}(\mathbf{m},\mathbf{n})\, \varepsilon^{\alpha}_{\mathbf{k}_1\sigma_1}\, \varepsilon^{\beta}_{\mathbf{k}_2\sigma_2} \exp\,[i(\mathbf{k}_1\cdot\mathbf{m}+\mathbf{k}_2\cdot\mathbf{n})], \quad \text{etc.} \tag{3.13.28}$$

Just as in section 3.13.3, crystal momentum is conserved; further, since here we are only concerned with two wave-vectors ($\mathbf{k}_1$ and $\mathbf{k}_2$) both of which may be restricted to lie within the BZ [see equation (3.2.23)], if $\mathbf{k}_1+\mathbf{k}_2 = \mathbf{K}$, the points $\mathbf{k}_1$ and $-\mathbf{k}_2$ are equivalent,§ and $-\mathbf{k}_2$ lies in the BZ so that we may always put $\mathbf{k}_1 = -\mathbf{k}_2$. The strain Hamiltonian may therefore be rewritten

$$H_s = \sum_{\substack{\mu\gamma\\ \mathbf{k}\sigma_1\sigma_2}} u_{\mu\gamma}\, V_{\mu\gamma}\begin{pmatrix}\sigma_1\,\sigma_2\\ \mathbf{k}\;-\mathbf{k}\end{pmatrix} q_{\mathbf{k}\sigma_1} q_{-\mathbf{k}\sigma_2} + \sum_{\substack{\mu\gamma,\mathbf{k}\\ \nu\lambda,\sigma_1\sigma_2}} u_{\mu\gamma}\, u_{\nu\lambda}\, V_{\mu\nu\gamma\lambda}\begin{pmatrix}\sigma_1\,\sigma_2\\ \mathbf{k}\;-\mathbf{k}\end{pmatrix} q_{\mathbf{k}\sigma_1} q_{-\mathbf{k}\sigma_2}. \tag{3.13.29}$$

As an example, we shall neglect $\tilde{\Phi}_4$, a case which we consider further in section 3.14. The full dynamical part of the Hamiltonian is then, to third order,

$$H = \sum_{\mathbf{k}\sigma} \hbar\omega(\mathbf{k}\sigma)\,[a^{\dagger}_{\mathbf{k}\sigma} a_{\mathbf{k}\sigma} + \tfrac{1}{2}] + \sum_{\substack{\mu\gamma\\ \mathbf{k}\sigma_1\sigma_2}} V_{\mu\gamma}\begin{pmatrix}\sigma_1\,\sigma_2\\ \mathbf{k}\;-\mathbf{k}\end{pmatrix} q_{\mathbf{k}\sigma_1} q_{-\mathbf{k}\sigma_2}\, u_{\mu\gamma}$$
$$+ \sum_{\substack{\mathbf{k}_1\mathbf{k}_2\mathbf{k}_3\\ \sigma_1\sigma_2\sigma_3}} V\begin{pmatrix}\sigma_1\,\sigma_2\,\sigma_3\\ \mathbf{k}_1\,\mathbf{k}_2\,\mathbf{k}_3\end{pmatrix} q_{\mathbf{k}_1\sigma_1} q_{\mathbf{k}_2\sigma_2} q_{\mathbf{k}_3\sigma_3} + \ldots. \tag{3.13.30}$$

‡ We have omitted here the expansion of Φ_0 because it does not enter into the dynamical part of the Hamiltonian; it is only this part with which we are concerned.

§ Vectors $\mathbf{k}$ and $\mathbf{k}+\mathbf{K}$ label the same phonon.

This is the central result which displays the correction terms to the harmonic Hamiltonian.

3.13.5 Thermal expansion

We have mentioned that the quasi-harmonic approximation predicts thermal expansion. Let us suppose on expansion that the lattice maintains the same structure. Then instead of the nine parameters $u_{\alpha\beta}$ we need consider only one, the lattice parameter a. For each value of the lattice parameter we expand Φ as a Taylor series, cutting it off at Φ_2. We then minimize the free energy

$$F = \Phi_0 + F_v \tag{3.13.31}$$

[F_v being given by equation (3.7.12)] with respect to the lattice parameter a.

The difference from the discussion of equilibrium in section 3.6 lies in our use of a different Φ for each a. Reference to equation (3.6.5) shows how this comes about. As in section 3.6 we put $\Delta\mathbf{l} = \mathbf{l} - \tilde{\mathbf{l}}$, etc., and see that Φ_0 includes terms from $\tilde{\Phi}_3$ cubic in the $\Delta\mathbf{l}$'s, while $\tilde{\Phi}_4$ contributes terms quartic in them; the contributions from Φ_3 and $\tilde{\Phi}_4$ to Φ_2 are linear in the $\Delta\mathbf{l}$'s and quadratic in the $\Delta\mathbf{l}$'s respectively.

We now have ω's that depend on a, and so $\partial F/\partial a = 0$ gives a different lattice parameter for each temperature. A coefficient of thermal expansion is thus obtainable.

This quasi-harmonic approximation is not internally consistent, as we remarked in section 3.13.1. Nevertheless, the method does predict the thermal expansion correct to fourth order in the expansion about the static equilibrium positions, as Leibfried and Ludwig (1961) show. Other thermodynamic data is discussed by Barron (1963).

The reader may recognize the quasi-harmonic approximation to be a refinement of Grüneisen's (1952) method; Grüneisen effectively used the quasi-harmonic approximation together with the assumption of a Debye spectrum. Since cubic crystals follow Grüneisen's law to a fair degree of accuracy, it is perhaps not out of place for us to derive it, although it gives us no information as to the character of the anharmonic forces.

We use the thermodynamic relations

$$p = -\left(\frac{\partial F}{\partial V}\right)_{\mathrm{T}}, \tag{3.13.32}$$

$$C_V = -T\left(\frac{\partial^2 F}{\partial T^2}\right)_V, \tag{3.13.33}$$

where V is the volume. Now we may verify that in the Debye theory, F has the form

$$F = \Phi_0 + Tf\left(\frac{\theta_D}{T}\right). \tag{3.13.34}$$

Suppose θ_D is independent of T, but depends on the volume. We have

$$p = -\frac{\partial \Phi_0}{\partial V} - T\frac{\partial}{\partial V} f\left(\frac{\theta_D}{T}\right) = -\frac{\partial \Phi_0}{\partial V} + T^2 \theta_D^{-1} \frac{\partial \theta_D}{\partial V}\frac{\partial f}{\partial T}$$

and

$$\left(\frac{\partial p}{\partial T}\right)_V = 2T\theta_D^{-1}\frac{\partial \theta_D}{\partial V}\frac{\partial f}{\partial T} + T^2 \theta_D^{-1}\frac{\partial \theta_D}{\partial V}\frac{\partial^2 f}{\partial T^2}.$$

From equations (3.13.32) and (3.13.33), this may be written as

$$\left(\frac{\partial p}{\partial T}\right)_V = -\theta_D^{-1}\frac{\partial \theta_D}{\partial V} C_V, \tag{3.13.35}$$

which is usually put into the form

$$\left(\frac{\partial p}{\partial T}\right)_V = \frac{\gamma C_V}{V}. \tag{3.13.36}$$

γ is known as Grüneisen's constant; one sees that it is defined as‡

$$\gamma = -\frac{V}{\theta_D}\bigg/\frac{\partial V}{\partial \theta_D} = -\frac{d(\ln \theta_D)}{d(\ln V)}. \tag{3.13.37}$$

The linear expansion coefficient α is one-third of the volume expansion coefficient, so that

$$\alpha = \frac{1}{3V}\left(\frac{\partial V}{\partial T}\right)_P = -\frac{1}{3V}\left(\frac{\partial p}{\partial T}\right)_V \bigg/ \left(\frac{\partial p}{\partial V}\right)_T. \tag{3.13.38}$$

Thus (3.13.36) gives the relation

$$\alpha = \frac{K\gamma C_V}{3V}, \tag{3.13.39}$$

where K is the compressibility.

We close this section by deriving a formula for thermal expansion based on first-order perturbation theory and which is sometimes used to check the anharmonic potentials.

‡ The above analysis, if carried out with the general form (3.7.12), would result in $\gamma = -d(\ln \omega_{\mathbf{k},\sigma})/d(\ln V)$, with γ assumed independent of $\mathbf{k}, \sigma$.

We note that

$$\left(\frac{\partial F}{\partial u_{\alpha\beta}}\right)_{\mathrm{T}} = -\frac{1}{\beta}\frac{\partial}{\partial u_{\alpha\beta}}\left[\ln \sum_i \mathrm{e}^{-\beta E_i}\right] = \sum_i \frac{\partial E_i}{\partial u_{\alpha\beta}} \mathrm{e}^{-\beta E_i} \Big/ \sum_i \mathrm{e}^{-\beta E_i} = \left\langle \frac{\partial E}{\partial u_{\alpha\beta}} \right\rangle_{\mathrm{T}}. \tag{3.13.40}$$

Here $\langle\ \rangle_{\mathrm{T}}$, as elsewhere, denotes a thermal average. It is to be noted that if $E_i = E_i^0 + \Delta_i$, Δ_i being the perturbation energy, then since $\mathrm{e}^{-\beta E_i} = \mathrm{e}^{-\beta E_i^0}(1-\beta\Delta_i+\ldots)$, to first order in Δ_i we have

$$\left\langle \frac{\partial \Delta}{\partial u_{\alpha\beta}} \right\rangle_{\mathrm{T}} = \sum_i \frac{\partial \Delta_i}{\partial u_{\alpha\beta}} \mathrm{e}^{-\beta E_i^0} \Big/ \sum_i \mathrm{e}^{-\beta E_i^0}. \tag{3.13.41}$$

We now take the Hamiltonian (3.13.29), taking it to only second order in the q's. Remembering the definitions (3.13.13), the energy for any phonon state $|\psi\rangle$ is then

$$\langle\psi|H|\psi\rangle = \sum_{\mathbf{k}\sigma} \hbar\omega(\mathbf{k}\sigma)\,[\langle\psi|a^\dagger_{\mathbf{k}\sigma}a_{\mathbf{k}\sigma}|\psi\rangle + \tfrac{1}{2}] + \sum_{\mu\gamma}\sum_{\mathbf{k}\sigma_1\sigma_2} V_{\mu\gamma}\begin{pmatrix}\sigma_1 & \sigma_2\\ \mathbf{k} & -\mathbf{k}\end{pmatrix} u_{\mu\gamma}$$
$$\times[\langle\psi|a_{\mathbf{k}\sigma_1}a_{-\mathbf{k}\sigma_2}|\psi\rangle + \langle\psi|a_{\mathbf{k}\sigma_1}a^\dagger_{\mathbf{k}\sigma_2}|\psi\rangle + \langle\psi|a^\dagger_{-\mathbf{k}\sigma_1}a_{-\mathbf{k}\sigma_2}|\psi\rangle$$
$$+\langle\psi|a^\dagger_{-\mathbf{k}\sigma_1}a^\dagger_{\mathbf{k}\sigma_2}|\psi\rangle]. \tag{3.13.42}$$

The terms in $u_{\mu\gamma}$ are taken for the perturbation, so that $|\psi\rangle$ is an exact state of the original harmonic Hamiltonian; thus all terms are zero unless $\sigma_1 = \sigma_2$. We also have

$$\langle\psi|a_{\mathbf{k}\sigma_1}a_{-\mathbf{k}\sigma_2}|\psi\rangle = \langle\psi|a^\dagger_{-\mathbf{k}\sigma_1}a^\dagger_{\mathbf{k}\sigma_2}|\psi\rangle = 0.$$

Remembering from section 3.7 that the number operator is

$$\hat{n}(\mathbf{k}\sigma) = a^\dagger_{\mathbf{k}\sigma}a_{\mathbf{k}\sigma} = a_{\mathbf{k}\sigma}a^\dagger_{\mathbf{k}\sigma} - 1, \tag{3.13.43}$$

we thus have

$$\langle\psi|H|\psi\rangle = \sum_{\mathbf{k}\sigma}\hbar\omega(\mathbf{k}\sigma)\,[\langle\psi|\hat{n}(\mathbf{k}\sigma)|\psi\rangle + \tfrac{1}{2}]$$
$$+\sum_{\substack{\mu\gamma\\ \mathbf{k}\sigma}} V_{\mu\gamma}\begin{pmatrix}\sigma & \sigma\\ \mathbf{k} & -\mathbf{k}\end{pmatrix}[\langle\psi|\hat{n}(\mathbf{k}\sigma)|\psi\rangle + \langle\psi|\hat{n}(-\mathbf{k}\sigma)|\psi\rangle + 1]\,u_{\mu\gamma}. \tag{3.13.44}$$

Now $\omega(\mathbf{k}\sigma)$ and $\hat{n}(\mathbf{k},\sigma)$ are defined for the static configuration and are independent of $u_{\mu\gamma}$. Hence

$$\frac{\partial}{\partial u_{\mu\gamma}}\langle\psi|H|\psi\rangle = \sum_{\mathbf{k}\sigma} V_{\mu\gamma}\begin{pmatrix}\sigma & \sigma\\ \mathbf{k} & -\mathbf{k}\end{pmatrix}[\langle\psi|\hat{n}(\mathbf{k}\sigma)|\psi\rangle + \langle\psi|\hat{n}(-\mathbf{k}\sigma)|\psi\rangle + 1]. \tag{3.13.45}$$

We see from (3.13.41) that the thermal average is taken with respect to the unperturbed solution. **k** and $-\mathbf{k}$ are then equivalent on taking the thermal average. Reference to section 3.7 therefore shows that if we define the occupation numbers

$$n(\mathbf{k}\sigma) = \frac{1}{\exp\left[\dfrac{\hbar\omega(\mathbf{k}\sigma)}{k_B T}\right]-1}, \tag{3.13.46}$$

then

$$\frac{\partial F_D}{\partial u_{\alpha\beta}} = \sum_{\mathbf{k}\sigma} V_{\alpha\beta}\begin{pmatrix}\sigma & \sigma\\ \mathbf{k} & -\mathbf{k}\end{pmatrix}[2n(\mathbf{k}\sigma)+1]. \tag{3.13.47}$$

The subscript D on F reminds us that we have as yet only differentiated the dynamical part of the free energy. We must also differentiate the static strain energy, which would be non-zero even if the crystal were purely harmonic. We define this energy using the elastic constants (defined about the static configuration). If we look at equation (A3.1.4) we can see that, defining for the moment generalized elastic moduli $c_{\alpha\beta\gamma}\delta$, the strain energy *per unit cell* will be of quadratic form‡:

$$U = \frac{\Omega}{2}\sum_{\substack{\alpha\beta\\ \gamma\delta}} c_{\alpha\beta,\gamma\delta}\, u_{\alpha\beta}\, u_{\gamma\delta} \tag{3.13.48}$$

and for equilibrium

$$0 = (\partial F/\partial u_{\alpha\beta})_T = \mathscr{V}\sum_{\gamma\delta} c_{\alpha\beta,\gamma\delta}\, u_{\gamma\delta} + \sum_{\mathbf{k}\sigma} V_{\alpha\beta}[2n(\mathbf{k},\sigma)+1]. \tag{3.13.49}$$

The strain with which we are here concerned is thermal strain, which we indicate by putting $u^T_{\beta\alpha}$. For cubic crystals the off-diagonal elements of u^T are zero, and from (3.13.48) we obtain, writing in the elastic moduli defined in section 3.1,

$$\frac{\partial u^T_{\alpha\alpha}}{\partial T} = \frac{-1}{V(c_{11}+2c_{12})}\sum_{\mathbf{k}\sigma} V_{\alpha\alpha}\begin{pmatrix}\sigma & \sigma\\ \mathbf{k} & -\mathbf{k}\end{pmatrix}\left(\frac{\partial n(\mathbf{k}\sigma)}{\partial T}\right). \tag{3.13.50}$$

3.13.6 Self-consistent phonon approximation

In the self-consistent phonon (SCP) approximation, first suggested by Born (1951), the Hamiltonian is harmonic but the force constants are the thermal averages, of the second derivative of the pair potential, taken with

‡ Equation (3.13.48) is only correct if, as in the crystals with which we are here concerned, the external strain induces no *internal* strain (internal strain occurs when the crystal consists of interpenetrating Bravais lattices which move relative to one another). Born and Huang (1954) show by what $c_{\alpha\beta,\gamma\delta}$ must be replaced if internal strain is present.

respect to the states these force constants generate. The SCP theory can thus be seen to be the analogue in phonon theory of the Hartree method. That self-consistency can be very important is shown by calculations on solid helium, when real phonon frequencies are obtained, in contrast with the quasi-harmonic approximation (Hooton, 1955).

The SCP method also resembles the Hartree method in that it is equivalent to a first-order variational calculation of the energy and we shall discuss it from this standpoint, largely following the derivation of Boccara and Sarma (1965) as presented by Gillis, Werthamer and Koehler (1968). We should also mention the diagrammatic discussion of Choquard (1967).

We begin by assuming the Hamiltonian to be of the form

$$H = -\sum_{\mathbf{l}} \frac{1}{2M} \frac{\partial^2}{\partial \mathbf{R}_{\mathbf{l}}^2} + \sum_{\mathbf{ll'}}{}' v(\mathbf{l}-\mathbf{l}'+\mathbf{u}_l-\mathbf{u}_{l'}), \tag{3.13.51}$$

where $\mathbf{l}$, $\mathbf{l}'$ represent the mean equilibrium positions and $\mathbf{u}_l = \mathbf{R}_l - \mathbf{l}$. We next take the harmonic trial Hamiltonian

$$H_{\mathrm{h}} = -\sum_{\mathbf{l}} \frac{1}{2M} \frac{\partial^2}{\partial \mathbf{R}_{\mathbf{l}}^2} + \frac{1}{2} \sum_{\mathbf{ll'}} (\mathbf{u}_l - \mathbf{u}_{l'})_\alpha \phi_{\alpha\beta}(\mathbf{l}-\mathbf{l}')(\mathbf{u}_l - \mathbf{u}_{l'})_\beta. \tag{3.13.52}$$

The density motion operator corresponding to this is

$$\rho_{\mathrm{h}} = \mathrm{e}^{-\beta H_{\mathrm{h}}} / \mathrm{Tr}\{\mathrm{e}^{-\beta H_{\mathrm{h}}}\} \tag{3.13.53}$$

yielding a trial free energy

$$F_{\mathrm{trial}} = \mathrm{Tr}\{\rho_{\mathrm{h}}(H + \beta^{-1} \ln \rho_{\mathrm{h}})\} = \langle H + \beta^{-1} \ln \rho_{\mathrm{h}} \rangle. \tag{3.13.54}$$

If we also write

$$F_{\mathrm{h}} = \langle H_{\mathrm{h}} + \beta^{-1} \ln \rho_{\mathrm{h}} \rangle \tag{3.13.55}$$

and introduce the matrix

$$\Lambda_{\alpha\beta}(\mathbf{l}-\mathbf{l}') = \langle (\mathbf{u}_l - \mathbf{u}_{l'})_\alpha (\mathbf{u}_l - \mathbf{u}_{l'})_\beta \rangle \tag{3.13.56}$$

we may rewrite equation (3.13.54) as

$$F_{\mathrm{trial}} = F_{\mathrm{h}} + \sum_{ll'}{}' \langle v(\mathbf{l}-\mathbf{l}'+\mathbf{u}_l-\mathbf{u}_{l'}) \rangle - \frac{1}{2} \sum_{\substack{\mathbf{ll'} \\ \beta\alpha}}{}' \Lambda_{\alpha\beta}(\mathbf{l}-\mathbf{l}') \phi_{\alpha\beta}(\mathbf{l}-\mathbf{l}'). \tag{3.13.57}$$

We now use the result that for any function $f(\mathbf{r})$

$$f(\mathbf{r}+\mathbf{u}) = \mathrm{e}^{\mathbf{u}\cdot\nabla} f(\mathbf{r}), \tag{3.13.58}$$

whereupon we can write

$$\langle v(\mathbf{l}-\mathbf{l}'+\mathbf{u}_l-\mathbf{u}_{l'}) \rangle = \langle \exp[(\mathbf{u}_l - \mathbf{u}_{l'}) . \partial/\partial\mathbf{R}] \rangle \, v(\mathbf{R})|_{\mathbf{R}=\mathbf{l}-\mathbf{l}'}. \tag{3.13.59}$$

On further invoking the property of harmonic oscillators that

$$\langle \mathrm{e}^{\mathbf{u}.\boldsymbol{\xi}} \rangle = \exp[\langle (\mathbf{u}.\boldsymbol{\xi})^2 \rangle / 2] \tag{3.13.60}$$

(see Chapter 5, p. 457) it is evident that

$$\langle v(\mathbf{l}-\mathbf{l}'+\mathbf{u}_l-\mathbf{u}_{l'})\rangle = \exp\left[\frac{1}{2}\sum_{\alpha\beta}\Lambda_{\alpha\beta}(\mathbf{l}-\mathbf{l}')\frac{\partial^2}{\partial R_\alpha\,\partial R_\beta}\right] v(\mathbf{R})|_{\mathbf{R}=\mathbf{l}-\mathbf{l}'}. \tag{3.13.61}$$

Inserting this result into equation (3.13.57) we see F_{trial} to be a function of Λ and ϕ, and minimizing the trial free energy with respect to these we find

$$0 = \frac{\delta F_{\text{trial}}}{\delta\Lambda_{\alpha\beta}(\mathbf{l}-\mathbf{l}')} = \frac{1}{2}\left\langle\frac{\partial}{\partial R_\alpha\,\partial R_\beta}v(\mathbf{l}-\mathbf{l}'+\mathbf{u}_l-\mathbf{u}_{l'})\right\rangle - \frac{1}{2}\phi_{\alpha\beta}(\mathbf{l}-\mathbf{l}'), \tag{3.13.62}$$

$$0 = \frac{\delta F_{\text{trial}}}{\delta\phi_{\alpha\beta}(\mathbf{l}-\mathbf{l}')} = \frac{\delta F_{\text{h}}}{\delta\phi_{\alpha\beta}(\mathbf{l}-\mathbf{l}')} - \tfrac{1}{2}\Lambda_{\alpha\beta}(\mathbf{l}-\mathbf{l}'). \tag{3.13.63}$$

Of these equations, the first shows the optimum choice of $\phi_{\alpha\beta}$ to be the thermal average of second derivatives of the potential while the second relates Λ to ϕ. By use of equation (3.13.61) we can write these more usefully in the form

$$\phi_{\alpha\beta}(\mathbf{l}) = (8\pi^3 \det\Lambda)^{-\frac{1}{2}}\int d\mathbf{u}\exp\left[-\frac{1}{2}\sum_{\alpha\beta}u_\alpha\Lambda^{-1}_{\alpha\beta}(\mathbf{l})\,u_\beta\, v_{\alpha\beta}(\mathbf{l}+\mathbf{u})\right], \tag{3.13.64}$$

$$\Lambda_{\alpha\beta}(\mathbf{l}) = \frac{2}{N}\sum_{\mathbf{k}\sigma}(1-e^{i\mathbf{k}.\mathbf{l}})\,\varepsilon^{*\alpha}_{\mathbf{k}\sigma}\,\varepsilon^{\beta}_{\mathbf{k}\sigma}(2M\omega_{\mathbf{k}\sigma})^{-1}\coth\tfrac{1}{2}\beta\omega_{\mathbf{k}\sigma}, \tag{3.13.65}$$

where the $\boldsymbol{\varepsilon}_{\mathbf{k}\sigma}$ and $\omega_{\mathbf{k}\sigma}$ are respectively the eigenvectors and eigenfrequencies of the dynamical matrix

$$D_{\alpha\beta}(\mathbf{k}) = \sum_{\mathbf{k}\sigma}(1-e^{i\mathbf{k}.\mathbf{l}})\,\phi_{\alpha\beta}(\mathbf{l}). \tag{3.13.66}$$

Equations (3.13.64) and (3.13.65) are to be solved self-consistently. For references of the application of this method to solid argon see Klein and co-workers (1970), and to solid helium see Chell (1970).

An alternative method of obtaining the frequencies in this same self-consistent phonon approximation is to sum the sub-series, the first three terms of which are shown in Figure 3.12 below, for the self-energy. One sees that only even derivatives of the potential are involved, a conclusion one can also draw from equation (3.13.61) above.

3.14 One-phonon Green functions

3.14.1 Definitions and general properties

When anharmonicity is present, one can no longer visualize the system as consisting of non-interacting quasi-particles or phonons. However, as we assumed in section 3.13, provided the anharmonicity is not too strong we can retain the phonon picture, as long as we now include the possibilities that phonons can collide, split or combine.

It is a little more convenient to consider modes, rather than phonons, using the operators $q_{\mathbf{k}\sigma}$ defined in equation (3.13.13), for, as we see from the Hamiltonian (3.13.30), the phonon operators always appear in pairs $a_{\mathbf{k}\sigma}+a^{\dagger}_{-\mathbf{k}\sigma}$ in the anharmonic contribution to H. Equations (3.2.13) and (3.2.15) show that when the phonons are non-interacting, $Q_{\mathbf{k}\sigma} \propto \mathrm{e}^{i\omega(\mathbf{k}\sigma)t}$. This time dependence is most simply modified by making the frequency complex; then $Q_{\mathbf{k}\sigma} \propto \mathrm{e}^{i\omega(\mathbf{k}\sigma)t}\,\mathrm{e}^{-\Gamma t}$. We can see that $\tau = \Gamma^{-1}$ will be a measure of the phonon lifetime. We might also expect a real frequency shift $\Delta(\mathbf{k}, \sigma)$.

To give the interacting quasi-particle picture more formal meaning, we may consider the one-particle Green function, as we did for electrons in the previous chapter. However, we cannot simply take over the formalism developed in Chapter 2 for there are important differences between the two cases. Firstly, the phonons are bosons, instead of Fermions, and, moreover, the particle number is not constant. Secondly, phonon energies are usually comparable to $k_B T$ and finite-temperature theory must be used from the outset.

We begin, by analogy with equation (2.7.70) for the one-electron Green function, by defining the one-phonon Green function for a particular eigenstate $|i\rangle$ of the system. We put

$$G(\mathbf{k}\sigma, \mathbf{k}'\sigma'; t) = \langle i|\,P[q_{\mathbf{k}\sigma}(t)\,q^{\dagger}_{\mathbf{k}'\sigma'}(0)]\,|i\rangle, \tag{3.14.1}$$

where, as usual, $q(t)$ is the Heisenberg operator

$$q(t) = \mathrm{e}^{iHt/\hbar}\,q(0)\,\mathrm{e}^{-iHt/\hbar} \tag{3.14.2}$$

and P is the Dyson chronological operator ordering earlier times to the right.

Just as we found for the Green functions of electrons in a periodic structure, G is zero unless $\mathbf{k} = \mathbf{k}'$, provided both $\mathbf{k}$ and $\mathbf{k}'$ are restricted to lie in the BZ and we shall assume this from now on. The proof is left to the reader (cf. section 2.7).

We now thermally average equation (3.14.1) and denoting the partition function by Z (for a discussion of the statistical density matrix, see Appendix 3.6) we have

$$\begin{aligned} G(\mathbf{k}\sigma\sigma' t) &= \frac{1}{Z}\sum_i \mathrm{e}^{-\beta E_i}\langle i|\,P[q_{\mathbf{k}\sigma}(t)\,q^{\dagger}_{\mathbf{k}\sigma'}(0)]\,|i\rangle \\ &= \frac{1}{Z}\sum_i \langle i|\,\mathrm{e}^{-\beta H}\,P[q_{\mathbf{k}\sigma}(t)\,q^{\dagger}_{\mathbf{k}\sigma'}(0)]\,|i\rangle \\ &= \langle P[q_{\mathbf{k}\sigma}(t)\,q^{\dagger}_{\mathbf{k}\sigma'}(0)]\rangle. \end{aligned} \tag{3.14.3}$$

It should be noted that the chemical potential $\mu = \partial F/\partial N$ is zero for the problem in question, since N, the number of quasi-particles, is fixed by the equilibrium condition. In the presence of the factor $e^{-\beta H}$, it is convenient to make the time purely imaginary and to introduce the variable

$$\tau = it \tag{3.14.4}$$

with P understood to operate on τ. Then

$$G(\mathbf{k}, \sigma, \sigma'; \tau) = \begin{cases} Z^{-1} \sum_i \langle i | e^{-(\beta - \tau/\hbar)H} q_{\mathbf{k}\sigma} e^{-\tau H/\hbar} q^{\dagger}_{\mathbf{k}\sigma'} | i \rangle, & \tau > 0, \\ Z^{-1} \sum_i \langle i | e^{-\beta H} q^{\dagger}_{\mathbf{k}\sigma'} e^{\tau H/\hbar} q_{\mathbf{k}\sigma} e^{-\tau H/\hbar} | i \rangle, & \tau < 0. \end{cases} \tag{3.14.5}$$

To see the simplifying nature of the transformation to imaginary times, we first note that equation (3.14.5) involves a trace, which is defined, for a general operator A, as

$$\mathrm{Tr}\{A\} = \sum_i \langle i | A | i \rangle, \tag{3.14.6}$$

the summation on the right-hand side being over a complete set of orthonormal functions. The trace has two very important properties. The first is that it has the same value no matter what complete set we sum over, provided the members of the set obey the correct boundary conditions. To see this, we use the completeness relation

$$\sum_f | f \rangle \langle f | = 1. \tag{3.14.7}$$

Evidently

$$\begin{aligned} \sum_i \langle i | A | i \rangle &= \sum_{if} \langle i | f \rangle \langle f | A | i \rangle \\ &= \sum_{if} \langle f | A | i \rangle \langle i | f \rangle = \sum_j \langle f | A | f \rangle. \end{aligned} \tag{3.14.8}$$

We use this result to prove the second property. We take A as a product of operators $A_1, A_2, \ldots, A_N$. On employing equation (3.14.8) we see that

$$\begin{aligned} \mathrm{Tr}\{A_1 \ldots A_N\} &= \sum_{if} \langle i | A_1 \ldots A_{N-1} | f \rangle \langle f | A_N | i \rangle \\ &= \sum_{if} \langle f | A_N | i \rangle \langle i | A_1 \ldots A_{N-1} | f \rangle \\ &= \sum_f \langle f | A_N A_1 \ldots A_{N-1} | f \rangle, \end{aligned} \tag{3.14.9}$$

that is

$$\mathrm{Tr}\{A_1 \ldots A_{N-1} A_N\} = \mathrm{Tr}\{A_N A_1 \ldots A_{N-1}\}. \tag{3.14.10}$$

This is referred to as the cyclic property of the trace.

Now suppose in equation (13.14.5) that we have $-\beta\hbar<\tau<0$ whereupon $\tau+\beta\hbar>0$. Using the cyclic property

$$G(\mathbf{k}\sigma\sigma';\tau+\beta\hbar) = Z^{-1}\operatorname{Tr}\{e^{\tau H/\hbar}q_{\mathbf{k}\sigma}e^{-\tau H/\hbar}e^{-\beta H}q^{\dagger}_{\mathbf{k}\sigma'}\}$$
$$= Z^{-1}\operatorname{Tr}\{e^{-\beta H}q^{\dagger}_{\mathbf{k}\sigma'}e^{\tau H/\hbar}q_{\mathbf{k}\sigma}e^{-\tau H/\hbar}\} \qquad (3.14.11)$$

or

$$G(\mathbf{k}\sigma\sigma';\tau+\beta\hbar) = G(\mathbf{k}\sigma\sigma';\tau) \quad (0>\tau>-\beta\hbar). \qquad (3.14.12)$$

This periodicity in the imaginary time direction allows representation by a Fourier series, provided τ is never allowed outside the limits $(-\beta\hbar, \beta\hbar)$. We have

$$G(\mathbf{k}\sigma\sigma';\tau) = \sum_{n=-\infty}^{\infty} G(\mathbf{k}\sigma\sigma';i\omega_n)\,e^{i\omega_n\tau}, \quad \omega_n = \frac{2\pi n}{\beta\hbar} \qquad (3.14.13)$$

with

$$G(\mathbf{k}\sigma\sigma';i\omega_n) = \frac{1}{2\beta\hbar}\int_{-\beta\hbar}^{\beta\hbar} G(\mathbf{k}\sigma\sigma';\tau)\,e^{-i\omega_n\tau}\,d\tau. \qquad (3.14.14)$$

As in section 2.7.6 of Chapter 2, we may introduce a spectral representation $A(\mathbf{k},\sigma\sigma',\omega)$ of G, which we choose to define through

$$G(\mathbf{k}\sigma\sigma';i\omega_n) = \frac{1}{\beta\hbar}\int_{-\infty}^{\infty}\frac{A(\mathbf{k}\sigma\sigma';\omega)}{\omega+i\omega_n}\,d\omega. \qquad (3.14.15)$$

A describes the spectrum of energies associated with a phonon of wave-vector $\mathbf{k}$.

3.14.2 *Results for harmonic crystal*

It is of interest to see what forms the above correlation functions take in the harmonic approximation, the starting-point of perturbation theory. It is readily shown that for this case

$$G_0(\mathbf{k}\sigma\sigma';\tau) = \delta_{\sigma\sigma'}\left[\frac{e^{-\tau\omega(\mathbf{k}\sigma)}}{1-e^{-\beta\omega(\mathbf{k}\sigma)}}+\frac{e^{\tau\omega(\mathbf{k}\sigma)}}{e^{\beta\omega(\mathbf{k}\sigma)}-1}\right], \qquad (3.14.16)$$

where we have set $\hbar = 1$, as we shall in the following discussion.

On taking the transform of equation (3.14.16) we find

$$G_0(\mathbf{k}\sigma\sigma';i\omega_n) = \frac{\delta_{\sigma\sigma'}}{\beta}\left[\frac{1}{\omega(\mathbf{k}\sigma)+i\omega_n}+\frac{1}{\omega(\mathbf{k}\sigma)-i\omega_n}\right]. \qquad (3.14.17)$$

This form, in which the temperature dependence is simply β^{-1}, shows that considerable simplicity results from the transformation to imaginary time. For future reference, we note that equation (3.14.17) may also be written as

$$G_0(\mathbf{k}\sigma\sigma'; i\omega_n) = \frac{\delta_{\sigma\sigma'}}{\beta}\frac{2\omega(\mathbf{k}\sigma)}{\omega^2+\omega_n^2}. \tag{3.14.18}$$

Finally, the spectral density function is readily seen to be

$$A(\mathbf{k}\sigma\sigma'; \omega) = \delta_{\sigma\sigma'}\{\delta(\omega-\omega(\mathbf{k}\sigma))-\delta(\omega+\omega(\mathbf{k}\sigma))\}, \tag{3.14.19}$$

for inserting this into equation (3.14.15) we immediately regain equation (3.14.17). This is an obvious result: we simply have delta function peaks at the phonon frequencies $\omega(\mathbf{k}\sigma)$.

3.14.3 *Effect of anharmonicity*

We expect the presence of phonon–phonon interactions to change G_0 in a fashion analogous to that in which electron–electron interactions were shown to change the one-electron Green function in Chapter 2. Then we expect that, provided the anharmonicity is not too strong, G will be dominated by simple poles, but these poles will be shifted off the imaginary axis by a real amount Γ. The δ-functions of equation (3.14.19) will be then broadened out into Lorentzian peaks, so that the spectral density takes the form

$$A(\mathbf{k}\sigma\sigma') = \delta_{\sigma\sigma'}\left\{\frac{2\Gamma(\mathbf{k}\sigma)}{[\omega-\Omega(\mathbf{k}\sigma)]^2+\Gamma(\mathbf{k}\sigma)^2}-\frac{2\Gamma(\mathbf{k}\sigma)}{[\omega+\Omega(\mathbf{k}\sigma)]^2+\Gamma(\mathbf{k}\sigma)^2}\right\}. \tag{3.14.20}$$

In Chapter 5, we shall see how the true phonon frequency Ω, and the inverse lifetime Γ of the mode, may be found from neutron-scattering experiments. For the present, we shall see how Ω and Γ arise from a perturbative treatment of anharmonicity.

3.14.4 *Dyson's equation and self energy*

We saw in Chapter 2 how the perturbation series for the one-electron Green function can be represented diagrammatically. A similar graphical method can be applied to calculate the one-phonon Green function. In each diagram for the Green function, a single mode represented by a line is seen entering the diagram. This mode then splits into other modes, each of which may also split, or combine with one or other modes. At each splitting or combination, the total $\mathbf{k}$ and ω are conserved. The process ends with modes combining to produce a single mode leaving the diagram. Representative diagrams are shown in Figure 3.12. Each line propagating between two vertices and labelled by $\mathbf{k}_i$, ω_i, σ_i is associated with a factor $G_0(\mathbf{k}_i\sigma_i\omega_i)$ and each vertex associated with a factor appropriate to whatever term Φ_i the vertex represents, in the expansion of the perturbing Hamiltonian,

$$H_1 = \Phi_s + \Phi_3 + \Phi_4 + \dots . \tag{3.14.21}$$

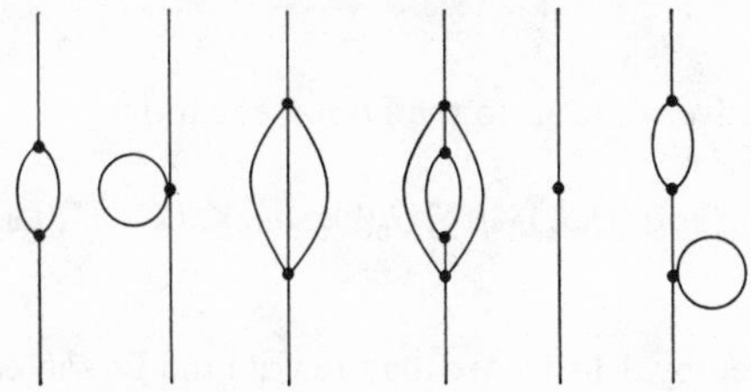

FIGURE 3.12. Representative diagrams for self-energy Σ, showing contributions of strain Hamiltonian [diagram (e)] and three- and four-phonon processes. Note that diagram (f) is reducible and does not contribute.

Here a vertex with i lines running from it arises from the term Φ_i. Thus, Figure 3.12(a) has interactions arising from Φ_3, (b) and (c) have interactions arising from Φ_4 and (d) has interactions arising from both Φ_3 and Φ_4. We may also obtain diagrams, like Figure 3.12(e), associated with Φ_s, the strain Hamiltonian, and representing forward scattering of the mode by the strain field. The full set of rules for translating the graphical terms into algebraic form is derived and listed in Appendix A3.7.

We note that Figure 3.12(f) is reducible, that is it may be separated into two by cutting a single internal line. This leads us to draw the series for G as in Figure 3.13, where Σ represents the sum of all irreducible diagrams.

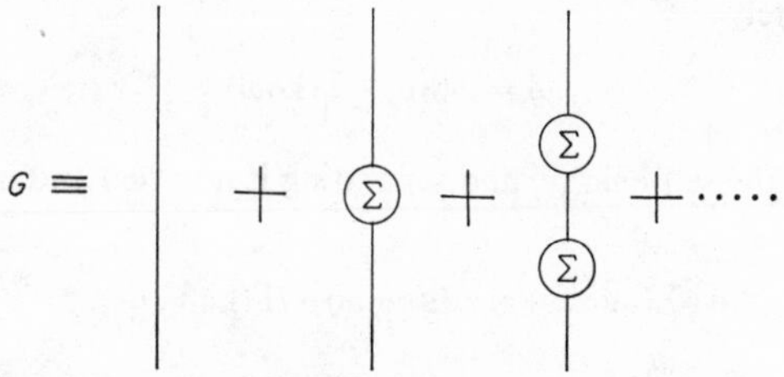

FIGURE 3.13. Diagrammatic series for Green function in terms of self-energy Σ.

Since crystal momentum and frequency are conserved at each vertex, each internal line shown is labelled by **k**, ω and we may write

$$G(\mathbf{k}\sigma\sigma';\, i\omega_n) = \delta_{\sigma\sigma'}\, G_0(\mathbf{k}\sigma;\, i\omega_n) - G_0(\sigma)\,\Sigma(\sigma\sigma')\,G_0(\sigma')\,\beta$$
$$+\beta^2 \sum_{\sigma''} G_0(\sigma)\,\Sigma(\sigma\sigma'')\,G_0(\sigma'')\,\Sigma(\sigma''\sigma')\,G_0(\sigma')+\ldots, \tag{3.14.22}$$

which is the iterative solution to the Dyson equation

$$G(\mathbf{k}\sigma\sigma';\, i\omega_n) = G_0(\mathbf{k}\sigma\sigma';\, i\omega_n) - \beta \sum_{\sigma''} G_0(\mathbf{k}\sigma;\, i\omega_n)\,\Sigma(\mathbf{k},\sigma\sigma'';\, i\omega_n)\,G(\mathbf{k}\sigma''\sigma';\, i\omega_n). \tag{3.14.23}$$

Since G_0 is given by 3.14.18, we may rewrite the Dyson equation as

$$\sum_{\sigma''}\{[\omega(\mathbf{k}\sigma)^2+\omega_n^2]\,\delta_{\sigma\sigma''}+2\omega(\mathbf{k}\sigma)\,\Sigma(\mathbf{k}\sigma\sigma'';\, i\omega_n)\}\,G(\mathbf{k}\sigma''\sigma') = \delta_{\sigma\sigma'}\,2\omega(\mathbf{k}\sigma)\,\beta^{-1}. \tag{3.14.24}$$

The modes σ' and σ'' may be coupled if they transform according to the same irreducible representation of the space group of the crystal (see section 3.5), in which case equation (3.14.20), in which A is diagonal in the σ's, will not be entirely correct. If such coupling is neglected, however, equation (3.14.24) yields

$$G(\mathbf{k}\sigma\sigma';\, i\omega_n) = \frac{\delta_{\sigma\sigma'}\,2\omega(\mathbf{k}\sigma)}{\beta[\omega(\mathbf{k}\sigma)^2+\omega_n^2]+2\omega(\mathbf{k}\sigma)\,\Sigma(\mathbf{k}\sigma i\omega_n)}. \tag{3.14.25}$$

We see that provided $\Sigma(\mathbf{k}) \ll \omega(\mathbf{k})$, the phonon frequencies may still be defined as the poles of G, given by $(i\omega_n \equiv \omega)$

$$\omega^2 = \omega(\mathbf{k}\sigma)^2+2\omega(\mathbf{k}\sigma)\,\Sigma(\mathbf{k}\sigma,\omega) \tag{3.14.26}$$

or approximately

$$\omega = \omega(\mathbf{k}\sigma)+\Sigma(\mathbf{k}\sigma\omega). \tag{3.14.27}$$

Σ is evidently the self-energy, and separating it into real and imaginary parts we can write

$$\Sigma(\mathbf{k}\sigma\sigma'\omega) = \Delta(\mathbf{k}\sigma\sigma'\omega) - i\Gamma(\mathbf{k}\sigma\sigma',\, i\omega). \tag{3.14.28}$$

It is plain that Δ gives the frequency shifts, $\Omega(\mathbf{k}\sigma)-\omega(\mathbf{k}\sigma)$ and Γ the inverse lifetime of the mode. The lowest order contributions to Σ, given by diagrams (a), (b) and (e) of Figure 3.12, are calculated explicitly in Appendix 3.7. It should be noted that, since in equation (3.14.24) Σ is only defined for values of the discrete variable $\omega_n = 2\pi n/\beta$, the values of $\Sigma(\mathbf{k}\sigma\omega)$ must be obtained by analytic continuation.

3.15 Molecular dynamics

We shall see in Chapter 5 that neutron diffraction data on the dispersion relations $\omega(\mathbf{q})$ for various solids require us to take anharmonic effects into account. The methods described above afford one route, but the calculations become complicated and an alternative method, namely the use of molecular dynamical machine computations, has recently been explored (Dickey and Paskin, 1969).

In particular, the method of molecular dynamics turns out to be a very useful way of exploring the temperature and volume dependence of phonon properties, since anharmonicity can, in essence, be included exactly.

The computations available were made with a Lennard-Jones potential, and results were obtained, at a number of volumes and for different temperatures, of average phonon properties and, in particular of pressure and energy, as well as correlation functions, frequency spectra and phonon dispersion curves.

It is the fact that, from neutron-diffraction studies, we have phonon-dispersion curves for the rare-gas solids, at different temperatures, with the volume held constant (Daniels and co-workers, 1967; Leake and co-workers, 1969) that has motivated this further theoretical study of anharmonicity.

The Newtonian equations of motion are solved on a computer for a finite number (864) of atoms, with periodic boundary conditions, for a Lennard-Jones potential of the form

$$\phi(r) = \varepsilon\left[\left(\frac{\sigma}{r}\right)^{12} - 2\left(\frac{\sigma}{r}\right)^{6}\right]. \tag{3.15.1}$$

The results can be expressed in terms of ε and σ in a universal way, and no specific choice is needed.

Figure 3.14 gives the temperature dependence of the thermodynamic properties at constant volume which are in a form relevant to solid Kr, with a lattice constant of 5.67 Å. Kr was chosen because its heavy mass validates the classical calculations.

Obviously, the slope of the energy–temperature curve is the specific heat at constant volume, which, classically, is $3k_B$ in the harmonic limit. It can be seen that these computations reproduce the harmonic value, with a very tiny curvature at high temperatures resulting from anharmonicity.

In the case of the pressure, one would expect a constant from the harmonic approximation, and the effect of anharmonicity is clear in these calculations.

The frequency spectrum was also calculated, and is compared with the harmonic approximation obtained from the work of Grindley and Howard (1965) in Figure 3.15. There is good agreement between the two calculations.

The dispersion curves for phonons were also calculated by introducing a small perturbation (a periodic modification of the velocities or positions of

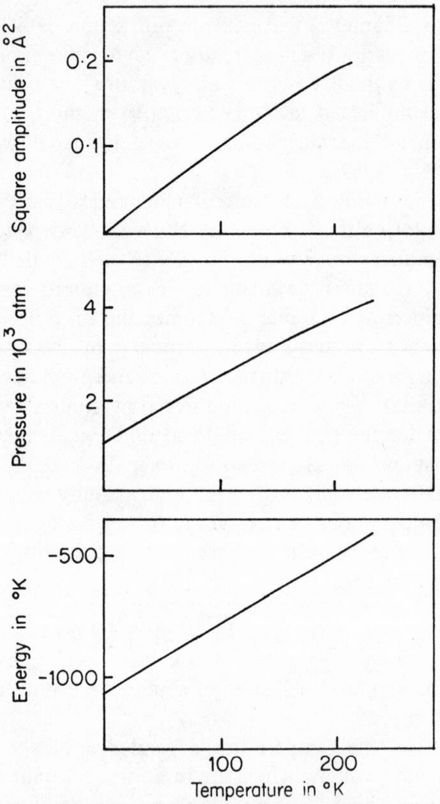

FIGURE 3.14. Temperature dependence of properties at constant volume. Relevant to solid Kr (after Dickey and Paskin).

each atom in the solid), with a particular wave vector, and examining the response of the system. Because of the background noise at high temperatures, this technique had to be combined with the usual Fourier analysis to extract the dispersion relations.

The results are shown in Figure 3.16, where the frequencies are shown relative to the low-temperature values. These are, to within the accuracy of the computations, identical with the harmonic results of Grindley and Howard.

Finally, these calculations give some valuable information on phonon lifetimes. As an illustration, we show in Figure 3.17 the time evolution of a

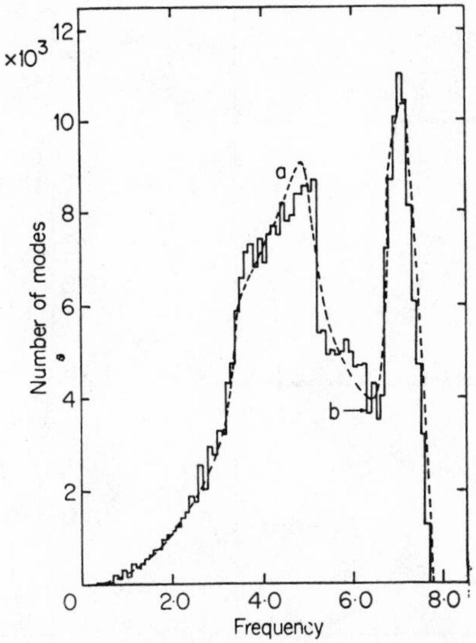

FIGURE 3.15. Frequency spectrum for Kr from (a) molecular dynamics, (b) harmonic approximation. Frequency is measured in reduced units, related to parameters in the Lennard-Jones potential (after Dickey and Paskin). See Grindley and Howard, 1965.

perturbed normal mode at three different temperatures. The reduction of the phonon lifetime as the temperature is increased is clearly evident. A lifetime τ can be calculated from the elementary formula for the ratio of the amplitudes A of adjacent maxima

$$A_1/A_2 = e^{-t/\tau}. \tag{3.15.2}$$

At high temperatures, τ is seen to be of the order of a few periods only.

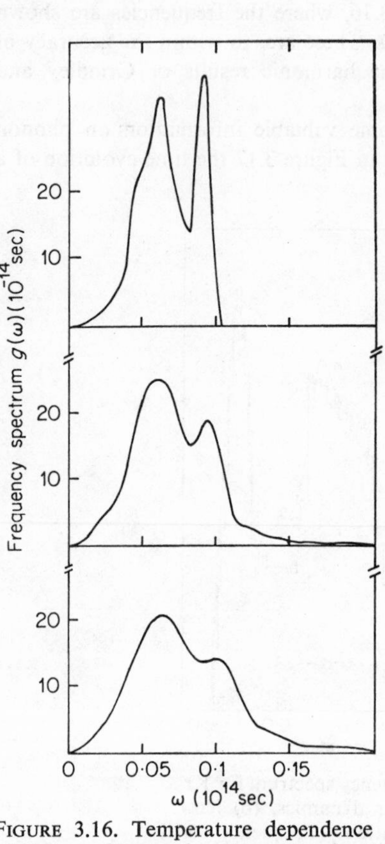

FIGURE 3.16. Temperature dependence of phonon frequencies. Top curve is at 2.5 K and others at 91 K and 186 K (after Dickey and Paskin).

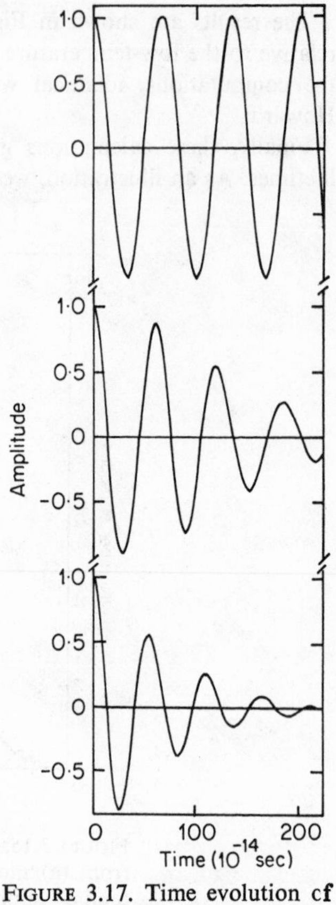

FIGURE 3.17. Time evolution cf perturbed normal mode at three temperatures, as in Fig. 3.16.

3.16 Phonon theory of ferroelectrics

We shall conclude this chapter by dealing with two further topics, how phonon theory can be used to help us to understand the properties of ferroelectric materials, and secondly to say a bit about crystal stability and its connection with the Morse relations of section 1.7 of Chapter 1. Though we shall not give attention to it here, Cochran's theory of ferroelectrics also involves consideration of crystal stability in a basic way (see Cochran, 1969), so that the topics are in fact related in this loose sense. In the theory of ferroelectrics we shall, at some stage, incorporate phonon interactions, and we shall do this using one-phonon Green functions. For this reason we have deferred the discussion until now. We shall assume that the reader is familiar with the basic facts of ferroelectricity and go straight into the discussion of the dielectric properties.

3.16.1 Dielectric properties

We shall describe below how ferroelectric behaviour may be interpreted in terms of phonon theory. It will be convenient to restrict ourselves to the non-polar phase above the Curie temperature and our discussion will be applicable to a material like $BaTiO_3$.

We introduce the frequency and wave-number-dependent susceptibility by $\chi(\mathbf{k}\omega)$ as follows. Applying an electric field $\mathscr{E}(\mathbf{r}t) = e^{i(\mathbf{k}.\mathbf{r}+\omega t)}$, the Fourier component $\mathscr{M}(\mathbf{k})$ of the polarization of the ferroelectric above the critical point may be written as

$$\mathscr{M}_\alpha(\mathbf{k}) = \sum_\beta \chi_{\alpha\beta}(\mathbf{k}\omega)\,\mathscr{E}_\beta, \tag{3.16.1}$$

provided the response is linear. This equation defines the susceptibility tensor, the components $\chi_{\alpha\beta}(\mathbf{k}\omega)$ of which will separately satisfy the properties of linear-response functions discussed in section 4.6 of the next chapter.

To calculate $\chi_{\alpha\beta}(\mathbf{k}\omega)$ explicitly, some model must be assumed and the simplest is the rigid-ion approach. Here, the electronic polarizability is ignored so that the dielectric constant is unity at optical frequencies but not at phonon frequencies. To find the polarization in terms of the ionic positions, we note that, in the non-polar phase, the polarization is zero when the ions are all at their equilibrium positions $\mathbf{R}_{li}$, i labelling the ith ion in the unit cell about the Bravais lattice point $\mathbf{l}$. Hence, when an ion of charge Z_i is displaced an amount $\mathbf{u}_{li}$, there is, in effect, a charge $-Z_i$ at $\mathbf{R}_{li}$, so that a dipole moment $Z_i\mathbf{u}_{li}$ has been created. We therefore have a distribution of dipole moments

$$\boldsymbol{\mu}(\mathbf{r}t) = \sum_{li} Z_i\,\mathbf{u}_{li}\,\delta(\mathbf{r}-\mathbf{R}_{li}) \tag{3.16.2}$$

with Fourier transform

$$\boldsymbol{\mu}(\mathbf{k}t) = \sum_{li} Z_i\,\mathbf{u}_{li}\,e^{i\mathbf{k}.\mathbf{R}_{li}}. \tag{3.16.3}$$

Using the expression

$$\mathbf{u}_{li}(t) = \frac{1}{(NM_i)^{\frac{1}{2}}}\sum_{\mathbf{q}\sigma} \boldsymbol{\varepsilon}_{i\sigma}(\mathbf{q})\,Q_\sigma(\mathbf{q},t)\,e^{i\mathbf{q}.\mathbf{R}_{li}}, \tag{3.16.4}$$

we then find

$$\boldsymbol{\mu}(\mathbf{k}t) = N^{\frac{1}{2}}\sum_\sigma \mathscr{M}_\sigma(\mathbf{k})\,Q_\sigma(\mathbf{q},t), \tag{3.16.5}$$

where

$$\mathcal{M}_\sigma(\mathbf{k}) = \sum_i \frac{Z_i}{M_i^{\frac{1}{2}}}\, \boldsymbol{\varepsilon}_{i\sigma}(\mathbf{k})\, e^{i\mathbf{K}.\mathbf{R}_i}. \tag{3.16.6}$$

Here

$$\mathbf{R}_i = \mathbf{R}_{\mathbf{l}i} - \mathbf{l} \tag{3.16.7}$$

and the reciprocal lattice vector $\mathbf{K}$ is given by

$$\mathbf{K} = \mathbf{k} + \mathbf{q} \tag{3.16.8}$$

with $\mathbf{q}$ restricted to the BZ. Since, by definition,

$$\mathcal{M}(\mathbf{k}) = \langle \boldsymbol{\mu}(\mathbf{k}) \rangle \tag{3.16.9}$$

equation (3.16.5), expressing $\boldsymbol{\mu}$ in terms of phonon coordinates Q, translates the problem of the dielectric behaviour of the ferroelectric into one in which we obtain the behaviour of the phonons under the influence of an electric field.

3.16.2 Harmonic approximation

We follow Cochran (1969) in starting from an equation for forced harmonic oscillators, namely

$$\ddot{Q}_\sigma(\mathbf{q}t) + \omega_\sigma^2(\mathbf{q})\, Q_\sigma(\mathbf{q}t) = \frac{\partial}{\partial Q_\sigma^*(\mathbf{q}, t)} (\boldsymbol{\mu}^* . \mathcal{E}) = N^{\frac{1}{2}} \mathcal{M}_\sigma^*(\mathbf{k}) . \mathcal{E}. \tag{3.16.10}$$

It is then a straightforward matter to show that

$$\mathcal{M}(\mathbf{k}) = N \sum_\sigma \frac{\mathcal{M}_\sigma(\mathbf{k})\, [\mathcal{M}_\sigma(\mathbf{k}) . \mathcal{E}]}{\omega_\sigma^2 - \omega^2}, \tag{3.16.11}$$

giving the real part of the susceptibility tensor as

$$\Omega \chi'_{\alpha\beta}(\mathbf{k}\omega) = \sum_\sigma \frac{\mathcal{M}_{\alpha\sigma}(\mathbf{k})\, \mathcal{M}_{\beta\sigma}^*(\mathbf{k})}{\omega_\sigma^2(\mathbf{k}) - \omega^2}. \tag{3.16.12}$$

Only the real part is obtained, since in applying the harmonic approximation all loss mechanisms have been neglected. However, the imaginary part χ'' of a response function is generally related to the real part by the Kramers–Krönig relations. We shall prove these in Chapter 7, but it is convenient to summarize the result at this point:

$$\chi'(\omega) = -\frac{1}{\pi} \int_{-\infty}^{\infty} \frac{\chi''(\omega')}{\omega - \omega'}\, d\omega' \tag{3.16.13}$$

and

$$\chi''(\omega) = \frac{1}{\pi} \int_{-\infty}^{\infty} \frac{\chi'(\omega')}{\omega - \omega'}\, d\omega'. \tag{3.16.14}$$

From equation (3.16.12), we now find

$$\chi''_{\alpha\beta}(\mathbf{k},\omega) = \frac{\pi}{2\Omega}\sum_{\sigma}\frac{\mathscr{M}_{\alpha\sigma}(\mathbf{k})}{\omega_{\sigma}(\mathbf{k})}\mathscr{M}^{*}_{\beta\sigma}(\mathbf{k})\{\delta(\omega-\omega_{\sigma}(\mathbf{k}))-\delta(\omega+\omega_{\sigma}(\mathbf{k}))\}. \tag{3.16.15}$$

It also follows directly from the use of the harmonic approximation in equation (3.16.5) that

$$\int_{-\infty}^{\infty}\langle\mu_{\alpha}(-\mathbf{k},0)\,\mu_{\alpha}(\mathbf{k},t)\rangle\,\mathrm{e}^{-i\omega t}\,dt$$

$$= N\pi\sum_{\sigma}\frac{|\mathscr{M}_{\alpha}(\mathbf{k})|^{2}}{\omega_{\sigma}(\mathbf{k})\,[1-\mathrm{e}^{-\beta\omega}]}\{\delta(\omega-\omega_{\sigma}(\mathbf{k}))-\delta(\omega+\omega_{\sigma}(\mathbf{k}))\} \tag{3.16.16}$$

and we see, on comparison with equation (3.16.15), that

$$\chi''_{\alpha\alpha}(\mathbf{k},\omega) = \frac{1}{\pi}(1-\mathrm{e}^{-\beta\omega})\int_{-\infty}^{\infty}\langle\mu_{\alpha}(-\mathbf{k},0)\,\mu_{\alpha}(\mathbf{k},t)\rangle\,\mathrm{e}^{-i\omega t}\,dt. \tag{3.16.17}$$

Actually, this relation is quite general, as we shall see in Chapter 4, section 4.6. Integrating over ω we obtain

$$\langle|\mu_{\alpha}(\mathbf{k})|^{2}\rangle = \frac{1}{\pi}\int_{-\infty}^{\infty}(1-\mathrm{e}^{-\beta\omega})^{-1}\,\chi''_{\alpha\alpha}(\mathbf{k},\omega)\,d\omega. \tag{3.16.18}$$

When the temperature is sufficiently high, we can replace $(1-\mathrm{e}^{-\beta\omega})^{-1}$ by $k_{\mathrm{B}}T/\omega$, whereupon we can see, when we use the Kramers–Krönig relation,

$$\frac{1}{\pi}\int_{-\infty}^{\infty}\frac{\chi''(\mathbf{k},\omega)}{\omega}\,d\omega = \chi(\mathbf{k},0) \tag{3.16.19}$$

that

$$\langle|\mu_{\alpha}(\mathbf{k})|^{2}\rangle = k_{\mathrm{B}}T\chi_{\alpha\alpha}(\mathbf{k},0). \tag{3.16.20}$$

3.16.3 General linear-response theory

To go beyond the harmonic approximation, we must obtain a general formula for the linear-response function. We write that part of the Hamiltonian containing the applied field as

$$H_{1} = \int\boldsymbol{\mu}(\mathbf{r})\,.\,\mathscr{E}(\mathbf{r}t)\,d\mathbf{r} \tag{3.16.21}$$

and it is our aim to calculate the response in the form [see equation (3.16.1)]

$$\Delta\mathscr{M}_{\beta}(\mathbf{r},t) = \sum_{\alpha}\int d\mathbf{r}'\int_{-\infty}^{t}\chi_{\alpha\beta}(\mathbf{r},\mathbf{r}',t-t')\,\mathscr{E}_{\alpha}(\mathbf{r}',t')\,dt'. \tag{3.16.22}$$

We should note, incidentally, that we are not at the moment necessarily assuming $\mathscr{M}_\beta = 0$ in equilibrium (as, in fact is true if the phase is non-polar) and that is why we write the left-hand side of equation (3.16.22) as $\Delta\mathscr{M}$ instead of $\mathscr{M}$ itself.

We shall utilize the density-matrix operator, which, as we show in Appendix 3.6, obeys the equation

$$\frac{\partial \rho}{\partial t} = \frac{i}{\hbar}[\rho, H + H_1(t)]. \tag{3.16.23}$$

Here H represents the unperturbed Hamiltonian and the system in equilibrium has the density-matrix operator ρ_0 obeying

$$0 = [\rho_0, H]. \tag{3.16.24}$$

Subject to the boundary $\rho(-\infty) = \rho_0$, equation (3.16.23) then has the integral solution

$$\rho = \rho_0 + \frac{i}{\hbar}\int_{-\infty}^{t} e^{st'} e^{iH(t'-t)/\hbar}[\rho(t), H_1(t')]\, e^{-iH(t'-t)/\hbar}\, dt' \tag{3.16.25}$$

as may be verified directly by differentiation. The factor e^{st} $(s \to 0)$ ensures convergence and corresponds to a perturbation that is strictly $e^{st} H_1(t)$. This can be interpreted to mean that $H_1(t)$ was switched very slowly on to the state $\rho(-\infty) = \rho_0$. The convergence factor will be dropped in the following, on the understanding that it can always be reinserted if necessary.

We now iterate to obtain

$$\rho = \rho_0 + \rho_1 + \rho_2 + \ldots, \tag{3.16.26}$$

where

$$\rho_1 = \frac{i}{\hbar}\int_{-\infty}^{t} e^{iH(t'-t)/\hbar}[\rho_0, H_1(t')]\, e^{-iH(t'-t)/\hbar}\, dt', \tag{3.16.27}$$

which is linear in H_1, ρ_2 being quadratic in H_1 and so on. We shall set H_1 equal to the right-hand side of equation (3.16.21) and because in defining χ we assume a linear response to $\mathscr{E}$, it suffices to terminate the series for ρ at ρ_1.

Now, by definition,

$$\Delta\mathscr{M}_\beta(\mathbf{r}, t) = \mathrm{Tr}\{\rho_1(t)\,\mu_\beta(\mathbf{r})\} = \frac{i}{\hbar}\int_{-\infty}^{t} \mathrm{Tr}\{[\rho_0, H_1(t')]\, e^{-iH(t'-t)/\hbar}\,\mu_\beta(\mathbf{r})\, e^{iH(t'-t)/\hbar}\}\, dt$$

$$= \frac{i}{\hbar}\mathrm{Tr}\left\{\int_{-\infty}^{t} dt'[\rho_0, H_1(t')]\,\mu_\beta(\mathbf{r}, t-t')\right\}$$

$$= \frac{i}{\hbar}\mathrm{Tr}\left\{\int_{-\infty}^{t} dt'\;\rho_0[H_1(t'), \mu_\beta(\mathbf{r}, t-t')]\right\} \tag{3.16.28}$$

where we have used the cyclic property of the trace and defined Heisenberg operators with respect to the unperturbed system. Inserting equation (3.16.21) for H_1 we now find

$$\Delta\mathscr{M}_\beta(\mathbf{r}t) = \frac{i}{\hbar}\sum_\alpha \int_{-\infty}^{t} dt' \int d\mathbf{r}'\, \mathscr{E}_\alpha(\mathbf{r}', t')\langle[\mu_\alpha(\mathbf{r}'), \mu_\beta(\mathbf{r}, t-t')]\rangle \quad (3.16.29)$$

with the thermal average carried out over the unperturbed system:

$$\langle A\rangle = \mathrm{Tr}\{\rho_0 A\}. \quad (3.16.30)$$

Comparison of equations (3.16.22) and (3.16.29) shows immediately that

$$\chi_{\alpha\beta}(\mathbf{r}, \mathbf{r}'; t) = \begin{cases} \frac{i}{\hbar}\langle[\mu_\alpha(\mathbf{r}'), \mu_\beta(\mathbf{r}t)]\rangle, & t>0, \\ 0, & t<0, \end{cases} \quad (3.16.31)$$

and so

$$\chi_{\alpha\beta}(\mathbf{k}, \mathbf{k}'; t) = \begin{cases} \frac{i}{\hbar}\langle[\mu_\alpha(-\mathbf{k})\, \mu_\beta(\mathbf{k}, t)]\rangle, & t>0, \\ 0, & t<0. \end{cases} \quad (3.16.32)$$

3.16.4 Response functions in terms of Green functions

We have seen earlier that, at least formally, one can deal with the effects of anharmonicity by using the one-phonon thermodynamic Green function $G_{\sigma\sigma'}(\mathbf{k}, \omega)$ and we therefore seek to relate the susceptibility to this quantity. With reference to equation (3.16.31), let us first define, for any two operators a and b,

$$\chi(t) = i\langle[a(0), b(t)]\rangle, \quad (3.16.33)$$

which we shall relate to the function

$$C(\tau) = \langle T\{a(0)\, b(\tau)\}\rangle, \quad (3.16.34)$$

where τ is an imaginary time, that is,

$$b(\tau) = \mathrm{e}^{H\tau}\, b\, \mathrm{e}^{-H\tau} \quad (3.16.35)$$

and T orders smaller τ's to the right. From equation (3.16.34) we see that C becomes the one-phonon Green function, on putting $a = q^\dagger_{\mathbf{k}\sigma}$ and $b = q_{\mathbf{k}\sigma}$.

Now, provided $0 \leqslant \tau \leqslant \beta$,

$$C(\tau-\beta) = \langle a(0)\, b(\tau-\beta)\rangle = \frac{1}{Z}\mathrm{Tr}\{\mathrm{e}^{-\beta H}\, a\, \mathrm{e}^{-\beta H}\, b(\tau)\, \mathrm{e}^{\beta H}\}$$

$$= \frac{1}{Z}\mathrm{Tr}\{a\, \mathrm{e}^{-\beta H}\, b(\tau)\}, \quad (3.16.36)$$

that is,

$$C(\tau-\beta) = C(\tau), \tag{3.16.37}$$

so that we can write

$$C(\tau) = \sum_{n=-\infty}^{\infty} C(i\omega_n)\,\mathrm{e}^{i\omega_n\tau}, \quad -\beta \leqslant \tau \leqslant \beta, \tag{3.16.38}$$

with

$$\omega_n \beta = 2\pi n. \tag{3.16.39}$$

Evidently

$$C(i\omega_n) = \frac{1}{\beta}\int_0^\beta C(\tau)\,\mathrm{e}^{-i\omega_n\tau}\,d\tau. \tag{3.16.40}$$

Taking the trace with respect to eigenstates of H, we have

$$\begin{aligned} C(i\omega_n) &= \frac{1}{\beta}\int_0^\beta \mathrm{Tr}\{\rho b(\tau)\,a(0)\}\,\mathrm{e}^{-i\omega_n\tau} \\ &= \frac{1}{\beta}\sum_{ij}\rho_i\int_0^\beta \langle i|\,\mathrm{e}^{H\tau}\,b\,\mathrm{e}^{-H\tau}|j\rangle\langle j|\,a\,|i\rangle\,\mathrm{e}^{-i\omega_n\tau}\,d\tau \\ &= \frac{1}{\beta}\sum_{ij}\rho_i\,b_{ij}\,a_{ji}\int_0^\beta \mathrm{e}^{(\varepsilon_i-\varepsilon_j-i\omega_n)\tau}\,d\tau, \end{aligned} \tag{3.16.41}$$

where we have written $\rho_i = \mathrm{e}^{-\beta\varepsilon_i}/Z$ and $b_{ij} = \langle i|\,b\,|j\rangle$, etc.

Equation (3.16.41) may also be written as

$$-C(i\omega_n) = \frac{1}{\beta}\sum_{ij}\frac{(\rho_i-\rho_j)}{\varepsilon_i-\varepsilon_j-i\omega_n}\,b_{ij}\,a_{ji}. \tag{3.16.42}$$

On the other hand, let us define

$$\chi(\omega) = i\int_0^\infty \mathrm{e}^{-i\omega t}\langle[a(0), b(t)]\rangle\,dt. \tag{3.16.43}$$

Then, since

$$\langle[a(0), b(t)]\rangle = \sum_{ij}(\rho_j-\rho_i)\,b_{ij}\,a_{ji}\,\mathrm{e}^{i(\varepsilon_i-\varepsilon_j)t}, \tag{3.16.44}$$

we find, on allowing an infinitesimal negative imaginary part to ensure convergence of the integral,

$$-\chi(\omega) = \sum_{ij}\frac{(\rho_j-\rho_i)\,b_{ij}\,a_{ji}}{\varepsilon_i-\varepsilon_j-\omega}. \tag{3.16.45}$$

Comparison with equation (3.16.42) shows that

$$\chi(\omega-i\varepsilon) = -\beta C(\omega-i\varepsilon), \quad \varepsilon\to+0, \tag{3.16.46}$$

where $C(\omega)$ is the analytic continuation of $C(i\omega_n)$ off the imaginary axis.

We now take a particular case of equation (3.16.46) by writing $a = \mu_\alpha(-\mathbf{k})$ and $b = \mu_\beta(\mathbf{k})$. We have then using equation (3.16.5)

$$C(\mathbf{k}, \tau) = \langle T\{\mu_\alpha(-\mathbf{k})\,\mu_\beta(\mathbf{k}, \tau)\}\rangle$$
$$= \sum_{\sigma\sigma'} \mathscr{M}_{\alpha\sigma'}(-\mathbf{k})\,\mathscr{M}_{\beta\sigma}(\mathbf{k})\,\langle T\{Q_\sigma(-\mathbf{k})\,Q_{\sigma'}(\mathbf{k}, \tau)\}\rangle. \quad (3.16.47)$$

That is, by making reference to equation (3.14.5),

$$C(\mathbf{k}, \tau) = -\sum_{\sigma\sigma'} \mathscr{M}_{\alpha\sigma'}(-\mathbf{k})\,\mathscr{M}_{\beta\sigma}(\mathbf{k})\frac{G_{\sigma\sigma'}(\mathbf{k}, \tau)}{[\omega_\sigma(\mathbf{k})\,\omega_{\sigma'}(\mathbf{k})]^{\frac{1}{2}}} \quad (3.16.48)$$

and hence

$$\chi(\mathbf{k}, \omega) = \beta \sum_{\sigma\sigma'} \mathscr{M}_{\alpha\sigma'}(-\mathbf{k})\frac{\mathscr{M}_{\beta\sigma}(\mathbf{k})\,G_{\sigma\sigma'}(\mathbf{k}, \omega)}{[\omega_\sigma(\mathbf{k})\,\omega_{\sigma'}(\mathbf{k})]^{\frac{1}{2}}}. \quad (3.16.49)$$

3.16.5 Generalization of rigid-ion model

The rigid-ion model used above can, in principle, be replaced by an exact formalism, by allowing the ionic charges Z_i to be replaced by an "apparent charge matrix" with elements $Z_{i\sigma}^{\alpha\beta}(\mathbf{k})$. In practice, it is difficult to see how to calculate these quantities reliably, except for the alkali halides, when, as we saw in an ealier section, we can use the shell model to represent polarizable atoms. Therefore, one generally adopts "apparent charges" Z_i, which may differ from the ionic charges somewhat.

It should finally be noted that, in any model, the dielectric susceptibility, as defined here, does not necessarily reduce to $[\varepsilon(0\omega)-1]/4\pi$ (ε being the usual dielectric function) as $\mathbf{k}\to 0$, but it will be convenient to postpone further discussion of the relation of $\chi(\mathbf{k}\omega)$ to $\varepsilon(\mathbf{k}\omega)$ until Chapter 5.

3.17 Crystal stability

Using Feynman's theorem, we may show that the first-order variation in the effective nuclear potential V due to nuclear displacements of the atoms from the crystal sites is given by

$$\Delta V = \Delta V_{\mathrm{nn}} + \int \rho(\mathbf{r})\,\Delta V_{\mathrm{en}}(\mathbf{r})\,d\mathbf{r}$$
$$= \Delta V_{\mathrm{nn}} + \int \rho(\mathbf{r})\,\mathbf{u}_l.\frac{\partial V(\mathbf{r}-\mathbf{l})}{\partial \mathbf{u}_l}\,d\mathbf{r}, \quad (3.17.1)$$

where V_{nn} is the nuclear–nuclear Coulomb energy, $\rho(\mathbf{r})$ is the perfect lattice density and V_{en} is the electron–nuclear interaction at $\mathbf{r}$. This may be rewritten in the form, for a Bravais lattice (the extension for two atoms per unit cell being evident)

$$\Delta V = \sum_{l} Z_l \mathbf{u}_l \cdot [\mathscr{E}_{\mathrm{n}}(\mathbf{l}) + \mathscr{E}_{\mathrm{e}}(\mathbf{l})], \tag{3.17.2}$$

where Z_l is the charge on the lth nucleus whose initial position is $\mathbf{l}$. $\mathbf{u}_l$ as usual is the displacement of the lth nucleus while $\mathscr{E}_{\mathrm{n}}(\mathbf{l})$ is the electric field at $\mathbf{l}$ due to all nuclei, and $\mathscr{E}_{\mathrm{e}}(\mathbf{l})$ is the electric field at each nucleus due to all the electrons. It is clear that, for equilibrium, the total electric field at each nucleus must vanish.

It is of interest to enquire what these equations have to say about the types of crystal structure that can be in equilibrium. It is an almost trivial result of inversion symmetry that these equations are automatically satisfied for any Bravais lattice having one atom per unit cell.

Our interest here, following Inglesfield and Johnson (1970), is to discuss the conditions which are imposed by (3.17.1) and (3.17.2) on structures with two or more atoms per unit cell.

It will be useful to divide crystals with two atoms per unit cell into two classes:

(i) those in which $\mathscr{E}_{\mathrm{n}} + \mathscr{E}_{\mathrm{e}}$ vanishes at every site, but $\mathscr{E}_{\mathrm{n}}$ and $\mathscr{E}_{\mathrm{e}} \neq 0$;

(ii) those in which $\mathscr{E}_{\mathrm{n}}$ and $\mathscr{E}_{\mathrm{e}}$ vanish separately.

To deal with two atoms per unit cell, we can start with a simple Bravais lattice and introduce a second identical interpenetrating Bravais lattice. Then the field $\mathscr{E}_{\mathrm{n}}(\mathbf{l})$ at one nucleus due to all the nuclei can be written as a sum of

(i) that due to nuclei in the same sub-lattice;

(ii) that due to nuclei on the other.

Contribution (i) vanishes, as remarked above, because of the inversion symmetry of the simple Bravais lattice.

If we distinguish the sites on the two sub-lattices A and B by $\mathbf{l}_a$ and $\mathbf{l}_b$, then it follows from Newton's third law that

$$Z_a \mathscr{E}_{\mathrm{n}}(\mathbf{l}_a) = -Z_b \mathscr{E}_{\mathrm{n}}(\mathbf{l}_b). \tag{3.17.3}$$

We further distinguish two classes of crystals, namely:

(i) those for which $\mathscr{E}_{\mathrm{n}}(\mathbf{l}_a) \neq 0$;

(ii) those for which $\mathscr{E}_{\mathrm{n}}(\mathbf{l}_a) = 0$.

The first class will not be discussed further since the structures are expected to depend on the properties of the particular atoms involved, because Z_a, etc. are involved. But in class (ii), it follows that, since $\mathscr{E}_{\mathrm{n}}(\mathbf{l}_a) = 0$, $\mathscr{E}_{\mathrm{e}}(\mathbf{l}_a) = 0$, and similarly for sub-lattice B. Thus, the equilibrium condition is no longer dependent on the Z's.

3.17.1 Critical points in electro-static potential

Evidently, to get the equilibrium positions, we need to locate points of zero electric field or minima in the electro-static potential.

While it is not possible to predict, on general grounds, the positions of all these minima, we have seen in section 1.7 that the Morse relations give the minimum number of critical points, within each unit cell, of different types in a periodic function. Further, symmetry considerations enable us to predict the positions of these critical points.

The periodic function in our case, of course, is the electro-static potential, and the critical points we require are the minima of this potential. An example of how the critical points are found is given in section 1.7 for a face-centred cubic lattice. Tables of the critical points are given for eight structures including simple cubic, face-centred cubic, body-centred tetragonal and hexagonal structures by Inglesfield and Johnson (1970) assuming a maximum of the electro-static potential at a nucleus.

3.17.2 Effect of electronic distribution

When we consider the introduction of sub-lattice B, whose sites $\mathbf{l}_b$ are located at the critical points in the potential of the first sub-lattice, it follows that $\mathscr{E}_{\mathrm{n}}(\mathbf{l}_a)$ and $\mathscr{E}_{\mathrm{n}}(\mathbf{l}_b)$ are zero.

However, the charge density $\rho(\mathbf{r})$ has the full symmetry of the crystal with two atoms per unit cell, and it is now necessary to find the conditions that both $\mathscr{E}_{\mathrm{e}}(\mathbf{l}_a)$ and $\mathscr{E}_{\mathrm{e}}(\mathbf{l}_b)$ are zero.

These conditions are simply stated. Both $\mathscr{E}_{\mathrm{e}}(\mathbf{l}_a)$ and $\mathscr{E}_{\mathrm{e}}(\mathbf{l}_b)$ will be zero if the point group symmetry of the A sub-lattice about the B site contains the inversion symmetry elements $\bar{1}$, $\bar{3}$, $\bar{4}$ or $\bar{6}$. From the tables of the critical points, it is clear where the second sub-lattice must be placed for equilibrium. (The reader may construct Table 1 of Inglesfield and Johnson, where the minimal set of critical points in a simple cubic lattice (CsCl, say) is summarized.)

To examine whether the equilibrium is stable or unstable, it is necessary to examine the second-order variation in the effective nuclear potential as done by Inglesfield and Johnson (1970), and we refer the reader to the original paper for the details.

CHAPTER 4

Spin and orbital magnetism

4.1 Introduction

The discussion of energy bands in Chapter 1 and phonons in the last chapter has been posed in a sufficiently general way to apply to any perfect crystal. Naturally enough, the accuracy of the approximation involved, the one-electron approximation as applied to excited states for example, or the harmonic treatment of lattice dynamics, is different for different types of crystals, but, qualitatively, the phenomena of electron wave and elastic wave propagation are the same in all crystalline media.

In contrast, when we turn to the problem of magnetism, some of the most striking phenomena are very specific; the ferromagnetism of Fe, Ni and Co being perhaps the most outstanding examples, or the most unusual antiferromagnetic state that exists in Cr (see section 4.12 for a detailed discussion). Why such magnetic properties occur in a localized region of the periodic table is, of course, an interesting question in itself. It is rather clear now that the nature of the d-electrons in Fe, Ni and Co and the detailed form of the Fermi surface in Cr are very relevant to the occurrence of such remarkable magnetic properties.

It is with such, rather rare, cooperative phenomena that we shall deal in the first (and major) part of this chapter. There are, it is true, interesting aspects of magnetism which are common to all crystals, and we shall deal with these towards the end of the chapter.

The reason why the rather rare cases of cooperative magnetism in the transition metals will be of prime interest is that their explanation not only requires a detailed study of their energy bands but also, crucially, the strength of electron–electron interactions relative to, say, the appropriate energy band width. Thus, we may hope to learn not only about the phenomena of ferromagnetism but also about ways that we can proceed to tackle the many-body problem.

Alas, this problem, in general terms, as will be clear from the discussion of Chapter 2, remains rather intractable and hence one must, at the outset, attempt to isolate what are likely to be the essential features of the problem.

Progress can be made in the study of correlation effects in two directions:

(i) Inclusion of the full Coulomb interactions between electrons in the uniform electron gas. Here, the powerful methods we now have available are probably giving us a relatively good description in the range of real metallic densities, though it is only fair to say that rigorous results are still restricted to the high-density limit.

However, this problem seems to have little or nothing to do with the ferromagnetism of Fe, Ni or Co, though a study of it was fruitful initially in leading to a definitive theory of the spin–density wave antiferromagnetism of Cr which we shall take up later in the chapter.

(ii) Inclusion of short-range intra-atomic interactions in narrow energy bands which can, at least in principle, simulate the d-bands in Fe, Ni and Co. This problem is much simpler to treat (though still sufficiently difficult that no exact solution is known) than the full Coulomb interaction and it seems, at least, to incorporate some of the main features of the d-electrons in the transition metals.

Going back to (i) for a moment, we shall see that, oversimplified as the model is, it has something to say about the conduction bands in simple (non-transition) metals and alloys. Thus, we shall pay particular attention to it when we come to discuss the wave-vector and frequency-dependent spin and orbital susceptibilities, which are accessible to observation using neutron scattering.

In spite of the fact that a lot of the most interesting magnetic phenomena arise in metals, we shall not neglect the well-established picture of Heisenberg, which is probably essentially correct for a whole variety of insulating magnetic materials, and which still affords the most direct route to an understanding of the main properties of spin waves, which we discuss in detail in section 4.3 below. This theory has indeed had great success in giving a quantitative description of the salient properties of magnetic insulators.

Notwithstanding the necessity for the simplifications discussed above, it will be helpful to start from the many-electron Hamiltonian for we can then display, albeit in an approximate way, how all the features we isolated above can arise. It has to be said, though, that the relative importance of the various contributions referred to above—the band motion, the intra-atomic interaction in narrow bands and the Heisenberg exchange interaction—is still a controversial matter. Nevertheless, as we shall see, we are at least in a position where we can construct a theory which will lead to the energy band theory for very large overlap of the atomic wave functions and to the correct atomic or Heitler–London picture when the atoms are far apart.

It is often true, in solid-state physics, that one or other of these pictures—band motion or Heitler–London description—is adequate in itself. As we indicated above, much of the fascination of the theory of ferromagnetism in metals lies in the fact that we have essential features from both limits which experiment demands we must incorporate in the theory. Thus, to explain the metallic conduction in Fe, Co and Ni we must have a band theory, whereas to understand the neutron results on Ni, which reveal spin waves, we must incorporate electron–electron interactions. This was clearly recognized in the work of Izuyama, Kim and Kubo (1963), in which the spin waves arise as a manifestation of collective behaviour resulting from the inclusion of electron–electron interactions in the band model.

As we shall see in detail below, such a picture can be connected directly with the simplest description of a spin wave which comes from the Heisenberg theory, where we think simply of the propagation of a misaligned spin through the lattice (see section 4.3 for full details). From the band theory as starting point, we can describe the spin wave as a bound state of an electron of one spin with a hole of opposite spin, the relative motion of the electron and the hole being such that, for most of the time, they reside on the same atom. Such a physical picture turns out to be not at all dissimilar to the Heisenberg case, though the starting points were quite different.

We turn now to analyse the many-body Hamiltonian into different contributions which we shall interpret physically. Since we cannot, at present, discuss the whole of the Hamiltonian, we shall have to be content with examining the physical situations which can arise when emphasis is placed on different terms in this Hamiltonian.

4.2 Simplification of Hamiltonian

The Hamiltonian for an electron in a solid has the form

$$H = \sum_i \left[\frac{1}{2m} p_i^2 + V(\mathbf{r}_i)\right] + \frac{1}{2}\sum_{i\neq j} \frac{e^2}{|\mathbf{r}_i - \mathbf{r}_j|}, \tag{4.2.1}$$

where $V(\mathbf{r})$ is the potential energy of the ion–electron interaction, while the last term represents the electron–electron interaction. While it is clear from Chapter 1 that a modified potential can be found which incorporates the effects of the electron–electron interactions in calculating the single-particle excitations, we shall also be concerned in this chapter with collective excitations, mainly spin waves, which can occur due to the electron–electron interactions.

4.2.1 Second quantized form of Hamiltonian

As in Chapter 3, where we discussed the quasi-particles associated with lattice vibrations, we shall find it helpful to work with the second quantized form of the Hamiltonian. If we represent the one-electron part by $h(\mathbf{r}_i)$, and the interaction term by $\frac{1}{2}\Sigma_{ij} v_{ij}$, then we find [cf. equations (2.5.16) and (2.5.24)]

$$H = \sum_{\mu\nu} \langle \mu | h | \nu \rangle a_\mu^\dagger a_\nu + \frac{1}{2} \sum_{\substack{\mu\nu \\ \sigma\tau}} \langle \mu\nu | v | \sigma\tau \rangle a_\mu^\dagger a_\nu^\dagger a_\tau a_\sigma, \tag{4.2.2}$$

where

$$\langle \mu | h | \nu \rangle = \int \phi_\mu^*(\mathbf{r})\, h(\mathbf{r})\, \phi_\nu(\mathbf{r})\, d\mathbf{r} \tag{4.2.3}$$

and

$$\langle \mu\nu | v | \sigma\tau \rangle = \int \phi_\mu^*(\mathbf{r})\, \phi_\nu^*(\mathbf{r}')\, v(\mathbf{r}-\mathbf{r}')\, \phi_\sigma(\mathbf{r})\, \phi_\tau(\mathbf{r}')\, d\mathbf{r}\, d\mathbf{r}'. \tag{4.2.4}$$

With Fermions, we have (cf. equation 2.5.9) the anticommutation relation

$$[a_\mu^\dagger, a_\nu]_+ = \delta_{\mu\nu}, \tag{4.2.5}$$

with all other anticommutators vanishing.

The complete set of functions ϕ will now be chosen as Bloch functions $\psi_\mathbf{k}^i$, i being as usual the band index. If σ is the spin variable, then corresponding to each Bloch function will be two creation or destruction operators, corresponding to spin up and spin down. Then, if $\varepsilon_\mathbf{k}^i$ are the one-electron eigenvalues $\langle \psi_\mathbf{k}^i | h | \psi_\mathbf{k}^i \rangle$ we may write

$$H = \sum_{\mathbf{k}i\sigma} \varepsilon_\mathbf{k}^i a_{\mathbf{k}i\sigma}^\dagger a_{\mathbf{k}i\sigma}$$

$$+ \frac{1}{2} \sum_{\mathbf{k}_1\mathbf{k}_2\mathbf{k}_1'\mathbf{k}_2'} \sum_{i_1 i_2 i_1' i_2'} \sum_{\sigma\sigma'} \langle \mathbf{k}_1 i_1 \mathbf{k}_2 i_2 | \frac{1}{r} | \mathbf{k}_1' i_1' \mathbf{k}_2' i_2' \rangle a_{\mathbf{k}_1 i_1 \sigma}^\dagger a_{\mathbf{k}_2 i_2 \sigma'}^\dagger a_{\mathbf{k}_2' i_2' \sigma'} a_{\mathbf{k}_1' i_1' \sigma}, \tag{4.2.6}$$

where

$$\langle \mathbf{k}_1 p_1 \mathbf{k}_2 p_2 | \frac{1}{r} | \mathbf{k}_1' p_1' \mathbf{k}_2' p_2' \rangle = e^2 \int \frac{\psi_{\mathbf{k}_1}^{*\,p_1}(\mathbf{r})\, \psi_{\mathbf{k}_2}^{*\,p_2}(\mathbf{r}')\, \psi_{\mathbf{k}_1'}{}^{p_1'}(\mathbf{r})\, \psi_{\mathbf{k}_2'}{}^{p_2'}(\mathbf{r}')}{|\mathbf{r}-\mathbf{r}'|}\, d\mathbf{r}\, d\mathbf{r}'. \tag{4.2.7}$$

To handle the sum over all bands is obviously impracticable. While for the 3d electrons in transition metals we would like to deal with the five sub-bands, it will be convenient to follow Kubo *et al.* and Hubbard and, in order to bring some of the essential features to light, to treat with an isolated s-band. Thus we shall consider the approximate form of the Hamiltonian

$$H = \sum_{\mathbf{k}\sigma} \varepsilon_\mathbf{k} a_{\mathbf{k}\sigma}^\dagger a_{\mathbf{k}\sigma} + \frac{1}{2} \sum_{\mathbf{k}_1\mathbf{k}_2\mathbf{k}_1'\mathbf{k}_2'\sigma\sigma'} \langle \mathbf{k}_1 \mathbf{k}_2 | \frac{1}{r} | \mathbf{k}_1' \mathbf{k}_2' \rangle a_{\mathbf{k}_1\sigma}^\dagger a_{\mathbf{k}_2\sigma'}^\dagger a_{\mathbf{k}_2'\sigma'} a_{\mathbf{k}_1'\sigma}. \tag{4.2.8}$$

4.2.2 *Wannier representation*

It will prove very convenient for emphasizing the atomic aspects of the problem to work now in a Wannier-function representation (or, approximately, for narrow bands, an atomic orbital picture). The Wannier function w was introduced in Chapter 1, and for present purposes may be defined through [cf. equation (1.8.3)]

$$\psi_{\mathbf{k}}(\mathbf{r}) = \frac{1}{N^{\frac{1}{2}}} \sum_i w(\mathbf{r}-\mathbf{R}_i)\, e^{i\mathbf{k}.\mathbf{R}_i}, \tag{4.2.9}$$

where $\mathbf{R}_i$ is a lattice vector. Using the result

$$\sum_{\mathbf{k}} e^{i\mathbf{k}.(\mathbf{R}_i-\mathbf{R}_j)} = N\delta_{ij}, \tag{4.2.10}$$

equation (4.2.9) may be inverted to give

$$w(\mathbf{r}-\mathbf{R}_i) = \frac{1}{N^{\frac{1}{2}}} \sum_{\mathbf{k}} \psi_{\mathbf{k}}(\mathbf{r})\, e^{-i\mathbf{k}.\mathbf{R}_i}. \tag{4.2.11}$$

Corresponding to this relation we define

$$a_{i\sigma}^{\dagger} = \frac{1}{N^{\frac{1}{2}}} \sum_{\mathbf{k}} a_{\mathbf{k}\sigma}^{\dagger}\, e^{-i\mathbf{k}.\mathbf{R}_i},$$

$$a_{i\sigma} = \frac{1}{N^{\frac{1}{2}}} \sum_{\mathbf{k}} a_{\mathbf{k}\sigma}\, e^{i\mathbf{k}.\mathbf{R}_i}. \tag{4.2.12}$$

It is easily shown by use of equation (4.2.10) that the $a_{i\sigma}^{\dagger}$ and $a_{i\sigma}$ obey the usual anticommutation relations for Fermion operators. We can think of $a_{i\sigma}^{\dagger}$ as creating a particle of spin σ on the ith site (that is, in the ith Wannier orbital).

It now follows that equation (4.2.8) can be written in the form

$$H = \sum_{ij}\sum_{\sigma} \varepsilon_{ij}\, a_{i\sigma}^{\dagger} a_{j\sigma} + \frac{1}{2} \sum_{\substack{ij\\kl}} \sum_{\sigma\sigma'} \langle ij|\frac{1}{r}|kl\rangle\, a_{i\sigma}^{\dagger} a_{j\sigma'}^{\dagger} a_{l\sigma'} a_{k\sigma}, \tag{4.2.13}$$

where

$$\varepsilon_{ij} = \int w^*(\mathbf{r}-\mathbf{R}_i)\, h(\mathbf{r})\, w(\mathbf{r}-\mathbf{R}_j)\, d\mathbf{r} \tag{4.2.14}$$

and

$$\langle ij|\frac{1}{r}|kl\rangle = e^2 \int \frac{w^*(\mathbf{r}-\mathbf{R}_i)\, w^*(\mathbf{r}'-\mathbf{R}_j)\, w(\mathbf{r}-\mathbf{R}_k)\, w(\mathbf{r}'-\mathbf{R}_l)}{|\mathbf{r}'-\mathbf{r}|}\, d\mathbf{r}\, d\mathbf{r}'. \tag{4.2.15}$$

The Hamiltonian can be split into contributions from a one-centre integral and two-centre integrals of various kinds. The decomposition is mathematically trivial and we only comment on the one-centre term, for which $i = k = j = l$; this is

$$\frac{1}{2}\sum_i \langle ii|\frac{1}{r}|ii\rangle \sum_{\sigma\sigma'} a^\dagger_{i\sigma} a^\dagger_{i\sigma'} a_{i\sigma'} a_{i\sigma} = \sum_i \langle ii|\frac{1}{r}|ii\rangle \sum_\sigma a^\dagger_{i\sigma} a_{i\sigma} a^\dagger_{i-\sigma} a_{i-\sigma} \quad (4.2.16)$$

since we cannot create two particles of the same spin on the same site—an expression of the exclusion principle through the anticommutation relations. Introducing [cf. equation (2.5.4)]

$$a^\dagger_{i\sigma} a_{i\sigma} = n_{i\sigma} \quad (4.2.17)$$

the number operator for electrons of spin σ on the ith site, the decomposition of H takes the form

$$H = \sum_{ij}\sum_\sigma \varepsilon_{ij} a^\dagger_{i\sigma} a_{j\sigma} + \frac{I}{2}\sum_i\sum_\sigma n_{i\sigma} n_{i-\sigma} + \frac{1}{2}\sum_{ij}{}' \sum_{\sigma\sigma'} \langle ij|\frac{1}{r}|ij\rangle n_{i\sigma} n_{j\sigma'}$$

$$-\frac{1}{2}\sum_{ij}{}' \sum_{\sigma\sigma'} J_{ij} a^\dagger_{i\sigma} a_{i\sigma'} a^\dagger_{j\sigma'} a_{j\sigma} + \sum_{ij}{}' \sum_{\sigma\sigma'} \langle ii|\frac{1}{r}|jj\rangle a^\dagger_{i\sigma} a^\dagger_{i\sigma'} a_{j\sigma'} a_{j\sigma}, \quad (4.2.18)$$

where

$$I = e^2 \int \frac{|w(\mathbf{r})|^2 |w(\mathbf{r}')|^2}{|\mathbf{r}'-\mathbf{r}|} d\mathbf{r}'\, d\mathbf{r} \quad (4.2.19)$$

$$J_{ij} = e^2 \int \frac{w^*(\mathbf{r}-\mathbf{R}_i)\, w(\mathbf{r}-\mathbf{R}_j)\, w^*(\mathbf{r}'-\mathbf{R}_j)\, w(\mathbf{r}'-\mathbf{R}_i)}{|\mathbf{r}'-\mathbf{r}|} d\mathbf{r}\, d\mathbf{r}'. \quad (4.2.20)$$

I is presumably greatest in magnitude, of the integrals concerned, if the electrons are truly localized, and J_{ij} is just the customary direct exchange integral. Of the other terms, since

$$n_i = \sum_\sigma n_{i\sigma} \quad (4.2.21)$$

is the number operator for electrons on site i irrespective of spin, the third term is

$$\frac{1}{2}\sum_{ij} \langle ij|\frac{1}{r}|ij\rangle n_i n_j \quad (4.2.22)$$

with no spin dependence and therefore unimportant for magnetism. It will be seen that the effect of the operators of the last term is to remove antiparallel spin pairs from one site to another previously vacant; this term will also play no part in our considerations.

Because the Heisenberg model gives us a very graphic way of understanding low-lying excitations which are collective rather than single-particle in character, we shall first tackle this problem, dealing later with the narrow energy-band picture, and finally with Coulombic interactions in a uniform electron gas.

4.3 Spin waves and Heisenberg Hamiltonian

4.3.1 Form of Hamiltonian

The integral I in the second term of the Hamiltonian of (4.2.18) is essentially the interaction energy between two spins on the same atom, and so if I is very large we can expect a state in which all atoms are singly occupied (taking N electrons in a system of N atoms). The wave function is then a determinant composed of the Wannier functions or, equivalently, the $\psi_{\mathbf{k}}$'s. Every state $\mathbf{k}$ in the BZ is singly occupied, and the system is an insulator. In terms of creation operators, the state is described by a linear combination of terms of the form

$$|\sigma_1 \ldots \sigma_N\rangle = a^\dagger_{1\sigma(1)}\, a^\dagger_{2\sigma(2)} \ldots a^\dagger_{N\sigma(N)}|0\rangle \tag{4.3.1}$$

which differ one from the other only in the distribution of spins $\sigma(i)$ on the lattice sites i. Between such states we find matrix elements of equation (4.2.18) given by (with $\varepsilon_{ii} = \varepsilon_0$)

$$\begin{aligned}\langle\sigma_1'(1)\ldots\sigma_N'(N)|\,H\,|\sigma_1(1)\ldots\sigma_N(N)\rangle \\ = \prod_{i=1}^{N} \delta_{\sigma'(i)\sigma(i)}\left[N\varepsilon_0 + \frac{1}{2}\sum_{ij}{}' \langle ij|\frac{1}{r}|ij\rangle\right] - \frac{1}{2}\sum_{ij}{}'\sum_{\sigma\sigma'} J_{ij} \\ \times\langle\sigma_1'\sigma\ldots\sigma_N'(N)|\,a^\dagger_{i\sigma}a_{i\sigma'}a^\dagger_{j\sigma'}a_{j\sigma}\,|\sigma_1(1)\ldots\sigma_N(N)\rangle.\end{aligned} \tag{4.3.2}$$

That is, the first three terms contribute only diagonal terms independent of spin configuration and the last term makes no contribution. As $I \to \infty$, therefore, the exchange term J_{ij} completely determines the states of the system. Since, as remarked, these states only differ in their spin configurations, it ought to be possible to write H solely in terms of spin operators, provided we allow it to operate only on functions described by (4.3.1) or, equivalently, a product

$$\sigma_1(1)\,\sigma_2(2)\ldots\sigma_N(N), \tag{4.3.3}$$

where $\sigma_i(i)$ is a spin function

$$\alpha \equiv \begin{pmatrix}1\\0\end{pmatrix} \quad \text{or} \quad \beta \equiv \begin{pmatrix}0\\1\end{pmatrix}.$$

We first observe that the operators of the exchange term are

$$\sum_{\sigma\sigma'} a^\dagger_{i\sigma} a_{i\sigma'} a^\dagger_{j\sigma'} a_{j\sigma} = a^\dagger_{i\uparrow} a_{i\uparrow} a^\dagger_{j\uparrow} a_{j\uparrow} + a^\dagger_{i\uparrow} a_{i\downarrow} a^\dagger_{j\downarrow} a_{j\uparrow} + a^\dagger_{i\downarrow} a_{i\downarrow} a^\dagger_{j\downarrow} a_{j\downarrow} + a^\dagger_{i\downarrow} a_{i\uparrow} a^\dagger_{j\uparrow} a_{j\downarrow}. \tag{4.3.4}$$

Now operating on functions defined in equation (4.3.1), $a^\dagger_\uparrow a_\uparrow$ leaves up spins unchanged and gives zero for down spins. This is equivalent to

$$\begin{pmatrix} 1 & 0 \\ 0 & 0 \end{pmatrix}$$

operating on α or β, since

$$\begin{pmatrix} 1 & 0 \\ 0 & 0 \end{pmatrix}\alpha = \begin{pmatrix} 1 & 0 \\ 0 & 0 \end{pmatrix}\begin{pmatrix} 1 \\ 0 \end{pmatrix} = \begin{pmatrix} 1 \\ 0 \end{pmatrix} = \alpha; \quad \begin{pmatrix} 1 & 0 \\ 0 & 0 \end{pmatrix}\beta = 0. \tag{4.3.5}$$

Similarly we obtain 2×2 matrices equivalent to other Fermion operator pairs:

$$a^\dagger_\uparrow a_\uparrow \equiv \begin{pmatrix} 1 & 0 \\ 0 & 0 \end{pmatrix}, \quad a^\dagger_\downarrow a_\downarrow \equiv \begin{pmatrix} 0 & 0 \\ 0 & 1 \end{pmatrix},$$

$$a^\dagger_\uparrow a_\downarrow \equiv \begin{pmatrix} 0 & 1 \\ 0 & 0 \end{pmatrix}, \quad a^\dagger_\downarrow a_\uparrow \equiv \begin{pmatrix} 0 & 0 \\ 1 & 0 \end{pmatrix}. \tag{4.3.6}$$

In terms of the Pauli spin matrices, introduced in Chapter 1,‡

$$\xi = \begin{pmatrix} 0 & 1 \\ 1 & 0 \end{pmatrix}, \quad \eta = \begin{pmatrix} 0 & -i \\ i & 0 \end{pmatrix}, \quad \zeta = \begin{pmatrix} 1 & 0 \\ 0 & -1 \end{pmatrix}: \tag{4.3.7}$$

we have

$$a^\dagger_\uparrow a_\uparrow \equiv \tfrac{1}{2}(\zeta+1), \quad a^\dagger_\downarrow a_\downarrow \equiv \tfrac{1}{2}(1-\zeta),$$

$$a^\dagger_\uparrow a_\downarrow \equiv \tfrac{1}{2}(\xi+i\eta), \quad a^\dagger_\downarrow a_\uparrow \equiv \tfrac{1}{2}(\xi-i\eta). \tag{4.3.8}$$

Thus from equation (4.3.4) the operator which exchanges spins of i and j sites is

$$\begin{aligned}\sum_{\sigma\sigma'} a^\dagger_{i\sigma} a_{i\sigma'} a^\dagger_{j\sigma'} a_{j\sigma} &\equiv \tfrac{1}{4}[(\zeta_i+1)(\zeta_j+1)+(\xi_i+i\eta_i)(\xi_i-i\eta_j)+(\xi_i-i\eta_i)(\xi_j+i\eta_j) \\ &\quad +(1-\zeta_i)(1-\zeta_j)] \\ &= \tfrac{1}{2}(\zeta_i\zeta_j+\eta_i\eta_j+\xi_i\xi_j+1)\end{aligned} \tag{4.3.9}$$

‡ ξ, η and ζ were referred to in Chapter 1 as σ_1, σ_2 and σ_3 respectively.

or, defining the Pauli spin operator,

$$\boldsymbol{\sigma} = \xi\hat{\mathbf{x}}+\eta\hat{\mathbf{y}}+\zeta\hat{\mathbf{z}} \tag{4.3.10}$$

$$\sum_{\sigma\sigma'} a_{i\sigma}^{\dagger}\, a_{i\sigma'}\, a_{j\sigma'}^{\dagger}\, a_{j\sigma} = \tfrac{1}{2}(\boldsymbol{\sigma}_i \cdot \boldsymbol{\sigma}_j + 1) \tag{4.3.11}$$

when operating on functions of the form (4.3.1) or (4.3.3). Neglecting diagonal elements, therefore, which are immaterial in determining the spin configuration, we have an equivalent Hamiltonian of the form

$$H_{\mathrm{H}} = -\frac{1}{4}\sum_{ij}{}' J_{ij}\, \boldsymbol{\sigma}_i \cdot \boldsymbol{\sigma}_j, \tag{4.3.12}$$

the familiar Heisenberg Hamiltonian, also sometimes termed the Dirac equivalent Hamiltonian.

Inspection of equation (4.2.20) for J_{ij} shows it to be positive, leading, as we shall see below, to ferromagnetism. We shall also see that to obtain antiferromagnetism we shall again use equation (4.3.12), except that J_{ij} must be chosen *negative*, and the question arises as to the way this change of sign can take place. We ought to remark that in the general Heitler–London method, of which the above is a restricted application, J_{ij} is not necessarily given by equation (4.2.20) when spins are antiparallel, and can in fact become negative. However, in the context of our present model a different answer must be given, as follows (compare Anderson, 1963).

In equation (4.2.18), we can interpret ε_{ij} $(i \neq j)$ as the energy required to take an electron from site j to site i, which we shall refer to as the hopping energy, as can be seen from the fact that $a_{j\sigma}$ annihilates an electron on site j and $a_{i\sigma}^{\dagger}$ creates an electron with the same spin on site i. If we allow I to become infinite, the energy involved in such hopping processes with one electron per atom, as we are assuming, becomes prohibitive, and the only contribution to the first term in equation (4.2.18) is from $\varepsilon_{ii} = \varepsilon_0$. Suppose, however, that I is very large, but finite. This suggests that we take the hopping Hamiltonian

$$H_{\mathrm{h}} = \sum_{ij}{}' \varepsilon_{ij} \sum_{\sigma} a_{i\sigma}^{\dagger}\, a_{j\sigma} \tag{4.3.13}$$

as a perturbation on a zeroth order function $|\xi_0\rangle$ which is a linear combination of terms of the form (4.3.1), the prime in equation (4.3.13) indicating that the terms $i = j$ are omitted. Standard Rayleigh–Schrödinger perturbation theory then gives a second-order energy

$$E^{(2)} = \sum_{\alpha} \frac{\langle \xi_0 | H_{\mathrm{h}} | \xi_\alpha \rangle \langle \xi_\alpha | H_{\mathrm{h}} | \xi_0 \rangle}{E_0 - E_\alpha}. \tag{4.3.14}$$

But from equation (4.3.13), if $\langle \xi_\alpha | H_{\mathrm{h}} | \xi_0 \rangle$ is non-zero, $|\xi_\alpha\rangle$ describes a state in which there is one atom which has an antiparallel spin pair. By definition the self-energy of this pair is I and so $E_0 - E \simeq -I$, because I is very large. Thus we find

$$E^{(2)} = -\sum_{\alpha} \frac{\langle \xi_0 | H_{\mathrm{h}} | \xi_\alpha \rangle \langle \xi_\alpha | H_{\mathrm{h}} | \xi_0 \rangle}{I} = -\frac{\langle \xi_0 | H_{\mathrm{h}}^2 | \xi_0 \rangle}{I}. \tag{4.3.15}$$

We can similarly handle higher-order terms in the series, yielding an expansion in inverse powers of I; we see that when operating on functions of the form (4.3.1) we can take an effective hopping Hamiltonian

$$H_{\mathrm{eff}} = H_{\mathrm{h}} - \frac{H_{\mathrm{h}}^2}{I} + \frac{H_{\mathrm{h}}^3}{I^2} \ldots. \tag{4.3.16}$$

The first-order term is the original hopping Hamiltonian, and so to lowest order in I^{-1} we need to examine H_{h}^2. A typical term of this operator is $\varepsilon_{ij}\, \varepsilon_{kl}\, a_{i\sigma}^\dagger\, a_{j\sigma}\, a_{k\sigma'}^\dagger\, a_{l\sigma'}$, which gives zero, when operating on a function describing a state with all sites singly occupied, unless $i = l$ and $j = k$. Thus on using the anticommutation relations, we find, since H_{h} gives zero contribution when operating on the functions (4.3.1),

$$\begin{aligned} H_{\mathrm{eff}} = -\frac{H_{\mathrm{h}}^2}{I} &= -\frac{1}{I} \sum_{ij}{}' \varepsilon_{ij}\, \varepsilon_{ji} \sum_{\sigma\sigma'} a_{i\sigma}^\dagger\, a_{i\sigma'}\, a_{j\sigma}\, a_{j\sigma'}^\dagger \\ &= \frac{1}{I} \sum_{ij}{}' |\varepsilon_{ij}|^2 \{ a_{i\uparrow}^\dagger\, a_{i\uparrow}\, a_{j\uparrow}^\dagger\, a_{j\uparrow} + a_{i\uparrow}^\dagger\, a_{i\downarrow}\, a_{j\downarrow}^\dagger\, a_{j\uparrow} \\ &\quad + a_{i\downarrow}^\dagger\, a_{i\uparrow}\, a_{j\uparrow}^\dagger\, a_{j\downarrow} + a_{i\downarrow}^\dagger\, a_{i\downarrow}\, a_{j\downarrow}^\dagger\, a_{j\downarrow} - a_{i\uparrow}^\dagger\, a_{i\uparrow} - a_{i\downarrow}^\dagger\, a_{i\downarrow} \} \end{aligned} \tag{4.3.17}$$

which, by using the correspondence (4.3.8), can be written as

$$H_{\mathrm{eff}} \equiv \frac{1}{2} \sum_{ij}{}' \frac{|\varepsilon_{ij}|^2}{I} \{ \boldsymbol{\sigma}_i . \boldsymbol{\sigma}_j - 1 \}. \tag{4.3.18}$$

There is thus a positive term in opposition to $-J_{ij}$ in equation (4.3.12) and so if $|\varepsilon_{ij}|^2/I$ is large enough we can use an *effective* J_{ij} which is negative.

4.3.2 Linear chain with ferromagnetic interaction

In discussing ferromagnets and antiferromagnets, we shall want to consider the excited states, as well as the ground state of the magnetic system. Only then shall we be able to calculate the thermodynamic properties of the magnet. In the Heisenberg model, we shall see that the way to do this is to set up a formalism similar to that we used in discussing phonons. This will lead us to quantized waves of magnetism, the spin waves, with their quanta referred to as *magnons*.

Consider then a linear chain of atoms with a spin of $\frac{1}{2}$ localized on each lattice point. We suppose the spins are coupled ferromagnetically, that the exchange interactions are short range, and we then write the Hamiltonian in the form, with $\mathbf{s} = (\hbar/2)\boldsymbol{\sigma}$,

$$H = -2J\sum_{ij} \mathbf{s}_i . \mathbf{s}_j, \tag{4.3.19}$$

where the sum goes over all nearest neighbour pairs, and J is positive, the strength of J being written in the form shown purely for convenience later. We can rewrite equation (4.3.19) in the form

$$H = -2J\sum_{ij}(s_i^z s_j^z - \tfrac{1}{4}) - J\sum_{ij}(s_i^+ s_j^- + s_i^- s_j^+), \tag{4.3.20}$$

which differs from equation (4.3.19) only in a shift in the zero of energy. Here we have written

$$s_i^{\pm} = s_i^x \pm i s_i^y, \tag{4.3.21}$$

which is often convenient in such problems. Now clearly the total z component of spin M is a good quantum number and so we can look for eigenstates with $M = N/2$, $N/2-1$, $N/2-2$, etc. If J is positive, the ground state will be that with all spins parallel. The state with $M = N/2$ is clearly

$$\psi_0 = \alpha_1 \alpha_2 \alpha_3 \ldots \alpha_N, \tag{4.3.22}$$

where α represents the spin state function of spin up. This state has energy $E_0 = 0$ because of the way we have chosen our zero of energy in going from (4.3.19) to (4.3.20).

Now let us look for states with $M = N/2-1$, i.e. with one "wrong" spin. There are N states with this value of M, corresponding to the N possible atoms to which the downward spin electron with state function β can be assigned. The basis states we must use to build up the eigenfunctions are

$$\phi_n = \alpha_1 \alpha_2 \ldots \alpha_{n-1} \beta_n \alpha_{n+1} \ldots \alpha_N, \tag{4.3.23}$$

and we must now consider the operation of the Hamiltonian on this state. The first term of (4.3.20) is such that its matrix elements between ϕ_n and ϕ_m are zero unless $n = m$. In this latter case, the matrix element is simply

$$J \times [\text{number of pairs of antiparallel spins on nearest neighbour sites}] = 2J.$$

The second term has off-diagonal matrix elements and these reflect the fact that the operator changes any $\alpha\beta$ nearest neighbour pair into $\beta\alpha$. Hence

$$H\phi_n = 2J\phi_n - J\phi_{n+1} - J\phi_{n-1}. \tag{4.3.24}$$

From the basis states ϕ_n we can construct an eigenstate

$$\psi_n = \sum_n b_n \phi_n \tag{4.3.25}$$

and then equation (4.3.24) leads to

$$Eb_n = 2Jb_n - Jb_{n+1} - Jb_{n-1} \tag{4.3.26}$$

as an equation for the energy E and the coefficients b_n. We want a wave-like solution representing the misaligned spin propagating through the lattice. If a is the spacing between the atoms in the linear chain, the desired solution is

$$b_n = e^{iqna} \tag{4.3.27}$$

which leads to the dispersion relation $E(q)$ for the spin waves

$$E(q) = 2J[1 - \cos aq]. \tag{4.3.28}$$

It is characteristic of spin waves in ferromagnets that the energy $E(q)$ is quadratic in the wave number for long wavelengths. With the usual periodic boundary conditions (see Chapter 1, section 1.4) we find the allowed values of q as

$$q = \frac{2\pi}{Na}[0, 1, 2, \ldots, N-1]. \tag{4.3.29}$$

Substituting for b_n gives

$$\psi = \sum_n e^{iqna} \phi_n. \tag{4.3.30}$$

Clearly this is a progressive wave, with wave vector q, representing the motion of the $\downarrow$ spin through the background of $\uparrow$ spins. This then is the formal mathematical expression of the idea of a spin wave.

4.3.3 *Three-dimensional case*

Before proceeding to the quantum-mechanical solution of the three-dimensional case, it will be helpful to make some remarks on a classical spin system. To do so, let us write down the Heisenberg equation of motion for a spin $\mathbf{S}_n$ on the nth site:

$$\dot{\mathbf{S}}_n = -\frac{i}{\hbar}[\mathbf{S}_n, H] = \frac{i}{\hbar}\left[\mathbf{S}_n, \sum_{n'n''}{}' J\mathbf{S}_{n'}.\mathbf{S}_{n''}\right]$$

$$= \frac{2}{\hbar}\sum_{n'} J\mathbf{S}_n \times \mathbf{S}_{n'}. \tag{4.3.31}$$

These equations of motion are identical in form to the classical equations for an array of spins. While we must not suppose that the quantum-mechanical and the classical averages of these equations are the same, the classical picture is valuable in giving insight into the theory of coupled spins. It has the further merit that at high temperatures it can yield quantitatively correct results as we shall see later.

It is easy to envisage spin waves in such a classical system. Every spin will precess at the same polar angle about the z-axis whereas the azimuthal angle will vary along the direction of propagation, as shown in Figure 4.1. It should be clear from this discussion that the spin-wave frequency $\omega(\mathbf{q})$ ($\equiv E(\mathbf{q})/\hbar$) will be the precessional frequency of the spins.

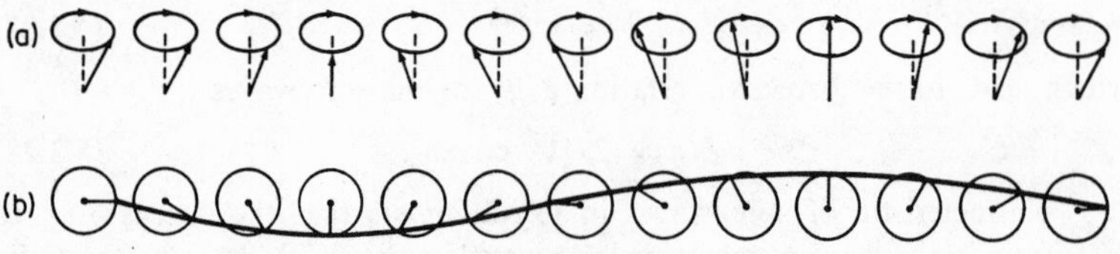

FIGURE 4.1. (a) Shows classical picture for spin wave associated with an array of precessing spins. (b) Same as (a), but seen from above.

While the treatment of a three-dimensional lattice is quite similar in principle to the above discussion of the linear chain, the problem is difficult when more than one spin wave is present because the spin waves interact and scatter one another. We shall ignore these complications and treat the spin waves as non-interacting. We again set up a formalism similar to that we used in discussing phonons. First we rewrite equation (4.3.20) in terms of "deviation operators" $\eta_{\mathbf{n}}$ representing the change in the total z component of spin on site $\mathbf{n}$, $S^z_{\mathbf{n}}$, from full saturation. Hence, if S is the magnitude of total spin,

$$\eta_{\mathbf{n}} = S - S^z_{\mathbf{n}}. \tag{4.3.32}$$

Replacing $-\frac{1}{4}$ by $-S^2$ in equation (4.3.20)—this is only a change in the zero of energy and is more convenient now we are treating the general case where S can be $\frac{1}{2}$, 1, etc.—the first term of equation (4.3.20) becomes

$$2JS \sum_{\mathbf{nm}} (\eta_{\mathbf{n}} + \eta_{\mathbf{m}}) - 2J \sum_{\mathbf{n,m}} \eta_{\mathbf{n}} \eta_{\mathbf{m}},$$

i.e.

$$2JSZ \sum_{\mathbf{n}} \eta_{\mathbf{n}} - 2J \sum_{\mathbf{n,m}} \eta_{\mathbf{n}} \eta_{\mathbf{m}}, \tag{4.3.33}$$

where Z is the number of nearest neighbours of each atom. Now from the known results of the spin angular momentum operators acting on eigenfunctions $|m\rangle$ of $S^z_{\mathbf{n}}$ with eigenvalues $m\hbar$ (see Rose (1957) or Feenberg and

Pake (1959))

$$S^{\pm}|m\rangle = \sqrt{[(S\mp m)(S+1\pm m)]}\,|m\pm 1\rangle \tag{4.3.34}$$

we get

$$S^{+}|\eta\rangle = (2S)^{\frac{1}{2}}\left\{1-\frac{\eta-1}{2S}\right\}^{\frac{1}{2}}\eta^{\frac{1}{2}}|\eta-1\rangle \tag{4.3.35}$$

and

$$S^{-}|\eta\rangle = (2S)^{\frac{1}{2}}\left\{1-\frac{\eta}{2S}\right\}^{\frac{1}{2}}(\eta+1)^{\frac{1}{2}}|\eta+1\rangle, \tag{4.3.36}$$

the states now being labelled by the "deviation" quantum number η which is an integer. We notice therefore that S^{+} acts like an annihilation operator and S^{-} like a creation operator for the deviations. Following Holstein and Primakoff (1940) we define Boson operators a_n, $a_n^{\dagger}$ such that

$$a_{\mathbf{n}}|\eta_{\mathbf{n}}\rangle = \sqrt{(\eta_{\mathbf{n}})}\,|\eta_{\mathbf{n}}-1\rangle,$$
$$a_{\mathbf{n}}^{\dagger}|\eta_{\mathbf{n}}\rangle = \sqrt{(\eta_{\mathbf{n}}+1)}\,|\eta_{\mathbf{n}}+1\rangle. \tag{4.3.37}$$

Thus

$$\eta_{\mathbf{n}} = a_{\mathbf{n}}^{\dagger}a_{\mathbf{n}},$$

$$S_{\mathbf{n}}^{+} = (2S)^{\frac{1}{2}}\left\{1-\frac{a_{\mathbf{n}}^{\dagger}a_{\mathbf{n}}}{2S}\right\}^{\frac{1}{2}}a_{\mathbf{n}},$$

$$S_{\mathbf{n}}^{-} = (2S)^{\frac{1}{2}}\,a_{\mathbf{n}}^{\dagger}\left\{1-\frac{a_{\mathbf{n}}^{\dagger}a_{\mathbf{n}}}{2S}\right\}^{\frac{1}{2}}. \tag{4.3.38}$$

In terms of these operators, the Hamiltonian is

$$H = 2JSZ\sum_{\mathbf{n}}a_{\mathbf{n}}^{\dagger}a_{\mathbf{n}} - 2J\sum_{\langle\mathbf{n},\mathbf{m}\rangle}a_{\mathbf{n}}^{\dagger}a_{\mathbf{n}}a_{\mathbf{m}}^{\dagger}a_{\mathbf{m}}$$

$$-2JS\sum_{\langle\mathbf{n},\mathbf{m}\rangle}\left[\left\{1-\frac{a_{\mathbf{n}}^{\dagger}a_{\mathbf{n}}}{2S}\right\}^{\frac{1}{2}}a_{\mathbf{n}}a_{\mathbf{m}}^{\dagger}\left\{1-\frac{a_{\mathbf{m}}^{\dagger}a_{\mathbf{m}}}{2S}\right\}^{\frac{1}{2}}\right.$$

$$\left.+a_{\mathbf{n}}^{\dagger}\left\{1-\frac{a_{\mathbf{n}}^{\dagger}a_{\mathbf{n}}}{2S}\right\}^{\frac{1}{2}}\left\{1-\frac{a_{\mathbf{m}}^{\dagger}a_{\mathbf{m}}}{2S}\right\}^{\frac{1}{2}}a_{\mathbf{m}}\right], \tag{4.3.39}$$

where the notation $\langle\mathbf{n},\mathbf{m}\rangle$ denotes a sum over near neighbours.

A problem arises from the introduction of the operators in equation (4.3.37). It can be seen most clearly by counting the number of states in the initial problem. This is the number of possible configurations $(2S+1)^N$ where N is the number of sites. However, in (4.3.37) we have imposed no constraint on the allowed values of η, whereas equation (4.3.32) shows that the maximum value is $2S$. This is equivalent to saying that the Hamiltonian (4.3.39) has an infinite number of eigenstates.

However, the result (4.3.39) is exactly equivalent to (4.3.31), provided the former operator is not allowed to act on states with $\eta > 2S$. It follows that these redundant states are not coupled to the physically admissible ones, as can be shown by explicit calculation of the matrix elements of H.

Our basic objective must be to diagonalize the Hamiltonian (4.3.39). So far, an exact solution has not been obtained but, even if this should prove possible, it would not reduce the problem to non-interacting spin waves which is such a fruitful concept in the theory. It is clear that such a situation can be regained from the Hamiltonian (4.3.39) provided that we neglect all but quadratic terms in the Boson operators. From the above discussion of the redundant states, it will be apparent that this approximation spoils the property that the (new spin-wave) Hamiltonian has zero matrix elements between physical and redundant states. Some justification of this procedure can be given for a ferromagnet by showing that the relevant matrix elements are small (see Marshall, 1955a). The Hamiltonian then becomes

$$H = 2JSZ\sum_{\mathbf{n}} a_{\mathbf{n}}^{\dagger} a_{\mathbf{n}} - 2JS\sum_{\mathbf{nm}} \{a_{\mathbf{n}} a_{\mathbf{m}}^{\dagger} + a_{\mathbf{n}}^{\dagger} a_{\mathbf{m}}\}. \tag{4.3.40}$$

Now let us define

$$b_{\mathbf{q}} = \frac{1}{\sqrt{N}}\sum_{\mathbf{n}} e^{-i\mathbf{q}.\mathbf{n}} a_{\mathbf{n}},$$

$$b_{\mathbf{q}}^{\dagger} = \frac{1}{\sqrt{N}}\sum_{\mathbf{n}} e^{i\mathbf{q}.\mathbf{n}} a_{\mathbf{n}}^{\dagger}, \tag{4.3.41}$$

i.e.

$$a_{\mathbf{n}} = \frac{1}{\sqrt{N}}\sum_{\mathbf{q}} e^{i\mathbf{q}.\mathbf{n}} b_{\mathbf{q}},$$

$$a_{\mathbf{n}}^{\dagger} = \frac{1}{\sqrt{N}}\sum_{\mathbf{q}} e^{-i\mathbf{q}.\mathbf{n}} b_{\mathbf{q}}^{\dagger}, \tag{4.3.42}$$

where there are N vectors $\mathbf{q}$ distributed uniformly over a unit cell of the reciprocal lattice, with density $\mathscr{V}/(2\pi)^3$.

From (4.2.5) we easily deduce

$$[b_{\mathbf{q}'}, b_{\mathbf{q}}^{*}] = \delta_{\mathbf{q},\mathbf{q}} \tag{4.3.43}$$

and all other commutators are zero. Substituting into (4.3.40) we get

$$H = 2JSZ\sum_{\mathbf{q}} b_{\mathbf{q}}^{*} b_{\mathbf{q}} - JSZ\sum_{\mathbf{q}} \gamma_{\mathbf{q}}(b_{\mathbf{q}} b_{\mathbf{q}}^{*} + b_{\mathbf{q}}^{*} b_{\mathbf{q}}), \tag{4.3.44}$$

where

$$\gamma_{\mathbf{q}} = \frac{1}{Z} \sum_{\mathbf{R}} e^{i\mathbf{q}.\mathbf{R}}. \tag{4.3.45}$$

The sum runs over the Z nearest neighbours. Making use of (4.3.43) we find

$$H = \sum_{\mathbf{q}} E_{\mathbf{q}} b_{\mathbf{q}}^* b_{\mathbf{q}} = \sum_{\mathbf{q}} E_{\mathbf{q}} n_{\mathbf{q}}, \tag{4.3.46}$$

where

$$E(\mathbf{q}) = 2JSZ\{1-\gamma_{\mathbf{q}}\}. \tag{4.3.47}$$

It is readily shown from equations (4.3.45) and (4.3.47) that for cubic lattices $E(\mathbf{q}) \propto \mathbf{q}^2$ for small $\mathbf{q}$. In equation (4.3.46) we can interpret $n_{\mathbf{q}}$ as the number of spin waves of wave-vector $\mathbf{q}$.

4.3.4 *Spontaneous magnetization*

To show how the magnetization varies with temperature in a ferromagnet in which the low-lying excitations are spin waves, we first use the fact, remarked on above, that the excitations are bosons. Neglecting the interactions between spin waves, we obtain the average number of quasi-particles, or magnons, in a state of wave-vector $\mathbf{q}$ as

$$\langle n_{\mathbf{q}} \rangle = \frac{1}{e^{\hbar\beta\omega_{\mathbf{q}}}-1}. \tag{4.3.48}$$

If $M_0 = N\mu$ is the total magnetic moment at $T=0$, N being the number of spins per unit volume, then

$$M = M_0 - \mu \sum_{\mathbf{q}} \langle n_{\mathbf{q}} \rangle. \tag{4.3.49}$$

Replacing the sum over $\mathbf{q}$ by an integration through the BZ we find

$$M_0 - M = \frac{\mu}{8\pi^3} \int_{\Omega_B} \frac{d\mathbf{q}}{e^{\hbar\beta\omega_{\mathbf{q}}}-1}. \tag{4.3.50}$$

Restricting ourselves to low temperatures, where our initial assumption of non-interacting spin waves is certainly valid, only the occupation numbers of the lowest excited states will be significant, and then we can use the quadratic form of $\omega_{\mathbf{q}}$ for small $\mathbf{q}$. We then find from equations (4.3.45) and (4.3.47), for cubic lattices

$$\omega_{\mathbf{q}} = 2JSa^2q^2 = \kappa q^2, \tag{4.3.51}$$

where a is the lattice constant.

We can integrate over the whole of space, because the Bose function is effectively zero except at small **q**, to obtain

$$M_0 - M = \frac{\mu}{8\pi^3} \int \frac{d\mathbf{q}}{e^{\hbar\beta\omega_\mathbf{q}} - 1} \tag{4.3.52}$$

or putting $\xi^2 = \beta\hbar\kappa q^2$ we have

$$M_0 - M = \frac{\mu}{8\pi^3} \int \frac{d\boldsymbol{\xi}}{(\beta\hbar\kappa)^{\frac{3}{2}}\,[e^{\xi^2} - 1]}. \tag{4.3.53}$$

Then $M_0 - M$ is seen to be proportional to $\beta^{-\frac{3}{2}}$ or $T^{\frac{3}{2}}$, the famous law first obtained by Bloch.

4.3.5 *Specific heat*

Similarly we can write the internal energy as

$$U = \sum_\mathbf{q} \frac{\hbar\omega_\mathbf{q}}{[e^{\hbar\beta\omega_\mathbf{q}} - 1]}. \tag{4.3.54}$$

Again, at low temperatures we can integrate through the whole of **q**-space, to obtain for the energy/unit volume

$$U = \frac{1}{8\pi^3} \int \frac{\hbar\omega_\mathbf{q}\, d\mathbf{q}}{(e^{\hbar\beta\omega_\mathbf{q}} - 1)}. \tag{4.3.55}$$

We see by analogy with the calculation of the magnetization that the extra factor $\omega_\mathbf{q} \propto q^2$ alters the temperature dependence by a factor T yielding

$$U \propto T^{\frac{5}{2}} \tag{4.3.56}$$

and hence a specific heat $C_v \propto T^{\frac{3}{2}}$.

We shall conclude this section with a brief discussion of the interactions which must occur between spin waves.

4.3.6 *Dynamical and kinematic interactions*

It can be seen that we may write the eigenfunction of states with one spin reversal as

$$|\psi\rangle = (2S)^{-\frac{1}{2}} S_\mathbf{k}^+ |0\rangle, \tag{4.3.57}$$

where $|0\rangle$ is the ground state of the system and

$$\mathbf{S}_\mathbf{k} = N^{-\frac{1}{2}} \sum_j e^{i\mathbf{k}.\mathbf{R}_j} \mathbf{S}_j. \tag{4.3.58}$$

With the Hamiltonian (4.3.19), including only nearest–neighbour interactions, the appropriate eigenvalue of equation (4.3.57) is, with applied field $\mathscr{H}$,

$$\varepsilon_{\mathbf{k}} = \mu\mathscr{H} + 2S[J(0) - J(k)]. \tag{4.3.59}$$

When there are many spin reversals, the nature of the Holstein–Primakoff procedure as followed through in section 4.3.3 makes it rather difficult to deal with the interactions which must take place between the spin waves (magnons) propagating through the crystal. The solution to this problem is due to Dyson (1956).

Dyson begins by generalizing equation (4.3.57) to the case of an arbitrary number $n = \sum n_{\mathbf{k}}$ of spin waves ($n_{\mathbf{k}}$ being the number with wave-vector $\mathbf{k}$) by writing

$$|n\rangle = \prod_{\mathbf{k}} (2S)^{-(\frac{1}{2})n_{\mathbf{k}}} (n_{\mathbf{k}}!)^{-\frac{1}{2}} (S_{\mathbf{k}}^{+})^{n_{\mathbf{k}}} |0\rangle \tag{4.3.60}$$

(with nearest neighbours only, for simplicity).

Applying the Hamiltonian (4.3.19), one then obtains

$$H|n\rangle = (E_0 + \sum_{\mathbf{k}} \varepsilon_{\mathbf{k}} n_{\mathbf{k}})|n\rangle + \sum_{m}^{\neq n} I_{nm}|m\rangle. \tag{4.3.61}$$

The term I_{nm} is a matrix element representing interactions between pairs of spin waves, and is termed by Dyson the "dynamical interaction". This distinguishes the interaction due to binary collision of spin waves from the "kinematical interaction" which is a reflexion of the non-orthogonality of the states $|n\rangle$ defined by equation (4.3.60). This non-orthogonality arises in turn from the fact that the number of reversed spins at any lattice point cannot exceed $2S$. Further, there are many more states defined by (4.3.60) than the total number $(2S+1)^N$ of independent states of a system of N spins S.

The kinematical interaction has no effect on the dynamics of the system—as, for example, the mutual scattering of two spin waves—but can affect the calculation of thermal averages. However, at low temperatures, $\sum_k n_{\mathbf{k}} \ll N$ and the $|n_{\mathbf{k}}\rangle$ are approximately orthogonal, so that in this régime the kinematical interaction has little effect on thermodynamic quantities.

Since the non-orthogonality of the states $|n\rangle$ inconveniences calculations, Dyson introduced the orthogonal "ideal states"

$$\prod_{\mathbf{k}} (n_{\mathbf{k}}!)^{-\frac{1}{2}} (a_{\mathbf{k}}^{\dagger})^{n_{\mathbf{k}}} |0\rangle, \tag{4.3.62}$$

where the $a_{\mathbf{k}}^{\dagger}$ are the oscillator operators defined previously. One must then replace H by an ideal Hamiltonian operating on these states to produce the same effect [see equation (4.3.61)] as H on the states $|n\rangle$.

However, we shall content ourselves with the above comments here, merely supplementing these by references to the work of Kubo, Oguchi (1960) and others, and by Dembinski and his colleagues (1964). A general reference is the article by Walker (1963) in the Rado–Suhl work but of course the original (but very difficult) paper is that of Dyson (1956).

4.4 Antiferromagnetism—spin wave theory

4.4.1 Physical picture

The simplest picture of an antiferromagnetic is that of two interpenetrating sub-lattices with ↑ spins on one and ↓ spins on the other. For example, in a body-centred cubic lattice, we have two interpenetrating simple cubic lattices.

This system can be treated quantitatively by starting from a Heisenberg Hamiltonian as above, with near-neighbour interactions, but with the exchange interaction J negative. Thus, we consider again the Hamiltonian

$$H = -J \sum_{ij} \mathbf{S}_i . \mathbf{S}_j, \tag{4.4.1}$$

the sum being over near neighbours. We then make the Holstein–Primakoff transformation analogous to equation (4.3.37). We have to have creation and annihilation operators referring to the ith atom on sub-lattice 1, and similarly for sub-lattice 2. Denoting these destruction operators by a_j^1, and a_j^2 for these two lattices, we now have, corresponding to equation (4.3.41),

$$b_{\mathbf{k}}^{(1)} = N^{-\frac{1}{2}} \sum_{\mathbf{l}} e^{i\mathbf{k}.\mathbf{l}} a_{\mathbf{l}}^{(1)}; \quad b_{\mathbf{k}}^{(1)*} = N^{-\frac{1}{2}} \sum_{\mathbf{l}} e^{-i\mathbf{k}.\mathbf{l}} a_{\mathbf{l}}^{(1)*} \tag{4.4.2}$$

for lattice 1 with lattice vectors $\mathbf{l}_1$, and similarly for sub-lattice 2

$$b_{\mathbf{k}}^{(2)} = N^{-\frac{1}{2}} \sum_{\mathbf{m}} e^{i\mathbf{k}.\mathbf{m}} a_{m}^{(2)}; \quad b_{\mathbf{k}}^{(2)*} = N^{-\frac{1}{2}} \sum_{\mathbf{m}} e^{-i\mathbf{k}.\mathbf{m}} a_{\mathbf{m}}^{(2)*}, \tag{4.4.3}$$

the **m**'s representing again the lattice vectors. The b's are now the anti-ferromagnetic magnon variables, in terms of which the Hamiltonian becomes, to lowest order,

$$H = 2NZJS^2 - 2JZS \sum_{\mathbf{k}} [\gamma_{\mathbf{k}}(b_{\mathbf{k}}^{(1)*} b_{\mathbf{k}}^{(2)*} + b_{\mathbf{k}}^{(1)} b_{\mathbf{k}}^{(2)}) + (b_{\mathbf{k}}^{(1)*} b_{\mathbf{k}}^{(1)} + b_{\mathbf{k}}^{(2)*} b_{\mathbf{k}}^{(2)})], \tag{4.4.4}$$

where $\gamma_{\mathbf{k}}$ is as in equation (4.3.45). With a centre of symmetry, $\gamma_{\mathbf{k}} = \gamma_{-\mathbf{k}}$.

Unlike the ferromagnon case, we cannot solve this trivially: we must diagonalize equation (4.4.4) by a suitable transformation. What we need is the Bogoliubov transformation, which involves new operators α and β defined through

$$\alpha_{\mathbf{k}} = u_{\mathbf{k}} b_{\mathbf{k}}^{(1)} - v_{\mathbf{k}} b_{\mathbf{k}}^{(2)*} \tag{4.4.5}$$

and

$$\beta_{\mathbf{k}} = u_{\mathbf{k}} b_{\mathbf{k}}^{(2)} - v_{\mathbf{k}} b_{\mathbf{k}}^{(1)*}, \tag{4.4.6}$$

where $u_{\mathbf{k}}$ and $v_{\mathbf{k}}$ are real, $u_{\mathbf{k}}^2 - v_{\mathbf{k}}^2 = 1$, but otherwise they are to be chosen to diagonalize the form (4.4.4).

Substituting in the Hamiltonian (4.4.4), the result is

$$H = 2NZJS(S+1) + \sum_{\mathbf{k}} \omega_{\mathbf{k}}(\alpha_{\mathbf{k}}^* \alpha_{\mathbf{k}} + \beta_{\mathbf{k}}^* \beta_{\mathbf{k}} + 1), \tag{4.4.7}$$

where $\omega_{\mathbf{k}}$ is given by

$$\omega_{\mathbf{k}}^2 = 4J^2 Z^2 S[1 - \gamma_{\mathbf{k}}^2]. \tag{4.4.8}$$

This result should be contrasted with the form (4.3.47) for ferromagnons, since $\omega_{\mathbf{k}}$ is evidently the dispersion relation.

As an example, if we take the long wavelength limit $ka \ll 1$, then for a simple cubic lattice

$$\omega_{\mathbf{k}} \doteqdot -4.3^{\frac{1}{2}} JSka. \tag{4.4.9}$$

The linear relation between ω and k is a basic characteristic of antiferromagnons.

Knowledge of these elementary excitations enables us to perform calculations of the temperature dependence of the sub-lattice magnetization, and also of the specific heat, along lines similar to those laid down in the previous section. The results are qualitatively that the sub-lattice magnetization decreases as $(T/\theta_c)^2$ for low temperatures, θ_c being of the order of the critical or Néel temperature. The specific heat on the basis of the excitations (4.4.9) is quite analogous to the phonon problem, and a T^3 law results.

4.4.2 Ground-state magnetization

For the calculation of thermodynamic properties, the excitations of the antiferromagnets are of prime interest. However, the magnetization at $T = 0$ is also of interest in that full saturation is not achieved, as we shall now discuss.

The magnetization in sub-lattice 1 is given immediately by

$$\mathscr{S}_z^{(1)} = \sum_l S_{lz}^{(1)} = NS - \sum_{\mathbf{k}} b_{\mathbf{k}}^{(1)*} b_{\mathbf{k}}^{(1)}. \tag{4.4.10}$$

Substituting for the b's in terms of the α's and β's and taking the expectation value, we find a departure from saturation given by

$$NS - \langle \mathscr{S}_z^{(1)} \rangle = \sum_{\mathbf{k}} v_{\mathbf{k}}^2$$

$$= \frac{1}{2} \sum_{\mathbf{k}} [(1-\gamma_{\mathbf{k}}^2)^{-\frac{1}{2}} - 1], \tag{4.4.11}$$

where the latter result follows from the choice $v_{\mathbf{k}}$ required to diagonalize the Hamiltonian. In this way, we easily find that the zero-point motion of the spin waves reduces the sub-lattice magnetization to about 85% of its maximum possible value. Applying perturbation theory, with the interaction between the spin waves as the perturbation, the sub-lattice magnetization has been shown by Davis (1960) to be 90% for the bcc lattice and 87% for the simple cubic lattice. The perturbation theory goes beyond linear spin-wave theory and should be more accurate, but we shall not give any details here.

4.4.3 Variational calculation of ground state

We consider instead the work of Thouless (1967), who discussed the ground-state wave function for a system with Hamiltonian

$$H = -\tfrac{1}{2} J \sum_{ij} \boldsymbol{\sigma}_i \cdot \boldsymbol{\sigma}_j, \tag{4.4.12}$$

where J is negative, the sum is over all pairs of nearest-neighbour sites and $\boldsymbol{\sigma}$ as usual is a Pauli spin operator. We follow Marshall (1955b) in setting up trial ground-state wave functions, although the conclusions drawn below are primarily based on the treatment of Thouless. Dividing the lattice into two sub-lattices A and B, the general wave function is specified by giving the amplitude of each arrangement of spin up or down among the various sites. Marshall showed that the ground-state wave function can be written in such a way that all amplitudes are real; positive if there are an even number of particles with $\uparrow$ spin on the A sub-lattice and negative if there are an odd number of particles with spin $\uparrow$ on the A sub-lattice. Furthermore, he showed that this ground state has zero total spin.

Let us assume that the amplitude of a particular configuration depends only on the number of particles with spin up on A, say p, and on q, the number of nearest-neighbour pairs of opposite spin; the sign is $(-1)^p$. This is then taken as a trial wave function for the ground-state energy. It cannot be used in an exact calculation of the variational energy except in one dimension (Appendix 4.1), but a satisfactory approximation can be obtained and the result is that the best wave function is strongly peaked in the neighbourhood of $p = N/4$, where N is the number of sites. Thus there is no preference for a particular spin direction on the A sub-lattice. This appears to contradict the physical picture outlined above, and led to the suggestion that a system of particles of spin $\frac{1}{2}$ with only nearest-neighbour interactions will be paramagnetic, rather than antiferromagnetic.

This is not the case, however, as we shall now show. Since the ground state of the system has zero total spin, neither sub-lattice can have a preferred direction for its total spin. A convenient measure of the sub-lattice magnetization is therefore given by the expectation value of $\boldsymbol{\sigma}_i . \boldsymbol{\sigma}_j$ when i and j are well separated. This quantity should be positive if i and j are on the same sub-lattice, and negative if they are on different sub-lattices. It is generally assumed that the expectation value of $\sigma_i^z \sigma_j^z$ in either spin-wave theory or perturbation theory gives a good approximation to this quantity, since $\sigma_i^x \sigma_j^x$ and $\sigma_i^y \sigma_j^y$ have vanishing expectation values for i and j well separated, in such theories. These theories single out a particular direction for the sub-lattice magnetization.

4.4.4 *Sub-lattice magnetization*

The wave function of Marshall, peaked up at $p = N/4$, gives no preference for spin up on either sub-lattice and so the expectation value of $\sigma_i^z \sigma_j^z$ vanishes when i and j are well separated. However, if we consider the other terms in $\frac{1}{2}\boldsymbol{\sigma}_i . \boldsymbol{\sigma}_j$, namely the operator

$$\tfrac{1}{2}(\sigma_i^x \sigma_j^x + \sigma_i^y \sigma_j^y) \tag{4.4.13}$$

which gives zero when the two spins are parallel, but exchanges the spins when one is $\uparrow$ and the other $\downarrow$, then this operator has an appreciable expectation value when i and j are well separated. Any wave function of the type discussed above will therefore give a large expectation value of $\boldsymbol{\sigma}_i . \boldsymbol{\sigma}_j$ alternating in sign. The point is that the sub-lattice magnetization is now in the xy-plane in this case.

Following Thouless, we can make an estimate of the degree of sub-lattice magnetization. The amplitude of a particular configuration, with p spins on the A sub-lattice and q nearest-neighbour pairs of opposite spin, is

$$(-1)^p \{W(p,q)\}^{-\frac{1}{2}} a(p,q), \tag{4.4.14}$$

where W is the number of configurations with a given p and q. The main contributions come from the neighbourhood of $p = p_0$, $q = q_0$, where a has its maximum. Since $p_0 = N/4$, where W also has a maximum, the dependence on q is the dominant feature and we neglect the dependence on p.

Further, guided by the linear chain, where nearest-neighbour spins can be parallel or antiparallel independently of one another, we take the amplitude to have the approximate form

$$(-1)^p (\cos\theta)^q (\sin\theta)^{\frac{1}{2}NZ-q}, \tag{4.4.15}$$

where θ is given by

$$\cos^2\theta = 2q_0/NZ \tag{4.4.16}$$

and Z is the number of nearest neighbours to any site.

Then we have

$$\begin{aligned}\tfrac{1}{2}\langle\sigma_i^x\sigma_j^x+\sigma_i^y\sigma_j^y\rangle &= \pm\frac{1}{2}\sum(\cos\theta)^{q_i'+q_j'}(\sin\theta)^{2Z-q_i'-q_j'}\\ &\quad\times(\cos\theta)^{2Z-q_i'-q_j'}(\sin\theta)^{q_i'+q_j'}\\ &= \pm 2^{2Z-1}(\cos\theta\sin\theta)^{2Z},\end{aligned} \tag{4.4.17}$$

where the sum goes over all possible arrangements of the spins which are neighbours to the sites i and j. The sign is positive or negative according to whether or not i and j are on the same sub-lattice and the factor $\frac{1}{2}$ occurs because $\sigma_i^x\sigma_j^x+\sigma_i^y\sigma_j^y$ gives zero when it acts on those configurations in which the spins on i and j are parallel. The quantities q_i' and q_j' are the numbers of nearest neighbours with spins opposite to the spins initially on i and j. The sub-lattice magnetization follows from equations (4.4.16) and (4.4.17) as

$$2^Z\left(\frac{2q_0}{NZ}\right)^{Z/2}\left(1-\frac{2q_0}{NZ}\right)^{Z/2}. \tag{4.4.18}$$

With the best variational values of q_0 this gives 87% for the simple cubic and 90% for the bcc lattice. These are strikingly close to the perturbation theory values of Davis. It turns out that the wave function also is very like that given by perturbation theory.

Thouless has made the interesting suggestion that an experimental test of these predictions might be possible, since the nuclear spins in solid He^3 may have an exchange interaction which is dominated by near-neighbour coupling. Finally, by analogy with the one-dimensional Bose system (Schultz, 1963), in which it is known that there is no long-range order, Thouless has suggested that the linear antiferromagnetic chain might not have true long-range order, but that the total spin of a group of n successive atoms may increase only as ln n. For other aspects of this problem, reference can be made to the book by Mattis (1964).

4.5 Strong correlations in narrow energy bands

While the Heisenberg model discussed above gives us a great deal of insight into magnetic problems, it clearly does not include all features of magnetic states. This is especially true for the metallic magnets, Fe, Ni and Co, in which there is high conductivity and we expect that the band theory developed in Chapter 1 will have some relevance. Attention has therefore been turning more recently towards a study of the many-body Hamiltonian (4.2.18), though it is clearly impracticable to treat it as it stands. In narrow energy-band theory, developed particularly by Gutzwiller (1963, 1964), Hubbard (1963, 1964) and Kanamori (1963), the description afforded by band theory, which is essentially improved by including not only the first term in (4.2.18) but also the second, is transcended by dealing with a total wave function which is not a single determinant of Bloch functions. Attention is thereby moved away from the direct exchange term in (4.2.18), which the Heisenberg model assumes to be dominant.

The model therefore on which the following discussion is based can be stated as one in which electrons move in a single band, built from Bloch functions, but in which the electrons interact when two electrons are in Wannier orbitals on the same atom. In a non-degenerate band, these two electrons obviously must have opposite spin. For this particular model, the ground state for a completely ferromagnetic configuration is not altered by the interaction. The restriction to one band is, of course, serious; we really should be treating five degenerate d-bands. The theory is therefore not outstandingly successful in solving the physical problems it purports to describe. However, it is very valuable as a many-body model of a magnet, from which further progress will surely stem.

4.5.1 One-band model with interactions

Thus, the Hamiltonian we shall work with, frequently referred to nowadays as the Hubbard Hamiltonian, is given by [see equation (4.2.18)]

$$H = \sum \varepsilon_{ij} a^{\dagger}_{i\sigma} a_{j\sigma} + \frac{I}{2} \sum_{i} \sum_{\sigma} n_{i\sigma} n_{i-\sigma}. \tag{4.5.1}$$

As we have said, electrons only interact when they have opposite spins and are sitting on the same atom. In the case of the Heisenberg model, direct exchange was assumed to dominate the problem, but the present model puts its trust in the importance of intra-atomic and intra-band interactions. If it should turn out in the future that, in magnetic problems, inter-atomic and inter-band effects are really large, it is obvious that H will have to be modified drastically.

We want to stress some features of the present problem, before proceeding to a detailed mathematical investigation of the Hamiltonian (4.5.1). It has already been implied, first of all, that the completely ferromagnetic configuration is described by a determinant of $\uparrow$ spin Bloch wave functions and is therefore uninfluenced by the electron interactions in this model.

However, two further points are worth stressing. The first is that, when we have that the total number of electrons N is equal to the total number of atoms, and at the same time the interaction constant I is much larger than the band width, then a determinant built up from Wannier orbitals which are singly occupied is almost an eigenfunction corresponding to the ground state. But there will be a very large number of wave functions of this kind, corresponding to 2^N ways of assigning the spins, and as the interaction I becomes very large, the energies of these different states will all tend to the same value. But we have seen that the fully ferromagnetic state will have the same energy for all I. It seems clear that, for sufficiently large I, this model will not have a ferromagnetic ground state, but that the ground state will have a smaller value of the total spin and will most probably be an anti-ferromagnetic configuration. Secondly, it is worth noting that the problem has symmetry between electrons and holes, which is simply seen if the Hamiltonian is rewritten in terms of creation and annihilation operators for holes, rather than electrons. We shall say more about this later.

4.5.2 Concepts arising from one-band model

It is worth stressing here, following Herring (1966), that two concepts are central to this model of a magnetic system. The first of these is very straightforward. When the interaction I is put to zero and one has a given number of electrons, there will exist a perfectly well-defined band width, W say. Inclusion of the interaction term in the Hamiltonian will alter the density of states at the Fermi level and also shift the Fermi level itself. It is then valuable to define an effective band width as

$$W_{\text{eff}} = W \times \text{ratio of density of states at Fermi level with and without interactions.} \tag{4.5.2}$$

Clearly then the effective band width depends on both the electron density and the strength of the interaction.

Secondly, it will sometimes be useful to think in terms of an effective interaction I_{eff}. Very crudely, we can think of it as related to the amount that energy bands for $\uparrow$ spin and $\downarrow$ spin electrons in a ferromagnetic are shifted relative to one another, ΔE say. Then we can think of the effective interaction I_{eff} as being given in order of magnitude by

$$\Delta E \sim I_{eff} \frac{\Delta N}{N_a}, \tag{4.5.3}$$

where ΔN is the excess of $\uparrow$ spins over $\downarrow$ spins and N_a is the total number of atoms.

In this language, we can rather confidently draw the conclusion that I_{eff} will be less than W, where W is the band width in the absence of the electron interactions. The basic physical reason for this is that it is always possible to keep two electrons from coming on to the same atom by paying the price in one-electron energy of the order of the band width multiplied by the fraction of lattice sites occupied by electrons.

After these preliminary remarks, we now turn to the mathematical treatment of the Hubbard Hamiltonian. Although we have spoken in terms of wave functions above, it will be more convenient in practice to seek the Green function and we next discuss the method we shall adopt to determine this quantity.

4.5.3 Green function and Hubbard Hamiltonian

As we saw in Chapter 2, the probability amplitude for a particle injected into the system in the one-particle state **k** at time $t = 0$ being found in the same state at time t is

$$\langle \Psi | a_{\mathbf{k}}(t)\, a_{\mathbf{k}}^{\dagger}(0) | \Psi \rangle = \langle a_{\mathbf{k}}(t)\, a_{\mathbf{k}}^{\dagger}(0) \rangle, \tag{4.5.4}$$

where Ψ is the exact ground state of the many-body system.

To deal with the dynamical properties, it will prove convenient to average the anticommutator

$$a_{\mathbf{k}}(t)\, a_{\mathbf{k}}^{\dagger}(0) + a_{\mathbf{k}}^{\dagger}(0)\, a_{\mathbf{k}}(t) = [a_{\mathbf{k}}(t), a_{\mathbf{k}}^{\dagger}(0)]_{+}. \tag{4.5.5}$$

We shall see that this average, essentially a single-particle Green function, contains most of the information we shall require on the consequences of the Hubbard Hamiltonian. However, to solve the equations of motion for the problem we shall need more general quantities, which we now define and examine.

The quantities we actually work with are defined in the following way:

$$\left.\begin{aligned} \langle\langle A; B \rangle\rangle_{\omega} &= -\frac{i}{2\pi} \int_{0}^{\infty} dt\, e^{i\omega t} \langle [A(t), B(0)]_{+} \rangle, \quad \operatorname{Im} \omega > 0, \\ \langle\langle A; B \rangle\rangle_{\omega} &= \frac{i}{2\pi} \int_{-\infty}^{0} dt\, e^{i\omega t} \langle [A(t), B(0)]_{+} \rangle, \quad \operatorname{Im} \omega < 0, \end{aligned}\right\} \tag{4.5.6}$$

where, in general, we are now dealing with a thermal average. Then $G_{AB}(\omega) = \langle\langle A; B\rangle\rangle_\omega$ is analytic in ω in the upper and lower half planes, with discontinuities or poles on the real axis. The discontinuity in $\langle\langle a_{\mathbf{k}}; a_{\mathbf{k}}^\dagger\rangle\rangle$ gives the number of particles in state $\mathbf{k}$:

$$\langle n_{\mathbf{k}}\rangle = \langle a_{\mathbf{k}}^\dagger a_{\mathbf{k}}\rangle = i \lim_{\epsilon\to 0} \int_{-\infty}^{\zeta} d\omega [G_{\mathbf{kk}}(\omega+i\varepsilon) - G_{\mathbf{kk}}(\omega - i\varepsilon)], \tag{4.5.7}$$

where ζ is the Fermi energy (see Problem 4.10). From this result we can see that $G_{\mathbf{kk}}(\omega)$, known for all N (the total particle number), determines the total energy E for the system, since, by definition,

$$\frac{\partial E}{\partial N} = \zeta. \tag{4.5.8}$$

To obtain an equation for G, we use the Heisenberg equation of motion

$$\dot{A} = \frac{i}{\hbar}[H, A]. \tag{4.5.9}$$

Then, taking $\operatorname{Im}\omega > 0$, for example,

$$\omega\langle\langle A; B\rangle\rangle_\omega = -\frac{i}{2\pi}\int_0^\infty \omega\, e^{i\omega t}\langle [A(t), B]_+\rangle\, dt$$

$$= -\frac{1}{2\pi} e^{i\omega t}\langle [A(t), B]_+\rangle \Big|_0^\infty + \frac{i}{2\pi}\int_0^\infty e^{i\omega t}\frac{d}{dt}\langle [A(t), B]_+\rangle\, dt, \tag{4.5.10}$$

that is

$$\omega\langle\langle A; B\rangle\rangle_\omega = \frac{1}{2\pi}\langle [A, B]_+\rangle + \langle\langle [H, A], B\rangle\rangle_\omega. \tag{4.5.11}$$

As we have implicitly assumed by our notation above, the one-particle states to which the annihilation operators a refer are most commonly chosen as plane-wave states, or Bloch states, with wave-vectors $\mathbf{k}$. However, in the problem to which we now address ourselves, the single-particle states are most conveniently taken as *localized* states, and the one-particle Green functions are of the form

$$G_{ij}^{\sigma\sigma'}(\omega) = \langle\langle a_{i\sigma}; a_{j\sigma'}^\dagger\rangle\rangle. \tag{4.5.12}$$

In Hubbard's treatment it is assumed that the system has translational invariance, that is, physical properties are invariant on translation through a lattice-vector $\mathbf{R}_i$ which excludes, for example, antiferromagnetism. It follows that the Green functions are diagonal in the spin indices, and we shall always set $\sigma = \sigma'$ in what follows. It may also be noted that the same assumption implies that $G^{\sigma}_{ij}(\omega)$ is a function only of $\mathbf{R}_i - \mathbf{R}_j$.

4.5.4 Isolated atom limit

In the limit of zero band width, $\varepsilon_{ij} = \varepsilon_0 \delta_{ij}$ and the Hubbard Hamiltonian becomes

$$H = \varepsilon_0 \sum_{i\sigma} n_{i\sigma} + \tfrac{1}{2} I \sum_{i\sigma} n_{i\sigma} n_{i-\sigma}. \tag{4.5.13}$$

Taking the commutator of this with $a_{i\sigma}$, we find

$$[a_{i\sigma}, H] = \varepsilon_0 a_{i\sigma} + I a_{i\sigma} n_{i-\sigma}. \tag{4.5.14}$$

Then the equation of motion (4.5.11) becomes

$$\omega G^{\sigma}_{ij}(\omega) = \frac{1}{2\pi}\delta_{ij} + \varepsilon_0 G^{\sigma}_{ij}(\omega) + I\Gamma^{\sigma}_{ij}(\omega), \tag{4.5.15}$$

where

$$\Gamma^{\sigma}_{ij}(\omega) = \langle\langle n_{i-\sigma} a_{i\sigma}; a^{\dagger}_{j\sigma}\rangle\rangle_{\omega}. \tag{4.5.16}$$

Now we make use of the identities

$$[n_{i-\sigma} a_{i\sigma}, a^{\dagger}_{j\sigma}]_{+} = \delta_{ij} n_{i-\sigma} \tag{4.5.17}$$

and

$$[n_{i\sigma}, H] = 0. \tag{4.5.18}$$

Then, using again the equation of motion, we obtain

$$\omega\Gamma^{\sigma}_{ij}(\omega) = \frac{1}{2\pi}\delta_{ij}\langle n_{i-\sigma}\rangle + \varepsilon_0 \Gamma^{\sigma}_{ij} + I\langle\langle n^2_{i-\sigma} a_{i\sigma}; a^{\dagger}_{j\sigma}\rangle\rangle_{\omega}. \tag{4.5.19}$$

At this stage we use the property $n^2_{i\sigma} = n_{i\sigma}$, and so the last term in equation (4.5.19) is simply $I\Gamma^{\sigma}_{ij}$. Thus

$$\Gamma^{\sigma}_{ij}(\omega) = \frac{1}{2\pi}\delta_{ij}\frac{\langle n_{i-\sigma}\rangle}{\omega - \varepsilon_0 - I}. \tag{4.5.20}$$

From the assumed translational invariance, $\langle n_{i\sigma}\rangle$ must be independent of i and σ, and therefore $\langle n_{i\sigma}\rangle = \frac{1}{2}n$ where n is the number of electrons per atom. Using this property in equation (4.5.20) and substituting in equation (4.5.15) we obtain the desired result

$$G^{\sigma}_{ij}(\omega) = \frac{1}{2\pi}\delta_{ij}\left\{\frac{1-\frac{1}{2}n}{\omega - \varepsilon_0} + \frac{\frac{1}{2}n}{\omega - \varepsilon_0 - I}\right\}. \tag{4.5.21}$$

Hence the density of states is given by [cf. equation (2.7.59)]

$$\rho_\sigma(\omega) = (1-\tfrac{1}{2}n)\,\delta(\omega-\varepsilon_0)+\tfrac{1}{2}n\delta(\omega-\varepsilon_0-I). \tag{4.5.22}$$

This shows that the system behaves as if it had two energy levels ε_0 and ε_0+I, containing $1-\frac{1}{2}n$ and $\frac{1}{2}n$ states per atom respectively. As electrons are added to the band, initially the Fermi energy will remain at $\mu = \varepsilon_0$ until the lower level is full. This happens when $n = 1$ and μ then jumps to ε_0+I whilst the remaining electrons are added.

4.5.5 Case when hopping is allowed

When hopping is allowed, the spin on a particular atom is the total spin of all the electrons on that atom, and, of course, the total spin at an atomic site will fluctuate in time.

But if we study the effect of the repulsive interactions acting at each site then if, at one instant, the spin on an atom was in a particular direction, say upwards, then it will tend to repel any downward spin electrons and in this way it will tend to persist.

If this repulsive force is strong enough, then the state of upward spin on an atom may persist for a time which is long compared to the hopping time between atoms. The spin can then be thought of as associated with a particular atom.

With the Hamiltonian (4.5.1), the equation of motion for $G_{ij}^\sigma(\omega)$ derived from equation (4.5.11) is

$$\omega G_{ij}^\sigma(\omega) = \frac{\delta_{ij}}{2\pi}+\sum_l \varepsilon_{il}G_{lj}^\sigma(\omega)+I\langle\langle n_{i-\sigma}a_{i\sigma};\, a_{j\sigma}^\dagger\rangle\rangle_\omega \tag{4.5.23}$$

rather than equation (4.5.15), and solving for $G_{ij}(\omega)$ evidently entails a knowledge of $\langle\langle n_{i-\sigma}a_{i\sigma};\, a_{j\sigma}^\dagger\rangle\rangle_\omega$. The equation of motion for this quantity is obtained in a similar fashion to that for $G_{ij}(\omega)$, and in fact the equation of motion reads

$$\omega\langle\langle n_{i-\sigma}a_{i\sigma};\, a_{j\sigma}^\dagger\rangle\rangle_\omega = \langle\langle [n_{i-\sigma}a_{i\sigma}, H];\, a_{j\sigma}^\dagger\rangle\rangle_\omega+\frac{1}{2\pi}\langle[n_{i-\sigma}a_{i\sigma}, a_{j\sigma}]_+\rangle. \tag{4.5.24}$$

If we go further and write down the equations of motion for the objects on the right-hand side, we obtain higher-order Green functions (cf. section 2.7.6). It is therefore clear that at some stage we have to make an approximate reduction of higher-order Green functions to lower-order ones, and in fact Hubbard chose to stop at equation (4.5.24), reducing quantities on the right-hand side in the manner first given by Zubarev (1960), which we illustrate below.

The example we take is that of the term

$$\langle\langle n_{i-\sigma} a_{l\sigma};\, a_{j\sigma}^{\dagger}\rangle\rangle_{\omega}.$$

By definition this is the transform (4.5.6) with $A = n_{i-\sigma} a_{l\sigma}$ and $B = a_{j\sigma}^{\dagger}$. Writing the anticommutator out, we have for the system in its ground state

$$\begin{aligned}\langle\Psi|\,[n_{i-\sigma}(t)\,a_{l\sigma}(t), a_{j\sigma}^{\dagger}(0)]_{+}\,|\Psi\rangle \\ = \langle\Psi|\,n_{i-\sigma}(t)\,a_{l\sigma}(t)\,a_{j\sigma}^{\dagger}(0)\,|\Psi\rangle + \langle\Psi|\,a_{j\sigma}^{\dagger}(0)\,a_{l\sigma}(t)\,n_{i-\sigma}(t)\,|\Psi\rangle\end{aligned} \tag{4.5.25}$$

where we have used the commutation relations to write the result in this form.

Now $|\Psi\rangle$ is the true ground state of the interacting system and in general will not be an eigenstate of the independent-particle number $n_{i-\sigma}$. The essential approximation we make is to replace the effect of $n_{i-\sigma}$ operating on $|\Psi\rangle$ by its ground-state expectation value

$$n_{i-\sigma}|\Psi\rangle \simeq \langle n_{i,-\sigma}\rangle|\Psi\rangle, \tag{4.5.26}$$

where $\langle n_{i-\sigma}\rangle$ is a pure number. Thus we obtain

$$\langle\Psi|\,[n_{i-\sigma}(t)\,a_{l\sigma}(t), a_{j\sigma}^{\dagger}(0)]_{+}\,|\Psi\rangle \simeq \langle n_{i-\sigma}\rangle\,\langle\Psi|\,[a_{l\sigma}(t), a_{j\sigma}^{\dagger}]_{+}\,|\Psi\rangle \tag{4.5.27}$$

and hence

$$\langle\langle n_{i-\sigma} a_{l\sigma};\, a_{j\sigma}^{\dagger}\rangle\rangle_{\omega} \simeq \langle n_{i-\sigma}\rangle\,\langle\langle a_{l\sigma};\, a_{j\sigma}^{\dagger}\rangle\rangle_{\omega}. \tag{4.5.28}$$

Similar treatment of the other terms of equation (4.5.24) leads to

$$\langle\langle a_{i-\sigma}^{\dagger} a_{l-\sigma} a_{i\sigma};\, a_{j\sigma}^{\dagger}\rangle\rangle_{\omega} \simeq \langle a_{i-\sigma}^{\dagger} a_{l-\sigma}\rangle\,\langle\langle a_{i\sigma};\, a_{j\sigma}^{\dagger}\rangle\rangle_{\omega}, \tag{4.5.29}$$

$$\langle\langle a_{l-\sigma}^{\dagger} a_{i-\sigma} a_{i\sigma};\, a_{j\sigma}^{\dagger}\rangle\rangle_{\omega} \simeq \langle a_{l-\sigma}^{\dagger} a_{i-\sigma}\rangle\,\langle\langle a_{i\sigma};\, a_{j\sigma}^{\dagger}\rangle\rangle_{\omega}. \tag{4.5.30}$$

These approximations yield the equation of motion

$$\begin{aligned}\omega\langle\langle n_{i-\sigma} a_{i\sigma};\, a_{j\sigma}^{\dagger}\rangle\rangle_{\omega} \simeq \frac{\langle n_{i-\sigma}\rangle\,\delta_{ij}}{2\pi} + (\varepsilon_0 + I)\,\langle\langle n_{i-\sigma} a_{i\sigma};\, a_{j\sigma}^{\dagger}\rangle\rangle_{\omega} \\ + \langle n_{i-\sigma}\rangle \sum_{l\neq i} \varepsilon_{il}\langle\langle a_{l\sigma};\, a_{j\sigma}^{\dagger}\rangle\rangle_{\omega},\end{aligned} \tag{4.5.31}$$

which is exact in the limit, already discussed, of widely separated atoms ($\varepsilon_{ij} = 0$ if $i \neq j$). Our assumption of translational invariance implies that the quantity $\langle n_{i-\sigma}\rangle$ in the above equation is independent of i and will in future be writtten as $\langle n_{-\sigma}\rangle$.

From equation (4.5.31) we find

$$\langle\langle n_{i-\sigma} a_{i\sigma};\, a_{j\sigma}^{\dagger}\rangle\rangle_{\omega} = \frac{\langle n_{-\sigma}\rangle}{\omega - \varepsilon_0 - I}\left\{\frac{\delta_{ij}}{2\pi} + \sum_{l\neq j} \varepsilon_{il}\langle\langle a_{l\sigma};\, a_{j\sigma}^{\dagger}\rangle\rangle_{\omega}\right\}, \tag{4.5.32}$$

which on substitution into equation (4.5.24) gives

$$\omega\langle\langle a_{i\sigma};a_{j\sigma}^{\dagger}\rangle\rangle_{\omega}=\varepsilon_0\langle\langle a_{i\sigma};a_{j\sigma}^{\dagger}\rangle\rangle_{\omega}+\left\{1+\frac{I\langle n_{-\sigma}\rangle}{\omega-\varepsilon_0-I}\right\}\times\left\{\frac{\delta_{ij}}{2\pi}+\sum_{l\neq i}\varepsilon_{il}\langle\langle a_{l\sigma};a_{j\sigma}^{\dagger}\rangle\rangle_{\omega}\right\}. \tag{4.5.33}$$

We solve this equation by writing

$$\langle\langle a_{i\sigma};a_{j\sigma}^{\dagger}\rangle\rangle_{\omega}=N^{-1}\sum_{\mathbf{k}}G^{\sigma}(\mathbf{k}\omega)\,e^{i\mathbf{k}.(\mathbf{R}_i-\mathbf{R}_j)} \tag{4.5.34}$$

and thereby obtain

$$\omega G^{\sigma}(\mathbf{k}\omega)=\varepsilon_0\,G^{\sigma}(\mathbf{k}\omega)+\left\{1+\frac{I\langle n_{-\sigma}\rangle}{\omega-\varepsilon_0-I}\right\}\left\{\frac{1}{2\pi N}+(E_{\mathbf{k}}-\varepsilon_0)\,G^{\sigma}(\mathbf{k}\omega)\right\}. \tag{4.5.35}$$

It follows that

$$G^{\sigma}(\mathbf{k}\omega)=\frac{1}{2\pi N}\frac{\omega-\varepsilon_0-I(1-\langle n_{-\sigma}\rangle)}{(\omega-E_{\mathbf{k}})(\omega-\varepsilon_0-I)+\langle n_{-\sigma}\rangle(\varepsilon_0-E_{\mathbf{k}})I}, \tag{4.5.36}$$

where

$$E_{\mathbf{q}}=\sum_j\varepsilon_j\,e^{i\mathbf{q}.\mathbf{R}_j}, \tag{4.5.37}$$

the dispersion relation in the absence of the interaction.

To find the propagation energies we seek the singularities on the real axis. These are given by

$$(\omega-E_{\mathbf{q}})(\omega-\varepsilon_0-I)+\langle n_{-\sigma}\rangle(\varepsilon_0-E_{\mathbf{q}})I=0. \tag{4.5.38}$$

This will have two roots, and if we replace $E_{\mathbf{q}}$ by ε_0 the atomic limit is regained. The general situation is sketched in Figure 4.2. Calling the two roots $\varepsilon_{\mathbf{k}\sigma}^{(1)}$ and $\varepsilon_{\mathbf{k}\sigma}^{(2)}$ we can write

$$G^{\sigma}(\mathbf{k}\omega)=\frac{1}{2\pi}\left\{\frac{A_{\mathbf{k}\sigma}^{(1)}}{\omega-\varepsilon_{\mathbf{k}\sigma}^{(1)}}+\frac{A_{\mathbf{k}\sigma}^{(2)}}{\omega-\varepsilon_{\mathbf{k}\sigma}^{(2)}}\right\}, \tag{4.5.39}$$

where from equation (4.5.36) we can show that the A's are positive definite and obey

$$A_{\mathbf{k}\sigma}^{(1)}+A_{\mathbf{k}\sigma}^{(2)}=1. \tag{4.5.40}$$

We also find, as shown in Figure 4.2,

$$\varepsilon_{\mathbf{k}\sigma}^{(1)}<\varepsilon_0+I(1-\langle n_{-\sigma}\rangle)<\varepsilon_{\mathbf{k}\sigma}^{(2)}. \tag{4.5.41}$$

From equation (4.5.39) it is evident that $\varepsilon_{\mathbf{k}\sigma}^{(1)}$ and $\varepsilon_{\mathbf{k}\sigma}^{(2)}$ may be interpreted as quasi-particle energies.

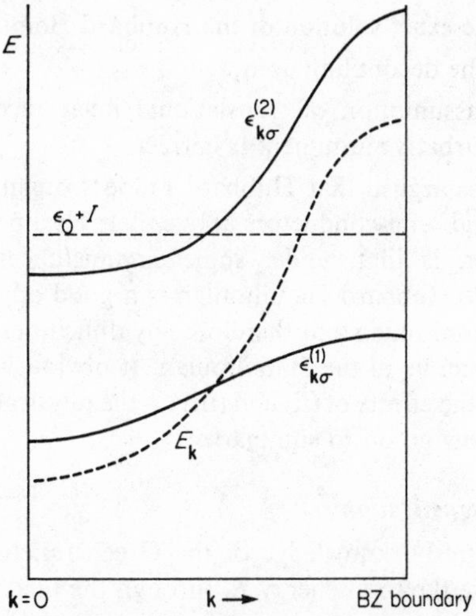

FIGURE 4.2. Showing splitting of band in Hubbard model. $E_{\mathbf{k}}$ is band dispersion relation in absence of interaction. $\varepsilon_{\mathbf{k}\sigma}^{(2)}$ and $\varepsilon_{\mathbf{k}\sigma}^{(1)}$ are the two roots of equation (4.5.38).

4.5.6 Volume enclosed by Fermi surface

Let us use equation (4.5.7) to evaluate $\langle n_{\mathbf{k}\sigma}\rangle$. Since

$$\frac{1}{2\pi i}\lim_{\delta\to 0}\left\{\frac{1}{\omega-\varepsilon+i\delta}-\frac{1}{\omega-\varepsilon-i\delta}\right\} = \delta(\omega-\varepsilon) \tag{4.5.42}$$

we find from (4.5.39)

$$\langle n_{\mathbf{k}\sigma}\rangle = A_{\mathbf{k}\sigma}^{(1)}\,\theta(\zeta-\varepsilon_{\mathbf{k}\sigma}^{(1)})+A_{\mathbf{k}\sigma}^{(2)}\,\theta(\zeta-\varepsilon_{\mathbf{k}\sigma}^{(2)}), \tag{4.5.43}$$

where $\theta(x) = 0$ for $x<0$, $= 1$ for $x>0$. If, as is in fact normally the case, $A_{\mathbf{k}\sigma}^{(1)}$ or $A_{\mathbf{k}\sigma}^{(2)}$ is non-zero when $\mathbf{k}$ crosses the Fermi surface, $\varepsilon_{\mathbf{k}\sigma}^{(1)} = \zeta$ or $\varepsilon_{\mathbf{k}\sigma}^{(2)} = \zeta$, there will be a discontinuous change in $\langle n_{\mathbf{k}\sigma}\rangle$ across this Fermi surface.

The volume enclosed by this surface, which is clearly defined by the discontinuity in $\langle n_{\mathbf{k}\sigma}\rangle$, is, in general, not conserved as we change the strength of the interactions. This is at variance with the result of Luttinger (1960) which we referred to in Chapter 2 that, to all orders in perturbation theory, this volume is unaffected by the interactions.

We are therefore faced with the question whether this apparent difficulty is

(i) present in the exact solution of the Hubbard Hamiltonian,

(ii) induced by the decoupling used,

(iii) due to the assumption of translational invariance, given that Luttinger's perturbative argument is correct.

Herring (1966) has argued that Hubbard's model ought to be appropriate to some narrow-band semi-conductors above their Néel points. Our point of view here, however, is that, under some circumstances, which we have described above, the Hubbard Hamiltonian is a good approximation to the true many-body Hamiltonian and therefore any difficulties that occur should not, in such a régime, lie in the Hamiltonian. It obviously is important, for the future, to study the effects of (ii) and (iii) on the physical predictions of the theory, which we now go on to summarize.

4.5.7 Energy of magnetization

As we saw previously, knowledge of the Green function G_{ii} is sufficient to allow us to derive the total energy E, through the relation

$$E = \int_0^N \zeta \, dN, \tag{4.5.44}$$

where the chemical potential ζ is the root of

$$N = \lim_{\delta \to 0} \sum_{\mathbf{k}\sigma} \int_{-\infty}^{\zeta} d\varepsilon [G^{\sigma}(\mathbf{k}\varepsilon + i\delta) - G^{\sigma}(\mathbf{k}\varepsilon - i\delta)]. \tag{4.5.45}$$

We now substitute Hubbard's Green function in this relation and this expression may then be simplified by using the result

$$A_{\mathbf{k}\sigma}^{(1)} = \frac{\partial \varepsilon_{\mathbf{k}\sigma}^{(1)}}{\partial \varepsilon_{\mathbf{k}}}, \tag{4.5.46}$$

which follows from equations (4.5.38) and (4.5.39). It is also convenient to replace the summation on $\mathbf{k}$ by N_a, the number of atoms in the crystal, multiplied by an integral over $\varepsilon_{\mathbf{k}}$, with a weighting factor $n_0(\varepsilon_{\mathbf{k}})$, which is the density of states of a single spin, per atom, in the noninteracting Bloch band. We find that if the Fermi surface is in the lower band then

$$\begin{aligned} \langle n \rangle = \frac{N}{N_a} &= \langle n_+ \rangle + \langle n_- \rangle \\ &= \int_{-\infty}^{\zeta} d\varepsilon_+^{(1)} \, n_0(\varepsilon_+^{(0)}) + \int_{-\infty}^{\zeta} d\varepsilon_-^{(1)} \, n_0(\varepsilon_-^{(0)}), \end{aligned} \tag{4.5.47}$$

where $\varepsilon_{\pm}^{(0)}$ is a function of $\varepsilon_{\pm}^{(1)}$, defined as the value of $\varepsilon_{\mathbf{k}}$ which gives $\varepsilon_{\pm}^{(1)}$ as the lower root of equation (4.5.38). The use of electron–hole symmetry enables the corresponding relation when the Fermi surface is in the upper band to be derived readily. For non-magnetic states, the two integrals in the above equation are equal.

We shall now consider two simple examples to illustrate these arguments.

Examples

(a) *I small: ground state is non-magnetic.* The defining equation for $\varepsilon_{\mathbf{k}}^{(1)}$ is then

$$\varepsilon_{\mathbf{k}} = \varepsilon_{\mathbf{k}}^{(1)} - \tfrac{1}{2}I\langle n\rangle + \dots. \tag{4.5.48}$$

This is just a "rigid band model", with constant energy shift $\frac{1}{2}I\langle n\rangle$, and it is clear that for a given $\langle n\rangle$ the value of the Fermi energy will exceed the value for non-interacting electrons by this same amount $\frac{1}{2}I\langle n\rangle$, to first order in the interaction. This is the same as the result of perturbation theory and so, to this order, Hubbard's total energy agrees with perturbation theory.

(*b*) *Case of square band.* Another example which has the merit that the total energy is readily evaluated is that of a square band. Thus, we define the band density states, with band width W, by

$$n_0(\varepsilon) = \begin{cases} \dfrac{1}{W} & \text{for } -\tfrac{1}{2}W \leqslant \varepsilon \leqslant \tfrac{1}{2}W, \\ 0 & \text{outside this range.} \end{cases} \tag{4.5.49}$$

Then, for a non-magnetic state, we obtain from the above equation for $\langle n\rangle$,

$$\begin{aligned} \zeta &= \tfrac{1}{2}W\langle n\rangle + \varepsilon_{\min}^{(1)} \\ &= \tfrac{1}{2}W\langle n\rangle + \tfrac{1}{2}\{I - \tfrac{1}{2}W - [(I+\tfrac{1}{2}W)^2 - IW\langle n\rangle]^{\frac{1}{2}}\}. \end{aligned} \tag{4.5.50}$$

The total energy may now be evaluated by elementary integration. The full result will not be given, but we note here that as $I \to \infty$

$$\frac{E}{N_{\mathrm{a}}} = W(-\tfrac{1}{2}\langle n\rangle + \tfrac{3}{8}\langle n\rangle^2). \tag{4.5.51}$$

Kanamori (1963) has also made a calculation, based on the Bethe–Goldstone method, for $\langle n\rangle \ll 1$, which yields

$$\frac{E}{N_{\mathrm{a}}} = W(-\tfrac{1}{2}\langle n\rangle + \tfrac{1}{2}\langle n\rangle^2 + \dots). \tag{4.5.52}$$

To this order, it would appear that this equation is exact, and hence Hubbard's method underestimates the energy.

4.5.8 Criterion for ferromagnetism

We can expect the condition for ferromagnetism, quite generally, to be more stringent than the Hartree–Fock condition. This is because Hartree–Fock theory is a better approximation when spins are parallel, for they all then "dig" a Fermi hole (cf. Figure 2.3) around themselves, than when some are antiparallel. An example of this, for a uniform electron gas, can be found in Problem 4.1.

It will be helpful at this stage to consider the Hartree–Fock solution in the present context. The Hartree–Fock Hamiltonian can be obtained by linearizing the interaction terms in the full Hamiltonian. For the Hubbard Hamiltonian (4.5.1), this amounts to replacing the term $n_{i\sigma}n_{i-\sigma}$ by $n_{i\sigma}\langle n_{i-\sigma}\rangle + n_{i-\sigma}\langle n_{i\sigma}\rangle$ where $\langle n_{i\sigma}\rangle$ is the average of the expectation value of $n_{i\sigma}$ over a canonical ensemble at some temperature T.

Thus, the Hartree–Fock Hamiltonian is

$$H_{\mathrm{HF}} = \sum_{ij}\sum_{\sigma} \varepsilon_{ij}\, a^{\dagger}_{i\sigma} a_{j\sigma} + I\sum_{i\sigma} n_{i\sigma}\langle n_{i-\sigma}\rangle. \tag{4.5.53}$$

If we consider only the class of solutions with translational invariance then we find

$$H_{\mathrm{HF}} = \sum_{\mathbf{k}}\sum_{\sigma} (\varepsilon_{\mathbf{k}} + In_{-\sigma})\, a^{\dagger}_{\mathbf{k}\sigma} a_{\mathbf{k}\sigma}, \tag{4.5.54}$$

which is clearly the Hamiltonian for non-interacting electrons with a slightly modified band structure, the energy of the $\mathbf{k}\sigma$ state being $\varepsilon_{\mathbf{k}} + In_{-\sigma}$. If $N(E)$ represents the density of states/atom for band structure $\varepsilon_{\mathbf{k}}$, then $\rho_{\sigma}(E)$ is given by

$$\begin{aligned}\rho_{\sigma}(E) &= N(E - In_{-\sigma})\\ &= N(E - In + In_{\sigma})\end{aligned} \tag{4.5.55}$$

since we must have

$$n_{\uparrow} + n_{\downarrow} = n. \tag{4.5.56}$$

Obviously we can therefore write

$$\begin{aligned}n_{\sigma} &= \int_{-\infty}^{\mu} \rho_{\sigma}(E)\, dE\\ &= \int_{-\infty}^{\mu} N(E - In + In_{\sigma})\, dE.\end{aligned} \tag{4.5.57}$$

If we solve these equations with $\sigma = \pm 1$, together with equation (4.5.56), we find that one solution is $n_{\uparrow} = n_{\downarrow} = \frac{1}{2}n$, and μ is determined by

$$\tfrac{1}{2}n = \int_{-\infty}^{\mu} N(E - \tfrac{1}{2}n)\, dE. \tag{4.5.58}$$

However, if the interaction I is sufficiently strong, it may be possible to find ferromagnetic solutions for which $n_\uparrow \neq n_\downarrow$. In this case, equations (4.5.57) must have two distinct solutions satisfying equation (4.5.56). The condition that this is so is simply

$$1 = IN(\mu - \tfrac{1}{2}In), \tag{4.5.59}$$

as is readily shown by expanding equation (4.5.57) in a series in $n_\sigma - \frac{1}{2}n$. The conclusion from this equation is that if for any E, $IN(E) > 1$ is satisfied, then for some n and μ the Hartree–Fock theory predicts that the system will become ferromagnetic.

It is a straightforward matter now to use the correlated solution derived above to see how this Hartree–Fock situation is altered. We have instead of equation (4.5.57) the result

$$n_\sigma = \int_{-\infty}^{\mu} N\{g(E, n - n_\sigma)\}\, dE, \tag{4.5.60}$$

where

$$g(E, n_{-\sigma}) = E - In_{-\sigma} - \frac{I^2 n_{-\sigma}(1 - n_{-\sigma})}{E - \varepsilon_0 - I(1 - n_{-\sigma})} \tag{4.5.61}$$

and the condition for ferromagnetism now becomes

$$-\tfrac{1}{2} = \int_{-\infty}^{\mu} \frac{\partial}{\partial n}[N\{g(E, \tfrac{1}{2}n)\}]\, dE. \tag{4.5.62}$$

Taking again as an example the density of states (4.5.49), the Hartree–Fock condition (4.5.59) is simply

$$I > W, \tag{4.5.63}$$

independent of n. For the correlated solution, from the electron–hole symmetry referred to above, we need only consider μ in the lower band when we find the condition for ferromagnetism as

$$1 < \frac{\frac{1}{4}I}{\sqrt{[(\frac{1}{2}I + \frac{1}{4}W)^2 - \frac{1}{4}nWI]}}. \tag{4.5.64}$$

Since μ is in the lower band, $n < 1$, and for this case the above condition cannot be satisfied for any I and W. Thus, we confirm that the Hartree–Fock condition is too stringent, since for this correlated case ferromagnetism is never possible even when the condition (4.5.63) is satisfied.

This is not to be taken to mean that ferromagnetism is never possible with the Hubbard solution. Indeed, he shows that, with two square bands symmetrically disposed about ε_0, ferromagnetism is possible. Thus, if we choose

$$N(E) = \begin{cases} \dfrac{1}{\delta} & \text{if } \varepsilon_0 - \frac{1}{2}W < E < \varepsilon_0 - \frac{1}{2}W + \frac{1}{2}\delta, \\ & \text{if } \varepsilon_0 + \frac{1}{2}W - \frac{1}{2}\delta < E < \varepsilon_0 + \frac{1}{2}W, \\ 0 & \text{otherwise,} \end{cases} \tag{4.5.65}$$

then it is a straightforward calculation to show that for small μ the condition for ferromagnetism is

$$\delta < \frac{\frac{1}{2}WI}{\sqrt{[(\frac{1}{2}I + \frac{1}{2}W)^2 - \frac{1}{4}nWI]}} \tag{4.5.66}$$

which can always be satisfied by making δ small enough. Since $\rho(\mu) = 1/\delta$, we can write, by analogy with the Hartree–Fock condition (4.5.59),

$$1 < I_{\text{eff}}\, \rho(\mu), \tag{4.5.67}$$

where (cf. the discussion of section 4.5.2)

$$I_{\text{eff}} = \frac{I\frac{1}{4}W}{\sqrt{[(\frac{1}{2}I + \frac{1}{4}W)^2 - \frac{1}{4}nWI]}}. \tag{4.5.68}$$

We see that I_{eff} is always less than I, the reduction being due to the weakening of the exchange interactions by correlation effects.

From equation (4.5.68), it is also clear that, even when I becomes very large, I_{eff} never becomes much greater than the band width W.

Thus, in summary we shall comment on the conditions favourable for ferromagnetism to occur. First, the Fermi energy must be in part of a band in which the density of states is rather greater than the mean density of states through the band. Also the high density of states peak in which the Fermi energy lies should be well away from the centre of gravity of the band. A high peak in the middle of a band appears to be ineffective in producing ferromagnetism.

4.5.9 Mott transition

We want only to consider one other consequence of the Hubbard theory. Mott, on numerous occasions, has stressed that, in a system of hydrogen atoms, as we take the atoms apart, and supposing we (formally) prevent molecular formation, then at some lattice spacing we shall go over from the conducting state predicted by band theory to an insulating state, this Mott transition being a consequence of the electron interactions.

We can say, straightaway, that the above Green function is not in accord with these ideas about the Mott transition, where we expect that when the band width is sufficiently small, the system will behave like an insulator, but for a band width greater than some critical value, the system behaves as a conductor, i.e. the transition from two bands with the lower one filled to a single half-filled band occurs.

Hubbard (1964b) has shown how these defects of the theory given above can be eliminated by the addition of scattering corrections and how the Mott transition can be included in the theory.

The physical motivation for these corrections was provided by Hubbard by making reference to an alloy model of the narrow band system (cf. Chapter 10, section 10.8). We consider the propagator (4.5.21), remembering that the main approximation used to derive this formula was that only interactions between band electrons on the same atom are important. The main physical effects which have been neglected concern the fact that the electrons are, all the time, propagating through the lattice.

The quantity $\varepsilon_{\mathbf{q}}-\varepsilon_0$ can be interpreted as due to the propagation of electrons between atoms and if we write equation (4.5.21) in the form

$$G_{ij}^{\sigma}(\omega) = \frac{1}{2\pi}\,\delta_{ij}\{F_0^{\sigma}(\omega)\}^{-1}, \tag{4.5.69}$$

where $F_0^{\sigma}(\omega)$ is then obtained by comparing equations (4.5.21) and (4.5.69) when $\frac{1}{2}n$ in equation (4.5.21) is replaced by $n_{-\sigma}$, we can see that F_0 describes the resonant properties of the atoms. Thus, using the ideas of the atomic limit (4.5.21), we can say that, if an electron of spin σ, propagating through the lattice, is incident on an atom with no electron present, then this atom absorbs the electron with resonant energy ε_0. If a $-\sigma$ spin electron is on this atom, the corresponding resonant energy is ε_0+I. Thus the Green function (4.5.69) can be thought of as describing the propagation of an electron, spin σ, through a random alloy consisting of two types of atom, with and without a $-\sigma$ spin present in proportion $n_{-\sigma}$, $1-n_{-\sigma}$ respectively, these $-\sigma$ spins considered fixed on the atoms.

Using this picture, the type of corrections to be applied to the Green function becomes clear. First, the quantity $F_0^{\sigma}(\omega)$ defined above does not take into account the damping of the σ spin electron waves which result from disorder scattering. Secondly, the $-\sigma$ spins are not fixed, but move about owing to the usual band motion and a correction must be applied to account for this. The extra terms introduced may be termed "scattering" and "resonance-broadening" corrections respectively.

The detailed inclusion of these corrections is complicated and we shall not therefore go into this quantitatively. But the main effect is to modify the function $F_0^{\sigma}(\omega)$ into a form which changes the analytic structure of the propagator. It is then found that $G^{\sigma}(\mathbf{k}\omega)$ has a branch cut along the real ω-axis for those values of ω for which $\rho^{\sigma}(E)>0$. The physical implication of this is that now the propagator describes pseudoparticles with finite lifetimes, corresponding to the damping of the electron wave by scattering.

It is then found that the Mott transition is indeed included in this refined theory. Following Hubbard (1964b) one can consider a specific non-ferromagnetic system in which there is a half-full s-band, with an unperturbed density of states having a parabolic form centred on ε_0 [cf. equation (10.5.1) of Chapter 10] with band width W. Then the pseudoparticle density of states is found to depend essentially on the ratio W/I. For small W/I, two bands occur, as shown in Figure 4.3, but at a critical value of this ratio, these

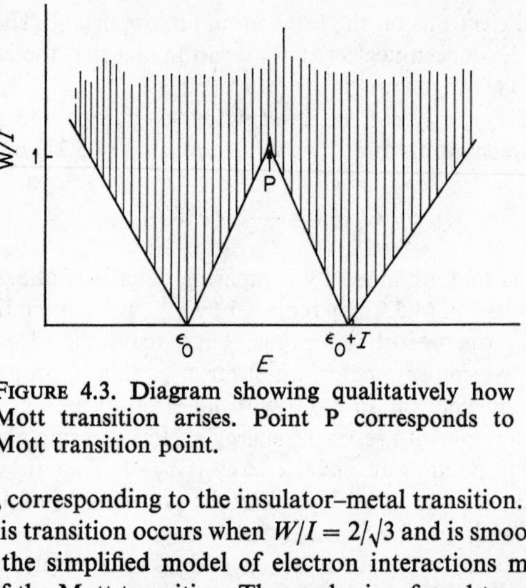

FIGURE 4.3. Diagram showing qualitatively how Mott transition arises. Point P corresponds to Mott transition point.

bands coalesce, corresponding to the insulator–metal transition. In the model of Hubbard, this transition occurs when $W/I = 2/\sqrt{3}$ and is smooth. However, corrections to the simplified model of electron interactions may well alter the character of the Mott transition. The reader is referred to a good deal of further work on the metal–insulator transition in *Reviews of Modern Physics*, Vol. 4, 1968 (see also Doniach, 1969), but we shall have to leave the discussion at this point.

4.5.10 Estimates of exchange splitting

Hubbard has later discussed how the one-band model may be extended to deal with orbital degeneracy. This has led him to make an estimate of the average exchange splitting in Ni. On the assumption that I is $\sim\frac{1}{2}W$ of the d-band, and assuming $W\sim 4$ eV, he has deduced this band splitting to be

~0.4 eV. Kanamori (1963), by different arguments, gets a somewhat smaller exchange splitting. But at least the order of magnitude seems now to be established. Having discussed at some length approximate solutions of the very specific Hamiltonian (4.5.1), we turn now to some more general considerations based on response theory.

4.6 Linear response theory

We begin by defining the dynamical susceptibility, which plays a central role in modern theories of magnetism (cf. Marshall and Lowde, 1968).

4.6.1 Dynamical susceptibility

Let us consider the application of a weak field to the system. We first suppose the field to be a localized impulse of the form

$$\mathscr{H}(\mathbf{r},t) = \mathscr{H}\,\delta(\mathbf{r}-\mathbf{r}')\,\delta(t). \tag{4.6.1}$$

Since the magnitude of the field is small, we take the response to be linear. We must therefore be able to write the induced magnetization as

$$\Delta\mathscr{M}_\beta(\mathbf{r},t) = \sum_\alpha \chi_{\alpha\beta}(\mathbf{r},\mathbf{r}',t-t')\,\mathscr{H}_\alpha \tag{4.6.2}$$

(α and β being Cartesian labels). The linear response function so defined will be termed the susceptibility. Because the induced magnetic moment must be zero until the field is applied,

$$\chi_{\alpha\beta}(\mathbf{r},\mathbf{r}',t) = 0, \quad t<0. \tag{4.6.3}$$

The general consequences of causality on the form of linear response functions will be discussed further in Chapter 7.

We shall in general consider a perfect lattice, with lattice vectors $\mathbf{R}$ (the results for a uniform electron gas are easily derived by inspection of the equations we then obtain). It is evident that the response at $\mathbf{r}+\mathbf{R}$ to a field of the form (4.6.1) is the same as the response at $\mathbf{r}$ to a field $\mathscr{H}\,\delta(\mathbf{r}-\mathbf{r}'+\mathbf{R})\,\delta(t)$. Thus

$$\chi_{\alpha\beta}(\mathbf{r}+\mathbf{R},\mathbf{r}',t) = \chi_{\alpha\beta}(\mathbf{r},\mathbf{r}'-\mathbf{R},t). \tag{4.6.4}$$

The generalization of equation (4.6.2) to fields of arbitrary spatial and temporal variation is easily effected, the result being

$$\Delta\mathscr{M}_\beta(\mathbf{r},t) = \sum_\alpha \int d\mathbf{r}' \int_{-\infty}^{t} \chi_{\alpha\beta}(\mathbf{r},\mathbf{r}',t-t')\,\mathscr{H}_\alpha(\mathbf{r}',t')\,dt'. \tag{4.6.5}$$

If we define the Fourier transforms

$$\Delta\mathcal{M}(\mathbf{k},t) = \int \Delta\mathcal{M}(\mathbf{r},t)\,\mathrm{e}^{-i\mathbf{k}.\mathbf{r}}\,d\mathbf{r}, \tag{4.6.6}$$

$$8\pi^3\mathscr{H}(\mathbf{k},t) = \int \mathscr{H}(\mathbf{r},t)\,\mathrm{e}^{-i\mathbf{k}.\mathbf{r}}\,d\mathbf{r} \tag{4.6.7}$$

and

$$\chi_{\alpha\beta}(\mathbf{k},\mathbf{k}',t) = \int \chi_{\alpha\beta}(\mathbf{r},\mathbf{r}',t)\,\mathrm{e}^{i\mathbf{k}'.\mathbf{r}'}\,\mathrm{e}^{-i\mathbf{k}.\mathbf{r}}\,d\mathbf{r}\,d\mathbf{r}' \tag{4.6.8}$$

it is evident that, on application of a field $\mathscr{H}(\mathbf{k}'t)\,\mathrm{e}^{i\mathbf{k}'.\mathbf{r}}$,

$$\Delta\mathcal{M}_\beta(\mathbf{k},t) = \sum_\alpha \int_{-\infty}^{t} \chi_{\alpha\beta}(\mathbf{k},\mathbf{k}',t-t')\,\mathscr{H}_\alpha(\mathbf{k}',t')\,dt'. \tag{4.6.9}$$

Now equation (4.6.4) implies that

$$\begin{aligned}
\chi_{\alpha\beta}(\mathbf{k},\mathbf{k}',t) &= \int d\mathbf{r} \sum_{\mathbf{R}} \mathrm{e}^{i\mathbf{k}'.\mathbf{R}} \int_\Omega d\mathbf{r}'\,\chi_{\alpha\beta}(\mathbf{r},\mathbf{r}'+\mathbf{R},t)\,\mathrm{e}^{i\mathbf{k}'.\mathbf{r}'}\,\mathrm{e}^{-i\mathbf{k}.\mathbf{r}} \\
&= \sum_{\mathbf{R}} \mathrm{e}^{i\mathbf{k}'.\mathbf{R}} \int_\Omega d\mathbf{r}' \int d\mathbf{r}\,\chi_{\alpha\beta}(\mathbf{r}-\mathbf{R},\mathbf{r}',t)\,\mathrm{e}^{i\mathbf{k}'.\mathbf{r}'}\,\mathrm{e}^{-i\mathbf{k}.} \\
&= \sum_{\mathbf{R}} \mathrm{e}^{i(\mathbf{k}'-\mathbf{k}).\mathbf{R}} \int_\Omega d\mathbf{r}'\,\mathrm{e}^{i\mathbf{k}'.\mathbf{r}'} \int d\mathbf{r}\,\chi_{\alpha\beta}(\mathbf{r},\mathbf{r}'\,t)\,\mathrm{e}^{-i\mathbf{k}.\mathbf{r}} \\
&= \Omega_B \sum_{\mathbf{K}} \delta(\mathbf{k}'-\mathbf{k}+\mathbf{K}) \int_\Omega d\mathbf{r}'\,\mathrm{e}^{i\mathbf{k}'.\mathbf{r}'} \int d\mathbf{r}\,\chi_{\alpha\beta}(\mathbf{r},\mathbf{r}',t)\,\mathrm{e}^{-i\mathbf{k}.\mathbf{r}},
\end{aligned} \tag{4.6.10}$$

that is,

$$\chi_{\alpha\beta}(\mathbf{k},\mathbf{k}',t) = \sum_{\mathbf{K}} \chi_{\alpha\beta}(\mathbf{k},\mathbf{k}+\mathbf{K},t)\,\delta_{\mathbf{k}',\mathbf{k}+\mathbf{K}}. \tag{4.6.11}$$

In general the fields we apply will be macroscopically varying, so that $\mathbf{k}$ is small. Then $\chi_{\alpha\beta}(\mathbf{k},\mathbf{k},t)$ will give the macroscopic response of the system to the field, while $\chi_{\alpha\beta}(\mathbf{k}+\mathbf{K},\mathbf{k},t)$ $(\mathbf{K}\neq 0)$ depends on the microscopic variation of the response within an individual cell. One should note that for spins completely localized at the lattice sites, as in the Heisenberg model, we will have

$$\chi_{\alpha\beta}(\mathbf{k}+\mathbf{K},\mathbf{k},t) = \chi_{\alpha\beta}(\mathbf{k},\mathbf{k},t) = \chi_{\alpha\beta}(\mathbf{k}+\mathbf{K},\mathbf{k}+\mathbf{K},t) \tag{4.6.12}$$

because the only material values of $\mathscr{H}(\mathbf{r})$ are those at the lattice sites, and $\mathscr{H}(\mathbf{R}) = \mathscr{H}\,\mathrm{e}^{i\mathbf{k}.\mathbf{R}} = \mathscr{H}\,\mathrm{e}^{i(\mathbf{k}+\mathbf{K}).\mathbf{R}}$.

Because of equation (4.6.11) we shall usually write χ as $\chi(\mathbf{k},t)$ below, on the understanding that in general $\mathbf{k}$ can be replaced by $\mathbf{k},\mathbf{k}'$.

If we take $\mathscr{H}(\mathbf{k}, t) = \mathscr{H}(\mathbf{k})\, e^{i\omega t}$ we have

$$\Delta\mathscr{M}_\beta(\mathbf{k}, t) = e^{i\omega t} \sum_{\alpha\beta} \chi_{\alpha\beta}(\mathbf{k}, \omega)\, \mathscr{H}_\alpha(\mathbf{k}), \tag{4.6.13}$$

where

$$\chi_{\alpha\beta}(\mathbf{k}, \omega) = \int_0^\infty e^{-i\omega t} \chi_{\alpha\beta}(\mathbf{k}, t)\, dt \tag{4.6.14}$$

and will be termed the dynamical susceptibility. In general it will be complex and to see the significance of $\operatorname{Im} \chi_{\alpha\beta}(\mathbf{k}, \omega)$ let us examine the response to a real field

$$\mathscr{H}(\mathbf{r}, t) = \mathscr{H} \cos(\mathbf{k}.\mathbf{r} + \omega t). \tag{4.6.15}$$

We find

$$\begin{aligned}
\Delta\mathscr{M}_\beta(\mathbf{r}, t) &= \tfrac{1}{2}[\Delta\mathscr{M}_\beta(\mathbf{k})\, e^{i(\mathbf{k}.\mathbf{r}+\omega t)} + \Delta\mathscr{M}_\beta(-\mathbf{k})\, e^{-i(\mathbf{k}.\mathbf{r}+\omega t)}] \\
&= \frac{1}{2} \sum_\alpha [\chi_{\alpha\beta}(\mathbf{k}, \omega)\, e^{i(\mathbf{k}.\mathbf{r}+\omega t)} + \chi_{\alpha\beta}(-\mathbf{k}, -\omega)\, e^{-i(\mathbf{k}.\mathbf{r}+\omega t)}]\, \mathscr{H}_\alpha \\
&= \frac{1}{2} \sum_\alpha \mathscr{H}_\alpha [\chi_{\alpha\beta}(\mathbf{k}, \omega) + \chi_{\alpha\beta}(-\mathbf{k}, -\omega)] \cos(\mathbf{k}.\mathbf{r} + \omega t) \\
&\quad + \frac{i}{2} \sum_\alpha \mathscr{H}_\alpha [\chi_{\alpha\beta}(\mathbf{k}, \omega) - \chi_{\alpha\beta}(-\mathbf{k}, -\omega)] \sin(\mathbf{k}.\mathbf{r} + \omega t).
\end{aligned} \tag{4.6.16}$$

This response $\Delta\mathscr{M}(\mathbf{r}, t)$ to a real field must itself be real and it therefore follows that

$$\begin{aligned}
\operatorname{Im} \chi_{\alpha\beta}(\mathbf{k}, \omega) + \operatorname{Im} \chi_{\alpha\beta}(-\mathbf{k}, -\omega) &= 0, \\
\operatorname{Re} \chi_{\alpha\beta}(\mathbf{k}, \omega) - \operatorname{Re} \chi_{\alpha\beta}(-\mathbf{k}, -\omega) &= 0,
\end{aligned} \tag{4.6.17}$$

that is,

$$\chi_{\alpha\beta}(\mathbf{k}, \omega) = \chi^*_{\alpha\beta}(-\mathbf{k}, -\omega). \tag{4.6.18}$$

We can now rewrite equation (4.6.16) as

$$\Delta\mathscr{M}_\beta(\mathbf{r}, t) = \sum_\alpha \mathscr{H}_\alpha [\operatorname{Re} \chi_{\alpha\beta}(\mathbf{k}, \omega) \cos(\mathbf{k}.\mathbf{r} + \omega t) + \operatorname{Im} \chi_{\alpha\beta}(\mathbf{k}, \omega) \sin(\mathbf{k}.\mathbf{r} + \omega t)]. \tag{4.6.19}$$

Hence $\operatorname{Im} \chi_{\alpha\beta}$ gives the out-of-phase response of the system and is the dissipative part of the susceptibility. A similar situation which arises in connection with the complex dielectric constant of a system will be discussed in Chapter 7.

We should mention here the Kramers–Krönig relations

$$\operatorname{Re}\chi_{\alpha\beta}(\mathbf{k},\omega)-\operatorname{Re}\chi_{\alpha\beta}(\mathbf{k},\infty)=\frac{2}{\pi}\int_0^\infty\frac{\omega'\operatorname{Im}\chi_{\alpha\beta}(\mathbf{k},\omega')}{\omega'^2-\omega^2}\,d\omega',$$

$$\operatorname{Im}\chi_{\alpha\beta}(\mathbf{k},\omega)=-\frac{2\omega}{\pi}\int_0^\infty\frac{\operatorname{Re}\left[\chi_{\alpha\beta}(\mathbf{k},\omega')-\chi_{\alpha\beta}(\mathbf{k},\infty)\right]}{\omega'^2-\omega^2}\,d\omega'. \tag{4.6.20}$$

These relations are derivable from explicit expressions for $\chi_{\alpha\beta}(\mathbf{k},\omega)$ given later, but we shall show in Chapter 7 that they are relations generally applicable to linear response functions.

4.6.2 *Relaxation of magnetization*

Suppose a steady field $\mathscr{H}(\mathbf{k})$ is applied for all times up to $t=0$, when it is switched off instantaneously. We define the relaxation factor R by writing the subsequent decay of magnetic moment as

$$\Delta\mathscr{M}_\beta(\mathbf{k},t)=\sum_\alpha\mathscr{H}_\alpha(\mathbf{k})\,R_{\alpha\beta}(\mathbf{k},t). \tag{4.6.21}$$

Since from equation (4.6.9) this is also

$$\Delta\mathscr{M}_\beta(\mathbf{k},t)=\sum_\alpha\int_{-\infty}^0\mathscr{H}_\alpha(\mathbf{k})\,\chi_{\alpha\beta}(\mathbf{k},t-t')\,dt' \tag{4.6.22}$$

it can be seen that

$$R_{\alpha\beta}(\mathbf{k},t)=\int_t^\infty\chi_{\alpha\beta}(\mathbf{k},t')\,dt',\quad t>0, \tag{4.6.23}$$

and

$$\chi_{\alpha\beta}(\mathbf{k},\omega)=-\int_0^\infty e^{-i\omega t}\frac{\partial R_{\alpha\beta}}{\partial t}(\mathbf{k},t)\,dt=R_{\alpha\beta}(\mathbf{k},0)-i\omega\int_0^\infty R_{\alpha\beta}(\mathbf{k},t)\,e^{-i\omega t}\,dt. \tag{4.6.24}$$

We shall find the subsidiary function R very useful in our discussions below.

4.6.3 *Susceptibility in terms of correlation functions*

We shall now obtain an expression for the dynamical susceptibility in terms of spin–correlation functions, under the assumption that the magnetic part of the Hamiltonian is

$$H_1=\int\boldsymbol{\mu}(\mathbf{r}')\,.\,\mathscr{H}(\mathbf{r}'t)\,d\mathbf{r}' \tag{4.6.25}$$

where $\boldsymbol{\mu}(\mathbf{r})$ is an operator representing the electronic magnetic-moment density:

$$\boldsymbol{\mu}(\mathbf{r}) = \sum_j \delta(\mathbf{r}-\mathbf{r}_j)\, \boldsymbol{\mu}_j, \tag{4.6.26}$$

$\boldsymbol{\mu}_j$ being the magnetic-moment operator of the jth electron.

We utilize the density matrix operator, as in section 3.16.3. Using arguments exactly like those given there, we find

$$\chi_{\alpha\beta}(\mathbf{r}, \mathbf{r}'; t) = \begin{cases} \dfrac{i}{\hbar}\langle[\mu_\alpha(\mathbf{r}'), \mu_\beta(\mathbf{r}, t)]\rangle, & t>0, \\ 0, & t<0, \end{cases} \tag{4.6.27}$$

when, by definition,

$$\chi_{\alpha\beta}(\mathbf{k}, \mathbf{k}'; t) = \begin{cases} \dfrac{i}{\hbar}\langle[\mu_\alpha(-\mathbf{k}), \mu_\beta(\mathbf{k}', t)]\rangle, & t>0, \\ 0, & t<0. \end{cases} \tag{4.6.28}$$

Here, the time dependence of the Heisenberg operators $\boldsymbol{\mu}$ is defined with respect to the unperturbed system, and the thermal average also refers to the unperturbed system.

In discussing frequency dependent properties it is convenient to introduce the Fourier transform

$$S_{\alpha\beta}(\mathbf{k}, \mathbf{k}'; \omega) = \frac{1}{2\pi}\int_{-\infty}^{\infty} \mathrm{e}^{-i\omega t}\langle\mu_\alpha(-\mathbf{k}')\, \mu_\beta(\mathbf{k}, t)\rangle\, dt, \tag{4.6.29}$$

which gives the frequency spectrum of spin correlations. Corresponding to equation (4.6.11),

$$S_{\alpha\beta}(\mathbf{k}, \mathbf{k}'; \omega) = \sum_{\mathbf{K}} S_{\alpha\beta}(\mathbf{k}, \mathbf{k}+\mathbf{K}; \omega)\, \delta_{\mathbf{k}',\mathbf{k}+\mathbf{K}} \tag{4.6.30}$$

and we shall usually simply write $S_{\alpha\beta}(\mathbf{k}, \omega)$. This is sometimes termed the *structure factor* or *power spectrum*, for reasons which will become evident in Chapters 5 and 6 respectively. It is to be observed that it obeys the relation

$$S_{\alpha\beta}(-\mathbf{k}, -\omega) = \mathrm{e}^{-\beta\omega}\, S_{\alpha\beta}(\mathbf{k}, \omega). \tag{4.6.31}$$

This equation can be regarded as a reflexion of the principle of detailed balancing which is proved in Appendix 4.2. By means of this condition equation (4.6.28) may be written as (with $\beta = \hbar/k_B T$ here)

$$\begin{aligned} \chi_{\alpha\beta}(\mathbf{k}, t) &= \frac{1}{i\hbar}\int_{-\infty}^{\infty} [S_{\alpha\beta}(\mathbf{k}, \omega) - S_{\beta\alpha}(-\mathbf{k}, -\omega)]\, \mathrm{e}^{i\omega t}\, d\omega \\ &= \frac{1}{i\hbar}\int_{-\infty}^{\infty} S_{\alpha\beta}(\mathbf{k}, \omega)\, (1-\mathrm{e}^{-\beta\omega})\, \mathrm{e}^{i\omega t}\, d\omega. \end{aligned} \tag{4.6.32}$$

We should like to express the relaxation factor in terms of $S_{\alpha\beta}(\mathbf{k}, \omega)$. It is evident that equation (4.6.23) can be written in the form

$$R_{\alpha\beta}(\mathbf{k}, t) = \frac{1}{\hbar i}\int_{t}^{\infty} dt' \int_{-\infty}^{\infty} S_{\alpha\beta}(\mathbf{k}, \omega)\,(1-\mathrm{e}^{-\beta\omega})\,\mathrm{e}^{i\omega t'}\,d\omega. \tag{4.6.33}$$

To perform the integral over t' we shall replace the upper limit by τ, after which we must take the limit $\tau \to \infty$. Then

$$R_{\alpha\beta}(\mathbf{k}, t) = \int_{-\infty}^{\infty} S_{\alpha\beta}(\mathbf{k}, \omega)\,\frac{(1-\mathrm{e}^{-\beta\omega})}{\hbar\omega}\,\mathrm{e}^{-i\omega t}\,d\omega$$

$$-\lim_{\tau\to\infty}\int_{-\infty}^{\infty} S_{\alpha\beta}(\mathbf{k}, \omega)\,\frac{(1-\mathrm{e}^{-\beta\omega})}{\hbar\omega}\,\mathrm{e}^{i\omega t}\,d\omega. \tag{4.6.34}$$

As $\tau \to \infty$ the integral involving $\mathrm{e}^{i\omega\tau}$ becomes significant about the origin only, where $(1-\mathrm{e}^{-\beta\omega})/\omega \to \beta$. Hence

$$\lim_{\tau\to\infty}\int_{-\infty}^{\infty} S_{\alpha\beta}(\mathbf{k}, \omega)\,\frac{(1-\mathrm{e}^{-\beta\omega})}{\omega}\,\mathrm{e}^{i\omega\tau}\,d\omega = \beta S_{\alpha\beta}(\mathbf{k}, t)\big|_{t=\infty}. \tag{4.6.35}$$

But

$$S_{\alpha\beta}(\mathbf{k}, t)\big|_{t=\infty} = \langle \mu_{\alpha}(-\mathbf{k})\,\mu_{\beta}(\mathbf{k}, \infty)\rangle = \langle \mu_{\alpha}(\mathbf{k})\rangle \langle \mu_{\beta}(\mathbf{k})\rangle \tag{4.6.36}$$

since in the limit of very large times $\mu_{\alpha}(-\mathbf{k})$ and $\mu_{\beta}(\mathbf{k}, t)$ will be uncorrelated. If we therefore define

$$S'_{\alpha\beta}(\mathbf{k}, \omega) = S_{\alpha\beta}(\mathbf{k}, \omega) - \delta(\omega)\,\langle \mu_{\alpha}(-\mathbf{k})\rangle \langle \mu_{\beta}(\mathbf{k})\rangle \tag{4.6.37}$$

(which, since the mean values $\langle \mu_{\alpha}(-\mathbf{k})\rangle \langle \mu_{\beta}(\mathbf{k})\rangle$ are subtracted out, refers to a spectrum of spontaneous fluctuations), equation (4.6.34) is just‡

$$R_{\alpha\beta}(\mathbf{k}, t) = \int_{-\infty}^{\infty} S'_{\alpha\beta}(\mathbf{k}, \omega)\,\frac{(1-\mathrm{e}^{-\beta\omega})}{\hbar\omega}\,\mathrm{e}^{i\omega t}\,dt. \tag{4.6.38}$$

Having now completed the discussion of the linear response theory in very general terms, we have reached the point at which models must be introduced to make further progress. The next two sections deal, respectively, with the Heisenberg model and with simple itinerant models. We shall, because of its simplicity and its numerous successes, use a molecular field approximation in dealing with Heisenberg systems.

‡ This is an expression of the fluctuation–dissipation theorem (qv).

4.7 Ground and low-lying excited states for Heisenberg systems

4.7.1 Weiss molecular field model

We shall now examine the main features of the magnetic behaviour of localized-spin systems at elevated temperatures and to do so again take the Heisenberg Hamiltonian

$$H = g\mu \sum_{\mathbf{R}} \mathbf{s}_{\mathbf{R}} \cdot \mathscr{H}(\mathbf{R}) - 2 \sum_{\mathbf{R}' \neq \mathbf{R}} J_{\mathbf{R}-\mathbf{R}'} \mathbf{s}_{\mathbf{R}} \cdot \mathbf{s}_{\mathbf{R}'} \tag{4.7.1}$$

in which we have included the influence of a weak external field in order to calculate the susceptibility. To take account of dipole–dipole effects and spin–orbit coupling, one would need to make J a tensor, but we shall treat the simpler scalar case here.

At present we shall solve within the molecular-field model. We write

$$\sum_{\mathbf{R}' \neq \mathbf{R}} J_{\mathbf{R}-\mathbf{R}'} \mathbf{s}_{\mathbf{R}} \cdot \mathbf{s}_{\mathbf{R}'} \simeq \sum_{\mathbf{R}' \neq \mathbf{R}} J_{\mathbf{R}-\mathbf{R}'} \mathbf{s}_{\mathbf{R}} \cdot \langle \mathbf{s}_{\mathbf{R}'} \rangle \tag{4.7.2}$$

so that equation (4.7.1) is approximated by

$$H \simeq g\mu \sum_{\mathbf{R}} \mathbf{s}_{\mathbf{R}} \cdot \mathscr{H}_{\text{eff}}(\mathbf{R}) \tag{4.7.3}$$

with an effective field given by

$$\mathscr{H}_{\text{eff}}(\mathbf{R}) = \mathscr{H}(\mathbf{R}) - \frac{2}{g\mu} \sum_{\mathbf{R}' \neq \mathbf{R}} J_{\mathbf{R}-\mathbf{R}'} \langle \mathbf{s}_{\mathbf{R}'} \rangle. \tag{4.7.4}$$

This effective field description gives a useful overall picture of the behaviour of the system at elevated temperatures, although it has certain limitations; in particular the spin waves we have already discussed are absent.

(a) *Below critical temperature.* The model Hamiltonian of equation (4.7.3) allows us to consider the energy of each spin separably, and this energy can be written as $s_z \mathscr{H}_{\text{eff}}$ if $\mathscr{H}_{\text{eff}}$ is in the z-direction. Hence

$$\langle s_z \rangle_{\mathscr{H}} = \sum_{s_z=-s}^{+s} s_z \exp(-\beta g\mu s_z \mathscr{H}_{\text{eff}}) \Big/ \sum_{s_z} \exp(-\beta g\mu s_z \mathscr{H}_{\text{eff}}). \tag{4.7.5}$$

We suppose β to be regarded as small and expand the right-hand side of the above equation in powers of β to yield

$$\langle s_z \rangle_{\mathscr{H}} = -(g\mu \mathscr{H}_{\text{eff}} \beta) \frac{s(s+1)}{3} + (g\mu \mathscr{H}_{\text{eff}} \beta)^2 \frac{s(s+1)}{45} (s^2+s+\tfrac{1}{2}) + \dots. \tag{4.7.6}$$

We first of all treat ferromagnets and accordingly assume $\langle s_{\mathbf{R}}^z \rangle_{\mathscr{H}=0}$ takes the same value for all $\mathbf{R}$. It is then possible to write

$$\langle s_{\mathbf{R}}^z \rangle_{\mathscr{H}} = \langle s_z \rangle + \Delta S_{\mathbf{R}}^z, \tag{4.7.7}$$

where, as usual, $\langle\ \rangle$ indicates an average in zero field. Substituting this into equation (4.7.4) we find

$$g\mu\mathscr{H}_{\text{eff}}(\mathbf{R}) = -2J(0)\langle s_z\rangle + \{g\mu\mathscr{H}(\mathbf{R}) - 2\sum_{\mathbf{R}'} J_{\mathbf{R}-\mathbf{R}'}\Delta S^z_{\mathbf{R}'}\}, \tag{4.7.8}$$

where $J(0)$ is the $\mathbf{k} = 0$ limit of

$$J(\mathbf{k}) = \sum_{\mathbf{R}}{}' J_{\mathbf{R}} e^{i\mathbf{k}.\mathbf{R}}. \tag{4.7.9}$$

To first order in ΔS equation (4.7.6) then becomes

$$\langle s_z\rangle + \Delta S^z_{\mathbf{R}} = \beta\frac{s(s+1)}{3}\{2J(0)\langle s_z\rangle - [g\mu\mathscr{H}(\mathbf{R}) - 2\sum_{\mathbf{R}'\neq\mathbf{R}} J_{\mathbf{R}-\mathbf{R}'}\Delta S^z_{\mathbf{R}'}]\}$$

$$+\beta^3\frac{s(s+1)}{45}(s^2+s+\tfrac{1}{2})\{-8J(0)^3\langle s_z\rangle^3 + 12J(0)\langle s_z\rangle^2$$

$$\times [g\mu\mathscr{H}(\mathbf{R}) - 2\sum_{\mathbf{R}'}{}' J_{\mathbf{R}-\mathbf{R}'}\Delta S^z_{\mathbf{R}'}]\}. \tag{4.7.10}$$

Putting $\mathscr{H} = 0$ we find the spontaneous magnetization is given by

$$\langle s_z\rangle = \tfrac{2}{3}\beta J(0)\, s(s+1)\langle s_z\rangle - \tfrac{8}{45}\beta^3 J(0)^3 s(s+1)(s^2+s+\tfrac{1}{2})\langle s_z\rangle^3, \tag{4.7.11}$$

which, excluding the trivial solution $\langle s_z\rangle = 0$, implies

$$\langle s_z\rangle^2 = \frac{\tfrac{2}{3}\beta J(0)\, s(s+1) - 1}{\tfrac{8}{45}\beta^3 J(0)^3 s(s+1)(s^2+s+\tfrac{1}{2})} \tag{4.7.12}$$

or, if we define T_c by

$$k_B T_c = \tfrac{2}{3}s(s+1)J(0), \tag{4.7.13}$$

$$\langle s_z\rangle^2 = \left(\frac{T_c}{T} - 1\right)\left(\frac{T}{T_c}\right)^3 \frac{5}{3}\frac{s^2(s+1)^2}{(s^2+s+\tfrac{1}{2})}, \tag{4.7.14}$$

showing that T_c is the Curie temperature within the present approximation. By abstracting from equation (4.7.10) the terms which vanish when $\mathscr{H} = 0$ we find

$$\Delta S^z_{\mathbf{R}} = \beta\frac{s(s+1)}{3}\{g\mu\mathscr{H}(\mathbf{R}) - 2\sum_{\mathbf{R}'} J_{\mathbf{R}-\mathbf{R}'}\Delta S^z_{\mathbf{R}'}\}\{-1 + \tfrac{4}{5}\beta^2(s^2+s+\tfrac{1}{2})J(0)^2\langle s_z\rangle^2\} \tag{4.7.15}$$

which, on using equations (4.7.13) and (4.7.14), becomes

$$\Delta S^z_{\mathbf{R}} = \frac{2T_c - 3T}{2J(0)\,T}\left\{g\mu\mathscr{H}(\mathbf{R}) - 2\sum_{\mathbf{R}'} J_{\mathbf{R}-\mathbf{R}'}\Delta S^z_{\mathbf{R}'}\right\} \tag{4.7.16}$$

and Fourier transform yields

$$\Delta S_z(\mathbf{k}) = \frac{2T_c - 3T}{2J(0)\,T}\,[g\mu\,\mathscr{H}(\mathbf{k}) - 2J(\mathbf{k})\,\Delta S_z(\mathbf{k})]. \tag{4.7.17}$$

The static susceptibility is therefore

$$\chi_{zz}(\mathbf{k}) = g^2\mu^2 \frac{3T - 2T_c}{2J(0)\,T\{[2(T_c - 3T)/T]\,[J(\mathbf{k})/J(0)] + 1\}}. \tag{4.7.18}$$

(b) *Behaviour near critical temperature* T_c. When $T \simeq T_c$,

$$\chi_{zz}(\mathbf{k}) \simeq \frac{g^2\mu^2}{2J(0)\{[2(T_c - T)/T_c] + [(J(0) - J(\mathbf{k}))/J(0)]\}}. \tag{4.7.19}$$

It is evident that when $T = T_c$, $\chi_{zz}(\mathbf{k})$ diverges as $k \to 0$. From equation (4.6.62) we see that the fluctuations in the magnetization increase without limit as $T \to T_c$. This is also true of their range, as we go on to show.

For small k, provided $J_\mathbf{R} = J_{-\mathbf{R}}$,

$$J(\mathbf{R}) = \sum_\mathbf{R}{}' J_\mathbf{R}\, e^{i\mathbf{k}.\mathbf{R}} = \sum_\mathbf{R} J_\mathbf{R}\left[1 - \frac{(\mathbf{k}.\mathbf{R})^2}{2!} + \frac{(\mathbf{k}.\mathbf{R})^4}{4!} + \ldots\right]. \tag{4.7.20}$$

We terminate this expression at the quadratic term for present purposes and spherically average to obtain

$$J(\mathbf{k}) = J(0) - \frac{k^2}{6} J^{(2)} + \ldots, \tag{4.7.21}$$

where

$$J^{(n)} = \sum_\mathbf{R} J_\mathbf{R}\, R^n. \tag{4.7.22}$$

No error is introduced by this procedure provided $J_\mathbf{R} = J_{|\mathbf{R}|}$. At small $\mathbf{k}$ equation (4.7.19) becomes

$$\chi_{zz}(\mathbf{k}) = \frac{\chi_c}{r_1^2(\kappa_1^2 + k^2)} \tag{4.7.23}$$

with

$$\chi_c = \frac{g^2\mu^2 s(s+1)}{3k_B T_c} \tag{4.7.24}$$

and

$$r_1^2 = \frac{J^{(2)}}{6J(0)}, \quad \kappa_1^2 = \frac{12J(0)}{J^{(2)}}\,\frac{T - T_c}{T}. \tag{4.7.25}$$

To examine the range of the fluctuations one must find $\langle s_0 s_{\mathbf{R}}\rangle$. Inspection of equation (4.6.32) will show that at sufficiently high temperatures, or provided $S'(\mathbf{k}, \omega)$ is sufficiently narrow in ω (which we shall later find to be the case for T near T_c) one can write, after integrating over t to obtain the static limit $\omega = 0$ (χ_0 being the susceptibility of N independent spins),

$$\chi_{\alpha\beta}(\mathbf{k}) \simeq \frac{\beta}{\hbar}\int_{-\infty}^{\infty} S'_{\alpha\beta}(\mathbf{k}, \omega)\, d\omega = \frac{3\chi_0}{Ns(s+1)}\langle s_\alpha(-\mathbf{k})\, s_\beta(\mathbf{k})\rangle \tag{4.7.26}$$

or, from equation (4.7.23),

$$\langle s_z(-\mathbf{k})\, s_z(\mathbf{k})\rangle = \frac{s(s+1)}{3T_c}\frac{T}{r_1^2(\kappa_1^2+\mathbf{k}^2)}. \tag{4.7.27}$$

This is true for small k only, but by means of equation (4.7.27) we can at least obtain the asymptotic form

$$\langle s_0^z s_{\mathbf{R}}^z\rangle = \frac{s(s+1)}{3T_c r_1^2} T \sum_{\mathbf{k}} \frac{e^{-i\mathbf{k}.\mathbf{R}}}{\kappa_1^2+k^2}, \quad R \to \infty. \tag{4.7.28}$$

To perform the sum over k we replace it by an integral in the usual way; the integration is elementary and yields

$$\langle s_0^z s_{\mathbf{R}}^z\rangle = \frac{Vs(s+1)}{12\pi}\frac{T}{T_c}\frac{e^{-\kappa_1 R}}{r_1^2 R}, \tag{4.7.29}$$

V being the volume. From equation (4.7.25) it is clear that $\kappa_1 \to 0$ as $T \to T_c$ and so as T_c is approached from below the spin correlations become longer and longer ranged.

4.7.2 Antiferromagnetic ordering

The modifications necessary to the above when our concern is with antiferromagnets will be illustrated in the simplest case. The spins sit on a Bravais lattice but whereas $\langle s_z\rangle$ has the same magnitude at every site the mean direction of every spin is opposite to that of its nearest neighbours. We therefore define a magnetic lattice, lattice vectors $\mathbf{R}$, on the sites of which sit all spins with direction in the same sense as that at the origin, whereas opposite spins sit upon sites $\mathbf{R}+\boldsymbol{\tau}$, $\boldsymbol{\tau}$ being a non-primitive translation in the magnetic lattice. Since a primitive translation $\mathbf{K}_p$ of the lattice reciprocal to the magnetic lattice is such that $\mathbf{K}_p . \boldsymbol{\tau} = \pi$ we have

$$\sum_{\mathbf{R}} J_{\mathbf{R}}\langle s_{\mathbf{R}}^z\rangle + \sum_{\mathbf{R}} J_{\mathbf{R}+\boldsymbol{\tau}}\langle s_{\mathbf{R}+\boldsymbol{\tau}}^z\rangle = \sum_{\mathbf{R}} J_{\mathbf{R}}\langle s_{\mathbf{R}}^z\rangle + \sum_{\mathbf{R}} J_{\mathbf{R}+\boldsymbol{\tau}}\langle s_{\mathbf{R}}^z\rangle\, e^{i\mathbf{K}_p.\boldsymbol{\tau}}$$

$$= \langle s_z\rangle \sum_{\mathbf{R}}\{J_{\mathbf{R}} + J_{\mathbf{R}+\boldsymbol{\tau}}\, e^{i\mathbf{K}_p.\boldsymbol{\tau}}\} = \langle s_z\rangle J(\mathbf{K}_p), \tag{4.7.30}$$

where $\langle s_z \rangle$ is the magnitude of $\langle s^z_{\mathbf{R}} \rangle$ and

$$J(\mathbf{k}) = \sum_{\mathbf{R}} \{J_{\mathbf{R}} + J_{\mathbf{R}+\boldsymbol{\tau}}\, e^{i\mathbf{k}.\boldsymbol{\tau}}\}\, e^{i\mathbf{k}.\mathbf{R}}. \tag{4.7.31}$$

The value of $\langle s_z \rangle$ is now obtained in a manner similar to that for the ferromagnet and equation (4.7.14) again holds, except that T_c is now the Néel temperature, and equation (4.7.13) is replaced by

$$k_B T_c = \tfrac{2}{3} s(s+1)\, J(\mathbf{K}_p). \tag{4.7.32}$$

We also note that equation (4.7.16) is replaced by

$$\Delta S^z_{\mathbf{R}} = \frac{2T_c - 3T}{2J(\mathbf{K}_p)\, T}\left\{g\mu\, \mathscr{H}(\mathbf{R}) - 2\sum_{\mathbf{R}'} [J_{\mathbf{R}-\mathbf{R}'}\, \Delta S^z_{\mathbf{R}'} + J_{\mathbf{R}-\mathbf{R}'+\boldsymbol{\tau}}\, \Delta S^z_{\mathbf{R}'+\boldsymbol{\tau}}]\right\}$$

$$\Delta S^z_{\mathbf{R}+\boldsymbol{\tau}} = \frac{2T_c - 3T}{2J(\mathbf{K}_p)\, T}\left\{g\mu\, \mathscr{H}(\mathbf{R}+\boldsymbol{\tau}) - 2\sum_{\mathbf{R}'} [J_{\mathbf{R}-\mathbf{R}'}\, \Delta S^z_{\mathbf{R}'+\boldsymbol{\tau}} + J_{\mathbf{R}-\mathbf{R}'+\boldsymbol{\tau}}\, \Delta S^z_{\mathbf{R}'}]\right\} \tag{4.7.33}$$

and Fourier inversion yields

$$\Delta S_z(\mathbf{k}) = \sum_{\mathbf{R}} \{\Delta S^z_{\mathbf{R}} + \Delta S^z_{\mathbf{R}+\boldsymbol{\tau}}\, e^{i\mathbf{k}.\boldsymbol{\tau}}\}\, e^{i\mathbf{k}.\mathbf{R}} = \frac{2T_c - 3T}{2J(\mathbf{K}_p)\, T}\{g\mu\, \mathscr{H}(\mathbf{k}) - 2J(\mathbf{k})\, \Delta S_z(\mathbf{k})\} \tag{4.7.34}$$

so that equations (4.7.17) and (4.7.18) are again true, except that $J(0)$ is replaced by $J(\mathbf{K}_p)$.

From equation (4.7.31) we note that, on introducing a new variable $\mathbf{q}$,

$$J(\mathbf{k}) = J(\mathbf{K}_p + \mathbf{q}) = \sum_{\mathbf{R}} \{J_{\mathbf{R}} + J_{\mathbf{R}+\boldsymbol{\tau}}\, e^{i\mathbf{K}_p.\boldsymbol{\tau}}\, e^{i\mathbf{q}.\boldsymbol{\tau}}\}\, e^{i\mathbf{q}.\mathbf{R}} \tag{4.7.35}$$

or, for small $\mathbf{q}$,

$$J(\mathbf{K}_p + \mathbf{q}) = \sum_{\mathbf{R}} J_{\mathbf{R}}\left[1 - \frac{(\mathbf{q}.\mathbf{R})^2}{2!} + \ldots\right] + \sum_{\mathbf{R}} J_{\mathbf{R}+\boldsymbol{\tau}}\, e^{i\mathbf{K}_p.\boldsymbol{\tau}}\left\{1 - \frac{[\mathbf{q}.(\mathbf{R}+\boldsymbol{\tau})]^2}{2!} + \ldots\right\}. \tag{4.7.36}$$

The spherical average is

$$J(\mathbf{K}_p + \mathbf{q}) = J(\mathbf{K}_p) - \tfrac{1}{6} J^{(2)} q^2 + \ldots, \tag{4.7.37}$$

where

$$J^{(n)} = \sum_{\mathbf{R}} J_{\mathbf{R}}\, R^n + e^{i\mathbf{K}_p.\boldsymbol{\tau}} \sum_{\mathbf{R}} J_{\mathbf{R}+\boldsymbol{\tau}} (\mathbf{R}+\boldsymbol{\tau})^n. \tag{4.7.38}$$

Equation (4.7.23) is then rewritten for antiferromagnets:

$$\chi_{zz}(\mathbf{K}_p + \mathbf{q}) = \frac{s(s+1)}{3T_c}\, \frac{g_2 \mu_2}{r_1^2(\kappa_1^2 + q^2)} \tag{4.7.39}$$

with T_c given by equation (4.7.32) and

$$r_1^2 = \frac{1}{6}\frac{J^{(2)}}{J(\mathbf{K}_p)}, \quad \kappa_1^2 = \frac{12J(\mathbf{K}_p)}{J^{(2)}}\frac{T-T_c}{T}. \tag{4.7.40}$$

The asymptotic form of equation (4.7.29) again obtains.

4.7.3 *Behaviour above critical temperature*

(a) *Curie–Weiss Law.* We now assume $\langle s_{\mathbf{R}}\rangle_{\mathscr{H}}$ to be small (in fact zero if $\mathscr{H} = 0$) whereupon we can use a linear theory to obtain the response to $\mathscr{H}_{\text{eff}}$, which will also be small. Moreover, since in equation (4.7.3) there is no term correlating spins on different atoms the susceptibility χ_0 by which the system responds to the effective field is simply that of independent spins and is given by equation (4.7.24) with T_c replaced by T. We then have

$$\mathscr{M}(\mathbf{R}) = \chi_0 \mathscr{H}_{\text{eff}}(\mathbf{R}) = \chi_0\left\{\mathscr{H}(\mathbf{R}) + \frac{2}{g^2\mu^2}\sum_{\mathbf{R}'\neq\mathbf{R}} J_{\mathbf{R}'-\mathbf{R}}\mathscr{M}(\mathbf{R}')\right\}. \tag{4.7.41}$$

Taking Fourier transforms,

$$\mathscr{M}(\mathbf{k}) = \chi_0\mathscr{H}(\mathbf{k}) + \frac{2\chi_0}{g^2\mu^2}J(\mathbf{k})\mathscr{M}(\mathbf{k}) \tag{4.7.42}$$

and

$$\chi_{\alpha\alpha}(\mathbf{k}) = \frac{\mathscr{M}(\mathbf{k})}{\mathscr{H}(\mathbf{k})} = \frac{g^2\mu^2 s(s+1)}{3k_B T - 2s(s+1)J(\mathbf{k})}. \tag{4.7.43}$$

We see that χ has the classical form $C/(T-T_c)$ and diverges as $\mathbf{k}\to\mathbf{k}_m$, where $J(\mathbf{k}_m)$ is the maximum value of $J(\mathbf{k})$, at a temperature given by

$$k_B T_c = \tfrac{2}{3}s(s+1)J(\mathbf{k}_m). \tag{4.7.44}$$

Comparing this with equations (4.7.13) and (4.7.32) we can see that if $J(\mathbf{k})$ takes on its maximum value at $\mathbf{k} = 0$ we can expect the system to be ferromagnetic below T_c whereas if the maximum is elsewhere we can expect an antiferromagnet of some kind.

(b) *Range of correlations near critical temperature.* Again, we can show that not only do the fluctuations of magnetization increase without limit as T_c is approached but the asymptotic form (4.7.29) again obtains, except that (specializing to a ferromagnet)

$$\kappa_1^2 = \frac{6J(0)}{J^{(2)}}\frac{T-T_c}{T}. \tag{4.7.45}$$

Hence, while the divergence of equation (4.7.27) as $\mathbf{k} \to 0$ is twice as fast below the critical temperature as above, fluctuations of the magnetization increase without limit in size and range as the critical temperature is approached from either above or below. After this rather specific discussion we shall go on to deal in more general terms with the approach to the critical temperature.

4.8 Approach to magnetic critical points

The transition from long-range magnetic order, either ferromagnetic, antiferromagnetic or ferrimagnetic, to a paramagnetic state is, of course, a phase transition, which is heralded, as we saw from molecular field theory by, for example, an increase in fluctuations, and a slowing down of their rate of decay (see Chapter 5), as the critical point is approached. Clearly we are speaking here of a cooperative phenomenon, and in this section we wish to discuss some rather general conclusions which can be obtained for magnetic phase transitions.

The "critical phenomena" occurring near the Curie or Néel temperatures have been very thoroughly investigated over a long period of time and a surprising degree of similarity has emerged between the magnetic phase transitions and other transitions of apparently quite different character. We shall restrict the discussion to equilibrium properties in this chapter.

The fundamental interest lies in the behaviour of thermodynamic quantities as the Curie (Néel) temperature T_c is approached. Thus, the object of the exercise is not to calculate T_c, though this is, of course, as we saw above, of considerable interest for particular materials in that the strength of the exchange interaction is involved, but to study the approach to the critical point.

In particular, we would like to describe how the magnetization $M(T, H)$ behaves. For ferromagnets, we shall attempt to describe:

(i) The spontaneous magnetization $M(T)$ as a function of temperature. For $T < T_c$, experiment shows that, for a few metallic and non-metallic ferromagnetics,

$$M \sim (T_c - T)^\beta \tag{4.8.1}$$

with $\beta \doteqdot 0.33 \pm 0.03$.

(ii) The behaviour of $M(T, H)$ as a function of H along the critical isotherm. For Ni and Gd the relation appears experimentally to take the form

$$|H \sim |M|^\delta, \quad T = T_c, \quad H \to 0. \tag{4.8.2}$$

There is some uncertainty in δ, but it appears that $\delta \geqslant 4$.

(iii) Initial susceptibility χ as T_c is approached. Experiment suggests that

$$\chi \sim (T-T_c)^{-\gamma} \tag{4.8.3}$$

with $\gamma = 1.33 \pm 0.02$ ($T \to T_c$ from above).

(iv) The specific heat C_H is found to behave as

$$C_H \sim (T-T_c)^{-\alpha}, \quad 0 \leqslant \alpha \leqslant 0.16 \quad (T \to T_{c+}, H = 0) \tag{4.8.4}$$

and

$$C_H \sim (T_c-T)^{-\alpha'}, \quad 0 \leqslant \alpha' < 0.2 \quad (T \to T_{c-}, H = 0). \tag{4.8.5}$$

(v) The pair correlation function for spins. As this is related intimately to neutron scattering, we shall defer discussion of this until Chapter 5.

Similar properties can be isolated for antiferromagnets but we shall not discuss these separately. In the following, therefore, we focus attention on ferromagnets.

The quantities α, β, γ, etc. discussed above are termed critical indices, and we shall start our discussion by seeing what information we can obtain by purely thermodynamic reasoning (cf. Kastelyn, 1971, or Fisher, 1967).

4.8.1 Thermodynamic relations between specific heats

The specific heat C_H at constant magnetic field is evidently given by

$$C_H = T\left(\frac{\partial S}{\partial T}\right)_H, \tag{4.8.6}$$

where S is the entropy. We can immediately write this in the form

$$C_H = T\left(\frac{\partial S}{\partial T}\right)_M + T\left(\frac{\partial S}{\partial M}\right)_T\left(\frac{\partial M}{\partial T}\right)_H. \tag{4.8.7}$$

But we have the thermodynamic relation

$$\left(\frac{\partial S}{\partial H}\right)_T = \left(\frac{\partial M}{\partial T}\right)_H \tag{4.8.8}$$

and from this it follows immediately that

$$\left(\frac{\partial S}{\partial M}\right)_T = \left(\frac{\partial M}{\partial T}\right)_H \Big/ \left(\frac{\partial M}{\partial H}\right)_T. \tag{4.8.9}$$

Hence we find, with C_M the specific heat at constant M,

$$C_H - C_M = \frac{T[(\partial M/\partial T)_H]^2}{(\partial M/\partial H)_T}. \tag{4.8.10}$$

Again, it can be shown by relating C_M to the fluctuation of the energy that if $H = 0$ then $C_M \geqslant 0$ and hence

$$C_H \geqslant T\left(\frac{\partial M}{\partial T}\right)_H^2 \Big/ \chi. \tag{4.8.11}$$

Now let us assume the forms discussed above, namely that for $T \leqslant T_c$

$$M \sim (T_c - T)^{\beta}, \tag{4.8.12}$$

$$\chi \sim (T_c - T)^{-\gamma'} \tag{4.8.13}$$

and

$$C_H \sim (T_c - T)^{-\alpha'}. \tag{4.8.14}$$

Hence the right-hand side of equation (4.8.11) behaves near the Curie point T_c as $(T_c - T)^{2(\beta-1)+\gamma'}$. If this diverges, that is if $2\beta + \gamma' - 2 < 0$, then C_H will diverge at least as fast and therefore

$$-\alpha' \leqslant 2\beta + \gamma' - 2 \tag{4.8.15}$$

or

$$\alpha' + 2\beta + \gamma' \geqslant 2. \tag{4.8.16}$$

This inequality is due to Rushbrooke (1963). In a similar way, Griffiths (1965) has shown that

$$\alpha' + \beta(1+\delta) \geqslant 2. \tag{4.8.17}$$

4.8.2 *Classical molecular field theories*

The Weiss molecular field treatment has been shown to lead to the results $\gamma' = 1$ [see equation (4.7.19)], $\beta = \frac{1}{2}$ [equation (4.7.14)] as T_c is approached. It is not difficult to show, in addition that $\alpha' = 0$ and $\delta = 3$. Hence we have immediately

$$\alpha' + 2\beta + \gamma' = 2 \tag{4.8.18}$$

and

$$\alpha' + \beta(1+\delta) = 2 \tag{4.8.19}$$

and the Rushbrooke and Griffiths inequalities are seen to become *equalities*.

Though we shall not go into any details, the 2- and 3-dimensional Ising model (see, for example, Fisher, 1967) appears also, at least to high numerical accuracy, to lead to similar conclusions: namely that the inequalities become equalities.

Thus, it is often conjectured that equations (4.8.18) and (4.8.19) are applicable quite generally, together with the results

$$\alpha' = \alpha \tag{4.8.20}$$

and

$$\gamma' = \gamma. \tag{4.8.21}$$

These are not established from first principles but appear to work!

4.8.3 *Scaling laws*

Widom (1965) and independently Kadanoff (1966) have put forward the idea that, near T_c, the range of the correlations is so great that local behaviour plays a negligible role and that the correlation length κ_c^{-1}, which diverges as $T \to T_c$, is the fundamental parameter, and the only one, entering the theory.

If this idea is correct, then there will be, essentially, a "Law of Corresponding States" between systems having the same lattice structure. We can express this more precisely as follows. Suppose we focus on the magnetization $M(H,T)$. Then, by integrating this with respect to H, we can obtain the free energy $F(H,T)$ and hence all thermodynamic quantities. The existence of a scaling law implies that, if $T - T_c = \varepsilon$, and we replace ε by $\lambda\varepsilon$, then near the critical point

$$\frac{M}{m(\lambda)} = f\left(\frac{H}{h(\lambda)}, \lambda\varepsilon\right). \tag{4.8.22}$$

If $\lambda \to 0$, we expect $m(\lambda)$ and $h(\lambda)$ to increase and we therefore assume

$$\left.\begin{aligned} m(\lambda) &= \lambda^{-u}, \\ h(\lambda) &= \lambda^{-v}, \end{aligned}\right\} \quad u, v > 0. \tag{4.8.23}$$

Then it is easy to show that, as a consequence of the above assumptions, we can write, since f must be such that (4.8.22) is independent of λ,

$$\frac{M}{|\varepsilon|^u} = f_{\pm}\left(\frac{H}{|\varepsilon|^v}\right), \tag{4.8.24}$$

where f_+ must be used for $\varepsilon > 0$ and f_- for $\varepsilon < 0$. It is clear that, if this relation holds near the critical point, then we have achieved the great simplification of reducing $M = M(H,T)$ to an unknown function of a single variable.

(a) *Identification of parameters with critical indices.* For $\varepsilon > 0$, i.e. $T > T_c$, $M = 0$ and therefore $f_+(0) = 0$. Similarly $f_-(0)$ must be a finite constant if M is to decrease with $|\varepsilon|$. Thus we can draw up Table 4.1, giving the four quantities we discussed earlier. These are expressed in the first column in terms of u and v and in terms of the critical indices in the second column. The consequences resulting from this are recorded in the third column.

TABLE 4.1

Physical quantity	Behaviour on approaching critical point: In terms of u and v	In terms of critical indices	Consequences
(i) Spontaneous magnetization	$M = \|\varepsilon\|^u f_\pm(0)$	$M \sim \|\varepsilon\|^\beta$	$u = \beta$
(ii) Magnetization M as function of H on critical isotherm	$M = \|H\|^{u/v}$	$M \sim \|H\|^{1/\delta}$	$v = \beta\delta$
(iii) Initial susceptibility	$\chi \sim \|\varepsilon\|^{\beta(1-\delta)} f'_\pm(0)$	$\chi \sim (T-T_c)^{-\gamma}$ or $(T_c-T)^{-\gamma'}$	$\beta(1-\delta) = -\gamma = -\gamma'$
(iv) Specific heat C_H at $H = 0$	$C_H \sim \|\varepsilon\|^{\beta+\beta\delta-2}$	$C_H \sim (T-T_c)^{-\alpha}$ or $(T_c-T)^{-\alpha'}$	$\alpha' = \alpha$; $\alpha' + \beta(1+\delta) = 2$

Given the "ansatz" (4.8.22), we get back a theory which turns out to have only two independent critical exponents, which we might choose as β and γ, say.

Various attempts have been made to support the "ansatz" (4.8.22), prominent work being that of Widom (1965) who assumed homogeneity of the free energy, cell scaling due to Kadanoff (1966) and the droplet model proposed by Fisher (1967). No convincing argument from first principles has, as yet, been given (see, however, Wilson, 1971).

4.8.4 Equation of state and its parametric representation

One of the consequences of the "scaling law" "ansatz" is that the magnetic equation of state takes the form

$$\frac{M}{|\varepsilon|^\beta} = f_\pm\left(\frac{H}{|\varepsilon|^{\beta\delta}}\right). \tag{4.8.25}$$

Experimental evidence exists, which gives strong support to this form very near to the critical point.

Schofield (1969) and Josephson (1969) have independently pointed out that it may be valuable to use "polar coordinates" in the H–T plane to represent this equation of state. The reason for this becomes clear if we refer to Figure 4.4. We choose a variable r related to the distance from T_c in the

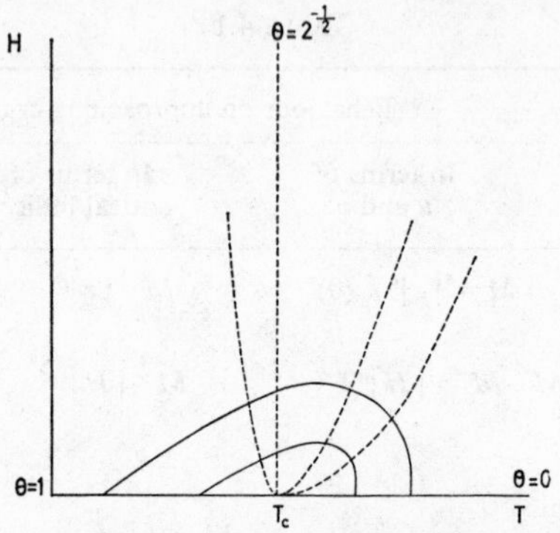

FIGURE 4.4. Schematic illustration of variables r and θ in parametric representation (4.8.32) of equation of state
———— curves of constant r
.... curves of constant θ

H–T plane, and θ, which varies continuously if we move from the coexistence line around the critical point and back to the coexistence line. Then, it would appear that the thermodynamic functions will be smooth functions of θ, but non-analytic functions of r as $r \to 0$.

We now choose the parametrization

$$H = r^{\beta\delta} h(\theta), \tag{4.8.26}$$

$$\varepsilon = rt(\theta), \tag{4.8.27}$$

which yields

$$\frac{H}{|\varepsilon|^{\beta\delta}} = \frac{h(\theta)}{[t(\theta)]^{\beta\delta}} \tag{4.8.28}$$

and hence, from equation (4.8.25),

$$M = r^{\beta} m(\theta), \tag{4.8.29}$$

where $m(\theta)$ is a regular, but unknown, function.

Use of molecular field theory. The value of such a parametrization becomes quite clear if we use as an example the molecular field theory.

It can then be shown (see Kasteleyn, 1971) that

$$\left(\frac{3\mu^2}{T_c}\right)^{\frac{1}{2}} \theta(1-\theta^2) = m(\theta)\,[1-2\theta^2] + \frac{T_c}{3\mu^2}\,[m(\theta)]^3 \tag{4.8.30}$$

which yields as a solution

$$m(\theta) = \left(\frac{3\mu^2}{T_c}\right)^{\frac{1}{2}} \theta. \tag{4.8.31}$$

Hence, in this case, the magnetization varies linearly with θ, and is, in fact, the "reduced magnetization". Experiments support the linear relation

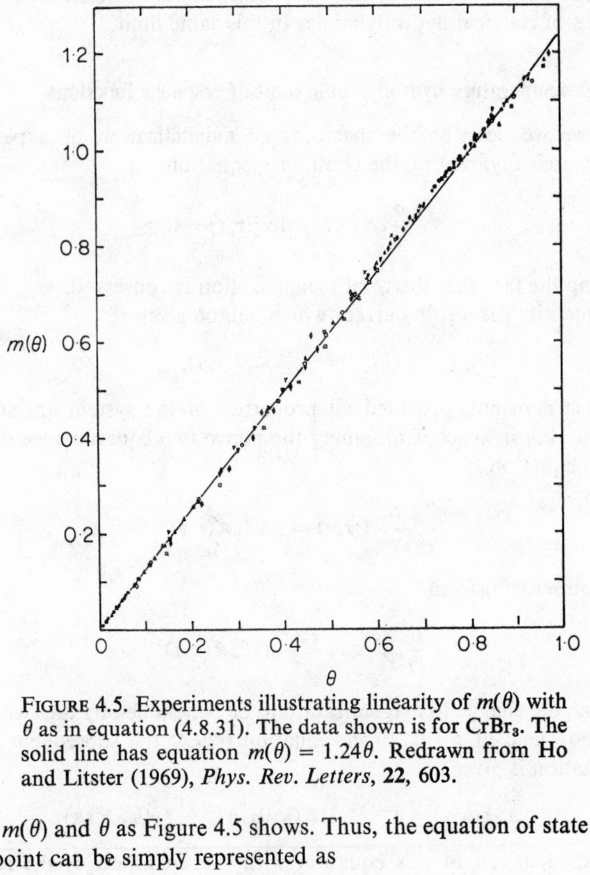

FIGURE 4.5. Experiments illustrating linearity of $m(\theta)$ with θ as in equation (4.8.31). The data shown is for $CrBr_3$. The solid line has equation $m(\theta) = 1.24\theta$. Redrawn from Ho and Litster (1969), *Phys. Rev. Letters*, **22**, 603.

between $m(\theta)$ and θ as Figure 4.5 shows. Thus, the equation of state near the critical point can be simply represented as

$$\left.\begin{aligned} H &= ar^{\beta\delta}\,\theta(1-\theta^2), \\ \varepsilon &= r(1-2\theta^2), \\ M &= gr^{\beta}\,\theta. \end{aligned}\right\} \tag{4.8.32}$$

No fundamental explanation has been offered but there is clearly important progress even at this early stage.

Having discussed critical phenomena in some detail, we shall now tackle the problem of gaining further information about the response functions introduced above. In the long wavelength–low frequency limit, this can be done in macroscopic terms, a situation quite analogous say to the way in which time-dependent correlation functions in liquids must connect with the equations of classical hydrodynamics in this same limit.

4.9 High-temperature hydrodynamic limit of response functions

Suppose we describe the macroscopic magnetization of a paramagnet hydrodynamically, writing the continuity equation

$$\frac{\partial}{\partial t}\mathcal{M}(\mathbf{r},t)+\operatorname{div}\mathbf{j}(\mathbf{r},t)=0, \tag{4.9.1}$$

expressing the fact that the total magnetization is conserved.

The quantity $\mathbf{j}$ is a spin current which will be given by

$$\mathbf{j}(\mathbf{r},t)=-\Lambda\nabla\mathcal{M}(\mathbf{r},t), \tag{4.9.2}$$

Λ being a constant, provided all properties of the system are sufficiently slowly varying in space. Combining the above two equations we obtain the diffusion equation

$$\frac{\partial}{\partial t}\mathcal{M}(\mathbf{r},t)=\Lambda\nabla^2\mathcal{M}(\mathbf{r},t) \tag{4.9.3}$$

or, in Fourier transform,

$$\frac{\partial}{\partial t}\mathcal{M}(\mathbf{k},t)=-\Lambda k^2\mathcal{M}(\mathbf{k},t). \tag{4.9.4}$$

Imagine the change in magnetization to be occasioned by the switching-off of a steady field at $t=0$. From equation (4.9.4) the subsequent decay of magnetization is given by

$$\Delta\mathcal{M}(\mathbf{k},t)=\mathrm{e}^{-\Lambda k^2 t}\Delta\mathcal{M}(\mathbf{k},0)=\mathrm{e}^{-\Lambda k^2 t}\chi(\mathbf{k})\mathcal{H}(\mathbf{k}) \tag{4.9.5}$$

and by comparison of this equation with the definition of the relaxation function in equation (4.6.21) we see that

$$R(\mathbf{k},t)=\chi(\mathbf{k})\,\mathrm{e}^{-\Lambda k^2|t|}, \tag{4.9.6}$$

where we have extended the definition of $R(\mathbf{k},t)$ to negative times.

Then

$$R(\mathbf{k},\omega)=\chi(\mathbf{k})\frac{\Lambda k^2}{\pi(\omega^2+\Lambda^2 k^4)}. \tag{4.9.7}$$

Because we began by assuming $\mathscr{M}(\mathbf{k},t)$ to be slowly varying, this is a small-$\mathbf{k}$ result, as is equation (4.9.6).

4.9.1 Diffusion equation for spin correlation functions

To see what the above implies concerning the spin correlation functions we note from equation (4.6.38) that for small ω

$$R_{\alpha\alpha}(\mathbf{k},\omega)=\beta S'_{\alpha\alpha}(\mathbf{k},\omega) \tag{4.9.8}$$

which, by definition of $S(\mathbf{k},\omega)$, gives for large times the asymptotic relation

$$\langle s_\alpha(-\mathbf{k})\,s_\alpha(\mathbf{k},t)\rangle=\frac{\chi(\mathbf{k})}{\beta g^2\mu^2}\mathrm{e}^{-\Lambda k^2|t|} \tag{4.9.9}$$

for sufficiently small $\mathbf{k}$. Evidently,

$$\frac{\partial}{\partial t}\langle s_\alpha(-\mathbf{k})\,s_\alpha(\mathbf{k},t)\rangle=-\Lambda k^2\langle s_\alpha(-\mathbf{k})\,s_\alpha(\mathbf{k},t)\rangle \tag{4.9.10}$$

and by Fourier inversion

$$\frac{\partial}{\partial t}\langle s_{\mathbf{0}}^\alpha\,s_{\mathbf{R}}^\alpha(t)\rangle=\Lambda\frac{\partial^2}{\partial\mathbf{R}^2}\langle s_{\mathbf{0}}^\alpha\,s_{\mathbf{R}}^\alpha(t)\rangle. \tag{4.9.11}$$

We can interpret this by imagining we put a unit of spin at the origin at $t=0$. If the spin system is disordered, as at high temperatures, one can expect the dominant process to be one in which the unit of spin jumps randomly from atom to neighbouring atom, just as a diffusing atom jumps from site to site. The same equation for the motion, the diffusion equation, should hold at large $\mathbf{r}$ and t, which is just the range of validity already ascribed to equation (4.9.11).

While there are in fact types of motion, other than diffusion, available to carry a spin away from the origin, all coherent motion such as that of spin waves is limited in range by interactions, so that at large enough $\mathbf{r}$ (small $\mathbf{k}$) only diffusion is available. In fact, as Bennett and Martin (1965) have pointed out, equation (4.9.11) can be arrived at in entirely formal fashion as $\mathbf{k},\omega\to 0$. We first note that since equation (4.6.2) shows $\chi_{\alpha\alpha}(\mathbf{k},t)$ to represent the response to a δ-function field it must be "well behaved"; it then follows from the definition (4.6.14) that $\chi_{\alpha\alpha}(\mathbf{k},\omega)$ is analytic in ω except on the real axis.

Hence the function

$$\frac{1}{k^2}\frac{\chi_{\alpha\alpha}(\mathbf{k},\omega)}{\chi_{\alpha\alpha}(\mathbf{k},0)-\chi_{\alpha\alpha}(\mathbf{k},\omega)}$$

is also analytic in ω except on the real axis. It is also even, because $\operatorname{Re}\chi_{\alpha\alpha}(\mathbf{k},\omega)$ is even and $\operatorname{Im}\chi_{\alpha\alpha}(\mathbf{k},\omega)$ is odd, as equation (4.6.18) shows. It now follows, by what amounts to an application of Cauchy's theorem, that we can write

$$\frac{k^{-2}\chi_{\alpha\alpha}(\mathbf{k},\omega)}{\chi_{\alpha\alpha}(\mathbf{k},0)-\chi_{\alpha\alpha}(\mathbf{k},\omega)} = \frac{1}{\pi}\int_{-\infty}^{\infty}\frac{\Lambda(\mathbf{k},\omega')}{\omega'^2-\omega^2}d\omega', \tag{4.9.12}$$

where Λ is some function analytic in ω'. From this equation it follows that

$$1-\frac{\chi_{\alpha\alpha}(\mathbf{k},\omega)}{\chi_{\alpha\alpha}(\mathbf{k},0)} = \left[1+\frac{k^2}{\pi}\int_{-\infty}^{\infty}\frac{\Lambda(\mathbf{k},\omega')}{\omega'^2-\omega^2}d\omega'\right]^{-1} \tag{4.9.13}$$

and taking imaginary parts by using the relation

$$\lim_{\epsilon\to 0}\int_{-\infty}^{\infty}\frac{d\omega'}{\pi}\frac{\Lambda(\mathbf{k},\omega')}{\omega'-(\omega+i\varepsilon)} = \frac{1}{\pi}⨍_{-\infty}^{\infty}d\omega'\frac{\Lambda(\mathbf{k},\omega')}{\omega'-\omega}+i\pi\Lambda(\mathbf{k},\omega) \tag{4.9.14}$$

we find

$$\operatorname{Im}\chi_{\alpha\alpha}(\mathbf{k},\omega) = \chi_{\alpha\alpha}(\mathbf{k})\frac{\Lambda(\mathbf{k},\omega)\,k^2}{\omega(1-k^2P)^2+\Lambda^2(\mathbf{k},\omega)\,k^4/\omega} \tag{4.9.15}$$

or, by equation (4.6.24), if $s(\mathbf{k}) = s(-\mathbf{k})$

$$R_{\alpha\alpha}(\mathbf{k},\omega) = \chi_{\alpha\alpha}(\mathbf{k})\frac{\Lambda(\mathbf{k},\omega)\,k^2}{\omega^2(1-k^2P)^2+\Lambda^2(\mathbf{k},\omega)\,k^4}. \tag{4.9.16}$$

Here P is the principal part of the integral in the right-hand side of equation (4.9.12). By comparison, with equation (4.9.7) we see that

$$\Lambda = \Lambda(0,0) \tag{4.9.17}$$

and processes other than diffusion are responsible for the dependence of Λ on $\mathbf{k}$ and ω.

4.9.2 Calculation of diffusion constant

To obtain an explicit expression for Λ it is convenient to use the auxiliary function F defined through

$$R_{\alpha\beta}(\mathbf{k},t) = \chi_{\alpha\beta}(\mathbf{k})\,F_{\alpha\beta}(\mathbf{k},t). \tag{4.9.18}$$

By the definition of R,

$$R_{\alpha\beta}(\mathbf{k},t) = R_{\beta\alpha}(-\mathbf{k},-t) \tag{4.9.19}$$

and so for F we have

$$F(\mathbf{k}, t) = F(\mathbf{k}, -t), \tag{4.9.20}$$

provided we restrict ourselves to substances for which $s(\mathbf{k}) = s(-\mathbf{k})$. Hence

$$\frac{dF(\mathbf{k}, 0)}{dt} = 0. \tag{4.9.21}$$

Since $R(\mathbf{k}, 0) = \chi(\mathbf{k})$ we also have $F(\mathbf{k}, 0) = 1$ and may therefore write

$$F(t) = 1 + \int_0^t \frac{dF}{dt}\, dt' = 1 + t\frac{dF}{dt}\bigg|_0^t - \int_0^t t' \frac{d^2 F(t')}{dt'^2}\, dt', \tag{4.9.22}$$

that is,

$$F(t) = 1 + \int_0^t (t - t')\frac{d^2 F(t')}{dt'^2}\, dt'. \tag{4.9.23}$$

We now argue that $\ddot{F}$ decays much faster in time than F. We rewrite the definition (4.6.38) as

$$R_{\alpha\beta}(\mathbf{k}, t) = g^2 \mu^2 \{s_\alpha(-\mathbf{k}), s_\beta(\mathbf{k}, t)\}, \tag{4.9.24}$$

where

$$\hbar\{A, B\} = \int_0^\beta \langle e^{\eta H} A\, e^{-\eta H} B\rangle\, d\eta - \beta\langle A\rangle\langle B\rangle. \tag{4.9.25}$$

The proof is simple. We note that $\int_0^\beta e^{-\eta\omega}\, d\eta = (1 - e^{-\beta\omega})/\omega$ so that equation (4.6.38) may be written as

$$\begin{aligned}\hbar R_{\alpha\beta}(\mathbf{k}, t) &= \int_0^\beta d\eta \int_{-\infty}^{\infty} S'_{\alpha\beta}(\mathbf{k}, \omega) \exp\,[i\omega(t + i\eta)]\, d\omega \\ &= g^2 \mu^2 \int_0^\beta d\eta [\langle s_\alpha(-\mathbf{k})\, s_\beta(\mathbf{k}, t + i\eta)\rangle - \langle s_\alpha(-\mathbf{k})\rangle \langle s_\beta(\mathbf{k})\rangle]\end{aligned} \tag{4.9.26}$$

from which equation (4.9.24) follows immediately. It also follows that

$$F_{\alpha\beta}(\mathbf{k}, t) = \frac{\{s_\alpha(-\mathbf{k}), s_\beta(\mathbf{k}, t)\}}{\{s_\alpha(-\mathbf{k}), s_\beta(\mathbf{k})\}}, \tag{4.9.27}$$

whence

$$\ddot{F}_{\alpha\alpha}(\mathbf{k}, t) = -\frac{\{\dot{s}_\alpha(-\mathbf{k}), \dot{s}_\alpha(\mathbf{k}, t)\}}{\{s_\alpha(-\mathbf{k}), s_\alpha(\mathbf{k})\}}. \tag{4.9.28}$$

In contrast to F, therefore, $\ddot{F}$ involves the correlation of microscopic quantities, for temperatures above the critical point, for in the absence of order the velocities $\dot{s}$ will have a correlation time of the order of that in which a spin moves from a neighbouring site by a flip-over motion. The uncertainty relation gives this time as the order of $\hbar/J$. Hence we can rewrite equation (4.9.23) as

$$F(t) \simeq 1 + t \int_0^\infty \frac{d^2F(t')}{dt'^2}\, dt' \tag{4.9.29}$$

for times not too short. Comparison with equation (4.9.6) now gives

$$\Lambda k^2 = -\frac{1}{2}\int_{-\infty}^{\infty} \ddot{F}(\mathbf{k}, t)\, dt. \tag{4.9.30}$$

Since $F(t)$ is even in t,

$$\ddot{F}(t) = \ddot{F}(0) + t^2 \left.\frac{d^4F}{dt^4}\right|_{t=0} + \dots. \tag{4.9.31}$$

It can be shown that

$$\left.\frac{d^4F}{dt^4}\right|_{t=0} < 0.$$

It is therefore consonant with our previous expressions to write

$$\ddot{F}(\mathbf{k}, t) = e^{-gt^2}\ddot{F}(\mathbf{k}, 0) \tag{4.9.32}$$

as an asymptotic expression in t, provided $F(t)$ is reasonably well behaved. Our earlier remarks on the decay of F imply $\sqrt{g^{-1}} \sim \hbar/J$, which we shall later see to be correct. Combination of equations (4.9.30) to (4.9.32) now gives

$$\Lambda k^2 = -\tfrac{1}{2}\ddot{F}(0) \int_{-\infty}^{\infty} e^{-g^2t^2}\, dt = -\frac{\ddot{F}(0)}{2}\sqrt{\frac{\pi}{g}}, \tag{4.9.33}$$

where

$$g = -\frac{1}{2}\left.\frac{d^4F}{dt^4}\right|_{t=0} \Big/ \ddot{F}(0). \tag{4.9.34}$$

This can also be written

$$g = \frac{1}{2}\frac{\{\ddot{s}_\alpha(-\mathbf{k}), \ddot{s}_\alpha(\mathbf{k})\}}{\{\dot{s}_\alpha(-\mathbf{k}), \dot{s}_\alpha(\mathbf{k})\}}. \tag{4.9.35}$$

Like $\{\dot{s}, \dot{s}\}$, $\{\ddot{s}, \ddot{s}\}$ involves correlations of microscopic quantities and will be much less affected by the temperature (through the long-range order) than $\chi = g^2\mu^2\{s, s\}$. In particular no pathological behaviour near the critical temperature T_c is to be expected. We will show this by more detailed examination of $F(0)$. From equation (4.6.38) we have, on using the detailed balance condition:

$$\frac{dR_{\alpha\beta}}{dt} = i\int_{-\infty}^{\infty} S'_{\alpha\beta}(\mathbf{k},\omega)\,(1-\mathrm{e}^{-\beta\omega})\,\mathrm{e}^{i\omega t}\,dt$$

$$= i\int_{-\infty}^{\infty} S'_{\alpha\beta}(\mathbf{k},\omega)\,\mathrm{e}^{i\omega t}\,dt - i\int_{-\infty}^{\infty} S'_{\alpha\beta}(-\mathbf{k},\omega)\,\mathrm{e}^{-i\omega t}\,dt$$

$$= ig^2\mu^2\{\langle s_\alpha(-\mathbf{k})\,s_\beta(\mathbf{k},t)\rangle - \langle s_\beta(\mathbf{k})\,s_\alpha(-\mathbf{k},-t)\rangle\}. \tag{4.9.36}$$

The cyclic property of the trace enables us to put this into the form

$$\hbar\dot{F}_{\alpha\beta}(\mathbf{k},t) = \frac{g^2\mu^2 i}{\chi_{\alpha\beta}(\mathbf{k})}\langle[s_\alpha(-\mathbf{k}), s_\beta(\mathbf{k},t)]\rangle, \tag{4.9.37}$$

whence

$$\hbar\ddot{F}_{\alpha\beta}(\mathbf{k},t) = \frac{g^2\mu^2 i}{\chi_{\alpha\beta}(\mathbf{k})}\langle[s_\alpha(-\mathbf{k}), \dot{s}_\beta(\mathbf{k},t)]\rangle. \tag{4.9.38}$$

Now with a Heisenberg Hamiltonian (cf. equation (4.3.31))

$$\dot{s}_{\mathbf{R}} = -\frac{1}{i\hbar}\left[\mathbf{s}_{\mathbf{R}}, \sum_{\mathbf{R}''\mathbf{R}'} J_{\mathbf{R}'-\mathbf{R}''}\mathbf{s}_{\mathbf{R}'}\,.\,\mathbf{s}_{\mathbf{R}''}\right] = \frac{2}{\hbar}\sum_{\mathbf{R}'} J_{\mathbf{R}'}\,s_{\mathbf{R}}\times s_{\mathbf{R}'+\mathbf{R}}, \tag{4.9.39}$$

that is,

$$\dot{s}^\alpha_{\mathbf{R}} = \frac{2}{\hbar}\sum_{\mathbf{R}'} J_{\mathbf{R}'}(s^\beta_{\mathbf{R}}\,s^\gamma_{\mathbf{R}+\mathbf{R}'} - s^\gamma_{\mathbf{R}}\,s^\beta_{\mathbf{R}+\mathbf{R}'}). \tag{4.9.40}$$

Using the commutation relations between the components of $s_{\mathbf{R}}$ and the results of section 4.7 we thus obtain equation (4.9.38) in the form ($\alpha=\beta$)

$$\ddot{F}_{\alpha\alpha}(\mathbf{k}) = \frac{2g^2\mu^2}{\hbar^2\chi_{\alpha\alpha}(\mathbf{k})}\sum_{\mathbf{R},\mathbf{R}'} J_{\mathbf{R}'}(1-\mathrm{e}^{i\mathbf{k}.\mathbf{R}'})\langle s^\beta_{\mathbf{R}}\,s^\beta_{\mathbf{R}+\mathbf{R}'} + s^\gamma_{\mathbf{R}}\,s^\gamma_{\mathbf{R}+\mathbf{R}'}\rangle. \tag{4.9.41}$$

If $T>T_c$, $\langle s_\alpha s_\alpha\rangle = \langle s_\beta s_\beta\rangle = \langle s_\gamma s_\gamma\rangle$ and so

$$\ddot{F}(\mathbf{k}) = \frac{2g^2\mu^2}{\hbar^2\chi(\mathbf{k})}\,\frac{2}{3}\sum_{\mathbf{R},\mathbf{R}'} J_{\mathbf{R}'}(1-\mathrm{e}^{i\mathbf{k}.\mathbf{R}'})\langle \mathbf{s}_{\mathbf{R}}\cdot\mathbf{s}_{\mathbf{R}+\mathbf{R}'}\rangle, \quad T>T_c, \tag{4.9.42}$$

which becomes, for small k,

$$\ddot{F}(\mathbf{k}) = \frac{2g^2\mu^2}{\hbar^2\chi(\mathbf{k})}\sum_{\mathbf{R}} J_{\mathbf{R}}(\mathbf{k}.\mathbf{R})^2\sum_{\mathbf{k}'}\mathrm{e}^{-i\mathbf{k}'.\mathbf{R}}\langle s_\alpha(-\mathbf{k}')\,s_\alpha(\mathbf{k}')\rangle. \tag{4.9.43}$$

If coupling is permitted between nearest neighbours only we obtain

$$\ddot{F}(\mathbf{k}) = \frac{2}{3\hbar}\frac{k^2 b^2}{\chi(\mathbf{k})}\sum_{\mathbf{k}'} J(-\mathbf{k}')\langle s_\alpha(-\mathbf{k}')\, s_\alpha(\mathbf{k}')\rangle, \tag{4.9.44}$$

b being the nearest-neighbour distance.

To estimate the variation of $F(0)$ with temperature above T_c (taking a ferromagnet to be definite) we use equation (4.7.26) and write

$$N\langle s_\alpha(-\mathbf{k}'), s_\alpha(\mathbf{k}')\rangle = \frac{s(s+1)}{3} N\left[1 - \frac{T_c}{T}\frac{J(\mathbf{k})}{J(0)}\right]. \tag{4.9.45}$$

As $T \to \infty$ equation (4.9.44) gives, since $\sum_{\mathbf{k}'} J(\mathbf{k}') = 0$

$$\frac{\chi(\mathbf{k})}{g^2\mu^2}\ddot{F}(\mathbf{k}) = \{\dot{s}_\alpha(-\mathbf{k}), \dot{s}_\alpha(\mathbf{k})\} = \beta\frac{4N}{27}\frac{J^2}{\hbar^2} Z s^2(s+1)^2 k^2 b^2, \tag{4.9.46}$$

where Z is the number of nearest neighbours. At general temperatures we find

$$\{\dot{s}_\alpha(-\mathbf{k}), \dot{s}_\alpha(\mathbf{k})\} \simeq 2\frac{s(s+1)}{9N\hbar^2} k^2 b^2 J(0)\frac{T}{T_c} w, \tag{4.9.47}$$

where

$$w = -1 + N^{-1}\sum_{\mathbf{k}} \frac{1}{1-(T_c/T)\,[J(\mathbf{k})/J(0)]}. \tag{4.9.48}$$

With nearest neighbour coupling only, w is determined by the structure and is in all cases the order of unity at the Curie point. Since at T_c equation (4.9.44) becomes, with $\beta = 1/k_B T$

$$\{\dot{s}_\alpha(-\mathbf{k}), \dot{s}_\alpha(\mathbf{k})\}_{T_c} \simeq \beta\frac{4wNJ^2}{27\,\hbar^2} Z^2 s^2(s+1)^2 k^2 b^2, \tag{4.9.49}$$

comparison of this with equation (4.9.46) shows $\{\dot{s}, \dot{s}\}$ to increase only by a factor wZ (about 3 or 4) as the temperature is lowered from infinity to the Curie point. A similar result can expected for $\{\ddot{s}, \ddot{s}\}$. Certainly, well above the critical temperature the value of Λ for $T = \infty$ should be a good approximation. The evaluation of g at infinite temperature is elementary but lengthy. Exact results for F^{iv} $(T = \infty)$ are given by Bennett and Martin (1965) but we quote here the approximate results given by Mori and Kawasaki (1962), obtained using a small k expansion, because of their simple form, namely

$$g = \frac{8Z\xi}{3} s(s+1)\left(\frac{J}{\hbar}\right)^2, \quad T = \infty, \tag{4.9.50}$$

with

$$\xi = 1 - \frac{39}{5Z^2}\left[1 + \frac{3}{26s(s+1)}\right]. \qquad (4.9.51)$$

The quantity ξ is close to unity; for example, if $s = 1$ and $Z = 8$, $\xi = 0.87$. From equations (4.9.4), (4.9.5) and (4.9.51) we now obtain (Mori and Kawasaki, 1962)

$$\Lambda = \frac{Jb^2}{\hbar}\sqrt{\left[\frac{\pi 2}{54\xi}s(s+1)\right]}, \quad T \gg T_c. \qquad (4.9.52)$$

An interesting method of calculating Λ, alternative to the above and yielding similar results, has been given by de Gennes (1958). The Gaussian approximation (4.9.6) is not directly used. Instead, de Gennes took explicit recognition of the fact that $F(t) = e^{-\Gamma t}$ must fail at short times because $F(0) = 0$. At large ω, therefore, $F(\omega) = R(\omega)/\chi$ should be greatly diminished below the Lorentzian of equation (4.9.7). A simple way of incorporating this fact into the theory is to use a truncated Lorentzian:

$$F(\mathbf{k}, \omega) = \begin{cases} \dfrac{1}{\pi}\dfrac{\Lambda k^2}{\omega^2 + \Lambda^2 k^4}, & \omega < \omega_c, \\ 0, & \omega > \omega_c. \end{cases} \qquad (4.9.53)$$

where for consistency ω_c must be much greater than the half-width Λk^2. To determine ω_c we use equation (4.9.53) to evaluate the derivatives

$$\frac{d^2 F(\mathbf{k}, 0)}{dt^2} = -\int_{-\infty}^{\infty} \omega^2 F(\mathbf{k}, \omega)\, d\omega, \quad \frac{d^4 F(\mathbf{k}, 0)}{dt^4} = \int_{-\infty}^{\infty} \omega^4 F(\mathbf{k}, \omega)\, d\omega, \qquad (4.9.54)$$

whereupon we find, provided $\omega_c \gg \Lambda k^2$,

$$\omega_c^2 = -3\frac{d^4 F(0)}{dt^0} \bigg/ \frac{d^2 F(0)}{dt^2} \qquad (4.9.55)$$

and

$$\Lambda^2 k^4 = -\frac{\pi^2}{12}\left(\frac{d^2 F(0)}{dt^2}\right)^3 \bigg/ \frac{d^4 F(0)}{dt^4}, \qquad (4.9.56)$$

a result similar to equation (4.9.33). A comparison of the high temperature results of Mori and Kawasaki and of de Gennes with experiment is made in Chapter 5.

4.9.3 Limits of validity

A few remarks on the consistency of some of the approximations and results above may conveniently be made at this point. First, equation (4.9.50) shows that $g^{-\frac{1}{2}} \sim \hbar/J$ so that the expression (4.9.4) is in agreement with our conclusion following equation (4.9.32). Secondly, for times less than the order of $\hbar/J$ the approximation of equation (4.9.5) cannot be made and the result $F(t) = \mathrm{e}^{-\Gamma t}$ must be invalid. We must therefore also expect the cut-off frequency of equation (4.9.55) to be roughly given by $\omega_c \sim J/\hbar$ and in fact equations (4.9.50) and (4.9.55) do give $\omega_c^{-1} \sim g^{-\frac{1}{2}} \sim \hbar/J$. Lastly, as mentioned previously, $\tau = \hbar/J$ is roughly the time in which a spin flips from one site to another. One ought therefore to have a diffusion coefficient given by

$$\Lambda = \frac{(\Delta x)^2}{\Delta t} \sim \frac{b^2}{\tau_c} \sim \frac{b^2 J}{\hbar}, \tag{4.9.57}$$

which is in fact in agreement with the result (4.9.56).

4.9.4 Critical diffusion

Let us now turn to a brief examination of critical diffusion, that is, diffusion when $T \simeq T_c$. Equations (4.9.28) and (4.9.33) indicate the small $\mathbf{k}$-dependence of $\Lambda(\mathbf{k})$ [a generalized diffusion parameter such as appears in equation (4.9.16)] to be entirely contained in $\chi^{-1}(\mathbf{k})$. Further, we have already discussed the temperature variations of $\{\dot{s}, \dot{s}\}$ and $\{\ddot{s}, \ddot{s}\}$ down from infinity to the critical point and concluded they show no dramatic changes. This leads us to suppose the essential $\mathbf{k}$ and T dependence of the diffusion parameter to be contained in

$$\Lambda_T \simeq \Lambda_\infty \frac{\chi_0}{\chi_T(\mathbf{k})}, \tag{4.9.58}$$

where Λ_∞ is the diffusion coefficient at $T = \infty$. This equation immediately tells us the diffusion coefficient goes to zero as $T \to T_c$, since $\chi \to \infty$ as $T \to T_c$ and $\mathbf{k} \to 0$. This is the basis of our earlier remark that $\chi(\mathbf{k}\omega)$ must narrow in $\mathbf{k}$ and ω as T_c is approached [cf. equations (4.9.4) and (4.9.5)]. In passing we note that this approach to zero by Λ has been termed "thermodynamic braking" by de Gennes (1957), who used the thermodynamics of irreversible processes in his discussion. From equation (4.9.11) we see the rate of change of $\langle s_\alpha(0)\, s_\alpha(\mathbf{r}, t)\rangle$ slows down as T_c is approached, this being associated with the influence of temperature on the spin dynamics. In contrast, de Gennes termed the slowing down of the rate of change of $\langle s_\alpha(-\mathbf{k})\, s_\alpha(\mathbf{k}, t)\rangle$ as $\mathbf{k} \to 0$ [see equation (4.9.10)], "kinematic braking". This latter slowing down is associated not with changes in the spin dynamics but rather with the fact that as $\mathbf{k} \to 0$ we are concerned with a constant of the motion.

We shall see in the next chapter how neutron diffraction studies confirm the prediction that $\Lambda = 0$ at $T = T_c$. On the other hand, the **k**-dependence of $\Lambda(\mathbf{k})$ is not correctly given by equation (4.9.58). Indeed, it has been shown by Kawasaki (1967) that such a **k**-dependence involves a contradiction in the theory. Let us insert the time-dependent form of equation (4.9.44) into (4.9.30). We obtain, to lowest order in **k**,

$$\Lambda k^2 = -\frac{g^2\mu^2}{\hbar^2 N}\frac{1}{\chi(\mathbf{k})}\sum_{\mathbf{R}} J_{\mathbf{R}}(\mathbf{k}.\mathbf{R})^2 \sum_{\mathbf{k}'} e^{-i\mathbf{k}'.\mathbf{R}} \int_{-\infty}^{\infty} \langle s_\alpha(-\mathbf{k}')\, s_\alpha(\mathbf{k}',t)\rangle\, dt. \tag{4.9.59}$$

Inserting the result (4.9.49) for $\langle s_\alpha(-\mathbf{k}')\, s_\alpha(\mathbf{k}',t)\rangle$, we have

$$\Lambda k^2 = \frac{2}{\hbar^2 N}\frac{k_B T}{\chi(\mathbf{k})}\sum_{\mathbf{R}} J_{\mathbf{R}}(\mathbf{k}.\mathbf{R})^2 \sum_{\mathbf{k}'} \frac{e^{-i\mathbf{k}'.\mathbf{R}}}{\Lambda k'^2}\chi(\mathbf{k}'), \tag{4.9.60}$$

or, if equation (4.9.58) is used for Λ on the right,

$$\Lambda k^2 = \frac{2}{\hbar}(\chi_0\,\chi(\mathbf{k})\,\Lambda_\infty)^{-1}\sum_{\mathbf{R}} J_{\mathbf{R}}(\mathbf{k}.\mathbf{R})^2 \sum_{\mathbf{k}'} \frac{e^{-i\mathbf{k}'.\mathbf{R}}}{k'^2}\chi^2(\mathbf{k}'). \tag{4.9.61}$$

If $\mathbf{k}'$ is small, $\chi(\mathbf{k}')$ is given by equation (4.7.23) and the sum over $\mathbf{k}'$ about the origin is $\sum_{\mathbf{k}'} [\chi_c^2/r_1^2(\kappa_1^2+k^2)]$, which diverges with κ_1^{-3} as the Curie temperature is approached. This internal inconsistency in the theory leads one to suspect pathological behaviour of $\chi(\mathbf{k})\ddot{F}(\mathbf{k},t)$ or, equivalently, $\{\dot{s}_\alpha(-\mathbf{k}), \dot{s}_\alpha(\mathbf{k},t)\}$ in the critical region so that the simple form (4.9.5) cannot be expected to be quantitative there.

Improved arguments have been given by Halperin and Hohenberg (1969). They formally generalize the theory by introducing, instead of Λq^2 in the hydrodynamic form, a characteristic frequency $\Gamma(q)$. This is defined precisely by

$$\int_{-\Gamma}^{\Gamma} F(q\omega)\, d\omega = \tfrac{1}{2}. \tag{4.9.62}$$

This depends on the correlation length κ_1^{-1}, and they find $\Gamma(q) \propto q^{\frac{5}{2}}$ in the limit $q \gg \kappa_1$, which is the critical region. Outside this region they write

$$\Gamma(q,\kappa_1) = \text{constant}\; q^{\frac{5}{2}} f\!\left(\frac{\kappa_1}{q}\right). \tag{4.9.63}$$

In the limit $q \ll \kappa_1$, $f(\kappa_1/q) \propto (\kappa_1/q)^{\frac{1}{2}}$ and to see when thermodynamic slowing down occurs, we must find how large κ_1/q has to be before this last form is valid. Résibois and Piette (1970) estimate $\kappa_1/q \gtrsim 2$ for this asymptotic form to hold. However, they use the kinetic equations of de Leener and Résibois, correct in the molecular field limit. Thus, since the critical indices in equilibrium situations are not correctly given, further work is still needed here.

4.10 Linear response functions in spin-wave regime

4.10.1 Equations of motion

We now return to the spin-wave problem, basing our discussion on the formalism constructed in the previous sections. We have seen that the spin-wave frequency will be the precession frequency of the spins, and we accordingly turn to further examination of the equations of motion for $s^{\pm}$, which in the spin-wave approximation obey

$$\dot{s}^{\pm}(\mathbf{k}) = i \mp \omega_{\mathbf{k}}\, s^{\pm}(\mathbf{k}), \tag{4.10.1}$$

where $\omega_{\mathbf{k}}$ is the spin-wave frequency. This follows on using the approximation

$$s^{+}(\mathbf{k}) \simeq \sqrt{(2Ns)}\, a_{\mathbf{k}}, \tag{4.10.2}$$

where $a_{\mathbf{k}}$ is the magnon annihilation operator introduced in section 4.3, the commutator of which we take with the Hamiltonian

$$H_0 = \sum_{\mathbf{k}} \hbar\omega_{\mathbf{k}}(a_{\mathbf{k}}^{\dagger} a_{\mathbf{k}} + \tfrac{1}{2}). \tag{4.10.3}$$

On the other hand, equation (4.3.31) gives the exact equation of motion

$$\dot{s}^{\pm}(\mathbf{k}) = \mp \frac{2i}{N} \sum_{\mathbf{q}} [J(\mathbf{q}) - J(\mathbf{k}-\mathbf{q})]\, s_z(\mathbf{q})\, s^{\pm}(\mathbf{k}-\mathbf{q}) \tag{4.10.4}$$

which can be rearranged to read

$$\begin{aligned}\dot{s}^{\pm}(\mathbf{k}) = {} & \mp 2i\langle s_z\rangle\, [J(0) - J(\mathbf{k})]\, s^{\pm}(\mathbf{k}) \\ & \mp \sum_{\mathbf{q}} [J(\mathbf{q}) - J(\mathbf{k}-\mathbf{q})]\, [s_z(\mathbf{q}) - \langle s_z(\mathbf{q})\rangle]\, s^{\pm}(\mathbf{k}-\mathbf{q}).\end{aligned} \tag{4.10.5}$$

Borrowing the language of the classical picture, we can say the second term on the right-hand side represents the torque arising from fluctuations in the z-component about the equilibrium value. If we neglect these fluctuations equation (4.10.1) is obtained with $\omega_{\mathbf{k}}$ given by

$$\omega_{\mathbf{k}}^{0} = 2\langle s_z\rangle\, [J(0) - J(\mathbf{k})], \tag{4.10.6}$$

which is the dispersion law for spin waves already obtained in section 4.3; it is now evident that it holds classically also.

4.10.2 Macroscopic treatment of spin waves

It is possible to arrive at the idea of spin waves in the long wavelength limit from macroscopic considerations. Suppose at time $t = 0$ the deviation from full equilibrium $\mathscr{M}(\mathbf{k}, 0)$ exists, as a result of either spontaneous fluctuation or the application of a field which has since been switched off. Now at $\mathbf{k} = 0$,

$\mathcal{M}$ is a constant of the motion, from which it follows that, for small $\mathbf{k}$, $\mathcal{M}(\mathbf{k}, t)$ will develop very slowly in time. On the other hand, microscopic processes will cause any small subregion to attain approximate internal equilibrium in a time $\sim \hbar/J$. We can therefore state that after this initial transient the system will to a large extent have lost the memory of how its magnetic state came to be prepared, and to a good approximation $(d/dt)\mathcal{M}(\mathbf{k}, t)$ will be determined entirely by $\mathcal{M}(\mathbf{k}, t)$—any possible variation in microscopic structure which does not affect the value of $\mathcal{M}(\mathbf{k}, t)$ will not affect $(d/dt)\mathcal{M}(\mathbf{k}, t)$ either. Now from equations (4.6.21) and (4.9.18) we know that if a previously steady field is switched off at $t = 0$

$$\mathcal{M}(\mathbf{k}, t) = F(\mathbf{k}, t)\mathcal{M}(\mathbf{k}, 0), \tag{4.10.7}$$

where, since only scalars appear in this equation, $\mathcal{M}$ must represent $\mathcal{M}_z$, $\mathcal{M}^+$ or $\mathcal{M}^-$, when F represents F_z, F^{+-} or F^{-+}. Since we have argued that the temporal development of the magnetization is independent of the preparation of the initial state after an initial transient lasting a time the order of $\hbar/J$, equation (4.10.7) must hold for any time, by which we mean

$$\mathcal{M}(\mathbf{k}, t+t_0) = F(\mathbf{k}, t_0)\mathcal{M}(\mathbf{k}, t_0), \tag{4.10.8}$$

which implies

$$F(\mathbf{k}, t+t_0) = F(\mathbf{k}, t_0)F(\mathbf{k}, t). \tag{4.10.9}$$

F must therefore be exponential in form:

$$F(\mathbf{k}, t) = e^{\alpha_{\mathbf{k}} t} \tag{4.10.10}$$

with

$$\alpha_{\mathbf{k}} = \dot{F}(\mathbf{k}, 0). \tag{4.10.11}$$

This is not precisely correct, for equation (4.10.8) only holds long after the initial transient, but if $\mathcal{M}$, and so F, is very slowly varying, it will be a good approximation, at low temperatures at least, where thermal fluctuations are minimal. It now follows that we can write

$$\frac{d\mathcal{M}^{\pm}(\mathbf{k}, t)}{dt} = \mp i\omega_{\mathbf{k}}\mathcal{M}^{\pm}(\mathbf{k}, t) \tag{4.10.12}$$

and

$$\frac{d\mathcal{M}_z(\mathbf{k}, t)}{dt} = 0, \tag{4.10.13}$$

where

$$i\omega_{\mathbf{k}} = \pm\alpha_{\mathbf{k}} = \frac{\{s^+(-\mathbf{k}), \dot{s}^-(\mathbf{k})\}}{\{s^+(\mathbf{k}), s^-(\mathbf{k})\}} = \dot{F}^{+-}(0). \tag{4.10.14}$$

From this expression it is easily shown that $\omega_{\mathbf{k}}$ is real. Equations (4.10.1) and (4.10.14) are obviously consistent with one another, and equation (4.10.12) describes the macroscopic manifestation of a spin wave.

Equation (4.10.10) gives

$$F^{+-}(\mathbf{k}, t) = e^{i\omega_{\mathbf{k}}t}, \quad F^{-+}(\mathbf{k}, t) = e^{-i\omega_{\mathbf{k}}t} \tag{4.10.15}$$

and

$$F_{zz}(\mathbf{k}, t) = 1. \tag{4.10.16}$$

It should also be noted that using the relation

$$F_{xx}(\mathbf{k}, t) = \tfrac{1}{2}\{F^{+-}(\mathbf{k}, t) + F^{-+}(\mathbf{k}, t)\} \tag{4.10.17}$$

equation (4.10.1) is equivalent to

$$F_{xx}(\mathbf{k}, t) = \cos \omega_{\mathbf{k}} t \tag{4.10.18}$$

and

$$F_{xx}(\mathbf{k}, \omega) = \tfrac{1}{2}[\delta(\omega - \omega_{\mathbf{k}}) + \delta(\omega + \omega_{\mathbf{k}})] \tag{4.10.19}$$

so that by equation (4.9.24) the spin waves give rise to delta functions in the frequency spectrum $S_{xx}(\mathbf{k}, \omega)$. This quantity is directly observable by neutron diffraction, as we shall see in the next chapter.

Equations (4.10.18) and (4.10.19) are of course completely consistent with the content of section 4.3, as can be seen by inserting (4.3.1) for $s^{\pm}(\mathbf{k})$ in the expression for $F^{+-}(\mathbf{k}, t)$ and $F^{-+}(\mathbf{k}, t)$, but, more generally, equation (4.10.1) includes a correction to $\omega_{\mathbf{k}}^0$ due to the fact that any spin wave moves in a background of other long wavelength spin waves. The resultant correction to order T^4 is given in Problem 4.5.

Since equation (4.10.10) strictly holds only after an initial transient depending on the preparation of $\mathscr{M}(\mathbf{k}, 0)$ we should write $\alpha_{\mathbf{k}} = \dot{F}(t)/F(t)$ rather than equation (4.10.11). Thus microscopic processes, which govern the motion during the transient, have been ignored. Before investigating consequences of the microscopic part of the motion, however, we shall calculate the time-independent susceptibility, which does not depend on the microscopic part.

4.10.3 Static field susceptibility

An exact expression for the perpendicular susceptibility in terms of the macroscopic spin-wave frequencies defined in equation (4.10.18) is easily written down. We have

$$\chi^{+-}(\mathbf{k}) = g^2 \mu^2 \{s^+(-\mathbf{k}), s^-(\mathbf{k})\} = \frac{g^2 \mu^2}{i\omega_{\mathbf{k}}} \{s^+(-\mathbf{k}), \dot{s}^-(\mathbf{k})\}, \tag{4.10.20}$$

that is,

$$\chi^{+-}(\mathbf{k}) = \dot{R}^{+-}(\mathbf{k}, 0)/i\omega_{\mathbf{k}}. \tag{4.10.21}$$

This becomes, on using equation (4.9.37)

$$\chi^{+-}(\mathbf{k}) = \frac{g^2 \mu^2}{\omega_{\mathbf{k}}} \langle [s^+(-\mathbf{k}), s^-(\mathbf{k})] \rangle \tag{4.10.22}$$

or, since

$$[s^+(\mathbf{k}'), s^-(\mathbf{k})] = 2s_z(\mathbf{k}+\mathbf{k}'), \tag{4.10.23}$$

$$\chi^{+-}(\mathbf{k}) = \frac{2g^2 \mu^2}{\omega_{\mathbf{k}}} \langle s_z(0) \rangle. \tag{4.10.24}$$

Calculation of $\chi_{zz}(\mathbf{k})$ requires care, as we shall see that in a sense it cannot be defined in the spin-wave region. We use the spin-wave expression.

$$s_z(\mathbf{k}) = Ns\delta_{\mathbf{k},0} - \sum_{\mathbf{q}} a_{\mathbf{q}}^\dagger a_{\mathbf{k}+\mathbf{q}} \tag{4.10.25}$$

to obtain, for $k \neq 0$,

$$\chi_{zz}(\mathbf{k}) = \frac{g^2 \mu^2}{N} \{s_z(-\mathbf{k}), s_z(\mathbf{k})\} = \frac{g^2 \mu^2}{N} \sum_{\mathbf{q},\mathbf{q}'} \{a_{\mathbf{k}+\mathbf{q}}^\dagger a_{\mathbf{q}}, a_{\mathbf{q}'}^\dagger a_{\mathbf{k}+\mathbf{q}'}\}. \tag{4.10.26}$$

For free spin waves

$$\begin{aligned}
\{a_{\mathbf{k}+\mathbf{q}}^\dagger a_{\mathbf{q}}, a_{\mathbf{q}'}^\dagger a_{\mathbf{k}+\mathbf{q}'}\} &= \int_0^\beta \langle a_{\mathbf{k}+\mathbf{q}}^\dagger a_{\mathbf{q}} \, \mathrm{e}^{-\eta H_0} a_{\mathbf{q}'}^\dagger a_{\mathbf{k}+\mathbf{q}'} \, \mathrm{e}^{\eta H_0} \rangle \, d\eta \\
&= \langle a_{\mathbf{k}+\mathbf{q}}^\dagger a_{\mathbf{q}} a_{\mathbf{q}'}^\dagger a_{\mathbf{k}+\mathbf{q}'} \rangle \int_0^\infty \exp[\eta(\omega_{\mathbf{k}+\mathbf{q}'}^0 - \omega_{\mathbf{q}'}^0)] \, d\eta \\
&= \langle a_{\mathbf{k}+\mathbf{q}}^\dagger a_{\mathbf{q}} a_{\mathbf{q}}^\dagger a_{\mathbf{k}+\mathbf{q}} \rangle \, \delta_{\mathbf{q}\mathbf{q}'} \{\exp[\beta(\omega_{\mathbf{q}+\mathbf{k}}^0 - \omega_{\mathbf{q}}^0)] - 1\}/(\omega_{\mathbf{k}+\mathbf{q}}^0 - \omega_{\mathbf{q}}^0).
\end{aligned} \tag{4.10.27}$$

Now it can be shown that‡

$$\langle a_{\mathbf{k}+\mathbf{q}}^\dagger a_{\mathbf{q}} a_{\mathbf{q}}^\dagger a_{\mathbf{k}+\mathbf{q}} \rangle = N_{\mathbf{k}+\mathbf{q}}(N_{\mathbf{q}}+1) \tag{4.10.28}$$

with

$$N_{\mathbf{q}}^{-1} = \exp[\beta(\omega_{\mathbf{k}}^0 + g\mu\mathscr{H})] - 1, \tag{4.10.29}$$

where we have included the effect of an external magnetic field on the magnon energy levels, for a reason which will soon become evident.

We take values of $\mathbf{k}$ for which $\omega_{\mathbf{k}}^0 \ll k_B T$ (which implies excitation of the magnons in considerable number). The important contributions to equation

‡ A simple way of evaluating such expressions is given in Chapter 6.

(4.10.27) will now occur for values of $\omega_{\mathbf{k}}^0$ less than $\sim k_B T$. Hence, since in section 4.3 we showed $\omega_{\mathbf{k}}^0 = \alpha k^2$ as $k \to 0$ we find

$$\omega_{\mathbf{k}+\mathbf{q}}^0 - \omega_{\mathbf{q}}^0 \ll k_B T \tag{4.10.30}$$

and equation (4.10.28) is then well approximated by

$$\{a_{\mathbf{k}+\mathbf{q}}^\dagger a_{\mathbf{q}}, a_{\mathbf{q}'}^\dagger a_{\mathbf{k}+\mathbf{q}'}\} = \delta_{\mathbf{q}\mathbf{q}'} \beta N_{\mathbf{k}+\mathbf{q}}(N_{\mathbf{q}}+1) \tag{4.10.31}$$

so that

$$\chi_{zz}(\mathbf{k}) = \frac{g^2 \mu^2}{N} \beta \sum_{\mathbf{q}} N_{\mathbf{k}+\mathbf{q}}(N_{\mathbf{q}}+1). \tag{4.10.32}$$

If $\mathscr{H} = 0$ and $\mathbf{k} \to 0$ the sum on the right-hand side diverges. However, if $\omega_{\mathbf{k}}^0$, $g\mu\mathscr{H} \ll k_B T$ we obtain the approximate form

$$\chi_{zz}(k) = \frac{\mathscr{V} g^2 \mu^2}{N 8\pi^2} \frac{k_B T}{\alpha^2 k} f\left(\frac{g\mu\mathscr{H}}{\omega_{\mathbf{k}}^0}\right), \tag{4.10.33}$$

where α is defined through

$$\omega_{\mathbf{k}}^0 = \alpha k^2 \tag{4.10.34}$$

and has been calculated in section 4.3, $\mathscr{V}$ is the volume, and

$$f(x) = \int_0^\infty \frac{t}{t^2+x} \ln \frac{(t+1)^2+x}{(t-1)^2+x}\, dt = \begin{cases} \dfrac{\pi}{\sqrt{x}}, & x \gg 1, \\ \pi^2, & x \ll 1. \end{cases} \tag{4.10.35}$$

We can therefore write the two limiting expressions

$$\chi_{zz}(\mathbf{k}) = \frac{\mathscr{V}}{N} \left(\frac{g\mu}{\alpha}\right)^{\frac{3}{2}} \frac{k_B T}{8\pi\sqrt{\mathscr{H}}} \quad (\omega_{\mathbf{k}}^0 \ll g\mu\mathscr{H} \ll k_B T), \tag{4.10.36}$$

$$\chi_{zz}(\mathbf{k}) = \frac{\mathscr{V}}{N} \left(\frac{g\mu}{\alpha}\right)^2 \frac{k_B T}{8k} \quad (g\mu\mathscr{H} \ll \omega_{\mathbf{k}}^0 \ll k_B T). \tag{4.10.37}$$

Equation (4.10.36) shows the magnetization to vary as $\mathscr{H}^{\frac{1}{2}}$, but in equation (4.10.37) we have an expression independent of $\mathscr{H}$ for negligible fields, with which we shall run into no trouble so long as we avoid setting $\mathbf{k} = 0$.

4.10.4 Diffusion in spin-wave region

Although in section 4.9 we gave physical justification for spin diffusion only in the high-temperature regime, we shall now show that the microscopic processes so far neglected in this section give rise to diffusion in the spin-wave region.

The discussion of F from equations (4.9.27) to (4.9.30) is applicable to $F_{zz}(k, t)$ in the low temperature region, and equation (4.10.16) is more accurately replaced by the asymptotic form

$$F_{zz}(\mathbf{k}, t) \simeq \mathrm{e}^{-\Gamma} \tag{4.10.38}$$

with

$$\Gamma = \tfrac{1}{2}\{s_z(-\mathbf{k}), s_z(\mathbf{k})\}^{-1} \int_{-\infty}^{\infty} \{\dot{s}_z(-\mathbf{k}), \dot{s}_z(\mathbf{k}, t)\, dt\} \tag{4.10.39}$$

where equation (4.9.27) has been used to define F.

Now the proof of the fluctuation dissipation theorem as expressed in equation (4.6.38) is independent of the particular nature of the operators involved in the definitions (4.6.27) and (4.6.29) (for a general concise proof see Chapter 6). We can therefore also write

$$\int_{-\infty}^{\infty} \mathrm{e}^{i\omega t}\{\dot{s}_\alpha(-\mathbf{k}), \dot{s}_\alpha(\mathbf{k}, t)\}\, dt = \frac{1-\mathrm{e}^{-\beta\omega}}{\hbar\omega} \int_{-\infty}^{\infty} \mathrm{e}^{i\omega t} \times [\langle \dot{s}_\alpha(-\mathbf{k})\, \dot{s}_\alpha(\mathbf{k}, t)\rangle - \langle \dot{s}_\alpha(-\mathbf{k})\rangle \langle \dot{s}_\alpha(\mathbf{k})\rangle]\, dt. \tag{4.10.40}$$

On taking the limit $\omega \to 0$ in this equation one can see that equation (4.10.39) can be written as

$$\Gamma(\mathbf{k}) = \frac{1}{2k_{\mathrm{B}}T}\{s_z(-\mathbf{k}), s_z(\mathbf{k})\}^{-1} \int_{-\infty}^{\infty} \langle \dot{s}_z(-\mathbf{k})\, \dot{s}_z(\mathbf{k}t)\rangle\, dt, \quad (\mathbf{k} \neq 0). \tag{4.10.41}$$

From equation (4.10.4) it can be shown that

$$\dot{s}_z(\mathbf{k}) = \frac{1}{iN} \sum_{\mathbf{q}} [J(\mathbf{q}) - J(\mathbf{k}-\mathbf{q})]\, s^+(\mathbf{q})\, s^-(\mathbf{k}-\mathbf{q}), \tag{4.10.42}$$

which becomes, on using the spin-wave approximation of equation (4.10.2),

$$\dot{s}_z(\mathbf{k}) = -2is \sum_{\mathbf{q}} [J(\mathbf{q}) - J(\mathbf{k}-\mathbf{q})]\, a_{\mathbf{q}}\, a^\dagger_{\mathbf{k}+\mathbf{q}}. \tag{4.10.43}$$

Equation (4.10.41) can now therefore be approximated by

$$\Gamma(\mathbf{k}) = \frac{\pi(2s)^2 g^2 \mu^2}{Nk_{\mathrm{B}}T\chi_{zz}(\mathbf{k})} \sum_{\mathbf{q}\mathbf{q}'} [J(\mathbf{q}) - J(\mathbf{k}-\mathbf{q})][J(\mathbf{q}') - J(\mathbf{k}-\mathbf{q}')] \times \langle a_{\mathbf{q}}\, a^\dagger_{\mathbf{q}-\mathbf{k}}\, a_{\mathbf{q}'-\mathbf{k}}\, a^\dagger_{\mathbf{q}'}\rangle\, \delta(\omega^0_{\mathbf{q}} - \omega^0_{\mathbf{q}'+\mathbf{k}}). \tag{4.10.44}$$

In the independent spin-wave approximation

$$\langle a_{\mathbf{q}}\, a^\dagger_{\mathbf{q}-\mathbf{k}}\, a_{\mathbf{q}'-\mathbf{k}}\, a_{\mathbf{q}'}\rangle = \delta_{\mathbf{q}\mathbf{q}'}\, N_{\mathbf{q}-\mathbf{k}}(N_{\mathbf{q}}+1) \tag{4.10.45}$$

so that

$$\Gamma(\mathbf{k}) = \frac{4\pi s^2 g^2 \mu^2}{Nk_B T\chi_{zz}(\mathbf{k})} \sum_{\mathbf{q}} [J(\mathbf{q}) - J(\mathbf{k}-\mathbf{q})]^2 N_{\mathbf{q}}(N_{\mathbf{k}+\mathbf{q}}+1)\,\delta(\omega^0_{\mathbf{q}} - \omega^0_{\mathbf{k}-\mathbf{q}}). \tag{4.10.46}$$

In the long-wavelength limit we put

$$J(0) - J(\mathbf{q}) \simeq \gamma q^2, \quad \gamma = \frac{Zb^2}{6} J, \tag{4.10.47}$$

in which case as $k \to 0$

$$J(\mathbf{q}) - J(\mathbf{k}-\mathbf{q}) \simeq 2\gamma \mathbf{k}.\mathbf{q}. \tag{4.10.48}$$

Now

$$\sum (\mathbf{k}.\mathbf{q})^2\, \delta(\mathbf{k}^2 - 2\mathbf{k}.\mathbf{q}) N_{\mathbf{q}}(N_{\mathbf{k}-\mathbf{q}}+1) = \frac{k\mathscr{V}}{4\pi^2} \int_{k/2}^{\infty} dq \int_0^1 d\xi\, \delta(k - 2q\xi) q^4 \xi^2 N_{\mathbf{q}}(N_{\mathbf{q}}+1)$$

$$= \frac{k^3 \mathscr{V}}{64\pi^2} \left\{ \exp\left[\beta\left(g\mu\mathscr{H} + \frac{\alpha k^4}{4}\right)\right] - 1 \right\}^{-1} \tag{4.10.49}$$

and so we finally reach an explicit expression for Γ, viz.

$$\Gamma(\mathbf{k}) = \frac{\eta g^2 \mu^2}{\chi_{zz}(\mathbf{k})} \frac{k^3}{\exp\{\beta[g\mu\mathscr{H} + (\alpha k^4/4)]\} - 1}, \tag{4.10.50}$$

where

$$\eta = \frac{\mathscr{V} s^2 \gamma^2}{N 4\pi\alpha^2} \simeq \frac{1}{16\pi} \frac{\mathscr{V}}{N}. \tag{4.10.51}$$

In the negligible field limit equations (4.10.50) and (4.10.51) give

$$\Gamma(\mathbf{k}) = \frac{2\alpha k^2}{\pi}. \tag{4.10.52}$$

Thus equation (4.9.6) holds for R_{zz}, with

$$\Lambda = \frac{2\alpha}{\pi} = \frac{2Z}{3\pi} b^2 s \left(\frac{J}{\hbar}\right); \tag{4.10.53}$$

a result due to Mori and Kawasaki (1962). Equation (4.9.5) shows that

$$\frac{d\mathscr{M}_z}{dt}(\mathbf{k}, t) = -\Lambda k_z^2 \mathscr{M}_z(\mathbf{k}, t), \tag{4.10.54}$$

the diffusion equation in **k**-space. This means that $\langle s_0^z s_R^z(t)\rangle$ also obeys the diffusion equation when **R** and t become very large, as we saw in equation (4.9.11).

We can also treat amplitude damping of spin waves. The statements made after equation (4.9.28) are no longer true of s_x and s_y or, equivalently, $s^{\pm}$, however. The microscopic quantities are rather $\dot{s}^{\pm} \pm i\omega_{\mathbf{k}} s^{\pm}$, representing the irregular fluctuations of the spin about a steady precessional motion. Our earlier arguments will hold, nevertheless, for the function

$$f^{+-}(\mathbf{k}, t) = \mathrm{e}^{-i\omega_{\mathbf{k}} t} F^{+-}(\mathbf{k}, t) \tag{4.10.55}$$

for which $df^{+-}/dt = 0$. Corresponding to equation (4.10.55) one finds for large times,

$$f^{+-}(\mathbf{k}, t) \simeq \mathrm{e}^{-\Gamma^{+-} t}, \tag{4.10.56}$$

where Γ^{+-} is now complex. The imaginary part is a frequency shift to be added to $\omega_{\mathbf{k}}$ as defined by equation (4.10.1). It is the shifted frequency that will appear in $S_{xx}(\mathbf{k}, \omega)$ as revealed by neutron diffraction.

Having discussed response functions in Heisenberg spin systems in considerable detail, we shall now begin the discussion of itinerant electron susceptibilities $\chi(\mathbf{k}\omega)$. Here, of course, the object must be to introduce the band structure into the calculation and then, later to deal with electron–electron interactions.

4.11 Frequency and wave-number-dependent susceptibility of non-interacting Bloch electrons

4.11.1 Calculation of induced current density

The method we shall employ in this section (cf. Hebborn and March, 1970) is to calculate the induced current density via the single particle density matrix ρ. As we have seen, this satisfies the Liouville equation

$$i\hbar \frac{\partial \rho}{\partial t} = [H, \rho]. \tag{4.11.1}$$

The Hamiltonian we shall use is that for Bloch electrons in a magnetic field $\mathscr{B}$, with corresponding vector potential $\mathscr{A}$, namely

$$H = \frac{1}{2m}\left(\mathbf{p} + \frac{e\mathscr{A}}{c}\right)^2 + V(\mathbf{r}) + \frac{e\hbar}{2mc}\boldsymbol{\sigma}.\mathscr{B}, \tag{4.11.2}$$

$V(\mathbf{r})$ being the periodic lattice potential. This will be assumed here to be independent of magnetic field, though the effect of including magnetic field dependence in $V(\mathbf{r})$ may well be interesting, as we discuss briefly in Chapter 10.

As indicated above, the theory is conveniently developed in terms of the Bloch eigenfunctions of the electrons in zero magnetic field. In this section, it will be notationally convenient to write explicitly

$$|\mathbf{k}\mu\rangle = u_{\mathbf{k}\mu}(\mathbf{r})\,e^{i\mathbf{k}.\mathbf{r}} = \psi_{\mathbf{k}\mu}(\mathbf{r}), \tag{4.11.3}$$

where the functions $u_{\mathbf{k}\mu}$ are of course periodic with the period of the lattice and the corresponding eigenvalues are denoted by $E_{\mathbf{k}\mu}$. As usual, we classify the zero-field electron states by wave-vector **k** and band index μ.

At this stage we linearize the Liouville equation with respect to $\mathscr{A}$ by writing

$$\rho = \rho_0 + \rho_1 \tag{4.11.4}$$

and we recall that the unperturbed density matrix ρ_0 satisfies

$$\rho_0|\mathbf{k}\mu\rangle = f_0(E_{\mathbf{k}\mu})|\mathbf{k}\mu\rangle \tag{4.11.5}$$

with $f_0(E_{\mathbf{k}\mu})$ as the Fermi–Dirac distribution function. Writing the Hamiltonian as

$$H = H_0 + H_1, \tag{4.11.6}$$

where H_0 is independent of $\mathscr{A}$ and H_1 is linear in the field (the second-order terms are not required in the calculation), we have

$$i\hbar\frac{\partial\rho_0}{\partial t} = [H_0, \rho_0] \tag{4.11.7}$$

for the zeroth-order Liouville equation and for the first-order equation

$$i\hbar\frac{\partial\rho_1}{\partial t} = [H_0, \rho_1] + [H_1, \rho_0]. \tag{4.11.8}$$

If we now take matrix elements between Bloch states $\langle\mathbf{k}\mu|$ and $|\mathbf{k}'\nu\rangle$ we find

$$i\hbar\langle\mathbf{k}\mu|\frac{\partial\rho_1}{\partial t}|\mathbf{k}'\nu\rangle = (E_{\mathbf{k}\mu} - E_{\mathbf{k}'\nu})\langle\mathbf{k}\mu|\rho_1|\mathbf{k}'\nu\rangle$$
$$- [f_0(E_{\mathbf{k}\mu}) - f_0(E_{\mathbf{k}'\nu})]\langle\mathbf{k}\mu|H_1|\mathbf{k}'\nu\rangle. \tag{4.11.9}$$

Making the assumption that $\langle\mathbf{k}\mu|\rho_1|\mathbf{k}'\nu\rangle$ has time dependence like $e^{-i\omega t+\delta t}$, where δ is a small positive infinitesimal corresponding to an adiabatic turning on of the perturbing potential, then we have

$$\langle\mathbf{k}\mu|\rho_1|\mathbf{k}'\nu\rangle = -\frac{[f_0(E_{\mathbf{k}\mu}) - f_0(E_{\mathbf{k}'\nu})]\langle\mathbf{k}\mu|H_1|\mathbf{k}'\nu\rangle}{\hbar\omega + E_{\mathbf{k}'\nu} - E_{\mathbf{k}\mu} + i\hbar\delta}. \tag{4.11.10}$$

The induced current may now be calculated from

$$\mathbf{j}_{\text{ind}}(\mathbf{r}t) = \text{Tr}\{\rho_1\mathbf{j}^0_{\text{op}}(\mathbf{r})\} + \text{Tr}\{\rho_0\mathbf{j}^1_{\text{op}}(\mathbf{r}t)\}, \tag{4.11.11}$$

where the current operator is given by (see Messiah (1969), vol. 2, p. 937)

$$\mathbf{j}^0_{\text{op}}(\mathbf{r}) = -\frac{e}{2m}[\mathbf{p}_e\,\delta(\mathbf{r}-\mathbf{r}_e)+\delta(\mathbf{r}-\mathbf{r}_e)\,\mathbf{p}_e]$$

$$-\frac{ie}{2m}[\delta(\mathbf{r}-\mathbf{r}_e)\,\mathbf{p}_e\times\boldsymbol{\sigma}-\mathbf{p}_e\times\boldsymbol{\sigma}\delta(\mathbf{r}-\mathbf{r}_e)],$$

$$\mathbf{j}^1_{\text{op}}(\mathbf{r}t) = -\frac{e^2}{mc}\mathscr{A}(\mathbf{r}t)\,\delta(\mathbf{r}-\mathbf{r}_e). \tag{4.11.12}$$

Here $\mathbf{r}_e$ and $\mathbf{p}_e$ are respectively position and momentum operators. In equation (4.11.12) the first bracket in $\mathbf{j}^0$ together with $\mathbf{j}^1$, gives the orbital current while the second bracket in $\mathbf{j}^0$ gives the spin current. The trace indicated in equation (4.11.11) will be taken with respect to the Bloch functions (4.11.3).

4.11.2 Orbital susceptibility

The total orbital current induced is now to be obtained from the relevant terms in equation (4.11.11). We have

$$\mathbf{j}(\mathbf{r}t) = \frac{e^2}{m^2c}\sum_{\mathbf{k}\mathbf{k}'\mu\nu}\frac{f_0(E_{\mathbf{k}\mu})-f_0(E_{\mathbf{k}'\nu})}{\hbar\omega+E_{\mathbf{k}'\nu}-E_{\mathbf{k}\mu}+i\hbar\delta}\langle\mathbf{k}\mu|\mathscr{A}.\mathbf{p}|\mathbf{k}'\nu\rangle\frac{\hbar}{i}\{\psi^*_{\mathbf{k}'\nu}\nabla\psi_{\mathbf{k}\mu}-\psi^*_{\mathbf{k}\mu}\nabla\psi_{\mathbf{k}'\nu}\}$$

$$-\frac{2e^2}{mc}\sum_{\mathbf{k}\mu} f_0(E_{\mathbf{k}\mu})\,\mathscr{A}(\mathbf{r}t)\,\psi^*_{\mathbf{k}\mu}\psi_{\mathbf{k}\mu}, \tag{4.11.13}$$

where a factor of 2 has been included to take account of the two spin directions.

The field $\mathscr{B}$ is assumed in the z-direction and the gauge is chosen so that div $\mathscr{A} = 0$. Choosing the time and space dependence of $\mathscr{B}$ such that

$$\mathscr{A} = \mathscr{A}(\mathbf{q}\omega)\exp[i(\mathbf{q}.\mathbf{r}-\omega t)] \tag{4.11.14}$$

we have

$$\mathscr{B} = \mathscr{B}\hat{\mathbf{z}},\quad \mathscr{A} = \mathscr{A}\hat{\mathbf{x}}\ \text{ and }\ \mathbf{q} = q\hat{\mathbf{y}}, \tag{4.11.15}$$

where $\hat{\mathbf{x}}$, $\hat{\mathbf{y}}$ and $\hat{\mathbf{z}}$ are unit vectors. The induced current in the x-direction will then be considered in order to obtain the diagonal part of the susceptibility tensor.

Taking the Fourier transform of equation (4.11.13) and using the above result for the magnetic field we find for that Fourier component of the current with the wave-vector $\mathbf{q}$ of the applied field

$$j_x(\mathbf{q}\omega) = -\frac{2e^2}{mc}\sum_{\mathbf{k}\mu} f_0(E_{\mathbf{k}\mu})\mathscr{A}(\mathbf{q}\omega)$$

$$-\sum_{\mathbf{k}\mathbf{k}'}\sum_{\mu\nu}\frac{2e^2\hbar^2}{m^2 c}\frac{[f_0(E_{\mathbf{k}\mu})-f_0(E_{\mathbf{k}'\nu})]}{\hbar\omega+E_{\mathbf{k}'\nu}-E_{\mathbf{k}\mu}+i\hbar\delta}\langle\mathbf{k}'\nu|e^{-i\mathbf{q}.\mathbf{r}}\frac{\partial}{\partial x}|\mathbf{k}\mu\rangle$$

$$\times\langle\mathbf{k}\mu|e^{i\mathbf{q}.\mathbf{r}}\frac{\partial}{\partial x}|\mathbf{k}'\nu\rangle\mathscr{A}(\mathbf{q}\omega). \tag{4.11.16}$$

The orbital susceptibility may be obtained directly from the relation

$$j_{\text{ind}}(\mathbf{q}\omega) = cq^2\chi_{\text{orb}}(\mathbf{q}\omega)\mathscr{A}(\mathbf{q}\omega) \tag{4.11.17}$$

with the result, first obtained by Szabó,

$$\chi_{\text{orb}}(q\omega) = -\frac{2e^2}{mc^2q^2}\sum_{\mathbf{k}\mu} f_0(E_{\mathbf{k}\mu})$$

$$-\frac{2e^2\hbar^2}{m^2c^2q^2}\sum_{\mathbf{k}\mathbf{k}'}\sum_{\mu\nu}\frac{[f_0(E_{\mathbf{k}\mu})-f_0(E_{\mathbf{k}'\nu})]}{\hbar\omega+E_{\mathbf{k}'\nu}-E_{\mathbf{k}\mu}+i\hbar\delta}$$

$$\times\langle\mathbf{k}\mu|e^{i\mathbf{q}.\mathbf{r}}\frac{\partial}{\partial x}|\mathbf{k}'\nu\rangle\langle\mathbf{k}'\nu|e^{-i\mathbf{q}.\mathbf{r}}\frac{\partial}{\partial x}|\mathbf{k}\mu\rangle. \tag{4.11.18}$$

Equation (4.11.18) then is the basic equation giving the orbital susceptibility in terms of the Bloch wave functions and eigenvalues. To make further progress analytically we shall consider free or nearly free electrons.

(a) *Evaluation for free electrons.* Calculating the matrix elements in equation (4.11.18) with the plane waves

$$|\mathbf{k}\mu\rangle = (2\pi)^{-\frac{3}{2}}\exp(i\mathbf{k}.\mathbf{r}) \tag{4.11.19}$$

we obtain

$$\left.\begin{aligned}\langle\mathbf{k}|e^{i\mathbf{q}.\mathbf{r}}\frac{\partial}{\partial x}|\mathbf{k}'\rangle &= \frac{ik'_x}{(2\pi)^2}\delta(\mathbf{k}'+\mathbf{q}-\mathbf{k})\\ \langle\mathbf{k}'|e^{-i\mathbf{q}.\mathbf{r}}\frac{\partial}{\partial x}|\mathbf{k}\rangle &= \frac{ik_x}{(2\pi)^3}\delta(\mathbf{k}-\mathbf{q}-\mathbf{k}')\end{aligned}\right\} \tag{4.11.20}$$

The susceptibility is then given by

$$\chi_{\text{orb}}(\mathbf{q}\omega) = -\frac{e^2}{mc^2q^2 4\pi^3}\int d\mathbf{k}\, f_0(E_{\mathbf{k}}) + \frac{e^2\hbar^2}{4\pi^3 m^2c^2q^2}\int d\mathbf{k}\frac{[f_0(E_{\mathbf{k}})-f_0(E_{\mathbf{k}-\mathbf{q}})]\,k_x^2}{\hbar\omega+E_{\mathbf{k}-\mathbf{q}}-E_{\mathbf{k}}+i\hbar\delta}. \tag{4.11.21}$$

The result (4.11.21) can equivalently be written, with N the particle density,

$$\chi_{\mathrm{orb}}(\mathbf{q}\omega) = -\frac{e^2}{mc^2q^2}\left\{N+\frac{2\hbar^2}{m}\sum_{\mathbf{k}}\frac{k_x^2[f_0(E_{\mathbf{k}})-f_0(E_{\mathbf{k}-\mathbf{q}})]}{E_{\mathbf{k}}-E_{\mathbf{k}-\mathbf{q}}-\hbar\omega-i\hbar\delta}\right\}. \tag{4.11.22}$$

An explicit expression for the real part of $\chi_{\mathrm{orb}}(\mathbf{q}\omega)$ is now readily obtained by direct calculation as

$$\operatorname{Re}\chi_{\mathrm{orb}}(\mathbf{q},\omega) = -\frac{\chi_{\mathrm{P}}}{8}\left\{1+\lambda^{-2}+\frac{3w^2}{\lambda^4}+\frac{1}{4\lambda^3}\left[1-\left(\lambda+\frac{w}{\lambda}\right)^2\right]^2\ln\left|\frac{1-\lambda-(w/\lambda)}{1+\lambda+(w/\lambda)}\right|\right.$$
$$\left.-\frac{1}{4\lambda^3}\left[1-\left(\lambda-\frac{w}{\lambda}\right)^2\right]^2\ln\left|\frac{1+\lambda-(w/\lambda)}{1-\lambda+(w/\lambda)}\right|\right\}, \tag{4.11.23}$$

where we have used the abbreviations $\lambda = q/2k_{\mathrm{f}}$, $w = \omega/2k_{\mathrm{f}}^2$ and χ_{P} is the Pauli susceptibility (see below). The imaginary part of equation (4.11.21) in the same notation is given by

$$\operatorname{Im}\chi_{\mathrm{orb}}(\mathbf{q}\omega) = \frac{\pi\chi_{\mathrm{P}}}{32\lambda^3}\left\{\left[1-\left(\lambda+\frac{w}{\lambda}\right)^2\right]^2\theta\left(1-\left(\lambda+\frac{w}{\lambda}\right)^2\right)\right.$$
$$\left.-\left[1-\left(-\lambda+\frac{w}{\lambda}\right)^2\right]^2\theta\left(1-\left(-\lambda+\frac{w}{\lambda}\right)^2\right)\right\}, \tag{4.11.24}$$

where $\theta(x) = 1$, $x \geqslant 0$, $\theta(x) = 0$, $x<0$.

(b) *Static orbital susceptibility and Landau diamagnetism.* We pause at this point to connect the discussion with the result for the orbital diamagnetism of free electrons first given by Landau (1930). Thus, if we first put $w = 0$ in equation (4.11.23) we find the result

$$\chi_{\mathrm{orb}}(q) = \chi_{\mathrm{L}}\frac{3}{8\lambda^2}\left[1+\lambda^2-\frac{(1-\lambda^2)^2}{2\lambda}\ln\left|\frac{1+\lambda}{1-\lambda}\right|\right], \tag{4.11.25}$$

where χ_{L} is the usual Landau diamagnetism. This, of course, is simply the long-wavelength limit of $\chi_{\mathrm{orb}}(q)$ and we show the relation as a function of q in curve 1 of Figure 4.6. The result appears first to have been given by Baltensperger (1966).

It is also relevant to remark that if we retain the Bloch wave functions in equation (4.11.18), and put $\omega = 0$, this is equivalent to the theory of the orbital susceptibility of Bloch electrons first given by Hebborn and Sondheimer (1960). To our knowledge, this theory has only been numerically evaluated in the limiting case of nearly free electrons (see, for example,

Misra and Roth, 1969) or in an extreme tight binding approximation. We turn therefore to discuss how the same approach developed above also yields the spin susceptibility.

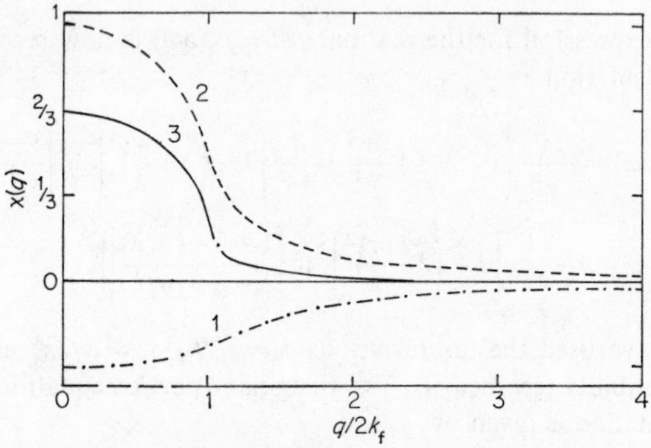

FIGURE 4.6. Wave number dependent susceptibilities for non-interacting electrons.

Curve 1: $\chi(q)$ from orbital diamagnetism.
Curve 2: $\chi(q)$ for spin only.
Curve 3: resultant susceptibility for free electrons.

4.11.3 Spin susceptibility

The spin susceptibility can be simply calculated by returning to equation (4.11.12) in which we saw earlier that the second bracket in $\mathbf{j}^0$ gives the spin current. It is then quite straightforward to show that the induced spin current is given by

$$j_{\text{spin}}(\mathbf{q},\omega) = \mathscr{A} q^2 \frac{e^2\hbar^2}{2m^2c} \sum_{\substack{\mathbf{k}\mathbf{k}'\\ \mu\nu}} \frac{[f_0(E_{\mathbf{k}\mu}) - f_0(E_{\mathbf{k}'\nu})]}{\hbar\omega + E_{\mathbf{k}'\nu} - E_{\mathbf{k}\mu} + i\hbar\delta} \times \langle \mathbf{k}\mu | e^{i\mathbf{q}\cdot\mathbf{r}} | \mathbf{k}'\nu \rangle \langle \mathbf{k}'\nu | e^{-i\mathbf{q}\cdot\mathbf{r}} | \mathbf{k}\mu \rangle. \tag{4.11.26}$$

Hence we find for the frequency- and wave-number-dependent spin susceptibility the result

$$\chi_{\text{spin}}(\mathbf{q}\omega) = \frac{e^2\hbar^2}{2m^2c^2} \sum_{\substack{\mathbf{k}\mathbf{k}'\\ \mu\nu}} \frac{[f_0(E_{\mathbf{k}\mu}) - f_0(E_{\mathbf{k}'\nu})]}{\hbar\omega + E_{\mathbf{k}'\nu} - E_{\mathbf{k}\mu} + i\hbar\delta} \times |\langle \mathbf{k}\mu | e^{i\mathbf{q}\cdot\mathbf{r}} | \mathbf{k}'\nu \rangle|^2. \tag{4.11.27}$$

Some calculations using energy-band theory have been made on this formula and we shall refer to these briefly again below. However, let us evaluate the formula explicitly for plane waves, as we did for the orbital contribution.

(a) *Evaluation for free electrons.* It is almost immediate that for plane waves equation (4.11.27) reduces to

$$\chi_{\text{spin}}(\mathbf{q}\omega) = \frac{e^2\hbar^2}{16\pi^3 m^2 c^2}\int d\mathbf{k}\,\frac{f_0(E_{\mathbf{k}})-f_0(E_{\mathbf{k}-\mathbf{q}})}{\hbar\omega+E_{\mathbf{k}-\mathbf{q}}-E_{\mathbf{k}}+i\hbar\delta}. \tag{4.11.28}$$

The real part of this, using the same variables λ and w employed in the orbital susceptibility discussion, is given by

$$\operatorname{Re}\chi_{\text{spin}} = \chi_{\text{P}}\left[\frac{1}{2}+\frac{1}{8\lambda}\left\{\left[1-\left(\frac{w}{\lambda}-\lambda\right)^2\right]\ln\left|\frac{1-(w/\lambda)+\lambda}{1+(w/\lambda)-\lambda}\right|\right.\right.$$
$$\left.\left.-\left[1-\left(\frac{w}{\lambda}+\lambda\right)^2\right]\ln\left|\frac{1-(w/\lambda)-\lambda}{1+(w/\lambda)+\lambda}\right|\right\}\right] \tag{4.11.29}$$

and the imaginary part is

$$\operatorname{Im}\chi_{\text{spin}} = \frac{\pi\chi_{\text{P}}}{8\lambda}\left\{\left[1-\left(\frac{w}{\lambda}-\lambda\right)^2\right]\theta\left(1-\left(\frac{w}{\lambda}-\lambda\right)^2\right)\right.$$
$$\left.-\left[1-\left(\frac{w}{\lambda}+\lambda\right)^2\right]\theta\left(1-\left(\frac{w}{\lambda}+\lambda\right)^2\right)\right\}. \tag{4.11.30}$$

These free electron results for the spin susceptibility appear to have been given first by Doniach (1967). Equation (4.11.30) will be used later in Chapter 5 when we come to discuss neutron-scattering experiments.

(b) *Static spin susceptibility and Pauli spin paramagnetism.* If we put w equal to zero, then we find for free electrons from equation (4.11.29) that

$$\chi_{\text{spin}}(q) = \chi_{\text{P}}\left[\frac{1}{2}+\left(\frac{1-\lambda^2}{4\lambda}\right)\ln\left(\frac{1+\lambda}{1-\lambda}\right)\right], \tag{4.11.31}$$

where χ_{P}, as we said earlier, is the Pauli spin susceptibility for free electrons. Of course, this is proportional to the density of states at the Fermi surface for a more general band structure (see Problem 4.2). The static q-dependent susceptibility given by equation (4.11.31) is shown in curve 2 of Figure 4.6.

At this stage, we wish to comment on the effect of electron interactions on the Pauli spin susceptibility. Below, we shall first do this from the Landau theory of Chapter 2, and for the Gell-Mann–Brueckner treatment of the high-density electron gas. Then we shall consider in some detail the discussion of Hamann and Overhauser (1966) which shows how the interactions affect not only the long-wavelength limit but also the q-dependent susceptibility.

4.11.4 Effect of interactions on spin susceptibility

It has been clear for many years that the spin susceptibility is rather sensitive to the effect of electron–electron interactions and we shall begin the discussion of this problem by giving the result of the Landau Fermi liquid theory discussed at length in Chapter 2. Naturally, the result will be expected to come out in terms of the Landau parameters we introduced earlier.

It will be convenient, in both this section and the next, to express the effect of the interactions in terms of a change in the gyromagnetic ratio, originally γ_0, to a new value, $\gamma(k_f)$ say.

(a) *Landau theory of spin susceptibility.* Then we can write for the induced moment $\mathscr{M}$ the result

$$\mathscr{M} = \mathrm{Tr}_\sigma \int d\mathbf{r}\, [\delta\rho(\mathbf{p}, \mathbf{r}\sigma)\, \gamma_0\, \boldsymbol{\sigma}]$$
$$= \chi\mathscr{H}, \tag{4.11.32}$$

where $\delta\rho$ is the change in the density matrix in a constant external field $\mathscr{H}$ while χ as usual is the volume susceptibility. In the presence of such an applied field and at absolute zero we have

$$\rho(\mathbf{p}\sigma) = \theta[\mu - E(\mathbf{p}\sigma)], \tag{4.11.33}$$

where

$$E(\mathbf{p}\sigma) = E^0(\mathbf{p}) + \delta E(\mathbf{p}\sigma) \tag{4.11.34}$$

with

$$\delta E(\mathbf{p}\sigma) = -\gamma\boldsymbol{\sigma}.\mathscr{H}. \tag{4.11.35}$$

This then yields the result

$$\chi = \frac{\gamma_0 m^* p_f}{\pi^2}\gamma. \tag{4.11.36}$$

To determine γ, we expand E in the form

$$E(\mathbf{p}\sigma) = E_{\mathbf{p}}^0 - \gamma_0\, \boldsymbol{\sigma}.\mathscr{H} + \int \frac{d\mathbf{p}'}{(2\pi)^3} \mathrm{Tr}_{\sigma'} f(\mathbf{p}\sigma, \mathbf{p}'\sigma')\, \delta p(\mathbf{p}', \sigma') \tag{4.11.37}$$

and for isotropic systems this yields the result

$$\gamma = \gamma_0 - \frac{\gamma m^* p_f}{2\pi^2} \int d\cos\theta\, \zeta(\theta). \tag{4.11.38}$$

Thus, it can be seen that γ depends only on the Landau B parameters, and indeed it is quickly shown that only B_0 enters, because of the orthogonality of the Legendre polynomials. Hence we find for the ratio of the susceptibility χ to that for a non-interacting system χ_p

$$\frac{\chi}{\chi_p} = \frac{m^*/m}{1+B_0} \tag{4.11.39}$$

which is the desired result of the Landau theory.

(b) *Gell-Mann–Brueckner theory*. In the Gell-Mann–Brueckner theory, a basic quantity entering the energy expression, as we shall see shortly in equation (4.11.43) below, is

$$\int_{\substack{k<k_f \\ |\mathbf{k}+\mathbf{q}|>k_f}} d\mathbf{k} \int_{-\infty}^{\infty} dt \exp\left[-|t|\left(\frac{q^2}{2}+\mathbf{q}.\mathbf{k}\right)+iqtu\right] \tag{4.11.40}$$

which will be denoted by $Q_q(u)$ and which is explicitly obtained from equation (4.11.40) as

$$Q_q(u) = 2\pi\left[1+\frac{1}{2q}(1-\tfrac{1}{4}q^2+u^2)\ln\frac{(1+\frac{1}{2}q)^2+u^2}{(1-\frac{1}{2}q)^2+u^2}\right.$$
$$\left. -u\tan^{-1}\frac{1+\frac{1}{2}q}{u} - u\tan^{-1}\frac{1-\frac{1}{2}q}{u}\right]. \tag{4.11.41}$$

When we switch on a magnetic field, then, as in the elementary treatment of Pauli spin susceptibility (cf. Problem 4.1) the electrons of $\uparrow$ spin occupy a Fermi sphere of radius k_f^+ and those of spin $\downarrow$ one of radius k_f^-. In the expression obtained for the correlation energy by Gell-Mann and Brueckner, in the approximation in which all ring diagrams are summed we simply have to replace the above function $Q_q(u)$ by

$$\tfrac{1}{2}[Q_q^+(u)+Q_q^-(u)] \tag{4.11.42}$$

where the + and − refer to the different Fermi wave numbers k_f^+ and k_f^-.

The correlation energy per particle in Rydbergs is then obtained immediately, in terms of the relative magnetization $\zeta = (N_+ - N_-)/N$, in an obvious notation, as

$$\varepsilon_{\text{corr}}(\zeta) = -\frac{3}{4\pi^5}\int_0^{\infty}\frac{dq}{q}\int_{-\infty}^{\infty} du \sum_{n=2}^{\infty}\frac{(-1)^n}{n}\left[\frac{Q_q^+(u)+Q_q^-(u)}{2}\right]^2\left(\frac{\alpha r_s}{\pi q^2}\right)^{n-2},$$

where

$$\alpha^3 = \left(\frac{4}{9\pi}\right). \tag{4.11.43}$$

To approximate this, we can replace Q_q by Q_0 and the upper limit of the q integration by unity, provided we then add a correction $\Delta(\zeta)$ to include a contribution from the second-order term arising from the range of q greater than unity. We then obtain

$$\varepsilon_{\text{corr}}(\zeta) = -\frac{3}{4\pi^5}\int_0^1 \frac{dq}{q}\int_{-\infty}^{\infty} du \sum_{n=2}^{\infty} \frac{(-1)^n}{n}\left[\frac{Q_0^+(u)+Q_0^-(u)}{2}\right]^n \left(\frac{\alpha r_s}{\pi q^2}\right)^{n-2} + \Delta(\zeta), \tag{4.11.44}$$

where

$$\Delta(\zeta) = \frac{3}{8\pi^5}\int_0^1 \frac{dq}{q}\int_{-\infty}^{\infty} du \left[\frac{Q_0^+(u)+Q_0^-(u)}{2}\right]^2 - \frac{3}{32\pi^6}\int \frac{d\mathbf{q}}{q^3}\int_{-\infty}^{\infty} du \left[\frac{Q_q^+(u)+Q_q^-(u)}{2}\right]^2. \tag{4.11.45}$$

Calculating these expressions to leading order in ζ we find

$$\varepsilon_{\text{corr}}(\zeta) = \varepsilon_{\text{corr}}(0) - \frac{\zeta^2}{6\pi^2}\left(\ln\frac{4\alpha r_s}{\pi} + \langle \ln R\rangle_{av}\right) + \frac{\zeta^2}{3\pi^3}(\ln 2 + \tfrac{1}{2}), \tag{4.11.46}$$

where

$$\langle \ln R\rangle_{av} = \int_{-\infty}^{\infty} du \frac{R(u)}{(1+u^2)^2}\ln R \Big/ \int_{-\infty}^{\infty} du \frac{R(u)}{(1+u^2)^2}, \tag{4.11.47}$$

the function R being simply

$$R(u) = 1 - u\tan^{-1}(1/u). \tag{4.11.48}$$

Adding the contribution $\varepsilon_{\text{corr}}(\zeta) - \varepsilon_{\text{corr}}(0)$ to the changes in kinetic and exchange energies to order ζ^2 (cf. Problem 4.1), the magnetic susceptibility can be calculated in the usual way.

It is interesting to note that the ratio of the paramagnetic spin susceptibility χ to the non-interacting Pauli value χ_{P}, when evaluated in this way, is not the same as for the specific heat ratio which we calculated in Chapter 2. This is just as for the Landau theory as we have seen already. The independent particle argument that both are proportional to the density of states at the Fermi surface is, of course, an oversimplification. As Brout and Carruthers (1964) stress, in addition to this effect, the gyromagnetic ratio γ_0 changes to a value $\gamma(k_{\text{f}})$ which they evaluate as

$$\frac{\gamma(k_{\text{f}})}{\gamma_0} = \left[1 - \left(\frac{9\pi}{4}\right)^{\frac{1}{3}}\frac{2}{\pi r_s}\int_{-1}^{+1}\frac{dx}{2(1-x)\{1+4(4/9\pi)^{\frac{1}{3}}(r_s/\pi)[1/2(1-x)]\}}\right]^{-1}. \tag{4.11.49}$$

If we multiply the ratio of specific heats by equation (4.11.49), we then get back the correct form at high density for the Pauli spin susceptibility as calculated above. Again, if we wish, we could obtain the small r_s form of the Landau parameter B_0 from the above theory. However, the treatment immediately above will not work at the interparticle spacing r_s appropriate to real metals, as we stressed earlier.

So far, we have dealt with the effect of interactions on the spin susceptibility when the applied field is constant in space. As we shall see below, it is important to know what happens away from $q = 0$. The discussion below is based principally on the work of Hamann and Overhauser (1966).

(c) *Interactions at* $\mathbf{q} \neq 0$ *and spin-density waves.* There is a generalization of the Hartree–Fock method which does not put electrons with opposed spins into the same space orbital. The idea is to consider the total density $\rho(\mathbf{r})$ at any point in the metal as composed of $\rho_\uparrow(\mathbf{r})$ and $\rho_\downarrow(\mathbf{r})$. If we doubly fill orbitals, $\rho_\uparrow(\mathbf{r}) \equiv \rho_\downarrow(\mathbf{r})$ at all points in space, and there is no spin density $\rho_\uparrow(\mathbf{r}) - \rho_\downarrow(\mathbf{r})$. But now we can insert the condition lost in the usual Hartree–Fock method that antiparallel spins can avoid one another by peaking up $\rho_\uparrow$ in a region where $\rho_\downarrow(\mathbf{r})$ is small. A way to do this is to write

$$\left.\begin{aligned} \rho_\uparrow(\mathbf{r}) &= \tfrac{1}{2}\rho(\mathbf{r}) + A\cos \mathbf{q}.\mathbf{r}, \\ \rho_\downarrow(\mathbf{r}) &= \tfrac{1}{2}\rho(\mathbf{r}) - A\cos \mathbf{q}.\mathbf{r}, \end{aligned}\right\}$$

giving back $\rho_\uparrow + \rho_\downarrow = \rho(\mathbf{r})$, but a spin density $\rho_\uparrow - \rho_\downarrow = 2A\cos \mathbf{q}.\mathbf{r}$. This idea of a spin-density wave crudely expressed in the above equation has been developed by Overhauser (1962) for a treatment of magnetism in an electron gas. It has been used in somewhat different form by many other workers in the fields of atomic and molecular physics (for example, it is related to the alternant molecular orbital method, though this makes a rather special choice of the wave-vector $\mathbf{q}$ of the spin-density wave).

Because, as we shall see below, it is unlikely that, in simple metals for which the electron gas model would be at all applicable, there are in fact spin-density wave states (or itinerant antiferromagnetism) we shall not give Overhauser's arguments in detail. We shall rather consider the result of Hamann and Overhauser on the influence of electron–electron interactions on the spin susceptibility $\chi_{\text{spin}}(q)$, as this, as we discuss in detail later in this section, will tell us about the stability of the non-magnetic state to magnetic deformations.

We need then to set up a formulation sufficiently general for the effect of interactions on $\chi(q)$ to be studied. We start out from the usual Hamiltonian for a homogeneous electron gas

$$H = \sum_i \frac{p_i^2}{2m} + \sum_{i<j} v(\mathbf{r}_i - \mathbf{r}_j) \tag{4.11.50}$$

and switch on a spatially varying magnetic field with wave vector **q**. Its effect is to add a perturbation H_1 to equation (4.11.50), which we can evidently write in the form

$$H_1 = -\mu_B \mathscr{H}_0 \sum_i (s_z)_i \cos \mathbf{q}.\mathbf{r}_i, \tag{4.11.51}$$

where $\mathscr{H}_0$ is the amplitude of the applied magnetic field. We can now calculate the matrix elements of equation (4.11.51) between plane-wave states, to obtain

$$\langle \mathbf{k}'\sigma' | H_1 | \mathbf{k}\sigma \rangle = -\tfrac{1}{2}\mu_B \mathscr{H}_0 (s_z)_{\sigma\sigma'} [\delta_{\mathbf{k}',\mathbf{k}+\mathbf{q}} + \delta_{\mathbf{k}',\mathbf{k}-\mathbf{q}}]. \tag{4.11.52}$$

Following Hamann and Overhauser (1966) we attempt to solve the problem in the Hartree–Fock approximation by assuming that the single-particle Hamiltonian has the form of equation (4.11.52) except that we can introduce a generalization that the δ-functions are multiplied by functions of **k**. Thus we can write down the matrix elements of the one-particle Hartree–Fock Hamiltonian H_{HF} as

$$\langle \mathbf{k}'\sigma' | H_{\mathrm{HF}} | \mathbf{k}\sigma \rangle = \varepsilon_{\mathbf{k}}\, \delta_{\mathbf{k}'\mathbf{k}}\, \delta_{\sigma'\sigma} + \tfrac{1}{2}(s_z)_{\sigma'\sigma} \{U[\mathbf{k}+(\mathbf{q}/2)]\, \delta_{\mathbf{k}',\mathbf{k}+\mathbf{q}} + U[\mathbf{k}-(\mathbf{q}/2)]\, \delta_{\mathbf{k}',\mathbf{k}-\mathbf{q}}\}. \tag{4.11.53}$$

Then, if we can show that this is indeed a self-consistent assumption, we have shown that the total effective field is redescribed in terms of $U(\mathbf{k})$, where the arguments introduced in U in equation (4.11.53) ensure that H_{HF} is a Hermitian operator.

We must next consider the states which represent the perturbed electrons when the field is switched on. Denoting the unperturbed states by $\phi_{\mathbf{k},\sigma}$ and the perturbed wave functions by $\psi_{\mathbf{k},\sigma}$, then, to first-order in U, we obtain readily by standard perturbation theory:

$$\psi_{\mathbf{k}} = \phi_{\mathbf{k},\sigma} - \frac{\tfrac{1}{2}(s_z)_{\sigma\sigma}\, U[\mathbf{k}+(\mathbf{q}/2)]}{\varepsilon_{\mathbf{k}+\mathbf{q}} - \varepsilon_{\mathbf{k}}} \phi_{\mathbf{k}+\mathbf{q},\sigma} - \frac{\tfrac{1}{2}(s_z)_{\sigma\sigma}\, U[\mathbf{k}-(\mathbf{q}/2)]}{\varepsilon_{\mathbf{k}-\mathbf{q}} - \varepsilon_{\mathbf{k}}} \phi_{\mathbf{k}-\mathbf{q},\sigma}. \tag{4.11.54}$$

Of course, in equations (4.11.53) and (4.11.54), $\varepsilon_{\mathbf{k}}$ is the single-particle energy, and we have next to show that, when we return to the interaction term in equation (4.11.50) and calculate its matrix elements, we find these are consistent with the assumption (4.11.53) for the Hartree–Fock Hamiltonian.

(i) *Matrix elements of interaction* v: Let us return to equation (4.11.50) and calculate the matrix elements of the interaction $v(\mathbf{r}_i - \mathbf{r}_j)$ with respect to the Hartree–Fock wave functions (4.11.54). We must remember only to work to first-order in U, as equation (4.11.54) is calculated correctly only to this approximation. When we sum over the spin states, the direct term makes no contribution while the exchange term gives

$$\langle \mathbf{k}'\sigma'|v|\mathbf{k}\sigma\rangle = -\delta_{\mathbf{k}',\mathbf{k}}\,\delta_{\sigma'\sigma}\sum_{k''<k_f} v(\mathbf{k}-\mathbf{k}'')$$

$$+\tfrac{1}{2}(s_z)_{\sigma'\sigma}\,\delta_{\mathbf{k}',\mathbf{k}+\mathbf{q}}\sum_{|\mathbf{k}''|<k_f}$$

$$\times\left\{\frac{U[\mathbf{k}''+(\mathbf{q}/2)]\,v(\mathbf{k}-\mathbf{k}'')}{\varepsilon_{\mathbf{k}''+\mathbf{q}}-\varepsilon_{\mathbf{k}''}}+\frac{U[\mathbf{k}''+(\mathbf{q}/2)]\,v(\mathbf{k}-\mathbf{k}''+\mathbf{q})}{\varepsilon_{\mathbf{k}''-\mathbf{q}}-\varepsilon_{\mathbf{k}''}}\right\}$$

$$+\tfrac{1}{2}(s_z)_{\sigma'\sigma}\,\delta_{\mathbf{k}',\mathbf{k}-\mathbf{q}}\sum_{|\mathbf{k}''|<k_f}$$

$$\times\left\{\frac{U[\mathbf{k}''-(\mathbf{q}/2)]\,v(\mathbf{k}-\mathbf{k}'')}{\varepsilon_{\mathbf{k}''-\mathbf{q}}-\varepsilon_{\mathbf{k}''}}+\frac{U[\mathbf{k}''-(\mathbf{q}/2)]\,v(\mathbf{k}-\mathbf{k}''-\mathbf{q})}{\varepsilon_{\mathbf{k}''+\mathbf{q}}-\varepsilon_{\mathbf{k}''}}\right\},$$

(4.11.55)

where $v(\mathbf{k})$ is the Fourier transform of the interaction. It is clear that this result is consistent with the form assumed for the Hartree–Fock Hamiltonian in equation (4.11.53) provided we write

$$\varepsilon_{\mathbf{k}} = \frac{k^2}{2m} - \sum_{|\mathbf{k}''|<k_f} v(\mathbf{k}-\mathbf{k}''), \tag{4.11.56}$$

the expression we would expect for the Hartree–Fock single particle energy, and

$$U\left(\mathbf{k}+\frac{\mathbf{q}}{2}\right) = -\mu_{\mathrm{B}}\mathscr{H}_0 + \sum_{|\mathbf{k}''|<k_f}$$

$$\times\left\{\frac{U[\mathbf{k}''+(\mathbf{q}/2)]\,v(\mathbf{k}-\mathbf{k}'')}{\varepsilon_{\mathbf{k}+\mathbf{q}}-\varepsilon_{\mathbf{k}}}+\frac{U[\mathbf{k}''+(\mathbf{q}/2)]\,v(\mathbf{k}-\mathbf{k}''+\mathbf{q})}{\varepsilon_{\mathbf{k}''-\mathbf{q}}-\varepsilon_{\mathbf{k}''}}\right\}.$$

(4.11.57)

It is quite clear then that, from equation (4.11.57), the total effective field is the applied field plus a self-consistent field contribution. Clearly, when v is zero, U is $-\mu_{\mathrm{B}}\mathscr{H}_0$.

(ii) *Integral equation for effective field U*: We shall next derive the basic integral equation for U, which, once the solution is known, will enable us to calculate the q-dependent susceptibility. In equation (4.11.57) we make a change of variables, by letting $\mathbf{k}+(\mathbf{q}/2)\to\mathbf{k}$, $\mathbf{k}''+(\mathbf{q}/2)\to\mathbf{k}'$ in the first term, and $\mathbf{k}''-(\mathbf{q}/2)\to\mathbf{k}'$ in the second term. If we also introduce n_k as the occupation number, i.e.

$$n_k = \begin{cases} 1, & k<k_f, \\ 0, & k>k_f, \end{cases} \tag{4.11.58}$$

then we find

$$U(\mathbf{k}) = -\mu_B \mathscr{H}_0 + \sum_{\mathbf{k}'} \frac{n_{\mathbf{k}'-(\mathbf{q}/2)} - n_{\mathbf{k}'+(\mathbf{q}/2)}}{\varepsilon_{\mathbf{k}'+(\mathbf{q}/2)} - \varepsilon_{\mathbf{k}'-(\mathbf{q}/2)}} v(\mathbf{k}-\mathbf{k}')\, U(\mathbf{k}'). \tag{4.11.59}$$

This, then is the integral equation determining the effective field U. Clearly, when the solution of equation (4.11.59) is known it is a straightforward matter to calculate the induced magnetization. For the spin density in the perturbed state is

$$\langle s_z(x)\rangle = \sum_{\substack{|\mathbf{k}|<k_f \\ \sigma\sigma'}} \psi^*_{\mathbf{k}\sigma'}(x)\,(s_z)_{\sigma'\sigma}\,\psi_{\mathbf{k}\sigma}(x)$$

$$= -\cos \mathbf{q}.\mathbf{x} \sum_{\mathbf{k}} \frac{n_{\mathbf{k}-(\mathbf{q}/2)} - n_{\mathbf{k}+(\mathbf{q}/2)}}{\varepsilon_{\mathbf{k}+(\mathbf{q}/2)} - \varepsilon_{\mathbf{k}-(\mathbf{q}/2)}} U(\mathbf{k}). \tag{4.11.60}$$

Now U is proportional to $\mathscr{H}_0$, and from equations (4.11.50) and (4.11.60) we can evidently write

$$\chi(q) = \mu_B^2 \sum_{\mathbf{k}} \frac{n_{\mathbf{k}-(\mathbf{q}/2)} - n_{\mathbf{k}+(\mathbf{q}/2)}}{\varepsilon_{\mathbf{k}+(\mathbf{q}/2)} - \varepsilon_{\mathbf{k}-(\mathbf{q}/2)}} u(\mathbf{k}), \tag{4.11.61}$$

where $u(\mathbf{k})$ is simply $-U/\mu_B \mathscr{H}_0$. From equation (4.11.59) $u(\mathbf{k})$ satisfies the equation

$$u(\mathbf{k}) = 1 + \sum_{\mathbf{k}'} \frac{n_{\mathbf{k}'-(\mathbf{q}/2)} - n_{\mathbf{k}'+(\mathbf{q}/2)}}{\varepsilon_{\mathbf{k}'+(\mathbf{q}/2)} - \varepsilon_{\mathbf{k}'-(\mathbf{q}/2)}} v(\mathbf{k}-\mathbf{k}')\, u(\mathbf{k}'). \tag{4.11.62}$$

Equations (4.11.61) and (4.11.62) appear to have been first derived by Wolff (1960).

In the case when the interaction v is zero, it is obvious from equation (4.11.62) that $u = 1$, and hence we regain the non-interacting spin susceptibility discussed in section 4.11.3. Thus, equation (4.11.62) now affords a tool for building into that theory an account of electron–electron interactions.

(iii) $\chi(q)$ *for short-range interactions*: Before going on to discuss the solution of the integral equation (4.11.62) for realistic interactions, it is worth deriving Wolff's (1960) result for the limit as the range of the interaction tends to zero. Thus, in equation (4.11.62) we can put

$$v(\mathbf{k}-\mathbf{k}') = \text{const}, \tag{4.11.63}$$

since

$$v(\mathbf{r}) \propto \delta(\mathbf{r}). \tag{4.11.64}$$

Hence, if we denote the non-interacting wave-number-dependent spin susceptibility (4.11.31) by $\chi_P\, g(q/2k_f)$ we find immediately [cf. equation (4.11.31)]

$$\chi(q) = \chi_P \frac{g(q/2k_f)}{1-\lambda g(q/2k_f)}, \tag{4.11.65}$$

where λ measures the strength of the interactions.

For positive λ, i.e. for repulsive interactions, we see that $\chi(q)$ is enhanced above $\chi_P\, g(q/2k_f)$ and these quantities are plotted in Figure 4.7 for various

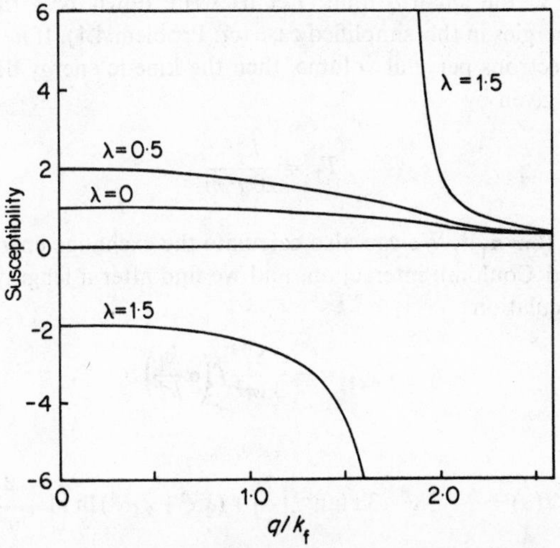

FIGURE 4.7. Showing effect of short-range interactions on spin susceptibility. Curves shown are, in fact, for delta function interactions, for several values of interaction strength λ. As the interaction λ is increased, the paramagnetic susceptibility initially increases. However, eventually an instability occurs, the curve shown for $\lambda = 1.5$ resulting from a transition to a ferromagnetic state. (After D. R. Hamann, thesis, 1966.)

values of the coupling parameter λ. For an interaction less than a certain value λ_c, the susceptibility is enhanced but still monotonically decreasing. For $\lambda > 1$, $\chi(q)$ is negative at small q, passes through a singularity and becomes well behaved at larger q.

(iv) $\chi(q)$ *for screened Coulomb interaction*: We now turn to a more realistic description of the screened Coulomb interaction. While, as we shall see, it is eventually important to consider a carefully calculated dynamic interaction, nevertheless many of the essential features of the problem are exposed by writing v as a screened Coulomb interaction with Fourier components

$$v(q) = \frac{4\pi e^2}{q^2 + \alpha^2 k_f^2}, \tag{4.11.66}$$

where the screening radius $(\alpha k_f)^{-1}$ is taken as a parameter at this stage. Thomas–Fermi screening discussed in Chapter 2 would yield $\alpha^2 k_f^2 = 4k_f/\pi a_0$.

One prime merit of this interaction is that some progress can be made analytically in the small q limit. Let us write down first the kinetic and exchange energies in this simplified case (cf. Problem 4.1). If $n_\uparrow$ is the number of $\uparrow$ spin electrons per unit volume, then the kinetic energy density of these electrons is given by

$$T_\uparrow = \frac{k_f^{+5}}{20\pi^2 m}, \tag{4.11.67}$$

where $k_f^+ = (6\pi^2 n_\uparrow)^{\frac{1}{3}}$. We can also calculate the exchange energy density for this screened Coulomb interaction, and we find after a lengthy but straightforward calculation

$$E_{\mathrm{ex}\uparrow} = -\frac{e^2 k_f^{+4}}{32\pi^3} f\left(\alpha \frac{k_f}{k_f^+}\right), \tag{4.11.68}$$

where

$$f(\alpha) = \tfrac{9}{4} - \tfrac{3}{8}\alpha^2 - 3\alpha \tan^{-1}\left(\frac{2}{\alpha}\right) + \left(\tfrac{1}{2}\alpha^2 + \tfrac{1}{24}\alpha^4\right) \ln\left(1 + \frac{4}{\alpha^2}\right). \tag{4.11.69}$$

Similar expressions hold for the downward spin electrons. Now if we transfer a small number of electrons δ from $\downarrow$ to $\uparrow$ spin states, the total energy will change. The change will be zero to first order in δ, but examining the second-order term it is a straightforward matter to show from equations (4.11.67) and (4.11.68) that there is a critical density at which the paramagnetic state becomes unstable with respect to infinitesimal ferromagnetic distortions. The condition is

$$a_0 k_f = \frac{1}{\pi}\left[1 - \frac{\alpha^2}{4} \ln\left(1 + \frac{4}{\alpha^2}\right)\right]. \tag{4.11.70}$$

For bare Coulomb interactions, i.e. $\alpha = 0$, the critical density is clearly

$$a_0 k_f = \frac{1}{\pi}. \tag{4.11.71}$$

It is interesting to note that, if we follow Bloch (1929), and calculate the energy of the completely polarized ferromagnetic state using equations (4.11.67) and (4.11.68) with $\alpha = 0$, we find that this state is stable with respect to the paramagnetic state when (cf. Problem 4.1)

$$a_0 k_f = \frac{5}{2(2^{\frac{1}{3}}+1)\pi} \approx \frac{1.1}{\pi}. \tag{4.11.72}$$

It is noteworthy therefore that instability tests are not by any means sufficient to determine the magnetic character of the ground state at any particular density.

(v) *Single-particle energy and effective mass.* To solve the integral equation for u, we must next calculate the single-particle energy $\varepsilon_{\mathbf{k}}$. We saw in equation (4.11.56) that this included a term from the interaction and, from a parallel calculation to that which yields equation (4.11.68) for the exchange energy, we find

$$\varepsilon_{\mathbf{k}} = \frac{k_f^2}{2m}\left\{y^2 + \frac{1}{2\pi a_0 k_f y}\left[4y + (1-y^2+\alpha^2)\ln\frac{(1+y)^2+\alpha^2}{(1-y)^2+\alpha^2} + 4y\tan^{-1}\frac{2\alpha}{1-y^2-\alpha^2}\right]\right\}, \tag{4.11.73}$$

where $y = k/k_f$. By differentiation with respect to k, we find for the effective mass m^*:

$$\frac{m^*}{m} = \frac{1}{1-(1/2\pi a_0 k_f)\{4-(2+\alpha^2)\ln[1+(4/\alpha^2)]\}}. \tag{4.11.74}$$

In the small q limit, we might suppose now intuitively that the denominator $[\varepsilon_{\mathbf{k}+(\mathbf{q}/2)} - \varepsilon_{\mathbf{k}-(\mathbf{q}/2)}]$ in the integral equation for u will be given by

$$\frac{k_f^2}{m} q\cos\theta_q \times \frac{m}{m^*} \tag{4.11.75}$$

and this turns out to be correct if we use equation (4.11.74) for m^*.

(vi) *Solution of integral equation in long-wave limit.* In the small q limit, we need also the occupation number difference $n_{\mathbf{k}+(\mathbf{q}/2)} - n_{\mathbf{k}-(\mathbf{q}/2)}$, in the integral equation (4.11.62), as well as the energy denominator (4.11.75). In this long-wave limit we may write

$$n_{\mathbf{k}+(\mathbf{q}/2)} - n_{\mathbf{k}-(\mathbf{q}/2)} \to -\delta(|\hat{\mathbf{k}}|-1)\,\mathbf{k}.\mathbf{q}. \tag{4.11.76}$$

Substituting equations (4.11.75) and (4.11.76) into the integral equation (4.11.62), we find

$$u(\mathbf{x}) = 1+\frac{1}{2\pi^2 a_0 k_f}\int d\mathbf{x}' \frac{-\mathbf{x}'.\mathbf{q}\delta(|\mathbf{x}'|-1)\,u(\mathbf{x}')}{(|\mathbf{x}-\mathbf{x}'|^2+\alpha^2)(\mathbf{x}'.\mathbf{q}/|\mathbf{x}'|)\,g(\alpha)}, \tag{4.11.77}$$

with

$$g(\alpha) = 1-\frac{1}{4\pi a_0 k_f}\left[4-(2+\alpha^2)\ln\left(1+\frac{4}{\alpha^2}\right)\right].$$

This integral equation, it turns out, only determines u for $|x| = 1$. But since the scalar product $\mathbf{x}'.\mathbf{q}$ goes out in equation (4.11.77), the $\mathbf{x}'$ integral has no preferred direction in space, except that defined by $\mathbf{x}$. Therefore the right-hand side of equation (4.11.77) is independent of the orientation of $\mathbf{x}$, so that $u(\mathbf{x})$ is a constant. Thus, it will suffice to find the density at which the electron gas is unstable in the long-wave limit, to determine the value of $1/a_0 k_f$ for which the homogeneous equation possesses a solution. Introducing spherical polar coordinates θ', ϕ', for the x' integration and completing the integration over x' and ϕ', we obtain then, putting $u(x) = 1$ in the right-hand side,

$$1 = \frac{1}{\pi a_0 k_f g(\alpha)}\int_{-1}^{1} d\xi \frac{1}{2+\alpha^2-2\xi}, \tag{4.11.78}$$

where $\xi = \cos\theta'$. This integral is elementary, and substituting for $g(\alpha)$ from equation (4.11.77) we find that for the critical density

$$a_0 k_f = \frac{1}{\pi}\left[1-\frac{\alpha^2}{4}\ln\left(1+\frac{4}{\alpha^2}\right)\right]. \tag{4.11.79}$$

This is the same result that we obtained for instability relative to strictly ferromagnetic deformations.

(vii) *Results of numerical solution of integral equation.* Hamann and Overhauser (1966) have solved the integral equation (4.11.62) numerically for various values of α and k_f. We shall summarize below their essential findings, which are also usefully illustrated in Figures 4.8 and 4.9.

(i) Taking $\alpha = 0.1$, it is found that for high densities χ is monotonically decreasing and in all respects is like the non-interacting gas susceptibility $\chi_0(q)$ cf equation (4.11.31).

(ii) As the density is decreased, a peak develops round $q = 2k_f$, as illustrated by Figure 4.8, with $\alpha = 0.1$ and $a_0 k_f = 0.39$. Actually, however, detailed calculation shows that the paramagnetic state remains stable at this density.

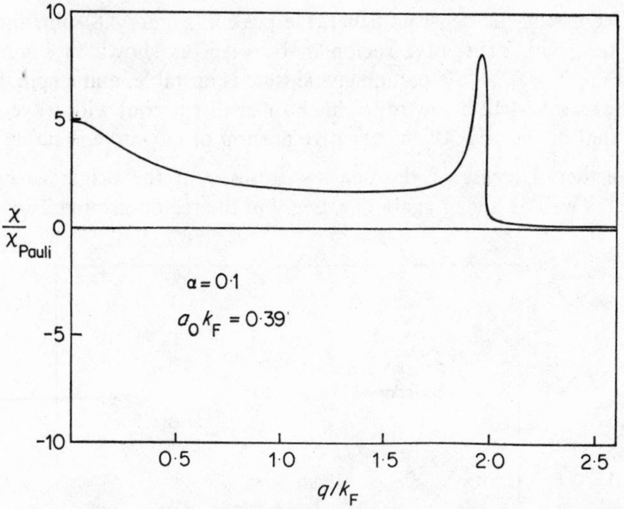

FIGURE 4.8. Illustrating enhancement of Pauli susceptibility for free electrons by switching on screened Coulomb interactions, as well as showing q dependence. Screening parameter α is too low to apply to realistic screened interaction. The peaking around $q = 2k_f$ is suppressed when more realistic screening is introduced. (After Hamann and Overhauser, 1966.)

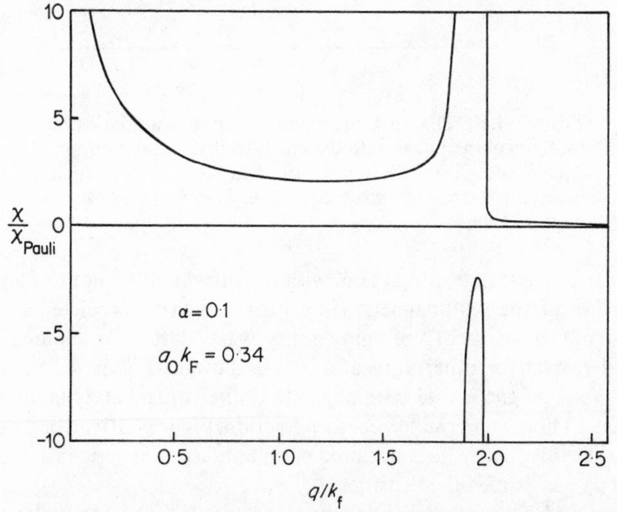

FIGURE 4.9. Same as Figure 4.8 but with a lower density. The instability around $2k_F$ now dominates $\chi(q)$. Again, the long-range of the interaction is artificial, rather than relevant to real metallic densities. (After Hamann and Overhauser, 1966.)

(iii) Decreasing the density further, the peak in Figure 4.8 splits into two singularities, with a negative region in between, as shown in Figure 4.9. Here with $a_0 k_f = 0.34$, the paramagnetic state is unstable, and a spin density wave (see section 4.12 below for a much fuller discussion) with wave vector close to that of the peak of the negative portion of the curve is stable.

(iv) Further decrease in the density causes χ at the origin to become negative, as well as giving again negative χ in the region around $2k_f$.

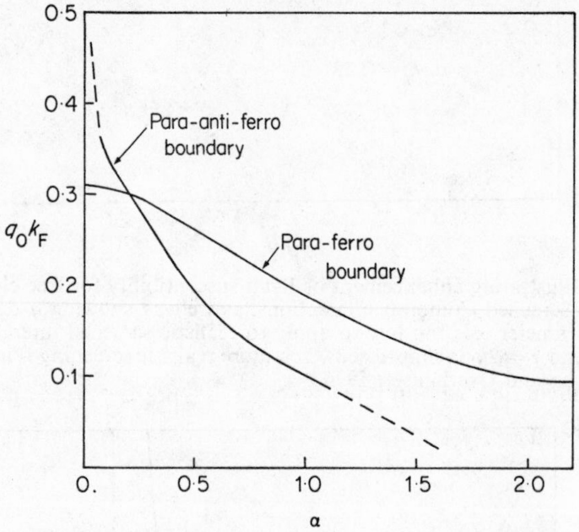

FIGURE 4.10. Diagram of magnetic phase boundaries in uniform electron gas with screened Coulomb interactions. The curves show the values of $a_0 k_f$ and α for which instabilities are first observed. (After Hamann and Overhauser, 1966.)

Thus, for a long-range interaction, with densities that are not too high, the susceptibility of the paramagnetic state varies much as predicted by Overhauser (1962) in his theory of spin-density wave states. To summarize the numerical results for other values of α, we show in Figure 4.10 the paramagnetic–ferromagnetic and paramagnetic–antiferromagnetic boundaries.

If we use a Thomas–Fermi screened interaction, Figure 4.10 can be employed to show that this theory does not predict an antiferromagnetic instability for the electron gas at metallic densities.

Hamann and Overhauser (1966) have also solved the integral equation numerically for an effective interaction calculated much more carefully from the dielectric screening theory discussed in Chapter 2. While the above results are confirmed, namely that in the region of metallic densities χ_{int} behaves like $\chi_{\text{non-int}}$, there is very substantial enhancement of the Pauli susceptibility by the interactions. We refer the reader to their paper for the further details.

4.11.5 Effect of Landau levels on transverse dynamical spin susceptibility of metals

We shall conclude the somewhat lengthy discussion of the q- and ω-dependent susceptibility by laying the foundations for treating the problem of spin waves (not to be confused with the spin-density waves just referred to) in non-ferromagnetic metals. This can be done either by Fermi liquid theory (Platzman and Wolff, 1967) or by using a random phase approximation (Edwards, 1969). We shall consider the results of the two methods in detail, in Chapter 7.

As we have seen above, the Pauli paramagnetism and the orbital diamagnetism are both consequences of placing a non-magnetic metal in a magnetic field. Though we have dealt fully with the diamagnetic effects, we have not so far considered how the existence of Landau levels can affect the dynamic spin susceptibility of the metal, and hence the spin waves.

In the presence of an applied field then, the one-electron energy levels for $\uparrow$ and $\downarrow$ spin are split by an energy which, neglecting electron interactions, is just the Zeeman energy $g\mu_B B$, where g is the spectroscopic splitting factor. The effect of electron interactions, as we have seen, is to increase the paramagnetic susceptibility and the energy splitting. Thus, with a non-magnetic metal in an applied field, we have a situation in which we have an excess of $\uparrow$ spins over $\downarrow$ spins. As we have seen to be the case in a ferromagnetic metal, there are then two types of excitation in which an $\uparrow$ spin flips to a $\downarrow$ spin. These are the independent-particle excitations and the spin-wave excitations.

The interest in this field has been greatly stimulated by the observation of spin-wave excitations in the alkali metals (Schultz and Dunifer, 1967; Dunifer, Schultz and Schmidt, 1968). A salient feature of their results is that the spin-wave dispersion law depends strongly on the angle between the direction of the applied field and that of the spin-wave propagation. These experiments have been interpreted by Platzman and Wolff using the Fermi liquid theory we discussed in Chapter 2, and we shall deal with its application to such spin-wave excitations in Chapter 7.

Random phase approximation. We turn now to the RPA treatment of Edwards (1969). The tool employed, as we have indicated, is the transverse dynamic spin susceptibility. This describes the spin response of the system to a transverse magnetic field, varying in space and time with wave-vector $\mathbf{q}$ and frequency ω. The work is essentially a generalization of the theory of Izuyama, Kim and Kubo (1963) to take account of the Landau levels.

The Hamiltonian on which the treatment is based is

$$H = \sum_i \left\{ \frac{[\mathbf{p}_i - (e/c)\mathscr{A}(\mathbf{r}_i)]^2}{2m} - g\sigma_i \mu_B \mathscr{H} \right\} + \tfrac{1}{2} I \sum_{i \neq j} \delta(\mathbf{r}_i - \mathbf{r}_j), \quad (4.11.80)$$

which is essentially Wolff's model of section 4.11.4 in the presence of a magnetic field.

(i) *Transverse dynamical susceptibility*: The excitations in which we are interested here, involving the reversal of a spin, can be investigated directly by calculating the transverse dynamic sysceptibility $\chi_{-+}(\mathbf{q}\omega)$. Omitting a factor $g^2 \mu_B^2$ for notational simplicity, $\chi_{-+}(q\omega)$ is given by (Izuyama and co-workers, 1963: compare the discussion of section 4.6.3)

$$\chi_{-+}(\mathbf{q}\omega) = \int_{-\infty}^{\infty} e^{-i\omega t} \langle\langle \sigma_-(\mathbf{q}t); \sigma_+(-\mathbf{q}) \rangle\rangle \, dt, \quad (4.11.81)$$

where

$$\langle\langle \sigma_-(\mathbf{q}t); \sigma_+(-\mathbf{q}) \rangle\rangle = \frac{i}{\hbar} \langle [\sigma_-(\mathbf{q}t), \sigma_+(-\mathbf{q})] \rangle \, \theta(t), \quad (4.11.82)$$

with $\theta(t) = 1$ for $t > 0$ and zero otherwise. It is to be understood that ω in the definition of χ_{-+} stands for $\omega - i\varepsilon$, where ε is a small positive quantity as usual, which we eventually let tend to zero. $\sigma_\pm(\mathbf{q})$ are the spin fluctuation operators and $\sigma_\pm(\mathbf{q}t)$ are the corresponding Heisenberg operators.

We now choose the x-axis so that $\mathbf{q} = (q_x, 0, q_z)$. Then, with c defined below,

$$\sigma_-(\mathbf{q}) = \sum_{k_y k_z m n} c^\dagger_{k_y, k_z + q_z, m \downarrow} \, c_{k_y k_z n \uparrow} \int_{-\infty}^{\infty} \exp(iq_x x) \, \xi_{mk_y}(x) \, \xi_{nk_y}(x) \, dx. \quad (4.11.83)$$

In this expression, we have used the Landau wave functions

$$\psi_{k_y k_z n\sigma} = \frac{1}{L} \xi_{nk_y}(x) \exp[i(k_y y + k_z z)] \chi_\sigma, \quad (4.11.84)$$

with corresponding energies

$$\varepsilon(k_z, n, \sigma) = \frac{\hbar^2 k_z^2}{2m^*} + (n + \tfrac{1}{2}) \hbar\omega_c - g\sigma\mu_B \mathscr{H}, \quad (4.11.85)$$

where the wave function is normalized in a cube of side L and χ_σ is a normalized spin function. The function $\xi_{nk_y}(x)$ is the nth normalized eigenfunction of a harmonic oscillator, centred on the point $x = -\hbar k_y / m^* \omega_c$, where

$\omega_c = eH/m^* c$ is the cyclotron frequency for an electron in the applied field H.

Also, with the Fermion creation operator $c^{\dagger}_{k_y k_z n\sigma}$ defined as creating an electron in the state described by $\psi_{k_y k_z n\sigma}$ of equation (4.11.84), the Hamiltonian can be expressed in the form

$$H = \sum_{k_y k_z n\sigma} \varepsilon(k_z, n, \sigma) N_{k_y k_z n\sigma} + \frac{I}{L^2} \sum_{\substack{k_{y1} k_{y2} k_{z1} k_{z2} \\ p_y p_z \\ n_1' n_2' n_1 n_2}} V\begin{pmatrix} n_1', k_{y1}-p_y; n_2', k_{y2}+p_y \\ n_1, k_{y1}; n_2, k_{y2} \end{pmatrix} c^{\dagger}_{k_{y1}-p_y, k_{z1}-p_z, n_1'\uparrow} \times c^{\dagger}_{k_{y2}+p_y, k_{z2}+p_z, n_2'\downarrow} c_{k_{y2} k_{z2} n_2 \downarrow} c_{k_{y1} k_{z1} n_1 \uparrow}, \tag{4.11.86}$$

where

$$V\begin{pmatrix} n_1' k_{y1}'; n_2' k_{y2}' \\ n_1, k_{y1}; n_2 k_{y2} \end{pmatrix} = \int_{-\infty}^{\infty} \xi_{n_1' k_{y1}'} \xi_{n_2' k_{y2}'} \xi_{n_1 k_{y1}} \xi_{n_2 k_{y2}} \, dx \tag{4.11.87}$$

and

$$N_{k_y k_z n\sigma} = c^{\dagger}_{k_y k_z n\sigma} c_{k_y k_z n\sigma}. \tag{4.11.88}$$

The expression for $\sigma_+(q)$ is obtained simply by changing round the $\uparrow$ and $\downarrow$ spins in equation (4.11.83). Using this expression (4.11.83) for $\sigma(q)$ and the fact that

$$\xi_{nk_y}(x) = \xi_n\left(x + \frac{\hbar k_y}{m^* \omega_c}\right), \tag{4.11.89}$$

it follows that

$$\langle\langle \sigma_-(\mathbf{q}t); \sigma_+(-\mathbf{q})\rangle\rangle = \sum_{k_y k_z n \mu} Q_{n,n+\mu}(q_x) \langle\langle \phi^{\mu}_{k_y k_z n}(q_z, t); \quad \sigma_+(-\mathbf{q})\rangle\rangle, \tag{4.11.90}$$

where

$$\phi^{\mu}_{k_y k_z n}(q_z) = \exp\left(\frac{-i\hbar k_y q_x}{m^* \omega_c}\right) c^{\dagger}_{k_y, k_z+q_z, n+\mu\downarrow} c_{k_y k_z n\uparrow} \tag{4.11.91}$$

and

$$Q_{m,n}(q_x) = \int_{-\infty}^{\infty} \exp(iq_x x) \, \xi_m(x) \, \xi_n(x) \, dx. \tag{4.11.92}$$

The summation over n and μ in equation (4.11.90) is such that n and $n+\mu$ range over all non-negative integers.

We now write down, as usual, the equation of motion for the Green functions on the right of equation (4.11.90), namely

$$i\hbar\frac{d}{dt}\langle\langle\phi^{\mu}_{k_y k_z n}(q_z, t); \sigma_+(-\mathbf{q})\rangle\rangle$$
$$= -\delta(t)\langle[\phi^{\mu}_{k_y k_z n}(q_z, t), \sigma_+(-\mathbf{q})]\rangle + \langle\langle[\phi^{\mu}_{k_y k_z n}(q_z, t), H]; \sigma_+(-\mathbf{q})\rangle\rangle. \tag{4.11.93}$$

The random phase approximation is now made, in order to simplify the last term of equation (4.11.93).

$N_{k_z n\sigma}$ denotes the occupation number of the Landau levels, and this means that only terms containing $N_{k_z' n' \sigma}$ are kept when the commutator is calculated, and these occupation numbers are replaced by $f_{k_z' n' \sigma}$, where

$$f_{k_z n\sigma} = f[\varepsilon(k_z, n, \sigma) - \sigma\Delta, \varepsilon_f] \tag{4.11.94}$$

with $f(\varepsilon, \eta)$ defined by

$$f(\varepsilon, \eta) = \left[\exp\left(\frac{\varepsilon-\eta}{k_B T}\right) + 1\right]^{-1}. \tag{4.11.95}$$

Here Δ is the exchange splitting between states with the same k_z and n, but different spin, namely $\Delta = I(N_\uparrow - N_\downarrow)$, where $N_\uparrow$ is the number of electrons per unit volume of $\uparrow$ spin.

Then the Fourier transform of the equation of motion may be written

$$[\varepsilon(k_z+q_z, n+\mu\downarrow) - \varepsilon(k_z, n\uparrow) + \Delta - \hbar\omega]\langle\langle\phi^{\mu}_{k_y k_z n}(q_z, \omega); \sigma_+(-\mathbf{q})\rangle\rangle$$
$$= (f_{k_z n\uparrow} - f_{k_z+q_z, n+\mu\downarrow})\, Q^*_{n,n+\mu}(q_x) + \frac{I}{L^2}(f_{k_z n\uparrow} - f_{k_z+q_z, n+\mu\downarrow})$$
$$\times \sum_{p_y p_z m \nu} V\begin{pmatrix} n, k_y; n+\mu, k_y \\ m, p_y; m+\nu, p_y \end{pmatrix} \exp\left[\frac{i\hbar(p_y - k_y)\, q_x}{m^* \omega_c}\right]$$
$$\times \langle\langle\phi^{\nu}_{p_y p_z m}(q_z, \omega); \sigma_+(-\mathbf{q})\rangle\rangle, \tag{4.11.96}$$

where

$$\langle\langle\phi^{\mu}_{k_y k_z n}(q_z, \omega); \sigma_+(-\mathbf{q})\rangle\rangle = \int_{-\infty}^{\infty} e^{-i\omega t}\langle\langle\phi^{\mu}_{k_y k_z n}(q_z, t); \sigma_+(-\mathbf{q})\rangle\rangle dt. \tag{4.11.97}$$

From equations (4.11.81), (4.11.90) and (4.11.97), we may write

$$\chi_{-+}(\mathbf{q}, \omega) = \sum_{k_y k_z n \mu} Q_{n,n+\mu}(q_x)\langle\langle\phi^{\mu}_{k_y k_z n}(q_z, \omega); \sigma_+(-\mathbf{q})\rangle\rangle \tag{4.11.98}$$

and we can now obtain the right-hand side from equation (4.11.96).

It is easy to show that the Green function $\langle\langle\phi^{\mu}_{k_y k_z n}(q_z, \omega); \sigma_+(-\mathbf{q})\rangle\rangle$ is independent of k_y, and then (4.11.96) can be solved. Using equations

(4.11.87) and (4.11.98) we may write equation (4.11.96) in the form

$$[\varepsilon(k_z+q_z, n+\mu\downarrow)-\varepsilon(k_z, n\uparrow)+\Delta-\hbar\omega]\langle\langle\phi^{\mu}_{k_y k_z n}(q_z,\omega);\sigma_+(-\mathbf{q})\rangle\rangle$$
$$=(f_{k_z n\uparrow}-f_{k_z+q_z, n+\mu\downarrow})\,Q^*_{n,n+\mu}(q_x)\left[1+\frac{I}{L^3}\chi_{-+}(\mathbf{q},\omega)\right]. \quad (4.11.99)$$

Dividing by $[1+(I/L^3)\chi_{-+}(\mathbf{q}\omega)]$, multiplying by $Q_{n,n+\mu}(q_x)$ and summing, it can be shown that

$$\chi_{-+}(\mathbf{q}\omega)=\Gamma(\mathbf{q}\omega)\left[1+\frac{I}{L^3}\chi_{-+}(\mathbf{q},\omega)\right], \quad (4.11.100)$$

where

$$\Gamma(\mathbf{q},\omega)=\sum_{k_y k_z n\mu}\frac{|Q_{n,n+\mu}(q_x)|^2(f_{k_z n\uparrow}-f_{k_z+q_z, n+\mu\downarrow})}{\varepsilon(k_z+q_z, n+\mu\downarrow)-\varepsilon(k_z, n\uparrow)+\Delta-\hbar\omega}. \quad (4.11.101)$$

Finally, we obtain then

$$\chi_{-+}(\mathbf{q},\omega)=\frac{\Gamma(\mathbf{q},\omega)}{1-(I/L^3)\,\Gamma(\mathbf{q},\omega)} \quad (4.11.102)$$

and this generalizes the result of Izuyama and co-workers (1963) by including the effect of Landau states.

Provided the correct exchange splitting is used, and also that $\omega_c = e\mathscr{B}/m^* c$, where $\mathscr{B}$ is the magnetic induction, equation (4.11.102) may also be used for ferromagnetic metals. We shall take up the application of this theory to spin–flip excitations in non-magnetic metals in Chapter 7, and make contact with the Fermi liquid theory there.

4.12 Spin-density waves and itinerant antiferromagnetism in chromium

The spin susceptibility $\chi_{\text{spin}}(q)$ given in equation (4.11.31) for free electrons tells us about the response of the system to a spin-density disturbance, and is relevant to the stability therefore of the paramagnetic state against a spin-density state or alternatively against a ferromagnetic deformation. In the later case, suppose that, as we switch on the electron–electron interactions, the quantity $\chi_{\text{spin}}(0)$, the usual Pauli susceptibility, tends to infinity at a critical value of the interaction strength. This then tells us that an infinitesimal magnetic field applied to the system creates a finite magnetic moment, indicating an instability against a spontaneous magnetization or a ferromagnetic state.

On the other hand, the simple form (4.11.31) which has a maximum at $q = 0$ and a kink around $q = 2k_f$, can be changed in another way, as we have seen above. For if we switch on a really long-range interaction, for example, an unscreened Coulomb interaction, then, as Overhauser has stressed, $\chi_{\text{spin}}(\mathbf{q})$ is dramatically changed around $q = 2k_f$, heralding an instability against a spin-density wave deformation. Though this instability exists for a very long-range interaction, we have argued that, in metals, the interaction which we obtained in Chapter 2 in the form $(e^2/r_{12})\exp(-qr_{12})$ is so strongly screened that, in an electron gas, certainly in the range of metallic densities (the mean inter-electronic spacing r_s measured in Bohr radii going from 2 to 5.5 for metals) the instability does not occur. However, though the *intraband* transitions which Overhauser was concerned with do not seem to lead to spin-density waves (itinerant antiferromagnetism) in simple metals, his work has led to an understanding of the itinerant antiferromagnetism in metallic chromium, through though an *interband* mechanism.

Indeed, building on Overhauser's work, Lomer (1962) pointed out that, in certain special cases in which particular Fermi surface properties obtain, Overhauser's idea can indeed lead to a lower energy state which can give the ground-state spin-density wave properties. Fedders and Martin (1966) have developed a very specific two-band model, which we shall discuss in some detail below, which brings out quantitatively many of the physical properties as Lomer had anticipated. A number of very striking observations confirm these ideas as relevant to Cr. Some of these involve alloys, however, and we shall defer the discussion of dilute Cr alloys to Chapter 10.

Following this discussion, we shall deal with almost ferromagnetic Pd, where, as we shall see, the criterion for ferromagnetism is almost exactly satisfied. In other words, the electron interactions are sufficiently strong so that they have a crucial effect on the magnetic properties. A new quasi-particle then emerges, the paramagnon.

(a) *Band structure and Fermi surface in chromium.* To understand Lomer's proposal in general terms, before turning to the specific model of Fedders and Martin (1966), we must briefly review the knowledge of the band structure and Fermi surface of bcc chromium.

The reciprocal lattice is fcc, with fundamental vectors $(\pi/a, \pi/a, 0)$, the cube edge being $2\pi/a$. The first Brillouin zone (BZ) is a rhombic dodecahedron, and the centre point Γ of the BZ and the point $H(\pi/a, 0, 0)$ will be of central importance for the argument.

There are, it turns out, two types of closed pockets of electrons, those centred at Γ being of particular importance in the present argument. There are two sets of holes, again closed pockets, the group centred at H being of particular relevance to use here.

This situation is represented in Figure 4.11, where two bands labelled a and b are drawn in the (100) plane of reciprocal space, the occupied part of each being shaded. It is clear from Figure 4.11 that the electron surface round Γ is of remarkably similar shape to the hole surface round H.

If we now go back to equation (4.11.61) with $u(k) = 1$ (the argument now is merely suggestive: we shall set up below the theory of the two-band model in some detail) we see that if the energy denominator could be made small over an appreciable region of **k**-space, then pathology could occur in $\chi_{\text{spin}}(q)$. This would herald an antiferromagnetic instability, with spin-density wave formation, of wave-vector **q**.

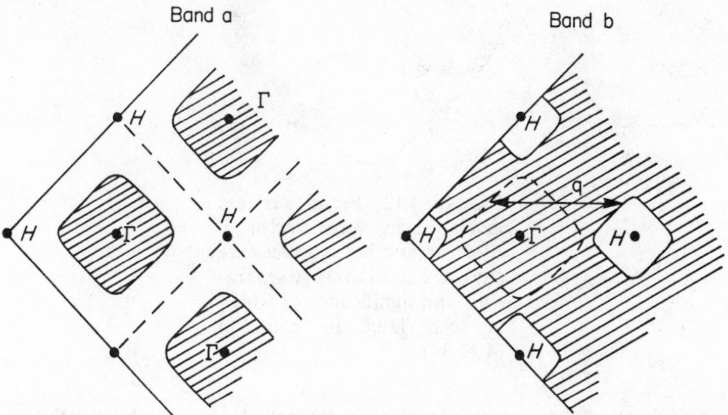

FIGURE 4.11. Fermi surface representative of Cr in (100) plane (after Lomer, 1962). Electron surface round Γ is seen to be similar in shape to hole surface round H. Choice of q as shown in the figure brings hole surface round H more or less into coincidence with electron surface round Γ.

As seen from Figure 4.11, a choice of **q** can be found which brings the hole surface round H more or less coincident with the electron surface round Γ, over quite a region of **k**-space, the situation after displacement **q** being shown in Figure 4.12.

Such a possibility of an instability in an intraband transition model, while not *impossible*, is extremely improbable (see below).

(b) *Fedders–Martin criterion for antiferromagnetism.* Fedders and Martin (1966) have established a criterion for the stability of the paramagnetic state against antiferromagnetism, by which they show that the screening of the Coulombic interaction in a homogeneous electron gas implies the absence of itinerant antiferromagnetism in such a system, in agreement with the work of Hamann and Overhauser discussed above.

However, by the same general methods, they show that it is possible for an interband mechanism such as that of Lomer, discussed above, to lead to an instability of the paramagnetic state.

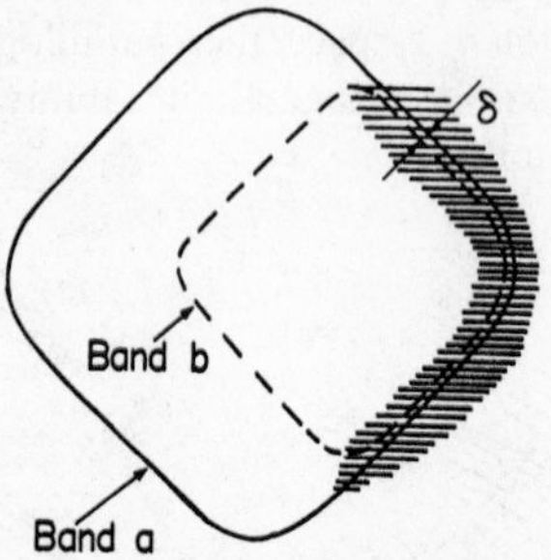

FIGURE 4.12. Fermi surfaces displaced by wave vector **q**. Electron and hole surfaces are shown. δ is an average separation, the significance of which is seen later in equation (4.12.34).

(i) *Theoretical framework*. As already remarked, the instability of a paramagnet is signalled by an infinity in the susceptibility χ. The non-local susceptibility may be seen, from equation (4.6.27), to be, in configuration space,

$$\chi_{ij}(\mathbf{rr}'\omega) = i\int_0^\infty dt\, e^{i\omega t}\langle[\mathscr{M}_i(\mathbf{r}t)\mathscr{M}_j(\mathbf{r}'0)]\rangle, \tag{4.12.1}$$

where $\mathscr{M}$ is the magnetic-moment density, written for present purposes as

$$\mathscr{M}_i(1) = \mu\sum_{\alpha\beta}\psi_\alpha^\dagger(1)\,\sigma_{\alpha\beta}^i\,\psi_\beta(1), \tag{4.12.2}$$

where μ is the magnetic moment of an electron, $(\sigma_{\alpha\beta})$ is the Pauli spin matrix, and the operator $\psi_\sigma(1)$ destroys a particle of spin σ at r_1 and time t_1 $(1 \equiv r_1, t_1)$:

$$\psi_\sigma(1) = \sum\phi_{\mathbf{k}\sigma}(\mathbf{r}t)\,a_{\mathbf{k}\sigma}. \tag{4.12.3}$$

The wave function $\phi_\mathbf{k}(\mathbf{r}, t)$ is an ordinary scalar describing an electron in the independent-particle state $\mathbf{k}\sigma$ and a is the annihilation operator for this state, as usual.

It is not possible to calculate χ directly, and so we first consider the off-diagonal susceptibility L or correlation function for electron–hole scattering, defined by

$$L(11', 22') = \frac{1}{2}\left\langle T\left(\sum_{\alpha\beta} \psi_\alpha^\dagger(1')\, \sigma_{\alpha\beta}^{(z)}\, \psi_\beta(1) \sum_{\gamma\delta} \psi_\gamma^\dagger(2')\, \sigma_{\gamma\delta}^{(z)}\, \psi_\delta(2)\right)\right\rangle, \quad (4.12.4)$$

where T is the time-ordering operator (see section 2.7.6). From equation (4.12.4), one obtains χ_{zz} by setting $1' = 1$ and $2' = 2$. It is only necessary to investigate this element, for in the paramagnetic state $\chi_{ij} \propto \delta_{ij}$ and $\chi_{xx} = \chi_{yy} = \chi_{zz}$.

If L is calculated in the Hartree–Fock approximation, the eigenstates are all Slater determinants of one-particle functions satisfying the Hartree–Fock equations of Appendix A1.1, the equation for L takes the form (cf. the Bethe–Salpeter equation of section 2.10.4), with $\beta = 1/k_B T$,

$$L(11', 22') = L_0(11', 22') + \beta \int d3\, d3'\, L_0(11', 33')\, v(33')\, L(3'3, 22'), \quad (4.12.5)$$

where L_0 is the off-diagonal susceptibility for non-interacting electrons and $v(33') = v(\mathbf{r}_3 - \mathbf{r}_3')$ is the Coulombic interaction between electrons. As we have discussed in Chapter 2, and earlier in this chapter, we can make the Hartree–Fock approximation more realistic by taking $v(\mathbf{r})$ as a screened Coulombic interaction, which we shall again do below (actually, when an interband mechanism is involved, the precise form of v is relatively unimportant, as we shall see). We might note that the exact integral equation for L is

$$L(11', 22') = L_0(11', 22') + \int d3\, d3'\, d4\, d4'\, L_0(11', 33')\, J(33', 44')\, L(44', 22'). \quad (4.12.6)$$

The electron–hole interaction kernel may be found to any given order in r_s (the mean interparticle spacing) by Green function and diagrammatic techniques. To lowest order, that is, order $1/r_s$, one obtains the Hartree–Fock approximation, which on comparison of equations (4.12.3) and (4.12.6) can be seen to be

$$J(33', 44') = \beta\delta(34')\, \delta(3'4)\, v(33'), \quad (4.12.7)$$

while the approximation to order $\ln r_s$ is

$$J(33'; 44') = \beta\delta(34')\, \delta(3'4)\, V_s(33'), \quad (4.12.8)$$

$V_s(33')$ being the dynamically screened potential v/ε, where ε is the RPA dielectric constant. The proofs of equations (4.12.6), (4.12.7) and (4.12.8) may be found in the work of Fedders and Martin (1966; see also Baym and Kadanoff, 1961; Baym, 1962).

Rather than working with L, we shall find it convenient to work with T, the scattering matrix (cf. Chapter 10, section 10.8) for electron–hole scattering, defined through

$$\int d2\,d2'\,T(11'\,22')\,L_0(22',33') = \int d2\,d2'\,J(11',22')\,L(22'\,33'). \tag{4.12.9}$$

Comparison of this equation with equation (4.12.6) shows that T obeys the integral equation

$$T(11',22') = J(11',22') + \int d3\,d3'\,d4\,d4'\,J(11',44')\,L_0(44',33')\,T(33',22'). \tag{4.12.10}$$

(ii) *Homogeneous electron gas.* Let us write L in momentum space. If we set $t_1' = t_1$ and $t_2' = t_2$ we can use the translational invariance of the system to write the Hartree–Fock equation (4.12.5) as (Fedders and Martin, 1966)

$$\langle \mathbf{p}|L(\mathbf{q},\omega)|\mathbf{p}'\rangle = \lim_{\omega_\nu\to\omega+i\epsilon} \beta\int_0^1 ds\, e^{\beta\omega_\nu s}\frac{1}{2}\Big\langle T\Big(\sum_{\alpha\beta}\psi_\alpha^\dagger(\mathbf{p}-\tfrac{1}{2}\mathbf{q},s)\,\sigma_{\alpha\beta}^{(z)}\,\psi_\beta(\mathbf{p}+\tfrac{1}{2}\mathbf{q},s)$$

$$\times\sum_{\gamma\delta}\psi_\gamma^\dagger(\mathbf{p}'+\tfrac{1}{2}\mathbf{q},0)\,\sigma_{\gamma\delta}^{(z)}\,\psi_\delta(\mathbf{p}'-\tfrac{1}{2}\mathbf{q},0)\Big)\Big\rangle, \tag{4.12.11}$$

where $s = it/\beta$ and $\omega_\nu = \pi\nu/-i\beta$, ν being an even integer. Then

$$\chi(\mathbf{q},\omega) = \frac{2\mu^2}{(8\pi^3)^2}\int d\mathbf{p}\,d\mathbf{p}'\,\langle\mathbf{p}|L(\mathbf{q},\omega)|\mathbf{p}'\rangle. \tag{4.12.12}$$

The independent-electron off-diagonal susceptibility L_0 is readily evaluated (averages of products of a and $a\dagger$ being evaluated by the trick used in section (4.10.3); note here, however, that the a's *anticommute*) and from equation (4.12.12) one obtains the equation,‡ with $f(x) = [\exp\beta(x-\mu)+1]^{-1}$,

$$[\omega-\Delta E(\mathbf{p},\mathbf{q})]\,\langle\mathbf{p}|L(q\omega)|\mathbf{p}'\rangle$$

$$= [f(E(\mathbf{p}+\tfrac{1}{2}\mathbf{q})-f(E(\mathbf{p}-\tfrac{1}{2}\mathbf{q}))]$$

$$\times\left[8\pi^3\,\delta(\mathbf{p}-\mathbf{p}')-\int\frac{d\mathbf{p}''}{8\pi^3}v(\mathbf{p}-\mathbf{p}'')\,\langle\mathbf{p}''|L(q\omega)|\mathbf{p}'\rangle\right], \tag{4.12.13}$$

where

$$\Delta E(\mathbf{p},\mathbf{q}) = E(\mathbf{p}+\tfrac{1}{2}\mathbf{q})-E(\mathbf{p}-\tfrac{1}{2}\mathbf{q}). \tag{4.12.14}$$

‡ The energies E may be taken to be the Hartree energies $E(k) = k^2/2m$ rather than the HF single particle energies, with negligible effect on the results (cf. Fedders and Martin, 1966).

Equivalently, T is the solution to the equation

$$\langle \mathbf{p}|T(\mathbf{q},\omega)|\mathbf{p}'\rangle = v(\mathbf{p}-\mathbf{p}') - \int \frac{d\mathbf{p}''}{(2\pi)^3} v(\mathbf{p}-\mathbf{p}'') \times \frac{\{f[E(\mathbf{p}''+\frac{1}{2}\mathbf{q})]-f[E(-\mathbf{p}''-\frac{1}{2}\mathbf{q})]\}}{\Delta E(\mathbf{p}'',\mathbf{q})-\omega} \langle \mathbf{p}''|T(\mathbf{q}\omega)|\mathbf{p}'\rangle. \tag{4.12.15}$$

If T has a pole, then so has $L(\omega)$ and the converse is also true.

It follows from the work of Hamann and Overhauser discussed above that below a certain critical temperature $T(\omega)$ has a pole at $q \doteqdot 2k_f$ when v is the unscreened interaction, and we shall not discuss that case further. On the other hand, it is worth while verifying the result of Hamann and Overhauser, when screening is included, by showing that no pole results when Thomas–Fermi screening is used.

If a pole in T results, but not otherwise, the finite term $v(\mathbf{p}-\mathbf{p}')$ on the right-hand side of equation (4.12.15) can be neglected and we obtain for ϕ the residue of T,

$$\phi(\mathbf{p}) = -\int \frac{d\mathbf{p}''}{8\pi^3} v(\mathbf{p}-\mathbf{p}'') \frac{\{f[E(\mathbf{p}''+\frac{1}{2}\mathbf{q})]-f[E(\mathbf{p}''-\frac{1}{2}\mathbf{q})]\}}{\Delta E(\mathbf{p}'',\mathbf{q})-\omega} \phi(\mathbf{p}''). \tag{4.12.16}$$

Now if $|\phi(x_m)|$ is the maximum value of the modulus of the solution of the homogeneous equation

$$\phi(x) = \int K(x,y)\,\phi(y)\,dy \tag{4.12.17a}$$

we have

$$|\phi(x_m)| = \left|\int K(x_m,y)\,\phi(y)\,dy\right| \leqslant \int |K(x_m,y)||\phi(y)|\,dy \leqslant |\phi(x_m)| \int |K(x_m,y)|\,dy, \tag{4.12.17b}$$

that is, if the homogeneous equation (4.12.17) has a solution. It now follows that a necessary condition for a pole in T to exist is

$$R(\mathbf{p},\mathbf{q}) = \left|\int \frac{d\mathbf{p}''}{8\pi^3} v(\mathbf{p}-\mathbf{p}'') \frac{f[E(\mathbf{p}''+\frac{1}{2}\mathbf{q})]-f[E(\mathbf{p}''-\frac{1}{2}\mathbf{q})]}{\Delta E(\mathbf{p}'',\mathbf{q})-\omega}\right| \geqslant 1. \tag{4.12.18}$$

We shall set $\omega = 0$ as the most favourable circumstance for the instability. Then setting

$$v(r) = \frac{e^2}{r} e^{-\xi r}, \tag{4.12.19}$$

one can show that for sufficiently small ξ, R has a maximum at $p = 0$ and $q = 2k_f$. R is then given by

$$R = \left(\frac{4}{9\pi}\right)^{\frac{1}{3}} \frac{r_s}{2\pi}\left\{\ln\left(\frac{2k_f}{\xi}\right)^2 - \mathscr{L}_2(-1) + \frac{1}{2}\mathscr{L}_2\left[-\left(\frac{\xi}{2k_f}\right)^2\right]\right\}, \quad \xi \leqslant 2k_f, \tag{4.12.20}$$

with $\mathscr{L}_2$ the dilogarithm

$$\mathscr{L}_2(x) = \Sigma \frac{x^n}{n^2}.$$

By setting $R = 1$ for a given r_s, we can find the maximum inverse screening length ξ_c for $R \geqslant 1$. Table 4.2 shows the Thomas–Fermi inverse screening length ξ_{TF} to be the greater for all values of r_s from 1 to 6. We therefore again confirm that the homogeneous electron gas, with Thomas–Fermi screening of the interparticle interaction, is stable against a transition to a state displaying itinerant antiferromagnetism. This result is unchanged when dynamical screening is included (see the earlier discussion of the work of Hamann and Overhauser as well as Fedders and Martin, 1966).

TABLE 4.2

r_s	1	2	3	4	5	6
ξ_{TF}/k_f	0.82	1.15	1.41	1.63	1.82	2.00
ξ_c/k_f	0.06	0.20	0.34	0.46	0.57	0.67

(iii) *Band model.* We investigate the band model in the same general way as we investigated the uniform electron gas. We shall find, however, that the possibility of interband transitions occasioned by the screened exchange interaction is all important and that the existence of an antiferromagnetic state depends on the band energies and rather critical matching of the shapes of the Fermi surfaces rather than on the exact nature of the interparticle interaction.

The independent-particle function L_0 may be evaluated in the manner described already for the case when the periodic potential $V(\mathbf{r}) = 0$. Then taking matrix elements with respect to the Bloch functions satisfying

$$-\frac{\hbar^2}{2m}\nabla^2\phi_{\mathbf{k}}^n + V(\mathbf{r})\,\phi_{\mathbf{k}}^n = E_n(\mathbf{k})\,\phi_{\mathbf{k}}^n, \tag{4.12.21}$$

n being the band index, we find L or, more conveniently, we can express the result in terms of the scattering matrix, as

$$\langle \mathbf{k}_1 n_1; \mathbf{k}_2 n_2 | T(\omega) | \mathbf{k}_3 n_3; \mathbf{k}_4 n_4 \rangle$$

$$= \langle \mathbf{k}_1 n_1; \mathbf{k}_2 n_2 | v | \mathbf{k}_3 n_3; \mathbf{k}_4 n_4 \rangle - \frac{1}{\Omega} \sum_{n_0 n_0'} \sum_{\mathbf{k}_0 \mathbf{k}_0'} \langle \mathbf{k}_1 n_1; \mathbf{k}_0 n_0 | v | \mathbf{k}_3 n_3; \mathbf{k}_0' n_0' \rangle$$

$$\times \frac{f_{n_0'}(\mathbf{k}_0') - f_{n_0}(\mathbf{k}_0)}{E_{n_0'}(\mathbf{k}_0') - E_{n_0}(\mathbf{k}_0) - \omega} \langle \mathbf{k}_0' n_0'; \mathbf{k}_2 n_2 | T(\omega) | \mathbf{k}_0 n_0; \mathbf{k}_4 n_4 \rangle, \quad (4.12.22)$$

where

$$\langle \mathbf{k}_1 n_1; \mathbf{k}_2 n_2 | v | \mathbf{k}_3 n_3; \mathbf{k}_4 n_4 \rangle = \Omega \int d\mathbf{r}_1\, d\mathbf{r}_2\, \phi_{\mathbf{k}_1}^{n_1 *}(\mathbf{r}_1)\, \phi_{\mathbf{k}_2}^{n_2 *}(\mathbf{r}_2)\, v(\mathbf{r}_{12})\, \phi_{\mathbf{k}_3}^{n_3}(\mathbf{r}_2)\, \phi_{\mathbf{k}_4}^{n_4}(\mathbf{r}_1) \quad (4.12.23)$$

and

$$f_n(\mathbf{k}) = \frac{1}{e^{\beta[E_n(\mathbf{k}) - \mu]} + 1}. \quad (4.12.24)$$

We shall neglect the umklapp processes induced by $v(\mathbf{k})$. To see the reason for this, we introduce the notation

$$\mathbf{k}_1 + \mathbf{k}_2 = [\mathbf{k}_1 + \mathbf{k}_2]_Z + [\mathbf{k}_1 + \mathbf{k}_2]_R, \quad (4.12.25)$$

where $\mathbf{k}_1$ and $\mathbf{k}_2$ both lie within the BZ, $[\mathbf{k}_1 + \mathbf{k}_2]_Z$ is the reduced part of the sum (also lying within the BZ) and $[\mathbf{k}_1 + \mathbf{k}_2]_R$ is a reciprocal lattice vector. Now equation (4.12.23) can be brought into the form

$$\langle \mathbf{k}_1 n_1, \mathbf{k}_2 n_2 | v | \mathbf{k}_3 n_3, \mathbf{k}_4 n_4 \rangle = \sum_{\mathbf{K}} v([\mathbf{k}_1 - \mathbf{k}_4]_Z + \mathbf{K}) \langle \mathbf{k}_1 n_1 | \mathbf{k}_4 + [\mathbf{k}_1 - \mathbf{k}_4]_R - \mathbf{K}, n_4 \rangle$$

$$\times \langle \mathbf{k}_2 n_2 | \mathbf{k}_3 + [\mathbf{k}_2 - \mathbf{k}_3]_R + \mathbf{K}, n_3 \rangle\, \delta_{[\mathbf{k}_3 - \mathbf{k}_2]_Z, [\mathbf{k}_1 - \mathbf{k}_4]_Z}, \quad (4.12.26)$$

where

$$\langle \mathbf{k} n | \mathbf{k}' + \mathbf{K}, n' \rangle = \int d\mathbf{r}\, \phi_{\mathbf{k}}^{n}(\mathbf{r})\, \phi_{\mathbf{k}'}^{n' *}(\mathbf{r})\, e^{i(\mathbf{k}' + \mathbf{K} - \mathbf{k}) \cdot \mathbf{r}}. \quad (4.12.27)$$

Because of the Kronecker delta in equation (4.12.26) we shall relabel $\mathbf{k}_1$, $\mathbf{k}_2$, $\mathbf{k}_3$ and $\mathbf{k}_4$ by

$$\mathbf{k}_1 = \mathbf{k} + \tfrac{1}{2}\mathbf{q}, \quad \mathbf{k}_2 = \mathbf{k}' - \tfrac{1}{2}\mathbf{q}, \quad \mathbf{k}_3 = \mathbf{k} - \tfrac{1}{2}\mathbf{q}, \quad \mathbf{k}_4 = \mathbf{k}' + \tfrac{1}{2}\mathbf{q}. \quad (4.12.28)$$

We shall neglect the umklapp processes because (a) the potential v behaves like

$$\frac{4\pi e^2}{|\mathbf{k} + \mathbf{K}|^2 + \xi^2}, \quad (4.12.29)$$

largest when the reciprocal lattice vector $\mathbf{K} = 0$ ($\mathbf{k}$ lying within the BZ and ξ being the order of the Thomas–Fermi inverse screening length); (b) the matrix elements of equation (4.12.27) are principally of interest when $n = n'$ and $|\mathbf{k}'-\mathbf{k}|$ is small compared with any reciprocal lattice vector. We there- therefore simplify our notation by defining

$$\langle \mathbf{k}n|\mathbf{k}'n'\rangle = \int d\mathbf{r}\, \phi_{\mathbf{k}}^{n}(\mathbf{r})\, \phi_{\mathbf{k}'}^{*n'}(\mathbf{r})\, e^{i([\mathbf{k}'-\mathbf{k}]_Z)\cdot\mathbf{r}} \tag{4.12.30}$$

and putting $v(\mathbf{k}-\mathbf{k}')$ for $v([\mathbf{k}'-\mathbf{k}]_Z)$. Equation (4.12.22) then becomes

$$\begin{aligned}
&\langle \mathbf{k}+\tfrac{1}{2}\mathbf{q}, n_1; \mathbf{k}'-\tfrac{1}{2}\mathbf{q}n_2|T(\omega)|\mathbf{k}-\tfrac{1}{2}\mathbf{q}, n_3; \mathbf{k}'+\tfrac{1}{2}\mathbf{q}, n_4\rangle \\
&= v(\mathbf{k}-\mathbf{k}')\langle \mathbf{k}+\tfrac{1}{2}\mathbf{q}, n_1|\mathbf{k}'+\tfrac{1}{2}\mathbf{q}; n_4\rangle \langle \mathbf{k}'-\tfrac{1}{2}\mathbf{q}, n_2|\mathbf{k}-\tfrac{1}{2}\mathbf{q}, n_3\rangle \\
&\quad -\frac{1}{\Omega}\sum_{\mathbf{k}''}\sum_{n_0,n_0'} v(\mathbf{k}-\mathbf{k}'')\langle \mathbf{k}+\tfrac{1}{2}\mathbf{q}, n_1|\mathbf{k}''+\tfrac{1}{2}\mathbf{q}, n_0'\rangle \langle \mathbf{k}''-\tfrac{1}{2}\mathbf{q}, n_0|\mathbf{k}-\tfrac{1}{2}\mathbf{q}, n_3\rangle \\
&\quad \times \frac{f_{n_0'}(\mathbf{k}''+\frac{1}{2}\mathbf{q})-f_{n_0}(\mathbf{k}''-\frac{1}{2}\mathbf{q})}{E_{n_0'}(\mathbf{k}''+\frac{1}{2}\mathbf{q})-E_{n_0}(\mathbf{k}''-\frac{1}{2}\mathbf{q})-\omega} \\
&\quad \times \langle \mathbf{k}''+\tfrac{1}{2}\mathbf{q}, n_0'; \mathbf{k}'-\tfrac{1}{2}\mathbf{q}, n_2|T(\omega)|\mathbf{k}''-\tfrac{1}{2}\mathbf{q}, n_0; \mathbf{k}'+\tfrac{1}{2}\mathbf{q}, n_4\rangle.
\end{aligned} \tag{4.12.31}$$

We assume inversion symmetry for the crystal, so that the matrix elements $\langle \mathbf{k}, n|\mathbf{k}'n'\rangle$ are real.

As previously, we set $\omega = 0$ as the most favourable circumstance for instability (if T has a pole at $\omega = 0$ for some temperature, it should have a pole for imaginary ω at a slightly lower temperature). We now use the earlier procedure, applied to the integral equation (4.12.31) and so we must examine when the kernel is largest. Here the kernel contains the positive-definite matrix

$$\begin{aligned}
\Gamma_{nn'}(\mathbf{k}'', \mathbf{q}) &= \Gamma_{n'n}(\mathbf{k}''-\mathbf{q}) \\
&= -\frac{f_n(\mathbf{k}''+\frac{1}{2}\mathbf{q})-f_{n'}(\mathbf{k}''-\frac{1}{2}\mathbf{q})}{E_n(\mathbf{k}''+\frac{1}{2}\mathbf{q})-E_{n'}(\mathbf{k}''-\frac{1}{2}\mathbf{q})}.
\end{aligned} \tag{4.12.32}$$

For simplicity we assume only two bands to be operative in this problem; that is, two bands which we label a and b are such that the denominator in equation (4.12.32) is small over a significant region of $\mathbf{k}$-space. The sum over n_0 and n_0' in equation (4.12.31) is then restricted to $n_0' = a$, $n_0 = b$ and $n_0' = b$, $n_0 = a$. Since the matrix elements between different bands will be small, we also keep only the two terms for which $n_1 = n_0'$ and $n_2 = n_0$. This eliminates the sum over bands. Applying equation (4.12.17), we now see that a necessary condition for itinerant antiferromagnetism is that at some value $\mathbf{q}_0$

$$R = \frac{1}{\Omega}\sum_{\mathbf{k}''} v(0)\gamma^2\Gamma_{ab}(\mathbf{k}'', \mathbf{q}_0) \geqslant 1. \tag{4.12.33}$$

Here we have written γ for the average of the quantity $\langle \mathbf{k}+\frac{1}{2}\mathbf{q}, n | \mathbf{k}''+\frac{1}{2}\mathbf{q}, n\rangle$ over the region for which Γ is unusually large.

It has been pointed out by Lomer (1962), as we remarked when we discussed the important features of the band structure of chromium for our present purposes above, that the matrix Γ of equation (4.12.32) is unusually large over an appreciable region, for this metal. Referring back to Figure 4.12, the occupied region of band a is similar in shape to that of band b and translation through $\mathbf{q}$ of band b shows the denominator of equation (4.12.32) to be small over the area of contact shown in Figure 4.12.

Since the energies E_n and E_n' are similar, the form of the numerator on the right-hand side of equation (4.12.32) implies that only a small volume near the Fermi surface will contribute appreciably to R and so we can rewrite the expression for R contained in equation (4.12.33) as

$$R \approx \frac{e^2 \gamma^2 \pi A}{2\xi_{\mathrm{TF}}^2 a^3 \varepsilon_{\mathrm{f}}(v_a + v_b)} \ln\left(\frac{\pi}{2\delta}\right), \tag{4.12.34}$$

where A is the area of close contact of the two Fermi surfaces (shaded in Figure 4.12), δ is an average separation of the surfaces also marked on that figure, v_a and v_b are velocities at the Fermi surfaces and $2a$ is the cube edge.

While the many approximations made in this section imply that the condition (4.12.33) must not be regarded as quantitatively precise, the essential properties of the band structure needed to make R large are evident from it. We can show the necessary condition for itinerant antiferromagnetism, namely $R > 1$, is satisfied for chromium, while it is not for a metal like potassium, where, as we shall see below, the de Haas–van Alphen effect experiments indicate a highly spherical Fermi surface. We shall say a little more about this theory of chromium in Chapter 10.

4.13 Paramagnons in nearly ferromagnetic systems

It has been known for some time that metallic Pd, with its d-band holes, has a very high paramagnetic susceptibility and comes near to satisfying the Stoner criterion (1938, 1939) $IN(E_{\mathrm{f}}) > 1$ for ferromagnetism (see also Wohlfarth, 1953).

Doniach and Engelsberg (1966), and independently Berk and Schrieffer (1966), have suggested that, in this case, new scattering corrections, to be included beyond a molecular field, or RPA treatment, will become very significant. In essence, some physical properties, and, in particular, the electronic specific heat, are then substantially enhanced above the value expected from elementary band theory, and the language used to express this nowadays is to say that the d-band holes are scattered by paramagnons.

We follow the short-range model used by Wolff (1960) and which we discussed in section 4.11.4. The effect of interactions in enhancing the spin susceptibility in a uniform electron gas was fully discussed in that section. We shall follow Doniach and Engelsberg in writing the short-range repulsions in the form

$$H_{\text{int}} = I\int d\mathbf{r}\, n_{\uparrow}(\mathbf{r})\, n_{\downarrow}(\mathbf{r}')\, \delta(\mathbf{r}-\mathbf{r}'), \tag{4.13.1}$$

where $n_{\uparrow}(\mathbf{r}) = \psi_{\uparrow}^{\dagger}(\mathbf{r})\psi_{\uparrow}(\mathbf{r})$ is the number density of upward-spin electrons.

To see the class of corrections which become important when the exchange enhancement is large, we consider the propagator for the spin fluctuations

$$\chi^{-+}(\mathbf{r}-\mathbf{r}', t-t') = i\langle T\{\sigma^{-}(\mathbf{r}, t)\,\sigma^{+}(\mathbf{r}'\, t)\}\rangle, \tag{4.13.2}$$

where $\sigma^{+}(\mathbf{r}) = \psi_{\uparrow}^{\dagger}(\mathbf{r})\psi_{\downarrow}(\mathbf{r})$. In the molecular field (RPA) approximation we can write

$$\chi^{-+}(\mathbf{k}, \omega_0) = \frac{\chi_0(\mathbf{k}, \omega_0)}{1-I\chi_0(\mathbf{k}, \omega_0)}, \tag{4.13.3}$$

where χ_0 is the usual spin susceptibility for non-interacting particles (see equation (4.11.27) above).

Using this RPA result, we can now calculate the correction $\langle H_{\text{int}}\rangle$ to the ground-state energy [see, for example, equation (2.7.101)] to obtain

$$\Delta E = I\langle n_{\uparrow}\rangle\langle n_{\downarrow}\rangle - i\int_c \frac{d\omega}{2\pi}\sum_{\mathbf{k}}\{\ln[1-I\chi_0(\mathbf{k}, \omega)] + I\chi_0(\mathbf{k}, \omega)\}, \tag{4.13.4}$$

where the contour is closed in the upper half-plane.

Writing out χ_0 explicitly we have

$$\chi_0(\mathbf{k}\omega_0) = \sum_{\mathbf{p}}\left[\frac{f_{\mathbf{p}\uparrow}(1-f_{\mathbf{p}+\mathbf{k}\downarrow})}{\varepsilon_{\mathbf{p}+\mathbf{k}\downarrow}-\varepsilon_{\mathbf{p}\uparrow}-\omega_0-i\delta} - \frac{f_{\mathbf{p}+\mathbf{k}\downarrow}(1-f_{\mathbf{p}\uparrow})}{\varepsilon_{\mathbf{p}+\mathbf{k}\downarrow}-\varepsilon_{\mathbf{p}\uparrow}-\omega_0+i\delta}\right], \tag{4.13.5}$$

where as usual, $f_{\mathbf{p}}$ represents the Fermi function and $E_p = (p^2/2m) - E_{\mathrm{f}}$.

At finite temperatures, equation (4.13.4) provides an approximation to the free energy, when the temperature-dependent Fermi functions are used.

Taking the functional derivative of ΔE with respect to $f_{\mathbf{p}\uparrow}$, we can get the quasi-particle self-energy correction $\sum(\mathbf{p}, \omega_0)$ on the energy shell $\varepsilon_p = \omega_0$ resulting from the spin fluctuations (cf. the Landau theory of Fermi liquids, section 2.10). The result

$$\sum_{\uparrow}(\mathbf{p}, \omega_0) = -iI^2\int\frac{d\mathbf{k}\,d\omega}{(2\pi)^4}\, G_{\downarrow}^0(\mathbf{p}+\mathbf{k}, \omega+\omega_0)\,\chi^{-+}(\mathbf{k}, \omega) \tag{4.13.6}$$

follows readily, where $G_{\downarrow}^0$ is the free-particle propagator.

If we now follow Doniach and Engelsberg and evaluate this formula using an expansion of $\chi^{-+}(\mathbf{k}\omega)$ appropriate when $\omega/E_f < (2|k|/k_f) \ll 1$, then it follows after some calculation that the effective mass correction at $T = 0$ is given by

$$\frac{m^*}{m} - 1 = -\lim_{\omega\to 0} \frac{\partial}{\partial\omega} \Sigma(k_f, \omega)$$

$$= 3IN(0) \ln\left\{1 + \frac{k_1^2}{k_f^2} \frac{IN(0)}{12[1 - IN(0)]}\right\}, \qquad (4.13.7)$$

where k_1 is a cut-off introduced because of the inadequacy of the above expansion for $\chi^{-+}(\mathbf{k}, \omega)$ for large k. This diverges as $IN(0)$ approaches unity, $N(0)$ being the density of states at the Fermi surface.

To proceed further the factor $[1 - IN(0)]$ can be determined from the measured susceptibility. To do so requires the calculation of the static susceptibility from ΔE above, when ΔE is expressed in terms of $E_{p\uparrow} = E_p - \mathscr{H}$, $E_{p\downarrow} = E_p + \mathscr{H}$, where $\mathscr{H}$ is the local field expressed in suitable units. Then the magnetization $\mathscr{M} = -\partial E/\partial\mathscr{H}$ is found to be such that

$$[1 - IN(0)]\, M = 2N(0)\mathscr{H} - i\Gamma^2 \int \frac{d\mathbf{k}\, d\omega}{(2\pi)^4} \frac{\partial\chi^0(k)}{\partial\mathscr{H}} \chi^{-+}(k). \qquad (4.13.8)$$

The first term on the right-hand side gives simply the exchange enhanced Pauli susceptibility. The second term can be shown to vanish, to the order to which m^*/m was calculated. This result indicates that the exchange (Stoner) enhancement factor usually calculated for Pd from the observed specific heat (which involves m^*/m) should be really calculated using the bare-band mass of the d-band holes for the unenhanced Pauli susceptibility.

The cut-off presents some problems, but an estimate can be made empirically. The exchange enhancement in Pd, calculated on the basis of the observed specific heat, is around 6.7. Using $m^*/m = 2$, from an APW calculation by Freeman, Furdyna and Dimmock (1966), gives $[1 - IN(0)] \sim 13$. The corresponding cut-off turns out to be $k_1/k_f \sim 0.6$.

We shall consider this effect in relation to PdFe alloys briefly later, in Chapter 10.

4.14 Ruderman–Kittel–Yosida interaction and magnetism in rare-earth metals

Let us suppose we introduce now a localized spin into a Fermi gas. The effect is to polarize the conduction electrons, in much the same way that we saw in Chapter 2 the introduction of a test charge caused a piling up of electrons to screen out its electrostatic field. The analogy goes further, in that the screening of the test charge led to long-range oscillations in the electron density at large distances, these reflecting the sharp Fermi edge. In the same way, the spin polarization induced in the conduction electron cloud has long-range oscillations, as we show below.

When we calculate the long-range interaction between two charged defects (see Chapter 10), we shall find that they interact essentially in accordance with an "electrostatic model", i.e. the interaction energy is just that of one charged defect sitting in the screened field of the other. This interaction therefore oscillates with distance.

A parallel situation arises when we calculate the interaction energy between two localized spins, embedded in a Fermi gas. This we shall now do, even though the details of the interaction in a metal will, in the end, depend on the detailed electronic band structure. The situation discussed above will exist for

(i) interaction between two nuclear spins via their hyperfine interaction with conduction electrons and

(ii) interaction between two magnetic ions via the d or f inner shells with conduction electrons.

We shall deal specifically below with the way this interaction is relevant in the study of the magnetism of the rare-earth metals.

The specific calculation we shall now make deals with the contact term $\mathbf{I}.\mathbf{S}$ of the hyperfine coupling between a nucleus of spin I and a conduction electron having a given state of the spin S. We follow largely the treatment of Kittel (1963).

Though we specialize later to plane waves, consider first electrons described by Bloch functions

$$\phi_{\mathbf{k}\sigma}(\mathbf{r}) = e^{i\mathbf{k}.\mathbf{r}}\, u_{\mathbf{k}\sigma}(\mathbf{r}) = \phi_{\mathbf{k}}(\mathbf{r})\,|\,s\rangle \tag{4.14.1}$$

normalized in unit volume. We need to calculate the perturbation in the electronic cloud created by the contact term discussed above. Thus, the perturbing Hamiltonian H' takes the form

$$H' = \lambda \sum_j \delta(\mathbf{r}_j - \mathbf{R}_n)\,\mathbf{S}_j . \mathbf{I}_n, \tag{4.14.2}$$

where λ measures the strength of the contact interaction.

Using the field operators introduced in Chapter 2, we can write

$$\psi(\mathbf{r}) = \sum_{\mathbf{k}\sigma} a_{\mathbf{k}\sigma}\,\phi_{\mathbf{k}\sigma}(\mathbf{r}); \quad \psi^{\dagger}(\mathbf{r}) = \sum_{\mathbf{k}\sigma} a^{\dagger}_{\mathbf{k}\sigma}\,\phi^{*}_{\mathbf{k}\sigma}(\mathbf{r}). \tag{4.14.3}$$

Then it is straightforward to express the Hamiltonian H' in the form

$$H' = \lambda \sum_{\substack{\mathbf{k}\mathbf{k}' \\ \sigma\sigma'}} \left[\int d\mathbf{r}\, \phi^*_{\mathbf{k}'\sigma'}\, \delta(\mathbf{r}-\mathbf{R}_n)\, \mathbf{S}.\mathbf{I}_n\, \phi_{\mathbf{k}\sigma}(\mathbf{r}) \right] a^\dagger_{\mathbf{k}'\sigma'}\, a_{\mathbf{k}\sigma}, \tag{4.14.4}$$

where $\mathbf{S}$ acts on the spin part of $\phi_{\mathbf{k}\sigma}$. Using equation (4.14.1), we then find

$$H' = \frac{1}{2}\sum_{\mathbf{k}\mathbf{k}'} e^{i(\mathbf{k}-\mathbf{k}').\mathbf{R}_n} J(\mathbf{k}'\mathbf{k})\, [I_n^+ a^\dagger_{\mathbf{k}'\downarrow} a_{\mathbf{k}\uparrow} + I_n^- a^\dagger_{\mathbf{k}'\uparrow} a_{\mathbf{k}\downarrow} + I_n^z (a^\dagger_{\mathbf{k}'\uparrow} a_{\mathbf{k}\downarrow} - a^\dagger_{\mathbf{k}'\downarrow} a_{\mathbf{k}\downarrow})], \tag{4.14.5}$$

where

$$J(\mathbf{k}'\mathbf{k}) = \lambda \int d\mathbf{r}\, \phi^*_{\mathbf{k}'}(\mathbf{r})\, \delta(\mathbf{r})\, \phi_{\mathbf{k}}(\mathbf{r}). \tag{4.14.6}$$

Hence $J(\mathbf{k}'\mathbf{k})$ is a constant, independent of $\mathbf{k}$ and $\mathbf{k}'$ for plane waves.

We can now proceed, as we have done a number of times previously, to set up the first-order wave functions when we find

$$|\mathbf{k}\uparrow\rangle = |\mathbf{k}\uparrow\rangle_0 + \sum_{\mathbf{k}'s}{}' |\mathbf{k}'s\rangle_0 \frac{\langle \mathbf{k}'s|\, H'\, |\mathbf{k}\uparrow\rangle}{\varepsilon_{\mathbf{k}} - \varepsilon_{\mathbf{k}'}}$$

$$= |\mathbf{k}\uparrow\rangle_0 + \sum_{\mathbf{k}'}{}' \frac{mJ}{k^2 - k'^2} (I_n^+ |\mathbf{k}'\downarrow\rangle_0 + I_n^z |\mathbf{k}'\uparrow\rangle_0) \tag{4.14.7}$$

for the wave function with the electron spin mainly $\uparrow$ and

$$|\mathbf{k}\downarrow\rangle = |\mathbf{k}\downarrow\rangle_0 + \sum_{\mathbf{k}'}{}' \frac{mJ}{k^2 - k'^2} (I_n^- |\mathbf{k}'\uparrow\rangle_0 - I_n^z |\mathbf{k}'\downarrow\rangle_0) \tag{4.14.8}$$

for the wave function with the electron spin pointing down.

For plane waves, it is then a matter of detail to show that this can be evaluated to yield for the total density of upward spins the result

$$\rho_\uparrow = \frac{k_{\mathrm{f}}^3}{6\pi^2} \left[1 - \frac{3mJI_n^z k_{\mathrm{f}}}{\pi} F(2k_{\mathrm{f}} r) \right], \tag{4.14.9}$$

where

$$F(x) = \frac{x \cos x - \sin x}{x^4}. \tag{4.14.10}$$

The net polarization is given at large r by the asymptotic form

$$\rho_\uparrow - \rho_\downarrow \simeq \frac{9\pi n^2}{4\varepsilon_{\mathrm{f}}} J I_n^z \frac{\cos 2k_{\mathrm{f}} r}{(k_{\mathrm{f}} r)^3}. \tag{4.14.11}$$

Thus, suppose now that a second magnetic moment I is added at a lattice point m at distance $\mathbf{r}$ from n. The moment at m will be perturbed by the spin polarization round the other moment at n and vice versa and we shall get an interaction between the localized moments via the conduction electrons. The second-order interaction between the localized moments can be written as

$$E''(\mathbf{r}) = \sum_{\substack{\mathbf{kk}' \\ ss'}} \frac{\langle \mathbf{k}\sigma | H' | \mathbf{k}'\sigma' \rangle \langle \mathbf{k}'\sigma' | H' | \mathbf{k}\sigma \rangle}{\varepsilon_{\mathbf{k}} - \varepsilon_{\mathbf{k}'}} \tag{4.14.12}$$

and we can evaluate this, summing over electron spin states (see Schiff, 1949), to obtain

$$E''(x) = \mathbf{I}_n . \mathbf{I}_m \frac{4J^2 mk_{\mathrm{f}}^4}{(2\pi)^3} F(2k_{\mathrm{f}} r). \tag{4.14.13}$$

The interaction is seen to be oscillatory, as we anticipated above. We have not gone into full details of the calculation here for we shall say something further about the interaction below, applied to rare-earth metals. Here, aside from the application to the interaction of two nuclei via their hyperfine interaction with the conduction electrons, this theory comes into its own, for the rare earths have very small $4f^n$ magnetic cores, immersed in conduction electrons from 6s–6p bands. Though a great deal of quantitative work has now been done, using the above approach, and building in realistic band structures, we shall conclude this discussion by presenting a simple approach to the rare earths, using a model of Elliott and Wedgwood (1963, 1964). While oversimplified in that it is based on free electrons, it shows up nicely some of the main features of the problem.

(a) *Rare-earth metals and indirect exchange.* We can summarize the observed types of spin ordering which occur in the rare-earth metals in terms of two types which may occur together;

$$\text{Type 1} \quad \langle S_n^z \rangle = MS \cos(\mathbf{q} . \mathbf{R}_n + \phi), \tag{4.14.14}$$

$\mathbf{R}_n$ being as usual a lattice point. Here $\mathbf{q}$ is in fact parallel to the c-axis and this gives a wave-like moment variation along this axis. This structure, to give specific examples, is found to occur in the high-temperature phase of Er and Tm.

$$\left. \begin{aligned} \text{Type 2} \quad \langle S_n^x \rangle &= M'S \cos(\mathbf{q} . \mathbf{R}_n), \\ \langle S_n^y \rangle &= M'S \sin(\mathbf{q} . \mathbf{R}_n), \end{aligned} \right\} \tag{4.14.15}$$

which is a spiral structure, the turn angle between spins in adjacent layers being qc. This is found in the high-temperature phase of Tb, Dy and Ho.

We allow the conduction electrons to interact with the localized magnetic ions as above, namely through a term of the form

$$H' = \frac{1}{N}\sum_{n} \lambda\delta(\mathbf{r}-\mathbf{R}_n)\,\mathbf{S}_n\,.\,\boldsymbol{\sigma}. \tag{4.14.16}$$

Then with a combination of equations (4.14.14) and (4.14.15), the unperturbed electron states of wave-vector $\mathbf{k}$ and energy $E(\mathbf{k})$ are coupled to those of $\mathbf{k}'$, by the matrix elements of equation (4.14.16). Since

$$\mathbf{S}\,.\,\boldsymbol{\sigma} = MS\sigma_z \cos(qz+\phi)+\tfrac{1}{2}M'\,S(\sigma_+\,\mathrm{e}^{-iqz}+\sigma_-\,\mathrm{e}^{iqz}), \tag{4.14.17}$$

only the matrix elements listed below are non-zero

$$\left.\begin{aligned} \langle \mathbf{k}\pm|H'|\mathbf{k}'\pm\rangle &= \pm\tfrac{1}{2}\lambda SM, \quad \mathbf{k}-\mathbf{k}' = \mathbf{K}\pm\mathbf{q},\\ \langle \mathbf{k}+|H'|\mathbf{k}'-\rangle &= \lambda SM', \quad \mathbf{k}-\mathbf{k}' = \mathbf{K}+\mathbf{q},\\ \langle \mathbf{k}-|H'|\mathbf{k}'+\rangle &= \lambda SM', \quad \mathbf{k}-\mathbf{k}' = \mathbf{K}-\mathbf{q}, \end{aligned}\right\} \tag{4.14.18}$$

where, as usual, $\mathbf{K}$ is any reciprocal lattice vector. The spin states of the conduction electrons $|\pm\rangle$ are defined relative to the z-axis.

From the first member of equation (4.14.18), it can be seen that magnetic ordering of type 1 mixes band states of the same spin polarization, and one obtains a two-by-two secular determinant for the new energy levels. The first-order correction to the wave function is evidently largest when $E(\mathbf{k}) = E(\mathbf{k}')$. As in customary nearly free-electron treatments, this leads to $\mathbf{k} = -\mathbf{k}'$, and hence, from equation (4.14.18), we have

$$\mathbf{k} = -\mathbf{k}' = \pm\tfrac{1}{2}(\mathbf{K}\pm\mathbf{q}) \equiv \pm\boldsymbol{\kappa}. \tag{4.14.19}$$

This leads to new zone boundaries in each spin sub-band, which in the reduced zone scheme ($\mathbf{K} = 0$) are evidently perpendicular to the c-axis at $\pm\frac{1}{2}\mathbf{q}$. From the first member of equation (4.14.18), the energy gap at these boundaries is just λSM.

For type 2 ordering, we see from the second and third members of equation (4.14.18) that band states of different spin polarization are mixed. Analogous arguments to those used above show that, from the second line of equation (4.14.18), strong mixing between states $|\frac{1}{2}(\mathbf{K}+\mathbf{q}),+\rangle$ and $|-\frac{1}{2}(\mathbf{K}+\mathbf{q}),-\rangle$ occurs, resulting in a new zone boundary. It should be emphasized that there is no boundary for the other sub-band at this point. From the last equation of (4.14.18), that has a zone boundary at $\pm\frac{1}{2}(\mathbf{K}-\mathbf{q})$. If we use the reduced zone scheme as before, the boundary is the plane $z = +\frac{1}{2}q$ for $\uparrow$ spins, and $z = -\frac{1}{2}q$ for $\downarrow$ spins. It is evident from equation (4.14.18) that the energy gap is $2\lambda SM'$ in each case.

In the neighbourhood of any boundary, it is a sufficient approximation to consider only that one boundary in calculating the perturbed energies. When both types 1 and 2 are present, a 4×4 secular determinant for the mixing of the states $|\mathbf{k}, \pm\rangle$, $|\mathbf{k}+2\boldsymbol{\kappa}, \pm\rangle$ results, where, from equation (4.14.19), $2\boldsymbol{\kappa} = \pm(\mathbf{K}\pm\mathbf{q})$ is the reciprocal lattice vector giving the new boundary.

The resulting energies are

$$E'(\mathbf{k}) = \tfrac{1}{2}[E(\mathbf{k})+E(\mathbf{k}+2\boldsymbol{\kappa})\pm\{[E(\mathbf{k})-E(\mathbf{k}+2\boldsymbol{\kappa})]^2+\lambda^2 S^2 M_{\pm}^2\}^{\frac{1}{2}}], \tag{4.14.20}$$

where

$$M_{\pm}^2 = M^2+2M'^2\pm 2M'(M^2+M'^2)^{\frac{1}{2}} \tag{4.14.21}$$

and the gap is given by

$$\Delta = \lambda S M^{\pm}. \tag{4.14.22}$$

For type 1 ordering, $M' = 0$ while for type 2, $M = 0$ and $M^{\pm} = 2M', 0$.

Though Elliott and Wedgwood go on to discuss the resistance of the rare-earth metals, occasioned by this change in the Fermi surface, and the main predictions of this theory have been verified experimentally (see, for example, Lee and Wilding, 1965), we want here only to point out the close relation with the Ruderman–Kittel interaction discussed above, and to discuss the theory of the way the turn angle varies with temperature in the heavy rare earths.

Introducing the structure factor $\mathscr{S}$ into the above theory, the energies (4.14.20) can be shown to become

$$2E'(\mathbf{k}) = E(\mathbf{k})+E(\mathbf{k}+2\boldsymbol{\kappa})\pm\{[E(\mathbf{k})-E(\mathbf{k}+2\boldsymbol{\kappa})]^2+\lambda^2 S^2 M_{\pm}^2|\mathscr{S}(\mathbf{K})|^2\}^{\frac{1}{2}}, \tag{4.14.23}$$

the energy gap evidently being given by

$$\Delta = \lambda S M_{\pm}|\mathscr{S}(\mathbf{K})|, \tag{4.14.24}$$

where, explicitly, the structure factor is

$$\mathscr{S}(\mathbf{K}) = \frac{1}{4}\sum_i \exp(i\mathbf{K}.\mathbf{R}_i), \tag{4.14.25}$$

the sum being over the position of the four atoms in each cell.

The argument by means of which we relate the above discussion to the Ruderman–Kittel interaction will now be sketched, complete details being found in the work of Elliott and Wedgwood.

The magnetic order to be expected in practice will be given, of course, by the values of M, M' and q which minimize the free energy. Since the entropy of the conduction electrons is of order $(k_B T/E_f)^2$ we can neglect it. The entropy of the localized spins depends only on M and M' and not on q. Thus, q can be found by minimizing the internal energy, rather than the free energy.

Restricting the discussion to type 2 ordering, a gap appears for one simple band only at $+\boldsymbol{\kappa}$ and in the other only at $-\boldsymbol{\kappa}$ [the solutions of equation (4.14.21) are $M_\pm = 2M'$ or 0]. Then, after some manipulation and summing the effect of the different gaps $M_\pm \lambda S|\mathscr{S}(\mathbf{K})|$ in the two sub-bands, the total energy E' is found to be

$$E' = \tfrac{3}{5}NE_f - \frac{3N\lambda^2 S^2}{8E_f}|\mathscr{S}(\mathbf{K})|^2(M^2+2M'^2)f\left(\frac{\kappa}{k_f}\right), \tag{4.14.26}$$

where N is the total number of conduction electrons and

$$f(x) = 1+\frac{1-x^2}{2x}\ln\left|\frac{1+x}{1-x}\right|. \tag{4.14.27}$$

Actually, in deriving equation (4.14.26) a number of approximations are made. First, an expansion in the band gap Δ has been made, the result (4.14.26) including the lowest-order term only. Secondly, though the theory on which equation (4.14.26) is based is that for type 2 ordering, the result shown is more general. The argument underlying this is that whereas for type 1 ordering, there are equal gaps $M_\pm = M$ at $\pm\boldsymbol{\kappa}$, these give the same result as one gap of $2M$, as long as the effect is small. In practice, it turns out that the metals having type 1 ordering have a small S, and since this ordering does not persist to low temperatures and large M, the effects are normally small. The theory of type 2 ordering can then be used to a reasonable approximation.

The connection between the result (4.14.26) and the Ruderman–Kittel interaction is now seen if we note that, apart from a constant factor, the function $f(x)$ in (4.14.27) is the Fourier transform of the oscillatory function F in equation (4.14.10). The second term in equation (4.14.26) is indeed just what one would then expect for a single new boundary at $\boldsymbol{\kappa}$. The total result is obtained by summing the contribution for all boundaries which is proportional to

$$\sum_{\mathbf{K}}|\mathscr{S}(\mathbf{K})|^2 f(|\mathbf{K}\pm\mathbf{q}|/2k_f). \tag{4.14.28}$$

The sum was calculated by Yoshida and Watabe (1962) who found the energy was minimized by a q along the c-direction such that $qc = 48°$, which is quite remarkably close to 51° observed in Tm and Er.

In general, without expanding in Δ it is only possible to solve for the total energy E' numerically. The results are conveniently expressed in terms of the unperturbed energy as

$$E'-E = -\frac{3N\Delta^2}{8E_f}\phi\left(\frac{\kappa}{k_f}, \Delta\right), \tag{4.14.29}$$

where $\phi(x,0) = f(x)$ of equation (4.14.27). The energy computed from equations (4.14.39) and (4.14.41), at $k_f = \kappa$, that is, $x = 1$, yields $\phi(1,\Delta) = 1$ and

$$\phi'(1,\Delta) = \ln(\Delta/16E_f), \tag{4.14.30}$$

giving the infinite slope of the Ruderman–Kittel formula only in the limit $\Delta = 0$.

(b) *Corrections to Ruderman–Kittel formula.* The results of the calculations of Elliott and Wedgwood for the correction $\phi(x,\Delta)-\phi(x,0)$ to the Ruderman–Kittel formula are shown in Figure 4.13, together with the positions of the

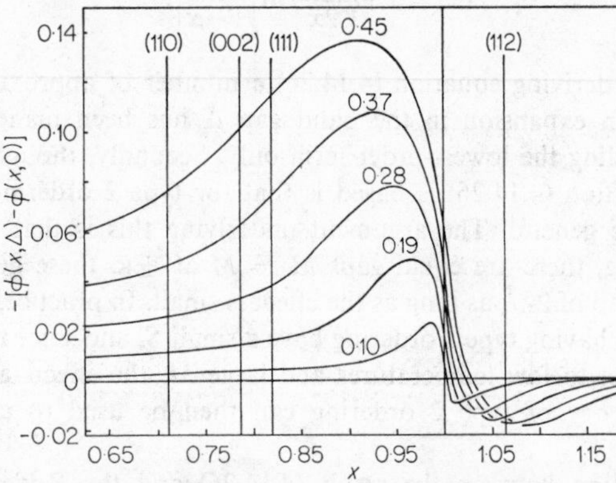

FIGURE 4.13. Correction $\phi(x,\Delta)-\phi(x,0)$ to Ruderman–Kittel interaction. The boundaries associated with reciprocal lattice vectors of hcp structure ($c/a = 1.57$) are also shown. Different curves are labelled with appropriate values of Δ. (From Elliott and Wedgwood, *Proc. Phys. Soc.*, **84**, p. 63, 1964. Copyright The Institute of Physics, by permission.)

boundaries associated with the reciprocal lattice vectors of the hexagonal close-packed lattice with $c/a = 1.57$ (see Yoshida and Watabe, 1957). The total energy is obtained by summing over all the superzone boundaries. This has been done exactly for the four **K**-values shown in Figure 4.13; and, for the rest, the deviation from the Ruderman–Kittel formula is negligible and f may be used instead of ϕ. The variation of this energy is shown in Figure 4.14, using the difference between the energy caused by gaps at **q** and at **q** = 0 such as would be caused by a ferromagnetic arrangement:

$$E'(\mathbf{q},\Delta) = E'(0,\Delta) = -\frac{3N\Delta^2}{8E_f}G(\mathbf{q}\,\Delta), \tag{4.14.31}$$

where

$$G(\mathbf{q}\Delta) = \frac{1}{2}\sum_{\mathbf{k}}|\mathscr{S}(\mathbf{K})|^2\left\{\phi\left[\frac{|\mathbf{q}+\mathbf{K}|}{2k_f},\Delta(\mathbf{K})\right]+\phi\left[\frac{|\mathbf{K}-\mathbf{q}|}{2k_f},\Delta(\mathbf{K})\right]\right.$$
$$\left.-2\phi\left[\frac{|\mathbf{K}|}{2k_f},\Delta(\mathbf{K})\right]\right\}. \tag{4.14.32}$$

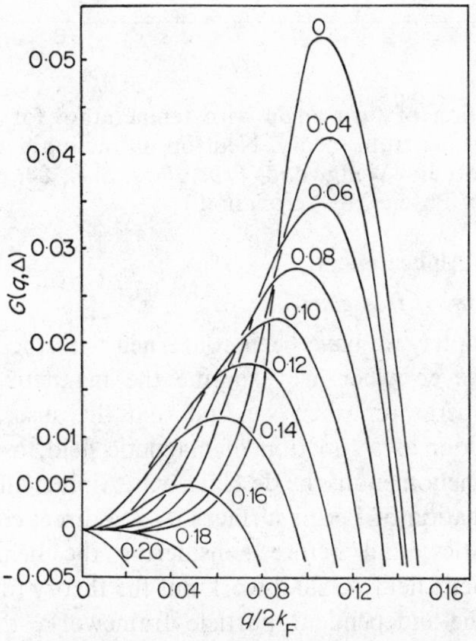

FIGURE 4.14. Variation of energy with superzone boundary gap. G is defined precisely in equation (4.14.32). Curves are labelled with values of $\Delta/2E_f$. (From Elliott and Wedgwood, *Proc. Phys. Soc.*, **84**, p. 63, 1964. Copyright The Institute of Physics, by permission.)

We finish by showing in Figure 4.15 the variation of the turn angle as a function of temperature for the heavy rare earths Tb–Tm. The solid curves are those calculated from the above theory while the experimental points are obtained from neutron diffraction data. The solid lines were calculated using values of M and M' from neutron diffraction, and a best fit is obtained with $E_f = 0.07$ or $\lambda = 0.55$). The agreement for the elements, Tm, Er and Ho with small values of S and Δ is remarkably good.

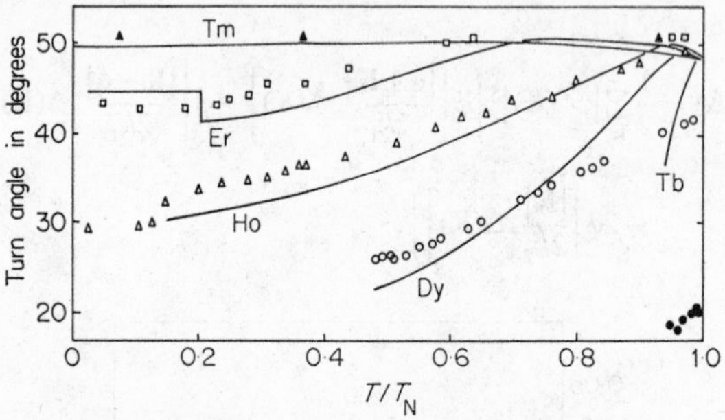

FIGURE 4.15. Variation of turn angle with temperature for heavy rare earths Tb–Tm. Solid curves are from theory. Neutron diffraction data shown for comparison. (From Elliott and Wedgwood, *Proc. Phys. Soc.*, **84**, p. 63, 1964. Copyright The Institute of Physics, by permission.)

4.15 De Haas–van Alphen effect

4.15.1 Diamagnetism of free electrons

So far in this chapter, we have been concerned with the field-independent susceptibility. However, when we examine the magnetic susceptibility of metals at low temperatures, then we find that the susceptibility shows a complicated behaviour as a function of magnetic field, oscillating with the field in fact. This phenomenon, the de Haas–van Alphen effect, is a powerful tool for the investigation of Fermi surfaces. It is a direct consequence of the Fermi–Dirac statistics, and before considering the behaviour of Bloch electrons in a magnetic field we shall work out the theory for a homogeneous electron gas. In the independent particle framework, this is an exactly soluble problem.

The classical diamagnetism of free electrons is identically zero and we shall start by briefly examining the reason for this. Classically, an electron moving with velocity v in a magnetic field $\mathscr{H}$ describes a helical path which, when projected on to a plane perpendicular to the field, is a circle of radius

$$r = \frac{mvc}{e\mathscr{H}} = \frac{v}{m\omega}, \tag{4.15.1}$$

where $\omega = e\mathscr{H}/2mc$ is the Larmor angular frequency. Thus a single orbit has a magnetic moment M given by Ampère's law

$$M = \text{Area} \times \text{current} = \pi r^2 e\omega/2\pi = cmv^2/2e\mathscr{H}. \tag{4.15.2}$$

This result, inversely proportional to $\mathscr{H}$, is not sensible when applied to all the electrons of the system and it becomes clear that we must consider the behaviour of the electrons near the boundary of the system. These electrons, as consideration of Figure 4.16 indicates, crawl round the walls in a direction

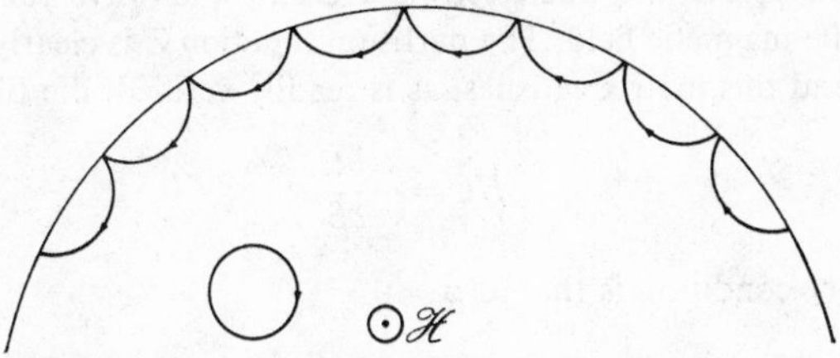

FIGURE 4.16. Illustrating effects of boundaries in classical diamagnetism.

opposite to that of the circular motion, and a detailed analysis (van Leeuwen, 1919) reveals that the magnetic moments arising from the two kinds of motion exactly cancel one another. That this must be so may be predicted thermodynamically. A magnetic field cannot do work on a system of electrons, the Lorentz force always being perpendicular to the velocity of the electron; in a reversible change at constant temperature the work done on a system is equal to the change in its free energy F and the magnetic moment of a system is given by

$$M = -\left(\frac{\partial F}{\partial \mathscr{H}}\right)_T \tag{4.15.3}$$

and thus $M = 0$. This argument can be made quite explicit by writing down the phase integral in terms of the vector potential $\mathscr{A}$ and moving the origin in momentum space. We shall see this quantitatively below.

(a) *Quantum-mechanical theory from Bloch density matrix.* As we have seen earlier, the Hamiltonian may be written for an electron in a magnetic field as

$$H = \left\{-\frac{\hbar^2}{2m}\nabla^2 + \frac{e\hbar}{mci}\mathscr{A}.\nabla + \frac{e^2\mathscr{A}^2}{2mc^2}\right\}, \tag{4.15.4}$$

where, for a constant field, we may choose

$$\mathscr{A} = \tfrac{1}{2}\mathscr{H}\times\mathbf{r}. \tag{4.15.5}$$

To obtain the magnetic moment we shall use equation (4.15.3) and since the classical partition function $Z(\beta)$, as we show below, will suffice to calculate the free energy in Fermi–Dirac statistics, we shall work with the Bloch density matrix

$$C(\mathbf{r}',\mathbf{r};\beta) = \sum_i \psi_i^*(\mathbf{r}')\,\psi_i(\mathbf{r})\,\mathrm{e}^{-\beta E_i}, \quad \beta = 1/k_\mathrm{B}T, \tag{4.15.6}$$

where E_i and ψ_i are the one-electron energies and wave functions of the electron in the magnetic field. The partition function Z is clearly given by the trace of C, and this matrix satisfies, as is readily verified, the Bloch equation

$$HC = -\frac{\partial C}{\partial \beta}. \tag{4.15.7}$$

The boundary condition is that at $\beta = 0$

$$C(\mathbf{r}'\mathbf{r}0) = \delta(\mathbf{r}'-\mathbf{r}) \tag{4.15.8}$$

which expresses the completeness condition on the eigenfunctions.

One of the advantages of working with C rather than the Landau wave functions ψ themselves is that we can treat an infinite crystal, thus avoiding the complications of the boundaries of the system. To find Z by explicit calculation of the E_i we would really have to take a large but finite volume and show that the susceptibility is independent of the shape and nature of the volume, as the volume is increased without limit.

Following Sondheimer and Wilson (1951), we are led to the form of the solution by the invariance properties of the Schrödinger equation with respect to a change of coordinates. The electrons are free to move through the whole of space and so no physical property can depend on our choice of origin. However, if $\mathbf{r}\to\mathbf{r}+\mathbf{a}$, $\mathscr{A}\to\mathscr{A}+\frac{1}{2}\mathscr{H}\times\mathbf{a} = \frac{1}{2}\nabla[\mathbf{r}.(\mathscr{H}\times\mathbf{a})]$. Thus, for the Schrödinger equation to remain invariant, every wave function must be multiplied by $\exp[-i\pi e\mathbf{r}.(\mathscr{H}\times\mathbf{a})/hc]$ and so

$$C(\mathbf{r}'+\mathbf{a},\mathbf{r}+\mathbf{a};\beta) = \exp[\pi ie(\mathbf{r}'-\mathbf{r}).(\mathscr{H}\times\mathbf{a})/hc]\,C(\mathbf{r}',\mathbf{r};\beta). \tag{4.15.9}$$

Hence C is of the form

$$C = \exp[-\pi ie\mathscr{H}.(\mathbf{r}'\times\mathbf{r})/hc]\,\mathscr{F}[\mathbf{r}-\mathbf{r}';\mathscr{H}.(\mathbf{r}-\mathbf{r}')]. \tag{4.15.10}$$

To find $\mathscr{F}$, we choose $\mathscr{H}$ such that the vector potential is

$$\mathscr{A} = (-\tfrac{1}{2}\mathscr{H}y, \tfrac{1}{2}\mathscr{H}x, 0). \tag{4.15.11}$$

This implies that the field $\mathscr{H} = \operatorname{curl}\mathscr{A}$ is in the z-direction. Equation (4.15.7) then becomes

$$\frac{\partial C}{\partial \beta} = \left[\frac{\hbar^2}{2m}\nabla^2 - \frac{e\hbar}{2mci}\mathscr{H}\left(\frac{x\partial}{\partial y}-\frac{y\partial}{\partial x}\right) - \frac{e^2\mathscr{H}^2}{8mc^2}(x^2+y^2)\right]C. \tag{4.15.12}$$

The solution of this equation when $\mathscr{H} = 0$ which satisfies the δ function condition (4.15.8) is readily seen to be

$$C(\mathbf{r}', \mathbf{r}; \beta) = \left(\frac{2\hbar^2 \beta\pi}{m}\right)^{-\frac{3}{2}} \exp\left(\frac{-m}{2\hbar^2 \beta}\{\mathbf{r}' - \mathbf{r}\}^2\right), \tag{4.15.13}$$

which can alternatively be obtained from the Dirac density matrix for free electrons (see Problem 2.4). This and equation (4.15.10) suggest that we try the form

$$C(\mathbf{r}', \mathbf{r}; \beta) = f(\beta) \exp\left\{-\frac{ie\mathscr{H}}{2\hbar c}(x'y - y'x) - g(\beta)\,[(x'-x)^2 + (y'-y)^2] - \frac{m}{2\hbar^2 \beta}(z-z')^2\right\}. \tag{4.15.14}$$

Substituting into equation (4.15.12), we find that this form does indeed satisfy the Bloch equation, provided that

$$\frac{df}{d\beta} = -\frac{2\hbar^2}{m} fg - \frac{f}{2\beta}; \quad \frac{dg}{d\beta} = \frac{e^2 \mathscr{H}^2}{8mc^2} - \frac{2\hbar^2}{m} g^2. \tag{4.15.15}$$

The solutions of these equations which recover equation (4.15.13) as $\mathscr{H} \to 0$ and give equation (4.15.8) as $\beta \to 0$ are

$$\left.\begin{aligned} f &= \left(\frac{\pi m}{2\beta}\right)^{\frac{1}{2}} \frac{e\mathscr{H}}{h^2 c} \operatorname{cosech}\left(\frac{e\hbar\mathscr{H}\beta}{2mc}\right), \\ g &= \frac{e\mathscr{H}}{4\hbar c} \coth\left(\frac{e\hbar\mathscr{H}\beta}{2mc}\right). \end{aligned}\right\} \tag{4.15.16}$$

$C(\mathbf{rr}, \beta)$ is independent of $\mathbf{r}$, and we find the classical partition function per unit volume to be

$$Z \equiv C(\mathbf{r}, \mathbf{r}; \beta) = \left(\frac{m}{2\pi\hbar^2 \beta}\right)^{\frac{3}{2}} \frac{\mu_B \mathscr{H} \beta}{\sinh(\mu_B \mathscr{H} \beta)}, \tag{4.15.17}$$

where $\mu_B = e\hbar/2mc$ as usual is the Bohr magneton.

(b) *Free energy*. For Boltzmann statistics, the free energy per unit volume is given by

$$F = -\rho k_B T \ln Z, \tag{4.15.18}$$

where ρ is the number density. Thus

$$M = \rho k_B T \frac{\partial \ln Z}{\partial \mathscr{H}} = -\rho\mu_B \left[\coth\left(\frac{\mu_B \mathscr{H}}{k_B T}\right) - \frac{k_B T}{\mu_B \mathscr{H}}\right]. \tag{4.15.19}$$

For small values of $\mu_B \mathscr{H}/k_B T$, this yields a susceptibility

$$\chi = -\rho\mu_B^2/3k_B T, \tag{4.15.20}$$

which is the correct high-temperature limit, but, of course, not applicable to a degenerate electron gas.

(i) *Application to degenerate gas*: In Fermi–Dirac statistics, the free energy per unit volume is given by

$$F = \rho\zeta - 2k_B T \sum_i \ln[1+e^{(\zeta-E_i)/k_B T}] = \sum_i F(E_i), \tag{4.15.21}$$

where ζ is the Fermi energy, which is to be calculated from the number of electrons by means of the condition

$$0 = \frac{\partial F}{\partial \zeta} = \rho - 2\sum_i f(E_i), \tag{4.15.22}$$

where f is the Fermi function $1/e^{(E_i-\zeta)/k_B T}+1$.
Let us define a quantity

$$\phi(\beta) = -2k_B T \int_0^\infty \ln[1+e^{(\zeta-E)/k_B T}]\, e^{-\beta E}\, dE, \tag{4.15.23}$$

which, being a Laplace transform, can be inverted to yield the result

$$\frac{\partial^2 f}{\partial E^2} = -2k_B T \frac{\partial^2}{\partial E^2} \ln[1+e^{(\zeta-E)\beta}] = \frac{1}{2\pi i}\int_{\sigma-i\infty}^{\sigma+i\infty} e^{Es} s^2 \phi(s)\, ds. \tag{4.15.24}$$

Since it is also readily verified that

$$F = \rho\zeta + \frac{1}{2\pi i}\int_{\sigma-i\infty}^{\sigma+i\infty} \frac{Z(-s)}{s^2}[s^2\phi(s)]\, ds, \tag{4.15.25}$$

we obtain the result

$$F = \rho\zeta + 2\int_0^\infty z(E)\frac{\partial f}{\partial E}\, dE, \tag{4.15.26}$$

where the quantity $z(E)$ is simply

$$z(E) = \frac{1}{2\pi i}\int_{c-i\infty}^{c+i\infty} e^{Es}\frac{Z(s)}{s^2}\, ds. \tag{4.15.27}$$

We can now calculate $z(E)$ from equation (4.15.17) by straightforward, though lengthy, methods. The result (see Sondheimer and Wilson, 1951) is

$$z(E) = \frac{(2\pi m)^{3/2}}{h^3}\left[\frac{8}{15\sqrt{\pi}}E^{5/2} - \frac{1}{3\sqrt{\pi}}(\mu_B \mathscr{H})^2 E^{1/2}\right.$$
$$+\frac{1}{\pi}(\mu_B \mathscr{H})^{5/2}\int_0^\infty\left(\frac{1}{y^{7/2}} - \frac{1}{6y^{3/2}} - \frac{1}{y^{5/2}\sinh y}\right)\mathrm{e}^{-yE/\mu_B\mathscr{H}}\,dy$$
$$\left.-2(\mu_B \mathscr{H})^{5/2}\sum_{r=1}^{\infty}\frac{(-1)^r}{(r\pi)^{5/2}}\cos\left(\frac{r\pi E}{\mu_B \mathscr{H}} - \frac{\pi}{4}\right)\right]. \tag{4.15.28}$$

Since in equation (4.15.26) at sufficiently low temperatures, $\partial f/\partial E$ acts as a delta function, it can be seen that equation (4.15.28) is essentially giving us the desired susceptibility. The second term in equation (4.15.28) gives us the steady Landau diamagnetism, which we referred to earlier in this chapter. The last term, showing clearly the oscillations in the susceptibility as a function of magnetic field, referred to above, represents then the de Haas–van Alphen effect.

However, what is really of the greatest importance in the theory of the de Haas–van Alphen effect is the way in which these oscillations depend on the shape of the Fermi surface. A semiclassical argument due to Onsager (1952) shows this dependence very clearly.

4.15.2 Onsager result for Bloch electrons

When the electrons move in a periodic potential an exact treatment is no longer practicable but fortunately the problem may be handled, as Onsager showed, by semi-classical theory. Before describing this, however, we might ask why it is that the classical result that a magnetic field can do no work on an electron does not carry over into quantum mechanics, as it obviously cannot do if a non-zero susceptibility is to be obtained. The answer is that the quantization of energy levels when a magnetic field is applied leads unavoidably to a small (except with truly colossal fields) energy change in the system, even though the mean density of states is unchanged.

Since therefore the quantization of levels in a magnetic field is at the heart of this problem, we begin by discussing the quantization rules of semi-classical theory.

(a) *Free electrons.* In the semi-classical treatment, we apply the Bohr–Sommerfeld method, which states that

$$\oint p\,dq = (n+\phi)\,h, \tag{4.15.29}$$

where p and q are canonically conjugate variables, the integration is taken round the semi-classical orbit, n is an integer and ϕ a phase factor ($0 \leqslant \phi < 1$).

In the presence of a magnetic field the Hamiltonian for free electrons is

$$H = \left(\mathbf{p} - \frac{e\mathscr{A}}{c}\right)^2 \Big/ 2m. \tag{4.15.30}$$

The canonical momentum $\mathbf{p}$ obtained from Hamilton's equations

$$\dot{x}_i = \frac{\partial H}{\partial p_i}, \quad \dot{p}_i = -\frac{\partial H}{\partial x_i} \tag{4.15.31}$$

is

$$\mathbf{p} = m\dot{\mathbf{r}} + \frac{e\mathscr{A}}{c} \tag{4.15.32}$$

and equation (4.15.29) becomes the line integral

$$\oint \left(m\dot{\mathbf{r}} + \frac{e\mathscr{A}}{c}\right) . d\mathbf{l} = (n+\phi)\, h. \tag{4.15.33}$$

From equation (4.15.30) we obtain Lorentz's equation

$$m\ddot{\mathbf{r}} = \frac{e}{c}(\dot{\mathbf{r}} \times \mathscr{H}) \tag{4.15.34}$$

and hence

$$m\dot{\mathbf{r}} = \frac{e}{c}(\mathbf{r} \times \mathscr{H}) = \mathscr{H}(y\hat{\mathbf{x}} - x\hat{\mathbf{y}})\frac{e}{c} \tag{4.15.35}$$

if $\mathscr{H}$ is in the z-direction.

From this equation we find

$$m \operatorname{curl} \dot{\mathbf{r}} = -\frac{2e}{c}\mathscr{H}, \tag{4.15.36}$$

while

$$\operatorname{curl} \mathscr{A} = \mathscr{H}. \tag{4.15.37}$$

Hence equation (4.15.33) becomes, from Stokes theorem,

$$(n+\phi)\, h = \int \left(\operatorname{curl} m\dot{\mathbf{r}} + e \operatorname{curl}\frac{\mathscr{A}}{c}\right) . d\mathbf{s} = -\frac{e}{c}\int \mathscr{H} . d\mathbf{s}; \tag{4.15.38}$$

that is,

$$\frac{e\mathscr{H}}{c} A = (n+\phi)\, h, \tag{4.15.39}$$

where A is the area of the orbit projected on the (x, y)-plane.

(b) *Bloch electrons.* We base our considerations here on the fact that, within a band, the number of states is very large and forms a quasi-continuum. We also note the group velocity result of section 1.10.1, which we write as

$$\dot{\mathbf{r}} = \frac{\partial E(\mathbf{k})}{\partial \mathbf{k}}, \tag{4.15.40}$$

where we now suppose $\mathbf{r}$ is the vector defining the semi-classical orbit. If $E(\mathbf{k})$ can be regarded as an effective Hamiltonian, equation (4.15.40) suggests that $\mathbf{r}$ and $\mathbf{k}$ can be regarded as canonically conjugate. This is consistent with the fact that, in the absence of an external perturbation

$$0 = \dot{\mathbf{k}} = -\frac{\partial E(\mathbf{k})}{\partial \mathbf{r}}. \tag{4.15.41}$$

We have, in fact, seen in section 1.10 when discussing the effective mass tensor that $E(\mathbf{k})$ does indeed play the part of an effective Hamiltonian.

Now we saw that for free electrons we could obtain the magnetic Hamiltonian by replacing $\mathbf{p}$ (canonically conjugate to $\mathbf{r}$) by $\mathbf{p}-e\mathscr{A}/c$. This suggests that we now replace $\mathbf{k}$ by $\boldsymbol{\kappa}-(e\mathscr{A}/c)$ in $E(\mathbf{k})$, where $\boldsymbol{\kappa}$ is the (now time-dependent) wave-vector for the Bloch electrons. This suggestion is strengthened by the following argument. If we can define a semi-classical orbit, the motion of the electron following out this orbit can be described by a wave-packet built up from Bloch orbitals. Let us see what is the wave-vector $\boldsymbol{\kappa}$ for a Bloch-like function making up this packet. We saw that, for free electrons, the Schrödinger equation remains invariant on translation $\mathbf{a}$ if the wave function is multiplied by $\exp[-ie\mathbf{r}.(\mathscr{H}\times\mathbf{a})/2\hbar c]$. In the presence of a periodic potential, we make the translation a lattice vector $\mathbf{R}$. Thus, supposing a $\mathbf{k}$-label still to exist

$$\psi_{\mathbf{k}}(\mathbf{r}+\mathbf{R}) = \psi_{\mathbf{k}}(\mathbf{r})\exp(i\mathbf{k}.\mathbf{R})\exp[-ie\mathbf{r}.(\mathscr{H}\times\mathbf{R})/2\hbar c]. \tag{4.15.42}$$

Now

$$\frac{e}{2c}\mathbf{r}.(\mathbf{R}\times\mathscr{H}) = \frac{e\mathscr{H}}{2c}(yR_x-xR_y) = -\frac{e}{c}\mathscr{A}(\mathbf{r}).\mathbf{R} \tag{4.15.43}$$

with the gauge of equation (4.15.5) and so

$$\psi_{\mathbf{k}}(\mathbf{r}+\mathbf{R}) = \psi_{\mathbf{k}}(\mathbf{r})\exp i\left[\mathbf{k}-\frac{e\mathscr{A}(\mathbf{r})}{\hbar c}\right].\mathbf{R}. \tag{4.15.44}$$

Thus, at any instant of time, we can say that the wave vector of a Bloch function with which the wave packet is built up is

$$\boldsymbol{\kappa} = \mathbf{k}-\frac{e}{c\hbar}\mathscr{A}(\mathbf{r}) \tag{4.15.45}$$

and the energy is

$$E(\boldsymbol{\kappa}) = E\left[\mathbf{k}-\frac{e}{\hbar c}\mathscr{A}(\mathbf{r})\right], \tag{4.15.46}$$

r, of course, being a position on the semi-classical orbit. Let us now regard $E(\boldsymbol{\kappa})$ as a Hamiltonian for semi-classical purposes, as was first postulated by Onsager and given formal justification by Wannier and others.

Hamilton's equations now take the form, with $\hbar = 1$,

$$\dot{\mathbf{r}} = \frac{\partial E(\boldsymbol{\kappa})}{\partial \mathbf{k}}, \quad \dot{\mathbf{k}} = -\frac{\partial E(\boldsymbol{\kappa})}{\partial \mathbf{r}} \tag{4.15.47}$$

and from the second of these we find

$$\dot{k}_x = \frac{\partial E}{\partial \boldsymbol{\kappa}}.\frac{\partial \mathscr{A}}{\partial x} = \dot{\mathbf{r}}\frac{\partial \mathscr{A}}{\partial x}, \quad \text{etc.} \tag{4.15.48}$$

On the other hand, the electron is in the state $\boldsymbol{\kappa} = \mathbf{k}-(e\mathscr{A}/c)$ and the crystal momentum varies as

$$\dot{\boldsymbol{\kappa}} = \dot{\mathbf{k}} - \frac{e\dot{\mathscr{A}}}{c}. \tag{4.15.49}$$

Thus

$$\dot{\kappa}_x = \dot{\mathbf{r}}.\frac{\partial \mathscr{A}}{\partial x} - \frac{e}{c}\dot{\mathscr{A}}_x, \quad \text{etc.,} \tag{4.15.50}$$

and if $\mathscr{A}$ does not depend explicitly on time we have

$$\dot{\mathscr{A}}_x = \frac{d\mathscr{A}_x}{dt} = \frac{\partial \mathscr{A}_x}{\partial x}\dot{x} + \frac{\partial \mathscr{A}_y}{\partial y}\dot{y} + \frac{\partial \mathscr{A}_z}{\partial z}\dot{z}. \tag{4.15.51}$$

From this it is readily seen that

$$\dot{\kappa}_x = \frac{e}{c}(\dot{\mathbf{r}} \times \operatorname{curl} \mathscr{A})_x, \quad \text{etc.,} \tag{4.15.52}$$

or

$$\dot{\boldsymbol{\kappa}} = \frac{e}{c}(\dot{\mathbf{r}} \times \mathscr{H}). \tag{4.15.53}$$

At this stage, we return to the Bohr–Sommerfeld quantization rules. Into these we insert $\mathbf{p} \equiv \mathbf{k}$ as from equation (4.15.40) the coordinate canonically conjugate to **r**. In view of equation (4.15.53) we have, corresponding to equation (4.15.33),

$$\int\left(+\boldsymbol{\kappa}\frac{+e\mathscr{A}}{c}\right).d\mathbf{l} = (n+\phi)h \tag{4.15.54}$$

and we arrive at equation (4.15.53) in just the same manner as before. We also observe that, on integrating equation (4.15.53),

$$\boldsymbol{\kappa} = \frac{e}{c}(\mathbf{r}\times\mathscr{H}), \tag{4.15.55}$$

where any constant of integration has been removed by a change of origin. We see that the orbit in **k**-space has the same shape as the orbit in **r**-space but is rotated through $\pi/2$. We also see from equation (4.15.55) that

$$\kappa = \frac{e}{c} r\mathscr{H} \tag{4.15.56}$$

and so the area in **k**-space is given by (inserting $\hbar$ explicitly)

$$A_k = \frac{e^2\mathscr{H}^2 A}{\hbar^2 c^2} = \frac{2\pi e}{c\hbar}\mathscr{H}(n+\phi). \tag{4.15.57}$$

The shape of the orbit is determined by the fact that $\boldsymbol{\kappa}$ must stay on a constant-energy surface, since the magnetic field does no work on the electron (apart from that arising from the quantization of the levels, as we have already remarked):

$$\frac{dE}{dt} = \dot{\boldsymbol{\kappa}}.\frac{dE}{d\boldsymbol{\kappa}} = \frac{e\mathbf{v}}{\hbar c}.(\dot{\mathbf{r}}\times\mathscr{H}) = 0. \tag{4.15.58}$$

Figure 4.17 shows three orbits in **k**-space (on a constant-energy surface).

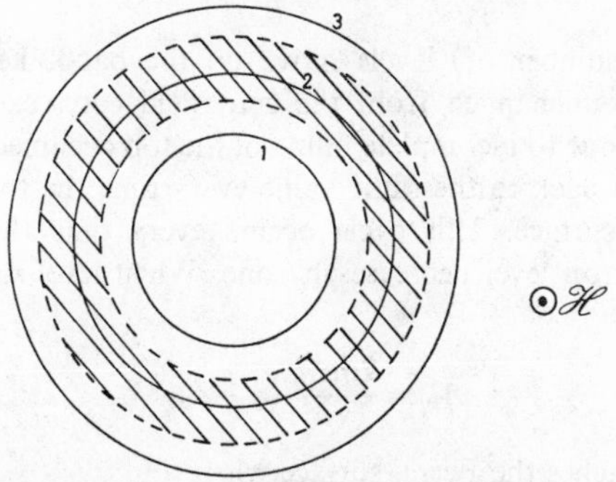

FIGURE 4.17. Orbits in **k**-space on constant-energy surfaces. Shaded area is area ΔA of equation (4.15.60).

The density of states in **k**-space is constant $n(\mathbf{k}) = 2/8\pi^3$ and to keep the mean density of states the same as in the absence of the field, the degeneracies of the quantized orbits must all be the same (adjacent orbits differ in area by $(2\pi e/c\hbar)\mathscr{H}$, independent of **k**). Taking a slice dk_z in the z-direction, the number of electrons going into a quantized orbit is, defining ΔA as

$$A_k = \frac{2\pi e}{c\hbar}\mathscr{H} \tag{4.15.59}$$

$$n(\mathbf{k})\,\Delta A\,dk_z = \frac{2\,dk_z}{8\pi^3}\Delta A = \frac{dk_z}{4\pi^3}\left(\frac{2\pi e}{\hbar c}\right)\mathscr{H}, \tag{4.15.60}$$

ΔA being the shaded area in Figure 4.18. One sees the degeneracy to be proportional to $\mathscr{H}$. Now looking at Figure 4.18, we see that as $\mathscr{H}$ is increased

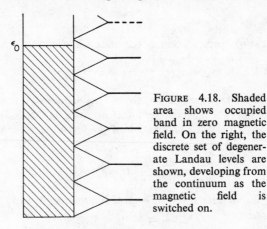

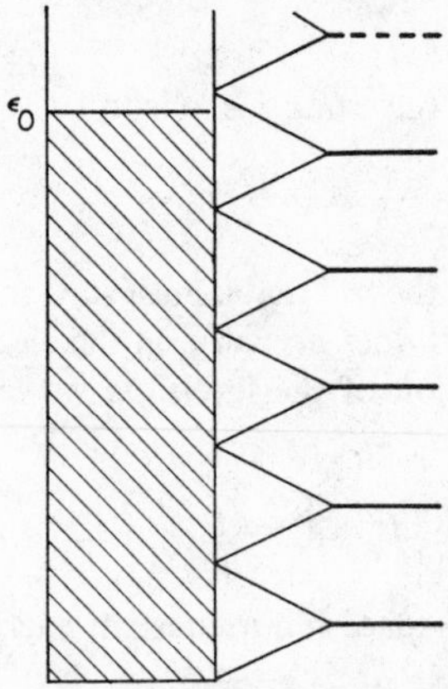

FIGURE 4.18. Shaded area shows occupied band in zero magnetic field. On the right, the discrete set of degenerate Landau levels are shown, developing from the continuum as the magnetic field is switched on.

the (very large number of) levels move up the band, keeping the mean density of states unchanged from the zero field case, except for a small change in energy due to incomplete filling of the top occupied level. However, the energy comes back to the same value every time the top level coincides with the Fermi surface. This cycle occurs every time the n of equation (4.15.60) for the top level decreases by one. When level n_i is at the Fermi surface

$$A_{k_f} = \frac{2\pi e\mathscr{H}}{c\hbar}(n_i+\phi). \tag{4.15.61}$$

The next level reaches the Fermi surface when

$$A_{k_f} = \frac{2\pi e}{c\hbar}(\mathscr{H}+\delta\mathscr{H})(n_i-1+\phi). \tag{4.15.62}$$

These equations may also be written as

$$\frac{1}{\mathscr{H}} = \frac{2\pi e}{c\hbar A_{k_f}}(n_i+\phi) \tag{4.15.63}$$

and

$$\frac{1}{\mathscr{H}} - \delta\left(\frac{1}{\mathscr{H}}\right) = \frac{2\pi e}{c\hbar A_{k_t}}(n_i - 1 + \phi). \tag{4.15.64}$$

The energy is therefore periodic in $1/\mathscr{H}$, with period

$$\delta\left(\frac{1}{\mathscr{H}}\right) = \frac{2\pi e}{c\hbar A_{k_t}}. \tag{4.15.65}$$

This period gives A_{k_t}, the area of the orbit at the Fermi surface. Let us recall, however, that all this is just for a slice of thickness dk_z. The total energy is obtained by integrating over dk_z; it should be noted that the susceptibility will also oscillate with $1/\mathscr{H}$. If we expand E as a Fourier series for each slice, we have

$$E = \sum_n \int A_n \cos\left(\frac{n\hbar c}{e} A_{k_t}/\mathscr{H}\right) dk_z. \tag{4.15.66}$$

Now the integrand oscillates very rapidly as a function of k_z, since for the fields used in practice

$$\frac{\hbar c}{2\pi e} A_{k_t}/\mathscr{H} \gg 1. \tag{4.15.67}$$

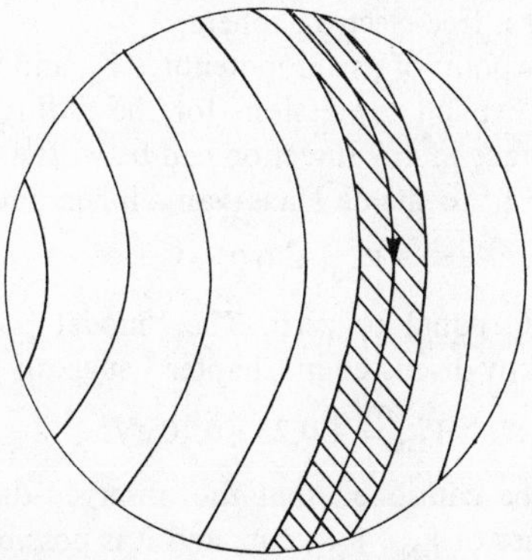

FIGURE 4.19. Showing Fermi surface, with orbits in **k**-space marked on it, perpendicular to k_z. Unshaded areas (referred to as a in text) are regions where A_k varies rapidly with k_z. Shaded region b shows extremal area when A_{k_f} is stationary as k_z is varied.

Thus the only appreciable contributions will come from regions where $A_{k_t}(k_z)$ is stationary with respect to k_z, and the period measures the extremal areas of the Fermi surface.

If we refer to Figure 4.19, in a region *a* the integrand of (4.15.66) varies very rapidly with k_z. The only appreciable contributions come from regions *b*.

The above considerations will hold only provided the temperature is sufficiently low that the spacing of the quantized energies is very much greater than the thermal energy.

4.15.3 Spherical nature of Fermi surfaces in alkali metals

It is not our purpose here to discuss the details of Fermi surfaces. However, because of its importance to Fermi liquid theory, we record that the alkali metals have much more nearly spherical Fermi surfaces than the other monovalent group, the noble metals. This is because, in the heavier alkalis, the d-bands are much further away from the Fermi level. In fact the d-bands in Cu, Ag and Au cause so much distortion in the Fermi surface that it contacts the zone boundary.

The conclusions from the de Haas–van Alphen effect experiments on the alkalis are that departures from a sphere are:

(i) about 1 in 1000 for Na and K,

(ii) about 1% for Rb,

(iii) several % for Cs.

Furthermore, for these four metals, the mean radius is, within experimental error, the same as for a free-electron sphere.

If we adopt the viewpoint of pseudopotentials V_K, and work only to second order in V_K/E_f, then explicit expressions for the radius and cross-sectional area of the Fermi surface in any direction can be written down. For Na, this proves to give a good fit to the de Haas–van Alphen data if one takes

$$|V_{110}| = 0.23 \text{ eV},$$

putting V_{200}, V_{211}, etc. equal to zero. The "model potential" calculation of Heine and Abarenkov discussed in Chapter 1 suggests that

$$V_{110} = +0.25 \pm 0.10 \text{ eV}.$$

However, it should be cautioned that the observed distortions are rather insensitive to the values of V_{200}, V_{211}, etc., and it is possible only to set upper limits on them.

The Fermi surface of Li has not, so far, been determined by the de Haas–van Alphen effect. However, there is evidence from positron annihilation experiments (Donaghy and Stewart, 1967; see Chapter 5 for details) that in the high-temperature bcc phase the Fermi surface is significantly distorted. In particular, it appears that k_{110} is some 5% greater than k_{100}.

CHAPTER 5

Interaction of particles with a crystal

5.1 Introduction

Having discussed the theoretical framework available for describing the behaviour of electrons, phonons and magnons in a crystal, we turn in this chapter to describe in detail the way in which the results of the theory may be compared with experiment. This comparison proves possible if one studies the scattering of X-rays, neutrons and electrons by the crystal. As we shall see, experiment not only provides information on the electron density in a crystal, but also the dispersion laws of phonons and magnons may be measured, together, at least in principle, with their lifetimes.

One may unify the treatments of the various kinds of scattering by working with the van Hove correlation function already introduced in Chapter 2, and we shall begin with a general discussion before proceeding to particular cases.

5.2 Differential cross-section in Born approximation

We now consider the scattering by a system of an incoming particle with wave function $e^{i\mathbf{k}\cdot\mathbf{r}}$. The particle may be specified as a neutron, but we shall see that photon scattering, under assumptions we shall emphasize later, also obeys the general form of the equations we shall derive.

It is our aim to obtain an expression for the out-going particle-current or flux. The classical relation between flux $\mathbf{F}$ and density ρ is

$$\operatorname{div}\mathbf{F} = -\frac{\partial\rho}{\partial t} \tag{5.2.1}$$

and can immediately be rewritten in quantum-mechanical terms by putting $\rho = |\psi|^2$, ψ being the wave function of the scattered particle, so that

$$-\operatorname{div}\mathbf{F} = \psi^*\frac{\partial\psi}{\partial t}+\psi\frac{\partial\psi^*}{\partial t}. \tag{5.2.2}$$

It is simplest to regard the scattering as a stationary state process; to do so we must remove the time derivatives from equation (5.2.2). With this aim, we write down the time-dependent Schrödinger equations for ψ and ψ^*, namely,

$$-\frac{\hbar^2}{2M}\nabla^2\psi + V(\mathbf{r})\psi = i\hbar\frac{\partial\psi}{\partial t} \tag{5.2.3}$$

and

$$-\frac{\hbar^2}{2M}\nabla^2\psi^* + V(\mathbf{r})\psi^* = \frac{\hbar}{i}\frac{\partial\psi^*}{\partial t}, \tag{5.2.4}$$

M being the mass of the scattered particle.

From these equations we obtain

$$\frac{\partial\rho}{\partial t} = \frac{\partial}{\partial t}|\psi|^2 = -\frac{i\hbar}{2M}(\psi\nabla^2\psi^* - \psi^*\nabla^2\psi), \tag{5.2.5}$$

i.e.

$$-\frac{\partial\rho}{\partial t} = \frac{i\hbar}{2M}\operatorname{div}(\psi\nabla\psi^* - \psi^*\nabla\psi). \tag{5.2.6}$$

By comparison of this equation with equations (5.2.1) and (5.2.2) we see that

$$\mathbf{F} = \frac{\hbar}{2Mi}(\psi^*\nabla\psi - \psi\nabla\psi^*) + \mathbf{A}, \tag{5.2.7}$$

where $\mathbf{A}$ is some vector whose divergence is zero. We in fact take $\mathbf{A} = 0$ to achieve correspondence with classical results: classically, $\mathbf{F} = \rho\mathbf{v}$, and the quantum mechanical operator corresponding to the momentum $M\mathbf{v}$ is $(\hbar/i)\nabla$; we therefore directly identify $\mathbf{F}$ as (cf. section 4.1.5)

$$\mathbf{F} = \frac{\hbar}{2Mi}(\psi^*\nabla\psi - \psi\nabla\psi^*). \tag{5.2.8}$$

We turn now to the quantal description of the entire system, scatterer plus scattered particle.

Let the scattering system have, in isolation, a Hamiltonian $H(q)$ with a complete set of eigenfunctions $\phi_n(q)$ and eigenvalues E_n. Then if we represent the interaction between the particle (coordinates $\mathbf{r}$) and scatterer as $V(\mathbf{r}, q)$ the Schrödinger equation for the entire system is

$$-\frac{\hbar^2}{2M}\frac{\partial^2}{\partial\mathbf{r}^2}\psi(\mathbf{r},q) + [H(q) + V(\mathbf{r},q)]\psi(\mathbf{r},q) = E\psi(\mathbf{r},q). \tag{5.2.9}$$

It is natural to insert into this expression an expansion of the form

$$\psi(\mathbf{r},q) = \sum_n \chi_n(\mathbf{r})\,\phi_n(q) \tag{5.2.10}$$

when we obtain

$$-\frac{\hbar^2}{2M}\sum_n \frac{\partial^2}{\partial \mathbf{r}^2}\chi_n(\mathbf{r})\,\phi_n(q)+\sum_n \chi_n(\mathbf{r})\,E_n\,\phi_n(q)+V(\mathbf{r}q)\,\psi(\mathbf{r},q) = E\sum_n \chi_n(\mathbf{r})\,\phi_n(q). \tag{5.2.11}$$

Taking the inner products of both sides with respect to $\phi_f(q)$, we obtain

$$-\frac{\hbar^2}{2M}\nabla^2\chi_f(\mathbf{r})+[E_f-E]\,\chi_f = -\langle\phi_f(q)|V|\psi(\mathbf{r},q)\rangle \tag{5.2.12}$$

and the general solution to this equation is (cf. section 1.11)

$$\chi_f(\mathbf{r}) = e^{i\mathbf{k}'.\mathbf{r}}-\frac{1}{4\pi}\int\frac{e^{ik'R}}{R}\langle\phi_f(q)|U|\psi(\mathbf{r}',q)\rangle\,d\mathbf{r}', \tag{5.2.13}$$

where we have written $\hbar^2 k'^2/2M = E-E_f$, $R = |\mathbf{r}-\mathbf{r}'|$ and $U = 2MV/\hbar^2$.

The experimental situation justifies our taking the asymptotic form of this equation. With the definition $\mathbf{k}' = k'\hat{\mathbf{r}}$, we have, when $r \geqslant r'$, $R \rightarrow (\mathbf{r}-\mathbf{r}').\hat{\mathbf{r}}$, and so $k'R \rightarrow k'r-\mathbf{k}'.\mathbf{r}'$; hence the part of equation (5.2.13) of interest becomes

$$\begin{aligned}\chi_f(\mathbf{r}) &= -\frac{1}{4\pi}\frac{e^{ik'r}}{r}\int e^{-i\mathbf{k}'.\mathbf{r}'}\langle\phi_f(q)|U(\mathbf{r}',q)|\,\psi(\mathbf{r}',q)\rangle\,d\mathbf{r}'\\ &= -\frac{1}{4\pi}\frac{e^{ik'r}}{r}\langle\mathbf{k}'f|U|\psi\rangle.\end{aligned} \tag{5.2.14}$$

We can see that this contribution to equation (5.2.13) represents the scattered particle, momentum $\mathbf{k}'$, leaving the scatterer in energy state E_f.

We obtain the contribution to the flux from equation (5.2.8). It is

$$\left(\frac{1}{4\pi}\right)^2\frac{\hbar}{i}\frac{1}{r^2}\left(\frac{ik'}{M}\right)|\langle\mathbf{k}'f|U|\psi\rangle|^2. \tag{5.2.15}$$

Suppose that the incoming particle found the scatterer in the state $\phi_i(q)$. The scattering system's energy change is then $E_f-E_i = \omega$ and we see that the total contribution to the flux by particles in an energy range $d\omega$ is

$$\left(\frac{1}{4\pi}\right)^2\frac{\hbar}{i}\frac{1}{r^2}\left(\frac{ik'}{M}\right)\sum_f|\langle\mathbf{k}'f|U|\psi\rangle|^2\,\delta(\omega+E_i-E_f)\,d\omega. \tag{5.2.16}$$

If, therefore, we define the differential cross-section per solid angle Ω by

$$d\sigma = \frac{\text{flux through area } r^2\, d\Omega}{\text{incident flux}}, \tag{5.2.17}$$

we have

$$(4\pi)^2 \frac{d^2\sigma}{d\omega\, d\Omega} = \frac{k'}{k} \sum_f |\langle \mathbf{k}' f | U | \psi \rangle|^2\, \delta(\omega + E_f - E_i). \tag{5.2.18}$$

The first Born approximation consists in substituting the initial state $e^{i\mathbf{k}.\mathbf{r}}\phi_i(q)$ for ψ on the right-hand side of equation (5.2.18), whereupon we obtain

$$\frac{d^2\sigma}{d\Omega\, d\omega} = \left(\frac{1}{4\pi}\right)^2 \frac{k'}{k} \sum_f |\langle \mathbf{k}' f | U | \mathbf{k} i \rangle|^2\, \delta(\omega + E_f - E_i). \tag{5.2.19}$$

The modification necessary to take account of thermal excitation of the scatterer may be made at once. If p_i is the Boltzmann probability of the particle finding the system in the state $|i\rangle$, we write

$$\frac{d^2\sigma}{d\Omega\, d\omega} = \left(\frac{1}{4\pi}\right)^2 \frac{k'}{k} \sum_{if} p_i |\langle \mathbf{k}' f | U | \mathbf{k} i \rangle|^2\, \delta(\omega + E_f - E_i). \tag{5.2.20}$$

Suppose now the interaction potential is a sum of pair potentials ($\mathbf{r}_i$ being the coordinates of the particles of the scatterer):

$$V(\mathbf{r}, q) = \sum_i V(\mathbf{r} - \mathbf{r}_i) \tag{5.2.21}$$

which is so for all particular situations we shall consider.

Then

$$\begin{aligned} \langle \mathbf{k}' f | V | \mathbf{k} i \rangle &= \langle f | \int V(\mathbf{r} - \mathbf{r}')\, e^{i(\mathbf{k}-\mathbf{k}').\mathbf{r}}\, d\mathbf{r} | i \rangle \\ &= \tilde{V}(\boldsymbol{\kappa}) \sum_i \langle f | e^{i\boldsymbol{\kappa}.\mathbf{r}_i} | i \rangle, \end{aligned} \tag{5.2.22}$$

where

$$\tilde{V}(\boldsymbol{\kappa}) = \int e^{-i\boldsymbol{\kappa}.\mathbf{r}}\, V(\mathbf{r})\, d\mathbf{r} \quad (\boldsymbol{\kappa} = \mathbf{k}' - \mathbf{k}). \tag{5.2.23}$$

The differential scattering cross-section may now be written as

$$(4\pi)^2 \frac{d^2\sigma}{d\omega\, d\Omega} = \frac{k'}{k} |\tilde{U}(\boldsymbol{\kappa})|^2\, S(\boldsymbol{\kappa}, \omega) \tag{5.2.24}$$

and $S(\boldsymbol{\kappa}, \omega)$ is the *structure factor* which we met earlier in Chapter 3 for the ground state, defined here by

$$S(\boldsymbol{\kappa}, \omega) = \sum_{if} p_i |\langle i | \sum_f e^{i\boldsymbol{\kappa}.\mathbf{r}_j} | f \rangle|^2\, \delta(\omega + E_f - E_i). \tag{5.2.25}$$

The intimate relation of the structure factor to the Van Hove function is readily found. Taking the Fourier transform of equation (5.2.25) with respect to time t, we get

$$S(\boldsymbol{\kappa}, t) = \sum_{if} p_i |\langle i| \sum_j e^{i\boldsymbol{\kappa}.\mathbf{r}_j} |f\rangle|^2 e^{i(E_i - E_f)t}, \tag{5.2.26}$$

which we write in the form

$$S(\boldsymbol{\kappa}, t) = \int g(\mathbf{r}_1 \mathbf{r}_2; t) e^{i\boldsymbol{\kappa}.(\mathbf{r}_1 - \mathbf{r}_2)} d\mathbf{r}_1 d\mathbf{r}_2. \tag{5.2.27}$$

$g(\mathbf{r}_1 \mathbf{r}_2; t)$ is the van Hove correlation function we introduced in Chapter 2.

As $t \to \infty$, the oscillations of $e^{i\omega t}$ become so rapid that any integral over ω picks out only terms for which $\omega = 0$. It follows that we can regard $\lim_{t\to\infty} e^{i\omega t} = 0$ unless $\omega = 0$, whereupon we see from equation (5.2.26) that

$$(4\pi)^2 \frac{d^2\sigma}{d\omega\, d\Omega}\bigg|_{\text{elastic}} = |\tilde{U}(\boldsymbol{\kappa})|^2 S(\boldsymbol{\kappa}, t)\bigg|_{t=\infty}. \tag{5.2.28}$$

In terms of Heisenberg operators, we can rewrite equation (5.2.26) in the alternative form

$$S(\boldsymbol{\kappa}, t) = \left\langle \sum_{ij} \exp\{i\boldsymbol{\kappa}.[\mathbf{r}_i(t) - \mathbf{r}_j(0)]\} \right\rangle, \tag{5.2.29}$$

where $\langle\ \rangle$ denotes a statistical average analogous to that of (5.2.26). Concerning the statistical average in equation (5.2.29), let us consider the average of the product of any two Heisenberg operators, A and B. As $t \to \infty$, we assume that the physical observables corresponding to $A(0)$ and $B(t)$ are uncorrelated. If this is so,

$$\lim_{t\to\infty} \langle A(0) B(t) \rangle = \langle A(0) \rangle \langle B(\infty) \rangle, \tag{5.2.30}$$

a result which will be used below.

5.3 Scattering of neutrons by nuclei

The scattering we are to describe in this section is due to the lattice vibrations alone and so it is convenient to discuss it prior to the scattering of particles by electrons, the theory of which must take some account of the effect of the lattice vibrations on the electrons.

Let us suppose a nucleus presents a deep and narrow well to a neutron. For small neutron energies, we may approximate the well by a delta function‡

$$V(\mathbf{r}) = a\delta(\mathbf{r}). \tag{5.3.1}$$

‡ For a discussion of the validity of this procedure, see Marshall and Lovesey (1971, p. 9).

By an analysis similar to that of the last section, the scattering of a particle by the potential $V(\mathbf{r})$ gives

$$\frac{d\sigma}{d\Omega} = \frac{2\pi}{M}\left|\int V(\mathbf{r})\, e^{i\boldsymbol{\kappa}.\mathbf{r}}\, d\mathbf{r}\right|^2 = b^2, \tag{5.3.2}$$

where the b defined here is termed the scattering length (cf. section 1.11.1, Chapter 1), independent of $\boldsymbol{\kappa}$, with the choice (5.3.1).

We see that in the Born approximation, if an assembly of scatterers all have the same scattering length, we may formally write the pair interaction between particles and scatterer as

$$V(\mathbf{r}) = \frac{2\pi b}{M}\, \delta(\mathbf{r}). \tag{5.3.3}$$

Let the scatterers be on sites $\mathbf{S}_n$. Then, from equation (5.2.24) we may write

$$\frac{d^2\sigma}{d\Omega\, d\omega} = b^2\left(\frac{k'}{k}\right) \sum_{\mathbf{S}_m \mathbf{S}_n} \int dt\, e^{-i\omega t} \langle \exp[-i\boldsymbol{\kappa}.\mathbf{S}_m(0)] \exp[i\boldsymbol{\kappa}.\mathbf{S}_n(t)]\rangle. \tag{5.3.4}$$

The scattering particles under consideration are the nuclei of a crystal, vibrating about the lattice points $\mathbf{m}$ and so we write

$$\mathbf{S}_m = \mathbf{m} + \mathbf{u}_m. \tag{5.3.5}$$

Then

$$\sum_{\mathbf{m},\mathbf{n}} \langle \exp[-i\boldsymbol{\kappa}.\mathbf{S}_m(0) \exp[i\boldsymbol{\kappa}.\mathbf{S}_n(t)]\rangle$$
$$= \sum_{\mathbf{m},\mathbf{n}} \langle \exp[-i\boldsymbol{\kappa}.\mathbf{u}_m(0)] \exp[i\boldsymbol{\kappa}.\mathbf{u}_n(t)]\rangle \exp[i\boldsymbol{\kappa}.(\mathbf{m}-\mathbf{n})]. \tag{5.3.6}$$

Now by a result obtained in Chapter 1, namely

$$\sum_{\mathbf{m},\mathbf{n}} \exp[i\boldsymbol{\kappa}.(\mathbf{m}-\mathbf{n})] = N \sum_{\mathbf{n}} \exp(i\boldsymbol{\kappa}.\mathbf{n}) = N\Omega_B \sum_{\mathbf{K}} \delta(\boldsymbol{\kappa}+\mathbf{K}), \tag{5.3.7}$$

the $\mathbf{K}$'s being reciprocal lattice vectors and N the number of cells, we can rewrite equation (5.3.4) in the case where the $\mathbf{u}_m$, $\mathbf{u}_n$ are zero as

$$\frac{d^2\sigma}{d\omega\, d\Omega} = b^2\, \delta(\omega)\, N\Omega_B \sum_{\mathbf{K}} \delta(\boldsymbol{\kappa}+\mathbf{K}). \tag{5.3.8}$$

This result, of course, ignores the effect of temperature. Let us now consider the modifications caused by the nuclear vibrations.

5.3.1 Elastic scattering and Debye–Waller factor

Using the theorem that, if operators A and B both commute with their commutator,

$$e^A e^B = e^{A+B} e^{[A,B]/2}, \tag{5.3.9}$$

the truth of which is readily demonstrated by expansion of the exponentials, we find

$$\langle \exp[-i\boldsymbol{\kappa}.\mathbf{u}_m(t)]\exp[i\boldsymbol{\kappa}.\mathbf{u}_n(0)]\rangle$$
$$= \langle \exp\{-i\boldsymbol{\kappa}.[\mathbf{u}_m(t)-\mathbf{u}_n(0)]\}\exp\{[\boldsymbol{\kappa}.\mathbf{u}_m(t), \boldsymbol{\kappa}.\mathbf{u}_n(0)]/2\}\rangle. \quad (5.3.10)$$

In accordance with equation (5.2.28) we obtain the elastic scattering by taking the limit $t\to\infty$. In this limit, $\mathbf{u}_m(t)$ and $\mathbf{u}_n(0)$ will be uncorrelated and so commute and we therefore obtain from equation (5.2.30)

$$\lim_{t\to\infty}\langle \exp[-i\boldsymbol{\kappa}.\mathbf{u}_m(t)]\exp[i\boldsymbol{\kappa}.\mathbf{u}_n(0)]\rangle = |\langle \exp(-i\boldsymbol{\kappa}.\mathbf{u}_m)\rangle|^2. \quad (5.3.11)$$

This expression is independent of n and m because of translational invariance. Then, combining equation (5.3.11) with equations (5.3.4) and (5.3.6) we find

$$\left.\frac{d^2\sigma}{d\Omega\, d\omega}\right|_{\text{elastic}} = b^2\,\delta(\omega)\,|d(\boldsymbol{\kappa})|^2 \sum_{\mathbf{K}} \delta(\boldsymbol{\kappa}+\mathbf{K})N\Omega_{\mathrm{B}}. \quad (5.3.12)$$

Thus, even when the lattice is vibrating, the elastic scattering consists only of Bragg peaks, multiplied however by the factor

$$|d(\boldsymbol{\kappa})|^2 = |\langle \exp(-i\boldsymbol{\kappa}.\mathbf{u}_m)\rangle|^2, \quad (5.3.13)$$

which is termed the Debye–Waller factor.

In the harmonic approximation, we can evaluate $d(\boldsymbol{\kappa})$ by using the theorem that for a three-dimensional harmonic oscillator

$$\langle \exp(i\boldsymbol{\xi})\rangle = \exp[-\tfrac{1}{2}\langle \xi^2\rangle]. \quad (5.3.14)$$

A proof of this can be found in the book by Messiah (1962), but we wish here to follow Ambegaokar, Conway and Baym (1964) in introducing the method of cumulants, which we shall later apply when considering the total scattering.

Given a set of variables $\{x_1, x_2, \ldots\}$ we define the cumulants $\langle x_i\rangle_c$, $\langle x_i x_j\rangle_c$, etc. in terms of expectation values as follows

$$\left.\begin{aligned}\langle x_i\rangle &= \langle x_i\rangle_c,\\ \langle x_i x_j\rangle &= \langle x_i x_j\rangle_c + \langle x_i\rangle_c\langle x_j\rangle_c,\\ \langle x_i x_j x_k\rangle &= \langle x_i x_j x_k\rangle_c + \langle x_i x_j\rangle_c\langle x_k\rangle_c + \langle x_i x_k\rangle_c\langle x_j\rangle_c,\\ &\quad + \langle x_j x_k\rangle_c\langle x_i\rangle_c + \langle x_i\rangle_c\langle x_j\rangle_c\langle x_k\rangle_c,\\ &\ldots\ldots\ldots\ldots\ldots\ldots,\end{aligned}\right\} \quad (5.3.15)$$

which in the case we require, when $x = x_i = x_j = \ldots$, become

$$\left.\begin{aligned}\langle x\rangle &= \langle x\rangle_c,\\ \langle x^2\rangle &= \langle x^2\rangle_c + \langle x\rangle_c^2,\\ \langle x^3\rangle &= \langle x^3\rangle_c + 3\langle x^2\rangle_c\langle x\rangle_c + \langle x\rangle_c^3,\\ &\ldots\ldots\ldots\ldots\end{aligned}\right\} \quad (5.3.16)$$

Expansions in cumulants are, in effect, "linked cluster" expansions as frequently used in the theory of liquids for example. One can readily show (see, for example, Fischer, 1964) that

$$\ln \langle e^x \rangle = \langle x \rangle_c + \frac{1}{2!}\langle x^2 \rangle_c + \frac{1}{3!}\langle x^3 \rangle_c + \dots$$
$$= \langle e^x \rangle_c - 1. \tag{5.3.17}$$

The property of cumulants which is of prime importance to us here is that, if the nth-order cumulant is zero, the expectation value of a product of n operators may be expressed entirely in terms of lower-order expectation values. In fact, in the harmonic approximation, it can be shown by direct calculation that

$$\langle u^n \rangle_c = 0 \quad (n > 2). \tag{5.3.18}$$

Now from equations (5.3.13) and (5.3.17) we find, using the fact that for a Bravais lattice $\langle(\boldsymbol{\kappa}.\mathbf{u})^n\rangle_c = 0$ if n is odd

$$d(\boldsymbol{\kappa}) = \exp\left[-\frac{1}{2!}\langle(\boldsymbol{\kappa}.\mathbf{u}_n)^2\rangle_c + \frac{1}{4!}\langle(\boldsymbol{\kappa}.\mathbf{u}_n)^4\rangle_c + \dots\right]. \tag{5.3.19}$$

Hence, in the harmonic approximation, we may write the Debye–Waller factor as

$$|d(\boldsymbol{\kappa})|^2 = e^{-2W}, \tag{5.3.20}$$

where

$$W = \tfrac{1}{2}\langle(\boldsymbol{\kappa}.\mathbf{u}_n)^2\rangle. \tag{5.3.21}$$

For a discussion of cumulants and linked cluster expansions applied to the thermodynamics of the anharmonic crystal, a general reference is Choquard (1967).

Debye model. We saw in Chapter 3 that the displacement $\mathbf{u}_m$ may be written, in terms of the phonon creation and annihilation operators $a^\dagger_{\mathbf{k}\sigma}$, $a_{\mathbf{k}\sigma}$ as

$$\mathbf{u}_m(t) = -i\sum_{\mathbf{q}}\left(\frac{\hbar}{2MN\omega(\mathbf{q}\sigma)}\right)^{\frac{1}{2}} \boldsymbol{\epsilon}_{\mathbf{q}\sigma}$$
$$\times\{\exp[-i(\mathbf{q}.\mathbf{m}+\omega(\mathbf{q}\sigma)t)]\,a^\dagger_{\mathbf{q}\sigma} + \exp[i(\mathbf{q}.\mathbf{m}+\omega(\mathbf{q}\sigma)t)]\,a_{\mathbf{q}\sigma}\}. \tag{5.3.22}$$

It then follows that, in the harmonic approximation,

$$\langle(\boldsymbol{\kappa}.\mathbf{u}_m(0))^2\rangle = \sum_{\mathbf{q}\sigma}\frac{\hbar}{2MN\omega(\mathbf{q}\sigma)}(\boldsymbol{\kappa}.\boldsymbol{\epsilon}_{\mathbf{q}\sigma})^2(g\langle n(\mathbf{q}\sigma)\rangle + 1), \tag{5.3.23}$$

where

$$2\langle n(\mathbf{q}\sigma)\rangle = \langle a_{\mathbf{q}\sigma}a^\dagger_{\mathbf{q}\sigma} + a^\dagger_{\mathbf{q}\sigma}a_{\mathbf{q}\sigma}\rangle - 1. \tag{5.3.24}$$

In the Debye model, we may take the three polarization branches to be degenerate, in which case we can always choose one of the polarization directions along $\boldsymbol{\kappa}$. Equation (5.3.23) then becomes

$$\frac{\kappa^2}{3}\langle u_n^2(0)\rangle = \sum_{\mathbf{q}} \frac{\hbar\kappa^2}{2NM}\left(\frac{2\langle n(q)\rangle+1}{\omega(q)}\right). \tag{5.3.25}$$

Two other assumptions of the Debye model are that $\omega = v_s q$ and that the spectrum is cut off at a maximum frequency ω_{D}, the Debye frequency, which is often written in terms of the Debye temperature θ_{D} by the relation $\hbar\omega_{\mathrm{D}} = k_{\mathrm{B}}\,\theta_{\mathrm{D}}$. We also have (see Chapter 3)

$$\langle n(q)\rangle = \frac{1}{\mathrm{e}^{\beta\omega(q)}-1} \quad \left(\beta = \frac{\hbar}{k_{\mathrm{B}}T}\right). \tag{5.3.26}$$

For low temperatures T, therefore, we obtain

$$\frac{2M}{3\hbar^2}\kappa^2\langle u_n^2\rangle = \frac{3\kappa^2}{2k_{\mathrm{B}}\,\theta_{\mathrm{D}}}\left[1+\frac{2\pi^2}{3}\left(\frac{T}{\theta_{\mathrm{D}}}\right)^2+\ldots\right] \quad (T \ll \theta_{\mathrm{D}}) \tag{5.3.27}$$

and more generally we find

$$\tfrac{1}{3}\kappa^2\langle u_n^2\rangle = \frac{3\hbar^2 T^2}{2Mk_{\mathrm{B}}\,\theta_{\mathrm{D}}^3}\int_0^{\theta_{\mathrm{D}}/T}\frac{x\mathrm{e}^x}{\mathrm{e}^x-1}\,dx \tag{5.3.28}$$

which at high temperatures tends to

$$\tfrac{1}{3}\kappa^2\langle u_n^2\rangle = \frac{3\hbar^2 T}{2Mk_{\mathrm{B}}\,\theta_{\mathrm{D}}^2}. \tag{5.3.29}$$

5.3.2 General formulation for total scattering

In Chapter 3, we described the phonons by means of the one-particle imaginary time Green function and so it is desirable to make a connection between this quantity and the neutron scattering. To do so, we shall adopt the method used by Ambegaokar, Conway and Baym (1965).

We note that if we define the ionic-density fluctuation operator as

$$\rho(\mathbf{k}, t) = \mathrm{e}^{i\mathbf{k}.\mathbf{S}_n(t)} \tag{5.3.30}$$

then the dynamical structure factor may be written as

$$S(\mathbf{k}, \omega) = \int_{-\infty}^{\infty} \mathrm{e}^{-i\omega t}\langle\rho(\mathbf{k}, t)\,\rho(-\mathbf{k}, 0)\rangle\,dt \tag{5.3.31}$$

and we therefore begin by discussing the quantity

$$C(\mathbf{k}, \tau) = \langle T\{\rho(-\mathbf{k}, 0)\,\rho(\mathbf{k}, \tau)\}\rangle, \tag{5.3.32}$$

where $\tau = it$ is the imaginary time. Referring to equations (3.16.34) and (3.16.38), we see that we can expand C as a Fourier series

$$\beta C(\mathbf{k}, \tau) = \sum_n C(\mathbf{k}, i\omega_n)\, e^{i\omega_n \tau} \quad (\beta\omega_n = 2\pi n) \tag{5.3.33}$$

in the region $-\beta \leqslant \tau \leqslant \beta$, with

$$C(\mathbf{k}, i\omega_n) = \int_0^\beta C(\mathbf{k}, \tau)\, e^{-i\omega_n \tau}\, d\tau. \tag{5.3.34}$$

Now we take $C(\mathbf{k}, \tau)$ to be obtained from the analytic continuation of the quantity

$$\mathscr{C}(\mathbf{k}, t) = \langle T\{\rho(-\mathbf{k}, 0)\, \rho(\mathbf{k}, t)\}\rangle \quad (t \text{ real}), \tag{5.3.35}$$

the analytic continuation proceeding from the positive real axis on to a strip of width β below the real axis, and from the negative real axis on to a strip of width β above the real axis. In view of equation (5.3.31) we can then write

$$2\pi C(\mathbf{k}, \tau) = \int_{-\infty}^{\infty} e^{-i\omega(i\tau)}\, S(\mathbf{k}, \omega)\, d\omega \quad (\tau > 0) \tag{5.3.36}$$

and equation (5.3.34) becomes

$$C(\mathbf{k}, i\omega_n) = \frac{1}{2\pi} \int_{-\infty}^{\infty} d\omega \int_0^\beta S(\mathbf{k}, \omega)\, e^{-(i\omega_n - \omega)\tau}\, d\tau. \tag{5.3.37}$$

Noting that $\beta\omega_n = 2\pi n$, the integration over τ yields

$$C(\mathbf{k}, i\omega_n) = \int_{-\infty}^{\infty} \frac{D(\mathbf{k}, \omega)}{i\omega_n - \omega}\, d\omega \tag{5.3.38}$$

with the spectral density function $D(\mathbf{k}, \omega)$ given by

$$D(\mathbf{k}, \omega) = (1 - e^{-\beta\omega})\, S(\mathbf{k}, \omega). \tag{5.3.39}$$

A direct connection between $S(\mathbf{k}, \omega)$ and $C(\mathbf{k}, i\omega_n)$ is effected by the relation (5.3.39), for we see from equation (5.3.38) that, since $S(\mathbf{k}, \omega)$ and hence $D(\mathbf{k}, \omega)$ is real,

$$D(\mathbf{k}, \omega) = 2 \operatorname{Im} C(\mathbf{k}, \omega - i\varepsilon), \tag{5.3.40}$$

where ε is a positive infinitesimal and $C(\mathbf{k}, \omega)$ is the analytic continuation of $C(\mathbf{k}, i\omega_n)$ as given by equation (5.3.38). Equations (5.3.39) and (5.3.40) are the basic results we have obtained here. In passing, it may be noted that these two equations may be regarded as a direct consequence of the fluctuation-dissipation theorem, which is discussed fully in section 6.11.3.

We shall shortly be ready to connect $S(\mathbf{k}, \omega)$ with the one-phonon Green functions. We must first observe, however, that from equations (5.3.17) and (5.3.19) it follows that

$$\ln\left\{\frac{\langle T\{\exp[-i\boldsymbol{\kappa}.\mathbf{u}_m(t)]\exp[i\boldsymbol{\kappa}.\mathbf{u}_n(0)]\}\rangle}{|d(\mathbf{k})|^2}\right\}$$
$$= \langle T[\{\exp[-i\boldsymbol{\kappa}.\mathbf{u}_m(t)]-1\}\{\exp[i\boldsymbol{\kappa}.\mathbf{u}_n(0)]-1\}]\rangle_c. \qquad (5.3.41)$$

Hence, by comparison with equation (5.3.32), we can see that

$$iC(\boldsymbol{\kappa}, t) = |d(\boldsymbol{\kappa})|^2 \sum_{\mathbf{m}} \exp[i\boldsymbol{\kappa}.(\mathbf{m}-\mathbf{n})]\exp(\langle T[\{\exp[-i\boldsymbol{\kappa}.\mathbf{u}_m(t)]-1\}$$
$$\times\{\exp[i\boldsymbol{\kappa}.\mathbf{u}_n(0)]-1\}]\rangle_c). \qquad (5.3.42)$$

In view of the factor $N^{-\frac{1}{2}}$ in equation (5.3.22), we expand the exponentials. The lowest-order term is readily seen to give the Bragg scattering of equation (5.3.12) and in the higher-order terms we are left with the inelastic scattering only.

Inelastic scattering. We shall write $S(\mathbf{k}, \omega)$ then in two parts, namely

$$S(\mathbf{k}, \omega) = S_0(\mathbf{k}, \omega) + S_1(\mathbf{k}, \omega), \qquad (5.3.43)$$

where $S_0(\mathbf{k}, \omega)$ gives the Bragg scattering, and $S_1(\mathbf{k}, \omega)$ contributes to the inelastic scattering only. Below we shall evaluate $S_1(\mathbf{k}, \omega)$ to lowest order.

(i) *Harmonic approximation*: We have seen that the harmonic crystal can be characterized by the condition (5.3.18), in which case equation (5.3.42) becomes

$$iC(\boldsymbol{\kappa}, t) = \exp(-2W)\sum_{m}\exp[i\boldsymbol{\kappa}.(\mathbf{m}-\mathbf{n})]\exp\{\langle T[\boldsymbol{\kappa}.\mathbf{u}_m(t)\,\boldsymbol{\kappa}.\mathbf{u}_n(0)]\rangle\}. \qquad (5.3.44)$$

Expanding the exponential, the lowest-order contribution to the inelastic scattering is given by

$$C_1(\boldsymbol{\kappa}, t) = \exp(-2W)\sum_{m}\exp[i\boldsymbol{\kappa}.(\mathbf{m}-\mathbf{n})]\langle T[\boldsymbol{\kappa}.\mathbf{u}_m(t)\,\boldsymbol{\kappa}.\mathbf{u}_n(0)]\rangle. \qquad (5.3.45)$$

To connect with the treatment of Chapter 3, it is convenient in this instance to define the one-phonon Green function as

$$G_{\alpha\beta}(\mathbf{k}, t) = -i\sum_{\mathbf{m}}\exp[+i\mathbf{k}.(\mathbf{m}-\mathbf{n})]\langle T\{u_{\mathbf{m}}^{\alpha}(t)\,u_{\mathbf{n}}^{\beta}(0)\}\rangle \qquad (5.3.46)$$

which differs only in minor detail from the definition (3.14.5) of the Green function. Comparing equations (5.3.45) and (5.3.46) we find

$$C_1(\mathbf{k}, t) = e^{-2W}\, i\sum_{\alpha\beta} G_{\alpha\beta}(\mathbf{k}, t)\, k_\alpha k_\beta. \qquad (5.3.47)$$

Defining now the spectral density $A_{\alpha\beta}$ by

$$G_{\alpha\beta}(\mathbf{k}, i\omega_n) = \int_{-\infty}^{\infty} d\omega \frac{A_{\alpha\beta}(\mathbf{k}, \omega)}{i\omega_n - \omega}, \tag{5.3.48}$$

we see from equations (5.3.39), (5.3.40) and (5.3.47) that

$$S_1(\mathbf{k}, \omega) = e^{-2W(\mathbf{k})} \sum_{\alpha\beta} A_{\alpha\beta}(\mathbf{k}, \omega)\, k_\alpha k_\beta (1 - e^{-\beta\omega})^{-1}. \tag{5.3.49}$$

From section 3.14.1 it can be seen that in the harmonic approximation

$$A_{\alpha\beta}(\mathbf{k}, \omega) = 2\pi \sum_{\sigma} \frac{\varepsilon_{\alpha\sigma}(\mathbf{k})\, \varepsilon_{\beta\sigma}(-\mathbf{k})}{2M\omega(\mathbf{k}\sigma)} \{\delta[\omega - \omega(\mathbf{k}\sigma)] - \delta[\omega + \omega(\mathbf{k}\sigma)]\}. \tag{5.3.50}$$

It is apparent that, to this order, only single-phonon processes are operative and the significance of the two terms of equation (5.3.50) is as follows. The first term describes the excitation of a phonon, of energy $\omega(\mathbf{q}\sigma)$, by a neutron, which loses momentum $\mathbf{k} = \mathbf{K} - \mathbf{q}$, where $\mathbf{q}$ is restricted to the BZ. The second term describes the destruction of a phonon of energy $\omega(\mathbf{q}\sigma)$ by a neutron which gains momentum $\mathbf{k} = \mathbf{K} + \mathbf{q}$.

We see that in the one-phonon approximation, inelastic neutron scattering reveals the entire phonon energy spectrum. The analysis is facilitated by considering only very slow neutrons, which cannot excite phonons but only absorb them.

(ii) *Linear approximation*: If we assume the response of the crystal to the neutrons to be linear and so the phonons created by a neutron propagate independently of one another, we obtain the same condition (5.3.18) which characterizes the harmonic approximation and we are again led to equation (5.3.49) for the inelastic scattering. However, the linear approximation differs from the harmonic approximation in one important respect. Although the interaction between the phonons created by a neutron are ignored, those interactions with the thermal phonons already present are assumed to be operative, allowing phonon decay, and $A_{\alpha\beta}(\mathbf{k}, \omega)$ is taken to be the true spectral density function for the crystal. If the anharmonicity is not too strong, we may replace the delta functions of equation (5.3.50) by Lorentzians to obtain, ignoring interactions between different modes σ

$$A_{\alpha\beta}(\mathbf{k}, \omega) = \sum_{\sigma=1}^{3} \varepsilon_{\alpha\sigma}(\mathbf{k})\, \varepsilon_{\beta\sigma}(-\mathbf{k})\, A_\sigma(\mathbf{k}\omega), \tag{5.3.51}$$

where

$$A_\sigma(\mathbf{k}\omega) = \frac{1}{2M\Omega(\mathbf{k}\sigma)} \left\{ \frac{\Gamma(\mathbf{k}\sigma)}{[\omega - \Omega(\mathbf{k}\sigma)]^2 + [\Gamma^2(\mathbf{k}\sigma)/4]} - \frac{\Gamma(\mathbf{k}\sigma)}{[\omega + \Omega(\mathbf{k}\sigma)]^2 + [\Gamma^2(\mathbf{k}\sigma)/4]} \right\}. \tag{5.3.52}$$

Here $\Omega(\mathbf{k}\sigma)$ is the true phonon frequency and $\Gamma(\mathbf{k}\sigma)/2$ the inverse lifetime of the mode.

It can be seen that, within the linear approximation, the peaks of the neutron scattering are symmetric, centred at the true phonon frequencies, and have half-widths directly proportional to the reciprocal of the phonon lifetime.

Because, in an anharmonic crystal, a phonon can always decay into several phonons, there is effectively an interaction between the one-phonon resonance and the diffuse multiphonon background. Thus, we ought to write the inelastic scattering cross-section $S_1(\mathbf{k}, \omega)$ in the form

$$S_1'(\mathbf{k}, \omega) + S_{\text{int}}(\mathbf{k}, \omega) + S_m(\mathbf{k}, \omega), \tag{5.3.53}$$

where S_1' is the cross-section for creation of one phonon, S_m represents the diffuse background and S_{int} represents interference between one-phonon and many-phonon processes. Like $S_1'(\mathbf{k}, \omega)$, $S_{\text{int}}(\mathbf{k}, \omega)$ will naturally be rapidly varying near the one-phonon scattering peak and in fact will contribute to it. This contribution, neglected in the linear approximation, leads to asymmetry in the one-phonon peaks. Expressions for this are given by Ambegaokar, Conway and Baym (1965) as summarized in Appendix 5.1, with an explicit estimate of the asymmetry for Pb. This shows that, for this metal, the asymmetry is likely to be small.

5.3.3 Experimental spectrum in terms of pair potentials

We shall now briefly summarize the results of the analysis of experimentally determined phonon spectra, particularizing to sodium, by use of pseudo-potential theory. The discussion here is clearly related to that given in Chapter 3, section 3.10. We also make some remarks about the range of the forces in Al.

Pseudo-potential theory was discussed in Chapter 1, and we can see that it leads naturally—though not inevitably (see in particular Chapter 7)—to the idea of pair interactions, essentially electrostatic in nature. In its simplest form it gives us the picture of "neutral pseudo-atoms", to use Ziman's term, each of which carries rigidly with it a charge distribution $\sigma(\mathbf{r})$ the sum of which, over all the ionic sites, gives the total distribution of the valence electrons. We must be a little careful, however, in calculating what electrostatic terms contribute to the pair interaction. As we have seen earlier, first-order perturbation theory immediately tells us that if we change an external potential $V_\lambda(\mathbf{r})$ by $[\partial V(r)/\partial\lambda]\,\delta\lambda$, the change in energy of a system of electrons of charge density $\rho(\mathbf{r})$ is $\delta\lambda\int[\partial V_\lambda(r)/\partial\lambda]\,\rho(\mathbf{r})\,d\mathbf{r}$; the changes in kinetic energy and electron–electron interaction energy cancel one another.

This will be true of the valence electrons when we change the external potential on them by movement of an ion. It therefore follows that the pair potential will be given by

$$\phi(\mathbf{R}) = \int \frac{Z(\mathbf{r}_1)\,\sigma(\mathbf{r}-\mathbf{R})}{|\mathbf{r}-\mathbf{r}_1-\mathbf{R}|}\,d\mathbf{r}_1\,d\mathbf{r} + \frac{Z^2}{|\mathbf{R}|}, \tag{5.3.54}$$

where $Z(\mathbf{r})$ is the ionic charge distribution such that $\int Z(\mathbf{r})\,d\mathbf{r} = Z$ and we have assumed that we may evaluate the ion–ion repulsion by taking the ions as point charges. To (5.3.54) we should add a term representing the short-range ion–ion exchange repulsion, usually represented by $A\,e^{-R/\lambda}$. The $Z(\mathbf{r})$ of equation (5.3.54) need not be regarded as the true ionic-charge distribution, but one simulating, through the integral $\int[Z(\mathbf{r}')\,d\mathbf{r}'/|\mathbf{r}-\mathbf{r}'|]$, the pseudo-potential $V_p(\mathbf{r})$ (taken to be $\mathbf{k}$-independent) acting on the valence electrons.

Now we can see that considerable information on the electronic properties of a metal can be found by an experimental determination of $\phi(\mathbf{R})$. We can check a pseudo-potential theory starting from first principles or, alternatively, increase the chances of success of the theory by making it semi-phenomenological in the sense that we choose, say, $Z(\mathbf{r}_1)$ to give the correct $\phi(R)$ before using $Z(\mathbf{r})$ to calculate other properties of the metal.

Now we have already seen in Chapter 3 that phonon spectra give us only $\partial\phi/\partial\mathbf{r}$ and $\partial^2\phi/\partial\mathbf{r}^2$ evaluated at the lattice points, giving us a variety of possibilities for the full curve of $\phi(r)$. However, another criterion may be involved to assist us in choosing the correct $\phi(r)$. From equation (5.3.54) we can see that its Fourier transform is

$$\tilde{\phi}(k) = \tilde{Z}(k)\,\sigma(k)/k^2 + Z^2/k^2. \tag{5.3.55}$$

Thus

$$\lim_{k\to 0} k^2\,\tilde{\phi}(k) = 0, \tag{5.3.56}$$

and, moreover, since in sodium and similar metals the σ must sum to a more or less constant distribution, we should have $\sigma(K)$ close to zero (for $\boldsymbol{\kappa}\neq 0$) and so $\tilde{\phi}(K)\,K^2$ should be close to Z^2 whenever $\mathbf{K}$ is a reciprocal lattice vector.

The above are the assumptions made by Cochran (1963) enabling him to sketch out a $\phi(r)$ consistent with experimentally determined phonon spectra of sodium. This is shown in Figure 5.1, and is seen to be quite short-range. This is in contrast to Al, to which we now turn.

Dispersion curves for phonons in Al. Foreman and Lomer (1957) showed that in a cubic crystal with one atom per unit cell, such as Al, phonons which propagate in the [100], [110] and [111] directions correspond to vibrations in which *entire planes* of atoms *normal* to the direction of propagation move together.

The polarization directions are determined by symmetry, and must be either parallel or perpendicular to the direction of propagation.

For each direction of propagation and mode of polarization separately, the equations of motion reduce to those for a linear chain of equally spaced identical particles, with *n*th neighbour harmonic forces.

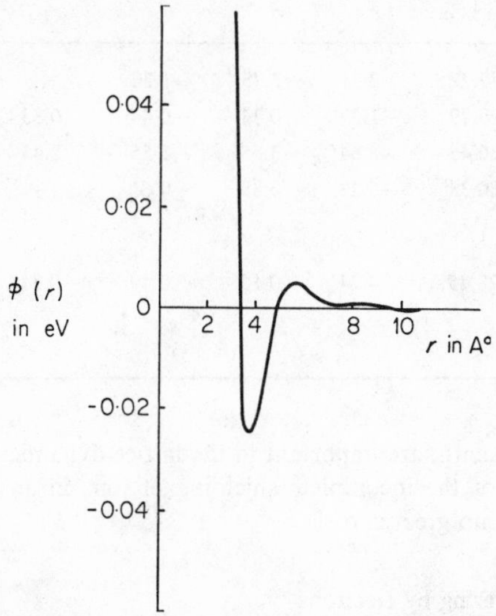

FIGURE 5.1. $\phi(r)$ for sodium (after Cochran, 1963).

Solution of these equations gives

$$M\omega^2 = \sum_{n=1}^{N} \phi_n(1-\cos n|\mathbf{q}|d), \tag{5.3.57}$$

M being the atomic mass, *d* is the distance between planes normal to $\mathbf{q}$, ϕ_n is the force on a given atom due to unit displacement of all of the atoms in the two planes located at a distance *nd* on either side of it, and *N* is large enough so that the forces due to more distant planes may be neglected.

The force and the displacement are both parallel to the polarization direction of the phonon.

The values of the interplanar force constants ϕ_n for Al at room temperature have been determined by Yarnell, Warren and Koenig (1965) from their neutron-scattering experiments, for propagation of phonons along the two directions in Al, and are shown in Table 5.1.

It is possible to conclude from the number of significant interplanar force constants that forces due to at least eighth nearest neighbours, and very

TABLE 5.1. Interplanar force constants ϕ_n in units of 10^3 dyne/cm for Al at 300°K

ϕ_n's: $n=$	1	2	3	4	5	6	7	8
L [100]	77.22	2.31	2.78	−1.30	—	—	—	—
T [100]	29.79	−1.37	0.34	−0.44	0.33	—	—	—
L [110]	30.45	48.64	−3.65	−1.55	2.43	0.08	0.77	—
T_2 [110] (polarized in [001] direction)	80.38	−5.29	−0.88	−0.61	—	—	—	—
T_1 [110] (polarized in [1$\bar{1}$0] direction)	28.45	4.24	0.15	0.49	−0.34	−0.28	—	—

probably to fifteenth, are important in the lattice dynamics of Al. In view of our discussion of the incomplete shielding of ions in an electron gas, this should come as no great surprise.

5.4 X-ray scattering by electrons

X-rays form, at the present time, the most important tool with which we are equipped to obtain information on electronic charge distributions from experiment. We have avoided the assertion that X-ray scattering *provides* us with the electronic charge distribution, for even this technique has severe limitations, as we shall see. As we pointed out in Chapter 1, the charge density is very basic in the many-body theory of electrons in crystals.

We shall not treat the X-ray scattering problem *ab initio*—for such treatments we refer the reader to other texts. Instead, we shall use the general formulation of scattering problems discussed already in section 5.2, the essential equation being equation (5.2.24). To be able to utilize this formula we must know $\tilde{V}(\boldsymbol{\kappa})$, the Fourier transform of the pair interaction of the scattered particle with an individual particle of the scatterer. Accordingly we shall find $|\tilde{V}(\boldsymbol{\kappa})|$ for the problem under discussion by considering the elastic scattering of photons by a single electron, as a simple example.

5.4.1 Interaction of photons with electrons

We have already seen (Chapter 4, equation 4.15.4) that the Hamiltonian for an electron in an electromagnetic field given by

$$\mathscr{E} = -\nabla V(\mathbf{r}) - \frac{1}{c}\frac{\partial \mathscr{A}}{\partial t}; \quad \mathscr{H} = \operatorname{curl} \mathscr{A}, \tag{5.4.1}$$

is‡

$$H = \frac{p^2}{2m} + eV(\mathbf{r}) - \frac{e}{2mc}(\mathbf{p}.\mathscr{A} + \mathscr{A}.\mathbf{p}), \tag{5.4.2}$$

which amounts to adding to the Hamiltonian for the usual Schrödinger equation the *magnetic Hamiltonian*,

$$H_{\mathrm{m}} = -\frac{e}{2mc}(\mathbf{p}.\mathscr{A} + \mathscr{A}.\mathbf{p}). \tag{5.4.3}$$

Following the procedure of section 5.2, we can see that the change in electron current **S** brought about by the incident radiation with vector potential $\mathscr{A}$ is given by

$$\psi^* H_m \psi - \psi H_{\mathrm{m}}^* \psi^* = -\frac{h}{2\pi i} \operatorname{div} \mathbf{S}, \tag{5.4.4}$$

where

$$\mathbf{S} = \frac{e}{mc}|\psi|^2 \mathscr{A}. \tag{5.4.5}$$

One readily sees that any change in $V(\mathbf{r})$ makes no contribution to **S**, except that ψ is not strictly the solution of the Schrödinger equation without the radiation field. We shall, however, neglect any perturbation of ψ arising from this field as being a second-order effect.

Let the incident radiation be in the z-direction, with

$$\mathscr{A} = (\mathscr{A}_x, 0, 0) = \mathscr{A}(\mathrm{e}^{ikz}, 0, 0). \tag{5.4.6}$$

Then the electric current is

$$e\mathbf{S} = e(S, 0, 0) = \frac{e^2}{mc}|\psi|^2 \mathrm{e}^{ikz}(\mathscr{A}, 0, 0). \tag{5.4.7}$$

This produces a secondary field $\mathscr{A}_{\mathrm{s}} = (\mathscr{A}_{\mathrm{s}}, 0, 0)$, where

$$\nabla^2 \mathscr{A}_{\mathrm{s}} + k^2 \mathscr{A}_{\mathrm{s}} = -\frac{4\pi e}{c} S = -\frac{4\pi e^2}{mc^2}|\psi|^2 \mathscr{A}\, \mathrm{e}^{ikz}. \tag{5.4.8}$$

‡ Here we have taken the non-relativistic limit of $[\mathbf{p} - (e\mathscr{A})/c]^2$.

The solution is found in the same way as we found that for $\chi_f(r)$ in section 5.2 (cf. equation 5.2.13). We put

$$\mathscr{A}_s = \frac{e^{ikr}}{r} f(\theta), \tag{5.4.9}$$

where

$$f(\theta) = \frac{e^2}{mc^2} \mathscr{A} \int e^{i\boldsymbol{\kappa}.\mathbf{r}} |\psi|^2 \, d\mathbf{r} \tag{5.4.10}$$

and

$$\kappa = |\boldsymbol{\kappa}| = 2k \sin \frac{\theta}{2}, \tag{5.4.11}$$

θ being the scattering angle.

Now the intensity I of the secondary radiation is

$$I = \frac{c}{4\pi} \mathscr{E}\mathscr{H} = \frac{\mathscr{E}^2}{4\pi} \tag{5.4.12}$$

and converting from $\mathscr{E}$ to $\mathscr{A}_s$, we obtain

$$\frac{I}{I_0} = (1 - \tfrac{1}{2} \sin^2 \theta) \left| \int e^{i\boldsymbol{\kappa}.\mathbf{r}} |\psi|^2 \, d\mathbf{r} \right|^2 \left(\frac{e^2}{mc^2} \right)^2, \tag{5.4.13}$$

where I_0 is the incident intensity.

We can see, from the general equation (5.2.24) and the subsequent considerations of that section, that this equation has the product form

$$|f(\boldsymbol{\kappa})| \, |V(\boldsymbol{\kappa})|^2; \quad f(\boldsymbol{\kappa}) = \int e^{i\boldsymbol{\kappa}.\mathbf{r}} |\psi|^2 \, d\mathbf{r}. \tag{5.4.14}$$

This represents the interaction between a single electron with the scattered photons; to treat the scattering by a system of many electrons we assume that the interaction with the photons is by pair interactions, any individual interaction being given by equation (5.4.14). We can now proceed to use the general formulation of section 5.2.

5.4.2 Scattering by all electrons of perfect lattice

We assume that the lattice points of the crystal are fixed. Some mention of temperature effects will be made later, but we should remark that temperature is an effect which an experimentalist usually corrects for, along with the semi-empirical corrections for such things as extinction and surface roughness, before presenting his results. The description of these corrections is outside our province, and so we shall assume here that we have "ideal" intensities presented to us by the experimentalist, to which the formulae below are directly applicable.

From equations (5.2.24) and (5.2.26) we can see that at zero temperature the elastic scattering is given by

$$\left.\frac{d\sigma}{d\Omega}\right|_{\text{elastic}} \propto |\tilde{V}(\boldsymbol{\kappa})|^2 S(\boldsymbol{\kappa}, t)\Big|_{t=\infty}, \tag{5.4.15}$$

where

$$S(\boldsymbol{\kappa}, \infty) = \sum_f \left| \langle i | \sum_j e^{i\boldsymbol{\kappa}\cdot\mathbf{r}_j} | f \rangle \right|^2 = \langle i | \left| \sum_j e^{i\boldsymbol{\kappa}\cdot\mathbf{r}_j} \right|^2 | i \rangle.$$

In other words, the elastic scattering is just

$$\left.\frac{d\sigma}{d\Omega}\right|_{\text{elastic}} \propto |V(\boldsymbol{\kappa})|^2 N^2 |f(\boldsymbol{\kappa})|^2, \tag{5.4.16}$$

where $f(\boldsymbol{\kappa})$ is the X-ray scattering factor (or "form-factor")

$$f(\boldsymbol{\kappa}) = \frac{1}{N}\int_{\mathscr{V}} \rho(\mathbf{r})\, e^{i\boldsymbol{\kappa}\cdot\mathbf{r}}\, d\mathbf{r} = \int_{\Omega} \rho(\mathbf{r})\, e^{i\boldsymbol{\kappa}\cdot\mathbf{r}}\, d\mathbf{r}. \tag{5.4.17}$$

Writing the charge density as a Fourier series, as in Chapter 1, equation (1.5.2),

$$\rho(\mathbf{r}) = \sum_{\mathbf{K}} \rho_{\mathbf{K}}\, e^{i\mathbf{K}\cdot\mathbf{r}} \tag{5.4.18}$$

we immediately see that

$$Nf(\boldsymbol{\kappa}) = \int \rho(\mathbf{r})\, e^{i\boldsymbol{\kappa}\cdot\mathbf{r}}\, d\mathbf{r} = \Omega_B \sum_{\mathbf{K}} \delta(\boldsymbol{\kappa}+\mathbf{K}) \int_{\Omega} \rho(\mathbf{r})\, e^{i\boldsymbol{\kappa}\cdot\mathbf{r}}\, d\mathbf{r} \tag{5.4.19}$$

so that

$$f(\boldsymbol{\kappa}) = \begin{cases} \Omega\rho_{\boldsymbol{\kappa}}, & \boldsymbol{\kappa} = \mathbf{K}, \\ 0, & \text{otherwise.} \end{cases} \tag{5.4.20}$$

Here we may note a fact we shall use later, that if a is any operator of the rotation group of the crystal,

$$\rho_{a\mathbf{K}} = \rho_{\mathbf{K}} \tag{5.4.21}$$

which readily follows from the fact that $\rho(\mathbf{r})$ has the rotational symmetry of the crystal:

$$\rho(\mathbf{r}) = \rho(a\mathbf{r}) = \sum_{\mathbf{K}} \rho_{\mathbf{K}}\, e^{ia^{\dagger}\mathbf{K}\cdot\mathbf{r}} = \sum_{\mathbf{K}} \rho_{a\mathbf{K}}\, e^{i\mathbf{K}\cdot\mathbf{r}}. \tag{5.4.22}$$

An expansion alternative to the Fourier series is one in terms of local distributions:

$$\rho(\mathbf{r}) = \sum_{\mathbf{R}} \sigma(\mathbf{r}-\mathbf{R}), \tag{5.4.23}$$

where the sum is over all lattice vectors. To show the validity of this equation, it is only necessary to assume this form and use it in the expression for the Fourier component $\rho_{\mathbf{K}}$:

$$\rho_{\mathbf{K}} = \frac{1}{\Omega}\int_{\Omega} \rho(\mathbf{r})\, e^{i\mathbf{K}.\mathbf{r}}\, d\mathbf{r} = \frac{1}{\Omega}\int \sum_{\mathbf{R}} \sigma(\mathbf{r}-\mathbf{R})\, e^{i\mathbf{K}.\mathbf{r}}\, d\mathbf{r} = \frac{1}{\Omega}\int_{\mathscr{V}} \sigma(\mathbf{r})\, e^{i\mathbf{K}.\mathbf{r}}\, d\mathbf{r}. \tag{5.4.24}$$

Thus the only requirements we make on $\sigma(\mathbf{r})$ are that its Fourier transform $\sigma(\boldsymbol{\kappa})$ becomes equal to $\rho_{\mathbf{K}}$ at each point for which $\boldsymbol{\kappa} = \mathbf{K}$.

(a) *Relative magnitudes of elastic and inelastic terms.* If it contributes significantly to the Fourier series (5.4.18), $\rho_{\mathbf{K}} \sim \rho_0$ and when $\boldsymbol{\kappa} = 0$, $d\sigma/d\Omega|_{\text{elastic}} \propto N^2$, where N is the number of electrons. We can obtain an estimate of the magnitude of the inelastic scattering most simply by use of the Hartree–Fock approximation. Referring to equation (5.2.24) and equation (2.7.38) of Chapter 2, we see that the *total* inelastic scattering is $\propto$

$$\sum_{p=1}^{N} \sum_{q=N+1}^{\infty} \left| \int \psi_p^* \psi_q e^{i\boldsymbol{\kappa}.\mathbf{r}}\, d\mathbf{r} \right|^2 < \sum_{p=1}^{\infty} \sum_{q=1}^{\infty} \left| \int \psi_p^* \psi_q e^{i\boldsymbol{\kappa}.\mathbf{r}}\, d\mathbf{r} \right|^2 = N, \tag{5.4.25}$$

the equality following since $\{\psi_q e^{i\boldsymbol{\kappa}.\mathbf{r}}\}$ forms a complete orthonormal set.

We see that the relative magnitude of significant elastic and inelastic scattering is of the order of $1/N$. The elastic scattering swamps the inelastic at the Bragg reflexions.

(b) *Effect of temperature.* Suppose we assume that the charge density can be decomposed into identical localized distributions $\sigma(\mathbf{r})$, one centred about each nucleus. Then, as we see in equation (5.2.24), the X-ray scattering can be assumed to be the resultant of scattering from these individual distributions. If we further assume that the local distributions $\sigma(\mathbf{r})$ *are carried unchanged by the nuclei as they vibrate*, that is we have "rigid ions" or "pseudo-atoms", we have a situation exactly analogous to that discussed in section 5.3, and by introduction of a scattering length dependent on the integral $\int \sigma(\mathbf{r})\, e^{i\boldsymbol{\kappa}.\mathbf{r}}\, d\mathbf{r}$ the considerations of section 5.4.2 apply to our assembly. In that section we saw that the sole effect of the lattice vibrations on the Bragg peaks was to reduce them by the Debye–Waller factor e^{-2W}.

The assumption we make here, that $\sigma(\mathbf{r})$ remains quite unchanged as the nucleus on which it is centred moves, is obviously a good one for the core electrons. Equally obviously, we have no guarantee that the valence electrons obey the criterion.‡ Such an assumption for the valence electrons can only constitute a first-order approximation.

A remark we might make is that our discussion above would indicate that the standard temperature correction made to experimental results might result in appreciable errors (perhaps a few per cent) in the estimation of the scattering factors of valence electrons. On the other hand, the experimentalists' correction, based on multiplication by a factor $e^{-2W(\kappa)}$ is semi-empirical.

5.4.3 Scattering by outer electrons

It is generally assumed that the charge-density of the core electrons in a crystal can be computed with adequate accuracy by the Hartree–Fock method. This being so, the importance of the X-ray scattering technique derives from (a) its value in crystal structure determination and (b) the information it gives on the charge density of the outer electrons. Our principal concern in this book is with the spatial distribution of electrons (b). At the present time, unfortunately, the technique has several severe limitations. First, the Fourier series for the charge density is very slowly convergent; second, since $\kappa = 4\pi \sin \theta/\lambda$ [equation (5.4.11)] we can obtain no information on the Fourier components $\rho_{\mathbf{K}}$ for which $|\mathbf{K}| > 4\pi/\lambda$; and third, much of the scattering intensity is due to the core electrons, and, especially in crystals built from the heavier atoms, the experimental errors are comparable with the magnitude of the scattering from the outer electrons. Figure 5.3, below, for the case of Fe, illustrates this last point very vividly, and errors in correcting for temperature effects can hardly be considered important when compared with other experimental errors. The first limitation is not one of principle, and can in any case be overcome by the method of local distributions which we shall later describe. The second limitation is one that we are forced to live with; no more will be said about it except that electron diffraction, with much shorter wavelengths, promises to provide us with information on higher-order reflexions when this technique is further developed. We shall briefly discuss electron diffraction in section 5.6. The third limitation may also yield to the ingenuity of experimentalists, eventually, and in view of our remark in Chapter 1 about the relation between electron density $\rho(\mathbf{r})$ and the one-body potential including many-body effects this would be an important area in which to make progress.

(a) *Crystalline field splitting.* Although the X-ray experiments lack the accuracy one would like to construct the charge density, the order of magnitude of the scattered intensity from the outer electrons can still give us valuable information if we work within the framework of a particular model.

‡ We saw in Chapter 3, section 3.10.5, that $\nabla\rho$ can be decomposed exactly into localized vectors which "move rigidly" with the nuclei.

The difficulty we immediately encounter is the choice of a suitable model. We require from such a model an expression for $\rho(\mathbf{r})$, based on sound physical reasoning, in which parameters appear which we can adjust by reference to X-ray scattering intensities. While in some cases, for example ionic crystals, obvious physical models suggest themselves, the problem in metals is really the most interesting in this context, the valence electrons being markedly modified by crystalline binding. Pseudopotential methods afford a useful approach for nearly free electron metals (see, for example, Ascarelli and Raccah, 1969, who discuss Al), but we want here to briefly consider the more interesting case of metallic Fe, though only qualitatively.

The aim of the treatment outlined below is to set up a model for the localized distribution $\sigma(\mathbf{r})$ introduced in equation (5.4.23). The simplest picture of all is to regard these localized distributions as exactly the same as those of isolated atoms and this, in fact, is the method adopted by crystallographers to determine crystal structures.

It was Weiss and de Marco (1958) who aroused interest in an atomic-like description of Fe. They made the usual assumption that the argon-like core is little perturbed from the free atom. Further, they assumed that the remaining electrons could be described by atomic orbitals, but with possibly different populations from the free atom, and therefore one had to consider 4p states in addition to 3d and 4s. 4p orbits are not split by the crystalline field so that the charge density in these orbits is spherically symmetric. The 3d orbits are split into E_g and T_{2g} orbitals,‡ which are linear combinations of atomic orbitals which span irreducible manifolds of the cubic group. The population of these can now, in principle, be determined from neutron diffraction results, in combination with the X-ray scattering. The original analysis of Weiss and Freeman assumed the unpaired spin distribution to have spherical symmetry, and the populations of E_g and T_{2g} orbitals came out to be about equal. However, Shull (1961) has subsequently shown experimentally that the unpaired spin density in the unit cell of Fe is not spherically symmetrical and the above populations should not be regarded as final.

Mook (1966) developed a similar treatment of Ni without assuming spherical symmetry. Again E_g and T_{2g} orbits are taken, and Mook finds that one must assume 19% of the 3d electrons to be in E_g orbitals—in contrast of the population of 40% demanded if spherical symmetry is to exist. Mook was enabled to examine this symmetry since he measured reflexions characterized by inequivalent reciprocal vectors with the same modulus.

We ought to note that Mook found the unpaired spin density going negative in certain outer regions of the unit cell (as did Shull for iron). This

‡ The understanding of the detailed notation here is not essential for what follows. However, the interested reader may consult Bethe (1929).

negative contribution was subtracted off before fitting with 3d orbitals, and may well be due to polarization of conduction electrons.

(b) *Method of local distributions.* We can either expand the charge density $\rho(\mathbf{r})$ in the crystal in a Fourier series, as we did in Chapter 1, or we can expand it in localized density distributions $\sigma(\mathbf{r})$ centred on each lattice site, as in equation (5.4.23).

Before we proceed to the mathematics, we shall list the advantages of the use of such an expansion, which are as follows:

(i) Since, as we have remarked already, the Fourier series for the charge density is very slowly convergent, the expansion in local distributions can be expected to converge rapidly.

(ii) The density we require is that of the outer electrons: the density of the core being localized, the natural expansion for the latter part of the density is in terms of local functions, leading us to expand the total density in the same manner.

(iii) The unobservable scattering factors for which $|\mathbf{K}| > 4\pi/\lambda$ are dominated by the charge density near the nuclei. We can choose localized functions both fitting the observable factors and behaving correctly near the nuclei off-setting to some extent the limitations of the experimental technique.

Let us now look at some of the mathematical aspects of the analysis in terms of local distributions, bearing in mind that while the total density at any point is readily computed using the series (5.4.23) the easier and more illuminating way of presenting results and comparing densities is by means of an expansion of the total density within the unit cell in terms of spherical harmonics.

First, let us recall that $\sigma(\mathbf{r})$ can be taken as any function

$$\sigma(\mathbf{r}) = \frac{1}{8\pi^3}\int \gamma(\boldsymbol{\kappa})\, e^{i\boldsymbol{\kappa}\cdot\mathbf{r}}\, d\mathbf{r} \tag{5.4.26}$$

for which

$$\gamma(\mathbf{K}) = \Omega\rho_{\mathbf{K}}. \tag{5.4.27}$$

From equation (5.4.21), we can see that $\sigma(\mathbf{r})$ can be taken to have the symmetry of the lattice—which is a natural choice and for example in iron the particular expansion in spherical harmonics begins,‡

$$\sigma(\mathbf{r}) = \sigma_s(r) + \sigma_g(r)g(\theta,\phi) + \dots, \tag{5.4.28}$$

‡ In the rest of this section much of the formalism will be particularized to cubic crystals, but can readily be generalized to other cases.

where g is a combination of spherical harmonics of order 4:

$$g(\theta,\phi) = \sum_{m=-4}^{4} c_m Y_{lm}(\theta,\phi) = \frac{x^4+y^4+z^4}{r^4} - \frac{3}{5}. \tag{5.4.29}$$

The s term must satisfy the requirement of integrating to Z, the number of atoms per unit cell:

$$\int \sigma_s(r)\, d\mathbf{r} = Z. \tag{5.4.30}$$

Although it can be shown that, apart from the requirement (5.4.30), the choice of $\sigma_s(r)$ is completely arbitrary because of the possibility of introducing angular terms as in (5.4.28), in fact, in practice, $\sigma_s(r)$ can be chosen as the dominant term in almost all cases.

We shall call lattice vectors $\mathbf{R}$ and $\mathbf{S}$ members of the same *equivalent set* S_n, with multiplicity μ_n, if there exists a member α of the rotation group of the crystal such that $\alpha\mathbf{R} = \mathbf{S}$. With this definition we may rewrite equation (5.4.23) as

$$\rho(\mathbf{r}) = \sum_n \frac{\mu_n}{g} \sum_\alpha \sigma(\mathbf{r}-\alpha\mathbf{R}^n), \tag{5.4.31}$$

where the first summation is over all equivalent sets, the second is over all members α of the rotation group of order g, and $\mathbf{R}^n$ is any member of the set S_n.

We now expand the *total* density within the unit cell as

$$\rho(\mathbf{r}) = \rho_s(r) + \rho_g(r) g(\theta,\phi) + \dots. \tag{5.4.32}$$

To find the s term from equation (5.4.31) we may conveniently choose the polar axis in the direction of $\mathbf{R}^n$, and we immediately find

$$\rho_s(r) = \frac{1}{2}\sum_n \mu_n \int_0^\pi \sigma(\mathbf{r}-\mathbf{R}^n) \sin\theta\, d\theta. \tag{5.4.33}$$

To similarly obtain the g term we remember that lattice harmonics are orthogonal, that

$$\int [g(\theta,\phi)]^2 \sin\theta\, d\theta\, d\phi = \frac{64\pi}{525} \tag{5.4.34}$$

and, if α is any member of the rotation group and $f(\mathbf{r})$ any function whatever,

$$\int_{\text{sphere}} f(\alpha\mathbf{r})\, g(\theta,\phi) \sin\theta\, d\theta\, d\phi = \int_{\text{sphere}} f(\mathbf{r})\, g(\theta,\phi) \sin\theta\, d\theta\, d\phi. \tag{5.4.35}$$

From this last equation we can see that

$$\begin{aligned} \int_{\text{sphere}} \sigma(\mathbf{r}-\alpha\mathbf{R}^n)\, g(\theta,\phi) \sin\theta\, d\theta\, d\phi &= \int \sigma(\alpha\mathbf{r}-\alpha\mathbf{R}^n)\, g(\theta,\phi) \sin\theta\, d\theta\, d\phi \\ &= \int_{\text{sphere}} \sigma(\mathbf{r}-\mathbf{R}^n)\, g(\theta,\phi) \sin\theta\, d\theta\, d\phi, \end{aligned} \tag{5.4.36}$$

where we have taken σ to have the symmetry of the lattice.

Hence

$$\rho_g(r) = \frac{525}{64\pi} \sum_n \mu_n \int_{\text{sphere}} \sigma(\mathbf{r} - \mathbf{R}^n) g(\theta\phi) \sin\theta \, d\theta \, d\phi. \qquad (5.4.37)$$

This can be simplified considerably if $\sigma(\mathbf{r})$ is spherically symmetrical (the conditions under which this choice can be made being discussed later).

The result is

$$\rho_g(r) = \frac{525}{32} \sum_n \mu_n g(\alpha_n, \beta_n) \int_0^{\pi} \sigma(|\mathbf{r} - \mathbf{R}^n|) P_4(\cos\theta') \sin\theta' \, d\theta', \qquad (5.4.38)$$

where the lattice vector $\mathbf{R}^n = (R_n, \alpha_n, \beta_n)$ has been chosen as polar axis and with respect to this axis we have written $\mathbf{r} = (r, \theta', \phi')$.

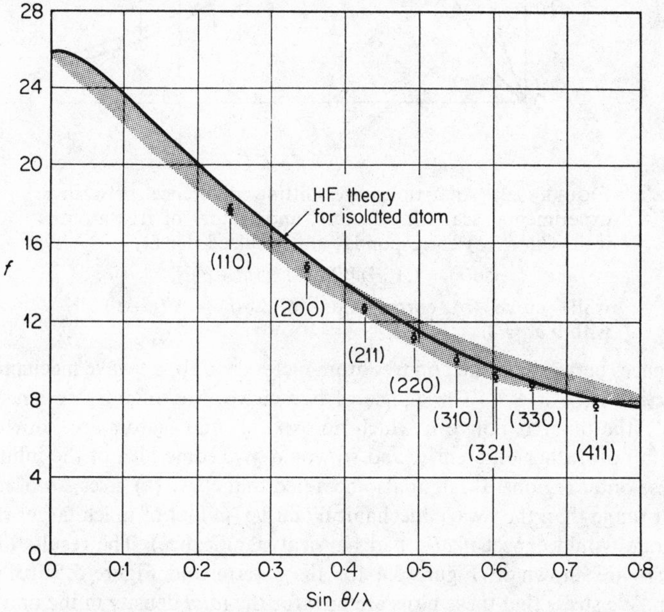

FIGURE 5.2. X-ray scattering from iron. Bragg reflexions are labelled by Miller indices. Hartree–Fock theory for an isolated atom is also shown (after Batterman and co-workers, 1961).

Jones, March and Tucker (1965) have used this expression and that of equation (5.4.33) to analyse the X-ray results of Batterman, Chipman and de Marco (1961) for Fe in the following way. Since the experimental results (Figure 5.2) lie close to the free-atom scattering curve Jones and co-workers wrote

$$\sigma(\mathbf{r}) = \sigma_0(r) + \Delta\sigma(r), \tag{5.4.39}$$

that is, they took a charge density made up of overlapping free atom distributions $\sigma_0(r)$ and estimated corrections to this on the basis of the experimental results, drawing for convenience analytic curves consistent with the experimental points as shown in Figure 5.3. The Fourier transforms of these curves gave estimates for $\Delta\sigma(r)$. They chose (a) a curve suggested solely by the

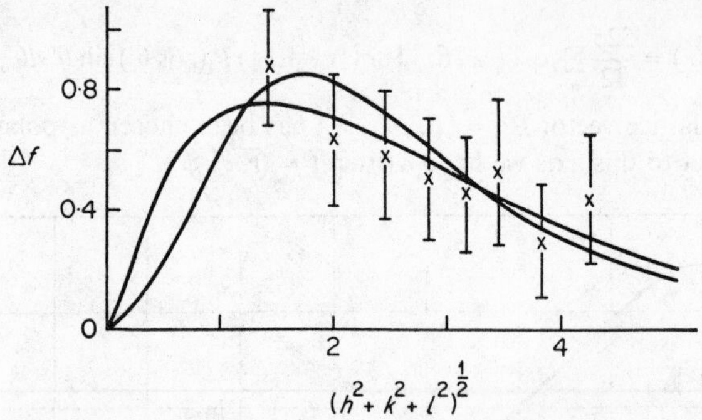

FIGURE 5.3. Analytic curves fitting difference between experimental scattering factors, and factors of free atom iron. Curve (a) corresponds to a localized density

$$\Delta\sigma(r) = (A/\pi^2)\,[(3a^2 - r^2)/(a^2 + r^2)^3],$$

while curve (b) corresponds to $\Delta\sigma(r) = A'(r - a')\,e^{-b'r}$ with $a'b' = 3$.

differences between crystal and free atom factors and (b) a "wave mechanical" correction $A(r-a)\,e^{-br}$. The choice of two curves was of interest, since the "tails" (the outer regions for which no experimental factors are known) of these fall off rather differently, and so would give some idea of the influence of these outer regions. It might also be noted that curve (a) gives $\Delta\sigma$ of much longer range than the "wave mechanical" curve (in fact of much larger range than one would expect if $\sigma(r)$ had physical significance). The results of the analysis are shown in Figure 5.4 for the s term and Figure 5.5 from the g term. We stress that these plots are now for the *total* density in the unit cell, obtained by allowing the localized distributions $\sigma(r)$ to overlap. It is clear from Figure 5.4 that the superposition of free atom densities is in fact quite a good first approximation, but that the corrections to this significantly increase the charge density at the unit cell boundary.

The g term is more sensitive to the choice of the localized distributions as Figure 5.5 shows. The best we can say from available data is that the superposition of free HF atoms gives a fair representation of this angular term.

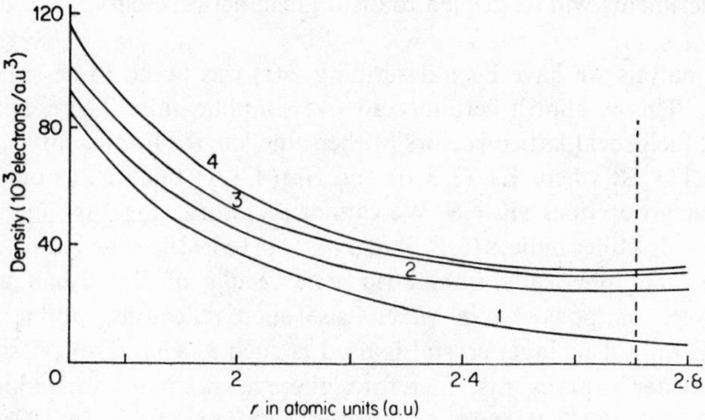

FIGURE 5.4. s term for total density within unit cell for Fe. Curve 1. Free atom density (Watson, 1952). Curve 2. Superposition of free atoms. Curve 3. Superposition plus correction using curve (a) of Figure 5.3. Curve 4. Same as curve 3 but using curve (b) of Figure 5.3. Dashed line is drawn at Wigner–Seitz sphere.

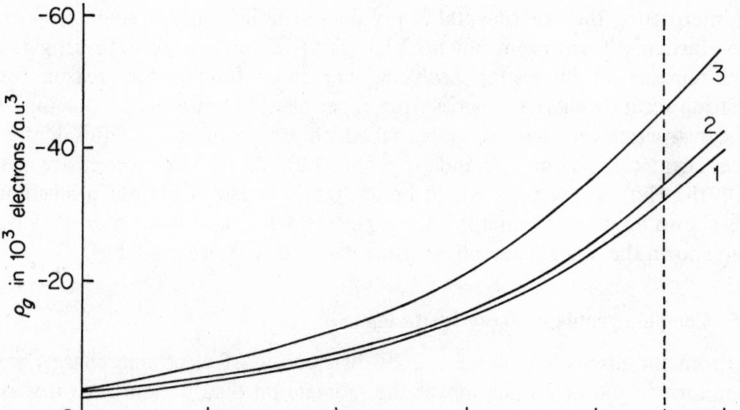

FIGURE 5.5. g term for total density within unit cell for Fe. Curve 1. Free atom superposition result. Curve 2. Superposition plus correction using curve (a) of Figure 5.3. Curve 3. Same as curve 2 but with curve (b) of Figure 5.3 correcting superposition result.

Better experiments will be needed to distinguish between curves 1, 2 and 3 in this case.

In the analysis we have been describing, $\sigma(\mathbf{r})$ was taken to be spherically symmetric. This is almost certainly an oversimplification. There exist non-equivalent reciprocal lattice vectors of the same length. For example, for bcc crystals $|\mathbf{K}| = |\mathbf{S}|$ where $\mathbf{K} \equiv (3, 3, 0)$ and $\mathbf{S} \equiv (4, 1, 1)$ but for no operator α of the cubic group does $\alpha\mathbf{K} = \mathbf{S}$. We cannot therefore expect the intensity of reflexions with Miller indices (3, 3, 0) and (4, 1, 1) to be the same, in which case $\sigma(\mathbf{r})$ cannot be spherically symmetric. The results of Batterman and co-workers were on powders, in which case such reflexions cannot be distinguished; only if a single crystal is used is such a separation possible. To analyse powder experiments, therefore, there is no point in making $\sigma(\mathbf{r})$ depart from spherical symmetry.‡ However, de Marco and Weiss (1965) have performed experiments on a single crystal of Fe. These experiments separate the (330)–(411) and (600)–(442) reflexions and show that an irreducible angular term of g symmetry must appear in the localized density $\sigma(\mathbf{r})$ in Fe. This term does not appreciably affect the analysis of the s term in the unit cell density shown in Figure 5.4. However, it does affect the g term somewhat, particularly towards the centre of the unit cell (see Tucker, Jones and March, 1965).

We close this section with a few remarks on the importance of this problem of interpreting the experimental X-ray data. The influence of angularity on the Hartree self-consistent potential has as yet never been fully investigated and remains an interesting problem. The more fundamental reason for desiring accurate charge densities from experiment is, however, the possibility of constructing the one-body potential which exactly reproduces this density (see Chapter 1, section 1.2, and Appendix 1.2). To be able to compare this with the Hartree potential would be of great interest. While our discussion has shown how we can interpret the experimental data, it has, unfortunately, also shown the inadequacy of experimental results as yet available.

5.5 Compton profile in X-ray scattering

From the discussion of the spatial distribution of electronic charge, we now turn to the determination of the momentum distribution. As early as 1933, Dumond extended the theory of Compton to deal with high-energy photons scattered by an assembly of non-interacting electrons, with a given momentum distribution. If we define a quantity $I(p)\,dp$ as the probability of finding an electron with momentum of magnitude between p and $p+dp$, the

‡ We are assuming here one atom per unit cell. For crystals such as Ge, the above statement would not be appropriate.

shape of the Compton line is conveniently represented in the reduced form

$$J(q) = \frac{1}{2}\int_q^\infty \frac{I(p)}{p}\,dp, \tag{5.5.1}$$

where $J(q)$ is proportional to the intensity of radiation scattered with a reduced wavelength between q and $q+dq$. More precisely

$$q = \frac{mc(\lambda-\lambda_0-\{2h\sin^2(\theta/2)\}/mc)}{2\lambda_0\sin(\theta/2)}, \tag{5.5.2}$$

where λ_0 and λ are respectively the wavelengths of the incident and scattered radiation and θ is the scattering angle.

Dumond's derivation of equation (5.5.1) was classical, but his result has been put on a firm quantum-mechanical footing by Kilby (1965) and Platzman and Tzoar (1965). Before giving a complete discussion, we shall first present a simpler argument. We use the "impulse approximation" which amounts to the assumption that the photon collides with a single electron, which instantaneously accepts all the energy loss ω and momentum transfer $\boldsymbol{\kappa}$. If ω is very large compared to the binding energy, the electron goes into a free-particle state $e^{i(\boldsymbol{\kappa}+\mathbf{k})\cdot\mathbf{r}}$ and

$$\omega = (\boldsymbol{\kappa}+\mathbf{k})^2, \tag{5.5.3}$$

where $\mathbf{k}$ is simply the difference between the free-particle momentum and the momentum transfer. It follows that we must assign a definite momentum, namely $\mathbf{k}$, to the electron before the collision. We can now argue that the probability of the photon colliding with an electron having this momentum is just $P(\mathbf{k})$, the momentum distribution, and so the Compton profile will simply be given by

$$\int P(\mathbf{k})\,\delta[\omega-(\mathbf{k}+\boldsymbol{\kappa})^2]\,d\mathbf{k}. \tag{5.5.4}$$

From this result, we can obtain Dumond's formula under certain conditions which we shall specify later. However, we shall first discuss the above result (5.5.4) in more detail.

5.5.1 Independent-electron approximation

We shall proceed to derive Dumond's result under the following six assumptions:

(i) The Hartree–Fock theory is applicable.

(ii) The electron binding energies are small compared with the energy transfer from the photons.

(iii) The spread in the momentum distribution is much less than the magnitude of the momentum transfer.

(iv) The energy transfer is only a small fraction of the total photon energy.

(v) The continuum states (i.e. the high-energy bands) are adequately described by plane-wave functions.

(vi) The momentum distribution is spherically symmetric.

Though we have made the assumption that the one-electron theory can be used at this stage, it will help in the later discussion to work in an occupation number formalism. Combining equations (5.2.24), (5.2.25) and (5.2.26), we obtain (putting $k'/k \simeq 1$ in accord with assumption (iii))

$$\frac{d^2\sigma}{d\Omega\, d\omega} = \left.\frac{d\sigma}{d\Omega}\right|_{\text{Thomson}} \frac{1}{2\pi}\int_{-\infty}^{\infty} dt\, e^{-i\omega t}\left\langle \sum_i e^{-i\mathbf{K}.\mathbf{r}_i} e^{iHt} \sum_j e^{i\mathbf{K}.\mathbf{r}_i} e^{-iHt}\right\rangle \tag{5.5.5}$$

which we now wish to write in terms of the creation and annihilation operators $a_{\mathbf{k}}^{\dagger}$ and $a_{\mathbf{k}}$ for the free-particle states $e^{i\mathbf{k}.\mathbf{r}}$. We note first that, in terms of the density operators $\rho_{\mathbf{k}}$, this may be written [cf. equation (5.3.31)]

$$\frac{d^2\sigma}{d\Omega\, d\omega} = \left.\frac{d\sigma}{d\Omega}\right|_{\text{Th}} \frac{1}{2\pi}\int_{-\infty}^{\infty} dt\, e^{-i\omega t}\langle \rho(-\boldsymbol{\kappa}, 0)\, \rho(\boldsymbol{\kappa}, t)\rangle \tag{5.5.6}$$

and in view of equation (2.5.22) we immediately obtain the desired expression as

$$\frac{d^2\sigma}{d\Omega\, d\omega} = \left.\frac{d\sigma}{d\Omega}\right|_{\text{Th}} \frac{1}{2\pi}\sum_{\mathbf{k}'\mathbf{k}} \int dt\, e^{-i\omega t}\langle a_{\mathbf{k}'}^{\dagger} a_{\mathbf{k}'+\boldsymbol{\kappa}}\, a_{\mathbf{k}+\boldsymbol{\kappa}}^{\dagger}(t)\, a_{\mathbf{k}}(t)\rangle. \tag{5.5.7}$$

Since $\boldsymbol{\kappa}$ is the momentum transfer, then by assumptions (iii) and (v) for all $\mathbf{k}$ for which the matrix elements of equation (5.5.7) are appreciable, $\mathbf{k}+\boldsymbol{\kappa}$ will pertain to a free-particle state and we can write

$$\langle a_{\mathbf{k}'}^{\dagger} a_{\mathbf{k}'+\boldsymbol{\kappa}}\, a_{\mathbf{k}+\boldsymbol{\kappa}}^{\dagger}(t)\, a_{\mathbf{k}}(t)\rangle \simeq \langle a_{\mathbf{k}'}^{\dagger} a_{\mathbf{k}'+\boldsymbol{\kappa}}\, a_{\mathbf{k}+\boldsymbol{\kappa}}^{\dagger}\, a_{\mathbf{k}}(t)\rangle\, e^{i\epsilon_{\mathbf{k}+\boldsymbol{\kappa}} t}. \tag{5.5.8}$$

By the same argument, $a_{\mathbf{k}'+\boldsymbol{\kappa}}$ annihilates a particle absent in the original state which must therefore be created by $a_{\mathbf{k}+\boldsymbol{\kappa}}^{\dagger}$ and this implies $\mathbf{k}' = \mathbf{k}$ if the matrix element is to be non-zero. Equation (5.5.8) now becomes

$$\langle a_{\mathbf{k}'}^{\dagger} a_{\mathbf{k}'+\boldsymbol{\kappa}}\, a_{\mathbf{k}+\boldsymbol{\kappa}}^{\dagger}(t)\, a_{\mathbf{k}}(t)\rangle \simeq \delta_{\mathbf{k}'\mathbf{k}}\langle a_{\mathbf{k}}^{\dagger} a_{\mathbf{k}}(t)\rangle\, e^{i\epsilon_{\mathbf{k}+\boldsymbol{\kappa}} t} \tag{5.5.9}$$

and

$$\frac{d^2\sigma}{d\Omega\, d\omega} = \left.\frac{d\sigma}{d\Omega}\right|_{\text{Th}} \frac{1}{2\pi}\sum_{\mathbf{k}} \int_{-\infty}^{\infty} dt\, e^{-i\omega t} e^{i\epsilon_{\mathbf{k}+\boldsymbol{\kappa}} t}\langle a_{\mathbf{k}}^{\dagger} a_{\mathbf{k}}(t)\rangle. \tag{5.5.10}$$

As yet, we have placed no restrictions on our description of the scatterer. However, at this point we introduce the assumption that independent-particle theory can be used. We represent the antisymmetrized product of one-particle wave functions by

$$|\psi\rangle = \int v_1(\mathbf{k}_1)\, v_2(\mathbf{k}_2) \ldots v_n(\mathbf{k}_n)\, a^\dagger_{\mathbf{k}_1} a^\dagger_{\mathbf{k}_2} \ldots a^\dagger_{\mathbf{k}_n} |0\rangle\, d\mathbf{k}_1 \ldots d\mathbf{k}_n, \qquad (5.5.11)$$

where

$$v_i(\mathbf{k}) = \frac{1}{8\pi^3} \int \psi_i(\mathbf{r})\, e^{i\mathbf{k}.\mathbf{r}}\, d\mathbf{r} \qquad (5.5.12)$$

is the momentum eigenfunction introduced in Chapter 1, section 1.9. Then we have

$$a_{\mathbf{k}} |\psi\rangle = \sum_{j=1}^{n} v_j(\mathbf{k}) \left(\prod_{i \neq j} \int d\mathbf{k}_i\, v_i(\mathbf{k}_i)\, a^\dagger_{\mathbf{k}_i} \right) |0\rangle \qquad (5.5.13)$$

so that, if ε_i represents a single-particle eigenvalue,

$$e^{iHt} a_{\mathbf{k}} |\psi\rangle = \sum_{j=1}^{n} v_j(\mathbf{k}) \exp\left(i \sum_{i \neq j} \varepsilon_i t\right) \left(\prod_{i \neq j} \int d\mathbf{k}_i\, v_i(\mathbf{k}_i)\, a^\dagger_{\mathbf{k}_i} \right) |0\rangle. \qquad (5.5.14)$$

It can now be seen that

$$\langle \psi | a^\dagger_{\mathbf{k}} a_{\mathbf{k}}(t) | \psi \rangle = \sum_{j=1}^{n} |v_j(\mathbf{k})|^2 e^{-i\varepsilon_j t}$$

$$= \sum_{j=1}^{n} P_j(\mathbf{k})\, e^{-i\varepsilon_j t}, \qquad (5.5.15)$$

where $P_j(\mathbf{k})$ is clearly the momentum distribution of the jth electron. Equation (5.5.10) is now

$$\frac{d^2\sigma}{d\Omega\, d\omega} = \left.\frac{d\sigma}{d\Omega}\right|_{\text{Th}} \sum_{j=1}^{n} \int P_j(\mathbf{k})\, \delta[\omega - (\mathbf{k}+\boldsymbol{\kappa})^2 - \varepsilon_i]\, d\mathbf{k}. \qquad (5.5.16)$$

We now incorporate assumption (ii) into the theory by neglecting ε_i, when we find

$$\frac{d^2\sigma}{d\Omega\, d\omega} = \left.\frac{d\sigma}{d\Omega}\right|_{\text{Th}} \int P(\mathbf{k})\, \delta[\omega - (\mathbf{k}+\boldsymbol{\kappa})^2]\, d\mathbf{k}. \qquad (5.5.17)$$

We can consider temperature effects to be included in this equation, the requisite modifications to the above argument being trivial.

The integral can be formally evaluated in the general case, but to obtain Dumond's formula we now invoke assumption (vi). Evaluating the angular integration, we find

$$\frac{1}{4\pi^2 \kappa} \int_0^\infty k\, dk \int_{|\kappa-k|}^{|\kappa+k|} P(p)\, p\, dp\, \delta(\omega + k^2) = \frac{1}{8\pi^2 \kappa} \int_{|\kappa-\sqrt{\omega}|}^{|\kappa+\sqrt{\omega}|} P(p)\, p\, dp. \qquad (5.5.18)$$

ω now represents the energy loss of the photon, and therefore the energy gain of an electron, the latter being approximately given by $(\kappa+q)^2 \approx \kappa^2(1+2k^{-1}q)$, where q is the reduced wavelength of the photon. Expressing $d^2\sigma/d\Omega\,d\omega = \mathscr{I}$ as a function of q, we then obtain

$$\mathscr{I}(q)\,dq = \left.\frac{d\sigma}{d\Omega}\right|_{\mathrm{Th}} \frac{dq}{4\pi^2}\int_q^{2\kappa+q} pP(p)\,dp. \tag{5.5.19}$$

By assumption (iii), we may replace the upper limit of integration by infinity. Then writing

$$2\pi^2 I(p) = p^2 P(p) \tag{5.5.20}$$

we obtain the desired result (5.5.1).

Assumptions (ii) to (vi) are confirmed in a typical example (cf. Kilby, 1965). Incident radiation of wavelength about 1Å is scattered through an angle $\theta \simeq 180°$. Then $\kappa = 4\pi/\lambda \sin(\theta/2)$ is approximately 6 atomic units and the energy transfer about 22 au. If we consider a light hydrogen-like atom with a binding energy of about 4 au and a root mean square momentum of about 3 au, $P(p)$ will be negligible beyond 2κ.

5.5.2 *Influence of correlations*

Only assumption (i) of the derivation above will be invalidated when correlation effects are strong. It is interesting to note in passing that the assumption that a particle of high momentum behaves as if free, while reasonable for Coulombic interactions, will not hold for more violent interactions. For example, the mean free path in a hard sphere gas will be approximately independent of momentum, so that the lifetime between collisions will be inversely proportional to the momentum.

Equation (5.5.19) does not depend for its validity on the independent-particle model. Since

$$P(\mathbf{k}) = \langle a_{\mathbf{k}}^\dagger a_{\mathbf{k}} \rangle \tag{5.5.21}$$

it is evident that equation (5.5.19) is obtained if we can neglect the time dependence of $\langle a_{\mathbf{k}}^\dagger a_{\mathbf{k}}(t)\rangle$. In Chapter 2, we saw that the decay in time of this quantity is related to the lifetime of a quasi-particle, the Fourier transform being a delta function if the electrons are free, and broadened out by correlation effects to a peak with breadth measured by the lifetime of the quasi-particle or imaginary part of the self-energy. In the presence of an external potential, we can see from equation (5.5.16) that, in the absence of correlations, we obtain rather a set of delta functions, each one of which we can expect correlations to broaden out. If this simple description of correlation effects is valid, the consequence is clear: $P(\mathbf{k})$ is seen with finite resolution.

For sufficiently high momentum transfer, the resolution will be good even if single-particle states are badly defined. Suppose the lifetimes of the particles in the single-particle states are very short and the delta functions are broadened in **k**-space over the entire volume enclosed by the Fermi surface of a metal, so that $\Delta\omega \simeq k_{\mathrm{f}}^2$. Noting the factor $\exp(i\varepsilon_{\mathbf{k}+\boldsymbol{\kappa}}t)$ in equation (5.5.10), the peak will be centred about $\omega = (\mathbf{k}+\boldsymbol{\kappa})^2$ with width $k_{\mathrm{f}}^2 \simeq \Delta\omega \simeq 2\boldsymbol{\kappa}.\Delta\mathbf{k}$. Since $\boldsymbol{\kappa}$ is large, $\Delta\mathbf{k}$ is small and we conclude that, within the assumptions made above, Dumond's formula holds when correlation effects are included. There is no point, however, in a finer resolution of energy than $\sim \hbar/\tau$, where τ is the lifetime of a quasi-particle in a definite state.

5.5.3 *Effect of binding energy*

The binding energies of the core electrons of a metal can have observable effects, and we now see how they modify our earlier theory. We take the Hartree–Fock method as adequate for the core electrons, and so refer to equation (5.5.16). We cannot now pass from this equation to equation (5.5.17), since the summation over the N electrons must include the delta function, and so we consider each term separately, in the form

$$\left|\int \psi_i(\mathbf{r})\, \mathrm{e}^{i(\mathbf{k}+\boldsymbol{\kappa}).\mathbf{r}}\, d\mathbf{r}\right|^2 \delta(\omega - \varepsilon_{\mathbf{k}} - \varepsilon_{\mathrm{B}})\, d\mathbf{k}. \tag{5.5.22}$$

Here we have put $\varepsilon_{\mathrm{B}} = -\varepsilon_i$, a positive quantity, and reversed the sign of ω so that it represents an energy loss. It is obvious that if ω is to be positive, no Compton scattering can occur unless $\omega > \varepsilon_{\mathrm{B}}$. When $\omega \simeq \varepsilon_{\mathrm{B}}$, we must have $\varepsilon_k \simeq 0$ of course when the final one-particle state is least like a plane wave, but we will ignore this additional complication.

We now write

$$\sum_{m=-l}^{+l} \left|\psi_{lm}(\mathbf{r})\, \mathrm{e}^{i(\mathbf{k}+\boldsymbol{\kappa}).\mathbf{r}}\, d\mathbf{r}\right|^2 = P_l(\mathbf{k}+\boldsymbol{\kappa}), \tag{5.5.23}$$

where we assume that the core electrons move in a spherically symmetric potential, so each level is $(2l+1)$-fold degenerate, and $P_l(\mathbf{k})$ is also spherically symmetric. Equation (5.5.22) becomes

$$\int P_l(\mathbf{k}+\boldsymbol{\kappa})\, \delta(\omega - k^2 - \varepsilon_{\mathrm{B}})\, d\mathbf{k} \tag{5.5.24}$$

and this integral can be carried out exactly as before, with ω now being replaced by $\omega + \varepsilon_{\mathrm{B}}$. We have as the integral replacing equation (5.5.18)

$$\frac{1}{8\pi^2 \kappa} \int_{|\kappa - \sqrt{(\omega+\varepsilon_{\mathrm{B}})}|}^{|\kappa + \sqrt{(\omega+\varepsilon_{\mathrm{B}})}|} P_l(p)\, p\, dp. \tag{5.5.25}$$

If we wish to continue to approximate as in section 5.5.1, we put $\omega \gg \varepsilon_B$, so that the lower limit becomes

$$\left|\kappa - \sqrt{\omega}\left(1+\frac{\varepsilon_B}{2\omega}\right)\right| \simeq \left|\kappa-(\kappa+q)\left(1+\frac{\varepsilon_B}{2\kappa^2}\right)\right|$$

$$\simeq \left|q+\frac{\varepsilon_B}{2\kappa}\right|. \tag{5.5.26}$$

Hence we finally obtain, with the same approximations as before,

$$J_l(q) = \int_{q+(\varepsilon_B/2\kappa)}^{\infty} \frac{I_l(p)}{p}\,dp. \tag{5.5.27}$$

The total Compton profile is, summing over occupied shells,

$$J(q) = \sum_l \int_{q+(\varepsilon_B/2\kappa)}^{\infty} \frac{I_l(p)}{p}\,dp. \tag{5.5.28}$$

As κ gets large, the original formula obtains since $\varepsilon_B/2\kappa$ may then be neglected.

In summary, if the Compton profile is zero for an energy transfer less than any core-binding energy for each core level equation (5.5.24) gives the dependence on ω and ε_B within the approximation that all continuum states are plane-waves, and equation (5.5.28) gives the Compton profile under the further assumption that ε_B is small compared with the energy transfer.

For the particular case of lithium, where we can take with fair accuracy for the 1s core $\psi_s(r) = e^{-\alpha r}$ ($\alpha = 2.69/a_0$), the integral in equation (5.5.23)

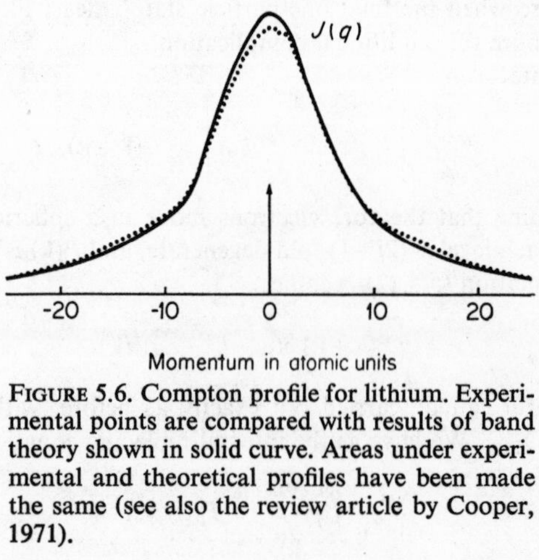

FIGURE 5.6. Compton profile for lithium. Experimental points are compared with results of band theory shown in solid curve. Areas under experimental and theoretical profiles have been made the same (see also the review article by Cooper, 1971).

can easily be evaluated analytically, but the resolution available with apparatus at the present time hardly warrants correction due to the 1s binding energy in lithium. There are, however, solids with observed Compton profiles showing binding-energy effects.

To show the degree of sophistication reached with this method, results of Cooper and colleagues (1970) for Li are shown in Figure 5.6. The agreement is seen to be very satisfactory, the theoretical curve being obtained from a momentum eigenfunction $v(\mathbf{p})$ for the conduction band of Li quite like that recorded in Table 1.7 of Chapter 1.

5.6 Scattering of electrons by electrons

The use of electrons instead of X-rays to investigate electronic structure has the advantage that higher-order reflexions may be investigated, since the electrons are readily obtainable at energies of 100 Kv, and so have much shorter wavelengths than the suitable X-ray radiation available.

We shall not go into detail about experimental technique of theory, but refer the interested reader to *Electron Microscopy of Thin Films* (Butterworth, 1965), since as yet experimental data are not available for our purposes. We merely note that, with appropriate relativistic cross-sections, we treat the electrons by nearly free electron theory—a single Bragg reflexion involving just two momentum eigenfunctions. As to what information we might expect to become available in future from electron scattering, because the 100 Kv electron will spend about 10^{-13} sec in each unit cell, we do not expect correlations between this electron and those of the metal to be of any importance, and the electron will just see the nuclear and Hartree potentials. Thus the scattering factors will be determined by the Fourier components of the series

$$V(\mathbf{r}) = \sum_{\mathbf{K}} \frac{Ze}{K^2} e^{i\mathbf{K}.\mathbf{r}} - e \sum_{\mathbf{K}} \frac{\rho_{\mathbf{K}}}{K^2} e^{i\mathbf{K}.\mathbf{r}}. \tag{5.6.1}$$

The presence of the factor $1/K^2$ makes the part of the electron scattering factor of interest fall off much more rapidly than the X-ray factor, and, moreover, it would appear that one cannot expect, for some time to come, great accuracy in the results of this method of experimental analysis; on the other hand, one can go to much higher-order reflexions than with X-rays, as already mentioned.

5.6.1 Positron annihilation

When positrons are introduced into metals, they rapidly thermalize and annihilate with conduction electrons in a time $\sim 10^{-13}$ sec (see Problem 5.1). Each individual interaction of an electron or positron with the radiation field can create only one photon.

The annihilation process must proceed therefore by an intermediate state in which the original positron–electron pair is still present: in addition one photon has been created.

The part of the Hamiltonian which can lead to a two-photon final state of total momentum $\hbar\mathbf{k}$ is, apart from proportionality constants,

$$\sum_{\mathbf{k}_1\mathbf{k}_2} a_{\mathbf{k}_1} b_{\mathbf{k}_2} \delta_{\mathbf{k}_1+\mathbf{k}_2,\mathbf{k}}, \tag{5.6.2}$$

where $a_{\mathbf{k}_1}$ and $b_{\mathbf{k}_2}$ respectively are plane-wave annihilation operators for the electron and positron. It is now convenient to introduce field operators $\psi(\mathbf{x}_1)$ and $\psi_p(\mathbf{x}_2)$ through

$$a_{\mathbf{k}_1} = \frac{1}{\sqrt{\mathscr{V}}} \int d\mathbf{x}_1 \, e^{-i\mathbf{k}_1\cdot\mathbf{x}_1} \psi(\mathbf{x}_1) \tag{5.6.3}$$

and

$$b_{\mathbf{k}_2} = \frac{1}{\sqrt{\mathscr{V}}} \int d\mathbf{x}_2 \, e^{-i\mathbf{k}_2\cdot\mathbf{x}_2} \psi_p(\mathbf{x}_2), \tag{5.6.4}$$

where $\mathscr{V}$ is the volume of the system. Then we can rewrite (5.6.2) in the form

$$\int d\mathbf{x} \, e^{-i\mathbf{k}\cdot\mathbf{x}} \psi(\mathbf{x}) \psi_p(\mathbf{x}). \tag{5.6.5}$$

This shows us that the essential quantity entering the theory of positron annihilation is the Fourier transform of the product of the wave functions of the positron and the electron.

If the positron wave function were quite flat, then it is clear that we could observe the Fourier transform of the electron wave function, or the momentum distribution. However, in general the positron wave function is not quite flat, and calculations of it, in simple metals, have been made by a number of authors. The positron, which is rapidly thermalized, essentially occupies the lowest state of an energy band, and, as we saw in Chapter 1, we have the valuable Wigner–Seitz method for calculating the wave function in this case.

The result obtained for a positron in Cu metal by Donovan and March (1956) is shown, as an example, in Figure 5.7. It will be seen that the positron avoids the ion-core region, and thus it largely cancels off the high-momentum components associated with the atomic-like oscillations of the electron wave functions. However, apart from annulling the effect of the high-momentum components, the result is to leave the Fourier transform of the electron wave function otherwise intact.

Angular correlation of photons. The above argument shows that the probability of finding two photons emitted with a certain angular deviation from 180° depends on the Fourier transform of the wave-function product. Also, as we shall verify quantitatively in the next section, the slowing down of positrons by conduction electrons leads to a time for thermalization which is very much shorter than the lifetime. Because thermal energies are very much smaller than the Fermi energy in a metal, the positron, as we assumed above, is in the state with $\mathbf{k} = 0$.

But if the positron wave function cuts out the high-momentum components, we are left with a rather plane-wave-like situation for the electrons. Thus, a useful approximation to the problem is that of a positron annihilating in a free electron gas.

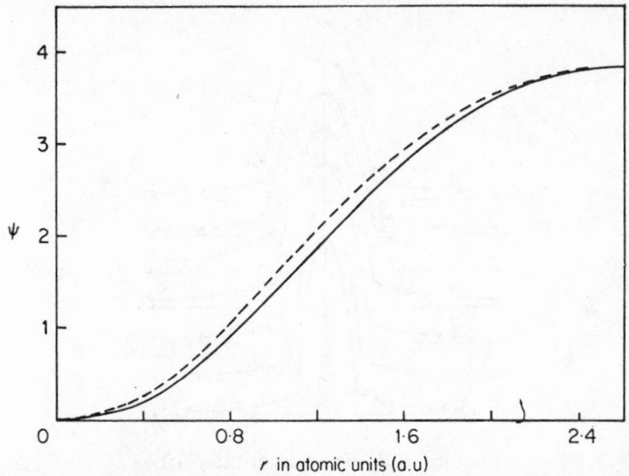

FIGURE 5.7. Positron wave function ψ_0 as function of radial distance r measured in atomic units, in Cu metal. Solid curve is for Hartree ion-core potential. Dashed curve is modification for Hartree–Fock ion core potential.

The two-photon angular correlation resulting from annihilation of a positron in an electron gas is simple because we are merely left with the probability that a momentum $\hbar\mathbf{k}$ is conveyed to the photons as, apart from a constant, the modulus squared of the Fourier transform of the wave-function product. But, for plane waves, this vanishes except for the positron–electron pair with momentum $\hbar\mathbf{k}$, in which case it is a constant independent of $\mathbf{k}$. The probability distribution function is therefore proportional to the momentum distribution $P(\mathbf{k})$, which is simply a constant for $k < k_f$ and zero otherwise.

The distribution function for the z component of momentum is therefore

$$P_z(k_z) = \iint P(\mathbf{k})\, dk_x\, dk_y = \begin{cases} 1 - \left(\dfrac{k_z}{k_f}\right)^2, & k_z < k_f, \\ 0, & k_z > k_f. \end{cases} \tag{5.6.6}$$

The angle θ between the photons is linearly related to k_z by $\theta = \hbar k_z/mc$, and hence the number of counts observed at angle θ should also be a parabolic function proportional to $1-(\theta/\theta_0)^2$ where the cut-off occurs at a value of θ_0 given by $\theta_0 = \hbar k_f/mc$.

Results obtained by Lang, DeBenedetti and Smoluchowski are shown in Figure 5.8 and confirm the general features of this primitive theory for

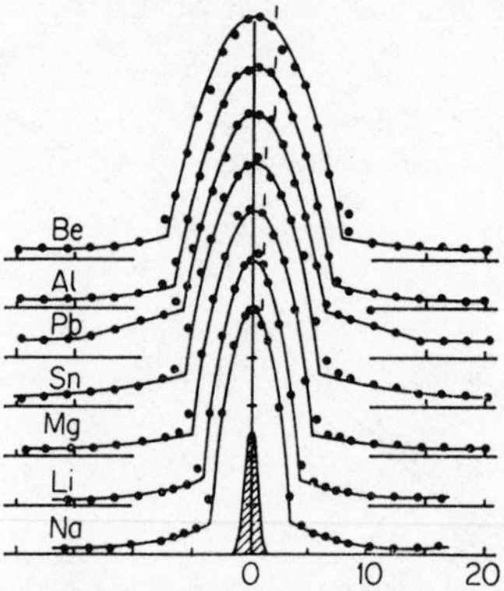

FIGURE 5.8. Experimentally observed angular distribution of annihilation radiation, for positrons introduced into the simple metals shown. Solid parabolas are computed from free-electron theory. Discontinuities are shown at calculated Fermi momenta. Shaded curve is the calculated resolution (after Lang, DeBenedetti and Smoluchowski, 1955).

simple metals. The cut-off angles agree well with the estimates given above.

Stewart has attempted to construct $P(k)$, assuming isotropy, by differentiation from the measured angular correlation. Very valuable results have come from the positron annihilation experiments and a full account is given in the work of Stewart and Roellig (1967).

5.6.2 Fast electron energy losses and plasma excitation

From the discussion of collective modes of the electron gas in a metal, it might be expected that plasmon modes could be excited by injection of electrons having sufficiently high energy. This proves indeed to be the case, the sharp peaks in the energy-loss spectrum of electrons shot at the material under investigation yielding the plasma frequencies. These energy-loss experiments also provide information on the dielectric constant.

To obtain the general formulae governing the results, we apply equation (5.2.24) to the scattering of the incoming electrons by a uniform gas. The probability P per unit time that the scattered electron transfers momentum $\mathbf{q}$ and energy ω to the electron gas is given by

$$P(\mathbf{q}, \omega) = 2\pi\left(\frac{4\pi e^2}{q^2}\right)^2 S(\mathbf{q}, \omega) \tag{5.6.7}$$

or, in view of the relation between the structure factor and the dielectric constant (see Chapter 2, equation 2.7.29),

$$P(\mathbf{q}, \omega) = -\frac{8\pi e^2}{q^2}\,\mathrm{Im}\left(\frac{1}{\varepsilon(\mathbf{q}, \omega)}\right). \tag{5.6.8}$$

From $P(\mathbf{q}, \omega)$ we can easily find $d^2\sigma/d\Omega\, d\theta$ and the energy transfer per unit time associated with scattering in a given direction.

From our previous discussion of the physical meaning of $S(\mathbf{k}, \omega)$ it is evident that undamped plasmons will appear as δ-function peaks in $P(\mathbf{q}, \omega)$, but a short independent argument based on equation (5.6.8) will now be given. We first note that the plasmon frequencies $\omega_{\mathbf{q}}$ are such that

$$\varepsilon(\mathbf{q}, \omega_{\mathbf{q}}) = 0, \tag{5.6.9}$$

i.e.

$$\varepsilon_1(\mathbf{q}, \omega_{\mathbf{q}}) = \varepsilon_2(\mathbf{q}, \omega_{\mathbf{q}}) = 0, \tag{5.6.10}$$

where ε_1 and ε_2 are the real and imaginary parts of ε respectively. We anticipate a result at this stage which we shall discuss fully in Chapter 6, that, as a consequence of the causality principle, certain response functions, among which is $\varepsilon^{-1}(\mathbf{q}, \omega)$, must be analytic in the upper complex half-plane. Then near $\omega = \omega_q$, which we assume to be real,

$$\begin{aligned} S(\mathbf{q}, \omega) &= \frac{-q^2}{4\pi^2 e^2}\,\mathrm{Im}\left[\frac{1}{\varepsilon(\mathbf{q}, \omega)}\right] \\ &= \frac{-q^2}{4\pi^2 e^2}\,\mathrm{Im}\left[\lim_{\eta\to 0}\frac{1}{\varepsilon_1(\mathbf{q}, \omega_{\mathbf{q}}) + (\partial\varepsilon_1/\partial\omega)|_{\omega_{\mathbf{q}}}(\omega - \omega_{\mathbf{q}} + i\eta)}\right] \\ &= \frac{q^2}{4\pi e^2}\,\frac{1}{\partial\varepsilon_1/\partial\omega}\bigg|_{\omega=\omega_{\mathbf{q}}}\delta(\omega - \omega_{\mathbf{q}}). \end{aligned} \tag{5.6.11}$$

In practice, damping occurs (so that we saw that the plasma frequencies turn out to be complex quantities), but this damping is not expected to be large as $\mathbf{q}\to 0$.

We ought to note that, at large angles, the scattering reveals an incoherent excitation region, the plasmons no longer being well defined. The significant contributions to $S(\mathbf{q}, \omega)$ then come mainly from single-pair and multiple-pair excitations. There is also a transition region in which the plasmons appear as broad peaks in $S(\mathbf{q}, \omega)$, until the single-pair excitations give way completely to the plasmons in the region we have discussed above. The frequency dependence of the dielectric constant will be referred to again in Chapter 7.

5.7 Scattering of neutrons by electrons

The technique of neutron scattering can be used to explore not only the phonon spectrum of a crystal, but also the spin-wave spectrum or, more generally, the microscopic magnetic structure of the crystal. This is so because, for example, the magnetic interaction of the electrons of a paramagnetic atom with a slow neutron has pertaining to it a scattering length comparable to that of the nuclear-neutron scattering already considered in section 5.3. We shall also find that it is possible to explore the details of the spin-density distribution within each cell of the crystal.

5.7.1 Magnetic Hamiltonian and matrix elements

The pair potential of equation (5.2.24) is now replaced by

$$H_{\mathrm{m}}^{j} = -\boldsymbol{\mu}^{j} . \mathscr{H}, \tag{5.7.1}$$

where $\boldsymbol{\mu}^j$ is the magnetic moment of the jth electron and $\mathscr{H}$ is the magnetic field associated with the magnetic moment $\boldsymbol{\mu}_{\mathrm{n}}$ of the neutron. We shall neglect the magnetization induced by orbital electronic motions at this stage; this is usually permissible but we shall discuss the modifications involved later.

The magnetic field of a magnetic moment $\boldsymbol{\mu}_{\mathrm{n}}$ of very small extent and localized at the origin, is given by

$$\begin{aligned} \mathscr{H} &= \operatorname{curl}\left(\frac{\boldsymbol{\mu}_{\mathrm{n}} \times \mathbf{r}}{r^3}\right) = -\operatorname{curl}\left[\boldsymbol{\mu}_{\mathrm{n}} \times \nabla\left(\frac{1}{r}\right)\right] \\ &= (\boldsymbol{\mu}_{\mathrm{n}} . \nabla) \nabla\left(\frac{1}{r}\right) - \boldsymbol{\mu}_{\mathrm{n}} \nabla^2\left(\frac{1}{r}\right) = \nabla\left(\boldsymbol{\mu}_{\mathrm{n}} . \nabla\frac{1}{r}\right) + 4\pi\delta(\mathbf{r})\, \boldsymbol{\mu}_{\mathrm{n}}. \end{aligned} \tag{5.7.2}$$

Thus

$$H_{\mathrm{m}} = -\boldsymbol{\mu} . \nabla\left[\boldsymbol{\mu}_{\mathrm{n}} . \nabla\left(\frac{1}{r}\right)\right] - 4\pi\delta(\mathbf{r})\, \boldsymbol{\mu}_{\mathrm{n}} . \boldsymbol{\mu}. \tag{5.7.3}$$

This is the form we shall use, but it should be remarked that H_m is often written as

$$H_m = H_{dipole} + H_{contact}, \tag{5.7.4}$$

where

$$H_{dipole} = \frac{\boldsymbol{\mu}.\boldsymbol{\mu}_n}{r^3} - \frac{3(\boldsymbol{\mu}.\mathbf{r})(\boldsymbol{\mu}_n.\mathbf{r})}{r^5} \tag{5.7.5}$$

and

$$H_{contact} = -\frac{8\pi}{3}\boldsymbol{\mu}.\boldsymbol{\mu}_n\,\delta(\mathbf{r}). \tag{5.7.6}$$

The first step in the calculation of the scattering is to evaluate the matrix elements between initial and final states. Let us write the states of the scatterer as $\langle i|$ and $|f\rangle$ and the states of the neutron as $e^{i\mathbf{k}.\mathbf{r}}|s_n\rangle$ and $e^{i\mathbf{k}'.\mathbf{r}}|s_{n'}\rangle$, $|s\rangle$ being a spin state. The scattering matrix element is therefore

$$\int d\mathbf{r}\,\langle s_n'|\langle f|e^{-i\mathbf{k}'.\mathbf{r}}\sum_j H_m^j(\mathbf{r}-\mathbf{r}_j)\,e^{i\mathbf{k}.\mathbf{r}}|i\rangle|s_n\rangle$$

$$= \int d\mathbf{r}\sum_j \langle s_n'|\langle f|e^{i\boldsymbol{\kappa}.\mathbf{r}_j}e^{i\boldsymbol{\kappa}.\mathbf{r}}H_m^j(\mathbf{r})|i\rangle|s_n\rangle \quad (\boldsymbol{\kappa} = \mathbf{k}-\mathbf{k}'). \tag{5.7.7}$$

We must now evaluate

$$\int e^{i\boldsymbol{\kappa}.\mathbf{r}}H_m(\mathbf{r})\,d\mathbf{r} = -\int e^{i\boldsymbol{\kappa}.\mathbf{r}}\boldsymbol{\mu}.\nabla\left[\boldsymbol{\mu}_n.\nabla\left(\frac{1}{r}\right)\right]d\mathbf{r} - 4\pi\boldsymbol{\mu}_n.\boldsymbol{\mu}$$

$$= \boldsymbol{\mu}.\boldsymbol{\kappa}\int i e^{i\boldsymbol{\kappa}.\mathbf{r}}\boldsymbol{\mu}_n.\nabla\left(\frac{1}{r}\right)d\mathbf{r} - 4\pi\boldsymbol{\mu}_n.\boldsymbol{\mu}$$

$$= \frac{4\pi}{\kappa^2}(\boldsymbol{\mu}.\boldsymbol{\kappa})(\boldsymbol{\mu}_n.\boldsymbol{\kappa}) - 4\pi\boldsymbol{\mu}.\boldsymbol{\mu}_e. \tag{5.7.8}$$

The right-hand side of equation (5.7.7) then becomes

$$-4\pi\langle s_n'|\langle f|\sum_j e^{i\boldsymbol{\kappa}.\mathbf{r}_j}[\boldsymbol{\mu}_n.\boldsymbol{\mu}^j - (\boldsymbol{\mu}^j.\boldsymbol{\kappa})(\boldsymbol{\mu}_n.\boldsymbol{\kappa})]|i\rangle|s_n\rangle$$

$$= -4\pi\langle s_n'|\boldsymbol{\mu}_n|s_n\rangle.\langle f|\sum_j e^{i\boldsymbol{\kappa}.\mathbf{r}_j}(\hat{\boldsymbol{\kappa}}\times[\boldsymbol{\mu}_j\times\hat{\boldsymbol{\kappa}}])|i\rangle$$

$$= -4\pi\langle s_n'|\boldsymbol{\mu}_n|s_n\rangle.\hat{\boldsymbol{\kappa}}\times\langle f|\boldsymbol{\mu}(\boldsymbol{\kappa})|i\rangle\times\hat{\boldsymbol{\kappa}}, \tag{5.7.9}$$

where we have introduced the Fourier transform $\boldsymbol{\mu}(\boldsymbol{\kappa})$ of the magnetic moment operator

$$\boldsymbol{\mu}(\mathbf{r}) = \sum_i \delta(\mathbf{r}-\mathbf{r}_j)\,\boldsymbol{\mu}^j. \tag{5.7.10}$$

5.7.2 *Total scattering*

The total scattering may be obtained by reference to the general formulation of section 5.2 with two additional considerations: first, we must take account of the initial polarization of the neutrons and, secondly, we must remember to sum over neutron spin states as well as the states of the scatterer. To take account of the polarization, we introduce the neutron spin density matrix $\langle s_n|\rho_n|s_n\rangle$, the scattering being directly proportional to this. We then find

$$\frac{d^2\sigma}{d\Omega\,d\omega} \propto \frac{k'}{k}\int \sum_{\substack{s_n s_n' \\ if}} \langle s_n|\rho_n|s_n\rangle\langle s_n|\boldsymbol{\mu}_n|s_n'\rangle.\langle i|\hat{\boldsymbol{\kappa}}\times[\boldsymbol{\mu}(-\boldsymbol{\kappa})\times\hat{\boldsymbol{\kappa}}]|f\rangle$$
$$\times\langle s_n'|\boldsymbol{\mu}_n|s_n\rangle.\langle f|\hat{\boldsymbol{\kappa}}\times[\boldsymbol{\mu}(\boldsymbol{\kappa},t)\times\hat{\boldsymbol{\kappa}}]|i\rangle\,e^{-i\omega t}\,dt. \qquad (5.7.11)$$

Now we have the relation

$$\sum_{s_n s_n'}\langle s_n|\rho_n|s_n\rangle\langle s_n|\mu_n^\alpha|s_n'\rangle\langle s_n'|\mu_n^\beta|s_n\rangle = \sum_{s_n}\langle s_n|\rho_n|s_n\rangle\langle s_n|\mu_n^\alpha\mu_n^\beta|s_n\rangle$$
$$= \mu_n^2\sum_{s_n}\langle s_n|\rho_n|s_n\rangle\,\delta_{\alpha\beta}, \qquad (5.7.12)$$

so that equation (5.7.11) becomes

$$\frac{d^2\sigma}{d\Omega\,d\omega} \propto \frac{k'}{k}\sum_{s_n}\langle s_n|\rho_n|s_n\rangle\int e^{-i\omega t}\langle(\hat{\boldsymbol{\kappa}}\times[\boldsymbol{\mu}(-\boldsymbol{\kappa})\times\boldsymbol{\kappa}]).(\boldsymbol{\kappa}\times[\boldsymbol{\mu}(\boldsymbol{\kappa},t)\times\hat{\boldsymbol{\kappa}}])\rangle\,dt. \qquad (5.7.13)$$

This expression may be simplified by noting that for any two vectors **a** and **b**

$$(\hat{\boldsymbol{\kappa}}\times[\mathbf{a}\times\hat{\boldsymbol{\kappa}}]).(\hat{\boldsymbol{\kappa}}\times[\mathbf{b}\times\hat{\boldsymbol{\kappa}}]) = \mathbf{a}.\mathbf{b}-(\mathbf{a}.\hat{\boldsymbol{\kappa}})(\mathbf{b}.\hat{\boldsymbol{\kappa}}) \qquad (5.7.14)$$

so that

$$\frac{d^2\sigma}{d\Omega\,d\omega} = A(\mathbf{k}',\mathbf{k})\int dt\,e^{-i\omega t}\langle\boldsymbol{\mu}(-\boldsymbol{\kappa}).\boldsymbol{\mu}(\boldsymbol{\kappa},t)-(\boldsymbol{\mu}(-\boldsymbol{\kappa}).\hat{\boldsymbol{\kappa}})(\boldsymbol{\mu}(\boldsymbol{\kappa},t).\hat{\boldsymbol{\kappa}})\rangle, \qquad (5.7.15)$$

where the definition of A is simply $8\pi(M\mu_n/h)^2(k'/k)$.

In terms of the time-dependent structure factor

$$S_{\alpha\beta}(\boldsymbol{\kappa},t) = \frac{1}{N}\langle\mu_\alpha(-\boldsymbol{\kappa})\,\mu_\beta(\boldsymbol{\kappa},t)\rangle \qquad (5.7.16)$$

equation (5.7.15) may be put into the form

$$\frac{d^2\sigma}{d\Omega\,d\omega} = A(\mathbf{k}',\mathbf{k})\sum_{\alpha\beta}(\delta_{\alpha\beta}-\hat{\kappa}_\alpha\hat{\kappa}_\beta)\int_{-\infty}^{\infty}S_{\alpha\beta}(\boldsymbol{\kappa},t)\,e^{i\omega t}\,dt, \qquad (5.7.17)$$

which is simply

$$\frac{d^2\sigma}{d\Omega\,d\omega} = A(\mathbf{k}',\mathbf{k})\sum_{\alpha\beta}(\delta_{\alpha\beta}-\hat{\kappa}_\alpha\hat{\kappa}_\beta)\,S_{\alpha\beta}(\boldsymbol{\kappa},\omega). \qquad (5.7.18)$$

5.7.3 *Elastic scattering*

In section 5.4, we anticipated the result that the elastic scattering gives us the unpaired spin distribution. This can now be shown very readily. The elastic scattering is [see equation (5.2.28)]

$$\left.\frac{d^2\sigma}{d\Omega\, d\omega}\right|_{\text{elastic}} = A(\mathbf{k}', \mathbf{k}) \sum_{\alpha\beta} (\delta_{\alpha\beta} - \hat{\kappa}_\alpha \hat{\kappa}_\beta)\, S_{\alpha\beta}(\boldsymbol{\kappa}, t)\bigg|_{t=\infty}, \tag{5.7.19}$$

where, in accordance with equation (5.2.30),

$$S_{\alpha\beta}(\boldsymbol{\kappa}, t)\big|_{t=\infty} = \frac{1}{N}\langle \mu_\alpha(-\boldsymbol{\kappa})\rangle \langle \mu_\beta(\boldsymbol{\kappa})\rangle \tag{5.7.20}$$

and represents Bragg scattering. The function $\langle \boldsymbol{\mu}(\mathbf{r})\rangle$ is the unpaired spin density and will be periodic in the crystal, in general (see, however, the remarks on Cr metal in Chapter 4, section 4.12):

$$\langle \boldsymbol{\mu}(\mathbf{r})\rangle = \sum_{\mathbf{K}} \boldsymbol{\mu}_{\mathbf{K}}\, \mathrm{e}^{-i\mathbf{K}.\mathbf{r}}. \tag{5.7.21}$$

Hence

$$\langle \boldsymbol{\mu}(\mathbf{k})\rangle = 8\pi^3 \sum_{\mathbf{K}} \delta(\mathbf{k}+\mathbf{K})\, \boldsymbol{\mu}_{\mathbf{K}}. \tag{5.7.22}$$

We thus obtain Bragg peaks just as for the other types of elastic scattering we have discussed in this chapter. The nuclear cross-section must be allowed for, of course, and the nuclear scattering subtracted off. This can be done in some cases by measuring the nuclear-scattering cross-section in a compound not showing magnetic effects. It can also be seen that if $\boldsymbol{\kappa}$ is parallel to the direction of magnetization, the magnetic scattering vanishes and only the nuclear contribution remains. Temperature may be allowed for by multiplication by the Debye–Waller factor.

5.7.4 *Scatterings from spin assemblies*

(a) *Measurement of magnon spectrum*. The magnon spectrum is visible in the inelastic scattering of neutrons. This can be seen from the formalism when $\boldsymbol{\mu}$ is expanded in the magnon creation and annihilation operators introduced in Chapter 4. Indeed, the required form of $S(\mathbf{k}, \omega)$ at low temperatures is already implicit in the theory developed in the last chapter. Using equations (4.6.38), (4.9.18) and (4.10.19) we find

$$\begin{aligned} S'_{xx}(\boldsymbol{\kappa}, \omega) &= \frac{\chi_{xx}(\boldsymbol{\kappa})\, \omega\beta}{2(1-\mathrm{e}^{-\beta\omega})} k_{\mathrm{B}} T F_{xx}(\boldsymbol{\kappa}, \omega) \\ &= \frac{\hbar\chi_{xx}(\boldsymbol{\kappa})\, \omega_{\boldsymbol{\kappa}}}{2(1-\mathrm{e}^{-\beta\omega_{\boldsymbol{\kappa}}})}[\delta(\omega-\omega_{\boldsymbol{\kappa}}) + \delta(\omega+\omega_{\boldsymbol{\kappa}})]. \end{aligned} \tag{5.7.23}$$

It is easy to see that S_{xx} obeys the detailed balancing condition (cf. Appendix 4.2).

We shall now illustrate how dispersion relations are measured by taking the specific example of the spin-wave spectrum. The interpretation of the experiment is, in principle, quite simple. From equation (5.7.23) we see that for a one-magnon process

$$\mathbf{k}'-\mathbf{k} = \mathbf{q}+\mathbf{K} \tag{5.7.24}$$

and

$$\frac{\hbar}{2M}(\mathbf{k}'^2-\mathbf{k}^2) = \pm\hbar\omega_{\mathbf{q}}, \tag{5.7.25}$$

where $\mathbf{q}$ is the magnon wave-vector. The presence of any reciprocal lattice-vector $\mathbf{K}$ on the right-hand side of equation (5.7.24) is a trivial consequence of the periodicity of $\omega_{\mathbf{q}}$ in the reciprocal lattice.

We begin by noting that Bragg reflexions (elastic scattering) enable us to use crystals as spectrometers, first to obtain a monochromatic beam of neutrons and second to separate out neutrons of a definite energy from all those scattered by the target.

For example, the positions of source and analyser, the former giving incident neutrons of a particular energy and the latter counting scattered neutrons of a particular energy, may be fixed while the crystal orientation is varied. There will be a sudden rise in the count of scattered neutrons when equations (5.7.24) and (5.7.25) are simultaneously satisfied and the orientation of the crystal will tell us $\mathbf{q}$. The general nature of the situation in $\mathbf{k}$-space is illustrated in Figure 5.9, the conditions for satisfying both equations (5.7.24) and (5.7.25) being indicated.

In such an experiment one obtains effects due both to magnons and to phonons, and these can be separated in the following manner. We have seen that, if a magnetic moment $\boldsymbol{\mu}$ is perpendicular to $\boldsymbol{\kappa} = \mathbf{k}'-\mathbf{k}$, it will not scatter a neutron into the direction of $\mathbf{k}'$. Thus, application of a magnetic field perpendicular to $\mathbf{k}'-\mathbf{k}$ results in a reduction of the magnetic scattering by a specimen whereas the nuclear scattering (reflecting the phonon spectrum) will be unchanged if the two types of scattering are independent. In fact, the phonon peak is enhanced as Figure 5.10, obtained by varying the crystal orientation as mentioned above, shows. The peaks M_1 and M_2 about the (200) direction correspond to the creation of magnons, while the peak P corresponds to the creation of a phonon. On application of a field, M_1 and M_2 clearly decrease, while P increases. This is because of "magneto-vibrational" scattering. Its origin is not difficult to find; when analysing the phonon scattering one should strictly take into account the magnetic scattering of the individual atoms as well as the nuclear scattering.

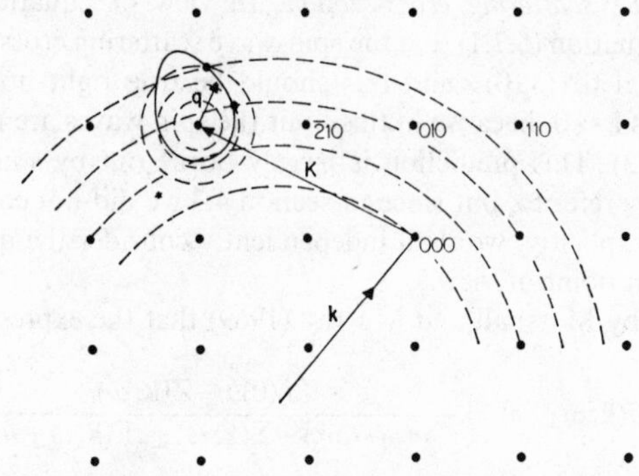

FIGURE 5.9. Inelastic scattering surface, shown as a solid curve, about the point ($\bar{2}10$) in reciprocal space for neutrons of wave-vector **k** incident on a target. On this surface equations (5.7.24) and (5.7.25) are simultaneously satisfied. The figure is applicable to both magnon scattering and phonon scattering. The dashed curves of larger radii represent constant-energy curves for the scattered neutron. Smaller radii curves are the same but for the annihilated phonon.

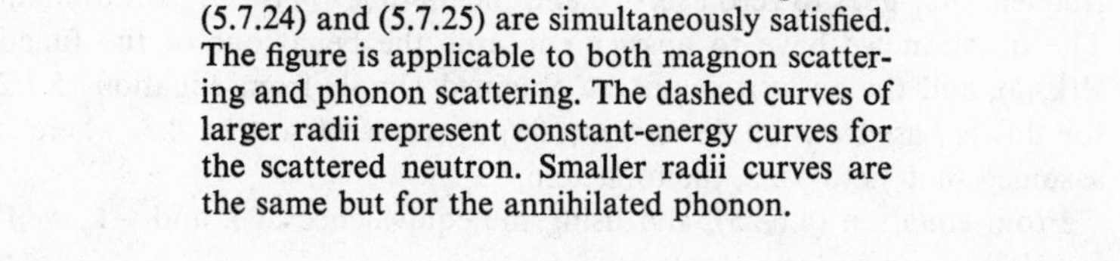

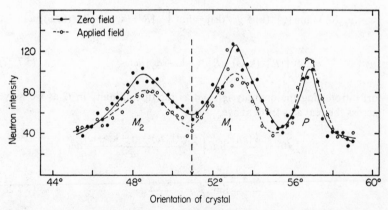

FIGURE 5.10. Experimental results showing magnon peaks M_1 and M_2 and phonon peak P. Applied field is perpendicular to $\mathbf{k}' - \mathbf{k}$. Reduction of magnon peaks is evident. However, phonon peak is enhanced, through magneto-vibrational scattering (after Brockhouse and Sinclair, 1960).

(b) *Spin-wave scattering cross-section.* In view of equation (5.7.23), we expect from equation (5.7.1) that the spin-wave scattering cross-section should be proportional to $\langle s_z(0)\rangle$ and this should be true right up to the critical temperature as $\mathbf{k}\to 0$, because in this limit the spin waves are non-interacting (see section 4.3). This prediction is largely borne out by some experiments we shall shortly refer to, but since in section 4.3 we did not consider elevated temperatures explicitly, we shall independently consider the question from a Green function point of view.

It is shown by Marshall and Murray (1969) that the expression

$$G(\mathbf{k}, \omega \pm i\varepsilon) = -\frac{\langle s_z(0)\rangle - Z(\mathbf{k}, \omega)}{\pi(\omega - \omega_k^0) - \Sigma(\mathbf{k}, \omega) \pm i\Gamma(\mathbf{k}, \omega \pm i\varepsilon)} \tag{5.7.26}$$

is a good approximation to the Green function at quite high temperatures. Here G is defined by equation (4.5.6) with $A = s_-(-\mathbf{k}), B = s_+(\mathbf{k})$. For reasons already stated, the frequency shift $\Sigma(\mathbf{k}, \omega)$ of the free spin-wave frequency ω_k^0 goes to zero as $\mathbf{k}\to 0$ and the lifetime $1/\Gamma(\mathbf{k}, \omega)$ goes to infinity. The question we have to answer concerns the behaviour of the function $Z(\mathbf{k}, \omega)$, and the answer cannot be obtained simply from equation (5.7.23), for this is based on the discussion of spin waves of section 4.3, where it is assumed that $\langle s_z(0)\rangle \simeq s$, the total spin.

From equation (4.6.23), and using the equivalence of $\mathbf{k}$ and $-\mathbf{k}$, we find for Heisenberg systems

$$\chi_{xx}(0) = \lim_{\mathbf{k}\to 0} [\chi^{+-}(\mathbf{k}) + \chi^{-+}(\mathbf{k})] = \lim_{\mathbf{k}\to 0} \int_{-\infty}^{\infty} [R^{+-}(\mathbf{k}, \omega) + R^{-+}(\mathbf{k}, \omega)]\, d\omega$$

$$= \lim_{\mathbf{k}\to 0} \int [R^{+-}(\mathbf{k}, \omega) + R^{+-}(-\mathbf{k}, -\omega)]\, d\omega \tag{5.7.27}$$

or, introducing the quantity $ig(\mathbf{k}, \omega)$ as the discontinuity in G as we move across the real axis from above,

$$\chi_{xx}(0) = \frac{N(g\mu)^2}{4} \lim_{\mathbf{k}\to 0} \int_{-\infty}^{\infty} \frac{g(\mathbf{k}, \omega) - g(-\mathbf{k}, -\omega)}{\omega}\, d\omega$$

$$= \frac{Ng^2\mu^2}{2} \lim_{\mathbf{k}\to 0} ⨍_{-\infty}^{\infty} \frac{g(\mathbf{k}, \omega)}{\omega}\, d\omega. \tag{5.7.28}$$

We began the derivation of this expression by using both R^{+-} and R^{-+} in order to make it clear that it is the principal part of the integral in equation (5.7.28) that must be utilized.

On the other hand, we find from equations (4.3.21) and (4.9.24) that for $s=\frac{1}{2}$

$$\chi_{xx}(0)=\frac{g^2\mu^2\beta}{N}[\langle s_x(0)\,s_x(0)\rangle]=\frac{g^2\mu^2\beta}{N}[\langle s^+(0)\,s^-(0)\rangle-\langle s_z(0)\rangle]. \tag{5.7.29}$$

It is also readily shown that

$$\frac{1}{N}\langle s^+(0)\,s^-(0)\rangle=\lim_{\mathbf{k}\to 0}\int\frac{g(\mathbf{k},\omega)}{e^{\beta\omega}-1}\,d\omega \tag{5.7.30}$$

and so, by inserting this equation into (5.7.29) and equating (5.7.29) and (5.7.28), we find

$$2\beta\langle s_0^z\rangle=2\lim_{\mathbf{k}\to 0}\left[\int_{-\infty}^{\infty}\frac{g(\mathbf{k},\omega)}{\omega}\,d\omega-\beta\int_{-\infty}^{\infty}\frac{g(\mathbf{k},\omega)}{e^{\beta\omega}-1}\,d\omega\right]. \tag{5.7.31}$$

Using equation (5.7.26) therefore

$$2\beta\langle s_0^z\rangle=\lim_{\mathbf{k}\to 0}\frac{1}{\pi}\int_{-\infty}^{\infty}d\omega\,\frac{[(2/\omega)-2\beta(e^{\beta\omega}-1)^{-1}]}{(\omega-\omega_{\mathbf{k}}^0-\Sigma)^2+\Gamma^2}$$
$$\times[\Gamma\{\langle s^z\rangle+C(\mathbf{k},\omega)\}+D(\mathbf{k},\omega)\,(\omega-\omega_{\mathbf{k}}^0-\Sigma)], \tag{5.7.32}$$

where

$$Z(\mathbf{k},\omega)=C(\mathbf{k},\omega)+iD(\mathbf{k},\omega) \tag{5.7.33}$$

and C and D are real functions. As $\mathbf{k}\to 0$, $\omega_{\mathbf{k}}^0+\Sigma$ and Γ tend to zero if $T<T_c$ and so the first term under the integral sign becomes $\delta(\omega)$ and we find

$$0=\lim_{\mathbf{k}\to 0}[2\beta C(\mathbf{k},0)]+\frac{2}{\pi}\lim_{\mathbf{k}\to 0}\int_{-\infty}^{\infty}d\omega\,D(\mathbf{k},\omega)\left[\frac{1}{\omega^2}-\frac{\beta}{\omega(e^{\beta\omega}-1)}\right]. \tag{5.7.34}$$

The analytic nature of G implies that Z is also analytic and so an application of Cauchy's theorem yields

$$C(\mathbf{k},0)=\frac{1}{\pi}\int_{-\infty}^{\infty}\frac{D(\mathbf{k},\omega)}{\omega}\,d\omega. \tag{5.7.35}$$

Inserting this equation into equation (5.7.34) we can show that

$$0=\lim_{\mathbf{k}\to 0}\int_{-\infty}^{\infty}d\omega\,D(\mathbf{k},\omega)\left[\frac{1}{\omega^2}+\frac{\beta}{\omega}-\frac{\beta}{\omega(e^{\beta\omega}-1)}\right]. \tag{5.7.36}$$

This must be true for all $\beta>1/k_B T_c$ and so we find

$$\lim_{\mathbf{k}\to 0}D(\mathbf{k},\omega)=0,\quad T<T_c, \tag{5.7.37}$$

which also implies

$$\lim_{\mathbf{k}\to 0}Z(\mathbf{k},\omega)=0,\quad T<T_c. \tag{5.7.38}$$

Since S_{xx} is proportional to R^{+-} and so to $g(\omega)$, we can see that the above result confirms that the scattering cross section is proportional to $\langle s_z \rangle$.

(c) *Spin diffusion at high temperatures.* In Chapter 4, section 4.9, it was argued that a diffusion picture of spin dynamics in Heisenberg systems should hold at high temperatures so that we could write

$$F(\mathbf{k}, \omega) = \frac{1}{\pi} \frac{\Lambda k^2}{(\hbar\omega)^2 + \Lambda^2 k^4}. \tag{5.7.39}$$

The corresponding scattering function $S(\mathbf{k}, \omega)$ can be obtained from equations (4.6.3) and (4.9.18) and has the same Lorentzian form, namely

$$S(\mathbf{k}, \omega) = \frac{\pi^{-1} \Lambda k^2 s(s+1)}{(\hbar\omega)^2 + (\Lambda k^2)^2}. \tag{5.7.40}$$

We show in Figure 5.11 experimental results of Windsor, Briggs and Kestigian (1968) on $RbMnF_3$. This substance has the simplest possible theoretical properties, being simple cubic in its magnetic structure, with, to fair accuracy, Heisenberg exchange interaction between nearest-neighbour Mn^{2+} ions only. This is shown by low-temperature spin-wave measurements on the antiferromagnetic phase (Windsor and Stevenson, 1966), supported

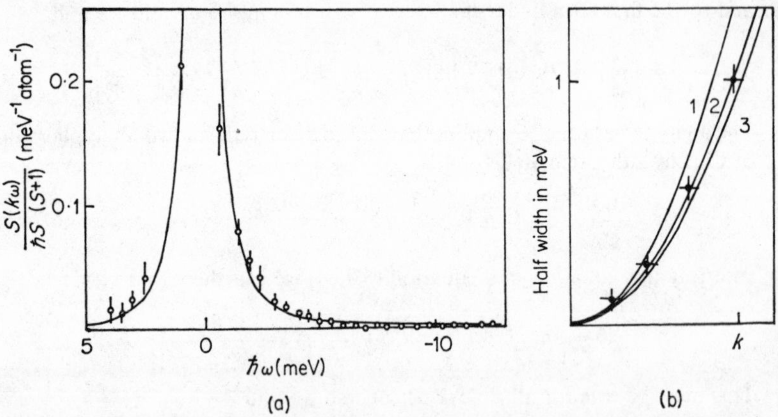

FIGURE 5.11. (a) Scattering-function $S(k\omega)$ against $\hbar\omega$ for neutron scattering from $RbMnF_3$ at $3.5T_N$, where T_N is the critical temperature. The solid curve is the Lorentzian function given by the theory of Bennett and Martin. (b) Shows half-width of observed Lorentzian spectral distribution functions for $RbMnF_3$. 1. Theory of Mori and Kawasaki. 2. Theory of Bennett and Martin. 3. Theory of de Gennes (after Windsor and co-workers, 1968).

by spin-wave measurements on the related salt $KMnF_3$ (Collins, Pickart and Windsor, 1960). That no significant distortion of the lattice occurs on transition from the antiferromagnetic to the paramagnetic phase (at a Néel temperature of $T_c = 83°K$) is shown by the X-ray studies of Tearney, Morazzi and Argyle (1966).

The Heisenberg Hamiltonian is thus found to be

$$H = \sum_{i<j} 2J\mathbf{s}_i \cdot \mathbf{s}_j \tag{5.7.41}$$

with $s = \frac{5}{2}$ and $J = 0.28$ meV; and the summation is to be taken over nearest-neighbour pairs.

An example of the results of the neutron-scattering experiments is shown in Figure 5.11 in comparison with the Lorentzian obtained by using the diffusion parameter Λ in equation (5.7.40) from the theory of Bennett and Martin (1965). Not only is the agreement good, but the experimental results also show evidence of a sharp drop in $S(\mathbf{k}, \omega)$ at $\omega \simeq \omega_c$, the cut-off frequency of the Lorentzian in de Gennes' theory (see section 4.9.2). In Figure 5.11 we show the experimental half-widths together with the theoretical curves Λk^2, Λ being calculated according to the various theories of Mori and Kawasaki (1962), de Gennes and Bennett and Martin. We should note that in order to obtain the experimental half-widths, Windsor and co-workers corrected for the distortion of experimental results from perfect Lorentzians because of the finite temperature.

We conclude this section by calling attention to the computer simulation experiments of Windsor (1967), which are useful for comparison with experiments like those above for larger values of $\mathbf{k}$. These simulation experiments were carried out for a classical spin system. We saw in Chapter 4, section 4.3, that the classical and quantum-mechanical equations of motion are formally identical and whereas, on the other hand, the classical and quantum-mechanical averaging processes are in general different, the temperatures in the present case are high enough to make a comparison of experimental results and classical calculations meaningful. We must refer the reader to the original paper for the details.

(d) *Critical scattering.* At the critical temperature T_c (Curie or Néel temperature), we expect from equation (4.9.8) that the width of the scattering function $S(\mathbf{k}, \omega)$ about $\omega = 0$ will vanish, because of the divergence in the susceptibility $\chi(\mathbf{k}, \omega)$. An experimental confirmation of this, for terbium, is shown in Figure 5.12 (Als-Nielsen and co-workers, 1967). On the other hand, the results are not in accord with the simple result discussed in section 4.2 that the $\mathbf{k}$ dependence of Λ [see equation (4.9.58)] near the critical temperature is given by $\Lambda_T \sim \Lambda_\infty \chi_0/\chi(\mathbf{k})$. Λ_T does vanish, but the "thermodynamical braking" (see section 4.9.4) sets in much closer to the critical temperature than simple theories might suggest. We discussed this point in Chapter 4, and for a related discussion of the experimental situation, reference can be made to Als-Nielsen (1970).

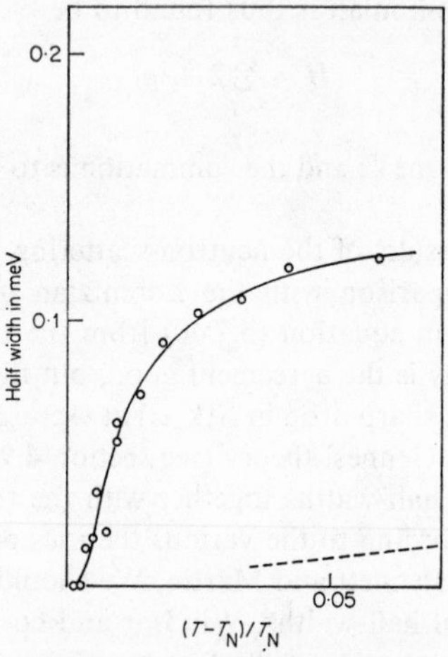

FIGURE 5.12. Shows vanishing width of $S(k\omega)$ as the critical temperature T_N of terbium is approached. Dashed line shows conventional theory, which is clearly quite inadequate over a wide temperature range (after Als-Nielsen and co-workers, 1967).

5.7.5 Orbital motion and neutron scattering

In many cases it appears that neutron scattering from metals is dominated by the spin contribution, and we discussed previously how the neutron scattering is related to the wave number and frequency-dependent spin susceptibility $\chi(\mathbf{q}, \omega)$. It seems clear that the orbital motion will play a role, however, and we shall outline the theory of neutron scattering when spin and orbital contributions are both included below, even though it is not clear that any experiments done to date allow one to check the orbital motion contribution.

Taking the one-electron approximation to the differential cross-section for neutron inelastic scattering per unit solid angle, per unit energy range, per unit volume of the specimen [see section 5.2 and especially equation (5.2.25)] it is readily verified that

$$\frac{d^2\sigma}{d\Omega\,d\omega} = \frac{k'}{k}\left(\frac{m_0}{2\pi\hbar^2}\right)^2 \sum_{\mathbf{q},\mathbf{q}'} f_0(\varepsilon_{\mathbf{q}})\,[1-f_0(\varepsilon_{\mathbf{q}'})]\,\delta(\omega+\omega_{\mathbf{q}}-\omega_{\mathbf{q}'})$$
$$\times \sum_{\substack{s_n,S_e\\ s_n',S_e'}} |\langle \mathbf{k}'s_n'\,S_e'\,\mathbf{q}'|V|\mathbf{q}S_e s_n\mathbf{k}\rangle|^2, \qquad (5.7.42)$$

where, to summarize, we have written m_0 here as the neutron mass, $\mathbf{k}$ and $\mathbf{q}$ as the wave vectors for neutron and electron respectively, s_n and S_e are neutron and electron spin coordinates respectively, $\varepsilon_q = \hbar\omega_q$, and for free electrons, which we shall be mainly treating in this section, $\varepsilon_q = \hbar^2 q^2/2m$.

Referring to equation (5.2.24) and making the one-electron approximation again, we want to relate the orbital and spin susceptibilities to the scattering function. For the interaction of a neutron having magnetic moment $\boldsymbol{\mu}_n$ the magnetic field given in equation (5.7.2) corresponds to a vector potential $\mathscr{A}$ with Fourier components

$$\mathscr{A}_{\mathbf{q}} = -\frac{4\pi i\boldsymbol{\mu}_n \times \mathbf{q}}{q^2} \qquad (5.7.43)$$

and the coupling between the electrons and the effective field is given by

$$H_e = -\frac{1}{c}\sum_{\mathbf{q}} \mathbf{j}_q \cdot \mathscr{A}_{-\mathbf{q}} \exp(i\mathbf{q}.\mathbf{R}). \qquad (5.7.44)$$

It should be noted that $\mathbf{q}.\mathscr{A}_{\mathbf{q}} = 0$, that is $\operatorname{div}\mathscr{A} = 0$.

In the first Born approximation we can evidently write

$$\langle \mathbf{k}'|V|\mathbf{k}\rangle = \int \exp(i\boldsymbol{\kappa}.\mathbf{r})\,H_e\,d\mathbf{r}, \quad \boldsymbol{\kappa} = \mathbf{k}-\mathbf{k}', \qquad (5.7.45)$$

and we can now work out the matrix elements of H_e for (a) orbital current and (b) spin current.

For the orbital part we have for the current $\mathbf{j}_q$ the result

$$\mathbf{j}_q = -\frac{e}{2m}[\mathbf{p}_e \exp(-i\mathbf{q}.\mathbf{r}_e) + \exp(-i\mathbf{q}.\mathbf{r}_e)\,\mathbf{p}_e], \qquad (5.7.46)$$

where $\mathbf{r}_e$ and $\mathbf{p}_e$ are respectively position and momentum operators, and hence the matrix elements are found to be

$$\langle \mathbf{k}'|H_e|\mathbf{k}\rangle = \frac{e}{2mc}\sum_{\mathbf{q}}\int d\mathbf{r}\exp(i\boldsymbol{\kappa}.\mathbf{r})\exp(i\mathbf{q}.\mathbf{r})\,\mathscr{A}_{-\mathbf{q}}$$
$$.\,[\mathbf{p}_e\exp(-i\mathbf{q}.\mathbf{r}_e) + \exp(-i\mathbf{q}.\mathbf{r}_e)\,\mathbf{p}_e]$$
$$= \frac{e}{2mc}\mathscr{A}_{\boldsymbol{\kappa}}\cdot[\mathbf{p}_e\exp(i\boldsymbol{\kappa}.\mathbf{r}_e) + \exp(i\boldsymbol{\kappa}.\mathbf{r}_e)\,\mathbf{p}_e]. \qquad (5.7.47)$$

For the spin part, we have for the current

$$\mathbf{j}_q = -\frac{ie}{2m}[\exp(-i\mathbf{q}.\mathbf{r}_e)(\mathbf{p}_e\times\boldsymbol{\sigma})+(\mathbf{p}_e\times\boldsymbol{\sigma})\exp(-i\mathbf{q}.\mathbf{r}_e)], \qquad (5.7.48)$$

from which the coupling between magnetic field and spins as given in equation (5.7.1) can be regained. For the matrix elements the result obtained by using equation (5.7.43) in (5.7.47) and adding the spin contribution is

$$\begin{aligned}\langle\mathbf{k}'|H_e|\mathbf{k}\rangle &= -\frac{4\pi i}{\kappa^2}\Big\{(\boldsymbol{\mu}_n\times\boldsymbol{\kappa})\frac{e}{2mc}[\exp(i\boldsymbol{\kappa}.\mathbf{r}_e)\mathbf{p}_e+\mathbf{p}_e\exp(i\boldsymbol{\kappa}.\mathbf{r}_e)]\\ &\quad -\frac{i\hbar e}{2mc}(\boldsymbol{\mu}_n\times\boldsymbol{\kappa}).(\boldsymbol{\sigma}\times\boldsymbol{\kappa})\exp(i\boldsymbol{\kappa}.\mathbf{r}_e)\Big\}\\ &= -\frac{2\pi ie}{mc}\frac{(\boldsymbol{\mu}_n\times\hat{\boldsymbol{\kappa}})}{|\boldsymbol{\kappa}|}[\exp(i\boldsymbol{\kappa}.\mathbf{r}_e)\mathbf{p}_e+\mathbf{p}_e\exp(i\boldsymbol{\kappa}.\mathbf{r}_e)]\\ &\quad -4\pi\mu[(\boldsymbol{\mu}_n.\hat{\boldsymbol{\kappa}})(\boldsymbol{\sigma}.\hat{\boldsymbol{\kappa}})-(\boldsymbol{\mu}_n.\boldsymbol{\sigma})]\exp(i\boldsymbol{\kappa}.\mathbf{r}_e), \qquad (5.7.49)\end{aligned}$$

where

$$\boldsymbol{\mu}_n = -\frac{\gamma e\hbar}{2m_0 c}\boldsymbol{\sigma}_n \quad\text{and}\quad \mu = \frac{e\hbar}{2mc} \qquad (5.7.50)$$

with $\gamma = 1.913$. If we write the classical electron radius e^2/mc^2 as r_0, then we find

$$\begin{aligned}\langle\mathbf{k}'|V|\mathbf{k}\rangle &= ir_0\gamma\frac{\hbar}{m_0}\pi\frac{1}{\kappa}(\boldsymbol{\sigma}_n\times\hat{\boldsymbol{\kappa}})[\exp(i\boldsymbol{\kappa}.\mathbf{r}_e)\mathbf{p}_e+\mathbf{p}_e\exp(i\boldsymbol{\kappa}.\mathbf{r}_e)]\\ &\quad +r_0\gamma\frac{\hbar^2}{m_0}\pi[(\boldsymbol{\sigma}_n.\hat{\boldsymbol{\kappa}})(\boldsymbol{\sigma}.\hat{\boldsymbol{\kappa}})-(\boldsymbol{\sigma}.\boldsymbol{\sigma}_n)]\exp(i\boldsymbol{\kappa}.\mathbf{r}_e). \qquad (5.7.51)\end{aligned}$$

Plane-wave case. In the case of plane waves, we can calculate the matrix elements $\langle\mathbf{q}'\mathbf{k}'|V|\mathbf{kq}\rangle$ by using the following results:

$$\left.\begin{aligned}\exp(i\boldsymbol{\kappa}.\mathbf{r}_e)\mathbf{p}_e+\mathbf{p}_e\exp(i\boldsymbol{\kappa}.\mathbf{r}_e) &= 2\exp(i\boldsymbol{\kappa}.\mathbf{r}_e)\mathbf{p}_e+\hbar\boldsymbol{\kappa}\exp(i\boldsymbol{\kappa}.\mathbf{r}_e),\\ [f(x),\mathbf{p}_e] &= i\hbar\frac{\partial}{\partial x}f(x),\\ \langle\mathbf{q}'|\exp(i\boldsymbol{\kappa}.\mathbf{r}_e)\mathbf{p}_e+\mathbf{p}_e\exp(i\boldsymbol{\kappa}.\mathbf{r}_e)|\mathbf{q}\rangle &= \hbar(2\mathbf{q}+\boldsymbol{\kappa})\,\delta_{\boldsymbol{\kappa},\mathbf{q}'-\mathbf{q}},\\ \langle\mathbf{q}'|\exp(i\boldsymbol{\kappa}.\mathbf{r}_e)|\mathbf{q}\rangle &= \delta_{\boldsymbol{\kappa},\mathbf{q}'-\mathbf{q}}.\end{aligned}\right\} \qquad (5.7.52)$$

The matrix element is then found to be

$$\langle \mathbf{q}'\mathbf{k}'|V|\mathbf{k}\mathbf{q}\rangle = \frac{\pi\hbar^2}{m_0} r_0\gamma \left[\frac{i}{\kappa}(\boldsymbol{\sigma}_{\mathrm{n}} \times \hat{\boldsymbol{\kappa}}).(\boldsymbol{\kappa}+2\mathbf{q})+(\boldsymbol{\sigma}_{\mathrm{n}}.\hat{\boldsymbol{\kappa}})(\boldsymbol{\sigma}.\hat{\boldsymbol{\kappa}})-(\boldsymbol{\sigma}_{\mathrm{e}}.\boldsymbol{\sigma}_{\mathrm{n}})\right] \delta_{\boldsymbol{\kappa},\mathbf{q}'-\mathbf{q}}. \tag{5.7.53}$$

Using the explicit forms

$$\left.\begin{aligned} \boldsymbol{\sigma}_{\mathrm{n}} &= 2\mathbf{s}_{\mathrm{n}}, \\ \boldsymbol{\sigma}_{\mathrm{e}} &= 2\mathbf{S}_{\mathrm{e}}, \end{aligned}\right\} \tag{5.7.54}$$

we find

$$\langle \mathbf{q}'\mathbf{k}'|V|\mathbf{k}\mathbf{q}\rangle = \frac{2\pi\hbar^2}{m_0} r_0\gamma \left\{\frac{i}{\kappa}(\mathbf{s}_{\mathrm{n}} \times \hat{\boldsymbol{\kappa}}).(\boldsymbol{\kappa}+2\mathbf{q})+2\mathbf{s}_{\mathrm{n}}.[\hat{\boldsymbol{\kappa}}(\mathbf{S}_{\mathrm{e}}.\hat{\boldsymbol{\kappa}})-\mathbf{S}_{\mathrm{e}}]\,\delta_{\boldsymbol{\kappa},\mathbf{q}'-\mathbf{q}}\right\}. \tag{5.7.55}$$

The first term in the curly bracket in equation (5.7.55) can be written as

$$\frac{i}{\kappa}\mathbf{s}_{\mathrm{n}}.[\hat{\boldsymbol{\kappa}} \times (\boldsymbol{\kappa}+2\mathbf{q})]$$

and hence we have

$$\begin{aligned} |\langle \mathbf{q}'\mathbf{k}'|V|\mathbf{k}\mathbf{q}\rangle|^2 = \left(\frac{2\pi\hbar^2}{m_0}\right)^2 (r_0\gamma)^2 \frac{1}{\kappa^2}\sum_{\alpha\beta} s_{\mathrm{n}}^{\alpha} s_{\mathrm{n}}^{\beta}[\boldsymbol{\kappa} \times (\boldsymbol{\kappa}+2\mathbf{q})]^{\alpha} \\ \times [\hat{\boldsymbol{\kappa}} \times (\boldsymbol{\kappa}+2\mathbf{q})]^{\beta} + 4\sum_{\alpha\beta} s_{\mathrm{n}}^{\alpha} s_{\mathrm{n}}^{\beta}[\kappa^{\alpha}(\mathbf{S}_{\mathrm{e}}.\boldsymbol{\kappa})-S_{\mathrm{e}}^{\alpha}] \\ \times [\hat{\boldsymbol{\kappa}}^{\beta}(\mathbf{S}_{\mathrm{e}}.\boldsymbol{\kappa})-S_{\mathrm{e}}^{\beta}]\,\delta_{\boldsymbol{\kappa},\mathbf{q}'-\mathbf{q}}. \end{aligned} \tag{5.7.56}$$

We now have to sum over spin indices, which we can do using equation (5.6.12). If ρ is the spin density for the initial state if the neutron beam is unpolarized we have

$$\begin{aligned} \sum_{s_{\mathrm{n}}} \langle s_{\mathrm{n}}|VV^*|s_{\mathrm{n}}\rangle\langle s_{\mathrm{n}}|\rho|s_{\mathrm{n}}\rangle = \left(\frac{2\pi\hbar^2}{m_0}\right)^2 (r_0\gamma)^2 \left\{\frac{1}{4\kappa^2}[\hat{\boldsymbol{\kappa}} \times (\boldsymbol{\kappa}+2\mathbf{q})]^2 \right. \\ \left. +[\hat{\boldsymbol{\kappa}}(\mathbf{S}_{\mathrm{e}}.\hat{\boldsymbol{\kappa}})-\mathbf{S}_{\mathrm{e}}]^2\right\} \delta_{\boldsymbol{\kappa},\mathbf{q}'-\mathbf{q}}. \end{aligned} \tag{5.7.57}$$

The sum over S_{e} is easily carried out for the first term. For the second term, with spin $\frac{1}{2}$, it can be shown (see Kittel, 1963, p. 383) that this yields the value $\frac{1}{2}$. Hence we can write for the differential scattering cross-section

$$\frac{d^2\sigma}{d\Omega\, d\omega} = \frac{k'}{k} S(\boldsymbol{\kappa}, \omega)\,\tfrac{1}{2}(r_0\gamma)^2, \tag{5.7.58}$$

where

$$S(\boldsymbol{\kappa}, \omega) = \sum_{\mathbf{q}\mathbf{q}'} f_0(\varepsilon_{\mathbf{q}})[1-f_0(\varepsilon_{\mathbf{q}'})] \times \delta(\omega+\omega_{\mathbf{q}}-\omega_{\mathbf{q}'}) \left\{\frac{1}{2\kappa^2}[\hat{\boldsymbol{\kappa}} \times (\boldsymbol{\kappa}+2\mathbf{q})]^2+1\right\} \delta_{\boldsymbol{\kappa},\mathbf{q}'-\mathbf{q}}. \tag{5.7.59}$$

We notice also that

$$[\hat{\boldsymbol{\kappa}} \times (\boldsymbol{\kappa}+2\mathbf{q})]^2 = \frac{1}{|\mathbf{q}'-\mathbf{q}|}[(\mathbf{q}'-\mathbf{q}) \times (\mathbf{q}'+\mathbf{q})]^2$$

$$= \frac{1}{|\mathbf{q}'-\mathbf{q}|}[2\mathbf{q}' \times \mathbf{q}]^2. \tag{5.7.60}$$

We can now write from equation (5.7.59), following Schneider (1970),

$$S_{\text{orb}}(\boldsymbol{\kappa}, \omega) = \sum_{\mathbf{q}\mathbf{q}'} f_0(\varepsilon_{\mathbf{q}})[1-f_0(\varepsilon_{\mathbf{q}'})]\,\delta(\omega+\omega_{\mathbf{q}}-\omega_{\mathbf{q}'}) \times \frac{1}{2\kappa^2}(\hat{\boldsymbol{\kappa}} \times 2\mathbf{q})^2\, \delta_{\boldsymbol{\kappa},\mathbf{q}'-\mathbf{q}} \tag{5.7.61}$$

and

$$S_{\text{spin}}(\boldsymbol{\kappa}, \omega) = \sum_{\mathbf{q}\mathbf{q}'} f_0(\varepsilon_{\mathbf{q}})[1-f_0(\varepsilon_{\mathbf{q}'})]\,\delta(\omega+\omega_{\mathbf{q}}-\omega_{\mathbf{q}'})\, \delta_{\boldsymbol{\kappa},\mathbf{q}'-\mathbf{q}}. \tag{5.7.62}$$

We wish finally to see how the neutron scattering described by $S(\boldsymbol{\kappa}, \omega)$ is related to the frequency and wave-number-dependent susceptibility discussed in Chapter 4, section 4.11. To see this relation we note the property of the scattering function (cf. Appendix 4.2):

$$S(\boldsymbol{\kappa}, \omega) = \exp(\beta\hbar\omega)\, S(-\boldsymbol{\kappa}, -\omega) \tag{5.7.63}$$

and it follows that

$$S(\boldsymbol{\kappa}, \omega) - S(-\boldsymbol{\kappa}, -\omega) = [1-\exp(-\beta\hbar\omega)]\, S(\boldsymbol{\kappa}, \omega). \tag{5.7.64}$$

Substituting for $S(\boldsymbol{\kappa}, \omega)$ from equation (5.7.59) shows that

$$S(\boldsymbol{\kappa}, \omega) - S(-\boldsymbol{\kappa}, -\omega) = -\sum_{\mathbf{q}} [f(\varepsilon_{\mathbf{q}+\boldsymbol{\kappa}}) - f(\varepsilon_{\mathbf{q}})] \left\{1+\frac{1}{2\kappa^2}[\hat{\boldsymbol{\kappa}} \times 2\mathbf{q}]^2\right\}$$

$$\times\, \delta(\omega+\omega_{\mathbf{q}}-\omega_{\boldsymbol{\kappa}+\mathbf{q}}). \tag{5.7.65}$$

By comparison with equation (4.11.18) for the orbital susceptibility and equation (4.11.28) for the spin susceptibility, we see that

$$\text{Im}\,[\tfrac{1}{2}\chi_{\text{orb}}(\boldsymbol{\kappa}, \omega) + \chi_{\text{spin}}(\boldsymbol{\kappa}, \omega)] = \frac{2\pi\mu^2}{\hbar} S(\boldsymbol{\kappa}, \omega)\,[\exp(-\beta\hbar\omega)-1]. \tag{5.7.66}$$

This is the fundamental relation between the neutron scattering function and the susceptibility. The presentation given above has followed closely that of Hebborn and March (1970). For further discussion the reader should consult Appendix 4.3.

5.7.6 Comparison of spin susceptibility with measured $S(\kappa, \omega)$ for nickel

For Ni, neutron experiments are available which in principle can be compared with equation (5.7.66). All available evidence points to the fact that the orbital contribution however is small and that the experiments can be interpreted usefully solely in terms of the spin susceptibility. The results are taken mainly from Windsor, Lowde and Allen (1969) and earlier references given there.

We shall consider the results for the magnetic inelastic scattering of neutrons from nickel above its Curie temperature T_c. This magnetic scattering cross-section was measured by Cable and co-workers (1967) for a single crystal of Ni at $1.6T_c$ over a wide range of wave vector $\mathbf{q}$ and in the energy transfer range between 0.02 and 0.12 eV. Using the form of $\chi_{\text{spin}}(\mathbf{q}, \omega)$ for free electrons at $T = 0$ given in equation (4.11.28), the exchange enhanced susceptibility (cf. equation 4.11.68) is given by

$$\chi_{\text{enhanced}}(\mathbf{q}, \omega) = \frac{\chi_{\text{spin}}(\mathbf{q}, \omega)}{1 - 2I_{\text{eff}}\,\chi_{\text{spin}}(\mathbf{q}, \omega)/g^2\mu^2}. \tag{5.7.67}$$

The magnitude of I_{eff} gives us, of course, a measure of the strength of the electron interactions in such a one-parameter theory.

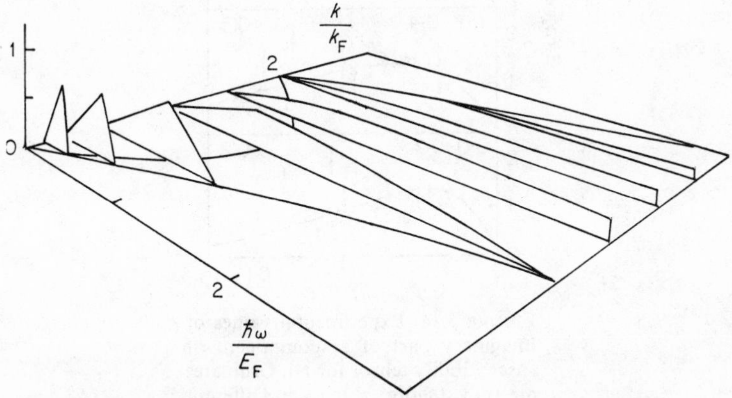

FIGURE 5.13. Calculation by Lowde and Windsor of Im χ_{spin} [see equation (4.11.30)] for free electrons.

Figure 5.13 shows the results calculated from equation (4.11.30) by Lowde and Windsor (1967) for the imaginary part of the spin susceptibility for free electrons. In Figure 5.14, absolute values of Im $\chi_{zz}(\mathbf{q}, \omega)/g^2\mu^2$ are shown, where χ_{zz} is a component of the susceptibility tensor, as deduced from neutron-scattering experiments on Ni at $1.6T_c$. The solid curves shown were calculated from free-electron theory with the parameters recorded in Table 5.2. The value of k_f turns out to be roughly what one would have expected, while the value of I_{eff} is in agreement, at least semi-quantitatively, with the estimates of Hubbard and other workers.

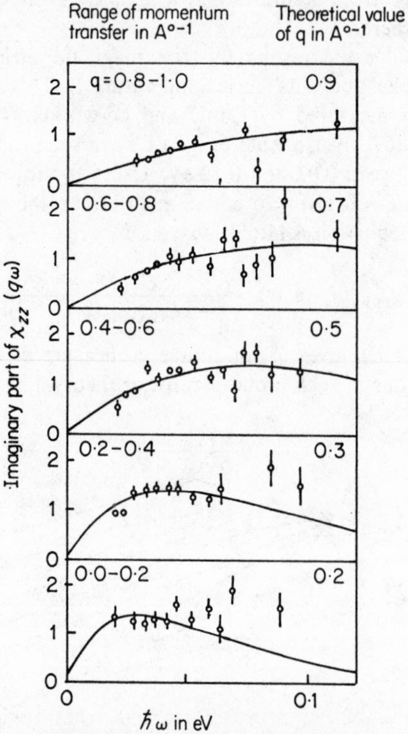

FIGURE 5.14. Experimental values of imaginary part of zz component of susceptibility tensor for Ni. Ordinates are Im $\chi_{zz}(q\omega)/g^2\mu^2$ in eV^{-1}. Different plots refer to different ranges of momentum transfer shown. Solid curves represent results given by equation (5.7.67), with χ_{spin} as in Figure 5.13. The parameters E_f, k_f and I_{eff} are shown in Table 5.2. The theory curves include thermal corrections (after Lowde and Windsor, 1967).

However, the estimated susceptibility for Ni, $\chi(00)$, calculated from free-electron theory with the constants given in Table 5.2 is 3×10^{-4} emu mole^{-1} which is only about one-third of the measured value for Ni at 1020° K. Since $\chi(00)$ is the small q limit of $\int \omega^{-1} \operatorname{Im} \chi \, d\omega$ by the Kramers–Kronig relations (4.6.20), we see that this latter quantity must be substantially greater at small q than the free-electron model predicts.

TABLE 5.2. Parameters for free electron fit to neutron scattering from Ni

Fitted parameters
$k_f = 1.3 \pm 0.2\ A^{-1}$
$E_f = 0.4 \pm 0.15\ A^{-1}$
$I_{\mathrm{eff}} = 0.4 \pm 0.2$ eV

The reasons seem clear; we must deal with a five 3d band model for Ni to get quantitative agreement. Refinements of the above model have been subsequently considered by Allen and co-workers (1968), who used the tight-binding approximation for the d-bands.

The main new points which then emerge from comparison with the neutron experiments are that, by inclusion of non-zero reciprocal lattice-vectors as an automatic consequence of using tight-binding rather than free-electron theory, rather good agreement can be obtained throughout the zone, provided the enhancement factor I_{eff} is increased somewhat from the value of 0.4 ± 0.2 eV suggested by Lowde and Windsor to a value of about $\frac{1}{2}$ eV.

Also, work on the response function for Ni (Windsor, Lowde and Allen, 1969) has extended the previous calculations into the ferromagnetic régime using the Hubbard Hamiltonian. The tight-binding approximation was again used and calculations were made at 0.5, 0.9, 1.1 and $1.9T_c$. Above T_c the distribution becomes that we discussed above, but one rather striking feature which emerges from this study is that at $1.1T_c$ the effect of spin waves on the scattering functions in the ferromagnetic state is still in evidence from the calculations. We should mention at this point, to conclude this discussion, the related work on critical and spin wave fluctuations in Ni by Minkiewicz and colleagues (1969).

5.7.7 Neutron scattering by ferroelectrics

(a) *Relation between dielectric susceptibility and structure factor.* We saw, in Chapter 3, how we can describe certain ferroelectrics like $BaTiO_3$ by assuming a rigid ion model, in terms of phonon theory. It is then possible to obtain expressions for the dielectric susceptibility $\chi(\mathbf{q},\omega)$ above the Curie temperature. We shall show that, in the one-phonon approximation for neutron scattering, $\chi(\mathbf{q},\omega)$ is directly proportional to $S_1(\mathbf{k},\omega)$, the dynamical structure factor with the Bragg peaks subtracted out. It therefore follows that neutron scattering by the nuclei of a crystal can provide important information on its ferroelectric behaviour. This has been emphasized especially by Cochran (1969), whose presentation we shall largely follow below.

(b) *Harmonic approximation.* The theory of section 5.3 is very simply extended to describe crystals with more than one atom per unit cell. For present purposes, it is convenient to define "form factors" by

$$F_\sigma(\mathbf{k}) = \sum_i (b_i/M_i^{\frac{1}{2}})\,\mathrm{e}^{-W(\mathbf{k})}[\mathbf{k}\,.\,\boldsymbol{\varepsilon}_{i\sigma}(\mathbf{k})]\,\mathrm{e}^{i\mathbf{K}.\mathbf{R}_i}, \tag{5.7.68}$$

where b_i is the scattering length for the nucleus at position $\mathbf{R}_i$ in the unit cell and $\mathbf{K}$ is a reciprocal lattice vector defined by

$$\mathbf{k} = \mathbf{q}+\mathbf{K} \tag{5.7.69}$$

with $\mathbf{q}$ restricted to the BZ. Using equation (5.3.50) we may then write

$$S_1(\mathbf{k},\omega) = N\pi \sum_\sigma (|F_\sigma(\mathbf{k})|^2/\omega)(1-\mathrm{e}^{-\beta\omega})^{-1}[\delta(\omega-\omega_\sigma(\mathbf{k}))+\delta(\omega+\omega_\sigma(\mathbf{k}))], \tag{5.7.70}$$

where the b_i are now absorbed into the structure factor.

By comparison with equation (3.16.5), we see that by considering the α-component in the direction of $\mathbf{k}$ we may rewrite the above equation as

$$S_1(\mathbf{k},\omega) = \frac{2\Omega|F(\mathbf{k},\omega)|^2}{|M_\alpha(\mathbf{k},\omega)|^2}(1-\mathrm{e}^{-\beta\omega})^{-1}\chi''_{\alpha\alpha}(\mathbf{k},\omega), \tag{5.7.71}$$

where both $F(\omega)$ and $M_\alpha(\omega)$ are zero unless $\omega = \pm\omega_\sigma(\mathbf{k})$, being defined respectively by

$$F(\mathbf{k},\omega) = \sum_{\sigma i} b_i \frac{\mathrm{e}^{-W(\mathbf{k})}}{M_i^{\frac{1}{2}}}[\mathbf{k}\,.\,\boldsymbol{\varepsilon}_{\sigma i}(\mathbf{k})]\,\mathrm{e}^{i\mathbf{K}.\mathbf{R}_i}\,\delta(\omega\pm\omega_\sigma(\mathbf{k})) \tag{5.7.72}$$

and

$$M_\alpha(\mathbf{k},\omega) = \sum_{\sigma i}\frac{Z_i}{M_i^{\frac{1}{2}}}\varepsilon_{i\sigma}^{\alpha}(\mathbf{k})\,\mathrm{e}^{i\mathbf{K}.\mathbf{R}_i}\,\delta(\omega\pm\omega_\sigma(\mathbf{k})) \tag{5.7.73}$$

with $\mathbf{k}$ again defined by equation (5.7.69). Equation (5.7.71) shows that χ'', the imaginary part of the susceptibility, is directly related to the structure factor, provided both quantities are calculated with the harmonic approximation.

(c) *Influence of anharmonicity*. By equation (3.16.17), the susceptibility is quite generally related to the phonon Green function $G_{\sigma\sigma'}(\mathbf{k}, \omega)$. Here we require the imaginary part of this relation [cf. equation (3.16.12)]:

$$\chi''_{\alpha\beta}(\mathbf{k}, \omega) = \frac{\beta}{2} \sum_{\sigma\sigma'} \frac{\mathscr{M}_{\sigma\alpha}(\mathbf{k})\, \mathscr{M}^*_{\sigma'\beta}(\mathbf{k})\, G''_{\sigma\sigma'}(\mathbf{k}, \omega)}{\Omega[\omega_\sigma(\mathbf{k})\, \omega_{\sigma'}(\mathbf{k}')]^{\frac{1}{2}}}. \tag{5.7.74}$$

On the other hand, if we recall that the spectral density is essentially the imaginary part of the Green function, we can express equation (5.3.49) for the scattering in the linear approximation in the form

$$S_1(\mathbf{k}, \omega) = N\beta(1 - e^{-\beta\omega})^{-1} \sum_{\sigma\sigma'} \frac{F_\sigma(\mathbf{k})\, F^*_{\sigma'}(\mathbf{k})\, G''_{\sigma\sigma'}(\mathbf{k}, \omega)}{[\omega_\sigma(\mathbf{k})\, \omega_{\sigma'}(\mathbf{k})]^{\frac{1}{2}}}, \tag{5.7.75}$$

the formal similarity of this with equation (5.7.74) being evident.

As in section 5.3, we may assume, provided the anharmonicity is not too great, that

$$G_{\sigma\sigma'}(\mathbf{k}, \omega) = 0, \quad \sigma \neq \sigma', \tag{5.7.76}$$

so that interaction between modes is neglected. Then we can take a Lorentzian form for the spectral density, in which case we can write

$$\chi''_{\alpha\alpha}(\mathbf{k}, \omega) = \frac{1}{\Omega} \sum_{\sigma} |\mathscr{M}_{\sigma\alpha}(\mathbf{k})|^2 \frac{2\omega \Gamma_\sigma(\mathbf{k})}{[\Omega^2_\sigma(\mathbf{k}) - \omega^2]^2 + 4\omega^2\, \Gamma^2_\sigma(\mathbf{k})}, \tag{5.7.77}$$

while [cf. equation (5.3.52)]

$$S_1(\mathbf{k}, \omega) = 2N(1 - e^{-\beta\omega})^{-1} \sum_{\sigma} \frac{|F_\sigma(\mathbf{k})|^2\, 2\omega \Gamma_\sigma(\mathbf{k})}{[\Omega^2_\sigma(\mathbf{k}) - \omega^2]^2 + 4\omega^2\, \Gamma^2_\sigma(\mathbf{k})}. \tag{5.7.78}$$

We therefore find a relation of the same form as (5.3.71), with the difference that the peaks in S_1 and χ'' are centred on the "quasi-harmonic" frequencies $\Omega_\sigma(\mathbf{k})$ and are Lorentzians of half-width $\Gamma_\sigma(\mathbf{k})$. The matter is not so simple when multiphonon contributions are included, but it is reasonable to expect, as in the analysis of neutron scattering in other contexts, that the above theory represents a useful first approximation. Further, while the present relations are nothing like as simple as those relating magnetic scattering and magnetic susceptibility (cf. section 5.6), we shall see that useful results emerge for ferroelectrics where only one phonon mode, the "ferroelectric mode", contributes significantly to the dielectric susceptibility. Before discussing the scattering further, however, it will be convenient to discuss the relation between $\chi(\mathbf{k}, \omega)$ and the dielectric constant.

(d) *Relation of* $\chi(\mathbf{k}, \omega)$ *to dielectric function*. The dielectric function $\varepsilon(\mathbf{k}, \omega)$ is usually defined at $\mathbf{k} = 0$ or at least for $k \ll K$, so that, from equation (5.7.69),

k and **q** are in fact identical. It is also assumed in defining ε that there is no macroscopic (or depolarizing) field; either it has been shorted out or the polarization is perpendicular to the wave-vector **q**. Then we have

$$\frac{P}{\mathscr{E}} = \frac{\varepsilon - 1}{4\pi}, \tag{5.7.79}$$

where P is the polarization and $\mathscr{E}$ the applied field. On the other hand, if the polarization is parallel to **q**, it is determined by

$$\frac{P}{\mathscr{E} - 4\pi P} = \frac{\varepsilon - 1}{4\pi} \tag{5.7.80}$$

or

$$\frac{P}{\mathscr{E}} = \frac{1}{4\pi}\left(1 - \frac{1}{\varepsilon}\right), \tag{5.7.81}$$

where $\mathscr{E}$ is again the applied field and $-4\pi P$ is the macroscopic field. Since χ has been defined as $P/\mathscr{E}$ in all cases, the limit $q \to 0$ depends on the direction of **q**. In fact, one sees immediately that, choosing **q** in the x-direction, the longitudinal susceptibility is given by

$$\chi_{xx}(\mathbf{q}, \omega) = \frac{1}{4\pi}\left[1 - \frac{1}{\varepsilon(\mathbf{q}, \omega)}\right], \quad \mathbf{q} = q\hat{\mathbf{x}}, \tag{5.7.82}$$

while the transverse susceptibility is given by

$$\chi_{yy}(\mathbf{q}, \omega) = \chi_{zz}(\mathbf{q}, \omega) = \frac{1}{4\pi}[\varepsilon(\mathbf{q}, \omega) - 1], \quad \mathbf{q} = q\hat{\mathbf{x}}. \tag{5.7.83}$$

We must now enquire what happens when **k** is not equal to **q**. To begin with, we note that in the model of a ferroelectric discussed in Chapter 3, the electric dipoles were assumed infinitesimal in extent and located at the lattice sites. Hence the only material values of the applied electric field $\mathscr{E}(\mathbf{q}, \mathbf{r}) = \mathscr{E}\, e^{i\mathbf{q}.\mathbf{r}}$ are the values $\mathscr{E}(\mathbf{R})$ at the lattice sites **R**. If the lattice were a Bravais lattice, then the response to $\mathscr{E}(\mathbf{q}+\mathbf{K}, \mathbf{r})$ would be exactly the same as the response to $\mathscr{E}(\mathbf{q}, \mathbf{r})$, because $e^{i\mathbf{K}.\mathbf{R}} = 1$ for a Bravais lattice, and $\chi(\mathbf{k}, \omega)$ would be periodic in the reciprocal lattice (these considerations are precisely similar to those applied to the Heisenberg model of a magnetic material in the previous chapter). However, a ferroelectric crystal does not constitute a Bravais lattice and $\chi(\mathbf{q}+\mathbf{K}) \neq \chi(\mathbf{q})$ in general. This lack of periodicity arises in equation (5.7.68) from the appearance of the factors $e^{i\mathbf{K}.\mathbf{R}_i}$. Nevertheless, if **K** corresponds to certain particular reciprocal lattice points such as $(2, 0, 0)$, we do obtain $e^{i\mathbf{K}.\mathbf{R}_i} = 1$ in an alkali halide structure and therefore

$\chi(\mathbf{k}) = \chi(\mathbf{q})$, at least in the rigid-ion model. Let us suppose, for example, that $\mathbf{q}$ is small, as shown in Figure 5.15(a). Clearly

$$\chi_{xx}(\mathbf{q}+\mathbf{K}, \omega) = \frac{1}{4\pi}\left(1 - \frac{1}{\varepsilon(\mathbf{q}, \omega)}\right), \tag{5.7.84}$$

while

$$\chi_{yy}(\mathbf{q}+\mathbf{K}, \omega) = \chi_{zz}(\mathbf{q}+\mathbf{K}, \omega)$$
$$= \frac{1}{4\pi}(\varepsilon(\mathbf{q}, \omega) - 1). \tag{5.7.85}$$

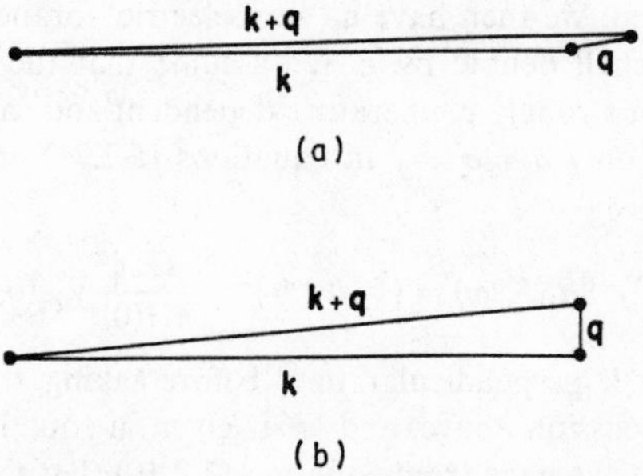

FIGURE 5.15. (a) Arrangement of reciprocal lattice vector **k** and phonon wave vector **q** for which neutron scattering cross-section is related to ferroelectric longitudinal susceptibility. (b) Arrangement for which scattering cross-section is related to transverse susceptibility. Note how the direction of **q** can markedly change even though **K**+**q** varies only slightly.

We may also note that, for a cubic crystal such as an alkali halide, when $\mathbf{q}$ is parallel to $(1, 0, 0)$ the eigenvectors $\boldsymbol{\varepsilon}_{i\sigma}(\mathbf{q})$, and thus the polarization, are constrained by symmetry to be purely transverse or purely longitudinal. Only the longitudinal optic mode contributes to $\chi_{xx}(\mathbf{q})$ as given by equations (5.7.84) and (5.7.85), and only the longitudinal optic and acoustic modes can contribute to $S_1(\mathbf{k}, \omega)$.

In contrast to the situation shown in Figure 5.16(*a*), when the relative directions of $\mathbf{k}$ and $\mathbf{q}$ are as shown in (*b*), we obtain, instead of equations (5.7.84) and (5.7.85),

$$\chi_{xx}(\mathbf{q}, \omega) = \chi_{zz}(\mathbf{q}, \omega) = \frac{1}{4\pi}[\varepsilon(\mathbf{q}, \omega) - 1] \tag{5.7.86}$$

and

$$\chi_{yy}(\mathbf{q}, \omega) = \frac{1}{4\pi}\left[1 - \frac{1}{\varepsilon(\mathbf{q}, \omega)}\right] \tag{5.7.87}$$

despite the fact that the position of **k** has changed but slightly.

(e) *Comparison with experiment*. While, within the rigid-ion model, the considerations of this section so far apply to any dielectric which can be described in terms of the theory of Chapter 3, section 3.2, based on phonon theory, it is only for ferroelectrics that the relations appear to be really useful. As already remarked, we then have a "ferroelectric" branch of the phonon modes, which we shall denote by f. We assume that the contributions of other modes are not strongly temperature dependent and can be allowed for. We therefore take only $\sigma = \sigma' = f$ in equations (5.7.71) and (5.7.73), from which we obtain

$$(2N\Omega)^{-1} S_1(\mathbf{K}, \omega) = (1 - e^{-\beta\omega})^{-1} \frac{\|F_f(\mathbf{K})|^2}{\mathcal{M}_{fx}^2(0)} \chi''_{xx}(0, \omega), \tag{5.7.88}$$

where we have set **k** perpendicular to **q** before taking the limit $\mathbf{q} \to 0$, in accord with the discussion above, and $\mathcal{M}$ is given in equation (3.16.6).

In the limit $q \to 0$, we have from equation (3.2.10) that the components of the displacements responsible for ferroelectric behaviour are

$$u_i = \frac{1}{(NM_i)^{\frac{1}{2}}} Q_f(0)\, \varepsilon_{if}(0), \tag{5.7.89}$$

while, by equation (3.16.5), the spontaneous polarization is given by

$$\mathscr{V} P = N^{\frac{1}{2}} Q_f(0) \mathcal{M}_f(0). \tag{5.7.90}$$

Hence equation (5.7.78) may be written

$$S_1(\mathbf{K}, \omega) = (1 - e^{-\beta\omega})^{-1} \frac{2N}{\Omega P^2} \left| \sum_i b_i e^{-W(\mathbf{K})} (\mathbf{K}.\mathbf{u}_i) e^{i\mathbf{K}.\mathbf{R}_i} \right|^2 \chi''_{xx}(0, \omega). \tag{5.7.91}$$

We next integrate over frequency, using the approximation that $(1 - e^{-\beta\omega})^{-1}$ can be replaced by $k_B T/\omega$. Then we find

$$S_1(\mathbf{K}) = \frac{Nk_B T}{\Omega P^2} \left| \sum_i b_i e^{-W(\mathbf{K})} (\mathbf{K}.\mathbf{u}_i) e^{i\mathbf{K}.\mathbf{R}_i} \right|^2 \chi_{xx}(0, 0) \tag{5.7.92}$$

with measurable quantities now appearing on both sides of the equation. In particular, since χ presumably obeys the Curie–Weiss law

$$\chi \propto \frac{1}{T - T_c}, \tag{5.7.93}$$

where T_c is the Curie temperature (a relation we derived from the magnetic susceptibility in the previous chapter), we also expect that

$$S_1(\mathbf{K}) \propto \frac{T}{T-T_c} \tag{5.7.94}$$

which has indeed been confirmed experimentally by Barker (1966) for $BaTiO_3$.

Finally, we ought to record here that, in some ferroelectric crystals, there are other elementary excitations, in addition to the phonons, which are responsible for the dielectric and scattering properties of the materials. For example, in KH_2PO_4 there are two positions of minimum potential energy in a hydrogen bond, one nearer each oxygen, and the proton can tunnel through the barrier which separates them. The ground state then splits into two levels, separated by some energy Ω. We can regard this two-level system as a (fictitious) spin system in a field Ω and if we add on some direct interaction between hydrogens, assuming one per unit cell, we can write down a Hamiltonian of the form

$$H = -2\Omega \sum_i X_i - \frac{1}{2} \sum_{ll'} J_{ll'} Z_l Z_{l'}, \tag{5.7.95}$$

where X_i and Z_i are components of the fictitious spin. If we further make the transformation

$$X_i \rightarrow S_i^0, \quad -2Z_i \rightarrow S_i^+ + S_i^- = S_i, \tag{5.7.96}$$

we can put the Hamiltonian in the form

$$H = -2\Omega N^{\frac{1}{2}} S^{(0)}(0) - \frac{1}{8} \sum_{\mathbf{q}} J(\mathbf{q}) S(\mathbf{q}) S(-\mathbf{q}) \tag{5.7.97}$$

just as we did for Heisenberg systems in Chapter 4, and the analysis to obtain the susceptibility and the scattering, etc. can proceed as for such systems. In fact, of course, the dielectric and scattering properties of a material such as KH_2PO_4 comes from mixed phonon and tunnelling modes, as has been worked out in some detail (for a review, see Cochran, 1969).

Appendixes

APPENDIX A1.1 HARTREE–FOCK EQUATIONS AND KOOPMAN'S THEOREM

The Thomas–Fermi theory discussed in section 1.2.1 gives us a means of calculating the self-consistent field in which particles move in atomic systems. Unfortunately though, it suffers from two important limitations:

(*a*) It is only valid when the potential field $V(\mathbf{r})$ varies but slowly over the de Broglie wavelength of a characteristic particle in the system.

(*b*) It assumes each particle moves in the same self-consistent field.

We now attempt, therefore, to transcend this approximation by appeal to the variational principle of quantum mechanics used in section 1.2.2 and elsewhere. Thus we form the quantity

$$E = \int \Psi^* H \Psi \, d\mathbf{r}, \tag{A1.1.1}$$

where Ψ is an approximate but normalized wave function of a form appropriate to describe the physical system under investigation. Clearly if Ψ were the exact ground-state wave function of our system, then E would be the ground-state energy. The variational principle states that E is stationary with respect to variation of Ψ, and indeed E is always an upper bound to the ground-state energy. It is possible to express (A1.1.1) in terms of the Dirac density matrix ρ of Chapter 2, section 2.7.5, or, what is entirely equivalent, we approximate the wave function by writing Ψ as a determinant of single-particle wave functions. Varying E with respect to the single-particle wave functions, then leads to Euler equations which are the famous Hartree–Fock equations. We derive these below using wave functions; a Dirac density matrix formulation of the Hartree–Fock theory is also possible but we shall not go into detail here.

A1.1.1 Euler equations of variation problem

Our interest here is restricted to the derivation of the Hartree–Fock equations assuming the special Coulomb form of the particle interactions. The general case can be given, but is not our concern in this appendix.

Then, explicitly, the Hamiltonian in (A1.1.1) may be written

$$H = \sum_i H_i + \frac{1}{2}\sum_{ij}\frac{e^2}{r_{ij}} \quad (i,j = 1, \ldots, N), \tag{A1.1.2}$$

where H_i includes the electronic kinetic energy operators and the electron–nuclear interactions.

We shall assume that we are dealing with a case where each single particle state of the Schrödinger wave function is doubly occupied, the ith state having then the wave functions $\psi_i \alpha$ or $\psi_i \beta$, where α and β are the usual spin functions. Then we write our approximation to the total wave function Ψ as

$$\Psi = (N!)^{-\frac{1}{2}} \det(\phi_1 \ldots \phi_N), \tag{A1.1.3}$$

where ϕ is a product of spin and space wave functions, or, more explicitly, in terms of the permutation operator P,

$$\Psi = \frac{1}{(N!)^{\frac{1}{2}}} \sum_P (\pm 1) P[\psi_1(\mathbf{r}_1) \ldots \psi_N(\mathbf{r}_N)\, \alpha(\sigma_1)\, \beta(\sigma_2) \ldots \alpha(\sigma_{N-1})\, \beta(\sigma_N)], \tag{A1.1.4}$$

where the space wave functions are equal in pairs, and the plus sign refers to even, and the minus sign to odd, permutations.

We handle the orthogonality and normalization conditions

$$\int \psi_i^*(\mathbf{r})\, \psi_j(\mathbf{r})\, d\mathbf{r} = 0 \quad (i \neq j) \tag{A1.1.5}$$

and

$$\int \psi_i^* \psi_i\, d\mathbf{r} = 1 \tag{A1.1.6}$$

by the customary use of Lagrange multipliers γ_{ij}. Then we must obtain the "best" single-particle orbitals ψ_i from the variational principle by requiring that

$$\begin{aligned} \delta E &= \delta \int \Psi^* H \Psi\, d\mathbf{r} \\ &= \int \Psi^* H \delta\Psi\, d\mathbf{r} + \int \delta\Psi^* H \Psi\, d\mathbf{r} = 0. \end{aligned} \tag{A1.1.7}$$

Now from (A1.1.4) it is readily shown that

$$\begin{aligned} \delta\Psi = \frac{1}{(N!)^{\frac{1}{2}}} \sum_P (\pm 1) P \Big[\sum_i \psi_1(\mathbf{r}_1) \ldots \psi_{i-1}(\mathbf{r}_{i-1})\, \psi_{i+1}(\mathbf{r}_{i+1}) \\ \times \ldots \psi_N(\mathbf{r}_N)\, \delta\psi_i(\mathbf{r}_i)\, \alpha_i(\sigma_i) \ldots \alpha_N(\sigma_N) \Big]. \end{aligned} \tag{A1.1.8}$$

Substituting in (A1.1.7), summing over spin and using (A1.1.5) and (A1.1.6) we then find

$$\sum_i \int \Bigg(\psi_i^*(r_1) \Bigg[\sum_j \int \psi_j^*(\mathbf{r}_2)\, H_2\, \psi_j(\mathbf{r}_2)\, d\mathbf{r}_2$$

$$+\frac{1}{2}\sum_{j,k}{}' \mathrm{e}^2 \int \frac{|\psi_j(\mathbf{r}_2)|^2 |\psi_k(\mathbf{r}_3)|^2}{r_{23}}\, d\mathbf{r}_2\, d\mathbf{r}_3$$

$$-\frac{1}{2} \sum_{\substack{jk \\ \text{parallel spins}}}{}' e^2 \int \frac{\psi_j^*(\mathbf{r}_2)\, \psi_k^*(\mathbf{r}_3)\, \psi_j(\mathbf{r}_3)\, \psi_k(\mathbf{r}_2)}{r_{23}}\, d\mathbf{r}_2\, d\mathbf{r}_3 + H_i$$

$$+\mathrm{e}^2 \sum_j \frac{|\psi_j(\mathbf{r}_2)^2|}{r_{12}}\, d\mathbf{r}_2 + \gamma_{ii} \Bigg]$$

$$-\Bigg\{ \sum_{\substack{j \\ \text{parallel spins}}} \psi_i^*(\mathbf{r}_1) \Bigg[e^2 \int \frac{\psi_i^*(\mathbf{r}_2)\, \psi_j(\mathbf{r}_2)\, d\mathbf{r}_2}{r_{12}} + \int \psi_i^*(\mathbf{r}_2)\, H_2\, \psi_j(\mathbf{r}_2) + \gamma_{ji} \Bigg] \Bigg\} \Bigg)$$

$$\times\, \delta\psi_i(\mathbf{r}_1)\, d\mathbf{r}_1$$

$$+\text{a similar expression in } \delta\psi_i^*(\mathbf{r}_1) = 0. \qquad \text{(A1.1.9)}$$

The Lagrange multipliers γ_{ji} are conveniently redefined through

$$\lambda_{ii} = \gamma_{ii} + \sum_j \int \psi_j^*(\mathbf{r}_2)\, H_2\, \psi_j(\mathbf{r}_2)\, d\mathbf{r}_2$$

$$+\frac{1}{2}\sum_{j,k}{}' e^2 \int \frac{|\psi_j(\mathbf{r}_2)|^2 |\psi_k(\mathbf{r}_3)|^2}{r_{23}}\, d\mathbf{r}_2\, d\mathbf{r}_3 - \frac{1}{2} \sum_{\substack{j,k \\ \text{parallel spins}}} e^2 \int \psi_i^*(\mathbf{r}_2)\, \psi_k^*(\mathbf{r}_3)$$

$$\times \frac{\psi_j(\mathbf{r}_3)\, \psi_k(\mathbf{r}_2)}{r_{23}}\, d\mathbf{r}_2\, d\mathbf{r}_3 \qquad \text{(A1.1.10)}$$

and

$$\lambda_{ij} = \gamma_{ij} + \sum_j \int \psi_i^*(\mathbf{r}_2)\, H_2\, \psi_j(\mathbf{r}_2)\, d\mathbf{r}_2. \qquad \text{(A1.1.11)}$$

At this point we now equate the coefficient of $\delta\psi_i(\mathbf{r}_1)$ to zero, and then we find

$$\Bigg[H_i + \sum_j e^2 \int \frac{|\psi_j(\mathbf{r}_2)|^2}{r_{12}}\, d\mathbf{r}_2 + \lambda_{ii} \Bigg] \psi_i - \sum_{\text{parallel spins}}{}' \Bigg[e^2 \int \frac{\psi_i^*(\mathbf{r}_2)\, \psi_i(\mathbf{r}_2)}{r_{12}}\, d\mathbf{r}_2 + \lambda_{ij} \Bigg] \psi_j = 0. \qquad \text{(A1.1.12)}$$

Taking suitable linear combinations of $\psi_1, \ldots, \psi_N$ to form new single-particle wave functions $\phi_1, \ldots, \phi_N$, equation (A1.1.12) can be brought into diagonal form. We use this below in equation (A1.1.13). This diagonalization is possible because the λ_{ij} are Hermitian.

Equations (A1.1.12) represent the Hartree–Fock equations for particles interacting by Coulomb forces. If we neglect the last term, which singles out those electrons with spins parallel to the one under discussion, and omit the term $i=j$ from the sum following H_i, we obtain the Hartree equations. These are simpler to interpret, for they tell us that each electron moves in an effective or average field due to the remaining electrons, calculated in the usual electrostatic manner from the charge distribution $-e|\psi_j|^2$ of the jth electron. This was, indeed, the way in which these equations were first written down by Hartree.

We see then the connection with the simpler assumption made earlier in section 1.2. There, the Hartree theory was further simplified, by assuming that each particle moved in the same potential field $V(\mathbf{r})$. However, when the one-electron wave functions have thereby been obtained, we can, of course, calculate the expectation value of H with respect to the determinant of these simpler (and orthogonal) one-electron functions. This "symmetrized" Hartree approximation is often useful.

On the other hand, aside from symmetrizing the potential, the previous discussion has neglected the "exchange" terms in (A1.1.12). These cannot be represented by any local potential $V(\mathbf{r})$, but imply either a velocity- or energy-dependent potential or, alternatively, the use of a potential matrix. While full use of the Hartree–Fock equations has been made in calculations on atoms, few such calculations exist for a solid.

However, an important result can be obtained, without actually solving the Hartree–Fock equations. It concerns the physical interpretation of the one-electron eigenvalues and we shall now show that these are intimately related to the low-lying excitations of the many-body system.

A1.1.2 Koopman's theorem

The philosophy we shall adopt is as follows. We consider a many-body wave function of determinantal form, in which the highest occupied ground-state orbital is replaced by the lowest unoccupied orbital. Let the eigenvalues of the corresponding one-electron equations be E_N and E_{N+1}. Strictly, making such a promotion would involve changing the lower-lying wave functions, because the field in which these electrons move is altered. This would involve complicated recalculation of all the one-electron functions in the determinant and no simple result emerges.

Nevertheless, it is often true on physical grounds that this rearrangement of the electrons which are not excited is small. If we neglect it, we can calculate the energy of the system with the original ground-state determinant and with the excited-state determinant. The answer is, as we show below, simply the difference $E_{N+1}-E_N$. Thus, the eigenvalues of the excited orbitals give us directly the excited states of the system.

Obviously, this method begins to fail when very highly excited states are considered. However, it suggests to us that the distribution of one-electron eigenvalues will be of fundamental importance in a solid.

While the Hartree–Fock method gives us a precise prescription for calculating the distribution of eigenvalues, and a *non-local* potential in a solid, evidence exists, especially in metals, that it is often not a good approximation, because of the neglect of electron correlations.

The essential mathematics behind Koopman's theorem may now be summarized as follows. Writing the Hartree–Fock equations formally as

$$H_{\mathrm{hf}}\phi_j = E_j\phi_j, \tag{A1.1.13}$$

we show first that the one-electron eigenvalue E_j (an element of the matrix λ_{ij} of equation (A1.1.12) brought into diagonal form) is the negative of the energy required to remove the electron in state described by the wave function ϕ_j from the solid.

If the Hamiltonian of the whole solid is H, then, on removal of the jth electron the new Hamiltonian is

$$H' = H-\left(-\frac{\hbar^2}{2m}\nabla_j^2+\sum_i{}'\frac{e^2}{r_{ij}}+V_j\right), \tag{A1.1.14}$$

where V_j is the ion-core potential. Then, the normalized wave function of the initial state is

$$\Psi = \frac{1}{\sqrt{N!}}\begin{vmatrix} \phi_1(\mathbf{r}_1) & \dots & \phi_{j-1}(\mathbf{r}_1) & \phi_j(\mathbf{r}_1) & \phi_{j+1}(\mathbf{r}_1) & \dots & \phi_N(\mathbf{r}_1) \\ \cdot & \cdot & \cdot & \cdot & \cdot & \cdot & \cdot \\ \phi_1(\mathbf{r}_N) & \dots & \dots & \dots & \dots & \dots & \phi_N(\mathbf{r}_N) \end{vmatrix}. \tag{A1.1.15}$$

Spin variables can be considered to be absorbed in the coordinates $\mathbf{r}_1, \dots, \mathbf{r}_N$. On the other hand, when the electron described by wave function ϕ_j is removed we may write the new wave function in the form

$$\Psi_j = \frac{1}{\sqrt{(N-1)!}}\begin{vmatrix} \phi_1(\mathbf{r}_1) & \dots & \phi_{j-1}(\mathbf{r}_1) & \phi_{j+1}(\mathbf{r}_1) & \dots & \phi_N(\mathbf{r}_1) \\ \cdot & \cdot & \cdot & \cdot & \cdot & \cdot \\ \phi_1(\mathbf{r}_N) & \dots & \dots & \dots & \dots & \phi_N(\mathbf{r}_N) \end{vmatrix}. \tag{A1.1.16}$$

We have argued earlier that the change in the ϕ's must be small and we shall neglect it. The error will tend to zero as the size of the system tends to infinity, provided the wave functions are of Bloch, and not localized, form.

The work done in removing the electron is given by

$$\int \Psi_j^* H' \Psi_j \, d\mathbf{r}' - \int \Psi^* H \Psi \, d\mathbf{r}, \tag{A1.1.17}$$

where $d\mathbf{r}'$ excludes integration over $d\mathbf{r}_j$. We have further that

$$\begin{aligned}\int \Psi_j^* H' \Psi_j \, d\mathbf{r}' = \sum_{i \neq j} \int \phi_i^*(\mathbf{r}_1) \left[-\frac{\hbar^2}{2m} \nabla_1^2 + V(\mathbf{r}_1) \right] \phi_i(\mathbf{r}_1) \, d\mathbf{r}_1 \\ + \frac{e^2}{2} \sum_{i,k \neq j} \left[\int \frac{|\phi_i(\mathbf{r}_1)|^2 |\phi_k(\mathbf{r}_2)|^2 \, d\mathbf{r}_1 \, d\mathbf{r}_2}{r_{12}} \right. \\ \left. - \int \frac{\phi_i^*(\mathbf{r}_1) \phi_k^*(\mathbf{r}_2) \phi_i(\mathbf{r}_2) \phi_k(\mathbf{r}_1) \, d\mathbf{r}_1 \, d\mathbf{r}_2}{r_{12}} \right].\end{aligned} \tag{A1.1.18}$$

Similarly $\int \Psi^* H \Psi \, d\mathbf{r}$ is obtained, the essential difference from (A1.1.18) being that the summations include the index j. Thus equation (A1.1.17) is

$$\begin{aligned}&-\left\{ \int \phi_j^*(\mathbf{r}_1) \left[-\frac{\hbar^2}{2m} \nabla_1^2 + V(\mathbf{r}_1) + \sum_i e^2 \int \frac{|\phi_i(\mathbf{r}_2)|^2}{r_{12}} d\mathbf{r}_2 \right] \phi_j(\mathbf{r}_1) \, d\mathbf{r}_1 \right. \\ &\left. - \sum_i e^2 \int \frac{\phi_i^*(\mathbf{r}_1) \phi_j^*(\mathbf{r}_2) \phi_i(\mathbf{r}_2) \phi_j(\mathbf{r}_1) \, d\mathbf{r}_1 \, d\mathbf{r}_2}{r_{12}} \right\} \\ &= - \int \phi_j^*(\mathbf{r}_1) H_{\mathrm{hf}} \phi_j(\mathbf{r}_1) \, d\mathbf{r}_1 \\ &= -E_j.\end{aligned} \tag{A1.1.19}$$

This proves Koopman's theorem. Also, in the Hartree–Fock approximation the energy required to take an electron from a Bloch state $\phi_{\mathbf{k}}$ to another Bloch state $\phi_{\mathbf{k}'}$ is simply $E(\mathbf{k}') - E(\mathbf{k})$.

APPENDIX A1.2
DENSITY-POTENTIAL RELATION AND ASSOCIATED VARIATIONAL PRINCIPLE IN ONE-BODY THEORY

In this appendix, we establish to all orders in perturbation theory the generalization of the Thomas–Fermi relation (1.2.5) between density and potential. The result (A1.2.8) below was first given by March and Murray (1961; see also 1960 for first-order theory). It will prove convenient here to work from the solution of the Bloch equation (A1.2.2), which was first given by Green (1952).

A1.2.1 Perturbative solution of Bloch equation

We work with the Bloch density matrix

$$C(\mathbf{r}\mathbf{r}_0\beta) = \sum_i \psi_i^*(\mathbf{r})\,\psi_i(\mathbf{r}_0)\,\mathrm{e}^{-\beta E_i}, \quad \beta = 1/k_B T \tag{A1.2.1}$$

where ψ_i and E_i are the wave function and energy of the ith state. The Bloch equation

$$H_\mathbf{r} C(\mathbf{r}\mathbf{r}_0\beta) = -\frac{\partial C}{\partial \beta} \tag{A1.2.2}$$

with boundary condition

$$C(\mathbf{r}\mathbf{r}_0 0) = \delta(\mathbf{r}-\mathbf{r}_0) \tag{A1.2.3}$$

is equivalent to the integral equation

$$C(\mathbf{r}\mathbf{r}_0\beta) = C_0(\mathbf{r}\mathbf{r}_0\beta) - \int_0^\beta d\beta_1 \int d\mathbf{r}_1\, C(\mathbf{r}\mathbf{r}_1\,\beta-\beta_1)\, V(\mathbf{r}_1)\, C_0(\mathbf{r}_1\mathbf{r}_0\beta_1), \tag{A1.2.4}$$

where C_0 is the free-particle result given in Chapter 4 [equation (4.15.13)]. We now proceed to iterate, by putting $C = C_0$ in the integral term as the first step. The resulting integral

$$\int_0^\beta d\beta_1\, C_0(\mathbf{r}\mathbf{r}_1\,\beta-\beta_1)\, C_0(\mathbf{r}_1\mathbf{r}_0\beta_1)$$

is readily evaluated using the relation that the Laplace transform of a convolution is the product of the Laplace transforms of the two separate factors. This same integral appears in all orders, and we can write down the final result as

$$C(\mathbf{r}\mathbf{r}_0\beta) = \sum_{j=0}^{\infty} C_j(\mathbf{r}\mathbf{r}_0\beta), \tag{A1.2.5}$$

where

$$C_j(\mathbf{r}\mathbf{r}_0\beta) = (2\pi\beta)^{-\frac{1}{2}} \int \prod_{l=1}^{j} \left[-\frac{d\mathbf{r}_l\, V(\mathbf{r}_l)}{2\pi}\right] \left(\sum_{l=1}^{j+1} s_l\right) \times \exp\left[-\frac{1}{2\beta}\left(\sum_{l=1}^{j+1} s_l\right)^2\right] \div \prod_{l=1}^{j+1} s_l \tag{A1.2.6}$$

with

$$s_l = |\mathbf{r}_l - \mathbf{r}_{l-1}|, \quad \mathbf{r}_{j+1} = \mathbf{r}.$$

A1.2.2 Perturbative solution for Dirac density matrix

We now make use of the Laplace transform relation between C and ρ (see section 4.15.1 of Chapter 4), namely

$$C(\mathbf{rr}_0\beta) = \beta\int_0^\infty \rho(\mathbf{rr}_0 E)\,\mathrm{e}^{-\beta E}\,dE. \tag{A1.2.7}$$

Straightforward calculation then leads to the final result. Writing $2E = k^2$,

$$\rho(\mathbf{rr}_0 k) = \sum_{j=0}^{\infty} \rho_j(\mathbf{rr}_0 k), \tag{A1.2.8}$$

where

$$\rho_j(\mathbf{rr}_0 k) = \frac{k^2}{2\pi^2}\left(-\frac{1}{2\pi}\right)^j \int\ldots\int d\mathbf{r}_1\, d\mathbf{r}_2 \ldots d\mathbf{r}_j\, V(\mathbf{r}_1)\, V(\mathbf{r}_2)\ldots V(\mathbf{r}_j)$$
$$\times\left\{\left[j_1\left(k\sum_{l=1}^{j+1} s_l\right)\right]\Big/\prod_{l=1}^{j+1} s_l\right\}. \tag{A1.2.9}$$

In equation (A1.2.9), $j_1(z)$ is the usual first-order spherical Bessel function defined by

$$j_1(z) = \frac{\sin z - z\cos z}{z^2}. \tag{A1.2.10}$$

Writing down the first two terms explicitly we find

$$\rho_0(\mathbf{rr}_0 k) = \frac{k^2}{2\pi^2}\frac{j_1(k|\mathbf{r}-\mathbf{r}_0|)}{|\mathbf{r}-\mathbf{r}_0|} \tag{A1.2.11}$$

which is the free-particle result, while the first-order term in $V(\mathbf{r})$ is given by

$$\rho_1(\mathbf{rr}_0 k) = -\frac{k^2}{2\pi^2}\int d\mathbf{r}_1 \frac{V(\mathbf{r}_1)}{2\pi}\frac{j_1(k|\mathbf{r}-\mathbf{r}_1| + k|\mathbf{r}_1-\mathbf{r}_0|)}{|\mathbf{r}-\mathbf{r}_1||\mathbf{r}_1-\mathbf{r}_0|}. \tag{A1.2.12}$$

This latter result is used in Chapter 10 to discuss the screening of an impurity in a metal.

A1.2.3 Sum of (ρ, V) series for slowly varying potentials

When the potential V varies only slowly in space, it proves possible to sum the perturbation series for $\rho(\mathbf{rr}k)$ to all orders. This may be done as follows. Taking the diagonal element of the first-order term, $\rho_1(\mathbf{rr}k)$, we have

$$\rho_1(\mathbf{rr}k) = -\frac{k^2}{2\pi^2}\int d\mathbf{r}_1 \frac{V(\mathbf{r}_1)}{2\pi}\frac{j_1(2k|\mathbf{r}-\mathbf{r}_1|)}{|\mathbf{r}-\mathbf{r}_1|^2} \tag{A1.2.13}$$

and the above assumption of slow variations is taken to imply that $V(\mathbf{r}_1)$ may be approximated by $V(\mathbf{r})$. The integration over $\mathbf{r}_1$ can then be carried out to yield

$$\rho_1(\mathbf{rr}k) = -\frac{kV(\mathbf{r})}{2\pi^2}. \tag{A1.2.14}$$

The higher-order terms may be calculated in a similar manner to give

$$\begin{aligned}\rho(\mathbf{rr}k) &= \frac{k^3}{6\pi^2}\left(1-\frac{3V}{k^2}+\frac{3V^2}{2k^4}-\frac{V^3}{2k^6}+\ldots\right)\\ &= \frac{1}{6\pi^2}[k^2-2V(\mathbf{r})]^{\frac{3}{2}}.\end{aligned} \tag{A1.2.15}$$

Thus, as expected, the Thomas–Fermi theory is regained, a factor of two for spin degeneracy relating (A1.2.15) to equation (1.2.5).

A1.2.4 Variational principle corresponding to Euler equation (A1.2.8)

The variational principle leading to the Euler equation (A1.2.8) has been constructed by Stoddart and March (1967). The idea is as follows. In the Thomas-Fermi theory, valid for slow variations of density and potential, the kinetic energy T is given in terms of the density $\rho(\mathbf{r})$ by

$$T = \frac{3h^2}{10m}\left(\frac{3}{8\pi}\right)^{\frac{2}{3}}\int d\mathbf{r}\{\rho(\mathbf{r})\}^{\frac{5}{3}}. \tag{A1.2.16}$$

It will be useful in what follows to introduce the displaced charge $\Delta = \rho-\rho_0$, where ρ_0 is the uniform density in the Fermi gas, to which a potential $V(\mathbf{r})$ is applied. Measuring the kinetic energy relative to that of the free Fermi gas, T_0, we obtain by expanding in Δ

$$T-T_0 = E_{\mathrm{f}}\int d\mathbf{r}\,\Delta+(E_{\mathrm{f}}/3\rho_0)\int d\mathbf{r}\,\Delta^2+O(\Delta^3), \tag{A1.2.17}$$

where E_{f} as usual is the Fermi energy. If now we cut this off at $O(\Delta^2)$, add a potential energy term and minimize the energy with respect to Δ, we obtain

$$\Delta = \text{const}\ V, \tag{A1.2.18}$$

which is the linearized form of the Thomas–Fermi (ρ, V) relation (1.2.5).

The object is to find the exact one-body form of $T-T_0$ in terms of Δ. We can either obtain T directly from $\rho(\mathbf{rr}_0 E)$ of equation (A1.2.8) as a functional of V, or we can ask for the variational principle which has the diagonal form of equation (A1.2.8) as its Euler equation. We adopt the latter procedure below.

A1.2.5 Generalized variational principle

We shall write the total energy in the form

$$E(\rho) = \int d\mathbf{r}\, t[\rho(\mathbf{r})] + \int d\mathbf{r}\, V(\mathbf{r})\, \rho(\mathbf{r}), \tag{A1.2.19}$$

where we are considering non-interacting particles in a potential field V, and t is the kinetic energy density. For convenience, we consider singly filled levels, and then, if k is the Fermi momentum, the gas density is

$$\rho_0 = k^3/6\pi^2 \tag{A1.2.20}$$

and

$$E(\rho) = E_0 + E(\Delta), \tag{A1.2.21}$$

where

$$E(\Delta) = \int F[\Delta(\mathbf{r})]\, d\mathbf{r} + \int \Delta(\mathbf{r})\, V(\mathbf{r})\, d\mathbf{r}. \tag{A1.2.22}$$

In equation (A1.2.22),

$$F[\Delta(\mathbf{r})] = T[\rho] - \frac{3(6\pi^2)^{\frac{2}{3}}}{10}\, \rho_0^{\frac{5}{3}}$$

and is a functional of the displaced charge density. The desired generalization of the variational principle is therefore to be sought as the minimization of $E(\Delta)$ with respect to density variations:

$$\delta E(\Delta) = \int d\mathbf{r}\, \delta F[\Delta(\mathbf{r})] + \int V(\mathbf{r})\, \delta\Delta(\mathbf{r})\, d\mathbf{r} = 0. \tag{A1.2.23}$$

Equation (A1.2.23) is equivalent to the exact relationship (A1.2.8) between density and potential and Stoddart and March (1967) show how these equations will give the functional F in the perturbational framework. Knowledge of this functional then completes the generalization of the Thomas–Fermi variational principle. The formulae, however, are complicated and we refer the reader to the original paper for the details.

APPENDIX A1.3 SPACE GROUPS

The purpose of this appendix is to supplement the discussion of space groups in the main text. In particular, we shall define the additional symmetry needed in discussing non-symmorphic crystals. As we shall see below, such crystals contain glide planes and screw axes.

We can divide lattices into types as follows. Every operator bringing an infinite crystal into coincidence with itself will be written $(\alpha|\mathbf{T})$; one applies a unitary coordinate transformation—a proper or improper rotation—and follows this by a translation **T**. The set {α} for a given crystal is, as we saw earlier, the *point group* G_p while the operators (α|**T**) constitute the space group. When α is the identity operator *E*, (*E*|**T**) represents a *pure*, or *primitive*, translation, **R**. The lattice generated by all **R**'s is the Bravais lattice introduced in Chapter 1. The fourteen types of Bravais lattice are shown in Figure A1.3.1. The Bravais lattice is invariant under the point group G_p of the

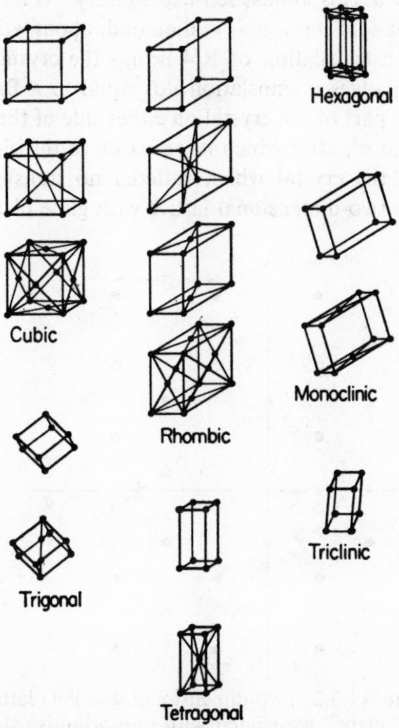

FIGURE A1.3.1. The fourteen types of Bravais lattice. Last diagram in extreme left column corresponds to a space lattice of the hexagonal system, and therefore should not be counted as a distinct Bravais lattice.

crystal, there being thirty-two possible point groups. However, the crystal itself need not be invariant under G_p, and will certainly not be if there exist operations (*A*|**T**) in which **T** is a non-primitive translation; that is, **T** is not a

Bravais lattice vector **R**. Where it is not, however, we shall later show that we may write

$$\mathbf{T} = \boldsymbol{\tau}(A) + \mathbf{R} \tag{A1.3.1}$$

where $\boldsymbol{\tau}(A)$ is some vector determined by A.

At this point, we must give definitions of screw axes and glide planes, since these are the symmetry elements for which $\boldsymbol{\tau}$ in equation (A1.3.1) is not zero.

The term *screw axis* is almost self-explanatory. As an example, we cite the existence of a four-fold screw axis in diamond, about which a rotation of 90° accompanied by a translation of $\mathbf{R}/4$ brings the crystal into itself. *A glide plane* similarly involves a translation not equal to a Bravais lattice vector: one translates the part of the crystal on either side of the glide plane through some displacement $\boldsymbol{\tau}'$, after which its reflexion in the glide plane is identical with the part of the crystal which suffered no translation. Figure A1.3.2 shows a diatomic two-dimensional lattice with glide plane g.

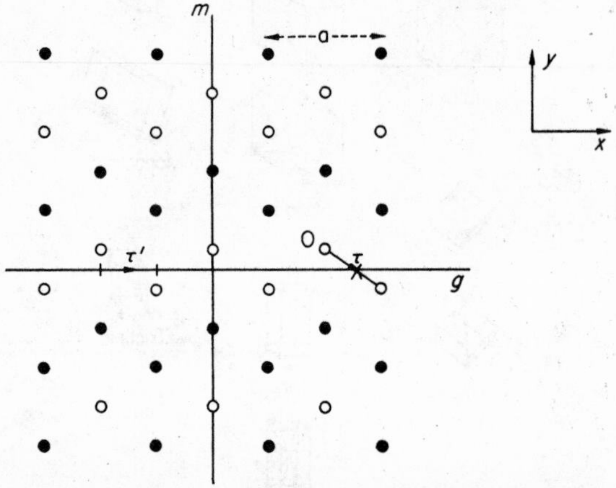

FIGURE A1.3.2. Two-dimensional diatomic lattice. g is a glide "plane" associated with non-primitive translation. m is a mirror plane. X is an inversion centre (axis of two-fold rotation C_2).

There are 73 *simple* or *symmorphic* space groups, being those which have no screw axes or glide planes; for these $\boldsymbol{\tau} = 0$ for every A in G_p. There are thus $230 - 73 = 157$ space groups which are *non-symmorphic*. Figure 1.4 shows an example of the diamond lattice, which has screw axes and glide planes and therefore has a non-symmorphic space group. We shall consider this lattice in detail in Volume 2.

It is worthwhile, at this stage, to recall that we can distinguish three groups, namely rotation, point and space groups.

(i) *Rotation group*. This is the group of pure rotations, which bring the crystal into itself.

(ii) *Point group*. This also consists of pure rotations, but these do not necessarily bring the crystal into itself without translation. The rotation group, when not identical with the point group, is a sub-group of it. However, for symmorphic lattices, these two groups are the same.

(iii) *Space group*. This consists of the totality of transformations which bring the crystal into itself. The point group and the rotation group are sub-groups of the space group.

If we take an eigenfunction $\psi_{\mathbf{k}}(\mathbf{r})$, of a Hamiltonian H invariant under every element of the space group G_s (which is the group consisting of all $(\alpha|\mathbf{T})$), we see that

$$\phi(\mathbf{r}) = (\alpha|\mathbf{T})\,\psi_{\mathbf{k}}(\mathbf{r}) = \psi_{\mathbf{k}}(\alpha\mathbf{r}+\mathbf{T}) \tag{A1.3.2}$$

is also an eigenfunction of H; to find the $\mathbf{k}$-vector with which we label it in accordance with Bloch's theorem, we move $\mathbf{r}$ through a Bravais lattice vector:

$$\phi(\mathbf{r}+\mathbf{R}) = \psi_{\mathbf{k}}(\alpha\mathbf{r}+\alpha\mathbf{R}+\mathbf{T}) = \psi_{\mathbf{k}}(\alpha\mathbf{r}+\mathbf{T})\,e^{i\mathbf{k}.\alpha\mathbf{R}} \tag{A1.3.3}$$

or

$$\phi(\mathbf{r}+\mathbf{R}) = e^{i\alpha^{\dagger}\mathbf{k}.\mathbf{R}}\,\phi(\mathbf{r}). \tag{A1.3.4}$$

Thus, whether the lattice is symmorphic or non-symmorphic, surfaces of constant energy have the symmetry of the point group of the lattice. The set of vectors $\alpha\mathbf{k}$, where α ranges over all elements of G_p, is known as the *star* of $\mathbf{k}$. However, the major problem is the degeneracy at a given $\mathbf{k}$, the degeneracy belonging to the star then following simply.

Attention is thus focused on the sub-group $G_p(\mathbf{k})$ of G_p; this will be called the point group of $\mathbf{k}$, and every element α in it is such that

$$\alpha\mathbf{k} = \mathbf{k}+\mathbf{K}, \tag{A1.3.5}$$

$\mathbf{K}$ being a reciprocal lattice vector. The corresponding space-group $G_s(\mathbf{k})$ consisting of all elements of the form $(\alpha|\mathbf{T})$ will be called the space group of $\mathbf{k}$. To obtain the degeneracy at $\mathbf{k}$, we must consider the irreducible representations of $G_s(\mathbf{k})$. If a function $\psi_{\mathbf{k}}$ lies in a manifold pertaining to a particular irreducible representation of $G_s(\mathbf{k})$, we shall say it *transforms according to this representation.*

These functions which we seek must, of course, obey Bloch's theorem and, as is evident from the main text, if the space group is symmorphic, the functions transform according to irreducible representations of the point group.

Earlier in this appendix, we asserted that all symmetry elements of a lattice may be written in the form $(A|\mathbf{T})$ where all the A's are operators belonging to the same point group. An essential characteristic of a point group is the coincidence of all its axes of rotation, whereas the axes of the symmetry elements g, m and X of Figure A1.3.2 do not have this characteristic. Let us show how the axes can be taken through the same origin, which is arbitrary, and so can, in any specific case, be chosen to make the symmetry operations as simple as possible. Suppose we have a symmetry operation $(A|\boldsymbol{\tau}')$, whose origin is displaced from our chosen one by $\mathbf{a}$, so the coordinate of a point $\mathbf{r}$ is, relative to it, $\mathbf{r}' = \mathbf{r}+\mathbf{a}$. Then‡

$$(A|\boldsymbol{\tau}')\mathbf{r}' = A\mathbf{r}'+\boldsymbol{\tau}' = A\mathbf{r}+A\mathbf{a}+\boldsymbol{\tau}' \tag{A1.3.6}$$

or

$$(A|\boldsymbol{\tau}')\mathbf{r} = A\mathbf{r}+A\mathbf{a}-\mathbf{a}+\boldsymbol{\tau}'. \tag{A1.3.7}$$

Now if we write the *same* symmetry operation in the form $(A|\boldsymbol{\tau})$ such that

$$(A|\boldsymbol{\tau})\mathbf{r} = A\mathbf{r}+\boldsymbol{\tau}, \tag{A1.3.8}$$

by comparison of (A1.3.8) and (A1.3.7) it is evident that

$$\boldsymbol{\tau}'+A\mathbf{a} = \mathbf{a}+\boldsymbol{\tau}. \tag{A1.3.9}$$

Let us check this by example from the two-dimensional lattice of Figure A1.3.2. We choose our origin at the atom marked O. Then relative to the inversion centre X, any point has coordinates

$$\mathbf{r}' = \mathbf{r}+\mathbf{a}, \quad \mathbf{a} = a(-\tfrac{1}{4}, \tfrac{1}{8}). \tag{A1.3.10}$$

Thus we can write the symmetry operator in question as $(C_2|\boldsymbol{\tau})$ where C_2 is now an inversion about O, with

$$\boldsymbol{\tau} = C_2\mathbf{a}-\mathbf{a} = a(\tfrac{1}{4}, -\tfrac{1}{8})-a(-\tfrac{1}{4}, \tfrac{1}{8}) = a(\tfrac{1}{2}, -\tfrac{1}{4}). \tag{A1.3.11}$$

It is easy to see from Figure A1.3.2 that this is indeed a correct result.

Equation (A1.3.11) shows us by what lateral displacement of axis one may turn a symmetry element $(C_4^n|\boldsymbol{\tau}_1)$ into an n-fold screw axis, and that such is always possible. One may make a similar statement about glide planes. To find the rule for multiplication of two operators $(A|\boldsymbol{\tau}_1)$ and $(A|\boldsymbol{\tau}_2)$, we successively apply transformations to the vector $\mathbf{r}$:

$$\mathbf{r}' = (B|\boldsymbol{\tau}_2)\mathbf{r} = B\mathbf{r}+\boldsymbol{\tau}_2. \tag{A1.3.12}$$

Applying the second transformation we then obtain

$$\mathbf{r}'' = (A|\boldsymbol{\tau}_1)\mathbf{r}' = A\mathbf{r}'+\boldsymbol{\tau}_1, \tag{A1.3.13}$$

‡ As no confusion can arise, we use the same notation for a symmetry operator acting on a vector as on a function.

that is,

$$(A|\boldsymbol{\tau}_1)(B|\boldsymbol{\tau}_2) = AB\mathbf{r}+A\boldsymbol{\tau}_2+\boldsymbol{\tau}_1. \tag{A1.3.14}$$

We can see that this is equivalent to

$$(A|\boldsymbol{\tau}_1)(B|\boldsymbol{\tau}_2) = (AB|\boldsymbol{\tau}_1+A\boldsymbol{\tau}_2). \tag{A1.3.15}$$

We can now prove the statement that any operator A is associated with but one non-equivalent non-primitive translation if any (we say $\boldsymbol{\tau}$ and $\boldsymbol{\tau}'$ are equivalent if $\boldsymbol{\tau}' = \boldsymbol{\tau}+\mathbf{R}$). Suppose $(A|\boldsymbol{\tau}_1)$ and $(A|\boldsymbol{\tau}_2)$ are elements of the space groups. Then there must exist an element $(C|\mathbf{t})$ such that

$$(C|\mathbf{t})(A|\boldsymbol{\tau}_1) = (A|\boldsymbol{\tau}_2). \tag{A1.3.16}$$

It is now obvious that $C = E$ and so $\mathbf{t} = \mathbf{R}$, E being associated with pure translations only. Thus (A1.3.16) becomes

$$(E|\mathbf{R})(A|\boldsymbol{\tau}_1) = (EA|\boldsymbol{\tau}_1+\mathbf{R}) = (A|\boldsymbol{\tau}_1+\mathbf{R}) = (A|\boldsymbol{\tau}_2). \tag{A1.3.17}$$

Hence we may write $\boldsymbol{\tau} \equiv \boldsymbol{\tau}(A)$, as in equation (A1.3.1).

This completes our discussion of the symmetry operators as such. The remaining problem, as we indicated in the main text, is to find their irreducible representations. This is dealt with in detail in GAII of Volume 2.

APPENDIX A1.4 FEYNMAN'S THEOREM

Feynman's theorem states that for any Hamiltonian H depending on a set of parameters λ, the ground state is such that

$$\frac{\partial E(\lambda)}{\partial \lambda} = \langle\psi_\lambda|\frac{\partial H}{\partial \lambda}|\psi_\lambda\rangle \Big/ \langle\psi_\lambda|\psi_\lambda\rangle, \tag{A1.4.1}$$

where E is the energy. To prove this we write

$$\psi_{\lambda'} = \psi_\lambda+(\lambda'-\lambda)\psi'_\lambda+\tfrac{1}{2}(\lambda'-\lambda)^2\psi''_\lambda+\ldots. \tag{A1.4.2}$$

Thus

$$\langle\psi_{\lambda'}|\psi_{\lambda'}\rangle = \langle\psi_\lambda|\psi_\lambda\rangle+(\lambda'-\lambda)\{\langle\psi_\lambda|\psi'_\lambda\rangle+\langle\psi'_\lambda|\psi_\lambda\rangle\}+\ldots \tag{A1.4.3}$$

so that if $\psi_{\lambda'}$ is normalized to the same value as ψ_λ

$$\langle\psi_\lambda|\psi'_\lambda\rangle+\langle\psi'_\lambda|\psi_\lambda\rangle = 0. \tag{A1.4.4}$$

Let us also write the Hamiltonian as

$$H(\lambda') = H(\lambda)+(\lambda'-\lambda)\frac{\partial H(\lambda)}{\partial \lambda}+\ldots. \tag{A1.4.5}$$

Then

$$E(\lambda') = E(\lambda) + (\lambda' - \lambda)\frac{\partial E(\lambda)}{\partial \lambda} + \ldots, \tag{A1.4.6}$$

where from (A1.4.2) and (A1.4.5) and with ψ_λ normalized to unity

$$\frac{\partial E(\lambda)}{\partial \lambda} = \langle \psi'_\lambda | H(\lambda) | \psi_\lambda \rangle + \langle \psi_\lambda | H(\lambda) | \psi'_\lambda \rangle + \langle \psi_\lambda | \frac{\partial H(\lambda)}{\partial \lambda} | \psi_\lambda \rangle. \tag{A1.4.7}$$

The first two terms give

$$E(\lambda)\{\langle \psi'_\lambda | \psi_\lambda \rangle + \langle \psi_\lambda | \psi'_\lambda \rangle\},$$

which is zero from equation (A1.4.4). Thus equation (A1.4.1) follows.

APPENDIX A1.5 GROUP VELOCITY FORMULA

We wish to prove that, for a local potential,

$$\frac{1}{\hbar}\frac{\partial E(\mathbf{k})}{\partial \mathbf{k}} = \frac{\langle \psi^* \mathbf{p} \psi \rangle}{m} \tag{A1.5.1}$$

and we shall also obtain a generalization in the presence of a non-local field. If the non-local potential is of the form $V(\mathbf{k}, \mathbf{r})$ it is most convenient to apply Feynman's theorem of Appendix A1.4 to the equation for $u_\mathbf{k}(\mathbf{r})$, the periodic part of a Bloch wave function, namely

$$-\frac{\hbar^2}{2m}\nabla^2 u_\mathbf{k}(\mathbf{r}) - \frac{i\hbar^2}{m}\mathbf{k}\,.\,\nabla u_\mathbf{k}(\mathbf{r}) + \frac{\hbar^2 k^2}{2m}u_\mathbf{k}(\mathbf{r}) + V(\mathbf{k}, \mathbf{r})\,u_\mathbf{k}(\mathbf{r}) = E(\mathbf{k})\,u_\mathbf{k}(\mathbf{r}). \tag{A1.5.2}$$

We start from the Hamiltonian

$$H = -\frac{\hbar^2}{2m}\nabla^2 + \frac{\hbar^2 k^2}{2m} - \frac{i\hbar^2}{m}\mathbf{k}\,.\,\nabla + V(\mathbf{k}, \mathbf{r}) \tag{A1.5.3}$$

and then we immediately obtain from Feynman's theorem

$$\frac{\partial E(\mathbf{k})}{\partial \mathbf{k}} = -\frac{i\hbar^2}{m}\langle u_\mathbf{k}^* \nabla u_\mathbf{k} \rangle + \frac{\hbar^2 \mathbf{k}}{m}\langle u_\mathbf{k}^* u_\mathbf{k} \rangle + \left\langle u_\mathbf{k}^* \frac{\partial V(\mathbf{k}, \mathbf{r})}{\partial \mathbf{k}} u_\mathbf{k} \right\rangle, \tag{A1.5.4}$$

provided $V(\mathbf{k}, \mathbf{r})$ is real. Putting $\psi_\mathbf{k}(\mathbf{r}) = u_\mathbf{k}(\mathbf{r})\,e^{i\mathbf{k}.\mathbf{r}}$, this can obviously be written

$$\frac{\partial E(\mathbf{k})}{\partial \mathbf{k}} = \frac{\hbar^2}{im}\langle \psi_\mathbf{k}^* \nabla \psi_\mathbf{k} \rangle + \left\langle \psi_\mathbf{k}^* \frac{\partial V(\mathbf{k}, \mathbf{r})}{\partial \mathbf{k}} \psi_\mathbf{k} \right\rangle. \tag{A1.5.5}$$

If V is independent of $\mathbf{k}$, we recover equation (A1.5.1). If the non-local potential is such that the Schrödinger equation is

$$-\frac{\hbar^2}{2m}\nabla^2 \psi_\mathbf{k} + \int V(\mathbf{r}, \mathbf{r}')\,\psi_\mathbf{k}(\mathbf{r}')\,d\mathbf{r}' = E(\mathbf{k})\,\psi_\mathbf{k}(\mathbf{r}), \tag{A1.5.6}$$

it is perhaps easier to adopt an alternative procedure rather than attempt reduction to a form involving $V(\mathbf{k},\mathbf{r})\,\psi_\mathbf{k}(\mathbf{r})$ instead of the integral. The equation for $u_\mathbf{k}(\mathbf{r})$ is now

$$-\frac{\hbar^2}{2m}\nabla^2 u_\mathbf{k}(\mathbf{r})-\frac{i\hbar^2}{m}\mathbf{k}.\nabla u_\mathbf{k}(\mathbf{r})+\int e^{-i\mathbf{k}.\mathbf{r}}\,V(\mathbf{r},\mathbf{r}')\,u_\mathbf{k}(\mathbf{r}')\,e^{i\mathbf{k}.\mathbf{r}'}\,d\mathbf{r}' = \left(E-\frac{\hbar^2 k^2}{2m}\right)u_\mathbf{k}(\mathbf{r}). \tag{A1.5.7}$$

Let us differentiate with respect to $\mathbf{k}$. We have

$$-\frac{i\hbar^2}{m}\nabla u_\mathbf{k}+\left[\frac{\hbar^2\mathbf{k}}{m}-\frac{\partial E(\mathbf{k})}{\partial\mathbf{k}}\right]u_\mathbf{k}-\frac{\hbar^2}{2m}\nabla^2\frac{\partial u_\mathbf{k}}{\partial\mathbf{k}}-\frac{\hbar^2 i\mathbf{k}}{m}.\nabla\frac{\partial u_\mathbf{k}}{\partial\mathbf{k}}$$
$$+\left[\frac{\hbar^2k^2}{2m}-E(\mathbf{k})\right]\frac{\partial u_\mathbf{k}}{\partial\mathbf{k}}-i\int\mathbf{r}V(\mathbf{r},\mathbf{r}')\,e^{-i\mathbf{k}.\mathbf{r}}\,\psi_\mathbf{k}(\mathbf{r}')\,d\mathbf{r}'$$
$$+\int V(\mathbf{r},\mathbf{r}')\,e^{-i\mathbf{k}.\mathbf{r}}\frac{\partial\psi_\mathbf{k}}{\partial\mathbf{k}}\,d\mathbf{r}'=0. \tag{A1.5.8}$$

We now premultiply by $u_\mathbf{k}^*$ and integrate over $\mathbf{r}$. From the first two terms we obtain

$$\frac{\hbar^2}{im}\int\psi_\mathbf{k}^*\,\nabla\psi_\mathbf{k}\,d\mathbf{r}-\frac{\partial E(\mathbf{k})}{\partial\mathbf{k}}, \tag{A1.5.9}$$

which is all that would remain if the potential were local. From the next three terms of (A1.5.8) we obtain, using the complex-conjugate of (A1.5.7)

$$-\int V^*(\mathbf{r},\mathbf{r}')\,\psi_\mathbf{k}^*(\mathbf{r}')\,e^{i\mathbf{k}.\mathbf{r}}\frac{\partial u_\mathbf{k}(\mathbf{r})}{\partial\mathbf{k}}\,d\mathbf{r}\,d\mathbf{r}'$$
$$=-\int V^*(\mathbf{r},\mathbf{r}')\,\psi_\mathbf{k}^*(\mathbf{r}')\frac{\partial}{\partial\mathbf{k}}\psi_\mathbf{k}(\mathbf{r})\,d\mathbf{r}\,d\mathbf{r}'+i\int\mathbf{r}V^*(\mathbf{r},\mathbf{r}')\,\psi_\mathbf{k}^*(\mathbf{r}')\,\psi_\mathbf{k}(\mathbf{r})\,d\mathbf{r}\,d\mathbf{r}'. \tag{A1.5.10}$$

Combining (A1.5.9) and (A1.5.10) with terms from the last two contributions to (A1.5.8), we finally obtain

$$\frac{\partial E(\mathbf{k})}{\partial\mathbf{k}}=\frac{\hbar^2}{im}\int\psi_\mathbf{k}^*\,\nabla\psi_\mathbf{k}\,d\mathbf{r}+\int V(\mathbf{r},\mathbf{r}')\,\psi_\mathbf{k}^*(\mathbf{r})\frac{\partial\psi_\mathbf{k}(\mathbf{r}')}{\partial\mathbf{k}}\,d\mathbf{r}\,d\mathbf{r}'$$
$$-\int V^*(\mathbf{r},\mathbf{r}')\,\psi_\mathbf{k}^*(\mathbf{r}')\frac{\partial\psi_\mathbf{k}(\mathbf{r})}{\partial\mathbf{k}}\,d\mathbf{r}\,d\mathbf{r}'$$
$$+i\int\mathbf{r}V^*(\mathbf{r},\mathbf{r}')\,\psi_\mathbf{k}^*(\mathbf{r}')\,\psi_\mathbf{k}(\mathbf{r})\,d\mathbf{r}\,d\mathbf{r}'-i\int\mathbf{r}V(\mathbf{r},\mathbf{r}')\,\psi_\mathbf{k}^*(\mathbf{r})\,\psi_\mathbf{k}(\mathbf{r}')\,d\mathbf{r}\,d\mathbf{r}' \tag{A1.5.11}$$

or, alternatively,

$$\frac{\partial E(\mathbf{k})}{\partial \mathbf{k}} = \frac{\hbar^2}{im}\int \psi_{\mathbf{k}}^* \nabla \psi_{\mathbf{k}}\, d\mathbf{r} + \int [V(\mathbf{r},\mathbf{r}') - V^*(\mathbf{r}',\mathbf{r})]\, \psi_{\mathbf{k}}^*(\mathbf{r}) \frac{\partial}{\partial \mathbf{k}} \psi_{\mathbf{k}}(\mathbf{r}')\, d\mathbf{r}\, d\mathbf{r}'$$

$$+ i \int [\mathbf{r} V^*(\mathbf{r},\mathbf{r}') - \mathbf{r}' V(\mathbf{r}',\mathbf{r})]\, \psi_{\mathbf{k}}^*(\mathbf{r}')\, \psi_{\mathbf{k}}(\mathbf{r})\, d\mathbf{r}\, d\mathbf{r}'. \qquad \text{(A1.5.12)}$$

If the integral potential operator is Hermitian so that $V(\mathbf{r},\mathbf{r}') = V^*(\mathbf{r}',\mathbf{r})$, this reduces to

$$\frac{\partial E(\mathbf{k})}{\partial \mathbf{k}} = \frac{\hbar^2}{im}\int \psi_{\mathbf{k}}^* \nabla \psi_{\mathbf{k}}\, d\mathbf{r} + i \int (\mathbf{r} - \mathbf{r}')\, V(\mathbf{r}',\mathbf{r})\, \psi_{\mathbf{k}}^*(\mathbf{r}')\, \psi_{\mathbf{k}}(\mathbf{r})\, d\mathbf{r}\, d\mathbf{r}'. \qquad \text{(A1.5.13)}$$

APPENDIX A1.6 TOPOLOGICAL BASIS OF MORSE RELATIONS

In this appendix we give a simplified account of the theory leading to the Morse relations, avoiding technical language as far as possible. These relations are concerned with the number of critical points possible for a single-valued function defined over a closed domain D, which is a region without boundaries but finite in every direction. It is first necessary to specify this domain for functions periodic in the reciprocal lattice.

A1.6.1 The domain D

We wish to map the values of $\mathbf{k}$ into a domain D, this mapping reflecting the periodicity of some function f. The mapping will therefore be such that $\mathbf{k}$ and $\mathbf{k}+\mathbf{K}$, $\mathbf{K}$ being any reciprocal lattice vector, become the same point in

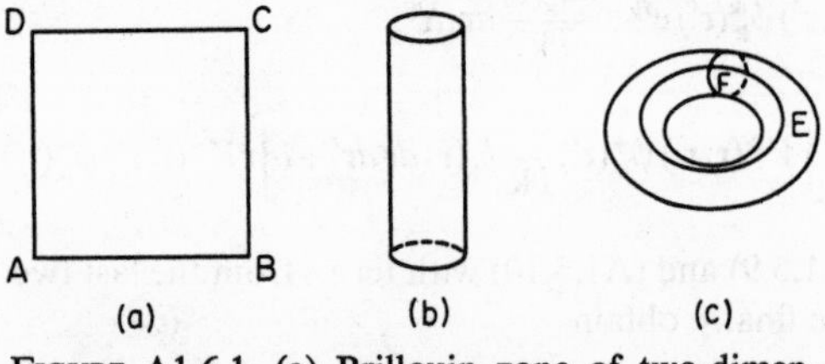

FIGURE A1.6.1. (a) Brillouin zone of two-dimensional lattice. (b) Cylinder formed by folding AD on to BC. (c) Torus formed from cylinder. Closed line E corresponds to lines AD and BC and closed line F corresponds to lines AB and CD.

D, but we require, of course, that all values of $\mathbf{k}$ within a BZ remain distinct when mapped into D. As an aid to the imagination, let us first consider a two-dimensional lattice. D is then the surface of a torus. The relation of the torus to the BZ can be clarified by reference to Figure A1.6.1. One may think of the torus as being formed by first folding the Brillouin Zone so that the boundary line AD coincides with BC (Figure A1.6.1b), thus forming a cylinder, and then bending the cylinder so that AB coincides with CD (Figure A1.6.1c). For a three-dimensional lattice, we have a similarly closed domain, but cannot properly express the closure pictorially, resort being necessary to a purely mathematical description. For our topological considerations this will be made by specifying the *connectivity numbers*. Preparatory to this we must say what we mean by *deformation* of a region of D. This implies change of shape or size, but we also include translation of the region in the idea, even if there is no other change. In addition, when we speak of a deformation we shall always assume that the deformation takes place in D; we never consider any deformation which requires a region to leave D in whole or part. Finally, regions not deformable into one another will be termed *distinct*.

A1.6.2 Connectivity numbers

The connectivity number R_n of a domain D is the number of distinct n-dimensional regions in D which have no boundaries and do not form boundaries of $(n+1)$ dimensional regions of D. We shall denote such regions as C_1^n, C_2^n, etc. For a torus, $R_0 = R_2 = 1$ and $R_1 = 2$. C^0 is any point on the torus, C^2 is the surface of the torus itself, and the two regions C_1^1 and C_2^1 are rings such as E and F in Figure A1.6.1. Such direct inspection is hardly possible for the three-dimensional case, except that we can immediately say $R_0 = R_3 = 1$, C^3 again being the domain D itself. We must therefore consider how the nature of the domain is ascertainable by more abstract considerations, and we shall first do this for two dimensions, so that our arguments may be checked by direct inspection.

Let us call the two variables in which f is periodic, ξ and η. Now on our domain we must be able to draw a closed line C of constant η over which ξ varies through its entire range. The closure expresses the equivalence of $\mathbf{k}$ and $\mathbf{k}+\mathbf{K}$. Let us suppose that such lines are drawn for each allowed value of η in the BZ. Let us now draw two closed lines B_1 and B_2 along each of which η varies over its whole range in the BZ. These lines are deformable into one another, for we can move each point of B_1 along a line C_η (of fixed η) until it coincides with a point of B_2 (see Figure A1.6.2). If, in defining D, we further specify that every line into which B_1 is deformable is one in which η varies over its whole range, it is evident that B_1 cannot be deformed into a line C_η. Moreover, B_1 cannot enclose a surface on D, for if it did, we could shrink it to a point.

By such considerations we establish that we require D to have just two distinct regions C_1^1 and C_2^1, one for each variable ξ and η. The two C^1's can be regarded as the mapping of the two primitive translations $\mathbf{K}_1$ and $\mathbf{K}_2$ on to the surface D of a torus when we map all allowed values of $\mathbf{k}$ in the BZ on to D. We can carry this argument over for the three-dimensional BZ, when we obtain $R_1 = 3$, and the three regions C^1 may be regarded as the mapping of the primitive translations $\mathbf{K}_1$, $\mathbf{K}_2$, $\mathbf{K}_3$ into D.

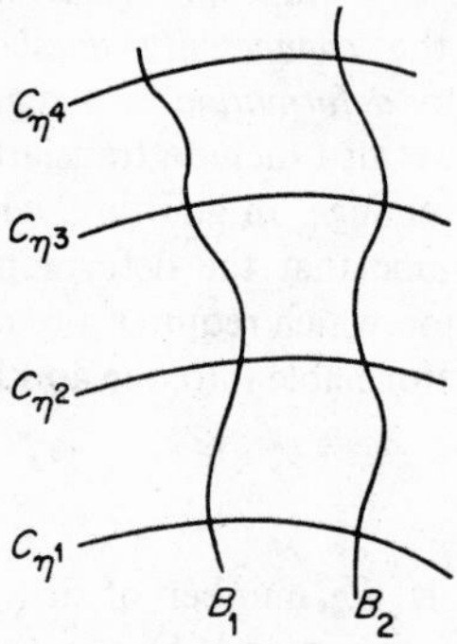

FIGURE A1.6.2. Lines indicated by C are parts of cycles along which ξ varies over its whole range. Figure indicates that deformation of B_1 into B_2 must be possible by moving each point along a line C_η of fixed η.

In three-dimensional situations, we also have to obtain R_2. Now a closed surface in the BZ must, of course, enclose a volume in the BZ, and any such volume becomes a volume in D under the mapping process. One can therefore see that surfaces must be found which are closed in D, but are *not* closed when we perform the inverse mapping of the values of the variables in D into the BZ, and it further becomes obvious that such closed surfaces must include C^1's. We therefore examine the nature of the mapping into D of a surface in the reciprocal lattice which includes primitive translations. If it includes one primitive translation only, it is fairly easy to see that it must still enclose a volume of D. The surfaces in question must therefore include two primitive translations, or, in D, two C^1's. Such a surface C_1^2 in D is that formed by taking all values of two variables ξ and η for fixed ζ. Although it has no boundaries, it does not enclose a volume of D for, if it did, it could be shrunk to a point, which is impossible, as it has regions C_ξ^1 and C_η^1 upon it. The similar surface containing lines C_ξ^1 and C_η^1 is not deformable into C_1^2, because C_η^1 is not deformable into C_ξ^1 or C_ζ^1. We also have a third distinct surface containing lines C_η^1 and C_ζ^1. There are no regions C^2 which cannot be deformed into one or other of these three surfaces, since, as already shown, any region C^2 must contain two C^1's not deformable into one another, and there are only three of these.

To summarize, for the three-dimensional domain,

$$R_0 = R_3 = 1, \quad R_1 = R_2 = 3. \tag{A1.6.1}$$

A1.6.3 Topological definition of critical points

Suppose that at a particular point $\mathbf{x}$, $f(\mathbf{x}) = \varepsilon$. Then we define the region $\{f < \varepsilon\}$ (which may be composed of two or more disconnected parts) as that region of D for which $f < \varepsilon$. Let us also take an n-dimensional region which lies entirely in $\{f < \varepsilon\}$ except that it contains $\mathbf{x}$. We then say the region lies entirely in $\{f < \varepsilon\} + \mathbf{x}$. If this region cannot be continuously deformed so as to lie entirely in $\{f < \varepsilon\}$, without its boundary passing over $\mathbf{x}$, under deformation always lying in $\{f < \varepsilon\} + \mathbf{x}$, then $\mathbf{x}$ is said to be a critical point of type n. The region will be denoted by Z_n when this happens. For these purposes a point has no boundaries and the boundaries of a line are its two ends.

We shall shortly show that the above topological definition for a function periodic in three variables is consistent with the analytic definition given in Section 1.7. Before doing so, let us consider a case readily represented pictorially in which $\mathbf{x}$ is at a local maximum of a doubly periodic function. Figure A1.6.3 shows the point $\mathbf{x}$ and the shaded area about it indicates part of the region $\{f < \varepsilon\}$. Z_n may be taken as the region enclosed by the line encircling $\mathbf{x}$ as shown. It is obvious that Z_n cannot be deformed so as to exclude $\mathbf{x}$ unless the boundary passes over $\mathbf{x}$.

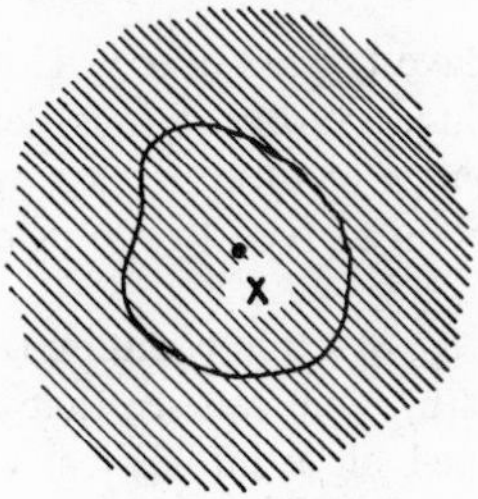

FIGURE A1.6.3. Point $\mathbf{x}$ enclosed by region $\{f < \varepsilon\}$.

Let us now consider all possibilities for the three-dimensional case. We take each type of critical point as analytically defined, and show that for every type the topological definition is equivalent.

Type 0. If $\mathbf{x}$ is a minimum, Z_n must be just the point $\mathbf{x}$, which cannot be continuously deformed into $\{f < \varepsilon\}$ for such a region does not exist about $\mathbf{x}$.

Type 1. If, referred to principal axes, we can write in the neighbourhood of $\mathbf{x}$,

$$f = \varepsilon + \sum_{i=1}^{3} \alpha_i x_i^2, \tag{A1.6.2}$$

and $\alpha_1 > 0$, $\alpha_2 > 0$, $\alpha_3 < 0$, the region $\{f < \varepsilon\}$ is a volume lying on both sides of $\mathbf{x}$ and narrowing to a point there, like a double cone (see Figure A1.6.4 and section 1.7.1). Any volume or surface lying in $\{f < \varepsilon\} + \mathbf{x}$ and including $\mathbf{x}$ can only do so by virtue of $\mathbf{x}$ lying on its boundary. Thus the volume or surface is deformable to lie in $\{f < \varepsilon\}$ (excluding $\mathbf{x}$) merely by slipping its

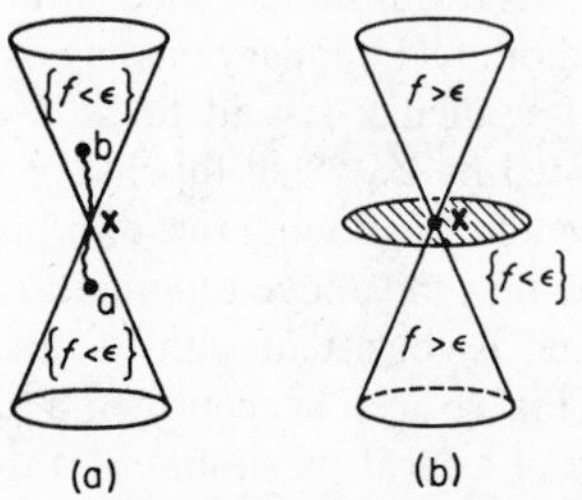

FIGURE A1.6.4. (a) For a critical point of type 1, $Z_n = Z_1$ is the line (a, b). (b) For a critical point of type 2, $Z_n = Z_2$ is the shaded disc.

boundary off $\mathbf{x}$—the boundary does not have to *pass over* $\mathbf{x}$. However, the line (a, b) shown in Figure A1.6.4(a) includes $\mathbf{x}$, but if it is deformed to exclude $\mathbf{x}$, one of its boundaries (i.e. a point a or b) must pass over $\mathbf{x}$. The existence of this line classifies the critical point as type 1 topologically.

Type 2. For a saddle-point where $\alpha_1 > 0$, α_2, $\alpha_3 < 0$ in equation (A1.6.2) we have the situation shown in Figure A1.6.4(b). One can see that any point, line or volume containing $\mathbf{x}$ may be slipped off $\mathbf{x}$ into $\{f < \varepsilon\}$ without its boundary passing over $\mathbf{x}$. This is not true of the disc-like region shown in Figure A1.6.4, however. Thus consistency of definitions is again established.

Type 3. If $\mathbf{x}$ is at a maximum, the whole volume immediately surrounding $\mathbf{x}$ is in $\{f < \varepsilon\}$ and so there are no regions Z_0, Z_1 or Z_2. There is, however, a region Z_3; any volume enclosing Z and part of the region $\{f < \varepsilon\}$ cannot be moved entirely into $\{f < \varepsilon\}$ without its boundary surface passing over $\mathbf{x}$.

Finally we remark that if $\mathbf{x}$ is at a point for which $\nabla f \neq 0$, any region of D, lying in $\{f < \varepsilon\} + \mathbf{x}$, can only include $\mathbf{x}$ in the sense of having it on its boundary, so that to move the region into $\{f < \varepsilon\}$ we merely slip the boundary off $\mathbf{x}$, it being unnecessary to make it pass over $\mathbf{x}$. Thus a point not defined as critical by the analytical definition is not defined as critical by the topological one either.

A1.6.4 Morse relations

These relate the numbers of regions Z_n to the connectivity numbers. To find them, let us take a parameter ε varying between the absolute minimum and maximum of $f(x)$. As ε is increased from its minimum we produce a region $\{f > \varepsilon\}$ which has connectivity numbers defined in an identical way to those we defined for the domain D in section A1.6.2.

Let us take a particular point $\mathbf{x}$ and look at the variation of the connectivity number of $\{f < \varepsilon\}$ as ε is varied so that the boundary of $\{f < \varepsilon\}$ passes over $\mathbf{x}$.

Let us take, therefore, $f(x) = \varepsilon$. Now if $\mathbf{x}$ is not at a critical point, the connectivity numbers of $\{f < \varepsilon\} + \mathbf{x}$ are the same as those of $\{f < \varepsilon\}$ since every region of $\{f < \varepsilon\} + \mathbf{x}$ is deformable into $\{f < \varepsilon\}$, while remaining during deformation in $\{f < \varepsilon\} + \mathbf{x}$, from the topological definition of a critical point.

On the other hand, suppose $\mathbf{x}$ is a critical point, so that there is associated with it a region as defined in the last section.

(i) Suppose the boundary of Z_n encloses a second region U_n lying entirely in $\{f < \varepsilon\}$; that is, the boundary of Z_n is also the boundary of a second region U_n. The combined region $Z_n + U_n$ has no boundary, for Z_n and U_n have a common boundary. Further, $Z_n + U_n$ does not enclose a region (of dimensions $n+1$) of $\{f < \varepsilon\} + \mathbf{x}$, for if it did, one could deform U_n continuously into Z_n while remaining in $\{f < \varepsilon\} + \mathbf{x}$ (so that the region enclosed collapsed), that is, a region lying entirely in $\{f < \varepsilon\}$ could be deformed to include $\mathbf{x}$ without its boundary crossing $\mathbf{x}$. For a moment let us suppose this to be possible. Then one can, by reversing the process of deformation, deform Z_n to lie entirely in $\{f < \varepsilon\}$ without its boundary passing over $\mathbf{x}$. But this conflicts with the definition of Z_n and hence the original supposition is invalid. The region $Z_n + U_n$ is thus a closed n-dimensional region C^n of $\{f < \varepsilon\} + \mathbf{x}$. It is not deformable into any region of $\{f < \varepsilon\}$ by analogous reasoning to the above. $\{f < \varepsilon\}$ may thus be called an "increasing region", for the connectivity number of $\{f < \varepsilon\}$ increases by one as the boundary of $\{f < \varepsilon\}$ passes over $\mathbf{x}$. We shall call the number of critical points for which this occurs N_+^n.

(ii) Suppose the boundary of Z_n does not enclose a region of $\{f < \varepsilon\}$. It is then a region C^{n-1} of $\{f < \varepsilon\}$ by definition, but not such a region for $\{f < \varepsilon\} + \mathbf{x}$, simply because it encloses Z_n. $\{f < \varepsilon\}$ may thus be called a "decreasing region", for the connectivity number R_{n-1} of $\{f < \varepsilon\}$ decreases by one as its boundary passes over $\mathbf{x}$. We shall call the number of critical points for which this occurs N_-^n.

Now it is obvious that

$$R_n = N_+^n - N_-^{n+1}; \quad N^n = N_+^n + N_-^n, \tag{A1.6.3}$$

which gives

$$N^n - R_n = N_-^n + N_-^{n+1}. \tag{A1.6.4}$$

If D is l-dimensional, $N_-^0 = N_-^{l+1} = 0$, there being no decreasing regions of dimension zero and no $(l+1)$ regions at all. Additionally, $N^1 = N_-^1$. Hence applying equation (A1.6.4),

$$N^0 - R_0 = N_-^0 + N_-^1 = N_-^1 \geqslant 0 \tag{A1.6.5}$$

and, further,

$$N^1 - R_1 = N_-^1 + N_-^2 = N^0 - R_0 + N_-^2, \tag{A1.6.6}$$

that is,

$$N^1 - N^0 \geqslant R_1 - R_0. \tag{A1.6.7}$$

From equations (A1.6.4) and (A1.6.6),

$$N^2 - R_2 = N_-^2 + N_-^3 = N^1 - R_1 - N^0 + R_0 + N_-^3, \tag{A1.6.8}$$

that is,

$$N^2 - N^1 + N^0 \geqslant R_2 - R_1 + R_0. \tag{A1.6.9}$$

We may proceed in similar fashion until we obtain the complete set of Morse relations, which are, for an l-dimensional region,

$$\left.\begin{aligned} N^0 &\geqslant R_0, \\ N^1 - N^0 &\geqslant R_1 - R_0, \\ N^2 - N^1 + N^0 &\geqslant R_2 - R_1 + R_0, \\ &\cdots\cdots\cdots \\ \sum_{i=0}^{l} (-1)^i N^i &= \sum_{i=0}^{l} (-1)^i R_i. \end{aligned}\right\} \tag{A1.6.10}$$

A1.6.5 Non-analytic critical points

As stated in the text, the functions to which our analysis is applied, the electron energy $E(\mathbf{k})$ or the squared frequency $\omega^2(\mathbf{k})$ for phonons (see Chapter 3), are not analytic when there are degenerate eigenvalues. This situation can be included in our discussion, however, as was first pointed out by Phillips (1956). Non-analytic minima and maxima are easily disposed of. If we look back at section A1.6.2, we can see that their type numbers in the topological classification are again 0 and 3 respectively, and no modification to our arguments is necessary. Modifications may, however, be required for non-analytic saddle-points, which we call *fluted* points. For example, the situation shown in Figure A1.6.4(a), where two regions in which $\{f < \varepsilon\}$ meet at a point, may be replaced by one in which n such regions meet at a point. In Figure A1.6.5 we have four such regions. There are now six lines C^1 with ends lying in different sections: from the figure they are the lines (a, b), (a, c), (a, d), (b, c), (b, d), (c, d). However, there are only $3(= n-1)$ C^1's not deformable one into the other without their boundaries passing *over* the critical point: for example, (b, c) is deformable so that it is contained in (a, b) and (a, c). Thus there are $n-1$ regions Z_1. Similarly if we obtain a point with n sectors $\{f > \varepsilon\}$ we find that there are $n-1$ regions Z_2.

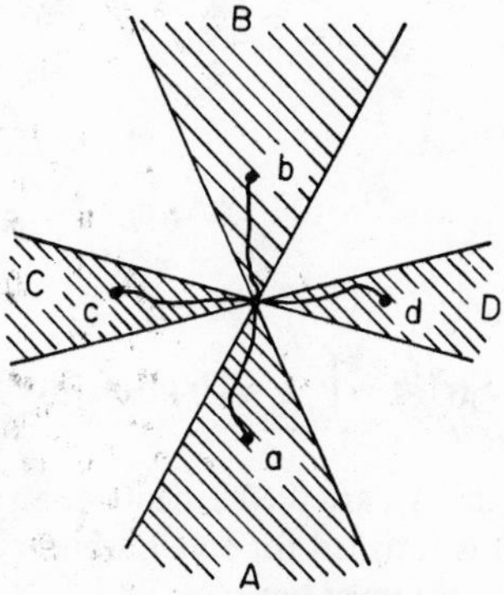

FIGURE A1.6.5. Generalization of Figure A1.6.4 to the case when four regions A, B, C, D for which $\{f<\varepsilon\}$ meet at a point. There are now six lines with ends lying in different regions A, B, C, D. However, only three of these lines are required, say (a,b), (a,c) and (a,d), to reproduce the Figure.

These points, as examination of the arguments of section A1.6.3 and A1.6.4 show, must be counted $n-1$ times in obtaining N^1 and N^2. For further arguments as to the different types of point possible, reference should be made to the work of Phillips (1956).

APPENDIX A 1.7 KORRINGA–KOHN–ROSTOKER METHOD

A1.7.1 Variational formulation

(i) *General equations.* We may readily verify directly that if we define

$$\Lambda \equiv \int_\Omega \psi^*(\mathbf{r})\, V(\mathbf{r})\, \psi(\mathbf{r})\, d\mathbf{r} - \int_\Omega \int_{\Omega'} \psi^*(\mathbf{r})\, V(\mathbf{r})\, G(\mathbf{r}, \mathbf{r}')\, V(\mathbf{r}')\, \psi(\mathbf{r}')\, d\mathbf{r}\, d\mathbf{r}' \tag{A1.7.1}$$

then

$$\delta\Lambda = 0 \tag{A1.7.2}$$

for small arbitrary deviations of ψ from the solution of the integral equation (1.11.25) of the main text.

Let us now apply the Rayleigh–Ritz technique. We insert a trial function of the form

$$\psi = \sum_{0}^{n} c_i \phi_i, \tag{A1.7.3}$$

whereupon we find

$$\Lambda = \sum_{i,j=0}^{n} c_i^* \Lambda_{ij} c_j \tag{A1.7.4}$$

with

$$\Lambda_{ij} = \int_{\Omega} \phi_i^*(\mathbf{r}) V(\mathbf{r}) \phi_j(\mathbf{r}) \, d\mathbf{r} - \int_{\Omega} \int_{\Omega'} \phi_i^*(\mathbf{r}) V(\mathbf{r}) G(\mathbf{rr}') V(\mathbf{r}') \phi_j(\mathbf{r}') \, d\mathbf{r} \, d\mathbf{r}'. \tag{A1.7.5}$$

It should be noted that the Λ_{ij} are the elements of an Hermitian matrix.

In order that (A1.7.2) is satisfied for small arbitrary variations of the c_i's we see from (A1.7.4) that we must have

$$\sum_{j=0}^{n} \Lambda_{ij} c_j = 0 \tag{A1.7.6}$$

which are compatible only if

$$\det|\Lambda_{ij}| = 0. \tag{A1.7.7}$$

Equation (A1.7.6), taken together with equation (A1.7.4), implies that

$$\Lambda(\psi, \mathbf{k}, E) = 0, \tag{A1.7.8}$$

which can also be seen immediately from equation (A1.7.1) if ψ is made to satisfy the integral equation.

On the other hand, the approximate energy E_t given by the trial function ψ_t is such that

$$\Lambda(\psi_t, \mathbf{k}, E_t) = 0. \tag{A1.7.9}$$

Now if we have

$$\psi_t = \psi + \varepsilon\chi, \tag{A1.7.10}$$

where ε is a small parameter, it must follow, because $\delta\Lambda = 0$, that

$$\Lambda(\psi_t, \mathbf{k}, E) = O(\varepsilon^2) \tag{A1.7.11}$$

and comparison with equation (A1.7.9) shows that

$$E - E_t = O(\varepsilon^2). \tag{A1.7.12}$$

Thus the error in the energy is of second order compared with that in the trial function.

(ii) *Muffin-tin potentials.* We now take the potential to be given by equations (1.11.28) and (1.11.29). To deal properly with the singularities of G, we must then use a limiting procedure in defining Λ. We take

$$\Lambda = \lim_{\varepsilon\to 0} \Lambda_\varepsilon, \tag{A1.7.13}$$

where

$$\Lambda_\varepsilon = \int_{r<r_i-2\varepsilon} d\mathbf{r}\, \psi^*(\mathbf{r})\, V(\mathbf{r}) \left[\psi(\mathbf{r}) - \int_{r'<r_i-\varepsilon} d\mathbf{r}'\, G(\mathbf{r}\mathbf{r}')\, V(\mathbf{r}')\, \psi(\mathbf{r}')\right]. \tag{A1.7.14}$$

Now we proved in the main text [cf. equations (1.11.35) and (1.11.36)] that provided $r<r'-\varepsilon$,

$$\psi(\mathbf{r}) - \int_{r'<r_i-\varepsilon} G(\mathbf{r}\mathbf{r}')\, V(\mathbf{r}')\, \psi(\mathbf{r}')\, d\mathbf{r}' = -\int_{r=r_i-\varepsilon} dS' \left[G(\mathbf{r}\mathbf{r}')\frac{\partial\psi(\mathbf{r}')}{\partial r'} - \psi(\mathbf{r}')\frac{\partial G(\mathbf{r}\mathbf{r}')}{\partial r'}\right]. \tag{A1.7.15}$$

The same procedure gives, if $r'>r_i-2\varepsilon$,

$$\int_{r<r_i-2\varepsilon} G(\mathbf{r}\mathbf{r}')\, V(\mathbf{r}')\, \psi^*(\mathbf{r})\, d\mathbf{r} = \int_{r=r_i-2\varepsilon} dS \left[\frac{\partial\psi^*(\mathbf{r})}{\partial r} G(\mathbf{r}\mathbf{r}') - \psi^*(\mathbf{r})\frac{\partial G(\mathbf{r}\mathbf{r}')}{\partial r}\right] \tag{A1.7.16}$$

and we can see that Λ_ε may be written

$$\Lambda_\varepsilon = \int_{r\leqslant r_i-2\varepsilon} dS \int_{r'<r_i-\varepsilon} dS' \left[\frac{\partial\psi^*(\mathbf{r})}{\partial r} - \psi^*(r)\frac{\partial}{\partial r}\right] \times \left[\psi^*(\mathbf{r}')\frac{\partial G(\mathbf{r}\mathbf{r}')}{\partial r'} - G(\mathbf{r}\mathbf{r}')\frac{\partial\psi(\mathbf{r}')}{\partial r'}\right]. \tag{A1.7.17}$$

One now inserts equation (1.11.30) for ψ and equation (1.11.45) for G into Λ_ε. When the limit $\varepsilon\to 0$ is taken, the coefficient of $c^*_{lm}\, c_{l'm'}$ is found to be

$$\Lambda_{lm,l'm'} = (L_l\, j_l - j_l')\, [(A_{lm,l'm'}\, j_{l'}' + \kappa\delta_{ll'}\, \delta_{mm'}\, n_{l'}') - A_{lm,l'm'}\, j_{l'} + \kappa\delta_{ll'}\, \delta_{mm'}\, n_{l'}\, L_{l'}] \tag{A1.7.18}$$

where the symbols have been explained in the main text [equations (1.11.47) and (1.11.48)]. Inserting these values of the Λ's into equation (A1.7.6), we obtain equation (1.11.46); that is the same result we obtained directly from the integral equation.

A1.7.2 Expressions for structure constants

Comparison of the two expressions

$$G(\mathbf{r}\mathbf{r}') = -\frac{1}{\Omega}\sum_{\mathbf{K}} \frac{\exp[i(\mathbf{K}+\mathbf{k}).(\mathbf{r}-\mathbf{r}')]}{\frac{1}{2}(\mathbf{k}+\mathbf{K})^2 - E} \tag{A1.7.19}$$

and [cf. equation (1.11.45)]

$$G(\mathbf{rr}') = \sum_{lm}\sum_{l'm'} [A_{lm,l'm'}\, j_l(\kappa r)\, j_{l'}(\kappa r') + \kappa \delta_{ll'}\,\delta_{mm'}\, j_l(\kappa r)\, n_l(\kappa r')]\, Y_{lm}(\theta\phi)\, Y^*_{l'm'}(\theta'\phi') \tag{A1.7.20}$$

yields explicit expressions for the A's when the individual terms $\exp[i(\mathbf{K}+\mathbf{k}).(\mathbf{r}'-\mathbf{r})]$ of the right-hand side of (A1.7.19) are expressed in spherical harmonics using Bauer's expansion. Specifically, we find

$$A_{lm,l'm'} = -\frac{(4\pi)^2}{\Omega} i^{(l-l')}[j_l(\kappa r) j_{l'}(\kappa r')]^{-1} \sum_{\mathbf{K}}$$

$$\times \frac{j_l(|\mathbf{K}+\mathbf{k}|r) j_{l'}(|\mathbf{K}+\mathbf{k}|r') Y_{lm}(\mathbf{k}+\mathbf{K}) Y_{l'm'}(\mathbf{k}+\mathbf{K})}{\frac{1}{2}(\mathbf{k}+\mathbf{K})^2 - E} - \kappa \delta_{ll'}\,\delta_{mm'} \frac{n_l(\kappa r')}{j_l(\kappa r)}. \tag{A1.7.21}$$

Here

$$Y_{lm}(\mathbf{p}) = Y_{lm}(\theta_p, \phi_p), \tag{A1.7.22}$$

where θ_p, ϕ_p are the polar angles of the vector $\mathbf{p}$. The sum on the right-hand side of equation (A1.7.21) is of course independent of the particular choice of r and r'.

From the Hermitian property

$$G(\mathbf{rr}') = G^*(\mathbf{r}'\mathbf{r}) \tag{A1.7.23}$$

it follows that

$$A_{lm,l'm'} = A^*_{l'm',lm} \tag{A1.7.24}$$

but, quite apart from this, the A's are not all linearly independent, as we shall see below. It follows that the use of equation (A1.7.21) is not the most efficient way to calculate the A's. Rather, they should be calculated from a smaller number of independent constants $D_{\lambda\mu}$, which we determine below.

From equation (A1.7.19), it is readily seen that $G(\mathbf{rr}')$ is a function of a single vector $\mathbf{r}'' = \mathbf{r}-\mathbf{r}'$. In order to make full use of this fact, we recall equations (1.11.27) and (1.11.40) of the main text and write

$$G(\mathbf{rr}') = G(\mathbf{r}'') = -\frac{1}{\Omega}\sum_{\mathbf{K}} \frac{\exp[i(\mathbf{K}+\mathbf{k}).\mathbf{r}'']}{\frac{1}{2}(\mathbf{K}+\mathbf{k})^2 - E}$$

$$= -\frac{1}{4\pi}\frac{\cos \kappa r''}{r''} + \sum_{\lambda''\mu''} D_{\lambda''\mu''}\, j_\lambda(\kappa r'') Y_{\lambda\mu}(\theta'', \phi''), \tag{A1.7.25}$$

where

$$D_{\lambda\mu} = -\frac{4\pi}{\Omega} i^{\lambda}[j_{\lambda}(\kappa r'')]^{-1} \sum_{\mathbf{K}} \frac{j_{\lambda}(|\mathbf{K}+\mathbf{k}|r'')\, Y^{*}_{\lambda\mu}(\mathbf{K}+\mathbf{k})}{\frac{1}{2}(\mathbf{K}+\mathbf{k})^2 - E} + \frac{\kappa}{4\pi}\delta_{\lambda 0}\,\delta_{\mu 0}\cot\kappa r'' \tag{A1.7.26}$$

and

$$D_{\lambda,-\mu} = (-1)^{\lambda} D^{*}_{\lambda\mu}. \tag{A1.7.27}$$

A1.7.3 Secular equation in effective potential form

We wish now to show that equation (1.11.84) can be transformed into the form

$$\det\left|\left(\frac{|\mathbf{k}-\mathbf{K}|^2}{2} - E\right)\delta(\mathbf{K},\mathbf{K}') + \Gamma_{\mathbf{KK}'}\right| = 0 \tag{A1.7.28}$$

and to determine the "effective potential" resulting from the KKR method. The argument follows Ziman (1965). It will be convenient to introduce a modified phase shift η_l' defined by

$$\cot\eta_l' = \cot\eta_l - \frac{n_l(\kappa r_i)}{j_l(\kappa r_i)}. \tag{A1.7.29}$$

Then equation (1.11.84) can be written, using equation (A1.7.21) for $A_{lm,l'm'}$, as

$$\det\left|\sum_{\mathbf{K}} \frac{F_{lm}(\mathbf{K})\,F_{l'm'}(\mathbf{K}')}{\frac{1}{2}(\mathbf{k}-\mathbf{K})^2 - E} - \kappa\cot\eta_l'\,\delta_{ll'}\,\delta_{mm'}\right| = 0, \tag{A1.7.30}$$

where

$$F_{lm}(\mathbf{K}) = \frac{4\pi}{\Omega^{\frac{1}{2}}} \frac{j_l(|\mathbf{k}-\mathbf{K}|r_i)}{j_l(\kappa r_i)} Y_{lm}(\mathbf{k}-\mathbf{K}). \tag{A1.7.31}$$

It should be noted that in using equation (A1.7.21) we have chosen $r = r' = r_i$. We now divide (A1.7.30) through, row and column, by $(\kappa\cot\eta_l')^{\frac{1}{2}}$, when we find

$$\det\left|\sum_{\mathbf{K}} \Phi_{lm}(\mathbf{K})\,\tilde{\Phi}_{l'm'}(\mathbf{K}) - \delta_{ll'}\,\delta_{mm'}\right| = 0, \tag{A1.7.32}$$

where the matrix $\Phi_{lm}(\mathbf{K})$, with lm labelling the rows and $\mathbf{K}$ the columns, is defined by

$$\Phi_{lm}(\mathbf{K}) = \frac{F_{lm}(\mathbf{K})}{[\kappa\cot\eta_l'(\frac{1}{2}|\mathbf{k}-\mathbf{K}|^2 - \mathrm{E})]^{\frac{1}{2}}} \tag{A1.7.33}$$

and $\tilde{\Phi}_{lm}(\mathbf{K})$ is the transpose of $\Phi_{lm}(\mathbf{K})$.

From (A1.7.32), we see that the matrix $\sum_{\mathbf{K}} \Phi_{lm}(\mathbf{K})\, \tilde{\Phi}_{l'm'}(\mathbf{K})$ has eigenvalues unity, and hence if α_{lm} is an element of an eigenvector, then

$$\sum_{l'm'} \sum_{\mathbf{K}} \Phi_{lm}(\mathbf{K})\, \tilde{\Phi}_{l'm'}(\mathbf{K})\, \alpha_{l'm'} = \alpha_{lm}. \tag{A1.7.34}$$

Let us now introduce another column vector with elements $\beta_{\mathbf{K}}$ defined by

$$\sum_{l'm'} \tilde{\Phi}_{l'm'}(\mathbf{K})\, \alpha_{l'm'} = \beta_{\mathbf{K}}. \tag{A1.7.35}$$

Substituting this in (A1.7.34) we have

$$\sum_{\mathbf{K}} \Phi_{lm}(\mathbf{K})\, \beta_{\mathbf{K}} = \alpha_{lm}. \tag{A1.7.36}$$

Multiplying by the matrix $\tilde{\Phi}_{lm}(\mathbf{K}')$ we find

$$\begin{aligned}\sum_{lm} \sum_{\mathbf{K}} \tilde{\Phi}_{lm}(\mathbf{K}')\, \Phi_{lm}(\mathbf{K})\, \beta_{\mathbf{K}} &= \sum_{lm} \tilde{\Phi}_{lm}(\mathbf{K}')\, \alpha_{lm} \\ &= \beta_{\mathbf{K}'}. \end{aligned} \tag{A1.7.37}$$

This shows that the matrix $\sum_{lm} \tilde{\Phi}_{lm}(\mathbf{K}')\, \Phi_{lm}(\mathbf{K})$ has eigenvalues unity and, reversing the argument used above, the secular determinant is

$$\det \left| \sum_{lm} \tilde{\Phi}_{lm}(\mathbf{K})\, \Phi_{lm}(\mathbf{K}') - \delta(\mathbf{K}, \mathbf{K}') \right| = 0. \tag{A1.7.38}$$

We can rewrite this determinant in terms of the $F_{lm}(\mathbf{K})$'s by use of equation (A1.7.33). On multiplying row and column by $(\frac{1}{2}|\mathbf{k}-\mathbf{K}|^2 - E)^{\frac{1}{2}}$ we then obtain

$$\det \left| \sum_{lm} \frac{F_{lm}(\mathbf{K})\, F_{lm}(\mathbf{K}')}{\kappa \cot \eta_l'} - \left(\frac{|\mathbf{k}-\mathbf{K}|^2}{2} - E \right) \delta(\mathbf{K}', \mathbf{K}) \right| = 0 \tag{A1.7.39}$$

which is of the required form (A1.7.28). From the definition (A1.7.31) of $F_{lm}(\mathbf{K})$ the matrix elements Γ of the effective potential are given by

$$\begin{aligned}\Gamma_{\mathbf{KK}'} &= \Gamma(\mathbf{k}-\mathbf{K}, \mathbf{k}-\mathbf{K}') \\ &= -\frac{4\pi}{\Omega\kappa} \sum_l \frac{(2l+1) \tan \eta_l'\, j_l(|\mathbf{k}-\mathbf{K}|r)\, j_l(|\mathbf{k}-\mathbf{K}'|r')}{j_l(\kappa r)\, j_l(\kappa r')} P_l(\cos \theta_{\mathbf{K},\mathbf{K}'}),\end{aligned} \tag{A1.7.40}$$

where $\theta_{\mathbf{K},\mathbf{K}'}$ is the angle between $\mathbf{k}-\mathbf{K}$ and $\mathbf{k}-\mathbf{K}'$.

APPENDIX A1.8 RELATIVISTIC APW METHOD

(i) *Variational principle.* In units for which $m = c = 1$ the relativistic Hamiltonian of equation (1.11.138) becomes

$$H = -\boldsymbol{\gamma} \cdot \mathbf{p} + \gamma_4 + V(\mathbf{r}). \tag{A1.8.1}$$

We seek to obtain a variational expression for the equation

$$H\Psi = E_T \Psi \tag{A1.8.2}$$

for solutions Ψ which obey Bloch boundary conditions and which are discontinuous across a sphere of radius r_i within the unit cell. We first note that for such functions

$$\int_\Omega \nabla.(\Phi^\dagger \boldsymbol{\gamma}.\Psi)\, d\mathbf{r} = \int_S (\Phi_i^\dagger \boldsymbol{\gamma}.\hat{\mathbf{n}}\Psi^{\mathrm{i}} - \Phi^{\mathrm{o}\dagger} \boldsymbol{\gamma}.\hat{\mathbf{n}}\Psi^{\mathrm{o}})\, dS, \tag{A1.8.3}$$

where the surface integration is over the sphere S of radius r_i and $\hat{\mathbf{n}}$ is the outward directed normal to that surface. As in the main text, superscripts i and o denote functions with arguments in the regions inside and outside the sphere S respectively. Using equation (A1.8.3) it is easy to show that

$$\int_\Omega \Phi^\dagger \boldsymbol{\gamma}.\nabla\Psi\, d\mathbf{r} = -\int_\Omega (\Psi^\dagger \boldsymbol{\gamma}.\nabla\Phi)^*\, d\mathbf{r} + \int_S (\Phi^{\dagger\mathrm{i}} \boldsymbol{\gamma}.\hat{\mathbf{n}}\Psi^{\mathrm{i}} - \Phi^{\mathrm{o}\dagger} \boldsymbol{\gamma}.\hat{\mathbf{n}}\Psi^{\mathrm{o}})\, dS. \tag{A1.8.4}$$

Since $\mathbf{p} \equiv -i\nabla$ and $\gamma_4 + mc^2$ is Hermitian, we now obtain

$$\int_\Omega \Phi^\dagger H\Psi\, d\mathbf{r} = \int_\Omega (\Psi^\dagger H\Phi)^*\, d\mathbf{r} + i\int_S (\Phi_i^\dagger \boldsymbol{\gamma}.\Psi^{\mathrm{i}}\hat{\mathbf{n}} - \Phi_{\mathrm{o}}^\dagger \boldsymbol{\gamma}.\hat{\mathbf{n}}\Psi^{\mathrm{o}})\, dS. \tag{A1.8.5}$$

Using this result, it can be verified after some calculation [cf. the non-relativistic discussion of Schlosser and Marcus (1963)] that equation (1.11.90) can be obtained by application of the variational principle to the expression (Loucks, 1965)

$$E_T \int_\Omega \Psi^\dagger \Psi\, d\mathbf{r} = \int_\Omega \Psi^\dagger H\Psi\, d\mathbf{r} + \frac{i}{2}\int_S (\Psi^{\mathrm{o}} + \Psi^{\mathrm{i}})^\dagger \boldsymbol{\gamma}.\hat{\mathbf{n}}(\Psi^{\mathrm{o}} - \Psi^{\mathrm{i}})\, dS. \tag{A1.8.6}$$

(ii) *Matrix elements.* We again follow Loucks (1965) in obtaining the secular equation resulting from the application of equation (A1.8.6).

We expand Ψ in terms of relativistic APWs (see equation (1.11.179)) and then we obtain the secular equation

$$\left| M\binom{n'm'}{nm} \right| = \left| H\binom{n'm'}{nm} - E_T Q\binom{n'm'}{nm} - S\binom{n'm'}{nm} \right| = 0, \tag{A1.8.7}$$

where

$$H\binom{n'm'}{nm} = \int_\Omega \Psi_{nm}^\dagger H\Psi_{n'm'}\, d\mathbf{r}, \tag{A1.8.8}$$

$$Q\binom{n'm'}{nm} = \int_\Omega \Psi_{nm}^\dagger \Psi_{n'm'}\, d\mathbf{r} \tag{A1.8.9}$$

and

$$S\begin{pmatrix} n'm' \\ nm \end{pmatrix} = -\frac{i}{2}\int_S (\Psi^o_{nm}+\Psi^i_{nm})^\dagger \boldsymbol{\gamma}.\mathbf{r}(\Psi^o_{n'm'}-\Psi^i_{n'm'})\,dS. \quad \text{(A1.8.10)}$$

If we take the eigenvalue E in equation (1.11.180) for the central field orbitals, to be identical with the eigenvalue $E(\mathbf{k})$, the integration of $(H-V)$ over the interior of the sphere of radius r_i is zero. In the outer region $(r>r_i)$ the eigenvalue is k_n^* as defined by equation (1.11.176), and so we can write

$$H\begin{pmatrix} n'm' \\ nm \end{pmatrix} - E_T\, Q\begin{pmatrix} n'm' \\ nm \end{pmatrix} = (k_{n'}^*-E_T)\, Q^0\begin{pmatrix} n'm' \\ nm \end{pmatrix}, \quad \text{(A1.8.11)}$$

where the superscript 0 on Q indicates the integration in equation (A1.8.9) is through the region $r>r_i$ only, instead of the entire unit cell. Since Ψ^o_{nm} is given by equation (1.11.177)

$$Q^0\begin{pmatrix} n'm' \\ nm \end{pmatrix} = \Omega\delta_{nn'}\,\delta_{mm'} - \frac{4\pi r_i^2 j_1(|\mathbf{k}_{n'}-\mathbf{k}_n|\,r_i)}{|\mathbf{k}_{n'}-\mathbf{k}_n|}\left(\frac{k_n^*+1}{2k_n^*}\right)^{\frac{1}{2}}\left(\frac{k_{n'}^*+1}{2k_{n'}^*}\right)^{\frac{1}{2}}$$

$$\times\left\{\left[1+\frac{\mathbf{k}_{n'}.\mathbf{k}_n}{(k_n^*+1)(k_{n'}^*+1)}\right]\delta_{mm'} + \frac{i\mathbf{k}_n\times\mathbf{k}_{n'}.\langle m|\boldsymbol{\sigma}|m'\rangle}{(\mathbf{k}_n^*+1)(\mathbf{k}_{n'}^*+1)}\right\}. \quad \text{(A1.8.12)}$$

To evaluate $S\binom{n'm'}{nm}$ from equation (A1.8.10) we take equation (1.11.178) for Ψ^i_{nm} and Ψ^o_{nm} from equation (1.11.177). Then

$$S\begin{pmatrix} n'm' \\ nm \end{pmatrix} = 4\pi r_i^2\left(\frac{k_n^*+1}{2k_n^*}\right)^{\frac{1}{2}}\left(\frac{k_{n'}^*+1}{2k_{n'}^*}\right)^{\frac{1}{2}}\sum_\kappa D_\kappa\begin{pmatrix} n'm' \\ nm \end{pmatrix} j_l(k_n r_i)$$

$$\times\left\{\left(\frac{k_n S_\kappa}{k_n^*+1}\right) j_{l'}(k_n r_i) - j_l(k_n r_i)\frac{f_\kappa(r_i)}{g_\kappa(r_i)}\right\}, \quad \text{(A1.8.13)}$$

where

$$D_\kappa\begin{pmatrix} n'm' \\ nm \end{pmatrix} = 4\pi\sum_\mu c(l\tfrac{1}{2}j;\,\mu-m;\,m)\,c(l\tfrac{1}{2}j;\,\mu-m';\,m')$$

$$\times\, Y_l^{(\mu-m')\dagger}(\hat{\mathbf{k}}_{n'})\,Y_l^{(\mu-m)}(\hat{\mathbf{k}}_n). \quad \text{(A1.8.14)}$$

The c's are defined in equation (1.11.184).

We finally convert our equations from ones for which $m=c$ to atomic units. In these units $c=137$ and, neglecting terms of second order in the fine structure constant $\alpha = 1/137$, we then obtain

$$M\begin{pmatrix} n'm' \\ nm \end{pmatrix} = (k_{n'}^2 - E)\,\Omega_{nn'}\,\delta_{mm'} + 4\pi r_i^2 \sum_{\kappa} D_{\kappa}\begin{pmatrix} n'm' \\ nm \end{pmatrix} j_l(k_n r_i)$$

$$\times \left\{ j_l(k_{n'} r_i) \left(c\frac{f_{\kappa}(r_i, E)}{g_{\kappa}(r_i, E)} - j_{l'}(k_{n'} r_i)\, k_{n'}\, S_{\kappa} \right\}, \qquad \text{(A1.8.15)}$$

where

$$\Omega_{nn'} = \Omega\delta_{nn'} - 4\pi r_i^2 \frac{j_1(|\mathbf{k}_{n'} - \mathbf{k}_n| r_i)}{|\mathbf{k}_{n'} - \mathbf{k}_n|}. \qquad \text{(A1.8.16)}$$

By applying the divergence theorem to plane waves in the region $r_i > r$, it is readily shown that

$$\Omega_{nn'}\,\delta_{mm'}(k_{n'}^2 - k_n^2) = 4\pi r_i^2 \sum_{\kappa} S_{\kappa}\, D_{\kappa}\begin{pmatrix} n'm' \\ nm \end{pmatrix}$$

$$\times \{ k_{n'} j_l(k_n r_i) j_{l'}(k_{n'} r_i) - k_n j_l(k_{n'} r_i) j_{l'}(k_n r_i) \} \qquad \text{(A1.8.17)}$$

which, when substituted into equation (A1.8.15) yields equation (1.11.191) of the main text.

APPENDIX A2.1 EQUIVALENCE OF HIGH-DENSITY PROCEDURES FOR ELECTRON GAS

In this appendix we shall examine more closely the relationships between the SCF, RPA and Sawada schemes. We should define what we mean by a random phase approximation; stemming from the work of Bohm and Pines mentioned in section 1.2, we extend the term to mean any approximation in which, given two momentum transfers $\mathbf{q}$ and $\mathbf{q}'$, we neglect the term unless $\mathbf{q} = \mathbf{q}'$. We shall show how the definition works in a specific case below.

Our particular aim will be to show why the SCF dielectric constant can be used to give the same correlation energy as the Sawada Hamiltonian, but it is also of interest to examine in general the relationship of the SCF equations of motion to the exact ones.

(a) *Equation of motion method.* The "equation of motion" procedure is typified by the way we sought the eigenfunctions of the Sawada Hamiltonian by means of the operator $\Omega_{\mathbf{k}}(\omega)$ (section 2.6.2). We shall further illustrate it here in obtaining the general equations of motion of the operator $a^{\dagger}_{\mathbf{k}+\mathbf{q}}\, a_{\mathbf{k}}$ and make comparisons with the SCF method and the Sawada method. The Heisenberg equation of motion is given by the commutator with the Hamiltonian. We can obtain this by straightforward use of the expressions given in section 2.5, but we shall follow here the elegant exposition of Ehrenreich and Cohen (1959) (to whom the proofs of equivalence we shall give are originally due) since it introduces some useful formalism.

If we introduce the field operator

$$\psi(\mathbf{x}) = \sum_{\mathbf{k}} a_{\mathbf{k}}\, e^{i\mathbf{k}.\mathbf{x}} \tag{A2.1.1}$$

which can be thought of as destroying a particle at position $\mathbf{x}$ whatever its momentum, then it is easily seen that equation (2.5.22) for the density operator may be written in configuration space as

$$\rho_{\text{op}}(\mathbf{x}) = \psi^{\dagger}(\mathbf{x})\, \psi(\mathbf{x}). \tag{A2.1.2}$$

Further, the density matrix operator may be defined as

$$\rho_{\text{op}}(\mathbf{x}'\mathbf{x}) = \psi^{\dagger}(\mathbf{x}')\, \psi(\mathbf{x}) \tag{A2.1.3}$$

and the Hamiltonian of equation (2.5.23) as

$$H = -\left(\frac{\hbar^2}{2m}\right) \int \psi^{\dagger}(\mathbf{x})\, \nabla^2 \psi(\mathbf{x})\, d\mathbf{x} + \frac{1}{2} \int \frac{e^2}{|\mathbf{x}'-\mathbf{x}|} \psi^{\dagger}(\mathbf{x}')\, \psi^{\dagger}(\mathbf{x})\, \psi(\mathbf{x})\, \psi(\mathbf{x}')\, d\mathbf{x}\, d\mathbf{x}'. \tag{A2.1.4}$$

We see here that the use of the field operator reveals a close formal correspondence with the Schrödinger representation. Moreover, if one evaluates the commutator $[\psi, H]$, one finds that the Heisenberg equation of motion yields

$$[\psi, H] = i\hbar \frac{\partial}{\partial t}\psi(\mathbf{x}) = \mathscr{H}(\mathbf{x})\,\psi(\mathbf{x}), \tag{A2.1.5}$$

where

$$\mathscr{H}(\mathbf{x}) = -\frac{\hbar^2}{2m}\nabla^2 + e^2 \int \frac{d\mathbf{x}'\,\rho(\mathbf{x}')}{|\mathbf{x}'-\mathbf{x}|}. \tag{A2.1.6}$$

From equation (A2.1.5) one finds that

$$\hbar i \frac{\partial}{\partial t}\rho_{\text{op}}(\mathbf{x}'\mathbf{x}) = \mathscr{H}(\mathbf{x}')\,\rho_{\text{op}}(\mathbf{x}'\mathbf{x}) - \rho_{\text{op}}(\mathbf{x}'\mathbf{x})\,\mathscr{H}(\mathbf{x}) \tag{A2.1.7}$$

and defining

$$\rho_{\text{op}}(\mathbf{k}, \mathbf{k}+\mathbf{q}) = a^{\dagger}_{\mathbf{k}+\mathbf{q}}\,a_{\mathbf{k}} \tag{A2.1.8}$$

it follows that

$$\begin{aligned} i\hbar \frac{\partial}{\partial t}\rho_{\text{op}}(\mathbf{k}, \mathbf{k}+\mathbf{q}) &= (\varepsilon_{\mathbf{k}} - \varepsilon_{\mathbf{k}+\mathbf{q}})\,\rho_{\text{op}}(\mathbf{k}, \mathbf{k}+\mathbf{q}) \\ &\quad + 4\pi e^2 \sum_{q'} \frac{1}{q'^2} \sum_{\mathbf{k}'} [\rho_{\text{op}}(\mathbf{k}', \mathbf{k}'+\mathbf{q}')\,\rho_{\text{op}}(\mathbf{k}+\mathbf{q}', \mathbf{k}+\mathbf{q}) \\ &\quad - \rho_{\text{op}}(\mathbf{k}, \mathbf{k}+\mathbf{q}-\mathbf{q}')\,\rho_{\text{op}}(\mathbf{k}', \mathbf{k}'+\mathbf{q})]. \end{aligned} \tag{A2.1.9}$$

We now examine how we can reduce this general equation to the approximate form (2.3.56). We recall first that the two approximations of the SCF method involved working in a time-dependent Hartree framework and linearizing the Hamiltonian. We can introduce the first approximation into equation (A2.1.9) by the Hartree factorization: if $\langle\ \rangle$ denotes an average, either over any eigenstate or an ensemble, we put

$$\langle AB \rangle = \langle A \rangle \langle B \rangle, \tag{A2.1.10}$$

A and B being any two operators. Writing

$$\langle \rho_{\text{op}}(\mathbf{k}, \mathbf{k}+\mathbf{q}) \rangle = \langle \mathbf{k} | \rho | \mathbf{k}+\mathbf{q} \rangle, \tag{A2.1.11}$$

thereby making contact with the notation of section 2.3, equation (A2.1.9) becomes

$$i\frac{\partial}{\partial t}\langle \mathbf{k}|\rho|\mathbf{k}+\mathbf{q}\rangle = (\varepsilon_{\mathbf{k}}-\varepsilon_{\mathbf{k}+\mathbf{q}})\langle \mathbf{k}|\rho|\mathbf{k}+\mathbf{q}\rangle$$

$$+4\pi e^2\sum_{q'}\frac{1}{q'^2}\sum_{\mathbf{k}'}(\langle \mathbf{k}'|\rho|\mathbf{k}'+\mathbf{q}'\rangle\langle \mathbf{k}+\mathbf{q}'|\rho|\mathbf{k}+\mathbf{q}\rangle$$

$$-\langle \mathbf{k}|\rho|\mathbf{k}+\mathbf{q}-\mathbf{q}'\rangle\langle \mathbf{k}'|\rho|\mathbf{k}'+\mathbf{q}'\rangle). \tag{A2.1.12}$$

To eliminate the sum over $\mathbf{q}'$ we make the "random phase approximation". We make any term in the sums zero unless $\mathbf{q}'=\mathbf{q}$. Then

$$i\frac{\partial}{\partial t}\langle \mathbf{k}|\rho|\mathbf{k}+\mathbf{q}\rangle = (\varepsilon_{\mathbf{k}}-\varepsilon_{\mathbf{k}+\mathbf{q}})\langle \mathbf{k}|\rho|\mathbf{k}+\mathbf{q}\rangle$$

$$+\frac{4\pi e^2}{q^2}(\langle \mathbf{k}+\mathbf{q}|\rho|\mathbf{k}+\mathbf{q}\rangle-\langle \mathbf{k}|\rho|\mathbf{k}\rangle)\sum_{\mathbf{k}'}\langle \mathbf{k}'|\rho|\mathbf{k}'+\mathbf{q}\rangle. \tag{A2.1.13}$$

Remembering that

$$\langle \mathbf{k}|\rho|\mathbf{k}\rangle = \langle a_{\mathbf{k}}^{\dagger}a_{\mathbf{k}}\rangle \tag{A2.1.14}$$

it is permissible, in the high-density limit, to write

$$\langle \mathbf{k}|\rho|\mathbf{k}\rangle = f_0(\varepsilon_{\mathbf{k}}) \tag{A2.1.15}$$

where $f_0(\varepsilon_{\mathbf{k}})$ is the Fermi–Dirac function as $T\to 0$. Equation (A2.1.13) is then precisely the same as equation (2.3.56). We can see that linearization is equivalent to the random phase approximation with the replacement of diagonal elements $\langle \mathbf{k}|\rho|\mathbf{k}\rangle$ by their values in the free-electron gas.

(b) *Sawada's equations of motion.* We have seen already that the equations for the eigenfrequencies of the SCF and Sawada procedures are the same, and we shall now see that the identity is complete as far as single-particle averages are concerned.

It is readily verified that the commutator $[V_s, d_{\mathbf{q}}^{\dagger}(\mathbf{k})]$ shown in equation (2.6.5) is obtained assuming the boson commutation rules

$$[d_{\mathbf{q}}^{\dagger}(\mathbf{k}), d_{\mathbf{q}'}^{\dagger}(\mathbf{k}')] = [d_{q}(\mathbf{k}), d_{\mathbf{q}'}(\mathbf{k}')] = 0,$$

$$[d_{\mathbf{q}}(\mathbf{k}), d_{\mathbf{q}'}^{\dagger}(\mathbf{k}')] = \delta_{\mathbf{q}\mathbf{q}'}\,\delta_{\mathbf{k}\mathbf{k}'}. \tag{A2.1.16}$$

It is evident that a form of RPA is used here; we ignore terms for which $\mathbf{q}\neq\mathbf{q}'$. We should further notice that, although $d_{\mathbf{q}}^{\dagger}(\mathbf{k})$ has been used only as a particle-hole pair operator, we may evaluate the commutator of $a_{\mathbf{k}+\mathbf{q}}^{\dagger}a_{\mathbf{k}}$ for both $|\mathbf{k}+\mathbf{q}|$, $k>k_{\mathrm{f}}$ or both $|\mathbf{k}+\mathbf{q}|$, $k<k_{\mathrm{f}}$ in exactly the same way as for the

particle-hole pair operators, and the commutator with V_s will be zero using the same approximations:

$$[H_s, a^\dagger_{\mathbf{k}+\mathbf{q}} a_{\mathbf{k}}] = \omega_{\mathbf{q}}(\mathbf{k})\, a^\dagger_{\mathbf{k}+\mathbf{q}} a_{\mathbf{k}} \begin{cases} |\mathbf{k}+\mathbf{q}|, & k > k_f \\ |\mathbf{k}+\mathbf{q}|, & k < k_f. \end{cases} \tag{A2.1.17}$$

From equation (A2.1.17), we obtain (ψ_s being the ground state of H_s, corresponding to energy E_0)

$$H_s(a^\dagger_{\mathbf{k}} a_{\mathbf{k}+\mathbf{q}})\psi_s = [E_0 - \omega_{\mathbf{q}}(\mathbf{k})](a^\dagger_{\mathbf{k}} a_{\mathbf{k}+\mathbf{q}})\psi_s. \tag{A2.1.18}$$

Now $\omega_{\mathbf{q}}(\mathbf{k})$ is positive, so that

$$(a^\dagger_{\mathbf{k}} a_{\mathbf{k}+\mathbf{q}})\psi_s = 0 \begin{cases} |\mathbf{k}+\mathbf{q}|, & k > k_f \\ |\mathbf{k}+\mathbf{q}|, & k < k_f \end{cases} \quad (\mathbf{q} \neq 0). \tag{A2.1.19}$$

One may readily verify that this implies

$$\langle a^\dagger_{\mathbf{k}} a_{\mathbf{k}+\mathbf{q}} \rangle = 0 \begin{cases} |\mathbf{k}+\mathbf{q}|, & k > k_f \\ |\mathbf{k}+\mathbf{q}|, & k < k_f \end{cases} \quad (\mathbf{q} \neq 0). \tag{A2.1.20}$$

Thus, in taking the average of equation (A2.1.17), we may include all terms of the form (A2.1.20) in the sum, and writing

$$\langle a^\dagger_{\mathbf{k}+\mathbf{q}} a_{\mathbf{k}} \rangle = \langle \mathbf{k} | \rho | \mathbf{k}+\mathbf{q} \rangle \tag{A2.1.21}$$

we obtain equation (2.3.56), as we shall see clearly below.

It is also illuminating to obtain the Sawada equations of motion directly from the exact equations (A2.1.9), in order to show that the use of random phase approximation (as defined at the beginning of this section), together with the replacement of the number operator $a^\dagger_{\mathbf{k}} a_{\mathbf{k}}$ by its eigenvalues when applied to the free-electron gas, leads to the restriction to particle–hole pairs in the commutator and so in the effective Hamiltonian.

We again remark that Sawada uses the random phase approximation, so that equation (A2.1.9) is reduced to

$$i\frac{\partial}{\partial t}\rho_{\text{op}}(\mathbf{k}, \mathbf{k}+\mathbf{q}) = (\varepsilon_{\mathbf{k}} - \varepsilon_{\mathbf{k}+\mathbf{q}})\,\rho_{\text{op}}(\mathbf{k}, \mathbf{k}+\mathbf{q})$$
$$+\frac{4\pi e^2}{q^2}[\rho_{\text{op}}(\mathbf{k}+\mathbf{q}, \mathbf{k}+\mathbf{q}) - \rho_{\text{op}}(\mathbf{k}, \mathbf{k})] \sum_{\mathbf{k}'} \rho_{\text{op}}(\mathbf{k}', \mathbf{k}'+\mathbf{q}). \tag{A2.1.22}$$

Now we know that

$$a^\dagger_{\mathbf{k}+\mathbf{q}} a_{\mathbf{k}} = \rho_{\text{op}}(\mathbf{k}, \mathbf{k}+\mathbf{q}) = d^\dagger_{\mathbf{q}}(\mathbf{k}), \quad k < k_f, \quad |\mathbf{k}+\mathbf{q}| > k_f. \tag{A2.1.23}$$

But since in the Sawada commutator [equation (2.6.5)] we find no negative terms in the $d_{\mathbf{q}}(\mathbf{k})$, it is evident that we must put

$$\rho_{\text{op}}(\mathbf{k}+\mathbf{q}, \mathbf{k}+\mathbf{q}) \equiv 0, \quad \rho_{\text{op}}(\mathbf{k}, \mathbf{k}) \equiv 1, \quad k < k_{\text{f}}, \quad |\mathbf{k}+\mathbf{q}| > k_{\text{f}}, \tag{A2.1.24}$$

in order to obtain equation (2.6.5), the commutator of $d_{\mathbf{q}}^{\dagger}(\mathbf{k})$ with the Sawada Hamiltonian. Equation (A2.1.24), we again note, will be a good approximation at high densities. Making the approximation (A2.1.24) for all values of $\mathbf{k}$ and $\mathbf{q}$ in equation (A2.1.22) we obtain

$$i\frac{\partial}{\partial t}\rho_{\text{op}}(\mathbf{k}, \mathbf{k}+\mathbf{q}) = (\varepsilon_{\mathbf{k}} - \varepsilon_{\mathbf{k}+\mathbf{q}})\,\rho_{\text{op}}(\mathbf{k}, \mathbf{k}+\mathbf{q}) + [f_0(\varepsilon_{\mathbf{k}+\mathbf{q}}) - f_0(\varepsilon_{\mathbf{k}})]\frac{4\pi e^2}{q^2}\sum_{\mathbf{k}'}\rho_{\text{op}}(\mathbf{k}', \mathbf{k}+\mathbf{q}), \tag{A2.1.25}$$

the average of which is just equation (2.3.56), the SCF equation. To obtain the Sawada equation, we remember that for both $|\mathbf{k}|$ and $|\mathbf{k}+\mathbf{q}|$ greater or less than k_{f}, equation (A2.1.25) reduces to equation (A2.1.17) from which equation (A2.1.19) is deduced. Taking account of this in the sum on the right-hand side of equation (A2.1.25) we obtain, for the purposes of operating on ψ_s,

$$\begin{aligned}
\sum_{\mathbf{k}}\rho_{\text{op}}(\mathbf{k}, \mathbf{k}+\mathbf{q}) &= \sum_{\mathbf{k}} a_{\mathbf{k}+\mathbf{q}}^{\dagger} a_{\mathbf{k}} = \sum_{\substack{k<k_{\text{f}}\\ |\mathbf{k}+\mathbf{q}|>k_{\text{f}}}} a_{\mathbf{k}+\mathbf{q}}^{\dagger} a_{\mathbf{k}} + \sum_{\substack{k>k_{\text{f}}\\ |\mathbf{k}+\mathbf{q}|<k_{\text{f}}}} a_{\mathbf{k}+\mathbf{q}}^{\dagger} a_{\mathbf{k}} \\
&= \sum_{\substack{k<k_{\text{f}}\\ |\mathbf{k}+\mathbf{q}|<k_{\text{f}}}} a_{\mathbf{k}+\mathbf{q}}^{\dagger} a_{\mathbf{k}} + \sum_{\substack{k<k_{\text{f}}\\ |\mathbf{k}-\mathbf{q}|>k_{\text{f}}}} a_{\mathbf{k}}^{\dagger} a_{\mathbf{k}-\mathbf{q}} = \sum_{\substack{k<k_{\text{f}}\\ |\mathbf{k}+\mathbf{q}|>k_{\text{f}}}} \{a_{\mathbf{k}+\mathbf{q}}^{\dagger} a_{\mathbf{k}} + a_{-\mathbf{k}}^{\dagger} a_{-\mathbf{k}-\mathbf{q}}\} \\
&= \sum_{\substack{k<k_{\text{f}}\\ |\mathbf{k}+\mathbf{q}|>k_{\text{f}}}} [d_{\mathbf{q}}^{\dagger}(\mathbf{k}) + d_{-\mathbf{q}}(-\mathbf{k})].
\end{aligned} \tag{A2.1.26}$$

We should emphasize that the two approximations involved above are intimately connected. The commutators (A2.1.16) above are approximations to the exact forms, easily verified,

$$\begin{aligned}
[d_{\mathbf{q}}(\mathbf{k}), d_{\mathbf{q}'}^{\dagger}(\mathbf{k}')] &= \delta_{\mathbf{k}+\mathbf{q},\mathbf{k}'+\mathbf{q}'}\, a_{\mathbf{k}}^{\dagger} a_{\mathbf{k}'} - \delta_{\mathbf{k}\mathbf{k}'}\, a_{\mathbf{k}'+\mathbf{q}'}^{\dagger} a_{\mathbf{k}+\mathbf{q}}, \\
[d_{\mathbf{q}}(\mathbf{k}), d_{\mathbf{q}'}(\mathbf{k}')] &= \delta_{\mathbf{k}+\mathbf{q},\mathbf{k}'}\, a_{\mathbf{k}}^{\dagger} a_{\mathbf{k}'+\mathbf{q}} - \delta_{\mathbf{k}'+\mathbf{q},\mathbf{k}}\, a_{\mathbf{k}'}^{\dagger} a_{\mathbf{k}+\mathbf{q}}.
\end{aligned} \tag{A2.1.27}$$

Now we argue that as $r_s \to 0$, ψ_s tends to a free-electron solution, in which case, with $k, k' < k_{\text{f}}$, $|\mathbf{k}+\mathbf{q}|$, $|\mathbf{k}'+\mathbf{q}| > k_{\text{f}}$,

$$\left.\begin{aligned}
a_{\mathbf{k}}^{\dagger} a_{\mathbf{k}'} |\psi_s\rangle &= 0, \\
a_{\mathbf{k}+\mathbf{q}} a_{\mathbf{k}'+\mathbf{q}}^{\dagger} |\psi_s\rangle &= 0
\end{aligned}\right\} \tag{A2.1.28}$$

unless $k = k'$ [cf. equation (A2.1.19)]. Since the approximations (A2.1.16) are used to evaluate commutators with V_s, already involving r_s as a parameter, the effect of replacing (A2.1.27) by (A2.1.16) will be an order r_s down on the approximation with which we are concerned, so that we use the boson-like commutators. This replacement was earlier referred to as the random-phase approximation, but we now see this and the other approximation leading to equation (A2.1.19) to be the same.

(c) *One-particle averages of equations of motion.* From what has been said above, one may gather that the SCF equations are, within the approximations common to the methods, the averages of the Sawada one-particle equations. It therefore follows that the Sawada dielectric constant is just $\varepsilon_{\text{SCF}}(\mathbf{k}\omega)$ and this is the reason why the use of ε_{SCF} in equation (2.4.1) gives the energy correct to order ln r_s and $(r_s)^0$. The Sawada scheme is, of course, more general, in that one can take many-particle averages, but we shall not require these.

APPENDIX A2.2 PERTURBATION THEORY AND DIAGRAMMATIC TECHNIQUES FOR GROUND STATE OF ELECTRON GAS

In this appendix we shall derive the rules needed for the detailed interpretation of Feynman diagrams as schematic representations of the various terms of perturbation expansions for the energy and one-particle Green function of a homogeneous electron gas in its ground state. We shall also expand on the discussion of the use of diagrams in the main text. For this purpose detailed knowledge of sections 1–4 of this appendix is not required, and the reader who is interested only in the application of these methods can turn immediately to section 4.2. We shall also briefly deal with the situation in which a non-uniform external potential exists.

A2.2.1 Time-development operator

The interaction representation of a time-dependent operator is given by

$$\tilde{q}(t) = e^{iH_0 t} q(0)\, e^{-iH_0 t}, \tag{A2.2.1}$$

where H_0 is the unperturbed Hamiltonian, and the full Hamiltonian is given by

$$H = H_0 + H_1. \tag{A2.2.2}$$

For most of this appendix, H_0 will be the kinetic energy operator, and H_1 the Coulombic interaction in an electron gas.

To obtain the original operator of the Heisenberg representation,

$$q(t) = e^{iHt} q(0) e^{-iHt}, \tag{A2.2.3}$$

we define the time-development operator $U(t, t_0)$, which is such that

$$e^{iH(t_0-t)} = e^{-iH_0 t} U(t, t_0) e^{iH_0 t_0}. \tag{A2.2.4}$$

One may then verify that

$$q(t) = U(0, t) \tilde{q}(t) U(t, 0). \tag{A2.2.5}$$

The time-development operator obeys the integral equation

$$U(t, t_0) = 1 - i \int_{t_0}^{t} \tilde{H}_1(t) U(t_1, t_0) dt_1 \tag{A2.2.6}$$

and, on iteration, we obtain the perturbative solution

$$U(t, t_0) = 1 - \sum_n (-i)^n \int_{t_0}^{t} dt_1 \int_{t_0}^{t_1} dt_2 \ldots \int_{t_0}^{t_{n-1}} dt_n \tilde{H}_1(t_1) \ldots \tilde{H}_1(t_n). \tag{A2.2.7}$$

It will be more suitable, in what follows, to convert all upper limits of integration to t. Consider the nth term

$$\int_{t>t_1>t_2\ldots>t_n} \tilde{H}_1(t_1) \ldots \tilde{H}_n(t_n) dt_1 \ldots dt_n.$$

This expansion remains unchanged on any of the $n!$ possible permutations of the t_1 to t_n, because these are merely dummy variables. We can therefore extend the range of every variable to t_1 if we divide by $n!$ and ensure that at every stage of the integration the correct time ordering is observed. We introduce the operator P to accomplish this, P ordering earlier times to the right. Then

$$U(t, t_0) = 1 - \sum_{n=1}^{\infty} (-i)^n \frac{1}{n!} \int_{t_0}^{t} dt_1 \ldots \int_{t_0}^{t} dt_n P[\tilde{H}_1(t_1) \ldots \tilde{H}_1(t_n)]. \tag{A2.2.8}$$

This is the expansion of U we shall apply, but before doing so we shall prove an important theorem. Suppose we insert any operator $\tilde{q}(t')$ inside the product $P[\tilde{H}_1(t_1) \ldots \tilde{H}_1(t_n)]$. The operator P is to be understood to apply to all time variables in an expression, not just variables of integration, and so orders $\tilde{q}(t')$ among the $\tilde{H}_1$'s according to the value of t' relative to $t_1 \ldots t_n$. If at any stage of the integration there are m times greater than t', there are $n!/[m!(n-m)!]$ possible combinations of the times in the two groups

$\{t_i > t'\}$ and $\{t_i < t'\}$. Hence, provided $t > t' > t_0$,

$$\frac{(-i)^n}{n!}\int_{t_0}^{t} dt_1 \ldots \int_{t_0}^{t} dt_n P[\tilde{H}_1(t_1)\ldots\tilde{H}_1(t_n)\tilde{q}(t')]$$

$$= \sum_{m=0}^{n}\frac{(-i)^n}{m!(n-m)!}\int_{t'}^{t} dt_1 \ldots \int_{t'}^{t} dt_m P[\tilde{H}_1(t_1)\ldots\tilde{H}_1(t_m)]\tilde{q}(t')$$

$$\times\int_{t_0}^{t'} dt_1' \ldots \int_{t_0}^{t'} dt_{n-m}' P[\tilde{H}_1(t')\ldots\tilde{H}_1(t_{n-m}')]$$

$$= \sum_{m=0}^{(m+p=n)}\sum_{p=0}\frac{(-i)^m}{m!}\int_{t'}^{t} dt_1 \ldots \int_{t'}^{t} dt_m P[\tilde{H}_1(t_1)\ldots\tilde{H}_1(t_m)]\tilde{q}(t')$$

$$\times\frac{(-i)^p}{p!}\int_{t_0}^{t'} dt_1' \ldots \int_{t_0}^{t'} dt_p' P[\tilde{H}_1(t_1')\ldots\tilde{H}_1(t_p')]. \qquad \text{(A2.2.9)}$$

The right-hand side is the nth-order term in the product $U(t,t')\tilde{q}(t')\,U(t',t_0)$ and so we may write

$$P[U(t,t_0)\tilde{q}(t')] = U(t,t')\tilde{q}(t')\,U(t',t_0) \quad (t > t' > t_0), \qquad \text{(A2.2.10)}$$

where it is to be understood, in defining the left-hand side, that $U(t,t_0)$ is to be expanded in the series (A2.2.8), and P orders $\tilde{q}(t')$ among the $\tilde{H}_1(t_1)$ to $\tilde{H}_1(t_n)$.

A special case of (A2.2.10) obtains when $\tilde{q}(t')$ is set equal to 1, when we evidently have

$$U(t,t_0) = U(t,t')\,U(t',t_0) \quad (t > t' > t_0), \qquad \text{(A2.2.11)}$$

which is also readily proved from the definition (A2.2.4) of U.

A2.2.2 Vacuum amplitude and ground-state energy

The vacuum amplitude‡ is defined by

$$R(t) = \langle\Phi_0|\,e^{iH_0t}e^{-iHt}\,|\Phi_0\rangle, \qquad \text{(A2.2.12)}$$

where Φ_0 is the ground state of the non-interacting system with Hamiltonian H_0.

By the definition of U, it follows that we also have

$$R(t-t_0) = \langle\Phi_0|\,U(t,t_0)\,|\Phi_0\rangle. \qquad \text{(A2.2.13)}$$

‡ The name "vacuum amplitude" is field theoretic in origin. It is appropriate here when we define creation operators $a_{\mathbf{k}}\dagger$ only for electrons excited above the Fermi sea, and creation operators $b_{\mathbf{k}}\dagger$ for holes in this sea; when Φ_0 does indeed behave just like the vacuum state, for $a_{\mathbf{k}}\Phi_0 = 0$, $b_{\mathbf{k}}\Phi_0 = 0$.

Its importance as far as we are concerned lies in the expression for the ground-state energy

$$E_0 = E_0^0 + i\frac{d}{dt}\ln R(t)\bigg|_{t\to\infty(1-i\eta)}, \tag{A2.2.14}$$

where E_0^0 is the ground-state energy of the non-interacting gas and η is a positive infinitesimal such that $t\eta \to \infty$ as $t\to\infty$. The form of the expression (A2.2.14) is suggested by an analogy from thermodynamics. We recall that the free energy is obtained by taking the logarithm of the partition function $\mathrm{tr}\{e^{-\beta H}\}$, where the trace is taken with respect to the eigenfunctions of any Hamiltonian we please, provided they obey the correct boundary conditions. We treat this aspect of finite-temperature theory elsewhere: at present we use a time variable rather than $\beta = 1/k_B T$.

Expanding Φ_0 in terms of the eigenstates Ψ_i of H, we have

$$\begin{aligned} R(t) &= e^{iE_0 t}\langle \Phi_0 | e^{-iHt} | \Phi_0\rangle = e^{iE_0 t}\sum_j \langle \Phi_0 | e^{-iHt} | \Psi_j\rangle \langle \Psi_j | \Phi_0\rangle \\ &= e^{iE_0 t}\sum_j e^{-iE_j t} |\langle \Psi_0 | \Psi_j\rangle|^2 \end{aligned} \tag{A2.2.15}$$

and

$$\frac{d}{dt}\ln R(t) = iE_0 - \frac{\sum_j iE_j e^{-iE_j t}|\langle \Phi_0|\Psi_j\rangle|^2}{\sum_j e^{-iE_j t}|\langle \Phi_0|\Psi_j\rangle|^2}. \tag{A2.2.16}$$

Putting $t = \tau(1-i\eta)$ and letting $\tau\to\infty$, the terms involving $\langle\Phi_0|\Psi_0\rangle$ dominate, provided of course

$$\langle \Phi_0 | \Psi_0\rangle \neq 0 \tag{A2.2.17}$$

and equation (A2.2.14) is obtained.

The condition (A2.2.17) is highly significant. $H_1 = H - H_0$ must genuinely be a perturbation on Φ_0; thus the proof given above does not allow for a phase change when the interaction is switched on.

A2.2.3 One-particle Green function

We now wish to express the Green function as the matrix element of some operator taken with respect to the ground state of the non-interacting system. Let us first re-express the Green function using operators in the interaction representation.

By definition the one-particle Green function for a homogeneous electron gas in its ground state is

$$G(\mathbf{k}, t-t_0) = -i\langle \Psi_0 | a_{\mathbf{k}}(t)\, a_{\mathbf{k}}^\dagger(t_0) | \Psi_0\rangle \quad (t>t_0) \tag{A2.2.18}$$

and, using equation (A2.2.5), this becomes

$$\begin{aligned} G(\mathbf{k}, t-t_0) &= -i\langle\Psi_0|\, U(0,t)\,\tilde{a}_{\mathbf{k}}(t)\, U(t,0)\, U(0,t_0)\,\tilde{a}^\dagger_{\mathbf{k}}(t_0)\, U(t_0,0)\,|\Psi_0\rangle \\ &= -i\langle\Psi_0|\, U(0,t)\,\tilde{a}_{\mathbf{k}}(t)\, U(t,t_0)\,\tilde{a}^\dagger_{\mathbf{k}}(t_0)\, U(t_0,0)\,|\Psi_0\rangle, \quad \text{(A2.2.19)} \end{aligned}$$

where we have also used equation (A2.2.11).

We now wish to transfer to matrix elements between Φ_0. As in the last section, a hint as to our procedure comes from thermodynamics, where the average of any operator A is given by the trace of $A\,\mathrm{e}^{-\beta H}$ divided by the partition function. We therefore proceed by noting that for any operator A

$$\begin{aligned} \langle\Phi_0|\, U(\tau,0)\, A\, U(0,\tau_0)\,|\Phi_0\rangle &= \langle\Phi_0|\,\mathrm{e}^{iH_0\tau}\,\mathrm{e}^{-iH\tau}\, A\,\mathrm{e}^{iH\tau_0}\,\mathrm{e}^{-iH_0\tau_0}|\,\Phi_0\rangle \\ &= \mathrm{e}^{iE_0{}^0(\tau-\tau_0)}\sum_i \langle\Phi_0|\,\mathrm{e}^{-iH\tau}\,|\Psi_i\rangle\,\langle\Psi_i|\, A\,\mathrm{e}^{iH\tau_0}|\,\Phi_0\rangle, \end{aligned} \quad \text{(A2.2.20)}$$

that is,

$$\begin{aligned} &\langle\Phi_0|\, U(\tau,0)\, A\, U(0,\tau_0)\,|\Phi_0\rangle \\ &\quad = \mathrm{e}^{iE_0{}^0(\tau-\tau_0)}\sum_{ij} \langle\Phi_0|\Psi_i\rangle\,\langle\Psi_i|\,A\,|\Psi_j\rangle\,\langle\Psi_j|\Phi_0\rangle\,\mathrm{e}^{-iE_i\tau}\,\mathrm{e}^{iE_j\tau_0}. \end{aligned} \quad \text{(A2.2.21)}$$

If we put $A = 1$ in this expression, we have by (A2.2.11)

$$\langle\Phi_0|\, U(\tau,\tau_0)\,|\Phi_0\rangle = \mathrm{e}^{iE_0{}^0(\tau-\tau_0)}\sum_i |\langle\Phi_0|\Psi_i\rangle|^2\,\mathrm{e}^{-iE_i\tau}\,\mathrm{e}^{iE_i\tau_0} \quad \text{(A2.2.22)}$$

and, dividing (A2.2.21) by (A2.2.22),

$$\frac{\langle\Phi_0|\, U(\tau,0)\, A\, U(0,\tau_0)\,|\Phi_0\rangle}{\langle\Phi_0|\, U(\tau,\tau_0)\,|\Phi_0\rangle} = \frac{\sum_{ij} \langle\Phi_0|\Psi_i\rangle\,\langle\Psi_j|\Phi_0\rangle\,\langle\Psi_i|\,A\,|\Psi_j\rangle\,\mathrm{e}^{-iE_i\tau}\,\mathrm{e}^{iE_j\tau_0}}{\sum_i |\langle\Phi_0|\Psi_i\rangle|^2\,\mathrm{e}^{-iE_i\tau}\,\mathrm{e}^{iE_i\tau_0}}. \quad \text{(A2.2.23)}$$

Now take the limits $\tau \to \infty(1-i\eta)$, $\tau_0 \to \infty(i\eta-1)$. Provided (A2.2.17) holds, only terms for which $i = j = 0$ survive, and

$$\lim_{\substack{\tau\to\infty(1-i\eta)\\ \tau_0\to\infty(i\eta-1)}} \frac{\langle\Phi_0|\, U(\tau,0)\, A\, U(0,\tau_0)\,|\Phi_0\rangle}{\langle\Phi_0|\, U(\tau,\tau_0)\,|\Phi_0\rangle} = \langle\Psi_0|\,A\,|\Psi_0\rangle. \quad \text{(A2.2.24)}$$

If we now set $A = U(0,t)\,\tilde{a}_{\mathbf{k}}(t)\, U(t,t_0)\,\tilde{a}^\dagger_{\mathbf{k}}(t_0)\, U(t_0,0)$, we see from equation (A2.2.19) that we have the expression

$$G(\mathbf{k}, t-t_0) = \left.\frac{-i\langle\Phi_0|\, U(\tau,t)\,\tilde{a}_{\mathbf{k}}(t)\, U(t,t_0)\,\tilde{a}^\dagger_{\mathbf{k}}(t_0)\, U(t_0,\tau)\,|\Phi_0\rangle}{\langle\Phi_0|\, U(\tau,\tau_0)\,|\Phi_0\rangle}\right|_{\substack{\tau\to\infty(1-i\eta)\\ \tau_0\to\infty(i\eta-1)}}, \quad (t > t_0). \quad \text{(A2.2.25)}$$

By using equation (A2.2.10) we can further compress the form of the numerator. We have, since $\tau > t > t_0 > \tau_0$,

$$\begin{aligned} U(\tau, t)\, \tilde{a}_{\mathbf{k}}(t)\, U(t, t_0)\, \tilde{a}_{\mathbf{k}}^{\dagger}(t_0)\, U(t_0, \tau) &= P[U(\tau, t)\, \tilde{a}_{\mathbf{k}}(t)\, U(t, \tau)\, \tilde{a}_{\mathbf{k}}^{\dagger}(t_0)] \\ &= P[U(\tau, \tau_0)\, \tilde{a}_{\mathbf{k}}(t)\, \tilde{a}_{\mathbf{k}}^{\dagger}(t_0)]. \qquad \text{(A2.2.26)} \end{aligned}$$

We may similarly handle G when $t < t_0$, and since for either time-ordering

$$G(\mathbf{k}, t - t_0) = -i\langle \Psi_0 | T[a_{\mathbf{k}}(t)\, a_{\mathbf{k}}^{\dagger}(t_0)] | \Psi_0 \rangle, \qquad \text{(A2.2.27)}$$

where

$$T = (-1)^p P, \qquad \text{(A2.2.28)}$$

p being the number of interchanges of Fermion operators necessitated by P, we can write

$$G(\mathbf{k}, t - t_0) = \left. \frac{-i\langle \Phi_0 | T[U(\tau, \tau_0)\, \tilde{a}_{\mathbf{k}}(t)\, \tilde{a}_{\mathbf{k}}^{\dagger}(t_0)] | \Phi_0 \rangle}{\langle \Phi_0 | U(\tau, \tau_0) | \Phi_0 \rangle} \right|_{\substack{\tau \to \infty(1 - i\eta) \\ \tau_0 \to \infty(i\eta - 1)}}, \qquad \text{(A2.2.29)}$$

where it is to be understood that T operates inside the integrals of the perturbation series for $U(\tau, \tau_0)$.

A2.2.4 Evaluation of matrix elements

(i) *Wick's theorem.* We shall now show how every matrix element in the perturbation series for $R(t)$ and $G(\mathbf{k}, t)$ can be expressed as sums of products of free-particle Green functions. The result we use to do so is called Wick's theorem. We shall not prove the theorem in its general form [for a statement of which see, for example, March, Young and Sampanthar (1967)] but only in the form we need here.

In stating the theorem it is convenient to use the idea of the *contraction* $\overline{AB}$ of two operators A and B. This will be defined as

$$\overline{AB} = \langle 0 | T(AB) | 0 \rangle, \qquad \text{(A2.2.30)}$$

where $|0\rangle$ is the vacuum state (state with no particles). For our purposes we may replace $|0\rangle$ by $|\Phi_0\rangle$, the ground state of the non-interacting system, as we shall see later. The usefulness of the symbol for a contraction can be seen from the following defining equality

$$\overbrace{A_1 A_2 A_3}^{} A_4 \;(\text{contractions } A_1 A_3,\ A_2 A_4) = \overline{A_1 A_3}\; \overline{A_2 A_4}. \qquad \text{(A2.2.31)}$$

Thus we can show contractions between a product of many operators without disturbing the order of the product. This order can, of course, be important; when we use the operator T for example.

We now state the theorem for a product $A_1 \ldots A_{2n}$ of creation and annihilation operators. It is

$$\langle 0|T(A_1 \ldots A_{2n})|0\rangle = \sum_{\text{all pairs}} (-1)^p (\overline{A_1 A_2 \ldots A_k} \ldots A_l \ldots \overline{A_m \ldots A_{2n}}), \tag{A2.2.32}$$

where the sum is over all possible products of n contractions and p is the number of permutations of Fermion operators necessary to obtain the elements of each contraction next to one another given the original ordering on the left.

Before we prove the theorem we should note that although we have written it in terms of $|0\rangle$, the vacuum state, we can replace $|0\rangle$ by $|\Phi_0\rangle$ without loss of validity. This is easily seen if we rewrite our operators according to the correspondence

$$a_{\mathbf{k}} \to b_{\mathbf{k}}^\dagger, \quad a_{\mathbf{k}}^\dagger \to a_{\mathbf{k}} \quad (k < k_f), \tag{A2.2.33}$$

where $b^\dagger$ and b are respectively hole creation and annihilation operators. We prove the theorem below for the vacuum state rather than $|\Phi_0\rangle$ only because the proof is a little simpler.

We introduce a second set of operators $B_1, \ldots, B_{2n}$, each being equal to one of the A's in a one-to-one correspondence, such that

$$T(A_1 A_2 \ldots A_{2n}) = (-1)^{p_1} B_1 B_2 \ldots B_{2n}. \tag{A2.2.34}$$

Thus the product $B_1 \ldots B_{2n}$ is already time-ordered, p_1 being the number of exchanges of Fermion operators needed to time order the product on the left.

If B_1 is a creation operator, we have $B_1^\dagger|0\rangle = 0$ and so

$$\langle 0|B_1 B_2 \ldots B_n|0\rangle = \langle 0|B_{2n}^\dagger \ldots B_2^\dagger B_1^\dagger|0\rangle^* = 0. \tag{A2.2.35}$$

Similarly

$$\overline{B_1 B_i} = 0 \tag{A2.2.36}$$

and we can see that equation (A2.2.32) is then trivially true because both left- and right-hand sides are identically zero.

The more difficult case is when B_1 is an annihilation operator. We shall use the method of induction, assuming the result true for $(n-1)$ pairs. For n pairs, we use the result‡

$$[B_1, B_2, \ldots, B_{2n}]_+ = \sum_{i=2}^{2n} (-1)^{i-1} [B_1, B_i]_+ \prod_{j \neq 1,2}' B_j$$

‡ We take the anticommutator under the assumption that the B_i are Fermion operators. The theorem is not, however, restricted to this case.

or since $B_1|0\rangle = 0$ when B_1 is an annihilation operator,

$$\langle 0|B_1, B_2 \dots B_{2n}|0\rangle = \sum_{i=2}^{2n}(-1)^{i-1}[B_1, B_i]_+ \langle 0|\prod_{j\neq 1,2} B_j|0\rangle. \quad \text{(A2.2.37)}$$

Because the anticommutator is a c-number, we also have

$$[B_1, B_i]_+ = \langle 0|[B_1, B_i]_+|0\rangle = \langle 0|B_1 B_i|0\rangle = \overline{B_1 B_i}, \quad \text{(A2.2.38)}$$

where we have again used the fact that B_1 is an annihilation operator. Hence equation (A2.2.37) becomes

$$\langle 0|B_1 B_2 \dots B_{2n}|0\rangle = \sum_{i=2}^{2n}(-1)^{i-1}\overline{B_1 B_2}\langle 0|T\Big(\prod_{j\neq 1,2} B_j\Big)|0\rangle. \quad \text{(A2.2.39)}$$

Noting that $i-1$ is the number of exchanges necessary to reorder the product $B_1 \dots B_{2n}$ so that B_1 and B_i are next to one another, we see that if the theorem holds for $2n-2$ operators, equation (A2.2.39) becomes

$$\langle 0|B_1 B_2 \dots B_{2n}|0\rangle = \sum_{\text{all pairs}} (-1)^p (B_1 B_2 \dots B_i \dots B_j \dots B_k \dots B_{2n}). \quad \text{(A2.2.40)}$$

On resubstituting the A's for the B's, the statement (A2.2.32) of the theorem obtains. Our proof by induction now follows immediately since the theorem is trivially true for $n = 1$.

(ii) *Linked-cluster expansion for vacuum amplitude.* We are now in a position to see the connection of the diagrams with the series for $R(t)$ and $G(\mathbf{k}, t)$. We first discuss $R(t)$.

Each term is of the form

$$\frac{(-i)^n}{n!}\int_{t_0}^{t} dt_1 \dots \int_{t_0}^{t} dt_n\, T[H_1(t_1) \dots H_1(t_n)] \quad \text{(A2.2.41)}$$

with

$$H_1(t) = \sum_{\mathbf{q}} V_{\mathbf{q}} \sum_{\mathbf{k}_1\mathbf{k}_2} a^\dagger_{\mathbf{k}_1+\mathbf{q}}(t)\, a^\dagger_{\mathbf{k}_2-\mathbf{q}}(t)\, a_{\mathbf{k}_2}(t)\, a_{\mathbf{k}_1}(t). \quad \text{(A2.2.42)}$$

It will be noticed that we have omitted the tildes over the time-dependent operators; in future it will be understood that the time dependence of all operators is taken with respect to that of the non-interacting system.

From equation (A2.2.42) we see that at time t there are two creations and two annihilations, which we draw as in Figure A2.2.1, the dashed horizontal line representing the interaction. The diagram is labelled to express the conservation of momentum as it appears in equation (A2.2.42). It must also be remarked that a line directed from a vertex represents a creation and a line directed to a vertex represents an annihilation.

If we now evaluate matrix elements of the expression (A2.2.4) with respect to Φ_0 by using equation (A2.2.32), each contraction will be of the form $\overline{a_{\mathbf{k}'}(t_1)\, a^{\dagger}_{\mathbf{k}'}(t_2)}$ and so we represent the contraction by a line arrowed from t_2 to

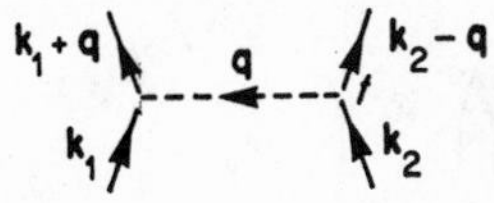

FIGURE A2.2.1. Scattering event at time t, showing conservation of momentum according to equation (A2.2.42).

t_1. Thus in second order we get, by joining two sections of the form of Figure A2.2.1, the associated graphs shown in Figure A2.2.2. It is evident from equation (A2.2.27) that each full line (termed a Fermion line) is associated with a factor G_0, but it is not yet necessary to discuss the exact correspondence between the graphs and the series for $R(t)$. At the moment, we wish to

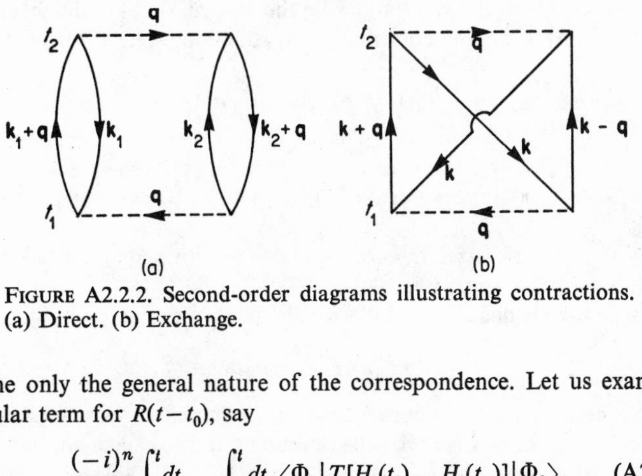

FIGURE A2.2.2. Second-order diagrams illustrating contractions. (a) Direct. (b) Exchange.

examine only the general nature of the correspondence. Let us examine a particular term for $R(t-t_0)$, say

$$\frac{(-i)^n}{n!}\int_{t_0}^{t} dt_1 \ldots \int_{t_0}^{t} dt_n \langle \Phi_0 | T[H_1(t_1) \ldots H_1(t_n)] | \Phi_0 \rangle. \qquad \text{(A2.2.43)}$$

We begin to make contractions, drawing lines between the vertices of the interaction lines, so we connect up figures such as shown in Figure A2.2.1. We pair an operator in $H_1(t_i)$ to one in $H_1(t_j)$, one in $H_1(t_j)$ to one in $H_1(t_k)$, and so on, until we come back again to t_i. Let us continue drawing until we have a diagram with each end of every interaction line connected to two Fermion lines. Now it is perfectly possible that the diagram includes only m

interaction lines, with $m<n$, so that all the times of the expression (A2.2.43) are not included. To include the other times, we must draw a second diagram linked in no way to the first by either Fermion lines or interaction lines. Hence a particular graph corresponding to a given term can consist of two or more completely unlinked parts. A simple example for $n = 6$ is shown in Figure A2.2.3. Suppose we have a graph consisting of k unlinked parts, with

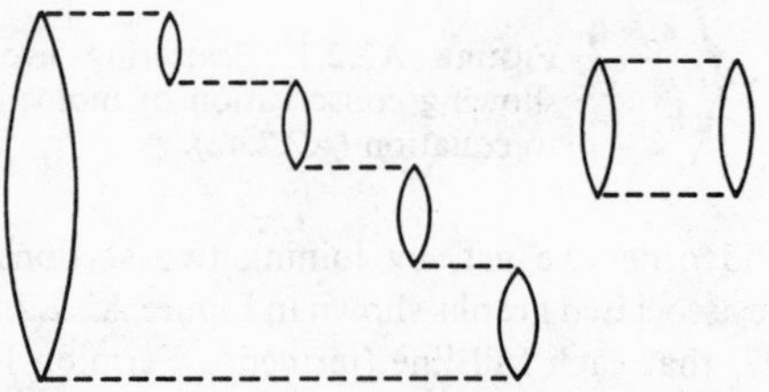

FIGURE A2.2.3. Seventh order graph with two unlinked parts.

m_1 interaction lines (and so m_1 times) in the first, m_2 times in the second, and so on. The corresponding contribution to equation (A2.2.43) is of the form

$$\frac{(-i)^n}{n!}\int dt_1^1 \dots \int dt_{m_1}^1 \langle T[H_1(t_1^1)\dots H_1(t_{m_1}^1)]\rangle_{\text{link}}$$

$$\times \int dt_1^2 \dots \int dt_{m_2}^2 \langle T[H_1(t_1^2)\dots H_1(t_{m_2}^2)]\rangle_{\text{link}} \times \dots, \qquad \text{(A2.2.44)}$$

where $\langle\ \rangle_{\text{link}}$ emphasizes the correspondence with a diagram all parts of which are joined in some way.

We may take it there are p_i links involving m_i times, so that

$$p_1 m_1 + p_2 m_2 + \dots + p_k m_k = n. \qquad \text{(A2.2.45)}$$

Now we need to know in how many ways n operators can be assigned to $p_1+p_2+\dots+p_k$ cells, where p_1 cells contain m_1 times, and so on, because the total contribution of terms like (A2.2.44) to (A2.2.43) is obtained by distributing the different times in all possible ways in the diagrams.

The number we require is, in fact,

$$F_k = \frac{n!}{p_1!(m_1!)^{p_1}\dots p_k!(m_k!)^{p_k}}, \qquad \text{(A2.2.46)}$$

being the number of ways we can distribute the n times among the various products $\langle\ \rangle_{\text{link}}$ of (A2.2.44).

Writing

$$\mathscr{L}_m = \frac{(-i)^m}{m!}\int_{t_0}^{t} dt_1 \ldots \int_{t_0}^{t} dt_m \langle 0|T[H_1(t_1)\ldots H_1(t_m)]|0\rangle_{\text{link}}, \tag{A2.2.47}$$

we obtain for (A2.2.44)

$$\frac{1}{p_1!}\mathscr{L}_{m_1}^{p_1}\frac{1}{p_2!}\mathscr{L}_{m_2}^{p_2}\ldots\frac{1}{p_k!}\mathscr{L}_{m_k}^{p_k}. \tag{A2.2.48}$$

Let us now label all possible terms of the form (A2.2.47) simply as $\mathscr{L}_1, \mathscr{L}_2, \ldots$ and look at the expression

$$\sum_{p_1}^{\infty}\frac{1}{p_1!}(\mathscr{L}_1)^{p_1}\sum_{p_2}^{\infty}\frac{1}{p_2!}(\mathscr{L}_2)^{p_2}\sum_{p_3}^{\infty}\frac{1}{p_3!}(\mathscr{L}_3)^{p_3}\ldots. \tag{A2.2.49}$$

This gives every possible product of the form (A2.2.48), and since the series for $R(t-t_0)$ is infinite, every product in it will be realized in the series. Moreover, it gives every possible product once only, and so is identical with $R(t-t_0)$. We finally note that

$$\sum_{p_1}\frac{1}{p_1!}(\mathscr{L}_1)^{p_1} = \exp(\mathscr{L}_1) \tag{A2.2.50}$$

so that

$$R(t-t_0) = \exp\{\mathscr{L}_1+\mathscr{L}_2+\ldots\}. \tag{A2.2.51}$$

Thus we see that to obtain $R(t-t_0)$ we need only draw *linked* diagrams, a considerable simplification. Furthermore, since the ground-state energy [see equation (A2.2.3)] involves $\ln R(t)$, we have obtained a direct perturbation series for the energy consisting of terms corresponding to linked diagrams only.

(iii) *Expansion of one-particle Green function.* The diagrams for $G(\mathbf{k}, t)$ may be drawn in similar fashion to those for $R(t)$, except that in the diagrams for the numerator of equation (A2.2.29) there are no interaction lines at the points corresponding to t and t_0. We shall therefore indicate the contractions containing $a_{\mathbf{k}}(t)$ and $a_{\mathbf{k}}^{\dagger}(t_0)$ by lines free at one end.

Just as in the case for $R(t)$, a graph may consist of two or more unlinked parts. Note, however, that if one of these parts contains one line free at one end, it must contain the other. This is because each interaction line has two ends, each of which must be connected to two Fermion lines. Those parts of graphs not linked to the lines with free ends will be termed *disconnected.*

We shall now prove that the denominator of equation (A2.2.29) is completely taken account of if we ignore all disconnected diagrams in the expansion of the numerator. The reason for this may be visualized as follows. We obtain all diagrams for the numerator by drawing every possible disconnected diagram and form other diagrams by drawing every possible disconnected diagram with the connected one. On the other hand, the denominator is the vacuum amplitude and so may be said to be represented by all disconnected diagrams. Let us show this in detail.

A typical term in the expansion of the numerator of equation (A2.2.29) is

$$\frac{(-i)^n}{n!}\int_{\tau_0}^{\tau} dt_1 \dots \int_{\tau_0}^{\tau} dt_n \langle 0|\, T[H_1(t_1)\dots H_n(t_n)\, a_{\mathbf{k}}(t)\, a_{\mathbf{k}}^{\dagger}(t_0)]\,|0\rangle. \tag{A2.2.52}$$

Consider now diagrams for this with disconnected parts. Denoting the decomposition by $\langle\;\rangle_{\text{con}}$ and $\langle\;\rangle_{\text{dis}}$, where $\langle\;\rangle_{\text{con}}$ emphasizes the correspondence with a connected diagram, we have

$$\frac{(-i)^n}{n!}\int dt_1 \dots \int dt_m \langle T[H_1(t_1)\dots H_1(t_m)\, a_{\mathbf{k}}(t)\, a_{\mathbf{k}}^{\dagger}(t_0)]\rangle_{\text{con}}$$
$$\times \int dt_{m+1} \dots \int dt_n \langle T[H_1(t_{m+1})\dots H_1(t_n)]\rangle_{\text{dis}} \tag{A2.2.53}$$

if there are m interaction lines in the connected part.

It is not hard to see that many diagrams give the same contribution; merely distributing the variables t_i among the brackets in a different way does not change the size of the contribution to G. The number of such possibilities is $n!/[m!\,(n-m)!]$ and the total contribution becomes

$$\frac{(-i)^m}{m!}\int dt_1 \dots \int dt_m \langle T[H_1(t_1)\dots H_1(t_m)\, a_{\mathbf{k}}(t)\, a_{\mathbf{k}}^{\dagger}(t_0)]\rangle_{\text{con}}$$
$$\times \frac{(-i)^{n-m}}{(n-m)!}\int dt_{m+1} \dots \int dt_n \langle T[H_1(t_{m+1})\dots H_1(t_n)]\rangle_{\text{dis}}. \tag{A2.2.54}$$

Now the number m of interaction lines in the connected part can vary from 0 to n and so, by summing over all terms, we find

$$1+\sum_{n=1}^{\infty}\sum_{m=0}^{n}\frac{(-i)^m}{m!}\int dt_1 \dots \int dt_m \langle T[H_1(t_1)\dots H_1(t_m)\, a_{\mathbf{k}}(t)\, a_{\mathbf{k}}^{\dagger}(t_0)]\rangle_{\text{con}}$$
$$\times \frac{(-i)^{n-m}}{m!}\int dt_{m+1} \dots \int dt_n \langle T[H_1(t_{m+1})\dots H_1(t_n)]\rangle_{\text{dis}}. \tag{A2.2.55}$$

This becomes, on changing the summation variables,

$$\sum_{n=0}^{\infty}\sum_{m=0}^{\infty}\frac{(-i)^m}{m!}\frac{(-i)^n}{n!}\int_{\tau_0}^{\tau}dt_1\ldots\int_{\tau_0}^{\tau}dt_m\langle T[H_1(t_1)\ldots H_1(t_m)\,a_{\mathbf{k}}(t)\,a_{\mathbf{k}}^{\dagger}(t_0)]\rangle_{\text{con}}$$

$$\times\int_{\tau_0}^{\tau}dt_1\ldots\int_{\tau_0}^{\tau}dt_n\langle T[H_1(t_1)\ldots H_1(t_n)]\rangle_{\text{dis}},\qquad \text{(A2.2.56)}$$

that is,

$$\langle 0|\,T\,[U(\tau,\tau_0)\,a_{\mathbf{k}}(t)\,a_{\mathbf{k}}^{\dagger}(t_0)]\,|0\rangle$$

$$=\langle 0|\,U(\tau,\tau_0)\,|0\rangle\left\{1+\sum_{n=1}^{\infty}\frac{(-i)^n}{n!}\int_{\tau_0}^{\tau}dt_1\ldots\int_{\tau_0}^{\tau}dt_n\langle T[H_1(t_1)\ldots a_{\mathbf{k}}(t)\,a_{\mathbf{k}}^{\dagger}(t_0)]\rangle_{\text{con}}\right\}$$

(A2.2.57)

which we required to show.

A2.2.5 Rules for diagrams

(i) *General correspondence.* Let us first examine limitations on the types of diagram we draw. One rule we can immediately comment on takes account of the vanishing of a contraction unless it contains both a creation and an annihilation operator. This is one purpose of drawing directed lines. An arrow leaving a vertex corresponds to a creation operator and an arrow running into a vertex corresponds to a destruction operator. We therefore draw only those diagrams in which the direction of an arrow along the line remains unchanged from vertex to vertex. Thus a line such as is shown in Figure A2.2.4 is not allowed.

FIGURE A2.2.4. Direction of arrow along line must be unchanged from vertex to vertex. Figure drawn is therefore invalid.

Next, while it is useful to regard the diagrams as plotted in time, the direction to later times being up the page, the principal feature of the term illustrated by a diagram is the nature of its contractions. Thus Figure A2.2.5(a) is entirely equivalent to Figure A2.2.5(b); the same contractions are illustrated and though the time ordering appears different, this difference disappears on integration. This is a special case of a general result. Since we integrate over all times (except t and t_0) in order to evaluate $G(\mathbf{k}, t-t_0)$, the ordering of the vertices is immaterial and we need only draw topologically distinct diagrams, that is those which cannot be distorted one into the other without the breaking or joining of lines, or changing the directions of the Fermion lines. Thus we need draw only Figure A2.2.5(a), which also represents Figures A2.2.5(b), (c), (d) and (e). If we do so, however, we must multiply by a factor to take account of the diagrams we have not drawn. For Figures A2.2.5(a) and (b) do correspond to different terms; in Figure A2.2.5(a) $a_{\mathbf{k}}(t)$ is contracted with a creation operator contained in $H_1(t_1)$, whereas in Figure A2.2.5(b) it is

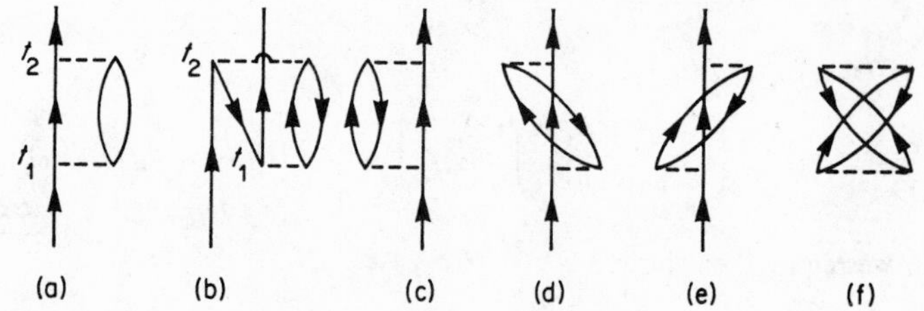

FIGURE A2.2.5. Diagrams (a)–(e) are topologically equivalent. Only topologically distinct diagrams need be drawn.

contracted with an operator contained in $H_1(t_2)$. Generalizing this result, we can say that each nth-order diagram represents $n!$ terms corresponding to the $n!$ ways in which we can arrange the order of the interaction lines in time. This cancels the factor $1/n!$ in equation (A2.2.57). We should realize, however, that for the closed diagrams representing $R(t-t_0)$ it is only the relative order that matters. As a very simple example, the closed loop of Figure A2.2.2(a) represents the same contractions between the two a's involved if we interchange H_1 and H_2. Thus there are only $(n-1)$ equal terms in the expansion of $\ln R(t)$ and a factor $1/n$ stays. The absolute order matters in the expansion for $G(\mathbf{k}, t-t_0)$ because of the presence of the two free lines.

As yet, we have taken account only of deformations affecting time order, as from Figure A2.2.5(a) to Figure A2.2.5(b). We have still to take account of diagrams such as Figures A2.2.5(d) and (e), obtained from Figure A2.2.5(a) by lateral deformation. That these correspond to different (although equal) terms can be seen by agreeing to associate the left end of every interaction line with the first and fourth operators of H_1 and the right end with every second and third operator [see equation (A2.2.42) and Figure A2.2.1]. If we draw topologically equivalent diagrams only, therefore, we must multiply by a factor accounting for this. We get a different diagram every time we reverse an interaction line, such as Figure A2.2.5(d) from (a), or Figure A2.2.5(c) from (d). It therefore follows that at nth order we multiply by 2^n. We can do this by associating every interaction line of the diagram with $-2iV_{\mathbf{q}}$; we have now also accounted for the factor $(-i)^n$ of the nth term of the right-hand side of equation (A2.2.57). However, this procedure must be modified for symmetric diagrams, such as Figure A2.2.2(a). From Figure A2.2.2(a) we get just one other diagram by distortion, Figure A2.2.5(f). This is easily generalized, with the conclusion that the factor 2^{n-1} is associated with symmetric diagrams; we again associate a factor $-2iV_{\mathbf{q}}$ with every interaction line, but divide by $\frac{1}{2}$ if the diagram is symmetric.

We have already seen that we associate a factor

$$\overbracket{a_{\mathbf{k}}\, a_{\mathbf{k}}^{\dagger}} = iG_0(\mathbf{k}, t-t_0) = \langle 0|\, T[a_{\mathbf{k}}(t)\, a_{\mathbf{k}}^{\dagger}(t_0)]\,|0\rangle \qquad \text{(A2.2.58)}$$

with every Fermion line, labelled by **k** and directed from t_0 to t and it remains for us, in enumerating factors associated with each diagram, to obtain the signs $(-1)^p$ of the terms in equation (A2.2.40).

We first note that, if we follow a Fermion line of a diagram for $R(t)$ from where it begins, that is from t_1 say, we shall go over vertices corresponding to t_2, t_3, etc., until eventually we come back to our starting point. We say we have traced out a closed Fermion loop. Figure A2.2.2(a) has two closed loops and Figure A2.2.2(b) one. Now each part of this loop between two times t_i and t_j corresponds to a contraction between operators $H_1(t_i)$ and $H_1(t_j)$ and if these H_1's are not next to one another we can order the product of equation (A2.2.47) until they are (with one exception we shall deal with later). Since each H_1 contains four Fermion operators, $(-1)^p = 1$ for this ordering. We now have the product

$$H(t_1)\, H(t_2)\, H(t_3) \ldots H(t_m) \qquad \text{(A2.2.59)}$$

for which an operator of $H(t_1)$ is contracted with an operator of $H_1(t_2)$ and so on. We shall suppose a destruction operator a of $H_1(t_1)$ is contracted with an $a^{\dagger}$ of $H(t_2)$ when an a of $H(t_2)$ is contracted with an $a^{\dagger}$ of $H(t_3)$ and so on. We next reorder the Fermion operators within each H_1 so that the creation operator to suffer contraction stands to the extreme left and the annihilation operator stands to the extreme right. It may be checked that this leaves $(-1)^p = 1$; an even number of exchanges is involved since we are dealing with the contractions of a Fermion loop. Finally, however, we must examine the contraction of the $a^{\dagger}$ of H_1 with the a of $H(t_m)$. To get $a^{\dagger}$ to the immediate right of the a (since we require the contraction in the form $\overbracket{a\, a^{\dagger}}$ in accordance with equation (A2.2.58)) involves $2n-1$ Fermion operator interchanges and we finally obtain the factor (-1) associated with each closed loop of Fermion lines. This conclusion is not changed if two operators of the same $H_1(t_i)$ are contracted together (see, in particular, our remarks on Figure A2.2.6 below).

Diagrams for $G(\mathbf{k}, t-t_0)$ can contain both closed loops and an open loop [see Figure A2.2.5(a), for example, with one closed loop]. We trace out the open loop if we begin from t_0. Our path must then lead to t; it cannot rejoin itself as this would mean three Fermion lines joining a vertex. Proceeding as before, we order the H_1's so that contractions are taken between neighbouring H_1's; the product then looks like

$$H_1(t_1)\, H_2(t_2) \ldots H_1(t_n)\, a(t)\, a^{\dagger}(t_0). \tag{A2.2.60}$$

The annihilation operator at the extreme right of $H_1(t_n)$ is now to be contracted with $a^{\dagger}(t_0)$ and the creation operator $a^{\dagger}(t_1)$ to the extreme left of $H_1(t_1)$ is to be contracted with $a(t)$. To obtain $a^{\dagger}(t_0)$ to the immediate right of $H_1(t_n)$ and $a^{\dagger}(t_1)$ to the immediate right of $a(t)$ involves an even number of Fermion exchanges.

Before we collect together the rules obtained in this section, we wish to clear up two further points. The first is concerned with simultaneous creation and annihilation of a Fermion at the same vertex, as in the forward scattering diagram of Figure A2.2.6(a). The loop of the diagram, consisting of a single

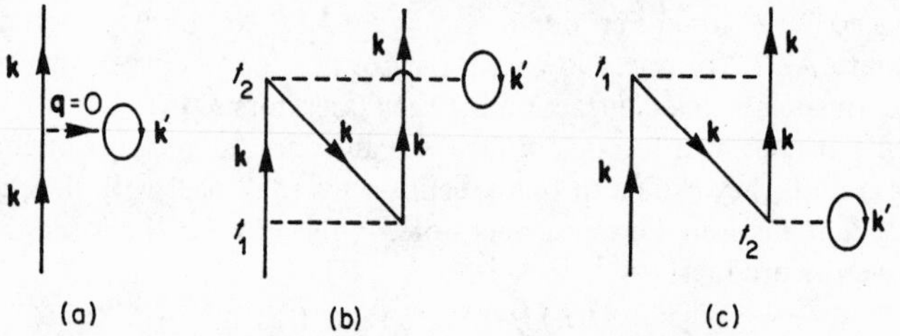

FIGURE A2.2.6. Three examples of diagrams including loops consisting of a single Fermion line.

Fermion line, corresponds to a contraction within a single H_1. One readily shows it is associated with a factor (-1) and so can be classified with other loops. Its peculiarity is that, in the expression $\langle 0|T[a(t_1)\, a^{\dagger}(t_1)]|0\rangle$, the operator T has no effect. However, the factor associated with it is easily obtained. Since the creation operators always stand to the left of the annihilation operators in $H_1(t_1)$, equation (A2.2.58) shows this loop to be associated with $-iG_0(\mathbf{k}', -0)$, i.e. we approach $t = 0$ along the negative axis.

The second point concerns the diagrams of Figures A2.2.6(b) and (c), the interaction lines of which cannot carry momentum, so that the Fermion lines are all labelled by $\mathbf{k}$. Because of the integration over times, the two diagrams evidently correspond to the same term, but it is useful here to show them separately because we then see that whether $t_2 > t_1$ or $t_1 > t_2$ we have both a particle line (a line directed upwards) and a hole line (a line directed downwards) labelled by the same $\mathbf{k}$ and propagating between the same times. The former has associated with it the factor $G_0(\mathbf{k}, t_2 - t_1)$ and the latter $G_0(\mathbf{k}, t_1 - t_2)$

and because of equation (2.8.5) of the main text, the contribution of such a diagram, referred to as anomalous, must therefore be zero. We shall now collect together the rules we have formulated.

(ii) *Detailed rules.* (1) For terms at nth order, draw all topologically inequivalent linked Feynman diagrams. Each diagram consists of n dashed lines (interaction lines) joining at each end a full line as shown in Figure A2.2.1. Such a junction is called a *vertex*. We join up the full lines and put on them arrows in such a way that at any part of the full line the arrow never changes direction. Thus there is an arrow pointing to, and an arrow pointing from, each vertex. For $\ln R(t)$, all full lines must be closed. For $G(\mathbf{k}, t-t_0)$ one full line must be free. Each part of a full line from vertex to vertex, or vertex to a free end, is termed a Fermion line.

Linked diagrams have no part entirely unconnected with another. Topologically distinct diagrams are all those which cannot be deformed one into another without breaking or joining of lines, or by reversal of direction of lines from vertex to vertex; the ends of free lines must be held fixed.

(2) Label each Fermion line with a momentum parameter $\mathbf{k}'$ and each interaction line with a momentum transfer parameter $\mathbf{q}'$, in accordance with conservation of this parameter at each vertex, as shown in Figure A2.2.1. Lines with free ends are to be labelled by $\mathbf{k}$. Label the interaction lines with times $t_1, t_2, \ldots, t_n$.

(3) Associate a factor $iG_0(\mathbf{k}', t_1-t_2)$ with each Fermion line labelled by $\mathbf{k}$ and running from t_2 to t_1. Lines linking the ends of the same interaction line ("non-propagating lines") are regarded as hole lines so that we must then have $k' < k_f$.

(4) Associate a factor $-2iV_{\mathbf{q}'}$ with each interaction line labelled by $\mathbf{q}'$.

(5) Associate a factor (-1) with each closed loop of Fermion lines [Figure A2.2.2(a) has two such loops, while Figures A2.2.5(a) and A2.2.6(a) have one].

(6) Associate a factor $\frac{1}{2}$ with each symmetric diagram, such as those of Figure A2.2.2.

(7) When calculating $\ln R(t)$, associate a factor n^{-1} with each diagram.

(8) Integrate over independent momenta and times with the same limits as those of the perturbation expansion. The factor associated with each momentum integration is $2/8\pi^3$ per unit volume. Integrate over independent momentum transfers $\mathbf{q}'$.

(iii) *Procedure for energy.* We shall now perform the integration over time necessary to obtain the ground-state energy from equation (A2.2.14) when $\ln R(t)$ is given by a diagrammatic series. We shall find that the nth-order integral in the expansion is most readily evaluated by separate integration over the $n!$ regions for which $t_1 > t_2 > \ldots > t_n$ (the labelling of the times by subscripts being arbitrary). These regions give identical results, with the consequent complete cancellation of the factor $(m!)^{-1}$ in the expansion (A2.2.47) and rule 7 above is inoperative. We must now, however, consider the graphs of Figures A2.2.7(a) and (b) separately, because the time ordering among the contractions is different.

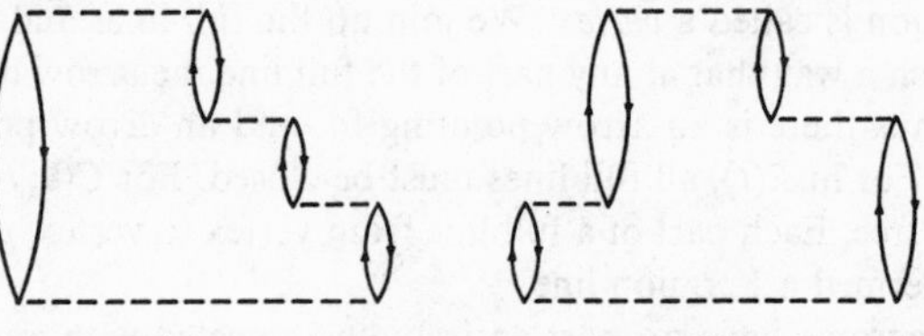

FIGURE A2.2.7. Fourth-order ring diagrams. Though these make the same contribution to the energy, the time ordering among the contractions is different. Therefore the contributions from the two diagrams must be counted separately.

Let us now take any nth-order diagram—the reader may wish to look at one of the diagrams of Figure A2.2.7—and take $t_n = t_4$. We remember that every line is associated with a factor of the form

$$iG_0(\mathbf{k}, t_1 - t_2) = \begin{cases} -\exp[i\varepsilon_{\mathbf{k}}(t_1 - t_2)], & t_1 < t_2, \quad k < k_{\mathrm{f}}, \\ \exp[-i\varepsilon_{\mathbf{k}}(t_1 - t_2)], & t_1 > t_2, \quad k > k_{\mathrm{f}}, \\ 0, \quad \text{otherwise}. \end{cases} \tag{A2.2.61}$$

Hence we first perform the integral

$$\int_0^{t_{n-1}} dt_n \exp(-i\Sigma_n t_n) = \frac{i\exp(-i\Sigma_n t_{n-1})}{\Sigma_n} - \frac{i}{\Sigma_n}, \tag{A2.2.62}$$

where we have put

$$\Sigma_n = \varepsilon_{\mathbf{k}_1{}^n} + \varepsilon_{\mathbf{k}_2{}^n} - \varepsilon_{\mathbf{k}_3{}^n} - \varepsilon_{\mathbf{k}_4{}^n}, \tag{A2.2.63}$$

$\mathbf{k}_1^n$ and $\mathbf{k}_2^n$ label particle lines and $\mathbf{k}_3^n$ and $\mathbf{k}_4^n$ hole lines; thus $k_1^n, k_2^n > k_{\mathrm{f}}$, and $k_3^n, k_4^n < k_{\mathrm{f}}$.

The integral over t_{n-1} yields (neglecting factors for the moment)

$$\int_0^{t_{n-2}} dt_{n-1} \left\{ \frac{\exp[-i(\Sigma_{n-1} + \Sigma_n)t_{n-1}] - \exp(-i\Sigma_{n-1}t_{n-1})}{\Sigma_n} \right.$$

$$= i\frac{\exp[-i(\Sigma_{n-1} + \Sigma_n)t_{n-2}]}{(\Sigma_{n-1} + \Sigma_n)\Sigma_n} - \frac{i}{\Sigma_{n-1}\Sigma_n}. \tag{A2.2.64}$$

We shall ignore the second of the two terms on the right-hand side and neglect all similar results from succeeding integrations also. We shall see the reason for this shortly. Then continuing the integrations we find, after the integration with respect to t_2, the term

$$\frac{\exp[(\Sigma_1+\Sigma_2+\ldots+\Sigma_n)]}{(\Sigma_2+\Sigma_3+\ldots+\Sigma_n)(\Sigma_3+\ldots+\Sigma_n)\ldots(\Sigma_n)} \tag{A2.2.65}$$

and this is, in fact, our final result, the integration over t_1 not being performed because of the differentiation in equation (A2.2.14). The exponential in equation (A2.2.65) is simply unity, because all the $\varepsilon_{\mathbf{k}}$'s in $\Sigma_1+\ldots+\Sigma_n$ appear twice with opposite signs, this arising because each $G_0(\mathbf{k}, t_i-t_j)$ involves two times of opposite sign. We now also see why we ignore terms such as the second on the right-hand side of equation (A2.2.62). In the final result, it appears as a term similar to (A2.2.65), but with some of the Σ's missing. Now each Σ for *intermediate* times will be positive, since $\varepsilon_{\mathbf{k}_1 n}, \varepsilon_{\mathbf{k}_2 n} > \varepsilon_{\mathbf{k}_3 n}, \varepsilon_{\mathbf{k}_4 n}$. In equation (A2.2.14) we take the limit $t\to\infty(1-i\eta)$ so that the exponential terms we get go to zero as $t\to\infty$.

As to the factors multiplying equation (A2.2.65), we have a factor i from each integration, as we saw in equation (A2.2.62). This gives a total factor i^{n-1}. From rule 4 of the last section we have a further factor $(-i)^n$; there is a factor i in equation (A2.2.14) and we finally arrive at the factor unity.

In order to complete a set of rules for obtaining the energy, we must see

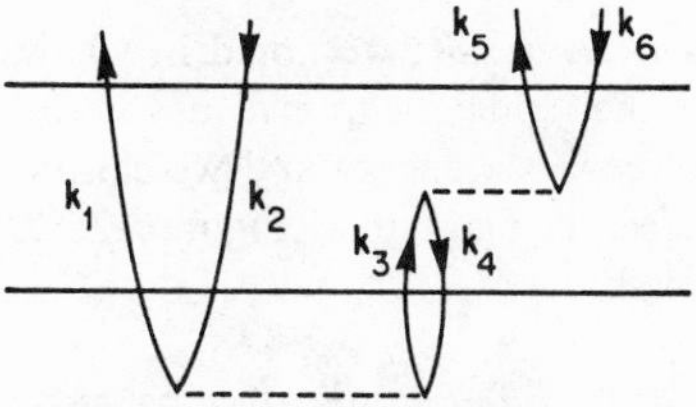

FIGURE A2.2.8. Illustrates procedure for calculating energy contribution from a given diagram. Significance of horizontal lines is discussed in text.

how to obtain the denominator of equation (A2.2.65) from the corresponding diagram. First take $(\Sigma_{n-1}+\Sigma_n)\Sigma_n$ with Figure A2.2.8 as an example. We have, with the labelling shown,

$$\Sigma_n = \varepsilon_{\mathbf{k}_1}+\varepsilon_{\mathbf{k}_3}-\varepsilon_{\mathbf{k}_2}-\varepsilon_{\mathbf{k}_4}, \tag{A2.2.66}$$

$$\Sigma_{n-1} = \varepsilon_{\mathbf{k}_4}+\varepsilon_{\mathbf{k}_5}-\varepsilon_{\mathbf{k}_3}-\varepsilon_{\mathbf{k}_6}, \tag{A2.2.67}$$

and therefore

$$\Sigma_n+\Sigma_{n-1} = \varepsilon_{\mathbf{k}_1}-\varepsilon_{\mathbf{k}_2}+\varepsilon_{\mathbf{k}_5}-\varepsilon_{\mathbf{k}_6}. \tag{A2.2.68}$$

Hence $\Sigma_n + \Sigma_{n-1}$ can be obtained by looking across the upper horizontal line of Figure A2.2.8 and taking a plus sign for $\varepsilon_{\mathbf{k}}$ if $\mathbf{k}$ labels a particle line but a minus sign if $\mathbf{k}$ labels a hole line. This result can be seen to be general. If we take such a horizontal line and raise it, its passing over an interaction line labelled by t_i corresponds to an integration over t_i; those lines labelled, say, by $\mathbf{k}_i$ and $\mathbf{k}_i'$ which end at this interaction line are no longer crossed by the horizontal line, which corresponds to the disappearance of $\varepsilon_{\mathbf{k}_i}$ and $\varepsilon_{\mathbf{k}'}$ from the sum $\Sigma_{i+1} + \Sigma_{i+2} + \ldots + \Sigma_n$ when Σ_i is added to it.

We now summarize the results for evaluating the ground-state energy. To obtain $i(d/dt)\ln R(t)|_{t\to\infty(1-i\eta)}$, we perform the following:

(1) We draw the nth-order linked Feynman diagrams that are topologically distinct or differ according to time order of the interactions.

(2) We associate a factor $2V_{\mathbf{q}_i}$ with each interaction line.

(3) We draw horizontal lines between each pair of vertices and associate with each the factor $(\sum \varepsilon_{\mathbf{k}_p} - \sum \varepsilon_{\mathbf{k}_n})^{-1}$, where $\mathbf{k}_p$ labels particle lines crossed by the horizontal line and $\mathbf{k}_{\mathrm{n}}$ labels hole lines similarly.

(4) We associate a factor (-1) with every hole line.

(5) We associate a factor (-1) with every Fermion loop.

(6) We associate a factor $\frac{1}{2}$ with each symmetric diagram.

(7) We integrate over momenta consistent with conservation of momentum at each vertex and similarly integrate over momentum transfer $\mathbf{q}'$, a factor $(4\pi^3)^{-1}$ being associated with each integration.

Gell-Mann and Brueckner result: we noted in Chapter 2 that Gell-Mann and Brueckner chose the "ring diagrams" as those giving the dominant correction to the Hartree–Fock energy and we can now see why this is so. Applying the above rules, the diagram of Figure A2.2.2(a) is found to contribute a term proportional to

$$\varepsilon_a^{(2)} = \int_{\substack{k_1,\,k_2<k_f \\ |\mathbf{k}_1+\mathbf{q}|,\,|\mathbf{k}_2+\mathbf{q}|>k_f}} \frac{d\mathbf{q}}{q^4}\, d\mathbf{k}_1\, d\mathbf{k}_2 \frac{1}{2\mathbf{q}\cdot(\mathbf{k}_1+\mathbf{k}_2+\mathbf{q})}, \tag{A2.2.69}$$

whereas Figure A2.2.2(b) (the exchange term) is

$$\varepsilon_{\mathrm{ex}}^{(2)} = \int_{\substack{k_1,\,k_2<k_f \\ |\mathbf{k}_1+\mathbf{q}|,\,|\mathbf{k}_2+\mathbf{q}|<k_f}} d\mathbf{q}\, d\mathbf{k}_1\, d\mathbf{k}_2 \frac{1}{q^2} \frac{1}{(\mathbf{k}_1-\mathbf{k}_2+\mathbf{q})^2} \frac{1}{2(\mathbf{k}_1+\mathbf{k}_2+\mathbf{q})\cdot\mathbf{q}}, \tag{A2.2.70}$$

$\varepsilon_a^{(2)}$ is divergent, while $\varepsilon_{\mathrm{ex}}^{(2)}$ is finite. The reason is easily found; in Figure A2.2.2(a) both interaction lines carry the same momentum transfer $\mathbf{q}$, giving rise to a factor $V_q^2 \propto 1/q^4$, whereas in Figure A2.2.2(b) the momentum transfers are $\mathbf{q}$ and $\mathbf{k}_1 - \mathbf{k}_2 + \mathbf{q}$. Similar results obtain at higher orders, it being easy to see that the most divergent terms come from those with the same momentum transfer at each vertex. It is these terms, therefore, which were isolated by Gell-Mann and Brueckner. The corresponding diagrams are readily recognized: they are the "ring diagrams". We ignore all diagrams but the ring diagram of a given order because the former diagrams give less divergent contributions. $\varepsilon_{\text{ex}}^{(2)}$, however, is kept, for it is independent of inter-particle spacing r_s, whereas the nth order term is formally proportional to r_s^{n-2}.

The Gell-Mann and Brueckner result is obtained in the main text by other means and so we shall not give any further details of their procedure, the mathematical device by which they carried out a formal summation of the infinite series of selected divergent terms being found in their original paper.

(iv) *Green function in Fourier transform.* The transformation to $\mathbf{k}, \omega$ space has been described in the main text and we may write down the rules for the diagrams almost immediately.

Worthy of special mention, however, are the non-propagating lines. In the rules of section (ii), these were treated as hole lines and as such give a factor -1 for $k<k_f$ and 0 for $k>k_f$. It is inconvenient to treat them as a special case here, however, and that is the reason for the integration convention described in rule 6 below. It is easily verified that

$$\lim_{\epsilon \to +0} \int_{-\infty}^{\infty} e^{i\omega\epsilon} G_0(\mathbf{k}, \omega)\, d\omega = \begin{cases} -1, & k<k_f \\ 0, & k>k_f \end{cases}. \tag{A2.2.71}$$

The rules are as follows:

(1) Draw all topologically distinct diagrams linked to a line or lines with two free ends, ignoring anomalous diagrams.

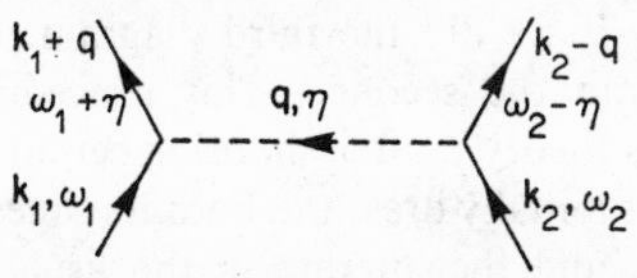

FIGURE A2.2.9. Single scattering event similar to Figure A2.2.1, except that time labelling is replaced by frequency, which is conserved at each vertex.

(2) Label each Fermion line with momentum and frequency indices $\mathbf{k}', \omega'$ and each interaction line with momentum and frequency transfer indices $\mathbf{q}', \eta'$, conserving momentum and frequency at each vertex, as shown in Figure A2.2.9.

(3) With each Fermion line labelled by $\mathbf{k}', \omega'$ associate a factor $iG_0(\mathbf{k}', \omega')$.

(4) Associate a factor $-2iV_{\mathbf{q}}$ with each interaction line labelled by **q**.

(5) Associate a factor -1 with each closed Fermion loop.

(6) Integrate over independent intermediate momenta $\mathbf{k}'$ and frequencies ω', performing the latter according to $(1/2\pi) \lim_{\epsilon\to+0} \int_{-\infty}^{\infty} d\omega\, e^{i\omega\epsilon}$; integrate over independent momentum transfers **q**. The factor associated with the momentum integration is $(4\pi^3)^{-1}$.

(v) *Dressing of skeletons and self-consistent perturbation theory.* As discussed in the main text, we can sum infinite sub-series by choosing a particular diagram and clothing it, so its Fermion lines and interaction lines are renormalized. We take the opportunity here to note that the RPA approximation contains the leading terms for the screened interaction for precisely the same reason as the ring diagrams contribute the dominant correction to the energy [see section iii(a)].

After dressing lines, the new set of diagrams form an infinite series which we may cut off at finite order, as in Figure A2.2.10. (We have not dressed

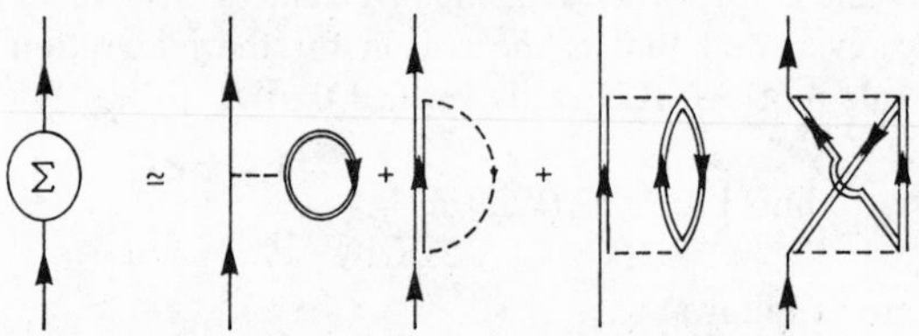

FIGURE A2.2.10. Dressing of skeletons. The diagrams for the self-energy are taken to second order and the Fermion lines are dressed.

interaction lines here—if we did the third diagram on the right-hand side would be absorbed into the second.) This procedure leads us to "self-consistent perturbation theory": after choosing certain "skeleton" diagrams, to clothe them we must initially dress the Fermion lines by an approximation suggested by intuition and then iterate in the usual fashion. The natural zeroth order approximation is, of course, the independent-particle approximation.

A difficulty here is to ensure that we obtain a Green function describing the correct number of particles, but this is readily overcome. We redefine the Hamiltonian as

$$H' = H - \mu N, \tag{A2.2.72}$$

where μ is the chemical potential and N the total number of particles. The interaction term is unaffected, while the kinetic energy term becomes

$$T' = \sum_{\mathbf{k}} (\varepsilon_{\mathbf{k}} - \mu)\, a_{\mathbf{k}}^{\dagger} a_{\mathbf{k}}. \tag{A2.2.73}$$

The free-particle propagator is now written as

$$G_0(\mathbf{k},\omega) = \frac{1}{\omega-(\varepsilon_{\mathbf{k}}-\mu)+i\delta_{\mathbf{k}}} \quad (\delta_{\mathbf{k}} \to \pm 0 \text{ as } \varepsilon_{\mathbf{k}} \gtrless \mu). \qquad \text{(A2.2.74)}$$

We carry μ through the calculation, adjusting it in the final expression to obtain the correct number of particles from the equation

$$N = \sum_{\mathbf{k}} \langle\psi_0|a_{\mathbf{k}}^{\dagger}a_{\mathbf{k}}|\psi_0\rangle = -2i\sum_{\mathbf{k}} G(\mathbf{k}, t \to +0). \qquad \text{(A2.2.75)}$$

A2.2.6 External potentials

(i) *External field in absence of Coulomb interaction.* In the absence of the Coulomb term, the Hamiltonian is given by

$$H = \sum_{\mathbf{k}} \varepsilon_{\mathbf{k}} a_{\mathbf{k}}^{\dagger} a_{\mathbf{k}} + \sum_{\mathbf{k}_1\mathbf{k}_2} a_{\mathbf{k}_1}^{\dagger} a_{\mathbf{k}_2} V_{\mathbf{k}_1-\mathbf{k}_2} \qquad \left(\varepsilon_{\mathbf{k}} = \frac{\hbar^2 k^2}{2m}\right), \qquad \text{(A2.2.76)}$$

$V_{\mathbf{k}}$ being the Fourier transform of the potential. We again apply the perturbation theory of section A2.2.1 with

$$H_1 = \sum_{\mathbf{k}_1\mathbf{k}_2} a_{\mathbf{k}_1}^{\dagger} a_{\mathbf{k}_2} V_{\mathbf{k}_1-\mathbf{k}_2}. \qquad \text{(A2.2.77)}$$

The diagrams are now particularly simple in structure, there being a single vertex for each time in the integration of the series for $\ln R(t)$ or $G(\mathbf{k}_1, \mathbf{k}_2, t)$ which we draw as Figure A2.2.11(a). As indicated, momentum is conserved at each vertex, as before, as is also the energy parameter if we take Fourier transforms on the times, η being non-zero if $V(\mathbf{r})$ is time-dependent.

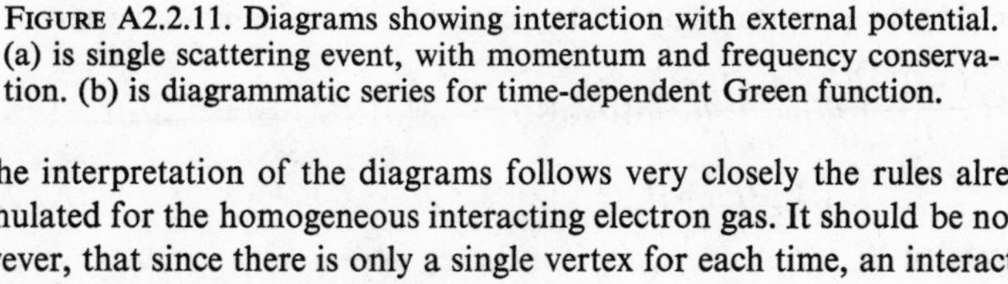

FIGURE A2.2.11. Diagrams showing interaction with external potential. (a) is single scattering event, with momentum and frequency conservation. (b) is diagrammatic series for time-dependent Green function.

The interpretation of the diagrams follows very closely the rules already formulated for the homogeneous interacting electron gas. It should be noted, however, that since there is only a single vertex for each time, an interaction line is associated with the factor $iV_{\mathbf{q}}$, not $2iV_{\mathbf{q}}$.

From the diagrammatic series, in Figure A2.2.11(b), there follows the Dyson equation

$$G(\mathbf{k}, \mathbf{k}_0, t-t_0) = G_0(\mathbf{k}, t-t_0)\,\delta_{\mathbf{k},\mathbf{k}_0} + \sum_{\mathbf{k}_1} \int_{-\infty}^{\infty} dt_1\, G_0(\mathbf{k}, t_1-t_0)\, V(\mathbf{k}-\mathbf{k}_1) \times G(\mathbf{k}_1, \mathbf{k}_0, t-t_1), \tag{A2.2.78}$$

and one sees that $-i\Sigma$ is just the external potential.

We should mention that there is a difficulty of principle in the discussion if $V(\mathbf{r})$ produces a non-spherical Fermi surface; we can see that if G_0 is defined relative to a spherical Fermi surface as hitherto, the Fermi surface remains spherical to all orders of perturbation theory. Further discussion is postponed, however, until we have reintroduced the particle interactions.

(ii) *External field applied to interacting gas.* We now take the perturbation to be

$$H_1 = H_{\text{ext}} + H_{\text{ee}} \tag{A2.2.79}$$

where H_{ext} and H_{ee} are respectively the perturbations due to the external potential and the electron–electron interactions. At present, we ignore the difficulty alluded to at the end of section (i). If we do so, the procedure is clear: we draw diagrams which for each time include both an electron–electron interaction line and an external interaction line; we then have three vertices for each time, as in Figure A2.2.12. It is rather restrictive, however,

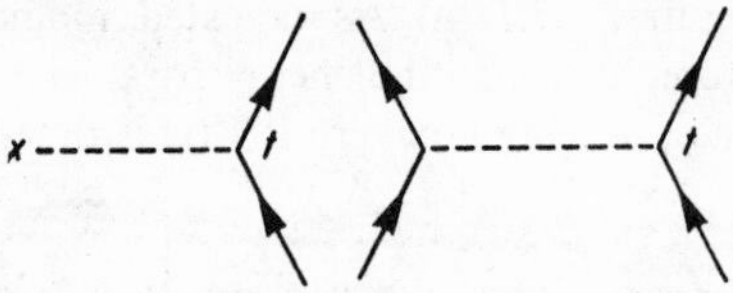

FIGURE A2.2.12. Shows at a particular time both an electron–electron scattering event and scattering from an external potential.

to have to pair internal and external interactions in this way and it is not hard to see that the practice is in fact unnecessary. From the nth-order term

$$\frac{(-i)^n}{n!} \int_{t_0}^{t} dt_1 \ldots \int_{t_0}^{t} dt_n\, T\{[H_{\text{ext}}(t_1) + H_{\text{ee}}(t_1)]\,[H_{\text{ext}}(t_2) + H_{\text{ee}}(t_2)] \ldots [H_{\text{ext}}(t_n) + H_{\text{ee}}(t_n)]\} \tag{A2.2.80}$$

we see that $H_{\text{ext}}(t_i)$ and $H_{\text{ee}}(t_i)$ can never appear together in any set of contractions and from this it follows that they will never appear in the same connected diagram. We may therefore consider the two sets of times, corresponding to external and internal interactions, as independent.

We now, therefore, draw all connected diagrams in which both types of interaction line can appear, but never associated with the same time. The only way in which the rules formulated in section 2.2.5 need modification is to associate a factor $-iV_{\mathbf{q}}$ with each interaction corresponding to the external potential.

A2.2.7 Redefinition of free-particle propagator

We recall from section A2.2.2 that a condition for the validity of perturbation theory is that the perturbed wave function must never be orthogonal to the unperturbed wave function. Forgetting interactions for a moment, it is evident that the perturbed many-body wave function *is* orthogonal to the free-electron wave function if the Fermi surface is deformed from its spherical shape by the presence of the external potential. This does not completely invalidate the discussion of A2.2.6(i), for we can regard the electrons as independent one-body systems and consider the perturbation theory to be one-body perturbation theory. We are thus in difficulties only near the Fermi surface. This difficulty is much more acute, however, when electron–electron interactions are considered as well, for the Fermi surface plays a dominant role in the quasi-particle theory discussed in Chapter 2. The orthogonality of the free-electron wave function and the perturbed wave function is now less obvious, but, instead, we can consider the behaviour of the Fermi surface as the interaction is switched on from zero. We suppose the system is in a "normal" state, and add a particle at the Fermi surface, with crystal momentum **k**, to the non-interacting system. We next switch on the interaction, whereupon the added particle becomes a quasi-particle with crystal momentum **k**; **k** remains a good quantum number of the same value because it is a constant of the motion with operator **P** independent of the interaction. It follows that, if we take the wrong Fermi surface, we must obtain an excited state after switching on the interaction. We see then that to obtain the correct ground state we must *start* with the correct Fermi surface S_f.

The way to overcome this difficulty has been pointed out by Nozières (1964). We require the unperturbed system to have the correct Fermi surface and a propagator of which the off-diagonal elements can be ignored. From this, it follows that we require the unperturbed Hamiltonian to be diagonal.

We can meet these requirements by writing, for the unperturbed propagator,

$$G_0'(\mathbf{k},\omega) = \frac{1}{\omega - \varepsilon'_{\mathbf{k}} + i\delta_{\mathbf{k}}} \quad (\delta_{\mathbf{k}} \to \pm 0, \text{ depending whether } \mathbf{k} \text{ is outside or inside } S_f). \tag{A2.2.81}$$

This can be obtained from the diagonal Hamiltonian

$$H_0' = \sum_{\mathbf{k}} \varepsilon_{\mathbf{k}}' a_{\mathbf{k}}^\dagger a_{\mathbf{k}}, \tag{A2.2.82}$$

where the $\varepsilon_{\mathbf{k}}'$'s are one-particle energies giving the correct Fermi surface S_f. To obtain the true Hamiltonian, we redefine the perturbation as

$$H_1' = H_1 - \sum_{\mathbf{k}} \lambda_{\mathbf{k}} a_{\mathbf{k}}^\dagger a_{\mathbf{k}}, \tag{A2.2.83}$$

where

$$\lambda_{\mathbf{k}} = \varepsilon_{\mathbf{k}}' - \varepsilon_{\mathbf{k}}. \tag{A2.2.84}$$

We shall now demonstrate a remarkable simplification. In the calculation of the self-energy and ground-state energy, the term involving $\lambda_{\mathbf{k}}$ on the right-hand side of equation (A2.2.83) may be ignored and the unperturbed propagator taken as

$$G_0'(\mathbf{k}, \omega) = \frac{1}{\omega - \varepsilon_{\mathbf{k}} + i\delta_{\mathbf{k}}} \quad (\delta_{\mathbf{k}} \to \pm 0 \text{ depending whether } \mathbf{k} \text{ is outside or inside } S_f). \tag{A2.2.85}$$

To show this, we note that the term in H_1' depending on $\lambda_{\mathbf{k}}$ will contribute parts to any diagram as shown in Figure A2.2.13(a) where we associate a factor

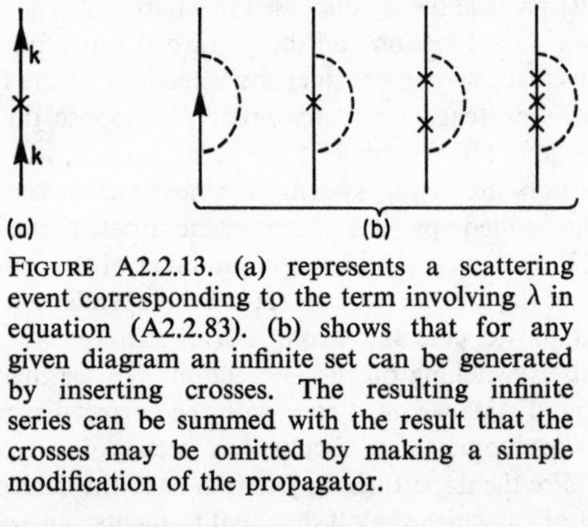

FIGURE A2.2.13. (a) represents a scattering event corresponding to the term involving λ in equation (A2.2.83). (b) shows that for any given diagram an infinite set can be generated by inserting crosses. The resulting infinite series can be summed with the result that the crosses may be omitted by making a simple modification of the propagator.

$-i\lambda_{\mathbf{k}}$ with the cross. Moreover, from any given diagram, we can generate an infinite set of other diagrams by inserting one, two, three, etc. crosses in any particular Fermion line, as shown in the example of Figure A2.2.13(b).

The Fermion line of the original diagram was associated with $G_0'(\mathbf{k}, \omega)$ and it can be seen from Figure A2.2.13 that the generated diagrams can be summed, with the result that we take the original diagrams only (those without crosses) with Fermion lines associated with

$$G_0(\mathbf{k}, \omega) = \frac{1}{\omega - \varepsilon'_{\mathbf{k}} + \lambda_{\mathbf{k}} + i\delta_{\mathbf{k}}} = \frac{1}{\omega - \varepsilon_{\mathbf{k}} + i\delta_{\mathbf{k}}}. \tag{A2.2.86}$$

A2.2.8 Rules for two-particle Green functions

The diagrams for the two-particle Green function have been illustrated in the main text. The factors associated with them may be derived along lines laid down for the single-particle Green function and the rules are in fact identical, except for a factor -1 when the line beginning at 1 ends at 4 (see Figure A2.2.14). It should also be noted that conservation of energy requires

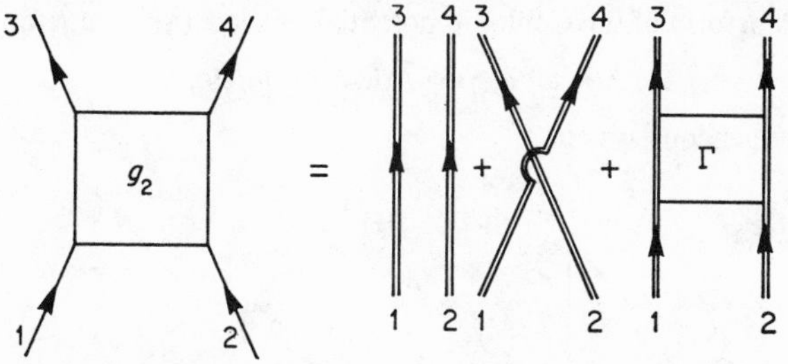

FIGURE A2.2.14. Diagrammatic representation of a two-particle Green function.

the δ-function $\delta(\omega_1 + \omega_2 - \omega_3 - \omega_4)$ when we consider G_2 in Fourier transform. This function becomes a product of two δ-functions for the two skeleton diagrams contributing to G_2, in which there is no interaction between the two particles. Γ in Figure A2.2.14 represents all repeated scatterings of two dressed particles.

APPENDIX A2.3 WARD IDENTITIES

A very interesting example of the use of diagrammatic techniques is afforded by the proof of the Ward identities, which relate limits of the vertex functions to derivatives of the self-energy. We shall consider only the Ward identities for the vertex function associated with the interaction of a scalar field, as discussed in the main text. The Ward identities for the vertex functions associated with interaction with a vector field can be found in the book by Nozières (1964).

(a) *Vertex functions.* In the main text, we were principally concerned with the sum Λ of all *proper* vertex parts (those irreducible with respect to internal interaction lines). Here we shall also be concerned with the sum of all vertex parts, including improper ones, which appear in the Ward identities for a system of particles interacting via short-range forces. We shall find it convenient to denote the latter sum, the vertex function, by Λ, where necessary distinguishing the *proper* vertex function by $\tilde{\Lambda}$.

Since Λ is reducible, we can insert any number of polarization parts into the interaction line of a given diagram and we then obtain the equivalence shown in Figure A2.3.1. The corresponding equation is

$$v(\mathbf{q})\,\Lambda(\mathbf{k},\omega;\mathbf{q},\eta) = V(\mathbf{q},\eta)\,\tilde{\Lambda}(\mathbf{k},\omega;\mathbf{q},\eta), \tag{A2.3.1}$$

where the factor $V(\mathbf{q}\eta)$ arises from the screened interaction line and $v_{\mathbf{q}}$ is the Fourier transform of the Coulomb potential. Since $V(\mathbf{q}\eta) = v_{\mathbf{q}}/\varepsilon(\mathbf{q}\eta)$, we have

$$\Lambda(\mathbf{k},\omega;\mathbf{q},\eta) = \tilde{\Lambda}(\mathbf{k},\omega;\mathbf{q},\eta)/\varepsilon(\mathbf{q},\eta) \tag{A2.3.2}$$

for a homogeneous system.

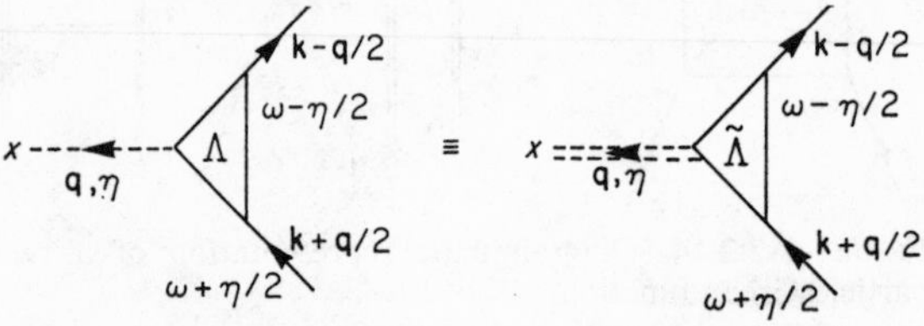

FIGURE A2.3.1. Relation between vertex function Λ and proper vertex function $\tilde{\Lambda}$.

(b) *Identities for homogeneous systems.* We shall discuss the origin of the Ward identities

$$\tilde{\Lambda}^0(\mathbf{k},\omega) = 1 + \frac{\partial \Sigma(\mathbf{k},\omega)}{\partial \omega}, \tag{A2.3.3}$$

$$\tilde{\Lambda}^\infty(\mathbf{k},\omega) = 1 + \frac{\partial \Sigma(\mathbf{k},\omega)}{\partial \omega} + \frac{\partial \Sigma(\mathbf{k},\omega)}{\partial \mu}, \tag{A2.3.4}$$

where the superscripts indicate the values of $\mathbf{r} = \mathbf{q}/\eta$ as $\mathbf{q}, \eta \to 0$. The equations hold for a homogeneous system of charged particles; if the forces are short range the tildes on the Λ are to be omitted.

(i) *First identity*: let us consider the derivative

$$\frac{\partial \Sigma(\mathbf{k},\omega)}{\partial \omega} = \lim_{\varepsilon\to 0} \frac{1}{\varepsilon}\left[\Sigma(\mathbf{k},\omega+\varepsilon) - \Sigma(\mathbf{k}\omega)\right]. \tag{A2.3.5}$$

The addition of ε to ω may be effected by adding ε to the energy parameter of every Fermion line of every diagram for the self-energy $\Sigma(\mathbf{k}\omega)$. Conservation of the energy parameter at each vertex ensures that the result is just $\Sigma(\mathbf{k}, \omega+\varepsilon)$. Now if we take any diagram for Σ with n Fermion lines to first order in ε, the diagram will be a sum of n terms, each corresponding to a diagram with only *one* of its internal lines associated with ε. One of these terms is shown graphically in Figure A2.3.2(a). Denoting the sum of all such

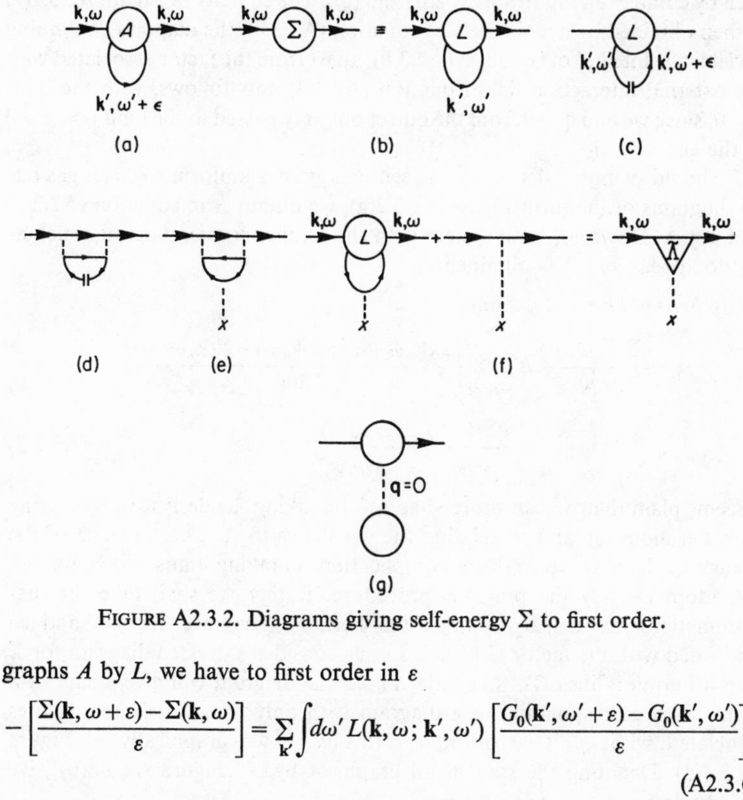

FIGURE A2.3.2. Diagrams giving self-energy Σ to first order.

graphs A by L, we have to first order in ε

$$-\left[\frac{\Sigma(\mathbf{k}, \omega+\varepsilon)-\Sigma(\mathbf{k}, \omega)}{\varepsilon}\right]=\sum_{\mathbf{k}'}\int d\omega' L(\mathbf{k}, \omega; \mathbf{k}', \omega')\left[\frac{G_0(\mathbf{k}', \omega'+\varepsilon)-G_0(\mathbf{k}', \omega')}{\varepsilon}\right]. \tag{A2.3.6}$$

As indicated in Figure A2.3.2(b), L is obtained by letting the line associated with ε be each internal line of Σ in turn.

Now

$$G_0(\mathbf{k}, \omega)=\frac{-1}{\frac{1}{2}k^2-\omega-i\delta} \tag{A2.3.7}$$

and hence

$$\frac{G_0(\mathbf{k}, \omega+\varepsilon)-G_0(\mathbf{k}, \omega)}{\varepsilon} = i^2\, G_0(\mathbf{k}, \omega+\varepsilon)\, G_0(\mathbf{k}, \omega). \qquad \text{(A2.3.8)}$$

It is now evident that the left-hand side of equation (A2.3.6) is obtained by cutting each internal line of Σ in turn and associating one-half of the cut line with $G_0(\mathbf{k}', \omega'+\varepsilon)$ and the other with $G_0(\mathbf{k}', \omega')$ as shown in Figure A2.3.2(c). But on cutting an internal line of Σ, one obtains a diagram of $\tilde{\Lambda}$ [as can be seen by comparing Figures A2.3.2(d) and (e)]. The equality of Figure A2.3.2(f) is then obtained, where, in view of equation (A2.3.8), the diagram containing L yields the integral of equation (A2.3.6), apart from the factor associated with the external interaction line. Equation (A2.3.3) now follows, with the limit $\mathbf{r} = 0$, since we had $\mathbf{q} = 0$ from the outset but only passed to the limit $\eta = \varepsilon \to 0$ at the end.

It should be noted that since the self-energy of a uniform electron gas has no diagrams of the form Figure A2.3.2(g), we obtain $\tilde{\Lambda}$ in equation (A2.3.3) and not Λ. However, for a system with short-range forces, diagrams such as (g) do appear and Λ is obtained.

(ii) *Second identity*: defining

$$\frac{\delta\Sigma(\mathbf{k}, \omega;\, \mu)}{\delta\mu} \equiv \lim_{d\mu\to 0}\left[\frac{\Sigma(\mathbf{k}, \omega+d\mu;\, \mu+d\mu)-\Sigma(\mathbf{k}, \omega;\, \mu)}{d\mu}\right]$$

$$= \frac{\partial\Sigma(\mathbf{k}, \omega)}{\partial\omega}+\frac{\partial\Sigma(\mathbf{k}, \omega)}{\partial\mu}, \qquad \text{(A2.3.9)}$$

it seems plain that we can proceed as before, taking the derivative by cutting each Fermion line and so relating the derivative to Λ. The presence of the change $d\mu$, however, introduces complications in taking limits and so we will not adopt entirely the previous procedure. Rather we shall take the diagrammatic series for Σ in which the Fermion lines are dressed, and so associated with the factor $G(\mathbf{k}', \omega')$. Let us consider a dressed diagram for Σ with n Fermion lines. To first order in $d\mu$, the diagram will give a sum of n terms, each corresponding to a diagram with only one of its internal lines associated with $d\mu$. One of these terms is shown graphically in Figure A2.3.3(a). Denoting the sum of all graphs A by $\mathscr{I}$ [Figure A2.3.3(b)], we have to first order in $d\mu$

$$[\Sigma(\mathbf{k}, \omega+d\mu;\, \mu+d\mu)-\Sigma(\mathbf{k}, \omega;\, \mu)]$$

$$= \sum_{\mathbf{k}'}\int d\omega'\, \mathscr{I}(\mathbf{k}, \omega;\, \mathbf{k}', \omega')\, [G(\mathbf{k}', \omega'+d\mu;\, \mu+d\mu)-G(\mathbf{k}', \omega';\, \mu)].$$

(A2.3.10)

Figure A2.3.3(c) shows that $\mathscr{I}$ is found by cutting each of the dressed lines of Σ in turn. Now we note that any diagram for ${}^0I$ as defined on p. 170, the superscript denoting zero total spin, contributes to Σ in precisely the same way as $\mathscr{I}$; the total spin is zero because the Fermion lines external to Σ are

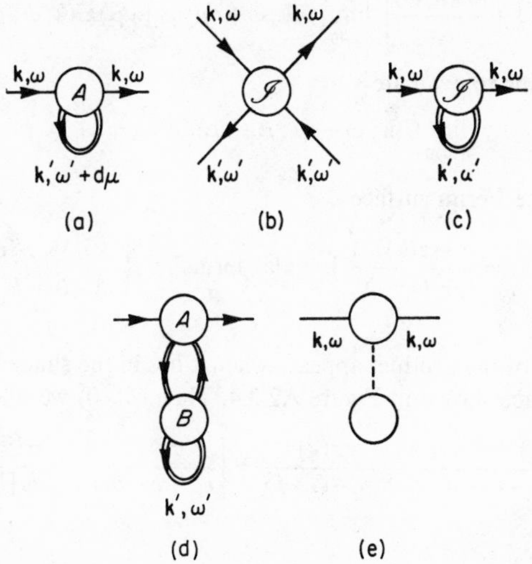

FIGURE A2.3.3. Diagrams leading to equation (A2.3.10) for self-energy Σ.

labelled with the same spin parameter and any diagram of the form of Figure A2.3.3(d) is contained in A, since the complex B simply contributes to the dressing of the Fermion line. Hence $\mathscr{I}$ is irreducible and, in fact,

$$\mathscr{I}(\mathbf{k}, \omega; \mathbf{k}', \omega') = {}^0I(\mathbf{k}, \omega; \mathbf{k}', \omega') = \lim_{\mathbf{q}, \eta \to 0} {}^0I(\mathbf{k}, \omega; \mathbf{k}', \omega'; \mathbf{q}, \eta) \tag{A2.3.11}$$

for short-range forces. For the electron system, we have

$$\mathscr{I}(\mathbf{k}, \omega; \mathbf{k}', \omega') = {}^0\tilde{I}(\mathbf{k}, \omega; \mathbf{k}', \omega') \tag{A2.3.12}$$

since no diagram of the form of Figure A2.3.3(e) contributes to Σ. Here $\tilde{I}$ is "proper", that is irreducible with respect to the internal interaction line, so that

$${}^0\tilde{I} = {}^0I - v_{\mathbf{q}}/2\pi i\mathscr{V}. \tag{A2.3.13}$$

Now since

$$G(\mathbf{k}, \omega) = \frac{-1}{\frac{1}{2}k^2 + \Sigma(\mathbf{k}, \omega) - \omega - i\delta}, \quad \begin{cases} \delta = +0, & k > k_{\mathrm{f}}, \\ \delta = -0, & k < k_{\mathrm{f}}, \end{cases} \tag{A2.3.14}$$

we have

$$-\lim_{d\mu\to 0}\left[\frac{G(\mathbf{k},\omega+d\mu;\mu+d\mu)-G(\mathbf{k},\omega;\mu)}{d\mu}\right]$$

$$=\left[1+\frac{\delta\Sigma(\mathbf{k},\omega)}{\delta\mu}\right]\lim_{d\mu\to 0}[G(\mathbf{k},\omega+d\mu;\mu+d\mu)\,G(\mathbf{k},\omega;\mu)] \quad \text{(A2.3.15)}$$

and so it remains to evaluate

$$\lim_{d\mu\to 0} G(\mathbf{k};\omega+d\mu;\mu+d\mu)\,G(\mathbf{k},\omega;\mu). \quad \text{(A2.3.16)}$$

Since near the Fermi surface

$$G(\mathbf{k},\omega)=\frac{-z(\mathbf{k})}{\varepsilon_{\mathbf{k}}-\omega-i\delta}+[\text{regular terms}],\qquad \begin{cases}\delta>0, & k_{\mathrm{f}}>0,\\ \delta<0, & k_{\mathrm{f}}<0,\end{cases} \quad \text{(A2.3.17)}$$

a singular part of the product appears when $\mathbf{k}$ lies in the shaded region about the Fermi surface shown in Figure A2.3.4. Then ($\delta>0$) we find

$$\frac{z(\mathbf{k})}{\varepsilon_{\mathbf{k}}+(\partial\varepsilon_{\mathbf{k}}/\partial\mu)\,d\mu-\omega-d\mu+i\delta}\,\frac{z(\mathbf{k})}{\varepsilon_{\mathbf{k}}-\omega-i\delta}=\left[\frac{z(\mathbf{k})}{\varepsilon_{\mathbf{k}}-\omega-i\delta}\right]^2-\frac{2\pi i z^2(\mathbf{k})\,\delta(\varepsilon_{\mathbf{k}}-\omega)}{d\mu[1-(\partial\varepsilon_{\mathbf{k}}/\partial\mu)]}. \quad \text{(A2.3.18)}$$

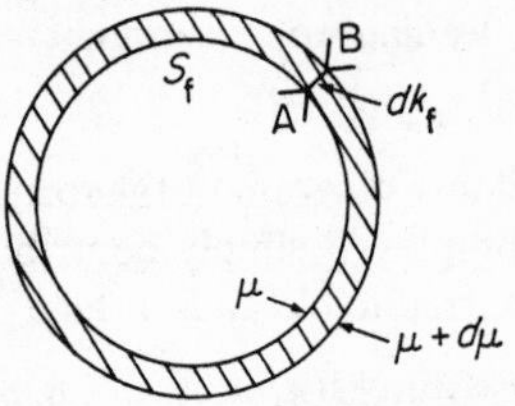

FIGURE A2.3.4. Part of $\mathbf{k}$ space near Fermi surface.

The first term on the right-hand side appears in the product $G^2(\mathbf{k},\omega)$. To take account of the fact that the second term only appears in the shaded regions of Figure A2.3.4, we multiply it by

$$d\varepsilon_{\mathbf{k}}\,\delta(\varepsilon_{\mathbf{k}}-\mu)=\frac{\partial\varepsilon}{\partial k}\,\delta(\varepsilon_{\mathbf{k}}-\mu)\,dk_{\mathrm{f}}=v_{\mathrm{f}}\,dk_{\mathrm{f}}\,\delta(\varepsilon_{\mathbf{k}}-\mu).$$

Since the energy of a quasi-particle at the Fermi surface is μ, we also find, referring to Figure A2.3.4, that

$$d\mu=\varepsilon(\mathbf{k}_{\mathrm{B}},\mu+d\mu)-\varepsilon(\mathbf{k}_{\mathrm{A}},\mu)=\frac{\partial\varepsilon_{\mathbf{k}}}{\partial\mu}\,d\mu+v_{\mathrm{f}}\,dk_{\mathrm{f}}. \quad \text{(A2.3.19)}$$

Thus we have

$$d\mu\left(1-\frac{\partial\varepsilon}{\partial\mu}\right) = v_{\mathrm{f}}\, dk_{\mathrm{f}} \tag{A2.3.20}$$

and hence we may finally write

$$\begin{aligned}\lim_{d\mu\to 0} G(\mathbf{k}, \omega+d\mu;\, \mu+d\mu)\, G(\mathbf{k}, \omega;\, \mu) &= G^2(\mathbf{k}, \omega;\, \mu) - 2\pi i z^2(\mathbf{k})\, \delta(\varepsilon_{\mathbf{k}}-\mu)\, \delta(\varepsilon_{\mathbf{k}}-\omega) \\ &= G^2(\mathbf{k}, \omega;\, \mu) + R^{\infty}(\mathbf{k}, \omega).\end{aligned} \tag{A2.3.21}$$

It follows that equation (A2.3.10) may be written in the form

$$\frac{\delta\Sigma(\mathbf{k}, \omega)}{\delta\mu} = \sum_{\mathbf{k}} \int d\omega'\, {}^0\tilde{I}(\mathbf{k}, \omega;\, \mathbf{k}', \omega') \left[G^2(\mathbf{k}', \omega') + R^{\infty}(\mathbf{k}', \omega') \left\{1+\frac{\delta\Sigma(\mathbf{k}', \omega')}{\delta\mu}\right\}\right] \tag{A2.3.22}$$

with ${}^0\tilde{I}$ replaced by ${}^0I$ if the force are short range.

On the other hand, it is evident from Figure A2.3.5 that Λ obeys the integral equation

$$\begin{aligned}\Lambda(\mathbf{k}, \omega;\, \mathbf{q}, \eta) = 1 + \sum_{\mathbf{k}'} \int d\omega'\, {}^0I(\mathbf{k}, \omega;\, \mathbf{k}', \omega';\, \mathbf{q}, \eta)\, G(\mathbf{k}'-\tfrac{1}{2}\mathbf{q}, \omega'-\tfrac{1}{2}\eta) \\ \times G(\mathbf{k}'+\tfrac{1}{2}\mathbf{q}, \omega'+\tfrac{1}{2}\eta)\, \Lambda(\mathbf{k}', \omega';\, \mathbf{q}, \eta).\end{aligned} \tag{A2.3.23}$$

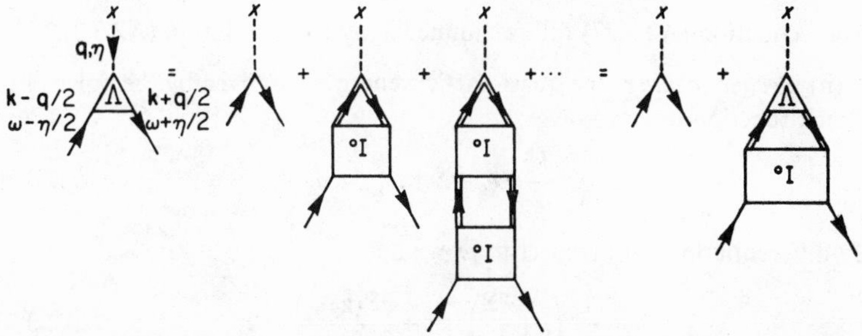

FIGURE A2.3.5. Diagrammatic representation of integral equation for vertex function Λ.

Similarly, applying diagrammatic analysis to the proper vertex function $\tilde{\Lambda}$, we find

$$\begin{aligned}\tilde{\Lambda}(\mathbf{k}, \omega;\, \mathbf{q}, \eta) = 1 + \sum_{\mathbf{k}'} \int d\omega'\, {}^0\tilde{I}(\mathbf{k}, \omega;\, \mathbf{k}', \omega';\, \mathbf{q}, \eta)\, G(\mathbf{k}'-\tfrac{1}{2}\mathbf{q}, \omega'-\tfrac{1}{2}\eta) \\ \times G(\mathbf{k}'+\tfrac{1}{2}\mathbf{q};\, \omega'+\tfrac{1}{2}\eta)\, \tilde{\Lambda}(\mathbf{k}', \omega';\, \mathbf{q}, \eta).\end{aligned} \tag{A2.3.24}$$

Since it can readily be shown that

$$\lim_{\mathbf{q}\to 0}\lim_{\eta\to 0} G(\mathbf{k}'-\tfrac{1}{2}\mathbf{q},\omega'-\tfrac{1}{2}\eta)\,G(\mathbf{k}'+\tfrac{1}{2}\mathbf{q},\omega'+\tfrac{1}{2}\eta) = G^2(\mathbf{k}',\omega')+R^{\infty}(\mathbf{k}',\omega'), \tag{A2.3.25}$$

comparison of equation (A2.3.24) with equation (A2.3.22) yields

$$1+\frac{\delta\Sigma(\mathbf{k},\omega)}{\delta\mu} = \tilde{\Lambda}^{\infty}(\mathbf{k},\omega) \tag{A2.3.26}$$

which was to be proved.

(c) *Identities at Fermi surface*. When $k = k_{\mathrm{f}}$ and $\omega = \mu$, equations (A2.3.3) and (A2.3.4) may be written as

$$z(\mathbf{k})\,\tilde{\Lambda}^{0}(\mathbf{k}) = 1 \tag{A2.3.27}$$

and

$$z(\mathbf{k})\,\tilde{\Lambda}^{\infty}(\mathbf{k}) = v_{\mathrm{f}}\frac{dk_{\mathrm{f}}}{d\mu}. \tag{A2.3.28}$$

For systems with short-range forces, the tildes are to be omitted.

(i) *First identity*: we repeat equation (2.10.72) of the main text:

$$z(\mathbf{k}) = \left.\frac{1}{1+[\partial\Sigma(\mathbf{k},\omega)/\partial\omega]}\right|_{\omega=\epsilon_{\mathbf{k}}} \tag{A2.3.29}$$

when equation (A2.3.27) follows immediately from equation (A2.3.3).

(ii) *Second identity*: the quasi-particle energy $\varepsilon_{\mathbf{k}}$ is given by the poles of the Green function or

$$\frac{k^2}{2}-\varepsilon_{\mathbf{k}}-\Sigma(\mathbf{k},\varepsilon_{\mathbf{k}}) = 0. \tag{A2.3.30}$$

By differentiating with respect to μ, we find

$$\frac{\partial\varepsilon_{\mathbf{k}}}{\partial\mu}\left(1+\frac{\partial\Sigma}{\partial\omega}\right)_{\omega=\epsilon_{\mathbf{k}}}+\frac{\partial\Sigma(\mathbf{k},\varepsilon_{\mathbf{k}})}{\partial\mu} = 0 \tag{A2.3.31}$$

and, using equation (A2.3.4), this becomes

$$\left(1-\frac{\partial\varepsilon_{\mathbf{k}}}{\partial\mu}\right)\left(1+\frac{\partial\Sigma}{\partial\omega}\right)_{\omega=\epsilon_{\mathbf{k}}} = \tilde{\Lambda}^{\infty}(\mathbf{k}). \tag{A2.3.32}$$

Multiplying through by $z(\mathbf{k})$, we obtain, in view of equation (A2.3.29),

$$1-\frac{\partial\varepsilon_{\mathbf{k}}}{\partial\mu} = z(\mathbf{k})\,\tilde{\Lambda}^{\infty}(\mathbf{k}). \tag{A2.3.33}$$

But from equation (A2.3.20)

$$1-\frac{\partial\varepsilon_{\mathbf{k}}}{\partial\mu}=v_{\mathrm{f}}\frac{dk_{\mathrm{f}}}{d\mu} \tag{A2.3.34}$$

which proves equation (A2.3.28).

(d) *Identities for inhomogeneous systems.* The Ward identities are readily generalized for a system of electrons moving in a periodic potential. As an example, we generalize equation (A2.3.4). The method of proof is largely the same except, of course, that the Fermi surface is not assumed to be spherical. The essential difference is that improper diagrams [Figure A2.3.3(e)] can now contribute to Σ (though not, of course, with any internal interaction line labelled by $\mathbf{q}=0$). Hence the proper vertex function $\tilde{\Lambda}$ is replaced by Λ and we obtain

$$\Lambda^{\infty}(\mathbf{k},\mathbf{k}';\omega)=\delta_{\mathbf{k},\mathbf{k}'}+\frac{\partial\Sigma(\mathbf{k},\mathbf{k}';\mu)}{\partial\omega}+\frac{\partial\Sigma(\mathbf{k},\mathbf{k}';\mu)}{\partial\mu}, \tag{A2.3.35}$$

where

$$\mathbf{k}'=\mathbf{k}+\mathbf{K}, \tag{A2.3.36}$$

$\mathbf{K}$ being a reciprocal lattice vector.

To obtain Λ^{∞} in terms of proper vertex functions, we note the equivalence shown in Figure A2.3.1, which now reads

$$\Lambda(\mathbf{k},\mathbf{k}';\mathbf{q},\eta)=\sum_{\mathbf{K}}\frac{\tilde{\Lambda}(\mathbf{k}-\frac{1}{2}\mathbf{K},\mathbf{k}'+\frac{1}{2}\mathbf{K},\mathbf{q}+\mathbf{K},\eta)}{\varepsilon(\mathbf{q},\mathbf{q}+\mathbf{K})}. \tag{A2.3.37}$$

Hence

$$\left(\frac{\partial}{\partial\omega}+\frac{\partial}{\partial\mu}\right)\Sigma(\mathbf{k},\mathbf{k}';\omega)+\delta_{\mathbf{k},\mathbf{k}'}=\sum_{\mathbf{K}}\frac{\tilde{\Lambda}(\mathbf{k}-\frac{1}{2}\mathbf{K},\mathbf{k}'+\frac{1}{2}\mathbf{K},\mathbf{K},0)}{\varepsilon(\mathbf{q},\mathbf{q}+\mathbf{K})}. \tag{A2.3.38}$$

As we saw in the main text, the charge distribution is obtained by closing the dressed line representing $G(\mathbf{k},\mathbf{k}';\omega)$ upon itself. Using equation (A2.3.15) we have, bearing in mind Figure A2.3.6, for the proper polarization part

$$\frac{\partial\rho(\mathbf{K})}{\partial\mu}=\sum_{\mathbf{K}'}\frac{\Pi(\mathbf{K},\mathbf{K}')}{\varepsilon(\mathbf{K}',0)}=\mathscr{F}(\mathbf{K},0), \tag{A2.3.39}$$

where we have used equation (2.11.55).

It should be emphasized that the above equations are not restricted to metals; the analysis holds equally well for insulators. We should finally remark that, in view of formal objections by Kohn and Luttinger (1960) to the use of ground-state perturbation theory on metals, the above Ward identities for crystals have been proved, using finite temperature theory, by Sham and Kohn (1966). At finite temperatures, the diagrams remain as illustrated in this appendix, the formalism being essentially the same as that applied to phonons in Chapter 3 and Appendix A3.7. For further details, recourse may be made to the book by Abrikosov, Gorkov and Dzaloshinski (1963).

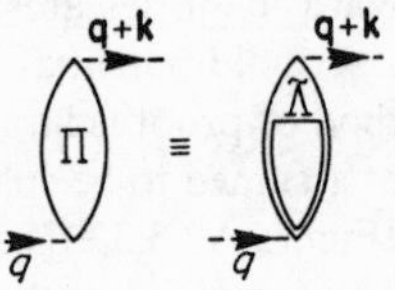

FIGURE A2.3.6. Diagram of proper polarization part Π showing relation between Π and $\tilde{\Lambda}$, the proper vertex part.

APPENDIX A2.4 GRADIENT EXPANSION OF ENERGY DENSITIES

The perturbation theory of Appendix A1.2 gave the Dirac density matrix $\rho(\mathbf{r},\mathbf{r}'E)$ as a functional of the potential $V(\mathbf{r})$. If now we go on to the diagonal, $\mathbf{r}'=\mathbf{r}$, then we have $\rho(\mathbf{r})$ as a functional of $V(\mathbf{r})$. As Stoddart and March (1967) have shown, this latter relation can be inverted to yield the one-body potential as a functional of the electron density. Then this expression can be reinserted into the full Dirac density matrix, which is thereby obtained as a functional of its diagonal element $\rho(\mathbf{r})$.

Clearly, since the kinetic energy density $t_{\mathbf{r}}[\rho]$ is simply given by

$$\text{kinetic energy} = \int t_{\mathbf{r}}[\rho]\,d\mathbf{r} = -\frac{1}{2}\int d\mathbf{r}[\nabla_{\mathbf{r}}^2\,\rho(\mathbf{r}\mathbf{r}'E_{\mathrm{f}})]_{\mathbf{r}'=\mathbf{r}} \qquad \text{(A2.4.1)}$$

we can obtain this, in principle, as a functional of the density $\rho(\mathbf{r})$, in the spirit of Thomas–Fermi theory. This programme has, as yet, proved possible to carry through only

(a) perturbatively from a uniform electron gas of density ρ_0,

(b) by gradient expansions,

the latter being clearly intimately connected with Thomas–Fermi theory. Our interest here is mainly in (b), though we shall derive the results starting from (a) (see Stoddart, Beattie and March, 1971).

If we write, as usual, the displaced charge $\Delta = \rho - \rho_0$ then the first-order relation between ρ and V can be inverted to yield

$$V(\mathbf{r}) = \frac{1}{2}\int d\mathbf{s}\,K(\mathbf{r}-\mathbf{s})\,\Delta(\mathbf{s}) + O(\Delta^2), \qquad \text{(A2.4.2)}$$

where

$$K(\mathbf{r}) = \frac{1}{(2\pi)^3} \int d\mathbf{q}\, e^{-i\mathbf{q}.\mathbf{r}}\, \tilde{K}(q, k_f) \qquad \text{(A2.4.3)}$$

and

$$\tilde{K}(q, k_f) = -\frac{2\pi^2}{k_f}\left[\frac{1}{2} + \frac{k_f}{2q}\left(1 - \frac{q^2}{4k_f^2}\right) \ln \left|\frac{q+2k_f}{q-2k_f}\right|\right]^{-1}, \qquad \text{(A2.4.4)}$$

the latter being closely connected with the Lindhard dielectric function of section 2.3.1. We now substitute this expression into equation (A1.2.10) and we then obtain, to second order in the displaced charge Δ,

$$\rho(\mathbf{rr}'E_f) = \rho_0(\mathbf{rr}'E_f) + \frac{1}{2}\int d\mathbf{s}\, A_1(\mathbf{rr}'\mathbf{s}E_f)\,\Delta(\mathbf{s})$$
$$+\frac{1}{4}\iint d\mathbf{s}_1\, d\mathbf{s}_2\, A_2(\mathbf{rr}'\mathbf{s}_1\,\mathbf{s}_2\, E_f)\,\Delta(\mathbf{s}_1)\,\Delta(\mathbf{s}_2) + \ldots, \qquad \text{(A2.4.5)}$$

where

$$A_1(\mathbf{rr}'\mathbf{s}E_f) = -\frac{k_f^2}{4\pi^3}\int d\mathbf{r}_1\, K(\mathbf{r}_1 - \mathbf{s})\, \frac{j_1(k_f|\mathbf{r}-\mathbf{r}_1| + k_f|\mathbf{r}_1 - \mathbf{r}'|)}{|\mathbf{r}-\mathbf{r}_1|\,|\mathbf{r}_1-\mathbf{r}'|}, \qquad \text{(A2.4.6)}$$

$$A_2(\mathbf{rr}'\mathbf{s}_1\,\mathbf{s}_2\, E_f) = -\frac{k_f^2}{4\pi^3}\int d\mathbf{r}_1\, W(\mathbf{r}_1\,\mathbf{s}_1\,\mathbf{s}_2)\, \frac{j_1(k_f|\mathbf{r}-\mathbf{r}_1| + k_f|\mathbf{r}_1 - \mathbf{r}'|)}{|\mathbf{r}-\mathbf{r}_1|\,|\mathbf{r}_1-\mathbf{r}'|}$$
$$+\frac{k_f^2}{8\pi^4}\iint d\mathbf{r}_1\, d\mathbf{r}_2\, K(\mathbf{r}_1 - \mathbf{s}_1)\, K(\mathbf{r}_2 - \mathbf{s}_2)$$
$$\times \frac{j_1(k_f|\mathbf{r}-\mathbf{r}_1| + k_f|\mathbf{r}_1-\mathbf{r}_2| + k_f|\mathbf{r}_2-\mathbf{r}'|)}{|\mathbf{r}-\mathbf{r}_1|\,|\mathbf{r}_1-\mathbf{r}_2|\,|\mathbf{r}_2-\mathbf{r}'|} \qquad \text{(A2.4.7)}$$

and

$$W(\mathbf{rs}_1\,\mathbf{s}_2) = -\frac{k_f^2}{8\pi^4}\int d\mathbf{s}\, d\mathbf{x}\, d\mathbf{y} K(\mathbf{r}-\mathbf{s})\, K(\mathbf{x}-\mathbf{s}_1)\, K(\mathbf{y}-\mathbf{s}_2)$$
$$\times \frac{j_1(k_f|\mathbf{s}-\mathbf{x}| + k_f|\mathbf{x}-\mathbf{y}| + k_f|\mathbf{y}-\mathbf{s}|)}{|\mathbf{s}-\mathbf{x}|\,|\mathbf{x}-\mathbf{y}|\,|\mathbf{y}-\mathbf{s}|}. \qquad \text{(A2.4.8)}$$

We now can use the result (A2.4.5) to obtain the kinetic energy of this system of non-interacting particles [and also the exchange energy: see equation (A2.4.23) below], when we find from equation (A2.4.1) that

$$t_r[\rho] = \frac{3}{10}(3\pi^2)^{\frac{2}{3}}\, \rho_0^{\frac{5}{3}} - \frac{1}{4}\int d\mathbf{s}\, K(\mathbf{r}-\mathbf{s})\,\Delta(\mathbf{r})\,\Delta(\mathbf{s}) + O(\Delta^3). \qquad \text{(A2.4.9)}$$

The use of this expression for the kinetic energy is clearly restricted to the linear response régime.

However, if we wish we can exploit the fact that the energy density is arbitrary with respect to the addition of a term of the general form $\nabla_r h(\mathbf{r})$.

Then it is readily shown that equation (A2.4.9) is equivalent to the result

$$t_r[\rho] = \frac{3}{10}(3\pi^2)^{\frac{2}{3}} \rho_0^{\frac{5}{3}} + \frac{1}{6}(3\pi^2)^{\frac{2}{3}} \rho_0^{-\frac{1}{3}} \Delta^2(\mathbf{r})$$
$$+\frac{1}{8}\int d\mathbf{r}' K(\mathbf{r}')\left[\Delta\left(\mathbf{r}+\frac{1}{2}\mathbf{r}'\right)-\Delta\left(\mathbf{r}-\frac{1}{2}\mathbf{r}'\right)\right]^2 + O(\Delta^3). \quad \text{(A2.4.10)}$$

The assumption made is that a local gradient series exists for the energy density and we write

$$t_r[\rho] = t_0(\rho(\mathbf{r})) + t_2^{(2)}(\rho(\mathbf{r}))(\nabla_r \rho(\mathbf{r}))^2 + O(r_0^{-4}), \quad \text{(A2.4.11)}$$

where, following Hohenberg and Kohn (1964), we put in the basic assumption that $\rho(\mathbf{r})$ is slowly varying by writing

$$\rho(\mathbf{r}) = \phi(\mathbf{r}/r_0) \quad \text{(A2.4.12)}$$

with r_0 large. It is in this way that the terms in equation (2.4.10) have been classified. Higher terms in the series (2.4.11) will involve more than two gradient operators acting on two or more $\rho(\mathbf{r})$'s and thus will be of order r_0^{-4} or higher order.

The coefficient $t_2^{(2)}$ in equation (A2.4.11) may now be obtained as a function of density from the terms in equation (A2.4.10) as follows. We obtain the energy density $t_r[\rho]$ to $O(\Delta^2)$ and to $O(r_0^{-2})$ using equations (A2.4.9) and (A2.4.10). From equation (A2.4.10) we may obtain t_r to $O(r_0^{-2})$ and $O(\Delta^2)$ by using

$$\Delta(\mathbf{r}+\tfrac{1}{2}\mathbf{r}') - \Delta(\mathbf{r}-\tfrac{1}{2}\mathbf{r}') = \mathbf{r}'.\nabla_r \rho(\mathbf{r}) + \ldots \quad \text{(A2.4.13)}$$

and we have

$$t_r[\rho] \simeq \frac{3}{10}(3\pi^2)^{\frac{2}{3}} \rho_0^{\frac{5}{3}} + \frac{1}{6}(3\pi^2)^{\frac{2}{3}} \rho_0^{-\frac{1}{3}} \Delta^2(\mathbf{r}) + \frac{1}{8}\int d\mathbf{r}' K(\mathbf{r}')\,[\mathbf{r}'.\nabla_r \rho(\mathbf{r})]^2. \quad \text{(A2.4.14)}$$

Similarly from equation (A2.4.11), using the result of the constant density limit (Thomas–Fermi result),

$$t_0(\rho(\mathbf{r})) = \tfrac{3}{10}(3\pi^2)^{\frac{2}{3}} \rho^{\frac{5}{3}}(\mathbf{r}); \quad \rho(\mathbf{r}) = \rho_0 + \Delta(\mathbf{r}), \quad \text{(A2.4.15)}$$

we have again, to $O(\Delta^2)$ and $O(r_0^{-2})$,

$$t_r[\rho] \simeq \tfrac{3}{10}(3\pi^2)^{\frac{2}{3}} \rho_0^{\frac{5}{3}} + \tfrac{1}{6}(3\pi^2)^{\frac{2}{3}} \rho_0^{-\frac{1}{3}} \Delta^2(\mathbf{r}) + t_2^{(2)}(\rho_0)\,[\nabla_r \rho(\mathbf{r})]^2 + \ldots. \quad \text{(A2.4.16)}$$

Thus we may equate the final terms shown in equations (A2.4.14) and (A2.4.16) and we find

$$t_2^{(2)}(\rho_0) = \frac{1}{24}\int d\mathbf{r}\, r^2 K(\mathbf{r}). \quad \text{(A2.4.17)}$$

But we can now make use of the assumption that $t_2^{(2)}$ and higher-order coefficients are functions rather than functionals of the density. Then knowing the coefficient (A2.4.17) as a function of ρ_0, or the Fermi momentum k_f, we can immediately obtain it as a function of $\rho(\mathbf{r})$ by the replacement of $k_f \to (3\pi^2 \rho(\mathbf{r}))^{\frac{1}{3}}$ in the right-hand side of equation (A2.4.17), with $K(\mathbf{r})$ given by equation (A2.4.3).

Though this procedure is straightforward, it has to be emphasized that it does not take full account of the singularity structure of $\tilde{K}(q, k_f)$ which is important for some purposes. In particular, the asymptotic behaviour of K is

$$K(\mathbf{r}) \sim \frac{\cos 2k_f r}{r^3}, \quad r \to \infty, \tag{A2.4.18}$$

arising from the singularity at $2k_f$ [compare equation (A2.4.4)]. In the above procedure, the function $\tilde{K}(q, k_f)$ is approximated by its low q expansion, which does not deal with behaviour like that in equation (A2.4.18). But if we use the small q form of $\tilde{K}(q, k_f)$, then equation (A2.4.17) yields

$$t_2^{(2)}(\rho_0) = \tfrac{1}{6}[\nabla_{\mathbf{q}}^2 \tilde{K}(q, k_f)]_{q=0}. \tag{A2.4.19}$$

This derivative exists and can be obtained from equation (A2.4.4) as $\pi^2/24k_f^3$. Using the replacement $k_f \to (3\pi^2 \rho(\mathbf{r}))^{\frac{1}{3}}$, we immediately regain the coefficient $\frac{1}{72}\rho^{-1}(\mathbf{r})$ referred to in the main text below equation (2.11.63).

The above argument has been improved by Hohenberg and Kohn (1964) who essentially use a result like equation (A2.4.10) and write

$$t_r[\rho] = \frac{3}{10}(3\pi^2)^{\frac{2}{3}}\{\rho(\mathbf{r})\}^{\frac{5}{3}} + \frac{1}{8}\int d\mathbf{r}'\, K(\mathbf{r}', \rho(\mathbf{r}))\left[\rho\left(\mathbf{r}+\frac{1}{2}\mathbf{r}'\right) - \rho\left(\mathbf{r}-\frac{1}{2}\mathbf{r}'\right)\right]^2, \tag{A2.4.20}$$

where from equations (A2.4.3) and (A2.4.4) we have

$$K(\mathbf{r}', \rho(\mathbf{r})) = \frac{1}{(2\pi)^3}\int d\mathbf{q}\, \mathrm{e}^{-i\mathbf{q}.\mathbf{r}'} \frac{(2\pi)^2}{2(3\pi^2)^{\frac{1}{3}}\rho^{\frac{1}{3}}(\mathbf{r})} \times \left[\frac{1}{2} + \frac{\{3\pi^2\rho(\mathbf{r})\}^{\frac{1}{3}}}{2q}\left(1 - \frac{q^2}{4\{3\pi^2\rho(\mathbf{r})\}^{\frac{2}{3}}}\right)\ln\left|\frac{q+2(3\pi^2)^{\frac{1}{3}}\rho^{\frac{1}{3}}(\mathbf{r})}{q-2(3\pi^2)^{\frac{1}{3}}\rho^{\frac{1}{3}}(\mathbf{r})}\right|\right]^{-1}. \tag{A2.4.21}$$

This formula is of interest, and retains the correct singularity structure of K, while still having the advantage of gradient expansions. However, it is usually important in getting the one-particle properties to calculate the kinetic energy from the wave functions essentially exactly.

But as Stoddart, Beattie and March (1971) emphasize, it is likely that the success of Dirac–Slater exchange theory means that a similar result for the exchange energy might converge very well. This result has been given by Stoddart and colleagues, and has the form

$$E_{\text{ex}}[\rho] = \int d\mathbf{r}\, \varepsilon_{\mathrm{r}}[\rho] \tag{A2.4.22}$$

with

$$\varepsilon_{\mathrm{r}}[\rho] = -\frac{3}{4}\left(\frac{3}{\pi}\right)^{\frac{1}{3}} \rho^{\frac{4}{3}}(\mathbf{r}) - \frac{1}{2}\int d\mathbf{r}'\, B(\mathbf{r}', \rho(\mathbf{r}))\left[\rho\left(\mathbf{r}+\frac{1}{2}\mathbf{r}'\right) - \rho\left(\mathbf{r}-\frac{1}{2}\mathbf{r}'\right)\right]^2, \tag{A2.4.23}$$

where the quantity B is, however, very complicated and we must refer to the original paper for details. Fortunately, however, $B(\mathbf{r})$ can be calculated once and for all in terms of its Fourier transform $\tilde{B}(q, k_{\mathrm{f}})$, exactly as for the kinetic energy in equations (A2.4.3) and (A2.4.4).

APPENDIX A2.5 CORRELATION ENERGY OF NON-UNIFORM ELECTRON GAS

We shall summarize here the way in which Ma and Brueckner (1968) calculate the quantity b_{c} in the form (2.11.85) for the correlation energy of an electron gas whose density varies slowly in space.

To calculate b_{c} to $O(e^2)$, we need to consider the diagrams (b) in Figure 2.44, on p. 205, labelled (1), (2) and (3). It does not seem practicable to calculate the diagrams for arbitrary k and then identify b_{c} with the coefficient of k^2 in the small k expansion. Thus, the integrand has to be expanded and the coefficient of k^2 picked out.

(a) *Diagrams* (1). The two diagrams have the same value. The contribution they make to $F(\mathbf{k}, 0)$ is

$$F^{(1)}(\mathbf{k}, 0) = 2 \times 2 \int dp\, G^2(p)\, \Sigma(p)\, G(p+\mathbf{k}), \tag{A2.5.1}$$

where

$$\int dp \equiv \int \frac{d\mathbf{p}\, d\varepsilon}{(2\pi)^4\, i}.$$

The Green function is given by

$$G(p) \equiv [\varepsilon - \varepsilon_p + \mu + i\eta \operatorname{sgn}(\varepsilon_p - \mu)]^{-1}, \quad \varepsilon_p \equiv p^2/2m, \quad \mu = p_{\mathrm{f}}^2/2m, \tag{A2.5.2}$$

and

$$\Sigma(p) \equiv -\int dq\, V_{\text{eff}}(q)\, G(p+q) \tag{A2.5.3}$$

with

$$V_{\text{eff}}(q) \equiv \frac{4\pi e^2}{q^2}\left[\frac{1}{\varepsilon(q)}-1\right], \tag{A2.5.4}$$

where $\varepsilon(q\omega)$ is the dielectric constant in the random-phase approximation. The quantity $G^2(p)$ in equation (A2.5.1) is precisely

$$G^2(p) \equiv -\frac{\partial G(p)}{\partial \varepsilon} \tag{A2.5.5}$$

which has a double pole, as required by momentum conservation, at

$$\varepsilon = \varepsilon_p - \mu - i\eta\, \operatorname{sgn}(\varepsilon_p-\mu).$$

We need to use the result

$$-\frac{\partial}{\partial \varepsilon} G(p) = -\frac{\partial}{\partial \mu} G(p) + 2\pi i \delta(\varepsilon)\, \delta(\mu-\varepsilon_p) \tag{A2.5.6}$$

which leads to the alternative form of equation (A2.5.1):

$$F^{(1)}(\mathbf{k},0) = -4\int dp\, \Sigma(p)\, G(p+\mathbf{k})\frac{\partial}{\partial \mu} G(p) + \Sigma(p_{\mathrm{f}},0)\frac{\partial}{\partial \mu} F^0(\mathbf{k},0), \tag{A2.5.7}$$

where

$$F^0(\mathbf{k},\omega) \equiv -2\int \frac{d\mathbf{p}}{(2\pi)^3}\frac{n_{\mathbf{p}+\mathbf{k}}-n_{\mathbf{p}}}{\omega-\varepsilon_{\mathbf{p}+\mathbf{k}}+\varepsilon_p+i\eta\, \operatorname{sgn}\omega}, \tag{A2.5.8}$$

$$n_p \equiv \theta(\mu-\varepsilon_p).$$

To find the k^2 term of $F^{(1)}(\mathbf{k},0)$, we expand $(\partial/\partial\mu)\, F^0(\mathbf{k},0)$ and $G(p+\mathbf{k})$:

$$\frac{\partial}{\partial \mu} F^{(0)}(\mathbf{k},0) = -\frac{m}{\pi^2 v_{\mathrm{f}}} + \frac{k^2}{p_{\mathrm{f}}^2}\frac{m}{12\pi^2 v_{\mathrm{f}}} + \dots, \quad v_{\mathrm{f}} = \frac{p_{\mathrm{f}}}{m}, \tag{A2.5.9}$$

$$\begin{aligned} G(p+\mathbf{k}) &= G(p) + \mathbf{k}\,.\,\nabla_p\, G(p) + \tfrac{1}{2}(\mathbf{k}\,.\,\nabla_p)^2\, G(p) + \dots \\ &= G(p) - \left(\frac{\mathbf{k}\,.\,\mathbf{p}}{m}+\frac{k^2}{2m}\right)\frac{\partial G(p)}{\partial \mu} \\ &\qquad + \frac{(\mathbf{k}\,.\,\mathbf{p})^2}{2m^2}\frac{\partial^2}{\partial \mu^2} G(p) + \dots . \end{aligned} \tag{A2.5.10}$$

By symmetry $(\mathbf{k}.\mathbf{p})^2$ can be replaced by $k^2p^2/3$, and therefore the k^2 term of $-G(p+\mathbf{k})\,\partial G(p)/\partial\mu$ is

$$\lim_{\delta\to 0}\left[-\frac{k^2}{2m}\frac{\partial}{\partial\mu}G(p+\delta)+\frac{k^2p^2}{6m^2}\frac{\delta^2}{\delta\mu^2}G(p+\delta)\right]\left[-\frac{\partial G(p)}{\partial\mu}\right]$$

$$=-\frac{k^2}{2m}\frac{1}{6}\frac{\partial^3}{\partial\mu^3}G(p)+\frac{2k^2}{3m}\varepsilon_p\frac{1}{24}\frac{\partial^4}{\partial\mu^4}G(p), \qquad \text{(A2.5.11)}$$

using the identity [for the proof see Ma and Brueckner (1968)]

$$\lim_{\delta\to 0}\frac{1}{(m-1)!}\left(-\frac{\partial}{\partial\mu}\right)^{m-1}G(p+\delta)\frac{1}{(n-1)!}\left(-\frac{\partial}{\partial\mu}\right)^{n-1}G(p)$$

$$=\frac{1}{(m+n-1)!}\left(-\frac{\partial}{\partial\mu}\right)^{m+n-1}G(p),\quad m \text{ and } n \text{ are integers } >0. \qquad \text{(A2.5.12)}$$

It follows that the k^2 term of $F^{(1)}(\mathbf{k},0)$ is given by

$$\frac{k^2}{m}\int dp\,\Sigma(p)\left[-\frac{1}{3}\frac{\partial^3}{\partial\mu^3}G(p)+\frac{1}{9}\varepsilon_p\frac{\partial^4}{\partial\mu^4}G(p)\right]-\frac{k^2\Sigma(p_{\mathrm{f}},0)}{24\pi^2v_{\mathrm{f}}\mu}\equiv b^{(1)}k^2. \qquad \text{(A2.5.13)}$$

(b) *Diagrams* (2) *and* (3). The k^2 term of diagram 2(b) in Figure 2.44(b) is given in Figure A2.5.1(a). A circle represents a factor $\mathbf{k}.\nabla$. The operator acts on the momentum variable of the Green function line on which the circle is drawn. The factor 2 comes, as usual, from the spin multiplicity.

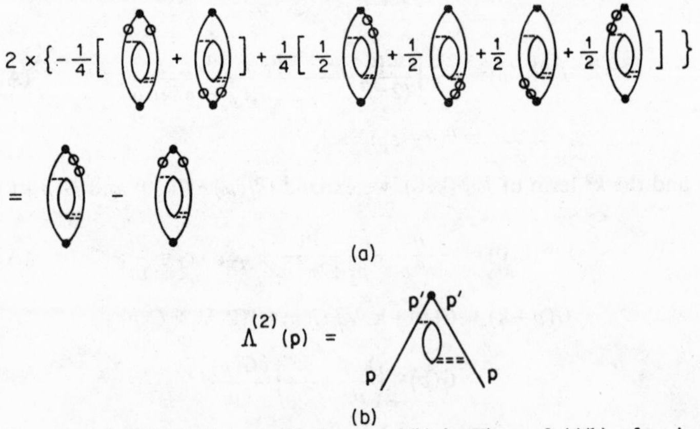

FIGURE A2.5.1. (a) k^2 term of diagram 2(b) in Figure 2.44(b) of main text. Circle represents the operator $\mathbf{k}.\nabla$. (b) Definition of vertex function $\Lambda^{(2)}(\mathbf{p})$.

The vertex function $\Lambda^{(2)}(p)$ is defined by Figure A2.5.1(b), that is,

$$\Lambda^{(2)}(p) = -\int dp'\, V_{\text{eff}}(p-p')\left[-\frac{\partial G(p')}{\partial \mu}\right]. \qquad \text{(A2.5.14)}$$

Using the identity (A2.5.12), the diagrams in Figure A2.5.1(a) contribute

$$\frac{k^2}{m}\int dp\, \Lambda^{(2)}(p)\left[\frac{1}{2}\frac{\partial^2}{\partial \mu^2}G(p) - \frac{1}{9}\varepsilon_p \frac{\partial^3}{\partial \mu^3}G(p)\right] \equiv b^{(2)}\, k^2. \qquad \text{(A2.5.15)}$$

Diagrams (3), namely the two labelled (3b) in Figure 2.44(b), contribute equally to the k^2 term. Just as with Figure A2.5.1(a), the contribution of the diagrams (3) to the k^2 term is shown in Figure A2.5.2(a). The factor 4 there

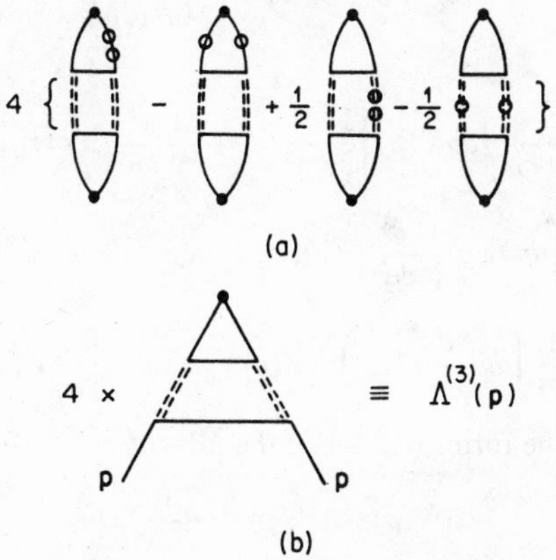

FIGURE A2.5.2. (a) Same as caption for Figure A2.5.1(a) except for modification described in text. (b) Definition of vertex function $\Lambda^{(3)}$ (**p**).

is due to each vertex having spin multiplicity 2. The last two terms of Figure A2.5.2(a) give

$$-\frac{1}{8}\int dq\left[\frac{\partial}{\partial \mu}F^{(0)}(q)\right]^2\{V(\mathbf{q})(\mathbf{k}\,.\,\nabla_{\mathbf{q}})^2\, V(\mathbf{q}) - [\mathbf{k}\,.\,\nabla_{\mathbf{q}}\, V(\mathbf{q})]^2\} \equiv b_1^{(3)}\, k^2, \qquad \text{(A2.5.16)}$$

where

$$F^0(\mathbf{q}\omega) \equiv 2\int dp\, G(p+q)\, G(p) \qquad \text{(A2.5.17)}$$

is also given by equation (A2.5.8). Each triangular loop gives a factor

$-\frac{1}{4}(\partial F^0/\partial \mu)$. $V(q)$ is defined by

$$V(q) = \frac{4\pi e^2}{q^2\,\varepsilon(q)}. \tag{A2.5.18}$$

Let the vertex function $\Lambda^{(3)}(p)$ be defined by Figure A2.5.2(b). The first two terms in Figure A2.5.2(a) are the same as those in Figure A2.5.1(a) except that $\Lambda^{(3)}$ replaces $\Lambda^{(2)}$. Thus their contribution $b_2^{(3)}\,k^2$ is found by substituting $\Lambda^{(3)}$ for $\Lambda^{(2)}$ in equation (A2.5.15). Finally b_c is obtained as

$$b_c = b^{(1)} + b^{(2)} + b_1^{(3)} + b_2^{(3)} \tag{A2.5.19}$$

which, in terms of the expansion for $b^{(1)}, b^{(2)}$, etc. resulting from the above procedure, can be written in the form

$$b_c = b' + b'' + b''' - \frac{\Sigma(p_f, 0)}{24\pi^2\, v_f\, \mu}, \tag{A2.5.20}$$

where

$$\left.\begin{aligned}
b' &= -\frac{1}{m}\frac{\partial}{\partial\mu}\int dp\,\Sigma(p)\left[\frac{1}{2}\frac{\partial^2}{\partial\mu^2}G(p) - \frac{1}{9}\frac{\partial^3}{\partial\mu^3}G(p)\,\varepsilon_p\right],\\
b'' &= \frac{1}{m}\int dp\,\Sigma(p)\frac{1}{6}\frac{\partial^3}{\partial\mu^3}G(p),\\
b''' &= -\frac{1}{24}\int dq\left(\frac{\partial}{\partial\mu}F^0(q)\right)^2\left\{V(q)\,\nabla_q^2\,V(q) - \left[\frac{d}{dq}V(q)\right]^2\right\}.
\end{aligned}\right\} \tag{A2.5.21}$$

In writing down the forms (A2.5.21), the identity

$$\Lambda^{(2)}(p) + \Lambda^{(3)}(p) = -\frac{\partial}{\partial\mu}\Sigma(p) \tag{A2.5.22}$$

has been employed. This can be proved by means of the methods of Appendix A2.3, equation (A2.5.22) being related to a Ward identity. All the principles involved in the calculation are now clear. The remaining evaluation is tedious, but straightforward. In particular, from the work of Quinn and Ferrell (1958), it can be shown that $\Sigma(p_f, 0)$ does not contribute to the order of equation (A2.5.21), while Ma and Brueckner show how the remaining integrals b', b'' and b''' can be explicitly evaluated, the desired result (2.11.86) being obtained in this way.

APPENDIX A2.6 DIELECTRIC FUNCTION OF SEMICONDUCTOR

We summarize here the model used by Penn (1962) to obtain an approximate expression for the wave-number-dependent dielectric constant of a semiconductor.

The expression we require for the dielectric function [cf. equation (2.3.63)] is

$$\varepsilon(\mathbf{q}) = 1-\left(\frac{4\pi e^2}{q^2}\right)\sum_{\mathbf{k},i,i'}|\langle\mathbf{k}i|\,e^{-i\mathbf{q}.\mathbf{r}}|\mathbf{k}+\mathbf{q},i'\rangle|^2\frac{n_{\mathbf{k}+\mathbf{q},i'}-n_{\mathbf{k}i}}{E_{\mathbf{k}+\mathbf{q},i'}-E_{\mathbf{k}i}}, \qquad \text{(A2.6.1)}$$

where

$$\langle\mathbf{k}i|\,e^{-i\mathbf{q}.\mathbf{r}}|\mathbf{k}+\mathbf{q},i'\rangle \equiv \int \psi^*_{\mathbf{k}i}(\mathbf{r})\,e^{-i\mathbf{q}.\mathbf{r}}\,\psi_{\mathbf{k}+\mathbf{q},i'}(\mathbf{r})\,d\mathbf{r} \qquad \text{(A2.6.2)}$$

and $n_{\mathbf{k}i}$ is the occupation number of the state $\mathbf{k}i$. The Bloch wave function having wave vector $\mathbf{k}$ and in band i satisfies the equation

$$H_0\psi_{\mathbf{k}i} = E_{\mathbf{k}i}\psi_{\mathbf{k}i} \qquad \text{(A2.6.3)}$$

where H_0 is the Hamiltonian in the absence of an external perturbation.

It is convenient for present purposes to express equation (A2.6.1) in the extended zone scheme and to write

$$\varepsilon(\mathbf{q}) = 1+\left(\frac{4\pi e^2}{q^2}\right)\sum_{\mathbf{k},\mathbf{K}}|\langle\mathbf{k}|\,e^{-i\mathbf{q}.\mathbf{r}}|\mathbf{k}+\mathbf{q}+\mathbf{K}\rangle|^2\frac{n_{\mathbf{k}+\mathbf{q}+\mathbf{K}}-n_{\mathbf{k}}}{E_{\mathbf{k}+\mathbf{q}+\mathbf{K}}-E_{\mathbf{k}}}, \qquad \text{(A2.6.4)}$$

where $\mathbf{K}$ is as usual a reciprocal lattice vector. The second term in equation (A2.6.4) represents virtual transitions, induced by the external perturbation, between the states $\mathbf{k}$ and $\mathbf{k}+\mathbf{q}+\mathbf{K}$. As usual, $\mathbf{K} = 0$ terms are called normal, while the $\mathbf{K}\neq 0$ contributions are the umklapp terms.

The first model for $\varepsilon(\mathbf{q})$ appears to be that introduced by Callaway (1959). He worked by analogy with the free-electron gas result (2.3.30), which it is useful to rewrite in terms of the plasma frequency ω_p given by equation (2.2.9). This is then

$$\varepsilon(\mathbf{q}) = 1+\tfrac{3}{8}(\hbar\omega_p/q_r E_f)^2\,[1+(q_r^{-1}-\tfrac{1}{4}q_r)\ln(|2+q_r|/|2-q_r|)], \qquad \text{(A2.6.5)}$$

where $q_{\text{reduced}} \equiv q_r = q/k_f$, and E_f is the Fermi energy.

Callaway modified this model by introducing an energy gap between the valence and conduction bands. The energy spectrum with the gap is simply

$$\left.\begin{aligned} E_{\mathbf{k}} &= \hbar^2k^2/2m, \quad k<k_f,\\ E_{\mathbf{k}} &= (\hbar^2k^2/2m)+E_g, \quad k>k_f, \end{aligned}\right\} \qquad \text{(A2.6.6)}$$

where E_g is a parameter, to be determined. For this model, equation (A2.6.4) leads to the result

$$\varepsilon(\mathbf{q}) = 1+\frac{3}{8}\left(\frac{\hbar\omega_p}{q_r E_f}\right)^2\left\{1+\frac{1}{2}\left(\frac{E_r}{q_r}\right)-\left(\frac{E_r}{q_r}\right)\left[\ln\left|1+2\left(\frac{q_r}{E_r}\right)-\left(\frac{q_r^2}{E_r}\right)\right|\right]\right.$$
$$\left.+\left(\frac{2-E_r}{2q_r}-\frac{q_r}{4}-\frac{E_r^2}{4q_r^3}\right)\ln\left|\frac{2+q_r+(E_r/q_r)}{2-q_r+(E_r/q_r)}\right|\right\}, \qquad \text{(A2.6.7)}$$

where $E_{\text{reduced}} \equiv E_r = E_g/E_f$. A plot of Callaway's model is given in Figure A2.6.1, where it is compared with the free-electron gas. There is one rather serious drawback to this model, however. The value of E_g which appears in Callaway's model ought to be determined by fitting the long-wave limit $\varepsilon(0)$

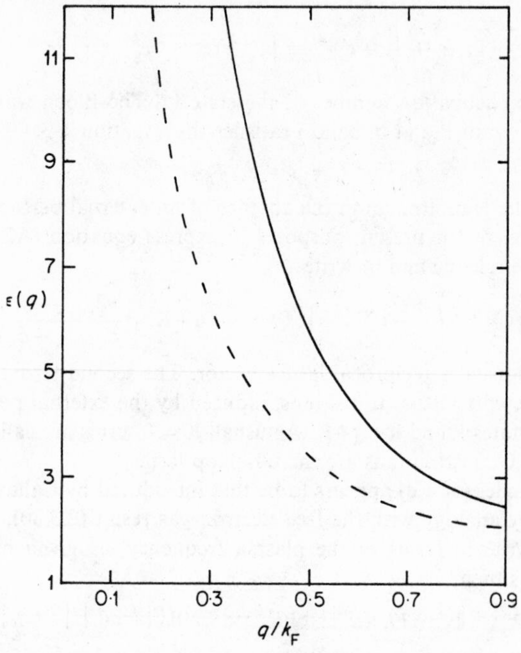

FIGURE A2.6.1. Dielectric function in Callaway's model of a semi-conductor (dashed curve). Solid curve gives result of free-electron model.

of equation (A2.6.7) to the measured static dielectric constant, but unfortunately the result (A2.6.7) gives an infinite value of $\varepsilon(0)$, independent of E_g. This fault of the model is due to the fact that the formation of standing waves at the BZ boundaries is not allowed for in the theory. Furthermore, the neglect of umklapp processes is also an oversimplification.

(a) *Penn's model.* For the reasons just outlined, Penn has proposed a more sophisticated model, namely the nearly free-electron model (cf. Chapter 1, section 1.9.2), isotropically extended to three dimensions. Specifically, the model is described by the energies and wave functions

$$E_{\mathbf{k}}^{\pm} = \tfrac{1}{2}\{E_{\mathbf{k}}^{0}+E_{\mathbf{k}'}^{0}\pm[(E_{\mathbf{k}}^{0}-E_{\mathbf{k}'}^{0})^{2}+E_{g}^{2}]^{\frac{1}{2}}\}, \tag{A2.6.8}$$

$$\psi_{\mathbf{k}} = (e^{i\mathbf{k}.\mathbf{r}}+\alpha_{\mathbf{k}}^{\pm}e^{i\mathbf{k}'.\mathbf{r}})/[1+(\alpha_{\mathbf{k}}^{\pm})^{2}]^{\frac{1}{2}}, \tag{A2.6.9}$$

where

$$\alpha_{\mathbf{k}}^{\pm} \equiv \tfrac{1}{2}E_{g}/(E_{\mathbf{k}}^{\pm}-E_{\mathbf{k}'}^{0}), \tag{A2.6.10}$$

$$E_{\mathbf{k}}^{0} \equiv \frac{\hbar^{2}k^{2}}{2m}, \quad \mathbf{k}' \equiv \mathbf{k}-2k_{\mathrm{f}}\hat{\mathbf{k}} \tag{A2.6.11}$$

and E_g is again a parameter. The superscripts + and − on $E_{\mathbf{k}}^{\pm}$ refer to the cases $k > k_{\mathrm{f}}$ and $k < k_{\mathrm{f}}$ respectively. Figure A2.6.2 shows that the corresponding density of states diverges at the top of the valence band and the bottom

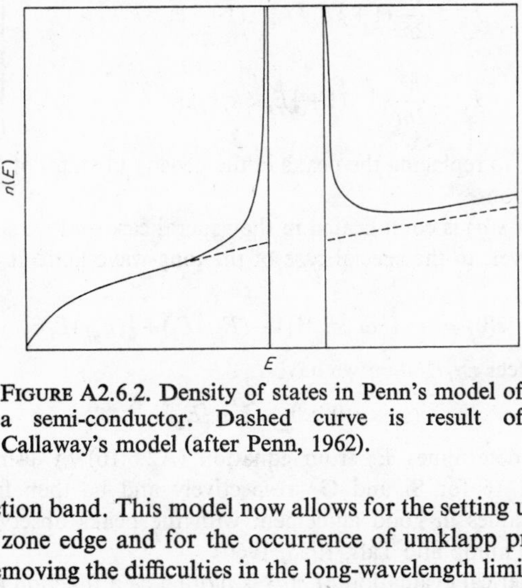

FIGURE A2.6.2. Density of states in Penn's model of a semi-conductor. Dashed curve is result of Callaway's model (after Penn, 1962).

of the conduction band. This model now allows for the setting up of standing waves at the zone edge and for the occurrence of umklapp processes. This will help in removing the difficulties in the long-wavelength limit, whereas for q/k_{f} greater than unity the model will give back the results of the Callaway scheme.

For the above model, equation (A2.6.4) can be written as

$$\varepsilon(\mathbf{q}) = 1+\frac{8\pi e^{2}}{q^{2}}\left[\sum_{\mathbf{k}} n_{\mathbf{k}}(1-n_{\mathbf{k}+\mathbf{q}})\frac{|\langle\mathbf{k}|e^{-i\mathbf{q}.\mathbf{r}}|\mathbf{k}+\mathbf{q}\rangle|^{2}}{E_{\mathbf{k}+\mathbf{q}}^{+}-E_{\mathbf{k}}^{-}} + \sum_{\mathbf{k}} n_{\mathbf{k}}(1-n_{\mathbf{k}'})\frac{|\langle\mathbf{k}|e^{-i\mathbf{q}.\mathbf{r}}|\mathbf{k}'\rangle|^{2}}{E_{\mathbf{k}'}^{+}-E_{\mathbf{k}}^{-}}\right], \tag{A2.6.12}$$

where $\mathbf{k}' \equiv \mathbf{k}+\mathbf{q}-[2k_f(\mathbf{k}+\mathbf{q})/|\mathbf{k}+\mathbf{q}|]$. The matrix elements turn out to be

$$\langle \mathbf{k}|e^{-i\mathbf{q}.\mathbf{r}}|\mathbf{k}+\mathbf{q}\rangle = (1+\alpha_{\mathbf{k}}^{-}\alpha_{\mathbf{k}+\mathbf{q}}^{+})/[1+(\alpha_{\mathbf{k}}^{-})^2]^{\frac{1}{2}}[1+(\alpha_{\mathbf{k}+\mathbf{q}}^{+})^2]^{\frac{1}{2}}, \tag{A2.6.13}$$

$$\langle \mathbf{k}|e^{-i\mathbf{q}.\mathbf{r}}|\mathbf{k}'\rangle = (\alpha_{\mathbf{k}'}^{+}+\alpha_{\mathbf{k}}^{-})/[1+(\alpha_{\mathbf{k}}^{-})^2]^{\frac{1}{2}}[1+(\alpha_{\mathbf{k}'}^{+})^2]^{\frac{1}{2}}. \tag{A2.6.14}$$

In order to evaluate the sums in equation (A2.6.12), Penn uses the approximations

$$\left.\begin{aligned}
E_{\mathbf{k}}^{-} &\approx \frac{\hbar^2 k^2}{2m} \quad (k/k_f < 1-\tfrac{1}{4}E_r),\\
E_{\mathbf{k}}^{-} &\approx \frac{\hbar^2}{2m}(1-\tfrac{1}{4}E_r)^2 k_f^2 \quad (1-\tfrac{1}{4}E_r < k/k_f < 1),\\
E_{\mathbf{k}}^{+} &\simeq \frac{\hbar^2}{2m}(1+\tfrac{1}{4}E_r)^2 k_f^2 \quad (1+\tfrac{1}{4}E_r > k/k_f > 1),\\
E_{\mathbf{k}}^{+} &\simeq \frac{\hbar^2}{2m}k^2 \quad (1+\tfrac{1}{4}E_r < k/k_f).
\end{aligned}\right\} \tag{A2.6.15}$$

These amount to replacing the peaks in the density of states of Figure A2.6.2 by delta functions.

The form of $\varepsilon(q)$ is complicated in the general case and we shall not quote it here. However, in the special case of the long-wave limit, it is then found that

$$\varepsilon(0) = 1+(\hbar\omega_p/E_g)^2[1-(E_g/4E_f)+\tfrac{1}{3}(E_g/4E_f)^2] \tag{A2.6.16}$$

and if we neglect E_g/E_f then we have

$$\varepsilon(0) \simeq 1+(\hbar\omega_p/E_g)^2. \tag{A2.6.17}$$

Penn now determines E_g from equation (A2.6.16) by using the values $\varepsilon(0) = 12$ and 16 for Si and Ge respectively and he then finds $E_g = 4.8$ and 4.2 eV, values in good agreement with the peaks observed in optical absorptivity (Philipp and Taft, 1958, 1960).

Penn's numerical evaluation of $\varepsilon(q)$ is plotted as a function of q in Figure A2.6.3 for Si, and also shown in a simple interpolation formula

$$\varepsilon(\mathbf{q}) = 1+(\hbar\omega_p/E_g)^2 F[1+(E_f/E_g)(q/k_f)^2 F^{\frac{1}{2}}]^{-2}, \tag{A2.6.18}$$

where $F \equiv 1-\frac{1}{4}(E_g/E_f) \simeq 1$. This formula gives back equation (A2.6.17) as q tends to zero, whereas for large q the effect of E_g can be neglected and the limit

$$\varepsilon(\mathbf{q}) \to 1+\left(\frac{\hbar\omega_p}{E_f}\right)^2\left(\frac{k_f}{q}\right)^2 \tag{A2.6.19}$$

is again correct. Though, of course, the model is only rough, and has replaced the often-complicated band structure of semi-conductors by an oversimplified model, it is gratifying that such a simple one-parameter model can give quite

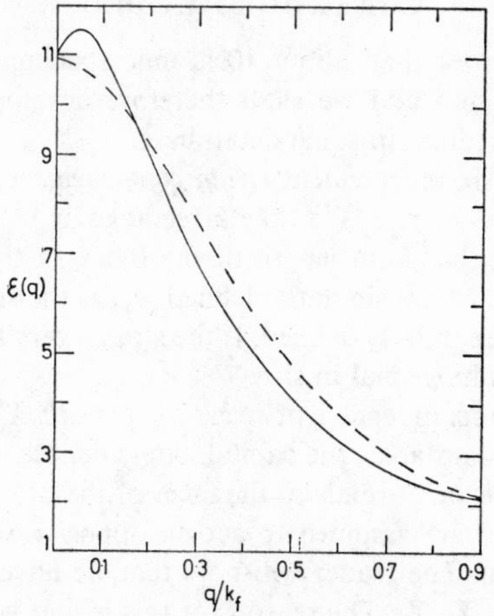

FIGURE A2.6.3. Dielectric function against wave number. Solid curve is numerical result from Penn model. Dashed curve is result of equation (A2.6.18) (after Penn, 1962).

sensible and useful results. Later workers have built on, and refined, Penn's results [see, for example, Jaros and Vinsome (1969), where other references can be found].

APPENDIX A3.1 ELASTIC CONSTANTS AND WAVE EQUATIONS IN CONTINUUM MODEL

For wavelengths greater than about 10 atomic spacings, the solid behaves as an elastic continuum and we shall therefore develop the theory of an anisotropic elastic medium in some detail here.

Let us define the six independent *strain components* e_{xx}, e_{yy}, e_{zz}, e_{xy}, e_{xz}, e_{yz} for Cartesian axes x, y, z. We take a rectangular block of the material; e_{xx} is the fractional change in length of the block in the x-direction when under stress; e_{yy} and e_{zz} are similarly defined. e_{xy} is the shear angle in the xy plane; e_{xz} and e_{yz} are similarly defined. If the strains vary from point to point, the block must be infinitesimal in size.

We also define the six independent *stress components* X_x, Y_y, Z_z, Y_z, Z_x, X_y. These are forces per unit area; the capital letters denote their directions, and the subscripts denote the normals to the faces of the block at which they are applied. Equal forces are assumed to act on opposite sides of the block to maintain equilibrium. The reader will note that we have not included three additional forces Z_y, X_z, Y_x. The reason for this is that we are not interested in forces producing pure rotations of the body. Thus we only consider forces for which the total torque is zero and then, as will be clear from Figure A3.1.1,

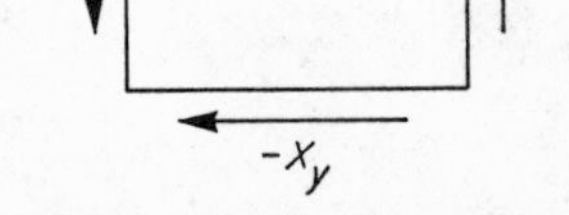

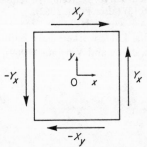

FIGURE A3.1.1. Shows that forces X_y and Y_x must be equal for no rotation of block to occur.

$X_y = Y_x$, etc. If the medium is elastic, the strains must be linear functions of the stresses, so that, for example,

$$e_{xx} = s_{11} X_x + s_{12} Y_y + s_{13} Z_z + s_{14} Y_z + s_{15} Z_x + s_{16} X_y, \qquad \text{(A3.1.1)}$$

the s's being constants. The relations between stresses and strains may be summarized in the matrix equation

$$\begin{pmatrix} e_{xx} \\ e_{yy} \\ e_{zz} \\ e_{yz} \\ e_{zx} \\ e_{xy} \end{pmatrix} = (s_{ij}) \begin{pmatrix} X_x \\ Y_y \\ Z_z \\ Y_z \\ Z_x \\ X_y \end{pmatrix}. \qquad \text{(A3.1.2)}$$

The s_{ij}'s are the elastic *compliance constants*. If we write the inverse matrix of (s_{ij}) as (c_{ij}), then

$$\begin{pmatrix} X_x \\ Y_y \\ Z_z \\ Y_z \\ Z_x \\ X_y \end{pmatrix} = (c_{ij}) \begin{pmatrix} e_{xx} \\ e_{yy} \\ e_{zz} \\ e_{yz} \\ e_{zx} \\ e_{xy} \end{pmatrix}. \qquad \text{(A3.1.3)}$$

The thirty-six *elastic moduli* c_{ij} can be reduced by considering the energy change δU per unit volume under infinitesimal strains of our block, which we now for convenience take as a cube of side L. Consider the work done under infinitesimal shear δe_{xy}. It is clear that although other forces might be required to produce this shear, only the force $X_y L^2$ contributes to the (first-order) energy change. In Figure A3.1.2 we see that the upper xy-face of the

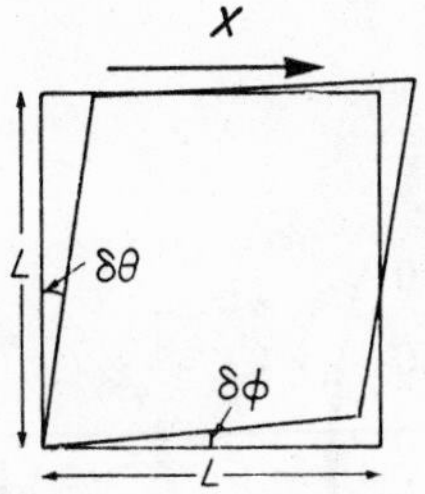

FIGURE A3.1.2. Shows shear of elastic block. From this the energy change follows as $X_y L^3(\delta\theta + \delta\phi)$.

block moves through a horizontal distance $L\,\delta\theta$, with energy change $X_y L^3\,\delta\theta$. There is also an energy change $X_y L^3\,\delta\phi$. Since $\delta e_{xy} = \delta\theta + \delta\phi$, the energy change per unit volume is $X_y\,\delta e_{xy}$. The energies involved in other infinitesimal deformations are similarly calculable and we have, for the energy change per unit volume,

$$\delta U = X_x\,\delta e_{xx} + Y_y\,\delta e_{yy} + Z_z\,\delta e_{zz} + Y_z\,\delta e_{yz} + Z_x\,\delta e_{xz} + X_y\,\delta e_{yx}. \tag{A3.1.4}$$

Thus, for example,

$$\frac{\partial U}{\partial e_{xx}} = X_x, \quad \frac{\partial U}{\partial e_{yy}} = Y_y. \tag{A3.1.5}$$

On further differentiation, we have immediately

$$\frac{\partial X_x}{\partial e_{yy}} = \frac{\partial Y_y}{\partial e_{xx}}. \tag{A3.1.6}$$

Now from equation (A3.1.3), we see that $\partial X_x/\partial e_{yy} = c_{12}$, $\partial Y_y/\partial e_{xx} = c_{21}$. Thus $c_{12} = c_{21}$; in general, we have

$$c_{ij} = c_{ji}. \tag{A3.1.7}$$

The matrix (c_{ij}) is therefore Hermitian. There are six diagonal elements and fifteen off-diagonal ones, making twenty-one elastic moduli in all.

This is still a large number, but, fortunately, for cubic crystals, we can reduce it to three. Application of a stress X_x cannot produce a shear for this would imply asymmetric behaviour of the crystal in the direction y or z. This must also apply to stresses Y_y and Z_z. One can also see that there would be such asymmetric behaviour if the stresses Z_x or X_y produced a shear in the x–y plane. Thus e_{yz} can only depend on Y_z, and

$$Y_z = c_{44}\,e_{yz}. \tag{A3.1.8}$$

Equivalence of axes means that $c_{66} = c_{55} = c_{44}$ and also that $c_{ij} = 0$ if $i \neq j$ and either i or j (or both) is greater than 3, where we have used the fact that c_{ij} is Hermitian. Hence X_x cannot depend on e_{xy}, e_{xz} or e_{yz}. Since its dependence on e_{yy} and e_{zz} must be the same,

$$X_x = c_{11}\,e_{xx} + c_{12}\,e_{yy} + c_{12}\,e_{zz}. \tag{A3.1.9}$$

In full, for cubic crystals

$$(c_{ij}) = \begin{pmatrix} c_{11} & c_{12} & c_{12} & 0 & 0 & 0 \\ c_{12} & c_{11} & c_{12} & 0 & 0 & 0 \\ c_{12} & c_{12} & c_{11} & 0 & 0 & 0 \\ 0 & 0 & 0 & c_{44} & 0 & 0 \\ 0 & 0 & 0 & 0 & c_{44} & 0 \\ 0 & 0 & 0 & 0 & 0 & c_{44} \end{pmatrix} \tag{A3.1.10}$$

In wave propagation the strains vary from point to point. Although we shall only write down the wave equations for cubic crystals, for the appropriate definitions of the strain components we can temporarily return to the general case. We write the position of any element of the medium as

$$\mathbf{r} = u\hat{\mathbf{x}} + v\hat{\mathbf{y}} + w\mathbf{z}, \tag{A3.1.11}$$

where $\hat{\mathbf{x}}$, etc. represent unit vectors in x, etc. directions. Under stress, the element goes to

$$\mathbf{r}' = (u+\delta u)\,\hat{\mathbf{x}} + (v+\delta v)\,\hat{\mathbf{y}} + (w+\delta w)\,\hat{\mathbf{z}}. \tag{A3.1.12}$$

Now on consideration of an infinitesimal block, sides δx, δy, δz, we can immediately write

$$e_{xx} = \frac{\partial u}{\partial x}, \quad e_{yy} = \frac{\partial v}{\partial y}, \quad e_{zz} = \frac{\partial w}{\partial z}. \tag{A3.1.13}$$

To similarly express the other strain components, consider Figure A3.1.2 where the rectangular block is now sheared through the finite angle $e_{xy} = \theta + \phi$. From the figure we can see that, with sides now δx and δy rather than L,

$$\left.\begin{aligned} &\theta = \frac{\delta u}{\delta y}, \quad \phi = \frac{\delta v}{\delta x}, \\ &\text{and} \\ &e_{xz} = \frac{\partial v}{\partial x} + \frac{\partial u}{\partial y}, \quad e_{xz} = \frac{\partial w}{\partial x} + \frac{\partial u}{\partial z}, \\ &e_{yz} = \frac{\partial w}{\partial y} + \frac{\partial v}{\partial z}. \end{aligned}\right\} \tag{A3.1.14}$$

To find the equations for elastic waves we proceed in the usual way. By considering a small block with sides δx, δy, δz and density ρ, the equation of motion in the x-direction is

$$\rho \frac{\partial^2 u}{\partial t^2} = \frac{\partial X_x}{\partial x} + \frac{\partial X_y}{\partial y} + \frac{\partial X_z}{\partial z}. \tag{A3.1.15}$$

For a cubic crystal, use of (A3.1.3) and (A3.1.10) gives us

$$\rho \frac{\partial^2 u}{\partial t^2} = c_{11} \frac{\partial e_{xx}}{\partial x} + c_{12}\left(\frac{\partial e_{yy}}{\partial y} + \frac{\partial e_{zz}}{\partial z}\right) + c_{44}\left(\frac{\partial e_{xy}}{\partial y} + \frac{\partial e_{xz}}{\partial z}\right). \tag{A3.1.16}$$

Using the definitions (A3.1.13) and (A3.1.14) of the strain components, this becomes

$$\rho \frac{\partial^2 u}{\partial t^2} = c_{11}\frac{\partial^2 u}{\partial x^2} + c_{44}\left(\frac{\partial^2 u}{\partial y^2} + \frac{\partial^2 u}{\partial z^2}\right) + (c_{12}+c_{44})\left(\frac{\partial^2 v}{\partial x\,\partial y} + \frac{\partial^2 w}{\partial x\,\partial z}\right). \tag{A3.1.17}$$

Equations for $\partial^2 v/\partial t^2$ and $\partial^2 w/\partial t^2$ may be obtained from this equation by symmetry.

It is to be noticed that, although these equations give no dispersion, the velocity of sound will vary with direction. We shall not solve (A3.1.17) for any direction here, but it finds application in section 3.6.

A typical sound velocity in a crystal is 5×10^5 cm/sec. With a wavelength of the order of 10 atomic spacings, i.e. $\sim 10^{-6}$ cm, we have a frequency of about 5×10^{11} c/s. Thus for all frequencies obtainable electronically at present, the continuum approach is adequate, and direct observation of the discrete nature of a crystal must be made with X-rays or slow neutrons, as is discussed in Chapter 5. Above a frequency of about 10^{12} c/s, the continuum approach begins to fail; nevertheless it is used, in the famous Debye model, to provide an approximate spectrum for lattice vibrations of all frequencies. Debye took the solid to be completely isotropic; we shall only briefly discuss such a medium, since the wave equations can be obtained from (A3.1.17) and are discussed in many elementary textbooks on the properties of matter.

A3.1.1 Isotropic elastic medium

The elastic medium propagates shear and compression waves without dispersion. Thus the frequency and wave number are linearly related and the travelling wave solutions may be written,

$$\mathbf{u}_{\mathbf{k}}(\omega, \mathbf{r}) = Q_{\mathbf{k}}\,\boldsymbol{\varepsilon}_{\mathbf{k}}\,e^{i(\omega t - \mathbf{k}.\mathbf{r})} \tag{A3.1.18}$$

with

$$\omega = ck, \tag{A3.1.19}$$

c being the velocity of sound, taking the value

$$c_t = \left(\frac{n}{\rho}\right)^{\frac{1}{2}} \tag{A3.1.20}$$

for shear waves, where n is the shear modulus and ρ the density, and the value

$$c_l = \left(\frac{K+4n/3}{\rho}\right)^{\frac{1}{2}} \tag{A3.1.21}$$

for compression waves, where K is the bulk modulus.

In (A3.1.18), $\boldsymbol{\varepsilon}_{\mathbf{k}}$ is a unit vector (the polarization vector) which is parallel to $\mathbf{k}$ for longitudinal (compression) waves, and perpendicular to $\mathbf{k}$ for transverse (shear) waves. Thus for each value of $\mathbf{k}$ there are three independent solutions of the wave equations, and two possible values of ω.

APPENDIX A3.2 CRYSTALS WITH MORE THAN ONE ATOM PER UNIT CELL

We now consider how to formulate the consequences of crystal symmetry quite generally. An argument as simple as sufficed for crystals with one atom per unit cell is no longer possible and instead of direct examination of the transformation properties of the matrix $D(\mathbf{k})$ through the force constants $\Phi_{\alpha\beta}$ it is probably easier to find how the polarization vectors transform. We shall in fact follow Chen (1967) in examining the transformations of the displacement $u_\alpha(\mathbf{k}, s)$ of the sth atom in direction α ($= x, y, z$),

$$u_\alpha(\mathbf{k}, s) = \left(\frac{M_s}{N}\right)^{\frac{1}{2}} \sum_{\mathbf{l}} u_\alpha(\mathbf{l}+\mathbf{s})\, \mathrm{e}^{-i\mathbf{k}.(\mathbf{l}+\mathbf{s})}, \tag{A3.2.1}$$

where M_s is the mass of the sth atom, and N is the number of unit cells. This equation is obtained by inversion of equation (3.3.2) and so

$$u_\alpha(\mathbf{k}, s) = \sum_\sigma Q_{\mathbf{k}\sigma}\, \varepsilon^{\alpha s}_{\mathbf{k}\sigma}, \tag{A3.2.2}$$

which shows that it suffices to consider $u_\alpha(\mathbf{k}, s)$ instead of the individual $\varepsilon^{\alpha s}_{\mathbf{k}\sigma}$, $u(\mathbf{k}, s)$ becoming proportional to $\varepsilon^{\alpha s}_{\mathbf{k}\sigma}$ when the mode σ is propagating alone.

Let us, therefore, apply the transformation $(a|\mathbf{t})$ to the crystal, where

$$(a|\mathbf{t})\,\mathbf{r} = a\mathbf{r}+\mathbf{t}, \tag{A3.2.3}$$

bearing in mind that $\mathbf{t}$ may be a non-primitive translation.

This transformation of the coordinate system causes $\mathbf{u}(\mathbf{l}+\mathbf{s})$ to be rotated, becoming $a\mathbf{u}(\mathbf{l}+\mathbf{s})$, and moved to the site $(a|\mathbf{t})\,(\mathbf{l}+\mathbf{s})$. Denoting new displacements by primes, we therefore write

$$u'_\alpha[(a|\mathbf{t})\,(\mathbf{l}+\mathbf{s})] = \sum_\beta a_{\alpha\beta}\, u_\beta(\mathbf{l}+\mathbf{s}) \tag{A3.2.4}$$

or

$$u'_\alpha(\mathbf{l}+\mathbf{s}) = \sum_\beta a_{\alpha\beta}\, u_\beta[(a|\mathbf{t})^\dagger\,(\mathbf{l}+\mathbf{s})]. \tag{A3.2.5}$$

Thus after transformation $u_\alpha(\mathbf{k}, s)$ becomes

$$u'_\alpha(\mathbf{k}, s) = \sum_{\alpha\beta} a_{\alpha\beta} \left(\frac{M_s}{N}\right)^{\frac{1}{2}} \sum_{l} u_\beta[(a|\mathbf{t})^\dagger\,(\mathbf{l}+\mathbf{s})]\, \mathrm{e}^{-i\mathbf{k}.(\mathbf{l}+\mathbf{s})}. \tag{A3.2.6}$$

We now change the summation over lattice vectors from $\mathbf{l}$ to $\mathbf{l}'$ where

$$\mathbf{l}+\mathbf{s} = (a|\mathbf{t})(\mathbf{l}'+\mathbf{s}'). \qquad \text{(A3.2.7)}$$

Then

$$u'_\alpha(\mathbf{k}, \mathbf{s}) = e^{-i\mathbf{k}.\mathbf{t}} \sum_{\alpha\beta} a_{\alpha\beta}\left(\frac{M_s}{N}\right)^{\frac{1}{2}} \sum_{\mathbf{l}'} u_\beta(\mathbf{l}'+\mathbf{s}')\, e^{-ia^\dagger\mathbf{k}.(\mathbf{l}'+\mathbf{s}')}. \qquad \text{(A3.2.8)}$$

Equation (A3.2.7) shows s and s' to be sites on which sit the same kind of atom: $M_s = M_{s'}$. It follows that equation (A3.2.8) may be rewritten as

$$u'_\alpha(\mathbf{k}, \mathbf{s}) = e^{-i\mathbf{k}.\mathbf{t}} \sum_{\beta} a_{\alpha\beta}\, u_\beta(a^\dagger\,\mathbf{k}, \mathbf{s}') \quad [\mathbf{s}' = (a|\mathbf{t})^\dagger\,\mathbf{s}]. \qquad \text{(A3.2.9)}$$

We now note that the collection of all possible primed polarization vectors must be the same as the unprimed ones, $(a|\mathbf{t})$ carrying the crystal into itself. We see further that

$$\omega^2(a\mathbf{k}, \sigma) = \omega^2(\mathbf{k}, \sigma) \qquad \text{(A3.2.10)}$$

which we have already anticipated, and also that there exists a matrix $A(a\mathbf{k})$, obtainable from equation (A3.2.9), such that

$$|\,\varepsilon_{a\mathbf{k},\sigma}\rangle = A(a\mathbf{k})\,|\,\varepsilon_{\mathbf{k}\sigma}\rangle. \qquad \text{(A3.2.11)}$$

Equation (3.2.15) then implies

$$D(a\mathbf{k})\,|\,\varepsilon_{a\mathbf{k},\sigma}\rangle = A(a\mathbf{k})\, D(\mathbf{k})\,|\,\varepsilon_{\mathbf{k}\sigma}\rangle. \qquad \text{(A3.2.12)}$$

This equation holds for all σ, from which it follows that

$$D(a\mathbf{k}) = A(a\mathbf{k})\, D(\mathbf{k})\, A^\dagger(a\mathbf{k}). \qquad \text{(A3.2.13)}$$

We know from Chapter 1 that the condition under which essential degeneracy (degeneracy implied by symmetry) can occur is

$$a^\dagger\,\mathbf{k} = \mathbf{k}+\mathbf{K} \qquad \text{(A3.2.14)}$$

($\mathbf{K}$ is a reciprocal lattice vector, possibly zero; $a^\dagger$ being an operator other than the identity). This is confirmed by writing the condition into equation (A3.2.9). From equation (A3.2.1)

$$u_\beta(a^\dagger\,\mathbf{k}, s') = u_\beta(\mathbf{k}+\mathbf{K}, s') = \left(\frac{M_s}{N}\right)^{\frac{1}{2}} \sum u_\beta(\mathbf{l}+\mathbf{s}')\, e^{-i\mathbf{k}.(\mathbf{l}+\mathbf{s}')}\, e^{-i\mathbf{K}.\mathbf{s}'}$$

$$= u_\beta(\mathbf{k}s')\, e^{-i\mathbf{K}.\mathbf{s}'}, \qquad \text{(A3.2.15)}$$

where we have used the fact that $\mathbf{l}.\mathbf{K} = 2\pi$.

Equation (A3.2.9) now reads

$$u'_\alpha(\mathbf{k}s) = e^{-i\mathbf{k}.\mathbf{t}}\, e^{i(\mathbf{k}-a^\dagger\mathbf{k}).\mathbf{s}'} \sum_{p} a_{\alpha\beta}\, u_\beta(\mathbf{k}s') \qquad \text{(A3.2.16)}$$

and so

$$|\,\varepsilon'_{\mathbf{k}\sigma}\rangle = A(\mathbf{k})\,|\,\varepsilon_{\mathbf{k}\sigma}\rangle, \tag{A3.2.17}$$

where

$$A^{ss'}_{\alpha\beta} = \mathrm{e}^{-i\mathbf{k}.\mathbf{t}}\,\mathrm{e}^{i(\mathbf{k}-a^{\dagger}\mathbf{k}).\mathbf{s}'}\,a_{\alpha\beta}\,\delta_{s's''} \quad [\mathbf{s}'' = (a\,|\,\mathbf{t})^{\dagger}\,\mathbf{s}]. \tag{A3.2.18}$$

In general,

$$\mathbf{t} = \mathbf{l} \quad \text{or} \quad \mathbf{t} = \mathbf{l}+\boldsymbol{\tau}, \tag{A3.2.19}$$

$\boldsymbol{\tau}$ being a non-primitive translation lying within the unit cell about the origin. Equation (3.2.16) shows that we need only consider A for $\mathbf{t} = 0$ and $\mathbf{t} = \boldsymbol{\tau}$, transformations for other $\mathbf{t}$ differing from A only by a phase factor. We should also note that, as in the case of electronic band structure, time-reversal effects can play a role [see, for example, Maradudin and Vosko (1968)].

APPENDIX A3.3 ADIABATIC APPROXIMATION

We consider here what approximations are involved in separating out the nuclear and electronic motions. As we shall see, the basic expansion parameter in such a separation is the ratio $(m/M)^{\frac{1}{4}}$, where m is the electronic and M the ionic mass.

We write the total Hamiltonian in the form

$$H = T_{\mathrm{el}}+V_{\mathrm{el}}+T_{\mathrm{ion}}+V_{\mathrm{ion}}+H_{\mathrm{int}} \tag{A3.3.1}$$

and abstract from this an electronic part

$$H_{\mathrm{el}} = T_{\mathrm{el}}+V_{\mathrm{el}}+H_{\mathrm{int}} \tag{A3.3.2}$$

depending parametrically (through H_{int}) on the ionic coordinates $\mathbf{R}$. We can now expand the solution Φ of the full Schrödinger equation

$$H\Phi = \mathscr{E}\Phi \tag{A3.3.3}$$

in terms of the solutions ψ of

$$H_{\mathrm{el}}\psi_n = E_n\psi_n. \tag{A3.3.4}$$

We accordingly write

$$\Phi = \sum_n \chi_n(\mathbf{R})\psi_n(\mathbf{R};\mathbf{r}) \tag{A3.3.5}$$

or, using the orthonormality of the ψ's,

$$(T_{\mathrm{ion}}+V_{\mathrm{ion}}+E_n)\chi_n+\sum_{n'} C_{nn'}\chi_{n'} = \mathscr{E}\chi_n, \tag{A3.3.6}$$

where

$$C_{nn'} = A_{nn'}+B_{nn'} \tag{A3.3.7}$$

with

$$\left.\begin{aligned} A_{nn'} &= -\frac{\hbar^2}{M}\sum_{\mathbf{R}_l}\int \psi_n^*(\mathbf{r})\frac{\partial}{\partial \mathbf{R}_l}\psi_{n'}(\mathbf{r})\,d\mathbf{r}\frac{\partial}{\partial \mathbf{R}_l}, \\ B_{nn'} &= -\frac{\hbar^2}{2M}\sum_{\mathbf{R}_l}\int \psi_n^*\frac{\partial^2}{\partial \mathbf{R}_l^2}\psi_{n'}\,d\mathbf{r}. \end{aligned}\right\} \tag{A3.3.8}$$

If there is no magnetic field the ψ's can be chosen to be real‡, from which it follows that $A_{nn} = 0$, since by differentiating the normalization condition we find

$$\int \psi_n^*\frac{\partial}{\partial \mathbf{R}_l}\psi_{n'}\,d\mathbf{r} + \int \psi_{n'}\frac{\partial}{\partial \mathbf{R}_l}\psi_n^*\,d\mathbf{r} = 0. \tag{A3.3.9}$$

The adiabatic approximation may evidently be obtained by dropping the non-zero elements of A and B. To see what this implies we follow Chester and Houghton (1959) in generating the Born–Oppenheimer expansion. We begin by writing

$$\mathbf{R}_l = \mathbf{l} + \lambda \mathbf{u}_l, \tag{A3.3.10}$$

where $\mathbf{l}$ is the lattice vector, i.e. the equilibrium position, and λ is an expansion parameter. We then have

$$H = \sum_n H^{(n)} \tag{A3.3.11}$$

with

$$\begin{aligned} H^{(0)} &= T_{\text{el}} + V_{\text{el}} + V_{\text{ion}}^{(0)} + H_{\text{int}}^{(0)}, \\ H^{(1)} &= \lambda V_{\text{ion}}^{(1)} + \lambda H_{\text{int}}^{(1)}, \\ H^{(2)} &= T_{\text{ion}} + \lambda^2 H_{\text{int}}^{(2)} + \lambda^2 V_{\text{ion}}^{(2)}, \\ H^{(n)} &= \lambda^n H_{\text{int}}^{(n)} + \lambda^n V_{\text{ion}}^{(n)}, \quad n > 2, \end{aligned} \tag{A3.3.12}$$

where $H_{\text{int}}^{(n)}$, etc. are to nth order in the displacements $\mathbf{u}_l$.

Now as $M \to \infty$, we must have $\lambda \to 0$; since the expansion parameter is dimensionless this suggests $\lambda = (m/M)^\alpha$ $(\alpha > 0)$. We can fix α by noting that if $\alpha = \frac{1}{4}$, $\lambda^2 H_{\text{int}}^{(2)}$ and $\lambda^2 V_{\text{ion}}^{(2)}$ are of the same order as T_{ion}, and to this order we have a harmonic Hamiltonian consistent with the assumption that the ions have an equilibrium configuration.

Writing

$$H_{\text{int}} = H_{\text{int}}^{0} + \lambda H_{\text{int}}^{(1)} + \lambda^2 H_{\text{int}}^{(2)} \ldots \tag{A3.3.13}$$

‡ We can always combine degenerate eigenfunctions to make them purely real (or purely imaginary). Then the purely real functions form a complete set.

in equation (A3.3.2), equation (A3.3.4) may be solved by standard perturbation theory. Next, on expanding each side of equation (A3.3.6) in powers of λ the expansion of ψ_n can be used to calculate expansions of $A_{nn'}$ and $B_{nn'}$, which are found to be of order λ^3 and λ^4 respectively; this means that the energy is adiabatic to order λ^6. Hence, in Born and Oppenheimer's expansion, non-adiabatic terms appear only at relatively high order, and so this expression may be unsuitable if these terms should, for any reason, become important. On the other hand $\lambda = (m/M)^{\frac{1}{4}}$ is never much greater than $\frac{1}{10}$ and so expansions in it may be expected to converge quickly.

APPENDIX A3.4 LATTICE DYNAMICS OF SHELL MODEL: TREATMENT OF COULOMB COEFFICIENTS

From equation (3.2.9), the equations of motion are

$$M_i \ddot{u}_\alpha(li) = \sum_{i'l'} \sum_{\beta} \Phi_{\alpha\beta}(li, l'i') \, u_\beta(l'i'), \tag{A3.4.1}$$

where M_i is the mass of the nucleus of type i. Then, as before, the dynamical matrix is

$$D_{\alpha\beta}(ii') = -\sum_{l'} \Phi_{\alpha\beta}(li, l'i') \exp\{i\mathbf{q}.[\mathbf{r}(l'i') - \mathbf{r}(li)]\}. \tag{A3.4.2}$$

Following the discussion of section 3.11.1, we divide D into two parts

$$D_{\alpha\beta}(ii') = B_{\alpha\beta}(ii') + C_{\alpha\beta}(ii'), \tag{A3.4.3}$$

where B and C are respectively bonding and Coulomb coefficients. It is readily shown that $B_{xx}(13) = B_{xx}(24) = -k$ and that the other bonding coefficients are determined by six independent force constants, $\Phi^{(B)}(li, l'i')$ and $\Phi^{(B)}(li, li')$ for $ii' = 12$, 34 and 14 when only near-neighbour interactions are included.

Following Cochran's notation, we write $D = B(12)$, $S = B(34)$ and $F = B(14)$, indicating core–core, shell–shell and core–shell bonding (see Figure 3.9).

A3.4.1 Coulomb coefficients and Ewald transformation

To deal with the Coulomb coefficients, we follow the treatment of Kellermann (1940) for NaCl. To handle the Coulomb interactions, let us first consider a lattice sum of the form

$$F_{\mathbf{k}}(\mathbf{r}) = \sum f(\mathbf{r} - \mathbf{R}_l) \exp(i\mathbf{k}.\mathbf{R}_l). \tag{A3.4.4}$$

This can be trivially rewritten as

$$F_{\mathbf{k}}(\mathbf{r}) = \exp(i\mathbf{k}.\mathbf{r}) \sum_{\mathbf{l}} f(\mathbf{r}-\mathbf{R}_{\mathbf{l}}) \exp[i\mathbf{k}.(\mathbf{R}_{\mathbf{l}}-\mathbf{r})]$$
$$= \exp(i\mathbf{k}.\mathbf{r}) \sum_{h} F_{\mathbf{k}}^{h} \exp(i\mathbf{K}_h.\mathbf{r}), \qquad \text{(A3.4.5)}$$

since we have a periodic function, with a Fourier expansion as shown. The Fourier coefficients $F_{\mathbf{k}}^{h}$ are given by

$$F_{\mathbf{k}}^{h} = \frac{1}{\Omega}\int f(\mathbf{r}) \exp[-i(\mathbf{K}_h+\mathbf{k}).\mathbf{r}]\, d\mathbf{r} = \frac{1}{\Omega}\phi(\mathbf{K}_h+\mathbf{k}), \qquad \text{(A3.4.6)}$$

where the integration is over the whole of space.

If we choose

$$f(r) = \frac{2}{\sqrt{\pi}} \exp(-\varepsilon^2 r^2) \qquad \text{(A3.4.7)}$$

then

$$\int_0^\infty f(r)\, d\varepsilon = \frac{1}{r}. \qquad \text{(A3.4.8)}$$

Determining ϕ in equation (A3.4.6) by Fourier transforming equation (A3.4.7), we find

$$F_{\mathbf{k}}(\mathbf{r}) = \frac{2}{\sqrt{\pi}} \sum_{\mathbf{l}} \exp[-\varepsilon^2(\mathbf{r}-\mathbf{R}_{\mathbf{l}})^2 + i\mathbf{k}.\mathbf{R}_{\mathbf{l}}]$$
$$= \frac{2\pi}{\Omega} \sum_{h} \frac{1}{\varepsilon^3} \exp[-\frac{1}{4\varepsilon^2}(\mathbf{K}_h+\mathbf{k})^2 + i(\mathbf{K}_h+\mathbf{k}).\mathbf{r}], \qquad \text{(A3.4.9)}$$

which is Ewald's transformation.

We now wish to evaluate that part of the sum in equation (A3.4.2) which comes from Coulomb interactions. Thus, we replace $\Phi(li, l'i')$ by $Z_{i'}Z_i F(r)$, where we choose $F(r) = 1/r$. Then we find

$$C_{\alpha\beta}(ii') = Z_{i'}Z_i \sum_{\mathbf{r}\to -\mathbf{r}_{i'i}} \sum_{\mathbf{l}} \frac{\partial^2}{\partial x_\alpha \partial x_\beta} F(\mathbf{r}-\mathbf{R}_l) \exp[i\mathbf{k}.(\mathbf{R}_i - \mathbf{R}_{i'})] \quad (i \neq i') \qquad \text{(A3.4.10)}$$

and

$$C_{\alpha\beta}(ii) = Z_i^2 \lim_{r\to 0} \left[\sum_{l} \frac{\partial^2}{\partial x_\alpha \partial x_\beta} F(\mathbf{r}-\mathbf{R}_l) \exp(i\mathbf{k}.\mathbf{R}_l) - \frac{\partial^2}{\partial x_\alpha \partial x_\beta} F(\mathbf{r}) \right]. \qquad \text{(A3.4.11)}$$

These sums can now be evaluated using the method described above, since $F(r)$ is evidently related to $f(r)$ through equation (A3.4.8). Numerical calculations of these Coulomb coefficients are given by Kellermann (1940), and Cochran (1959) has made the necessary modifications to apply to the Ge structure.

For Ge, all the other Coulomb coefficients can be derived from $C(11)$ and $C(12)$. It is convenient to introduce new dimensionless functions of $\mathbf{q}$, C_1 and C_2 such that

$$(Z^2 e^2/v)\, C_1 = C(11) = C(22) = C(33)$$
$$= C(44) = -C(13) = -C(24) \qquad \text{(A3.4.12)}$$

and

$$(Z^2 e^2/v)\, C_2 = C(12) = C(34) = -C(14)$$
$$= -C(32). \qquad \text{(A3.4.13)}$$

Here v is simply $2r_0^3$, where the two atoms in the unit cell are at $(0,0,0)$ and $-(r_0/2, r_0/2, r_0/2)$. In the limit $\mathbf{q}\to 0$, $C_1 = C_2$.

APPENDIX A3.5 PROOF OF LYDDANE–SACHS–TELLER RELATION

The argument is a long wavelength approximation. With optical modes, the positive ions vibrating in opposition to the negative ions lead us to a local time-dependent dipole moment, which, when transverse, will interact strongly with electromagnetic radiation.

If we have N_0 unit cells per unit volume, the macroscopic polarization $\mathscr{P}$ is

$$\mathscr{P} = N_0(e^*\,\mathbf{s} + \alpha\mathscr{E}_l). \qquad \text{(A3.5.1)}$$

$\mathbf{s}$ is the relative displacement of the positive and negative ions, while e^* is the effective charge on the ions. α is the electronic polarizability and is multiplied by the local field $\mathscr{E}_l$ which must be distinguished from the real field $\mathscr{E}$. The Lorentz formula

$$\mathscr{E}_l = \mathscr{E} + \frac{\mathscr{P}}{3\varepsilon_0} \qquad \text{(A3.5.2)}$$

relates the local field to $\mathscr{E}$, ε_0 being the free-space permittivity. Putting (A3.5.1) and (A3.5.2) together we find

$$\mathscr{P} = \frac{N_0[e^*\,\mathbf{s} + \alpha\mathscr{E}]}{[1-(N_0\,\alpha/3\varepsilon_0)]}. \qquad \text{(A3.5.3)}$$

Now to bring in the high-frequency dielectric constant, we note that at very high frequencies the ions cannot follow the applied field and $\mathbf{s}\to 0$. In this limit we can write

$$\mathscr{P} = (\varepsilon_\infty - 1)\,\varepsilon_0\,\mathscr{E}. \tag{A3.5.4}$$

From the above equations, putting $\mathbf{s} \to 0$ to get ε_∞, we can express α in terms of ε_∞ and hence we find, in general,

$$\mathscr{P} = \tfrac{1}{3}N_0\,e^*(2+\varepsilon_\infty)\,\mathbf{s} + (\varepsilon_\infty - 1)\,\varepsilon_0\,\mathscr{E}. \tag{A3.5.5}$$

A3.5.1 Equations of motion of ions

If we denote by K the strength of the restoring forces, and by $\mathbf{u}_+$ and $\mathbf{u}_-$ the displacements of positive and negative ions respectively, then we have

$$M_+ \frac{d^2\mathbf{u}_+}{dt^2} = -K[\mathbf{u}_+ - \mathbf{u}_-] + e^*\,\mathscr{E}_l, \tag{A3.5.6}$$

$$M_- \frac{d^2\mathbf{u}_-}{dt^2} = K[\mathbf{u}_+ - \mathbf{u}_-] - e^*\,\mathscr{E}_l. \tag{A3.5.7}$$

In terms of the reduced mass, we obtain

$$M_r \frac{d^2\mathbf{s}}{dt^2} = -K\mathbf{s} + e^*\,\mathscr{E}_l. \tag{A3.5.8}$$

Substituting for $\mathscr{E}_l$ we find

$$M_r \frac{d^2\mathbf{s}}{dt^2} = -M_r\,\omega_0^2\,\mathbf{s} + \tfrac{1}{3}e^*(2+\varepsilon_\infty)\,\mathscr{E}, \tag{A3.5.9}$$

where

$$\omega_0^2 = \frac{K}{M_r} - \frac{N_0\,e^{*2}(2+\varepsilon_\infty)}{9M_r\,\varepsilon_0}. \tag{A3.5.10}$$

It is convenient to write the displacement $\mathbf{s}$ in a different scale of length, namely

$$\mathbf{w} = \sqrt{(N_0 M_r)}\,\mathbf{s}. \tag{A3.5.11}$$

Then we find

$$\frac{d^2\mathbf{w}}{dt^2} = -\omega_0^2\,\mathbf{w} + \sqrt{\left(\frac{N_0}{M_r}\right)}\,\tfrac{1}{3}e^*(2+\varepsilon_\infty)\,\mathscr{E} \tag{A3.5.12}$$

and

$$\mathscr{P} = \sqrt{\left(\frac{N_0}{M_r}\right)}\,\tfrac{1}{3}e^*(2+\varepsilon_\infty)\,\mathbf{w} + (\varepsilon_\infty - 1)\,\varepsilon_0\,\mathscr{E}. \tag{A3.5.13}$$

When the applied field is static, $d^2\mathbf{w}/dt^2 = 0$ and we can find $\mathbf{w}$ in terms of $\mathscr{E}$. We get in this case

$$\mathscr{P} = [\varepsilon_s - 1]\,\varepsilon_0\,\mathscr{E} \tag{A3.5.14}$$

and

$$\varepsilon_s - \varepsilon_\infty = \frac{N_0 e^{*2}(2+\varepsilon_\infty)^2}{9M_r \omega_0^2 \varepsilon_0}. \tag{A3.5.15}$$

Hence the displacement **w** satisfies the equation of motion

$$\frac{d^2\mathbf{w}}{dt^2} = -\omega_0^2 \mathbf{w} + \omega_0 \sqrt{[\varepsilon_0(\varepsilon_s - \varepsilon_\infty)]}\,\mathscr{E} \tag{A3.5.16}$$

and

$$\mathscr{P} = \omega_0 \sqrt{[\varepsilon_0(\varepsilon_s - \varepsilon_\infty)]}\,\mathbf{w} + (\varepsilon_\infty - 1)\,\varepsilon_0 \mathscr{E}. \tag{A3.5.17}$$

At this stage, we enquire as to the nature of **w** in the absence of any applied field. To make progress it is useful to write

$$\mathbf{w} = \mathbf{w}_t + \mathbf{w}_l, \tag{A3.5.18}$$

where $\operatorname{div} \mathbf{w}_t = 0$ and $\operatorname{curl} \mathbf{w}_l = 0$.

Since, with no real charge ρ, Maxwell's equation gives us

$$\operatorname{div} \mathscr{D} = 0, \tag{A3.5.19}$$

we find

$$\operatorname{div} \mathscr{E} + \frac{\omega_0}{\varepsilon_\infty} \sqrt{\left(\frac{\varepsilon_s - \varepsilon_\infty}{\varepsilon_0}\right)} \operatorname{div} \mathbf{w} = 0. \tag{A3.5.20}$$

Solution of this equation yields

$$\mathscr{E} = -\frac{\omega_0}{\varepsilon_\infty} \sqrt{\left(\frac{\varepsilon_s - \varepsilon_\infty}{\varepsilon_0}\right)} \mathbf{w}_l. \tag{A3.5.21}$$

Hence we can write from equations (A3.5.16) and (A3.5.21)

$$\frac{d^2\mathbf{w}_t}{dt^2} = -\omega_0^2 \mathbf{w}_t \tag{A3.5.23}$$

and

$$\frac{d^2\mathbf{w}_l}{dt^2} = -\frac{\varepsilon_s}{\varepsilon_\infty}\omega_0^2 \mathbf{w}_l. \tag{A3.5.24}$$

If we assume $\mathbf{w}_t$ and $\mathbf{w}_l$ are proportional to the plane wave $e^{i\mathbf{q}\cdot\mathbf{r}}$, then the conditions that they are solenoidal and irrotational respectively mean that $\mathbf{w}_t$ is at right angles to **q** and $\mathbf{w}_l$ is parallel to **q**. Thus, the transverse mode $\mathbf{w}_t$ oscillates with frequency ω_0^2 and the longitudinal mode with a higher frequency $\omega_{l0}^2 = (\epsilon_s/\varepsilon_\infty)\,\omega_0^2$. This completes the proof of equation (3.11.31).

APPENDIX A3.6 STATISTICAL DENSITY MATRIX

Given that an unperturbed system is in thermal equilibrium with its surroundings, it is well known that the thermal average of any operator Q is given by the formula

$$\langle Q\rangle = \sum_i (e^{-\beta\varepsilon_i}/Z)\langle\phi_i^* Q\phi_i\rangle, \tag{A3.6.1}$$

where $\beta = 1/k_B T$, Z is the partition function and the ϕ_i are eigenfunctions of the Hamiltonian with eigenvalues ε_i. If we define the density-matrix operator

$$\rho = e^{-\beta H}/Z, \tag{A3.6.2}$$

equation (A3.6.1) may also be written as

$$\langle Q\rangle = \sum_i \langle i|\rho|i\rangle\langle i|Q|i\rangle \tag{A3.6.3}$$

or, since ϕ_i is an eigenfunction of ρ,

$$\langle Q\rangle = \sum_i \langle i|\rho|j\rangle\langle j|Q|i\rangle = \sum_i \langle i|\rho Q|i\rangle. \tag{A3.6.4}$$

We shall write this as

$$\langle Q\rangle = \mathrm{Tr}\{\rho Q\}, \tag{A3.6.5}$$

where the trace of any operator A is defined as

$$\mathrm{Tr}\{A\} = \sum_i \langle i|A|i\rangle. \tag{A3.6.6}$$

The great merit of the form (A3.6.5) arises from the fact that the value of the trace is unchanged if we take it with respect to any orthonormal set of functions obeying the correct boundary conditions. As discussed in Chapter 3 (p. 300), if ψ_i represents another orthonormal set,

$$\mathrm{Tr}\{A\} = \sum_i \langle\phi_i^* A\phi_i\rangle \tag{A3.6.7}$$

$$= \sum_j \langle\psi_j^* A\psi_j\rangle. \tag{A3.6.8}$$

This property means that it is unnecessary to know the eigenvalues of H to calculate $\langle Q\rangle$.

We may also remind the reader that by using the completeness relation it is simple to prove the cyclic property of the trace (see equation (3.14.10)).

Equation (A3.6.5) may be extended to situations where the system is not in equilibrium—for example, when a time-dependent perturbation is present. The statistical average of Q at time t will then be written

$$\langle Q\rangle = \mathrm{tr}\{\rho(t)\,Q\}. \tag{A3.6.9}$$

During periods of time in which the system may be considered to be isolated, $\rho(t)$ satisfies a formally very simple equation of motion. We do not wish to discuss the finer points of statistical mechanics in deriving this equation, but the following method of arriving at it may be considered to be reasonably satisfactory.

Let us suppose the system S is surrounded by a very large number of systems coupled to it in such a way that these latter systems can be considered to be a heat-bath of the kind usually imagined in discussions of the foundations of statistical mechanics. We also suppose that the assembly as a whole forms an isolated system to which we can ascribe a wave function $\Psi(q,x)$ obeying the Schrödinger equation

$$H_t \Psi = \hbar i \frac{\partial \Psi}{\partial t}, \qquad \text{(A3.6.10)}$$

where

$$H_t = H(x) + H_1(q) + H_2(q,x). \qquad \text{(A3.6.11)}$$

Here x denotes the dynamical variables of S, and q the dynamical variables of the rest of the ensemble, so that $H_2(q,x)$ describes the interaction of the system S with the rest. It is evident that the average of an operator Q depending only on the variables x can be written as

$$\langle Q \rangle = \int \Psi^*(q,x)\, Q(x)\, \Psi(q,x)\, dq\, dx. \qquad \text{(A3.6.12)}$$

On defining the density matrix

$$\rho(x',x) = \int \Psi^*(q,x')\, \Psi(q,x)\, dq \qquad \text{(A3.6.13)}$$

equation (A3.6.12) becomes

$$\langle Q \rangle = \int [Q(x)\, \rho(x',x)]_{x'=x}\, dx. \qquad \text{(A3.6.14)}$$

We next expand $\rho(x',x)$ in the set $\{\psi_i\}$ of orthonormal time-independent functions:

$$\rho(x',x;t) = \sum_{ij} \rho_{ij}(t)\, \psi_i^*(x')\, \psi_j(x) \qquad \text{(A3.6.15)}$$

whereupon equation (A3.6.14) becomes

$$\langle Q \rangle = \sum_{ij} \rho_{ij}(t) \langle i | Q | j \rangle. \qquad \text{(A3.6.16)}$$

This may immediately be put into the form (A3.6.9), where the operator $\rho(t)$ is such that

$$\langle j | \rho | i \rangle = \rho_{ij}. \qquad \text{(A3.6.17)}$$

Equation (A3.6.1) must evidently arise from (A3.6.16) in the special case where S is in thermal equilibrium with the rest of the ensemble. However, we are concerned here only to obtain an equation of motion for ρ. We note from equation (A3.6.15) that

$$\rho_{mn}(t) = \int \psi_n^*(x)\,\rho(x',x;t)\,\psi_m(x')\,dx'\,dx \tag{A3.6.18}$$

and therefore

$$\langle n|\frac{\partial \rho}{\partial t}|m\rangle = \frac{\partial \rho_{mn}}{\partial t} = \int \psi_n^*(x)\frac{\partial \rho(x',x)}{\partial t}\psi_m(x')\,dx'\,dx. \tag{A3.6.19}$$

Using equations (A3.6.10) and (A3.6.13) we find

$$\hbar i\frac{\partial \rho(x',x)}{\partial t} = \int \Psi^*(qx')\,H_t(q,x)\,\Psi(q,x)\,dq - \int [H_t(q,x')\,\Psi(q,x')]^*\,\Psi(q,x)\,dq. \tag{A3.6.20}$$

Referring now to equation (A3.6.11), since H_t is Hermitian the terms in $H_1(q)$ vanish, and, further, if S is isolated, $H_2(q,x)$ is zero. We are left with just

$$\begin{aligned}\hbar i\frac{\partial \rho(x',x)}{\partial t} &= \int \Psi^*(q,x')\,H(x)\,\Psi(q,x)\,dq - \int [H(x')\,\Psi(q,x')]^*\,\Psi(q,x)\,dq \\ &= \sum_{ij} \rho_{ij}\{\psi_i^*(x')\,H(x)\,\psi_j(x) - [H(x')\,\psi_i(x')]^*\,\psi_j(x)\}.\end{aligned} \tag{A3.6.21}$$

Hence, by equation (A3.6.19),

$$\begin{aligned}\hbar i\frac{\partial \rho_{mn}}{\partial t} &= \sum_j \rho_{mj}\langle n|H|j\rangle - \sum_i \rho_{in}\langle i|H|m\rangle \\ &= \sum_j [\langle n|H|j\rangle\langle j|\rho|m\rangle - \langle n|\rho|j\rangle\langle j|H|m\rangle],\end{aligned} \tag{A3.6.22}$$

that is,

$$i\langle n|\frac{\partial \rho}{\partial t}|m\rangle = \langle n|[H,\rho]|m\rangle \tag{A3.6.23}$$

and so

$$\hbar i\frac{\partial \rho}{\partial t} = [H,\rho], \tag{A3.6.24}$$

the desired equation of motion. If the density matrix is stationary,

$$0 = [H,\rho] \tag{A3.6.25}$$

satisfied by ρ as given by equation (A3.6.2).

The above discussion for a many-particle system may also be applied to the case where the system is made up of independent particles. For example, the density matrix for independent electrons may be written

$$\rho(\mathbf{r}', \mathbf{r}) = \sum_{ij} f_{ij}\,\psi_i^*(\mathbf{r}')\,\psi_j(\mathbf{r}), \tag{A3.6.26}$$

where the ψ_i are single-particle functions. Defining the one-particle operator f such that

$$f_{ij} = \langle j|f|i\rangle, \tag{A3.6.27}$$

the average of any one-particle operator may be written

$$\langle Q\rangle = \operatorname{tr}\{fQ_s\}, \tag{A3.6.28}$$

where the trace is over a set of independent particle states. The operator f obeys (A3.6.24), but in equilibrium its form must correspond to the Fermi–Dirac distribution, that is,

$$f(H) = \frac{1}{\mathrm{e}^{\beta(H-\zeta)}+1}, \tag{A3.6.29}$$

where ζ is chosen such that

$$\operatorname{tr}\{f(H)\} = N, \tag{A3.6.30}$$

N being the total number of particles.

Returning to ρ, we should finally note that various thermodynamic functions can be written in terms of it. In particular, the partition function Z is just

$$Z = \operatorname{tr}\{\mathrm{e}^{-\beta H}\} \tag{A3.6.31}$$

and the Helmholtz free energy is

$$F = -k_\mathrm{B} T \ln Z = \operatorname{tr}\{\rho H + k_\mathrm{B} T \rho \ln \rho\}. \tag{A3.6.32}$$

Thus if the system is in equilibrium we may write

$$\rho = \frac{1}{Z}\mathrm{e}^{-\beta H} = \mathrm{e}^{\beta(F-H)}. \tag{A3.6.33}$$

The entropy S also finds a formally simple expression in terms of ρ. This is [cf. Landau and Lifshitz (1958)]

$$S = -k_\mathrm{B}\operatorname{tr}\{\rho \ln \rho\}. \tag{A3.6.34}$$

APPENDIX A3.7 DIAGRAMMATIC TECHNIQUES FOR INTERACTIONS INVOLVING PHONONS

As we saw in the main text, it is convenient when dealing with temperature-dependent or thermodynamic Green functions to introduce an imaginary time τ $(= it)$. We then define the phonon Green function as

$$\mathscr{G}(\mathbf{k}, \sigma, \sigma'; \tau) = \langle T\{q_{\mathbf{k}\sigma}(\tau)\,q^\dagger_{\mathbf{k}\sigma'}(0)\}\rangle, \tag{A3.7.1}$$

where $q_{\mathbf{k}0}$ is the phonon-mode operator defined in equation (3.13.13), and

$$q(\tau) = e^{H\tau} q\, e^{-H\tau}. \qquad \text{(A3.7.2)}$$

The time-ordering operator orders operators labelled by smaller τ's to the right, and, since the q's are boson operators, $T = P$ here (cf. equation (3.14.5)).

The average of equation (A3.7.1) is a *thermal* average, i.e.

$$\langle P\{q_{\mathbf{k}\sigma}(\tau)\, q^{\dagger}_{\mathbf{k}\sigma'}(0)\}\rangle = \operatorname{tr}\{\rho P[q_{\mathbf{k}\sigma}(\tau)\, q^{\dagger}_{\mathbf{k}\sigma'}(0)]\} \qquad \text{(A3.7.3)}$$

and so in this appendix we exemplify diagrammatic techniques applied at elevated temperatures.

We begin by defining the development operator $U(\tau_1, \tau_2)$ such that

$$e^{(\tau_2-\tau_1)H} = e^{-\tau_1 H_0}\, U(\tau_1 \tau_2)\, e^{\tau_2 H_0}. \qquad \text{(A3.7.4)}$$

Hence

$$A(\tau) = U(0,\tau)\, \tilde{A}(\tau)\, U(\tau, 0), \qquad \text{(A3.7.5)}$$

where

$$\tilde{A}(\tau) = e^{H_0 \tau} A\, e^{-H_0 \tau}. \qquad \text{(A3.7.6)}$$

U may be written as

$$U(\tau_1 \tau_2) = 1 + \sum_n \left(-\frac{1}{\hbar}\right)^n \frac{1}{n!} P \int_{\tau_2}^{\tau_1} d\tau_1 \ldots \int_{\tau_2}^{\tau_1} d\tau_n\, \tilde{H}_1(\tau_1) \ldots \tilde{H}_1(\tau_n) \qquad \text{(A3.7.7)}$$

in which form it has the property (cf. equation (A2.2.11))

$$P[U(\tau_1, \tau_3)\, \tilde{A}(\tau_2)] = U(\tau_1 \tau_2)\, \tilde{A}(\tau_2)\, U(\tau_2 \tau_3) \quad (\tau_1 > \tau_2 > \tau_3). \qquad \text{(A3.7.8)}$$

If $A \equiv 1$ we have

$$U(\tau_1 \tau_3) = U(\tau_1 \tau_2)\, U(\tau_2 \tau_3) \quad (\tau_1 > \tau_2 > \tau_3). \qquad \text{(A3.7.9)}$$

Since from equation (A3.7.4) we have

$$e^{-\beta H} = e^{-\beta H_0}\, U(\beta, 0) \qquad \text{(A3.7.10)}$$

and from equations (A3.7.8) and (A3.7.9), together with the definition of U,

$$\begin{aligned} U(\beta,0) q(\tau)\, U(\beta,0)\, U(0,\tau) \tilde{q}(\tau)\, U(\tau,0) &= U(\beta,\tau) \tilde{q}(\tau)\, U(\tau,0) \\ &= P[U(\beta,0) \tilde{q}(\tau)] \quad (\beta > \tau > 0), \end{aligned} \qquad \text{(A3.7.11)}$$

we can now rewrite equation (A3.7.1) in the form (treating $\tau < 0$ similarly)

$$\begin{aligned} \mathscr{G}(\mathbf{k}, \sigma\sigma'; \tau) &= \langle U(\beta, 0\rangle_0^{-1} \langle P[U(\beta,0)\, \tilde{q}_{\mathbf{k}\sigma}(\tau) \tilde{q}^{\dagger}_{\mathbf{k}\sigma}(0)]\rangle_0 \\ &= \operatorname{tr}\{e^{-\beta H_0} P[U(\beta,0)\, \tilde{q}_{\mathbf{k}\sigma}(t)\, q^{\dagger}_{\mathbf{k}\sigma}(0)\}/\operatorname{tr}\{e^{-\beta H_0}\, U(\beta,0)\}. \end{aligned} \qquad \text{(A3.7.12)}$$

From now on we will drop the tildes from τ-dependent operators, assuming all such operators are in the interaction representation, so that their τ-dependence is according to the harmonic Hamiltonian.

A3.7.1 Perturbation expansion and thermodynamic form of Wick's theorem

From the form of the anharmonic terms as shown in equation (3.13.22) of the main text, it will be seen that, on insertion of equation (A3.7.7) into equation (A3.7.12), a typical term of the second factor of equation (A3.7.12) has the form

$$\int d\tau_1 \dots d\tau_n \operatorname{tr}\{e^{-\beta H_0} P[q_{\mathbf{k}\sigma}(\tau) q_{\mathbf{k}'\sigma'}(0) q_{\mathbf{k}_1\sigma_1}(\tau_1) q_{\mathbf{k}_2\sigma_2}(\tau_2) \dots]\}. \quad \text{(A3.7.13)}$$

When treating the electron gas in its ground state we used Wick's theorem to write an analogous expression as a product of independent particle functions. We shall see that this is possible here also. We use the theorem that for a set of operators A_i, which are linear combinations of creation and annihilation operators (as are the q's),

$$\langle T(A_1 A_2 \dots A_{2n})\rangle_0 = \sum_{\text{all pairs}} (-1)^p (\overbrace{A_1 A_2 \dots A_k \dots A_l \dots A_m \dots A_n}), \quad \text{(A3.7.14)}$$

where

$$\overline{AB} = \langle T(AB)\rangle_0$$

and

$$\overline{A_1 A_2 A_3 A_4} = \overline{A_1 A_3}\,\overline{A_2 A_4}, \quad \text{etc.} \quad \text{(A3.7.15)}$$

We prove equation (A3.7.14) only for the case we require, that is, when the A's are boson operators. Then $(-1)^p = 1$ (p being the number of exchanges of *Fermion* operators needed to bring contracted operators next to one another). The basic method of proof is unchanged however.

We first note that for any operator Q, the cyclic property of the trace implies that

$$\begin{aligned}\langle [a_{\mathbf{k}}, Q]\rangle_0 &= \operatorname{tr}\{(e^{-\beta H_0}/Z_0)[a_{\mathbf{k}}, Q]\} = \operatorname{tr}\{(e^{-\beta H_0}/Z_0)(a_{\mathbf{k}} Q - Q a_{\mathbf{k}})\} \\ &= \langle a_{\mathbf{k}} Q\rangle_0 - \frac{1}{Z_0}\operatorname{tr}\{a_{\mathbf{k}} e^{-\beta H_0} Q\} \\ &= (1 - e^{-\beta\epsilon_{\mathbf{k}}})\langle a_{\mathbf{k}} Q\rangle_0 \end{aligned} \quad \text{(A3.7.16)}$$

and, similarly,

$$\operatorname{tr}\{e^{-\beta H_0}[a_{\mathbf{k}}^\dagger, Q]\} = (1 - e^{\beta\epsilon_{\mathbf{k}}}) \operatorname{tr}\{e^{-\beta H_0} a_{\mathbf{k}}^\dagger Q\}. \quad \text{(A3.7.17)}$$

We apply these results to the product $B_1 \dots B_{2n}$, for which we assume

$$P(B_1 B_2 \dots B_{2n}) = B_1 B_2 \dots B_{2n}. \quad \text{(A3.7.18)}$$

At the moment we also take B_1 to be either a creation or an annihilation operator. Then by equation (A3.7.16) or (A3.7.17),

$$\langle B_1 B_2 \dots B_{2n}\rangle_0 = \langle [B_1, B_2 B_3 \dots B_{2n}]\rangle_0 (1 - e^{\pm\beta\epsilon_{\mathbf{k}}})^{-1}. \quad \text{(A3.7.19)}$$

In $e^{\pm\beta\epsilon_{\mathbf{k}}}$, the plus sign holds if B_1 is a creation operator and the minus sign if it is an annihilation operator. Using the result

$$[B_1, B_2 B_3 \ldots B_{2n}] = \sum_{i=2}^{2n} [B_1, B_i] \prod_{j\neq 1,i}' B_j, \tag{A3.7.20}$$

equation (A3.7.19) becomes

$$\langle B_1 B_2 \ldots B_{2n} \rangle_0 = (1 - e^{\pm\beta\epsilon_{\mathbf{k}}})^{-1} \sum_{i=2}^{2n} [B_1, B_i] \left\langle \prod_{j\neq 1,i}' B_j \right\rangle_0. \tag{A3.7.21}$$

Since the commutator is a *c*-number we can put

$$[B_1, B_i] = \langle [B_1, B_i] \rangle_0 = (1 - e^{\pm\beta\epsilon_{\mathbf{k}}}) \langle B_1 B_i \rangle_0 \tag{A3.7.22}$$

and equation (A3.7.21) becomes

$$\langle B_1 B_2 \ldots B_{2n} \rangle_0 = \sum_{i=2}^{2n} \overline{B_1 B_i} \left\langle \prod_{j\neq 1,i}^{2n}{}' B_j \right\rangle_0. \tag{A3.7.23}$$

It is now evident that the result (A3.7.14) may be proved by induction. From equation (A3.7.23) if it is true for $2n-2$ boson operators it is true for $2n$ boson operators. Then since it is true if $n = 1$, it is true for all n. Note also that while we have supposed B_1 to be either a creation operator or an annihilation operator, the same result obtains for either and hence is also applicable to a linear combination of the two.

A3.7.2 Graphical representation

To represent the series for $\mathcal{G}$ graphically we put a dot (vertex) for each time τ, and term Φ_i of the Hamiltonian (see section 3.13.3), and draw lines between the vertices, each line running from τ_i to τ_j being associated with a Green function $\mathcal{G}_0(\mathbf{k}, \tau_i - \tau_j)$. Since the term Φ_3, for example, of $H_1(\tau_i)$ is associated with a product of three q's all labelled with the same time τ_i, there will be three lines to the vertex associated with τ_i and Φ_3. Every line is labelled with $\mathbf{k}$, σ, and because of the condition stated after equation (3.13.23) of the main text the total $\mathbf{k}$ must be conserved, to within a reciprocal lattice vector, at each vertex. Every line must run between two vertices associated with a Φ_i, except the lines corresponding to the Green function arising from the factors $q_{\mathbf{k}\sigma}$ and $q^{\dagger}_{\mathbf{k}\sigma'}$ of equation (A3.7.12). We thus obtain the diagrams illustrated in Figure 3.12 of the main text.

A3.7.3 Elimination of disconnected diagrams

Apart from diagrams like Figure 3.12, however, we can also obtain disconnected diagrams, as in Figure A3.7.1, for example, which is associated with the product

$$\mathcal{G}(\mathbf{k}\tau\tau_1)\,\mathcal{G}(\mathbf{k}_1\,\tau_1\,\tau_2)\,\mathcal{G}(\mathbf{k}_2\,\tau_1\,\tau_2)\,\mathcal{G}(\mathbf{k}_3\,\tau_3\,\tau_4)\,\mathcal{G}(\mathbf{k}_4\,\tau_3\,\tau_4)\,\mathcal{G}(\mathbf{k}_5\,\tau_3\,\tau_4)\,\mathcal{G}(\mathbf{k}\tau_2\,0).$$

However the denominator $\mathrm{tr}\{e^{-\beta H_0} U(\beta, 0)\}$ is represented entirely by disconnected diagrams, and just as in Appendix A2.5 we can show that the contribution of the disconnected diagrams to the numerator exactly cancels the denominator.

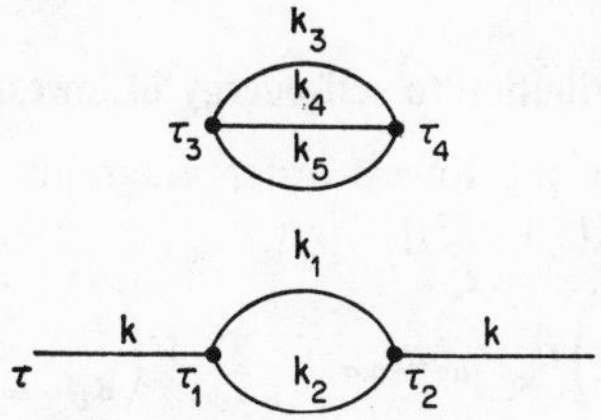

FIGURE A3.7.1. Simple diagram with a disconnected part.

A3.7.4 Rules for the interpretation of the graphs

The rules we give are applicable to $\mathscr{G}(\mathbf{k}\sigma\sigma', i\omega_n)$. The Fourier transformation is similar to that given for the electron Green function, and other matters of detail—for example, the equivalence of terms corresponding to topologically equivalent diagrams—not covered above, may also be discussed in a way similar to that of Appendix A2.5.

For the nth term (cf. Cowley, 1966):

1. Draw all topologically distinct diagrams with n vertices.

2. Associate momenta $\mathbf{k}_i$, frequencies ω_n and branch indices σ_i with the lines such that the external lines are labelled by $\mathbf{k}, \omega_n$, while crystal momentum and frequency is conserved at each vertex.

3. With each phonon line associate a factor $\mathscr{G}_0(k_i\,\sigma_i;\, i\omega_n^i)$.

4. Multiply by the expansion coefficients in H_1 appropriate to each vertex.

5. Multiply by $(-\beta)^n$.

6. Multiply by the number of ways of pairing the modes in the diagram. This essentially comes from symmetry of H_1 in the $(\mathbf{k}, \omega, \sigma)$'s and is a combinatorial factor, a product (a) × (b) × (c) where we have

(a) the number of topologically equivalent diagrams for a *fixed* arrangement of vertices;

(b) the number of different labellings $\mathbf{k}_i\,\sigma_i$ at each vertex for each pairing scheme;

(c) the number of ways of permuting 3- and 4-phonon vertices.

7. Sum over independent momenta $\mathbf{k}_i$, frequencies ω_n and branch indices σ_i.

We illustrate these rules by a calculation to lowest order of the phonon frequency shift Δ and inverse lifetime Γ.

A3.7.5 Example: Contribution to self-energy of lowest-order diagrams

Let us now consider the lowest-order diagrams from the perturbation [given by (3.13.22) and (3.13.27)]

$$H_1 = \sum_{\substack{\alpha\beta\mathbf{k} \\ \sigma\sigma'}} V_{\alpha\beta}\begin{pmatrix} \sigma & \sigma' \\ \mathbf{k} & -\mathbf{k} \end{pmatrix} u_{\alpha\beta} q_{\mathbf{k}\sigma} q_{-\mathbf{k}\sigma'} + \sum_{\substack{\mathbf{k}_1\mathbf{k}_2\mathbf{k}_3 \\ \sigma_1\sigma_2\sigma_3}} V\begin{pmatrix} \sigma_1 & \sigma_2 & \sigma_3 \\ \mathbf{k}_1 & \mathbf{k}_2 & \mathbf{k}_3 \end{pmatrix} q_{\mathbf{k}_1\sigma_1} q_{\mathbf{k}_2\sigma_2} q_{\mathbf{k}_3\sigma_3}$$
$$+ \sum_{\substack{\mathbf{k}_1\mathbf{k}_2\mathbf{k}_3\mathbf{k}_4 \\ \sigma_1\sigma_2\sigma_3\sigma_4}} V\begin{pmatrix} \sigma_1 & \sigma_2 & \sigma_3 & \sigma_4 \\ \mathbf{k}_1 & \mathbf{k}_2 & \mathbf{k}_3 & \mathbf{k}_4 \end{pmatrix} q_{\mathbf{k}_1\sigma_1} q_{\mathbf{k}_2\sigma_2} q_{\mathbf{k}_3\sigma_3} q_{\mathbf{k}_4\sigma_4}. \quad \text{(A3.7.24)}$$

The first term here is that of thermal strain, the lowest-order diagram being as in Figure A3.7.2(a). The third term gives Figure A3.7.2(b). The second term contains no linear part, and its lowest contribution involves two different times [Figure A3.7.2(c)]. The contribution to Σ from Figure A3.7.2(a) is just

$$\frac{2}{\hbar}\sum_{\alpha\beta} V_{\alpha\beta}\begin{pmatrix} \sigma & \sigma \\ \mathbf{k} & -\mathbf{k} \end{pmatrix} u_{\alpha\beta}$$

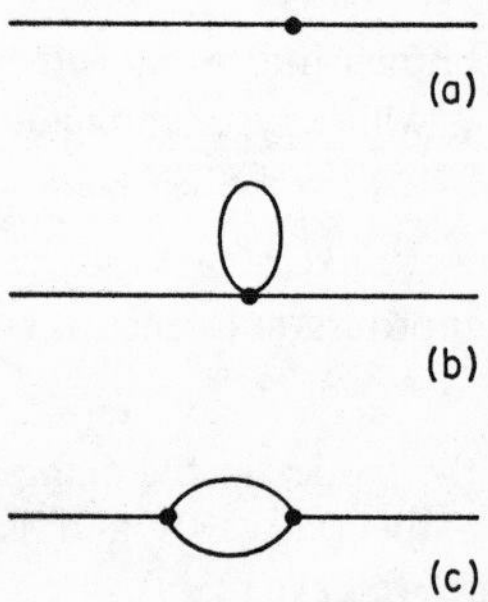

FIGURE A3.7.2. Lowest-order diagrams for self-energy. (a) Thermal strain contribution. (b) Linear term in four phonons. (c) Second-order term.

because the factor $(-\beta)$ from rule 5 of p. 625 cancels with the $(-1/\beta\hbar)$ in the definition to Σ to give $1/\hbar$; the factor 2 comes from rule 6 and is the number of pairing schemes. The contribution from Figure 3.7.2(b) is

$$\frac{12}{\hbar}\sum_{\mathbf{k}_1\sigma_1}\sum_{n_1} V\begin{pmatrix} \sigma & \sigma' & \sigma_1 & \sigma_1 \\ \mathbf{k} & -\mathbf{k} & \mathbf{k}_1 & -\mathbf{k}_1 \end{pmatrix} \mathscr{G}_0(\mathbf{k}\sigma_1; i\omega_{n_1}). \quad \text{(A3.7.25)}$$

The $1/\hbar$ arises as in (3.11.21). The factor 12 (rule 6) arises since in (3.13.22), $\mathbf{k}$ can equal $\mathbf{k}_1$, $\mathbf{k}_2$, $\mathbf{k}_3$ or $\mathbf{k}_4$; and $-\mathbf{k}$ can equal any of the $\mathbf{k}_i$ apart from $\mathbf{k} = \mathbf{k}_i$. This gives twelve possibilities, and the complete symmetry of the V in the $(\mathbf{k}\sigma)$'s means that we need not write down these twelve terms separately.

The summation over n_1, though easily done directly, will serve to illustrate a more general method. Consider the integral (see Cowley, 1966)

$$\int_C \frac{f(i\omega)\, d(i\omega)}{[e^{i\beta\hbar\omega}-1]}, \tag{A3.7.26}$$

where the contour is taken as a circle at infinity in the complex plane. Provided $f(i\omega)$ is a reasonable function, going to zero sufficiently rapidly as $(i\omega)\to\infty$, the integral round the contour is zero. Now the integral is also equal to $2\pi i$ times the sum of residues R at the poles. The simple zeros of the denominator $(e^{i\beta\hbar\omega}-1)$ occur whenever $\omega = 2\pi n_1/\beta\hbar$, with residues $1/\beta\hbar$. If the poles of $f(i\omega)$ occur at $i\omega_p$, then we have

$$0 = \frac{1}{\beta\hbar}\sum_n f(i\omega_n) + \sum_p \frac{R[f(i\omega_p)]}{[e^{i\beta\hbar\omega_p}-1]}. \tag{A3.7.27}$$

In (A3.7.25), we have $f(i\omega_n) = \mathscr{G}_0(\mathbf{k}\sigma_1; i\omega_n)$; $\mathscr{G}_0$ has poles at $i\omega_n = \pm\omega(\mathbf{k}\sigma)$ with residues of $-1/\beta\hbar$ and $1/\beta\hbar$. The summation is then

$$-\beta\hbar\left\{-\frac{1}{\beta\hbar}[e^{\beta\hbar\omega(\mathbf{k}_1\sigma_1)}-1]^{-1}+\frac{1}{\beta\hbar}[e^{-\beta\hbar\omega(\mathbf{k}_1\sigma_1)}-1]^{-1}\right\} = 2n(\mathbf{k}_1\sigma_1)+1, \tag{A3.7.28}$$

where $n(\mathbf{k}\sigma)$ is defined in (3.13.46). Thus (3.11.25) becomes

$$\frac{12}{\hbar}\sum_{\mathbf{k}_1\sigma_1} V\begin{pmatrix}\sigma & \sigma' & \sigma_1 & \sigma_1\\ \mathbf{k} & -\mathbf{k} & \mathbf{k}_1 & -\mathbf{k}_1\end{pmatrix}[2n(\mathbf{k}_1\sigma_1)+1]. \tag{A3.7.29}$$

One may also verify that the contribution of Figure A3.7.2(c) is

$$-\frac{18\beta}{\hbar}\sum_{\substack{\mathbf{k}_1\mathbf{k}_2\\ \sigma_1\sigma_2}}\sum_{n_1} V\begin{pmatrix}\sigma & \sigma_1 & \sigma_2\\ \mathbf{k} & -\mathbf{k}_1 & -\mathbf{k}_2\end{pmatrix} V\begin{pmatrix}\sigma' & \sigma_1 & \sigma_2\\ -\mathbf{k} & \mathbf{k}_1 & \mathbf{k}_2\end{pmatrix}$$
$$\times\, \mathscr{G}_0(\mathbf{k}_1\sigma_1; i\omega_n)\,\mathscr{G}_0(\mathbf{k}_2\sigma_2; i\omega_n - i\omega_{n_1}). \tag{A3.7.30}$$

We again sum with the help of (A3.7.27). This is left to the reader. The result is

$$-\frac{18}{\hbar^2}\sum_{\substack{\mathbf{k}_1\sigma_1\\ \mathbf{k}_2\sigma_2}} V\begin{pmatrix}\sigma & \sigma_1 & \sigma_2\\ \mathbf{k} & -\mathbf{k}_1 & -\mathbf{k}_2\end{pmatrix} V\begin{pmatrix}\sigma' & \sigma_1 & \sigma_2\\ -\mathbf{k} & \mathbf{k}_1 & \mathbf{k}_2\end{pmatrix}$$
$$\times\left(\frac{n_1+n_2+1}{\omega_1+\omega_2+i\omega_n}+\frac{n_1+n_2+1}{\omega_1+\omega_2-i\omega_n}+\frac{n_2-n_1}{\omega_1-\omega_2+i\omega_n}+\frac{n_1-n_2}{\omega_2-\omega_1+i\omega_n}\right). \tag{A3.7.31}$$

[Here we abbreviate $\omega(\mathbf{k}_i\sigma_i)$ to ω_i and $n(\mathbf{k}_i\sigma_i)$ to n_i.]

It is pertinent to note that physical properties can be obtained by the analytic continuation of $\mathscr{G}(\mathbf{k}\sigma\sigma'; i\omega_n)$, from the points $(i\omega_n)$, over the complex plane: we have already assumed this in discussing the physical meaning of Σ [cf. equation (3.14.25)]. Now in arriving at (A3.7.31) we encounter terms $e^{\beta\hbar i\omega_n}$ which are put equal to one because of the definition of ω_n. This *must* be done before considering ω_n as a continuous variable, or incorrect results will be obtained. The analytic continuation is obtained by putting $i\omega_n = \omega + i\varepsilon$ $(\varepsilon \to 0)$ and using the result

$$\lim_{\varepsilon\to 0_+} \frac{1}{x+i\varepsilon} = P\left(\frac{1}{x}\right) - i\pi\delta(x). \tag{A3.7.32}$$

APPENDIX A4.1 GROUND-STATE OF LINEAR ANTIFERROMAGNETIC CHAIN

In this appendix we amplify our discussion in the text of the variational treatment of an antiferromagnet with the Hamiltonian

$$H = \frac{J}{2}\sum_{ij}(\boldsymbol{\sigma}_i . \boldsymbol{\sigma}_j - 1), \tag{A4.1.1}$$

where the summation is taken over all nearest neighbours, J is a *positive* exchange integral, and

$$\boldsymbol{\sigma} = \xi\hat{\mathbf{x}} + \eta\hat{\mathbf{y}} + \zeta\hat{\mathbf{z}}. \tag{A4.1.2}$$

ξ, η, ζ are the Pauli spin matrices

$$\xi = \begin{pmatrix} 0 & 1 \\ 1 & 0 \end{pmatrix}, \quad \eta = \begin{pmatrix} 0 & -i \\ i & 0 \end{pmatrix}, \quad \zeta = \begin{pmatrix} 1 & 0 \\ 0 & -1 \end{pmatrix}. \tag{A4.1.3}$$

To illustrate this variational procedure, carried out by Marshall (1955) on cubic lattices, it will be sufficient to take as an example the linear chain, since formally identical results are obtained. The linear chain has been treated by Kasteleijn (1952), following earlier work by Slater (1930) and Hulthen (1938), and this appendix largely follows Kasteleijn's procedure.

The Hamiltonian for the linear chain of N spins is

$$H = \frac{J}{2}\sum_{i=1}^{N}(\boldsymbol{\sigma}_i . \boldsymbol{\sigma}_{i+1} - 1). \tag{A4.1.4}$$

If the x and y components of $\boldsymbol{\sigma}$ are neglected, one obtains the Hamiltonian of the Ising model with nearest-neighbour interactions:

$$H = \frac{J}{2}\sum_{i=1}^{N}(\zeta_i \zeta_{i+1} - 1). \tag{A4.1.5}$$

Now in operating on the spin functions $\alpha = \begin{pmatrix} 1 \\ 0 \end{pmatrix}$, $\beta = \begin{pmatrix} 0 \\ 1 \end{pmatrix}$ one can easily verify that

$$\begin{aligned} \xi\alpha = \beta, \quad \eta\alpha = i\beta, \quad \zeta\alpha = \alpha, \\ \xi\beta = \alpha, \quad \eta\beta = -i\alpha, \quad \zeta\beta = -\beta, \end{aligned} \tag{A4.1.6}$$

so that the ground state of the Hamiltonian of (A4.1.5) is evidently one in which all spins on even sites ($i = 2, 4, 6, \ldots$) are aligned in the z-direction ("up-spins"), and all spins on odd sites ($i = 1, 3, 5, \ldots$) are antiparallel to these. Observing that the isotropic Hamiltonian of (A4.1.5) does not show this division into sub-lattices, Kasteleijn introduced anisotropy by taking a Hamiltonian of the form

$$H = \frac{J}{2}\sum_{i=1}^{N} [(1-a)(\xi_i\,\xi_{i+1} + \eta_i\,\eta_{i+1}) + (\zeta_i\,\zeta_{i+1} - 1)] \tag{A4.1.7}$$

and calculated the value of a at which a transition to a sub-lattice division of spin alignment in the z-direction obtained.

In view of Thouless' remarks, reported in the main text (p. 338), we shall not deal with this aspect of Kasteleijn's work, and we shall keep $a = 0$.

A4.1.1 Form of wave function

One can expand the wave function Ψ in terms of functions, each representing a particular configuration of spins, formed by products of N α's and β's. These configurations are characterized by the numbers p, q and r. p is the number of up-spins at "even" places, q the number of antiparallel pairs, and r the number of up-spins at "odd" places. These three numbers do not completely characterize a particular configuration, but both Marshall and Kasteleijn assumed that configurations described by the same set of numbers p, q and r appear with the same amplitudes in the expansion of the ground-state wave function Ψ, so that the amplitudes depend only on local-spin ordering. It is actually somewhat more convenient to use, instead of p and r, numbers m, n defined by

$$\left.\begin{aligned} 2p &= m+n, \\ 2r &= m-n, \end{aligned}\right\} \tag{A4.1.8}$$

the advantage being that the functions produced when the Hamiltonian acts upon a particular configuration are characterized by the same parameter m, only n and q changing.

We therefore write

$$\Psi = \sum_{m,n,q} p(m,n,q) \sum_k \Phi_k(m,n,q). \tag{A4.1.9}$$

The functions Φ describe particular configurations, the distinction between the $\Omega(m,n,q)$ different configurations with the same parameters m, n and q being made with the subscript k. Note that the Φ's are orthonormal:

$$\langle \Phi_k(m,n,q) |\, \Phi_{k'}(m',n',q') \rangle = \delta_{mm'}\,\delta_{nn'}\,\delta_{qq'}\,\delta_{kk'}. \tag{A4.1.10}$$

We now examine the effect of the Hamiltonian on the Φ's. We shall follow Kasteleijn in using a Hamiltonian in which the coordinate system for the odd sites has been rotated through 180° relative to the system for even sites: we take

$$H = \frac{J}{2}\sum_i [(\zeta_i\,\zeta_{i+1}-1)-(\xi_i\,\xi_{i+1}+\eta_i\,\eta_{i+1})] = \sum_i H_i. \qquad \text{(A4.1.11)}$$

The reason for this change will be seen later in section A4.1.2. (It is, of course, not essential, merely convenient.)

From (A4.1.6) one can see that operating on a particular spin pair in a configuration,

$$H_i\,\alpha_i\,\alpha_{i+1} = 0, \quad H_i\,\beta_i\,\beta_{i+1} = 0,$$
$$H_i\,\alpha_i\,\beta_{i+1} = -J(\alpha_i\,\beta_{i+1}+\beta_i\,\alpha_{i+1}), \quad H_i\,\beta_i\,\alpha_{i+1} = J(\beta_i\,\alpha_{i+1}+\alpha_i\,\beta_{i+1}). \qquad \text{(A4.1.12)}$$

Hence, in the result of operating with H on $\Phi_k(m,n,q)$, we see that parallel spins contribute a zero term, so that the original function occurs q times with coefficient $-J$. Of the other terms, where spins are interchanged, $m = p+r$ does not change, whereas $n = p-r$ and q do. In the following table we show the changes of n and q when all possible configurations of 4-spins have their inner pair changed by the operation of H. Capital letters denote spin functions for even sites, and small letters those for odd sites.

Spin configurations	$A\beta A\beta$ $\beta A\beta A$	$\alpha B\alpha B$ $B\alpha B\alpha$	$A\beta A\alpha$ $\alpha A\beta A$ $B\beta A\beta$ $\beta A\beta B$	$\alpha B\alpha A$ $A\alpha B\alpha$ $\beta B\alpha B$ $B\alpha A\alpha$	$B\beta A\alpha$ $\alpha A\beta B$	$\beta B\alpha A$ $A\alpha B\beta$
Parameter changes	$n\to n-2$ $q\to q-2$	$n\to n+2$ $q\to q-2$	$n\to n-2$ $q\to q$	$n\to n+2$ $q\to q$	$n\to n-2$ $q\to q+2$	$n\to n+2$ $q\to q+2$

Let us now suppose that, in $\Phi_k(mnq)$, the configuration in the above table responsible for the transition $n\to n+\delta$, $q\to q+\Delta$ is present in the *fraction* $F^k_{\delta,\Delta}(m,n,q)$. Now for example, a function $\Phi_l(m,n+2,q+2)$ can only be formed from a $\Phi_k(m,n,q)$ by interchange of the inner pair of one of the configurations $A\alpha B\beta$ or $\beta B\alpha A$, giving $A\beta A\beta$ or $\beta A\beta A$; since the total number of the latter two configurations in $\Phi_l(m,n+2,q+2)$ is $NF^l_{-2,-2}(m,n+2,q+2)$ (if a transition is made only one way with a certain frequency, it is made in the opposite direction with equal frequency), $\Phi_l(m,n+2,q+2)$ can be formed in $NF^l_{-2,-2}(m,n+2,q+2)$ ways. If we define the normalized function

$$\Psi(m,n+2,q+2) = \sum_k \Phi_k(m,n,q)/\Omega^{\frac{1}{2}}(m,n,q) \qquad \text{(A4.1.13)}$$

it follows that $\Phi_l(m, n+2, q+2)$ occurs in $H\Psi(m, n, q)$ with the coefficient $-JNF^l_{2,-2}(m, n+2, q+2)$ divided by the normalizing factor:

$$H\Psi(m,n,q) = -qJ\Psi(m,n,q) - NJ \sum_{\delta,\Delta,l} \frac{F^l_{-\delta,-\Delta}(m, n+\delta, q+\Delta)}{[\Omega(m,n,q)]^{\frac{1}{2}}} \Phi_l(m, n+\delta, q+\Delta). \tag{A4.1.14}$$

The summations are over

$$\delta = \pm 2, \quad \Delta = 0, \pm 2, \quad = 1, 2, \ldots, \Omega(m, n+\delta, q+\Delta). \tag{A4.1.15}$$

The ground-state energy is now estimated by writing $\Psi = \sum_{m,n,q} p(m,n,q) \times \Psi(m,n,q)$, and minimizing

$$\frac{\langle\Psi|H|\Psi\rangle}{\langle\Psi|\Psi\rangle} = \frac{\sum\limits_{\substack{m,n,q\\ m',n',q'}} p(m,n,q)p(m',n',q')\langle\Psi(m,n,q)|H|\Psi(m',n',q')\rangle}{\sum\limits_{\substack{m,n,q\\ m',n',q'}} p(m,n,q)p(m',n',q')\langle\Psi(m,n,q)|\Psi(m,n,q)\rangle}. \tag{A4.1.16}$$

From equation (A4.1.3) and (A4.1.6) we can see that

$$\langle\Psi(m,n,q)|\Psi(m',n',q')\rangle = \delta_{m'n'}\delta_{nn'}\delta_{qq'} \tag{A4.1.17}$$

so that using (A4.1.14), equation (A4.1.16) becomes

$$-\frac{\langle\Psi|H|\Psi\rangle}{\langle\Psi|\Psi\rangle} = qJ + NJ \sum_{m,n,q} p(m,n,q) \times \frac{\sum\limits_{\delta,\Delta} F_{\delta,\Delta}(m,n,q)p(m,n+\delta,q+\Delta)}{\sum\limits_{m,n,q}|p(m,n,q)|^2}[\Omega(m,n,q)\Omega(m,n+\delta,q+\Delta)]^{-\frac{1}{2}}. \tag{A4.1.18}$$

Here we have made the transformations $\delta \to -\delta$, $\Delta \to -\Delta$ and introduced the definition

$$F_{\delta,\Delta}(m,n,q) = \frac{\sum\limits_k F^k_{\delta,\Delta}(m,n,q)}{\Omega(m,n,q)}. \tag{A4.1.19}$$

A4.1.2 Minimization procedure

Our reason for choosing the Hamiltonian in the form (A4.1.11) should now be stated. We wish, in the limit $N \to \infty$, to change from the discrete set m, n and q to a continuum to ease the analysis. Now the only difference in the right-hand side of (A4.1.18) that would appear if we were to take the original Hamiltonian of equation (A4.1.4) would be to change the sign of the second term. It is then evident that, $F_{\delta,\Delta}$ being real and positive, maximization of this right-hand side would be achieved by taking $p(m,n,q)$ and

$p(m, n+\delta, q+\Delta)$ opposite in sign (although of comparable magnitude). We have ensured continuity by transformation of the Hamiltonian.

Let us now introduce the quantities

$$\left.\begin{aligned} F_{\delta,\Delta}(m,n,q) &= \bar{F}_{\delta,\Delta}(m,n,q)\,[\Omega(m,n,q)/\Omega(m,n+\delta,q+\Delta]^{\frac{1}{2}}, \\ s &= [m-(N/2)]/(N/2), \quad m = (N/2)(1+s), \\ s' &= n/(N/2), \qquad\qquad\quad n = (N/2)\,s', \\ x &= [q-(N/2)]/(N/2), \quad\ q = (N/2)(1+x). \end{aligned}\right\} \tag{A4.1.20}$$

The quantities s, s', x become continuous variables in the limit, and that is why they are introduced.

We also define

$$p(m,n,q) = \exp[N\omega(s,s',x)] \tag{A4.1.21}$$

when, converting summations to integrations, we obtain

$$-\frac{\langle\Psi|H|\Psi\rangle}{\langle\Psi|\Psi\rangle} = qJ$$
$$+\frac{NJ\int \exp[2N\omega(s,s',x]\sum_{\delta,\Delta} F_{\delta\,\Delta}(s,s',x)\exp\{2\delta[(\partial\omega/\partial s')+2\Delta(\partial\omega/\partial x)]\}\,ds\,ds'\,dx}{\int \exp[2N\omega(s,s',x)]\,ds\,ds'\,dx}. \tag{A4.1.22}$$

The introduction of the exponential in equation (A4.1.21) is really no more than a formal indication that we assume we can proceed statistically in close analogy with the procedures of statistical thermodynamics (and the procedures used to handle the general Ising model).

We shall briefly remind the reader of the kind of considerations involved. Suppose m is significant in a range about a mean $\bar{m}$, the measure of this range being $\sqrt{(\Delta m)^2}$ [the average being taken using the probability amplitudes $p(m,n,q)$]. Now if the system is very large, we can divide into it a number (say $\mathcal{N}$) of sub-systems each macroscopic in size, and each having a parameter m_i describing it. Obviously

$$m = \sum_{i=1}^{\mathcal{N}} m_i,$$

and for the mean values,

$$\bar{m} = \sum_{i=1}^{\mathcal{N}} \overline{m_i}.$$

We can also say that we can let $N \to \infty$ by keeping the sub-systems of the same size, but multiplying them so that $\mathcal{N} \to \infty$; m will roughly increase in proportion to $\mathcal{N}$. Let us now look at $\overline{(\Delta m)^2}$ (corresponding to the mean-square fluctuation of a quantity in statistical physics). We have

$$\overline{(\Delta m)^2} = \overline{\sum_i (\Delta m_i)^2} = \sum_{i=1}^{\mathcal{N}} \overline{(\Delta m_i)^2} + \sum_{i \neq j}^{\mathcal{N}} \overline{\Delta m_i \, \Delta m_j}. \qquad \text{(A4.1.23)}$$

If the sub-systems are large enough, Δm_i and Δm_j will not sensibly depend on one another, and we can write

$$\overline{(\Delta m)^2} = \sum_{i=1}^{\mathcal{N}} \overline{(\Delta m_i)^2}.$$

Thus $\overline{(\Delta m)^2}$ is proportional to $\mathcal{N}$ or N, and the significant range of m is directly proportional to $\sqrt{1/N}$:

$$\frac{\sqrt{\overline{(\Delta m)^2}}}{\bar{m}} \propto \sqrt{\frac{1}{N}}.$$

The range of significant values of s, s' and x, therefore (these values depending on the proportions m/n and q/N), tends to zero as $N \to \infty$. This being so, we can calculate the integrals of equation (A4.1.22) assuming

$$\frac{\partial \omega}{\partial s} = \frac{\partial \omega}{\partial x} = 0 \qquad \text{(A4.1.24)}$$

and we can also take $\sum_{\delta,\Delta} F_{\delta,\Delta}(s, s', x)$ out from under the integral sign.

Then we obtain an estimate of the energy from

$$\begin{aligned} \frac{\langle \Psi | H | \Psi \rangle}{\langle \Psi | \Psi \rangle} &= -\max \Big[qJ + NJ \sum_{\delta,\Delta} F_{\delta,\Delta}(s, s', x) \Big] \\ &= -NJ \max \Big[\tfrac{1}{2}(1+x) + \sum_{\delta,\Delta} F_{\delta,\Delta}(s, s', x) \Big]. \qquad \text{(A4.1.25)} \end{aligned}$$

It remains to calculate the $F_{\delta,\Delta}(s, s', x)$. We shall shortly see that it is the $\mathbf{F}_{\delta,\Delta}$ that are calculable directly: from the definition of $\mathbf{F}_{\delta,\Delta}(m, n, q)$ in (A4.1.19) one sees that it is the mean fraction of configurations responsible for the transition $n \to n+\delta$, $q \to q+\delta$. We accordingly first show how $F_{\delta,\Delta}$ may be obtained from $\mathbf{F}_{\delta,\Delta}$. The relation of one to the other is given in equation (A4.1.20), but cannot be applied immediately because we do not know the values of the Ω's. To obtain this relationship in a workable form, we note that in the $\Omega(m,n,q)$ functions $\Phi_k(m,n,q)$ the quartets $A\beta A\beta$ and $\beta A\beta A$ occur together $\Omega(m,n,q)\bar{F}_{-2,-2}(m,n,q)$ times, and give, on interchange of inner pairs, chains with parameters $(m,n-2,q-2)$. Reversing each step, $\Omega(m,n,q)\bar{F}_{-2,-2}(m,n,q)$ must also be the number of times configurations $A\alpha B\beta$ and $\beta B\alpha A$ occur in the functions $\Phi_l(m,n-2,q-2)$:

$$\Omega(m,n,q)\,N\bar{F}_{-2,-2}(m,n,q) = \Omega(m,n-2,q-2)\,N\bar{F}_{2,2}(m,n-2,q-2).$$

Evidently, in general,

$$\Omega(m,n,q)\,N\bar{F}_{\delta,\Delta}(m,n,q) = \Omega(m,n+\delta,q+\Delta)\,N\bar{F}_{-\delta,-\Delta}(m,n+\delta,q+\Delta) \tag{A4.1.26}$$

and neglecting terms of order $1/q$ we find that

$$F_{\delta,\Delta}(s,s',x) = \bar{F}_{\delta,\Delta}(s,s',x)\,[\bar{F}_{-\delta,-\Delta}(s,s',x)/\bar{F}_{\delta,\Delta}(s,s',x)]^{\frac{1}{2}} \tag{A4.1.27}$$

or

$$F_{\delta,\Delta}(s,s',x) = [\bar{F}_{\delta,\Delta}(s,s',x)\,\bar{F}_{-\delta,-\Delta}(s,s',x)]^{\frac{1}{2}}. \tag{A4.1.28}$$

We now go on to the calculation of $\bar{F}_{\delta,\Delta}(s,s',x)$. The first step is to find the probabilities of any pair, in a chain of given m, n and q, being one of the possibilities $A\alpha$, $A\beta$, $B\alpha$, $B\beta$, αA, αB, βA or βB. We denote these probabilities by a, b, c, d, e, f, g and h. Since there are $N/2$ even places and $N/2$ odd ones, $a+b+c+d=\frac{1}{2}$; further $a+b$ is the probability of an even spin being A, i.e. $a+b=p/N$ and, similarly, $a+c=r/N$. Also, $2(b+c)=(b+c+f+g)=q/N$, the probability of a pair being antiparallel. Thus we find

$$\left.\begin{aligned} a &= (1+2s-x)/8, & b &= (1+2s'+x)/8,\\ c &= (1-2s-x)/8, & d &= (1-2s'+x)/8. \end{aligned}\right\} \tag{A4.1.29}$$

To compute $\bar{F}_{\delta,\Delta}$, we have to find, for example, the probability of four spins having configurations $A\beta A\beta$. The chance of βA being an inner pair is b, the chance then of having $A\beta$ as left-hand pair is $b/(b+d)$, and the chance that $A\beta$ is the right-hand pair when the third spin is A is $b/(b+a)$. The probability of $A\beta A\beta$ therefore being $b^3/[(b+d)(b+a)]$, we can compute $\bar{F}_{-2,-2}$, the mean function of configurations responsible for the transition $n\to n-2$, $q\to q-2$. We obtain

$$\bar{F}_{-2,-2}(s,s',x) = \frac{2b^3}{(b+d)(b+a)} = \frac{(1+2s'+x)^3}{16(1-s+s')(1+s+s')}. \tag{A4.1.30}$$

We can similarly calculate the other $\bar{F}_{\delta,\Delta}$'s and use equation (A4.1.28) to obtain the $F_{\delta,\Delta}$'s. The final results are

$$F_{-2,-2}(s,s',x) = F_{2,2}(s,s',x) = \frac{1+2s'+x}{16}\sqrt{\frac{[(1+x)^2-4s'^2]\,[(1-x)^2-4s^2]}{[(1-s)^2-s'^2]\,[(1+s)^2-s'^2]}}, \tag{A4.1.31}$$

$$F_{-2,0}(s,s',x) = F_{2,0}(s,s',x) = \frac{[(1+x)^2-4s'^2]\,[1-x]}{8\sqrt{\{[(1-s)^2-s'^2]\,[(1+s)^2-s'^2]\}}}, \tag{A4.1.32}$$

$$F_{-2,2}(s,s',x) = F_{2,-2}(s,s',x) = \frac{(1-2s'+x)}{16}\sqrt{\frac{[(1+x)^2-4s'^2]\,[(1-x)^2-4s^2]}{[(1-s)^2-s'^2]\,[(1+s)^2-s'^2]}}. \tag{A4.1.33}$$

A4.1.3 Results for energy and wave function

Inserting equations (A4.1.31)–(A4.1.33) into equation (A4.1.25) we obtain as our estimate of the energy

$$E = -NJ\max\left\{\tfrac{1}{2}(1+x)+\sqrt{\frac{(1+x^2)\,[(1+x)^2-4s'^2]\,[(1-x)^2-4s]}{16[(1-s)^2-s'^2]\,[(1+s)^2-s'^2]}} + \frac{(1-x)\,[(1+x)^2-4s'^2]}{4\sqrt{\{[(1-s)^2-s'^2]\,[(1+s)^2-s'^2]\}}}\right\}. \tag{A4.1.34}$$

We can see by inspection that $s = 0$, when we obtain

$$\begin{aligned} E &= -NJ\max f(s',x) \\ &= -NJ\max\{\tfrac{1}{2}(1+x)+[(1-x^2)\,[(1+x)^2-4s'^2]^{\frac{1}{2}} \\ &\quad +(1-x)\,[(1+x)^2-4s'^2]]/4(1-s'^2)\}. \end{aligned} \tag{A4.1.35}$$

To obtain the minimum of this expression, we must have

$$\left.\begin{aligned} \partial f/\partial s' = 0, \quad \partial^2 f/\partial s'^2 < 0, \\ \partial f/\partial x = 0, \quad \partial^2 f/\partial x^2 < 0. \end{aligned}\right\} \tag{A4.1.36}$$

This is achieved by

$$\begin{aligned} s = s' = 0 \\ x = \tfrac{1}{3}(\surd 2 - 1). \end{aligned} \tag{A4.1.37}$$

If $s = s' = 0$, $m = 0$ while $m = N/2$, and so

$$p = N/4, \quad r = N/4. \tag{A4.1.38}$$

There is no sub-lattice magnetization in the z-direction, but as we saw in the text, for the trial wave function taken, this magnetization *does* obtain in the x–y plane.

APPENDIX A4.2 DETAILED BALANCING

Equation (4.2.24) shows that the structure factor $S(\mathbf{k}, \omega)$ can be regarded as a transition probability, and so for a system in equilibrium must obey the principle of detailed balancing. In brief, this principle states that within a system in equilibrium any particular process must have a reverse process occurring with equal frequency. Consider for example the two energy levels illustrated in Figure A4.2.1 which in thermal equilibrium will have the relative Boltzmann

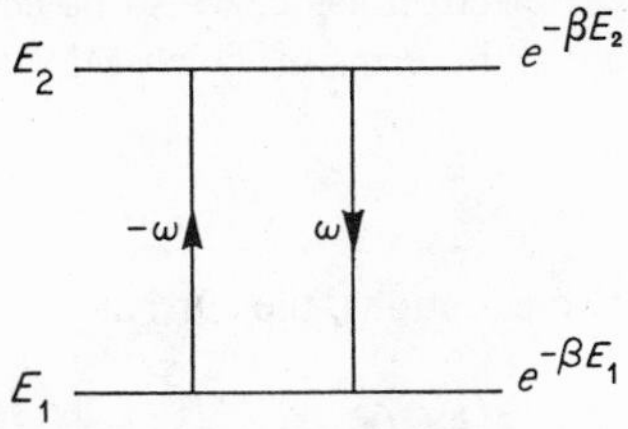

FIGURE A4.2.1. Showing two energy levels with equilibrium populations.

occupation probabilities $e^{-\beta E_1}$ and $e^{-\beta E_2}$. Assume $p(-\omega)$ is the probability of an individual upward transition. If n_1 is the population of the lower level, the total probability is $n_1 p(-\omega)$. Similarly, the total probability of a downward transition is $n_1 p(\omega) e^{-\beta\omega}$. But the principle of detailed balancing says these are equal, and hence

$$p(-\omega) = p(\omega) e^{-\beta\omega}. \tag{A4.2.1}$$

Referring again to the structure factor, we can now immediately see what the principle of detailed balancing implies for this quantity, if we note that the reverse of a process with momentum change $\mathbf{k}$ and energy change ω must involve a change of sign $\mathbf{k}$ as well as ω. The condition is

$$S(\mathbf{k}, \omega) = e^{\beta\omega} S(-\mathbf{k}, -\omega). \tag{A4.2.2}$$

From this it is also evident that if $T = 0\,(\beta = 1/k_B T = \infty)$

$$S(\mathbf{k}, -\omega) = 0 \quad (\omega > 0, T = 0) \tag{A4.2.3}$$

which is also obvious from the fact that a system in its ground state cannot lose energy.

APPENDIX A4.3 RELATION BETWEEN PERMEABILITY AND DIELECTRIC FUNCTIONS OF A UNIFORM ELECTRON GAS

In Chapter 2, section 2.3.2, we implicitly restricted ourselves to the longitudinal dielectric function ε_l. We shall show here that the discussion of the orbital susceptibility of the uniform gas in section 4.11 can be posed equally well in terms of the transverse dielectric function ε_t and ε_l.

We begin by noting that Maxwell's equation relating the magnetic and electric fields to the current in a uniform gas is (using the MKS system)

$$\varepsilon_0 c^2 \nabla \times \mathscr{B} = \mathbf{j} + \varepsilon_0 \frac{\partial \mathscr{E}}{\partial t}. \tag{A4.3.1}$$

On the other hand, the electron gas can also be considered as a medium with electric displacement $\mathscr{D}$ in terms of which Maxwell's equation is

$$\varepsilon_0 c^2 \nabla \times \mathscr{B} = \varepsilon_0 \frac{\partial \mathscr{D}}{\partial t}. \tag{A4.3.2}$$

By comparing these two equations, the current may be written

$$\mathbf{j} = (\varepsilon - 1)\, \varepsilon_0 \frac{\partial \mathscr{E}}{\partial t} \tag{A4.3.3}$$

or if $\mathscr{E} \propto e^{i\omega t}$

$$\mathbf{j} = i\omega(\varepsilon - 1)\, \varepsilon_0 \mathscr{E}. \tag{A4.3.4}$$

It should be noted that in the above equation ε in general is a tensor. If $\mathscr{E}$ varies in space as $e^{i\mathbf{q}\cdot\mathbf{r}}$ we also may write

$$\frac{\mathscr{D}}{\varepsilon_0} = \varepsilon_t \mathscr{E}_t + \varepsilon_l \mathscr{E}_l, \tag{A4.3.5}$$

where $\mathscr{E}_l$ is parallel to $\mathbf{q}$ while $\mathscr{E}_t$ is transverse to the direction of propagation. Examination of section 2.3.2 soon reveals that it is ε_l that is dealt with there.

As an alternative to the use of the transverse dielectric function, part of the current can be ascribed to the presence of a magnetic moment density $\mathscr{M}$ and then we have

$$\mathbf{j} = \sigma \mathscr{E} + \nabla \times \mathscr{M}, \tag{A4.3.6}$$

where σ is the conductivity. In terms of the dielectric function, we find by comparison with equation (A4.3.4) that this can also be written

$$\mathbf{j} = i\omega(\varepsilon_l - 1)\, \varepsilon_0 \mathscr{E} + \nabla \times \mathscr{M}. \tag{A4.3.7}$$

However, viewing ε as a tensor in equation (A4.3.4), we have

$$\begin{aligned} \mathbf{j} &= i\omega(\varepsilon_t - 1)\, \varepsilon_0 \mathscr{E}_t + i\omega(\varepsilon_l - 1)\, \varepsilon_0 \mathscr{E}_l \\ &= i\omega(\varepsilon_l - 1)\, \varepsilon_0 \mathscr{E} + i\omega(\varepsilon_t - \varepsilon_l)\, \varepsilon_0 \mathscr{E}_t. \end{aligned} \tag{A4.3.8}$$

Comparison with equation (A4.3.7) now gives

$$\nabla \times \mathscr{M} = i\omega(\varepsilon_t - \varepsilon_l)\, \varepsilon_0 \mathscr{E}_t. \tag{A4.3.9}$$

We next use Maxwell's equation

$$-\frac{\partial \mathscr{B}}{\partial t} = \operatorname{curl} \mathscr{E} \tag{A4.3.10}$$

when we obtain

$$-i\omega\mathscr{B} = \mathbf{q} \times \mathscr{E} \tag{A4.3.11}$$

and

$$-i\omega\mathbf{q} \times \mathscr{B} = -q^2 \mathscr{E}_t. \tag{A4.3.12}$$

Equation (A4.3.9) then becomes

$$\mathbf{q} \times \mathscr{M} = \frac{\omega^2}{q^2}(\varepsilon_t - \varepsilon_l)\,\varepsilon_0\,\mathbf{q} \times \mathscr{B} \tag{A4.3.13}$$

and hence, since

$$\mathscr{M} = \varepsilon_0 c^2(\mathscr{B} - \mathscr{H}) = \varepsilon_0 c^2(1 - \mu^{-1})\mathscr{B}, \tag{A4.3.14}$$

where μ is the permeability, we find

$$1 - \mu^{-1} = \frac{\omega^2}{c^2 q^2}(\varepsilon_t - \varepsilon_l), \tag{A4.3.15}$$

which is the central result of this Appendix.

We may further note that the susceptibility χ is conventionally defined through the relation

$$\mathscr{M} = \chi\mathscr{H}. \tag{A4.3.16}$$

Now the "magnetic current" is given by

$$\mathbf{j}_{\text{mag}} = \operatorname{curl} \mathscr{M} \tag{A4.3.17}$$

and so on writing

$$\mathscr{B} = \operatorname{curl} \mathscr{A} \tag{A4.3.18}$$

we find for a single Fourier component

$$\mathbf{j}_{\text{mag}} = q^2 \chi \mathscr{A}/\mu. \tag{A4.3.19}$$

This is to be compared with the definition (4.11.17). In the static case the two definitions agree, for

$$\mathscr{E} = -\nabla\phi - \frac{\partial \mathscr{A}}{\partial t} \tag{A4.3.20}$$

and so if $\nabla\phi = 0$ we see from equation (A4.3.6) that $\mathbf{j}_{\text{mag}}$ is the total current when $\omega = 0$. As Rajagopal and Jain (1972) point out, this relates χ_{orb} and μ by $\chi_{\text{orb}} = \varepsilon_0 c^2[\mu - 1]$ in this case. The ambiguity is resolved, however, by calculating the neutron scattering.

APPENDIX A5.1 INELASTIC SCATTERING OF NEUTRONS BY ANHARMONIC CRYSTALS

According to the discussion of Chapter 3, the true phonon frequency, including the frequency shift due to anharmonicity, is characterized by a Lorentzian peak in the spectral density $A_{ij}(k\omega)$. We shall see below that such a Lorentzian peak in $A_{ij}(k\omega)$ leads to an asymmetric peak in the neutron-scattering cross-section. While the maximum of this peak is not at the true phonon frequency, we shall see that, using frequency at the maximum plus details of the asymmetry, the true phonon frequency can be extracted. This can then be compared with the frequency shift from the harmonic frequency, which is clearly not an observable.

That such peaks, with finite widths, can be obtained from neutron measurements is evident from the work of Brockhouse, Larsson and co-workers. The theory outlined below is based on the work of Ambegaokar, Conway and Baym (1965).

As seen in Chapter 5, the spectral density $A_{ij}(k\omega)$ can be written for a harmonic crystal in terms of the phonon frequencies $\omega(\mathbf{k})$ and the polarization vectors $\boldsymbol{\varepsilon}(k\sigma)$. The precise form is

$$A_{ij}(k\omega) = 2\pi \sum_{\sigma=1}^{3} \varepsilon_i(\mathbf{k}\sigma)\, \varepsilon_j(-\mathbf{k}\sigma) \frac{1}{2M\omega(\mathbf{k}\sigma)} \{\delta[\omega-\omega(\mathbf{k}\sigma)] - \delta[\omega+\omega(\mathbf{k}\sigma)]\}. \tag{A5.1.1}$$

Broadening the δ-function peaks and allowing a frequency shift may both be incorporated, as in Chapter 3, by using a Lorentzian function, and then, with no coupling between the modes, we find

$$A_{ij}(\mathbf{k}\omega) = \sum_{\sigma=1}^{3} \varepsilon_i(\mathbf{k}\sigma)\, \varepsilon_j(-\mathbf{k}\sigma)\, A_\sigma(\mathbf{k}\omega), \tag{A5.1.2}$$

where $A_\sigma(\mathbf{k}\omega)$ is given by equation (5.3.52). A standard result for the exact spectral density defined through equation (3.14.15) may be written in terms of field operators as

$$A(\mathbf{k}\omega) = i \int d\mathbf{r} \int_{-\infty}^{\infty} dt\, \mathrm{e}^{-i\mathbf{k}.\mathbf{r}}\, \mathrm{e}^{i\omega t} [\langle\langle\{\psi(x)\,\psi^\dagger(0) + \psi^\dagger(0)\,\psi(x)\}\rangle\rangle]. \tag{A5.1.3}$$

The corresponding spectral density function A_{ij} of equation (5.3.48) can hence be shown from equation (A5.1.3) to satisfy the sum rule

$$\int_{-\infty}^{\infty} \frac{d\omega}{2\pi} \omega A_{ij}(\mathbf{k}\omega) = \frac{\delta_{ij}}{M}, \tag{A5.1.4}$$

which is also obeyed by the approximate A_σ of equation (A5.1.2).

To relate A_σ to $S(\mathbf{k}\omega)$ we next note that after the neutron is scattered by the crystal the final state can be written as

$$\sum_n a_n |n\rangle, \tag{A5.1.5}$$

where $|n\rangle$ denotes a state where n phonons have been created. The scattering cross-section is then proportional to the absolute value of the square of the expression (A5.1.5).

In the harmonic case the states $|n\rangle$ and $|m\rangle$ are orthogonal if $m \neq n$, but this is not true in the anharmonic crystal. Hence the inelastic scattering cross-section $\sigma(\mathbf{k}\omega)$ will have the form

$$\sigma(\mathbf{k}\omega) = \sigma_1(\mathbf{k}\omega) + \sigma_{\mathrm{i}}(\mathbf{k}\omega) + \sigma_{\mathrm{b}}(\mathbf{k}\omega), \tag{A5.1.6}$$

where σ_1 is the one-phonon cross-section, σ_{i} the interference term between the one-phonon and many-phonon processes, while σ_{b} represents the diffuse background (i.e. creation of more than one phonon). σ_1 and σ_{i} are peaked near the phonon frequency.

Corresponding to this, we shall write the scattering function in the form

$$S(\mathbf{k}\omega) = S_{1\mathrm{i}}(\mathbf{k}\omega) + S_{\mathrm{b}}(\mathbf{k}\omega), \tag{A5.1.7}$$

where $S_{1\mathrm{i}}$ represents the sum of one-phonon and interference terms, and satisfies

$$\int_{-\infty}^{\infty} \frac{d\omega}{2\pi} \omega S_{1\mathrm{i}}(\mathbf{k}\omega) = [d(\mathbf{k})]^2 \frac{k^2}{2M}, \tag{A5.1.8}$$

where

$$d(\mathbf{k}) = \langle \mathrm{e}^{i\mathbf{k}.\mathbf{r}-\mathbf{r}_0} \rangle. \tag{A5.1.9}$$

We shall discuss both this sum rule and the sum rule on the full scattering function $S(\mathbf{k}\omega)$ below.

To obtain $S_{1\mathrm{i}}$, we pick out those processes responsible for the peaks in the scattering. Following Ambegaokar and co-workers, we select processes in which the kinematical conservation conditions are restrictive, i.e. those containing one intermediate phonon. We work with the quantity $C(\mathbf{k}\omega)$ of the main text and write for the corresponding contributions to C (see also van Hove, 1961)

$$C_{1\mathrm{i}}(\mathbf{k}\omega) = [d(\mathbf{k})]^2 \sum_{\alpha} \bar{R}_\alpha(\mathbf{k}\omega)\, G_{\alpha\beta}(\mathbf{k}\omega)\, R_\beta(\mathbf{k}\omega), \tag{A5.1.10}$$

where

$$C_{1\mathrm{i}}(\mathbf{k}\omega) = \int_{-\infty}^{\infty} \frac{dz}{2\pi} S_{1\mathrm{i}}(\mathbf{k}z) \frac{(1-\mathrm{e}^{-\beta z})}{\omega - z}. \tag{A5.1.11}$$

In equation (A5.1.10) the factor $[d(\mathbf{k})]^2$ ensures that the sum rule (A5.1.8) is satisfied, $G_{\alpha\beta}$, the one-phonon Green function in the presence of interactions, describes the propagation of the intermediate phonon, and $\bar{R}_\alpha$ and R_β respectively describe the interaction of the neutron with the lattice as it enters and as it leaves the solid.

To find these quantities, we expand the exponentials of the thermal average in equation (5.3.42) and, noting that $\sum_{\mathbf{l}} e^{-i\mathbf{k}.(\mathbf{l}-\mathbf{u}_l)}$ is the neutron interaction operator, we describe the creation of one (dressed) phonon by the neutron by

$$\sum_{\alpha} \bar{R}_\alpha(\mathbf{k}\omega_m)\, G_{\alpha\beta}(\mathbf{k}\omega_m)$$

$$= \sum e^{-i\mathbf{k}.(\mathbf{l}-\mathbf{m})} \int_0^{-i\beta} dt\, e^{i\omega_m t} \langle e^{-i\mathbf{k}.\mathbf{u}_l(t)}\, u_m^\beta(0)\rangle_c \tag{A5.1.12}$$

while

$$\sum_{\beta} G_{\alpha\beta}(\mathbf{k}\omega_m)\, R_\beta(\mathbf{k}\omega_m)$$

$$= \sum_{l} e^{-i\mathbf{k}.(\mathbf{l}-\mathbf{m})} \int_0^{-i\beta} dt\, e^{i\omega_m t} \langle u_l^\alpha(t)\, e^{i\mathbf{k}.\mathbf{u}_m}(0)\rangle_c, \tag{A5.1.13}$$

the subscript c indicating that a cumulant average is being taken. The quantity $S_{1\mathrm{i}}$ is related to A_{ij} by

$$S_{1\mathrm{i}}(k\omega) = 2d(\mathbf{k})^2 \frac{\mathrm{Im}\,[R_i(\mathbf{k}\omega - i\varepsilon)\, G_{ij}(\mathbf{k}\omega - i\varepsilon)\, R_j(\mathbf{k}\omega - i\varepsilon)]}{1 - e^{-\beta\omega}}, \tag{A5.1.14}$$

where

$$G_{ij}(\mathbf{k}\nu) = \int_{-\infty}^{\infty} \frac{d\omega}{2\pi} \frac{A_{ij}(\mathbf{k}\omega)}{\nu - \omega} \tag{A5.1.15}$$

and $R_i(\mathbf{k}z)$ is found to be given by

$$R_i(\mathbf{k}z)\, G_{ij}(kz) = \int_{-\infty}^{\infty} \frac{d\omega}{2\pi} \frac{B_j(\mathbf{k}\omega)}{z - \omega}, \tag{A5.1.16}$$

where $B_j(\mathbf{k}\omega)$ is defined by

$$B_j(\mathbf{k}\omega) = i \sum_{\mathbf{k}} \exp(-i\mathbf{k}.\mathbf{K} - \mathbf{G})$$

$$\times \int_{-\infty}^{\infty} dt \exp(i\omega t) \langle \{\exp[i\mathbf{k}.\mathbf{u}_{\mathbf{K}}(t)], u_{\mathbf{G}i}(0)\}\rangle_c. \tag{A5.1.17}$$

Using finally the spectral representation of R in the form

$$R_i(\mathbf{k}z) = k_i + \int_{-\infty}^{\infty} \frac{d\omega}{2\pi} \frac{r_i(\mathbf{k}\omega)}{z - \omega}, \tag{A5.1.18}$$

we can write $S_{1i}(\mathbf{k}\omega)$ in the form

$$S_{1i}(\mathbf{k}\omega) = [d(\mathbf{k})]^2 \sum_{\sigma=1}^{3} \frac{1}{2M\Omega_\sigma} \times \left[\frac{a_\sigma(\mathbf{k}\omega)\,\Gamma_\sigma(\mathbf{k}) + b_\sigma(\mathbf{k}\omega)\,(\omega-\Omega_\sigma)}{(\omega-\Omega_\sigma)^2+\frac{1}{4}\Gamma_\sigma^2} - \frac{a_\sigma(\mathbf{k}\omega)\,\Gamma_\sigma(\mathbf{k}) + b_\sigma(\mathbf{k}\omega)\,(\omega+\Omega_\sigma)}{(\omega+\Omega_\sigma)^2+\frac{1}{4}\Gamma_\sigma^2}\right]$$

$$= \sum_{\sigma=1}^{3} S_{1i}^{\sigma}(\mathbf{k}\omega). \tag{A5.1.19}$$

The factor $[d(\mathbf{k})]^2$ has been introduced to again satisfy the sum rule (A5.1.8).

Here $a_\sigma(\mathbf{k}\omega)$ and $b_\sigma(\mathbf{k}\omega)$ are related to the polarization vectors $\boldsymbol{\varepsilon}$ and the function $\mathbf{r}(\omega)$ in (A5.1.18) by

$$a_\sigma(\mathbf{k}\omega) = \left[\left\{\mathbf{k}+P\int\frac{d\omega'}{2\pi}\frac{\mathbf{r}(\omega')}{\omega-\omega'}\right\}.\,\boldsymbol{\varepsilon}(\mathbf{k}\sigma)\right]^2 - \tfrac{1}{4}[\mathbf{r}(\mathbf{k}\omega).\,\boldsymbol{\varepsilon}(\mathbf{k}\sigma)]^2 \tag{A5.1.20}$$

and

$$b_\sigma(\mathbf{k}\omega) = 2[\mathbf{r}(\mathbf{k}\omega).\,\boldsymbol{\varepsilon}(\mathbf{k}\sigma)]\left[\mathbf{k}+P\int\frac{\mathbf{r}(\omega')}{\omega-\omega'}\frac{d\omega'}{2\pi}\right].\,\boldsymbol{\varepsilon}(\mathbf{k}\sigma). \tag{A5.1.21}$$

We now look at $S_{1i}(\mathbf{k}\omega)$ near a resonance, and clearly the second part is unimportant there. Near the resonance, then, $S_{1i}^{\sigma}(\mathbf{k}\omega)$ has a factor

$$\frac{\Gamma a + (\omega-\Omega_\sigma)b}{(\omega-\Omega_\sigma)^2+\frac{1}{4}\Gamma^2}. \tag{A5.1.22}$$

The observed maximum ω_{m} in the neutron scattering experiment will be related to the parameters introduced above by

$$\omega_{\mathrm{m}} \simeq \Omega_\sigma + \frac{\Gamma b}{8a}\left(1-\frac{b^2}{16a^2}\right) \tag{A5.1.23}$$

provided $b \lesssim a$. (a and b are taken as constants, rather than slowly varying in ω.)

Furthermore, the frequencies at half-height of the resonance peak are given by

$$\omega_{\frac{1}{2}} = \omega_{\mathrm{m}} + \frac{\Gamma b}{8a}\left(1-\frac{b^2}{16a^2}\right) \pm \frac{\Gamma}{2}\left(1+\frac{b^2}{16a^2}\right). \tag{A5.1.24}$$

This shows that the curve, as expected, is not symmetric about ω_{m}. If $\Gamma_{\pm}$ represent the two half-widths of the asymmetric peak, then these can be measured directly from the resonance and

$$\Gamma_+ - \Gamma_- = 2(\omega_m - \Omega_\sigma). \tag{A5.1.25}$$

Thus, the phonon frequency Ω_σ can be extracted from the maximum and the half-widths of the resonance peak. This tells us nothing about the frequency shift of the harmonic phonon frequency, but allows us to get out the true phonon frequency (maximum in the spectral density) from the neutron measurements. Ambegaokar and co-workers take the data of Brockhouse for Pb at 425°K in the (100) direction at the edge of the BZ and find $(b/a) \sim -0.2$. The corresponding asymmetry is small.

Sum rules. From the definition of S we have

$$\frac{\partial}{\partial t}\int_{-\infty}^{\infty} d\omega\, e^{i\omega t} S(\mathbf{k}\omega) = \frac{1}{\hbar N}\langle \rho_{\mathbf{k}}(0)\frac{\partial}{\partial t}\rho_{-\mathbf{k}}(t)\rangle, \tag{A5.1.26}$$

which can be written as

$$\int_{-\infty}^{\infty} d\omega\, \omega S(\mathbf{k}\omega) = -\frac{1}{\hbar^2 N}\langle \rho_{\mathbf{k}}(0)\,[\rho_{-\mathbf{k}}(0), H]\rangle. \tag{A5.1.27}$$

Expressing the right-hand side explicitly in terms of the eigenfunctions of the target Hamiltonian H we have

$$\begin{aligned}\langle \rho_{\mathbf{k}}[\rho_{-\mathbf{k}}, H]\rangle &= \frac{1}{Z}\sum_i e^{-\beta\varepsilon_i}\langle i|\,\rho_{\mathbf{k}}[\rho_{-\mathbf{k}}, H]\,|i\rangle \\ &= \frac{1}{Z}\sum_i e^{-\beta\varepsilon_i}\sum_j (\varepsilon_i - \varepsilon_j)|\langle i|\,\rho_{\mathbf{k}}|j\rangle|^2, \end{aligned} \tag{A5.1.28}$$

where Z is the partition function.

The right-hand side is just the thermal average of the energy lost by the target on receiving momentum $\hbar\mathbf{k}$. The energy gained by the scattered particle is simply $\hbar k^2/2M$, and so we have from equations (A5.1.27) and (A5.1.28)

$$\int_{-\infty}^{\infty} d\omega\, \omega S(\mathbf{k}\omega) = \frac{\hbar k^2}{2M}, \tag{A5.1.29}$$

which is the sum rule we wished to prove.

Finally we wish to indicate the proof of the sum rule (A5.1.8).

Taking the limit $|\omega|\to\infty$ in equation (A5.1.11) we find

$$C_{1i}(\omega) = \frac{1}{\omega}\int_{-\infty}^{\infty}\frac{dz}{2\pi}S_{1i}(\mathbf{k}z)(1-\mathrm{e}^{-\beta z}) + \frac{1}{\omega^2}\int_{-\infty}^{\infty}\frac{dz}{2\pi}zS_{1i}(\mathbf{k}z)(1-\mathrm{e}^{-\beta z})+\dots. \tag{A5.1.30}$$

On the other hand, for large ω the quantity $R(\mathbf{k}\omega)$ takes the form

$$\mathbf{R}(\mathbf{k}\omega) = \mathbf{k}+\frac{1}{\omega^2}\int_{-\infty}^{\infty}\frac{dz}{2\pi}z\mathbf{r}(z)+\dots \tag{A5.1.31}$$

and

$$G_{ij}(\mathbf{k}\omega) = \frac{1}{\omega^2}\int_{-\infty}^{\infty}\frac{dz}{2\pi}zA_{ij}(\mathbf{k}\omega). \tag{A5.1.32}$$

Hence we have the asymptotic result

$$R_i\,G_{ij}\,R_j\to\frac{k^2}{M\omega^2}. \tag{A5.1.33}$$

Comparing equations (A5.1.33), (A5.1.30) and (A5.1.10) we find

$$\int_{-\infty}^{\infty}\frac{d\omega}{2\pi}S_{1i}(\mathbf{k}\omega)(1-\mathrm{e}^{-\beta\omega}) = 0 \tag{A5.1.34}$$

and

$$\int_{-\infty}^{\infty}\frac{d\omega}{2\pi}\omega S_{1i}(\mathbf{k}\omega)(1-\mathrm{e}^{-\beta\omega}) = 2\int_{-\infty}^{\infty}\frac{d\omega}{2\pi}\omega S_{1i}(\mathbf{k}\omega) = |d(\mathbf{k})|^2\frac{k^2}{M}, \tag{A5.1.35}$$

in the last step use having been made of the detailed balancing condition of Appendix 4.2. Thus, equation (A5.1.8) is established.

Problems

Chapter 1

1. Prove that if $M = M_1 + M_2$, where M and M_1 are invariant manifolds, M_2 is also an invariant manifold.

2. Prove that if ϕ lies entirely in an irreducible manifold M of a group G, then any other vector ψ lying entirely in M may be expanded solely in terms of the functions generated by operating on ϕ with every operation of G.

3. Obtain all irreducible matrix representations for the group of the point W in the BZ of the fcc lattice (see p. 24, Figure 1.5, and Appendix A1.3).

Find also the linear combinations of plane-waves (with **k** vectors for the three inequivalent points W) realizing these matrix representations.

4. On the assumption that for a band of a bcc lattice, overlap between Wannier functions is negligible except for those centred on nearest neighbours, obtain an estimate of the energy-width of the band in terms of the Fourier coefficient $E_{\mathbf{R}1}$, $\mathbf{R}_1$ being a vector between nearest neighbours [see equations (1.4.21) and (1.8.21)].

5. Using the result of Problem 1.4 and the group-velocity formula [equation (1.10.13)], obtain a numerical estimate of the width of the 1s band in lithium. Use equation (1.8.24) for the Bloch wave function and take the core function ϕ as a Slater function $\sqrt{(\alpha^3/\pi)}\exp(-\alpha r)$. ($\alpha = 2.7$ in atomic units (au), and the nearest-neighbour distance is 5.719 au.)

The fact that all the information we require is contained in the overlap of ϕ's on nearest neighbours renders this procedure open to criticism. Why? Equation (1.8.24) is the basic equation of the "tight-binding" method, in which one takes $\psi_{\mathbf{k}}(\mathbf{r})$ as the trial wave function in a variational calculation. Write down the formula for the second-order overlap energy by this procedure to see if it is susceptible to the same criticism.

6. Show that there is no first-order change in energy (except by the mean value of the potential) when we perturb free electrons by a periodic potential unless **k** lies at a high symmetry point on the BZ, when resort can be made to degenerate perturbation theory. Exemplify the latter statement by considering the perturbation at the point W. (The results of Problem 1.3 are of relevance here, though they are not essential.) What does this suggest as to the magnitude of band gaps?

7. Estimate the form of the band structure of the 2s band of Li by (i) using a Seitz wave function $u_0(r)\exp(i\mathbf{k}.\mathbf{r})$, (ii) orthogonalizing a single plane wave to the core function given in Problem 1.5.

Take the mean value of the potential as $V_0 = -0.500$ au, the core energy as $E_0 = -1.883$ au, the volume of the unit cell as $\Omega = 153$ au and $E(0) = -0.343$ au (the bottom of the band energy).

8. Show that the Bloch eigenfunctions $\psi_{\mathbf{k}}$ for a given $\mathbf{k}$ vector span an invariant manifold of the full point group of the crystal, but that this manifold is, in general, reducible.

9. Construct the BZ for bismuth. The bismuth structure is based on a rhombohedral lattice in which there are three equal crystallographic axes inclined to each other at the same angle α, with $\alpha = 57° \, 16'$ (see Callaway, 1958).

10. Obtain the minimal set (or minimal sets) of critical points for a bcc lattice.

11. Calculate the Wigner–Seitz wave function for $\mathbf{k} = 0$ for high-density metallic hydrogen, taking the field within the Wigner–Seitz cell as that of a proton.

[*Hint*: Write lengths in units of the Wigner–Seitz radius r_W, and expand the radial wave function about the point $r = r_W$ in a Taylor series.]

To determine the Fermi energy according to the method of Bardeen (1938), one must solve the p state ($l = 1$) equation

$$\frac{1}{r_W}\frac{d^2\chi_1}{d\rho^2} + \left(\varepsilon_0 + \frac{2}{\rho} - \frac{2}{r_W\,\rho^2}\right)\chi_1 = 0,$$

where $\varepsilon_0 = r_s E(\mathbf{k} = 0)$, $\rho = r/r_W$, subject to the condition that χ_1/ρ^2 is finite at the origin. The mean Fermi energy is then given by

$$\langle E\rangle = \frac{2.21}{r_W^2}\alpha,$$

where

$$\alpha = \frac{\frac{1}{3}\chi_0^2(1)}{\int_0^1 \chi_0^2(\rho)\,d\rho}\left(\frac{1}{\chi_1}\frac{d\chi_1}{d\rho} - 1\right)_{\rho=1}.$$

Here $\chi_0 = r\psi_0$, where ψ_0 is the Wigner–Seitz wave function for $\mathbf{k} = 0$.

Hence show that

$$\alpha = 1 - \tfrac{13}{2100}r_W^2 - \tfrac{1}{420}r_W^3 + O(r_W^4).$$

Show that this procedure is a special case of the $\mathbf{k}.\mathbf{p}$ method.

12. The additional density of states due to a scattering potential in a Fermi gas may be written

$$\Delta n(E) = \frac{2}{\pi}\sum_l (2l+1)\frac{d\eta_l}{dE},$$

where η_l is the phase shift of the lth partial wave.

Assuming the problem is dominated by d-wave scattering, and that a d-wave resonance occurs, described by

$$\tan\eta_2 = \frac{\frac{1}{2}W}{E_0 - E}$$

show that

$$\Delta n(E) \simeq \frac{10}{\pi}\frac{\frac{1}{2}W}{(E_0-E)^2 + (\frac{1}{2}W)^2},$$

that is, the atomic level has become broadened, with a width W. What would be the lifetime of such a virtual bound state?

13. Ashcroft (1963) determined the Fermi surface of Al from the nearly free-electron model.

Set up the appropriate 4×4 secular equation, assuming, around the point $W = (\pi/a)(2, 0, 1)$, only the plane-waves with wave-vectors $\mathbf{k}-\mathbf{K}_n$, with

$$\mathbf{K}_1 = 0, \quad \mathbf{K}_2 = (\pi/a)(\bar{2}, 2, \bar{2}),$$

$$\mathbf{K}_3 = (\pi/a)(\bar{4}, 0, 0), \quad \mathbf{K}_4 = (\pi/a)(\bar{2}, \bar{2}, \bar{2})$$

are involved. With the form

$$|(\{\mathbf{k}-\mathbf{K}\}^2 - E)\,\delta_{\mathbf{KK}'} + V_{\mathbf{KK}'}| = 0,$$

where $\mathbf{K}$, $\mathbf{K}'$ are the vectors given above, the pseudo-potential $V_{\mathbf{KK}'}$ as determined by Ashcroft corresponds to

$$V_{\mathbf{K}-\mathbf{K}'} = \begin{cases} 0.0179 \text{ Ry} & \text{when } \mathbf{K}-\mathbf{K}' \equiv (111), \\ 0.0562 \text{ Ry} & \text{when } \mathbf{K}-\mathbf{K}' \equiv (200). \end{cases}$$

Construct the Fermi surface from these values.

14. In the free-electron model of a conjugated molecule, the chemists replace the molecule by a network of straight lines joining adjacent atoms and the valence electrons are supposed to move entirely freely along these various one-dimensional paths.

Apply this model literally to the case of a simple cubic lattice. Using the usual continuity conditions on the wave functions, Bloch's theorem and the conservation of current at any junction, show that the dispersion relation $E(\mathbf{k})$ is given by

$$3 \cos ql = \cos k_x l + \cos k_y l + \cos k_z l,$$

where $q^2 = 8\pi^2 mE/h^2$ and l is the lattice spacing. Briefly discuss the relation between this band structure given by the above 'network model' and that given for the simple cubic lattice by the tight-binding approximation.

As a more difficult example, apply the network model to the π-electrons in graphite (see also Chapter 7) and show that the density of states has a zero value in the middle of the band (see Coulson, 1954).

Chapter 2

1. Assuming that the virial theorem can be applied to a uniform electron gas such that the electrons exert a pressure p on the walls of the "container", of volume $\mathscr{V}$, show that the virial theorem can be re-expressed in the form

$$2T + V = -r_s\, dE/dr_s.$$

Hence

(a) Obtain separately the kinetic and potential energies T and V respectively from the Gell-Mann–Brueckner result for the total energy.

(b) Make a similar calculation using the Wigner interpolation formula for the correlation energy.

2. Lindemann's law of melting states that a lattice will melt when the amplitude of vibration becomes a certain fraction of the interatomic spacing. Using the same criterion for the Wigner lattice, estimate when this will melt.

Give another argument from the balance between the kinetic energy of localization and potential energy.

3. Set up the energy expression for the ground state of an inhomogeneous electron gas, including exchange. Hence generalize the Thomas–Fermi relation between density and potential to include exchange. This new relation is the Thomas–Fermi–Dirac relation.

4. Construct the first-order density matrix for plane waves by

(a) Direct summation;

(b) By solution of the Bloch equation and then performing an inverse Laplace transform.

Use the Wigner orbitals of the low-density electron gas to calculate the occupation numbers of plane-wave states in a low-density régime.

5. Integrate equation (2.6.18) analytically and hence derive the k^2 term in the plasmon dispersion relation. The result is

$$\omega^2 = \omega_p^2 + \tfrac{3}{5}v_f^2 k^2 + O(k^4),$$

where v_f is the Fermi velocity.

6. The second-order density matrix for the ground state of a homogeneous system of Fermions can be written in the following form:

$$\Gamma(\mathbf{x}_1' \mathbf{x}_2' \mid \mathbf{x}_1 \mathbf{x}_2) = \sum_{\substack{k_1<k_2 \\ l_1<l_2}} a_{k_1k_2l_1l_2} \frac{1}{\sqrt{2}} \begin{vmatrix} \psi_{k_1}^*(\mathbf{x}_1') & \psi_{k_2}^*(\mathbf{x}_1') \\ \psi_{k_1}^*(\mathbf{x}_2') & \psi_{k_2}^*(\mathbf{x}_2') \end{vmatrix} \frac{1}{\sqrt{2}} \begin{vmatrix} \psi_{l_1}(\mathbf{x}_1) & \psi_{l_2}(\mathbf{x}_1) \\ \psi_{l_1}(\mathbf{x}_2) & \psi_{l_2}(\mathbf{x}_2) \end{vmatrix},$$

where $\mathbf{x} \equiv (\mathbf{r}\sigma)$ and the space part of $\psi_{\mathbf{k}}$ is $e^{i\mathbf{k}.\mathbf{r}}$.

Show that the coefficients $a_{k_1k_2k_1k_2}$ must satisfy the conditions

(a) They are real;

(b) $\sum_{k_1<k_2} a_{k_1k_2k_1k_2} = \frac{1}{2}N(N-1)$; N being the total number of electrons;

(c) $0 \leqslant a_{k_1k_2k_1k_2} \leqslant 1$ (all k_1, k_2). These are the Pauli conditions on the second-order density matrix.

7. Take the Fourier transform of the Fermi hole pair function

$$g(r) = \left\{1 - \frac{9}{2}\left[\frac{j_1(k_f r)}{k_f r}\right]^2\right\}$$

to obtain the structure factor $S(k)$ as

$$S(k) = \begin{cases} \frac{3}{4}\left(\frac{k}{k_f}\right) - \frac{1}{16}\left(\frac{k}{k_f}\right)^3 & \text{for } k \leqslant 2k_f, \\ 1 \quad \text{for } k > 2k_f. \end{cases}$$

Interpret the asymptotic behaviour of $g(r)$ in terms of contributions from $S(k)$ at small k and at $k = 2k_f$.

8. To include the electron interactions in a homogeneous electron gas, the total wave function may be written in terms of the Fourier components of density $\rho_{\mathbf{k}}$ as

$$\psi = D\exp\left[-\sum_{k} d(k)\,\rho_{\mathbf{k}}\,\rho_{\mathbf{k}}^{*}\right],$$

where D is a determinant of plane waves, and $d(k)$ was determined variationally by Gaskell (1962), who found that for small r_s and for long wavelengths, with k in units of k_f,

$$d(k) \sim \frac{r_s^{\frac{1}{2}}}{k^2}\frac{1}{2}\left[\frac{4}{3\pi}\left(\frac{4}{9\pi}\right)^{\frac{1}{3}}\right]^{\frac{1}{2}}.$$

Then the pair function $g(r)$ takes the approximate form

$$g(r) \doteqdot \left\{1-\frac{3}{2}\int_0^\infty dk\,k\left[1-\frac{S_{ni}(k)}{1+4d\,S_{ni}(k)}\right]\right\}\frac{\sin kr}{r},$$

where S_{ni} is the non-interacting result given in the previous problem. Show that the leading term in $S(k)$ for small k is k^2, and that, therefore, the Coulomb interactions suppress the k term in the Fermi hole. Show explicitly that for small k

$$S(k) = \tfrac{1}{4}\hbar k^2(\pi\rho m e^2)^{-\frac{1}{2}},$$

where ρ is the average number density of the electrons.

9. Show that, for non-interacting Fermions, the function $F(k, t)$ obtained by a Fourier transform of ω in the van Hove function $S(k, \omega)$ is given, for $k > 2k_f$, by

$$F(k, t) = \frac{3\exp(-i\hbar k^2 t/2m)}{(\hbar k/m)\,k_f\,t}\,j_1\!\left(\frac{\hbar k k_f t}{m}\right).$$

From the long time behaviour of this expression, discuss the frequencies at which pathology occurs in $S(k, \omega)$ for $k > 2k_f$.

10. Prove that the spectral density function defined in section 2.7 obeys the sum rule

$$\int [A^-(\mathbf{k}\omega) + A^+(\mathbf{k}\omega)]\,d\mathbf{k}\,d\omega = 1.$$

11. Show that the linear response function $\mathscr{F}$ is given in terms of the second functional derivative of the energy with respect to the external potential V:

$$\mathscr{F}(1, 2) = \frac{\delta^2 E}{\delta V(1)\,\delta V(2)}$$

and hence obtain the diagrammatic series for $\mathscr{F}$ from that for E in the presence of an external field.

12. Use the one-body perturbation theory of Appendix A1.2 to second order in the potential $V(r)$ to obtain the dielectric function $\varepsilon(q)$ including exchange for small q.

Hence show that the Dirac–Slater exchange energy is regained.

13. Prove that, for a homogeneous system of N Fermions, the compressibility K is related to the chemical potential μ, at $T = 0$, by

$$K^{-1} = N\rho\left(\frac{\partial\mu}{\partial N}\right)_{\Omega}, \qquad \text{(P2.13.1)}$$

where ρ is the density and Ω the volume.

[*Hint*: Write the ground-state energy in the form $E = f(\rho)\,\Omega$ and use the usual thermodynamic relation for the pressure in terms of E, at $T = 0$.]

14. Using the methods of Appendix A2.3, show that the dynamical structure factor $S(\mathbf{q}t)$, given by

$$S(qt) = \frac{1}{\Omega}\langle\phi_0\,|\,P\{\rho_{\mathbf{q}}(t)\,\rho_{-\mathbf{q}}(0)\,|\,\phi_0\rangle, \tag{P.2.14.1}$$

where ϕ_0 is the many-body ground state and P is Dyson's chronological operator, in Fourier transform satisfies

$$\lim_{\mathbf{q}\to 0}\lim_{\omega\to 0} S(\mathbf{q}\omega) = 2\pi i \sum_{\mathbf{k}} \delta(\mu-\varepsilon_{\mathbf{k}})\, v_{\mathbf{k}} \frac{dk_{\mathrm{f}}}{d\mu}, \tag{P2.14.2}$$

where $v_{\mathbf{k}}$ is the quasi-particle group velocity at the Fermi surface, and $\varepsilon_{\mathbf{k}}$ the quasi-particle energy. Show that it then follows that

$$\lim_{\mathbf{q}\to 0}\lim_{\omega\to 0} S(\mathbf{q}\omega) = -2\pi i \frac{dN}{d\mu} = -N\rho K 2\pi i, \tag{P2.14.3}$$

where we have used the result of the previous problem.

Using the relation between $S(q)$ and $\varepsilon(q)$ given in the main text, show that, at small q,

$$\varepsilon(q) = 1+\frac{4\pi e^2}{q^2}\frac{1}{\Omega}\frac{dN}{d\mu}. \tag{P2.14.4}$$

Chapter 3

1. Show that the velocity of sound in the long-wavelength limit for a monatomic linear chain with near-neighbour interactions is given by $a(\alpha/m)^{\frac{1}{2}}$, where a is the atomic spacing, α the force constant and m the mass of the particles. Using the full dispersion relation, show that the density of states is given by

$$g(\omega) = \frac{2}{\pi a}\left\{\frac{4\alpha}{m}-\omega^2\right\}^{-\frac{1}{2}}.$$

Indicate the relation of this to the van Hove singularities discussed in Chapter 1.

Also obtain the form of $g(\omega)$ for the diatomic linear chain, showing how the three critical points arise in this case.

2. The diamond structure consists of two interpenetrating fcc lattices displaced with respect to each other by one-quarter of the body diagonal.

To make a rough theory, assume that only nearest-neighbour forces operate, and that we consider a general set of coupling constants limited only by symmetry and invariance conditions.

Show that

(a) There are five distinct coupling constant matrices required;

(b) These can be written in terms of two parameters only;

(c) The frequencies are determined by a secular equation of the general form.

$$\begin{pmatrix} b_{11}-(A-\omega^2)^2 & b_{12} & b_{13} \\ b_{12}^* & b_{22}-(A-\omega^2)^2 & b_{23} \\ b_{13}^* & b_{23}^* & (A-\omega^2)^2 \end{pmatrix} = 0.$$

Solve this secular equation for the case when $\mathbf{q} = (q, 0, 0)$ and plot the dispersion curves, labelling the TO, LO, LA and TA branches. Compare the results with the realistic dispersion curves shown in Figure 3.1.

3. Using equation (3.9.4), show that the Debye temperature θ_D may be expressed in terms of the moments of the frequency spectrum as

$$\frac{1}{\theta_D^3} = 2\left(\frac{k_B}{\hbar\omega}\right)^3 \sum_{n=0}^{\infty} \sum_{\nu=0}^{n} \frac{\mu_{2\nu} P_{2n}^{2\nu} p_{2n}^2}{4n+1}.$$

Here the moments μ_n are defined by

$$\mu_n = \int_0^{\omega_L} \omega^n g(\omega)\, d\omega,$$

where $g(\omega)$ is the frequency spectrum. The quantities p_n are defined by writing the Legendre polynomial $P_n(x)$ in the form

$$P_n(x) = \sum_{\nu=0}^{\infty} p_n^\nu x^\nu.$$

4. Show from the first-order perturbation theory result (see Appendix 1.2) for the Bloch density matrix, namely

$$C(\mathbf{r}\mathbf{r}_0\, \beta) = C_0(\mathbf{r}\mathbf{r}_0\, \beta) - (2\pi\beta)^{-\frac{3}{2}} \int d\mathbf{r}_1\, V(\mathbf{r}_1)\,(|\mathbf{r}-\mathbf{r}_1| + |\mathbf{r}_1-\mathbf{r}_0|) \times \frac{\exp[-(1/2\beta)(|\mathbf{r}-\mathbf{r}_1| + |\mathbf{r}_1-\mathbf{r}_0|)^2]}{2\pi|\mathbf{r}-\mathbf{r}_1||\mathbf{r}_1-\mathbf{r}_0|}, \qquad \text{(P3.4.1)}$$

that the corresponding Green function is given by

$$G(\mathbf{r}\mathbf{r}_0\, E) = G_0(\mathbf{r}\mathbf{r}_0\, E) - \int \frac{d\mathbf{r}_1\, V(\mathbf{r}_1)}{4\pi^2} \frac{\exp[i\sqrt{(2E)}\,(|\mathbf{r}-\mathbf{r}_1| + |\mathbf{r}_1-\mathbf{r}_0|)]}{|\mathbf{r}-\mathbf{r}_1||\mathbf{r}_1-\mathbf{r}_0|},$$

where $C_0(\mathbf{r}\mathbf{r}_0\, \beta) = (2\pi\beta)^{-\frac{3}{2}} \exp(-|\mathbf{r}-\mathbf{r}_0|^2/2\beta)$ and correspondingly

$$G_0(\mathbf{r}\mathbf{r}_0\, E) = \frac{1}{2\pi|\mathbf{r}-\mathbf{r}_0|} \exp[i\sqrt{(2E)}\,|\mathbf{r}-\mathbf{r}_0|].$$

Also derive from equation (P3.4.1) an expression for $\partial\rho(\mathbf{r}\mathbf{r}_0\, E)/\partial E$ to first order in V. Hence, from equation (3.10.55), derive the response function F to first order in V.

5. The pair potential $\phi(\mathbf{X})$ between ions in an degenerate electron gas can be written

$$\phi(\mathbf{X}) = \int \sigma(\mathbf{r})\, V_b(\mathbf{r}-\mathbf{X})\, d\mathbf{r},$$

where $V_b(\mathbf{r})$ is the bare-ion potential and $\sigma(\mathbf{r})$ is the localized charge density [see equation (3.10.69)]. The charge displaced by the bare-ion potential is simply, in Fourier transform and in a linear theory:

$$k^2 V_b(\mathbf{k})\, [\varepsilon^{-1}(\mathbf{k}) - 1],$$

where $\varepsilon(\mathbf{k})$ is the dielectric function including exchange and correlation, given by equation (2.11.5).

Show, by adding the ion–ion interaction on to $\phi(\mathbf{X})$ above, that the total pair interaction $\phi_{\text{total}}(\mathbf{X})$ is given by

$$\phi_{\text{total}}(\mathbf{X}) = \int \frac{k^2 V_b^2(\mathbf{k})}{\varepsilon(\mathbf{k})} e^{i\mathbf{k}.\mathbf{X}}\, d\mathbf{k}.$$

6. In the theory of the electron–ion interaction, we encounter matrix elements of the gradient of the bare–ion potential (see Section 3.10.4). Since the potential $V_i(\mathbf{r})$ entering the Hartree equations is the bare–ion potential plus a screening potential $-V_e(\mathbf{r})$ we first evaluate

$$\mathbf{M}_i(\mathbf{kk}') = \int_\Omega \psi^*_{\mathbf{k}'}(\mathbf{r})\, \nabla V_i(\mathbf{r})\, \psi_{\mathbf{k}}(\mathbf{r})\, d\mathbf{r},$$

where $V_i(\mathbf{r})$ is a local potential from which the periodic potential is built up by placing one of these on each lattice site. Assuming $V_i(\mathbf{r})$ vanishes outside the atomic polyhedron Ω, show that

$$\begin{aligned}\boldsymbol{\varepsilon}.\mathbf{M}_i(\mathbf{k}'\mathbf{k}) = \frac{\hbar^2}{4m}\int_\Omega \{&\psi^*_{\mathbf{k}'}\, \nabla(e^{i\mathbf{k}.\mathbf{r}}\,\boldsymbol{\varepsilon}.\nabla u_{\mathbf{k}}) - (\nabla\psi^*_{\mathbf{k}'})\, e^{i\mathbf{k}.\mathbf{r}}\,\boldsymbol{\varepsilon}.\nabla u_{\mathbf{k}} \\ &+ [\nabla(e^{-i\mathbf{k}'.\mathbf{r}}\,\boldsymbol{\varepsilon}.\nabla u^*_{\mathbf{k}'})]\,\psi_{\mathbf{k}} - (e^{-i\mathbf{k}'.\mathbf{r}}\,\boldsymbol{\varepsilon}.\nabla u^*_{\mathbf{k}'})\,\nabla\psi_{\mathbf{k}}\}.d\mathbf{s} \\ &+ \tfrac{1}{2}[E(\mathbf{k}) - E(\mathbf{k}')]\int_\Omega e^{i(\mathbf{k}-\mathbf{k}').\mathbf{r}}[u^*_{\mathbf{k}'}(\boldsymbol{\varepsilon}.\nabla u_{\mathbf{k}}) - (\boldsymbol{\varepsilon}.\nabla u^*_{\mathbf{k}})\, u_{\mathbf{k}}]\, d\mathbf{r},\end{aligned}$$

provided

$$\left[-\frac{\hbar^2}{2m}\nabla^2 + V_i(\mathbf{r})\right]\psi_{\mathbf{k}} = E(\mathbf{k})\,\psi_{\mathbf{k}}.$$

Following Bardeen (1937), we next take $\psi_{\mathbf{k}}$ to be given by the Seitz wave function

$$\psi_{\mathbf{k}} \doteqdot u_0(\mathbf{r})\, e^{i\mathbf{k}.\mathbf{r}},$$

where $u_0(\mathbf{r})$ is the wave function at $\mathbf{k} = 0$. Show that it then follows that

$$\boldsymbol{\varepsilon}.\mathbf{M}_i(\mathbf{k}'\mathbf{k}) = -i\boldsymbol{\varepsilon}.(\mathbf{k}'-\mathbf{k})\left[\frac{V_i(r_0) - E_0}{N}\right] G(|\mathbf{k}'-\mathbf{k}|\, r_0),$$

where r_0 is the radius of the Wigner–Seitz sphere and G is the shape factor defined by $G(x) = 3(\sin x - x\cos x)/x^3$. In the above equation $|u_0(\mathbf{r}_0)|^2$ has been approximated by unity.

To obtain the total matrix element of the Hartree potential, we must add on

$$\mathbf{M}_e(\mathbf{k}'\mathbf{k}) = \int \psi^*_{\mathbf{k}'}\, \nabla V_e(\mathbf{r})\, \psi_{\mathbf{k}}(\mathbf{r})\, d\mathbf{r},$$

where $V_e(\mathbf{r})$ is the potential due to the electrons. Assuming these electrons are uniformly distributed through a sphere of radius r_0 with total charge Z^* evaluate M_e by taking $\psi_{\mathbf{k}}$ and $\psi_{\mathbf{k}'}$ to be given by single plane-waves or OPW's, to finally obtain the total Bardeen matrix element

$$i\boldsymbol{\varepsilon}.\mathbf{M}(\mathbf{k}+\mathbf{q}, \mathbf{k}) = \boldsymbol{\varepsilon}.\mathbf{q}\left[\frac{4\pi Z^* e^2}{\Omega q^2} + \frac{V_i(r_0) - E_0}{N}\right] G(\mathbf{k}r_0).$$

What would be the effect of taking the $\psi_{\mathbf{k}}$ and $\psi_{\mathbf{k}'}$ to be Seitz wave functions, in the evaluation of $\mathbf{M}_e$?

7. Prove that the relation (3.11.1) follows if only near-neighbour interactions are included.

8. Show that the specific heat c_v may be obtained from the internal energy $E(T)$ given in an independent phonon model by

$$E(T) = \sum_{\mathbf{k}\sigma}^{3+N} \hbar\omega(\mathbf{k}\sigma)\left\{\frac{1}{\exp[\hbar\omega(\mathbf{k}\sigma)/k_B T]-1}+\tfrac{1}{2}\right\},$$

using the expansion

$$\frac{x}{e^x-1} = 1-\frac{x}{2}-\sum_{n=1}^{\infty}(-1)^n B_{2n}\frac{x^{2n}}{(2n)!} \quad (|x|<2\pi),$$

where the B's are the Bernoulli numbers,

$$\left.\begin{array}{l} B_2 = \tfrac{1}{6}, \quad B_4 = \tfrac{1}{30}, \quad B_6 = \tfrac{1}{42}, \quad B_8 = \tfrac{1}{30}, \\ B_{10} = \tfrac{5}{66}, \quad B_{12} = \tfrac{691}{2730}, \quad B_{14} = \tfrac{7}{6}, \quad \ldots, \end{array}\right\}$$

as

$$\frac{c_v(T)}{3rNk_B} = 1-\sum_{n=1}^{\infty}(-1)^n B_{2n}\frac{(1-2n)}{(2n)!}\left(\frac{\hbar\omega_L}{k_B T}\right)^{2n} u_{2n}.$$

The u_{2n} are the dimensionless moments of the frequency spectrum.

Discuss the radius of convergence of this expansion and show that typically it converges for $T > 50°$K. [In practice, with a few low-order moments, the series is useful only for $T > 80°$K. For more rapidly convergent series, see Maradudin, Montroll, and Weiss (1963).]

Chapter 4

1. Using the kinetic energy and exchange energy for electrons in a homogeneous gas, show that a transition occurs from a paramagnetic state to a ferromagnetic state as the electron density is lowered. Explain qualitatively what must be done to correct this theory.

2. The change in the nuclear magnetic resonance frequency when the ion is in a metal, compared with its value in a non-metallic environment, is termed the Knight shift and can be expressed in terms of the Pauli spin susceptibility χ_s, multiplied by a factor which is due to the wave-function enhancement at the nucleus. To obtain the requisite expression by deriving the formula for the fractional change in the frequency, $\mathscr{H}$ being the magnetic field,

$$K = \frac{\Delta\mathscr{H}}{\mathscr{H}} \tag{P4.2.1}$$

with

$$\Delta\mathscr{H} = \frac{8\pi}{3}\chi_s\,\mathscr{H}\Omega\langle|\psi(0)|^2\rangle_s \tag{P4.2.2}$$

(Ω being the volume of the unit cell: and $\langle\ \rangle_s$ denoting the average over the Fermi surface) first calculate the magnetic moment density at a nucleus at the origin due to the surrounding valence electrons, neglecting the orbital contribution. The resulting integral is indeterminate if $\psi(0) \neq 0$ but can be evaluated by drawing a sphere round the origin, thus splitting the integral into two parts. Equation (2) is found by ignoring the contribution from the electrons outside the sphere.

3. Show that the spin-lowering operator S^- commutes with S^2 and the Heisenberg Hamiltonian.

Relate these properties to the invariance of the Heisenberg Hamiltonian under rotation.

4. Under what circumstances can the exchange interaction $J(k)$ be regarded as spherical?

5. To evaluate the spin-wave frequency [see equation (4.10.14)]

$$i\omega = \frac{\{S^+, \dot{S}^-\}}{\{S^+, S^-\}} \tag{P4.5.1}$$

to order T^4, show that, using the spin-wave approximation,

$$S_{\mathbf{q}}^{\dagger} = \sqrt{(2NS)}\, a_{\mathbf{q}}, \tag{P4.5.2}$$

$$S_{\mathbf{q}}^{0} = NS\delta_{\mathbf{q},0} - \sum_{\mathbf{p}} a_{\mathbf{p}}^{\dagger}\, a_{\mathbf{p}+\mathbf{q}} \tag{P4.5.3}$$

and evaluating the averages over the creation and annihilation operators as in equation (4.10.27), we obtain, to lowest order,

$$\omega_{\mathbf{k}} \simeq 2SJ(0,\mathbf{k}) + \frac{2}{N}\sum_{\mathbf{q}} n_{\mathbf{q}}[J(\mathbf{q},\mathbf{k}+\mathbf{q}) - J(0,\mathbf{k})]. \tag{P4.5.4}$$

By taking the small q limit of $n_{\mathbf{q}}$ and using the relation

$$\frac{\partial^2}{\partial \mathbf{k}^2} J(\mathbf{k}) = -b^2 J(\mathbf{k}), \tag{P4.5.5}$$

where b is the near-neighbour distance, show that equation (P4.5.4) may be approximated by

$$\omega_{\mathbf{k}} \simeq 2\left(1 - \frac{\varepsilon}{2S}\right) SJ(0,\mathbf{k}); \tag{P4.5.6}$$

with

$$\varepsilon = \left(\frac{b^2}{3N}\right)\sum_{\mathbf{q}} n_q q^2 = 2\pi b^2\left(\frac{\mathscr{V}}{N}\right)\zeta\left(\frac{5}{2}\right)\left(\frac{k_B T}{4\pi D}\right)^{\frac{5}{2}} \tag{P4.5.7}$$

and

$$D = 2S \lim_{q\to 0} q^{-2} J(0,\mathbf{q}) = 2Sa^2 J, \tag{P4.5.8}$$

where a is the lattice constant and ζ is the Riemann zeta function.

The result (P4.5.6) for $\omega_{\mathbf{k}}$ agrees with that of Keffer and Loudon, using the Dyson spin-wave theory (Mori and Kawasaki, 1962).

6. Consider a two-dimensional square lattice with three atoms per unit cell, so that provided the periodic potential is small the nearest energy band is completely occupied and there are occupied portions of second, third and fourth bands. When the periodic potential is exactly zero it is natural to represent the Fermi "surface" in the extended zone scheme, when it is a circle in $\mathbf{k}$-space containing three electronic states per atom; this circle defines the orbital motion of electrons under the influence of a magnetic field perpendicular to the system. On the other hand, when the potential is non-zero, continuous electron orbits, about Fermi surfaces of the second, third and fourth bands, are found only in the reduced zone

scheme. Show that these Fermi surfaces for the free-electron model may be constructed by drawing a circle, containing three states per atom, about every lattice point, and that a point **k** lying within at least one circle corresponds to an occupied state of the lowest band, a point lying in at least two circles corresponds to an occupied state of the second band, and so on. This is Harrison's construction.

Show that the orbit round the Fermi surface of the second band is a "hole orbit", that is, the direction of the electron's motion round the direction of the magnetic field is in the opposite sense to that of a free electron.

Apply similar considerations to the three-dimensional case of aluminium.

7. Show that the Hartree–Fock equation obeyed by the fractional susceptibility $L(11'22')$ is given by equation (4.12.5).

8. If E is the energy of an electron in zero magnetic field, the energy of an electron with magnetic moment $\pm\mu_B$ in a field is $E\mp\mu_B\mathscr{H}$. Show that the magnetic moment per unit volume of an electron gas is

$$M = \mu_B\int[f_0(E-\mu_B\mathscr{H})-f_0(E+\mu_B\mathscr{H})]\,n(E)\,dE, \tag{P4.8.1}$$

f_0 being the Fermi function and $n(E)$ the density of states. Show that the chemical potential ζ is determined by the condition

$$n = \int[f_0(E-\mu_B\mathscr{H})+f_0(E+\mu_B\mathscr{H})]\,n(E)\,dE, \tag{P4.8.2}$$

where n is the number of electrons per unit volume.

Defining the Fermi–Dirac functions $F_s(\xi)$ by

$$F_s(\xi) = \int_0^\infty \frac{x^s\,dx}{e^{x-\xi}+1}, \tag{P4.8.3}$$

show that, for a parabolic band,

$$M = \frac{3}{4}n\mu_B\left(\frac{k_B T}{\zeta_0}\right)^{\frac{3}{2}}[F_{\frac{1}{2}}(\xi+\alpha)-F_{\frac{1}{2}}(\xi-\alpha)] \tag{P4.8.4}$$

and that the chemical potential is to be determined from the equation

$$1 = \frac{3}{4}\left(\frac{k_B T}{\zeta_0}\right)^{\frac{3}{2}}[F_{\frac{1}{2}}(\xi+\alpha)+F_{\frac{1}{2}}(\xi-\alpha)], \tag{P4.8.5}$$

where

$$\alpha = \mu_B\mathscr{H}/k_B T \quad\text{and}\quad \xi = \zeta/k_B T. \tag{P4.8.6}$$

In the Stoner theory of collective electron ferromagnetism, the assumption is made that the energy includes now a term porportional to the spontaneous magnetization, and given by $\pm k_B\,\theta'\sigma$; $\sigma = M/n\mu_B$, the sign depending on whether the electron spin is antiparallel or parallel to M.

Show that the equations (P4.8.1) and (P4.8.2) for M and n are modified by replacing α by $\alpha+\alpha'$ where $\alpha' = \sigma\theta'/T$. Hence show that we can write these equations as

$$F_{\frac{1}{2}}(\xi+\alpha+\alpha') = \tfrac{2}{3}(1+\sigma)(\zeta_0/k_B T)^{\frac{3}{2}} \tag{P4.8.7}$$

and

$$F_{\frac{1}{2}}(\xi-\alpha-\alpha') = \tfrac{2}{3}(1-\sigma)(\zeta_0/k_B T)^{\frac{3}{2}}. \tag{P4.8.8}$$

Using the property that for large x

$$F_{\frac{1}{2}}(x) = \frac{2}{3}x^{\frac{3}{2}}\left[1+\frac{\pi^2}{8x^2}+\dots\right] \qquad \text{(P4.8.9)}$$

eliminate ξ between equations (P4.8.7) and (P4.8.8) and hence prove that

$$2(\alpha+\alpha') = y_+ - y_- - \tfrac{1}{12}\pi^2(y_+^{-1}-y_-^{-1})+\dots, \qquad \text{(P4.8.10)}$$

where

$$y_+ = (1+\sigma)^{\frac{2}{3}}\,\zeta_0/k_B T, \quad y_- = (1-\sigma)^{\frac{2}{3}}\,\zeta_0/k_B T. \qquad \text{(P4.8.11)}$$

Hence show that when $T = 0$ and $\mathscr{H} = 0$

$$2\sigma_0\, k_B\, \theta'/\zeta_0 = (1+\sigma_0)^{\frac{2}{3}}-(1-\sigma_0)^{\frac{2}{3}}, \qquad \text{(P4.8.12)}$$

where σ_0 is the relative magnetization $M/n\mu_B$ at $T = 0$. Show from this result that a necessary condition for the existence of spontaneous magnetization is that

$$k_B\, \theta'/\zeta_0 > \tfrac{2}{3} \qquad \text{(P4.8.13)}$$

and that if

$$\tfrac{2}{3} < k_B\, \theta'/\zeta_0 < 2^{-\frac{1}{3}} \qquad \text{(P4.8.14)}$$

the relative magnetization is such that $0<\sigma_0<1$.

Chapter 5

1. Treating the positron as a static impurity, use the first-order equation (2.3.14) to show that the density $\rho(0)$ at the positron of electrons in an electron gas at high density (small r_s) is related to the mean density ρ_0 by

$$[\rho(0)-\rho_0]/\rho_0 = 2.46\, r_s/a_0.$$

When Hartree calculations on this system are carried out as a function of r_s, the results over a range of r_s can be fitted by the approximate formula

$$[\rho(0)-\rho_0]/\rho_0 = 2.46(r_s/a_0)+0.8(r_s/a_0)^2+0.6(r_s/a_0)^3.$$

What is wrong with this Hartree formula as r_s tends to infinity?

How much does the positron lifetime vary in going from Cs to Al, given that the lifetime is inversely proportional to the electron density at the positron?

2. Integrate the radial wave equation for the positron wave function in Cu, using one of the finite-difference approximations of Fox and Goodwin (1949). Use the Hartree–Fock–Slater potential given by Herman and Skillman (1963).

3. In relation to the Compton line of Li, it is of interest to estimate the mean kinetic energy per conduction electron, defined by

$$\mathcal{T} = \int_0^\infty \frac{p^2}{2m} I(p)\, dp,$$

where $I(p)\, dp$ is clearly the probability of finding momentum of magnitude between p and $p+dp$.

Use the virial theorem, together with the following information, to estimate $\mathcal{T}$ (see Kilby, 1963):

(i) Experimental values of the binding energy of metallic Li show that the total energy per atom is 1.6 eV less than that of the free atom.

(ii) The mean kinetic energy of the 2s electron in the free atom is 5.4 eV.

(iii) The kinetic energy of the 1s electrons is unchanged on binding.

4. Calculate the probability of an electron having a momentum between p and $p+dp$ from the Thomas–Fermi theory, by drawing the (r, p) plane, and noting that for each value of r there corresponds a value $p_f(r)$, where $p_f(r)$ is the maximum momentum. Using the same approach, but given in a ferromagnet like Fe the densities of upward and downward spin electrons, $\rho_\uparrow(\mathbf{r})$ and $\rho_\downarrow(\mathbf{r})$, show how the momentum distributions of the two separate spin densities can be obtained.

5. Prove the sum rule, for the dynamical structure factor:

$$\int_{-\infty}^{\infty} \frac{d\omega}{2\pi} \omega S(k\omega) = \frac{\hbar k^2}{2M}.$$

6. Show that if neutrons are scattered by the nuclei of a single isotope, the scattering length $\langle b \rangle$ determining the coherent cross-section is

$$\langle b \rangle = \left\{ \frac{I+1}{2I+1} b_+ + \frac{I}{2I+1} b_- \right\}$$

and the incoherent scattering is $\propto \langle |b|^2 \rangle - |\langle b \rangle|^2$, where

$$\langle |b|^2 \rangle = \left\{ \frac{I+1}{2I+1} |b_+|^2 + \frac{I}{2I+1} |b_-|^2 \right\}.$$

Here I is the spin of the nuclei, and b_+ and b_- are respectively the scattering lengths for the states of spin $I+\frac{1}{2}$ and $I-\frac{1}{2}$ of the nucleus plus the neutron.

7. Derive the contribution to the differential scattering cross-section for neutrons from a vibrating crystal composed of magnetic ions. Express the result in terms of the dynamical structure factor for phonons.

8. In connection with the theory of cumulants, prove that

$$\langle u^n \rangle_c = 0 \quad (n > 2)$$

for a harmonic oscillator.

9. The network model, set out in Problem 14 of Chapter 1, bunches the charge distribution in a crystal along particular directions and is therefore a model of extreme angularity.

Discuss the X-ray scattering factor for a bcc crystal, when the above model is represented by a localized charge distribution

$$\sigma(\mathbf{r}) = Ar^2 \exp(-ar) [\delta(x)\,\delta(y) + \delta(y)\,\delta(z) + \delta(z)\,\delta(x)].$$

Show, in particular, that the (110) reflexion has a scattering factor which differs from that of the corresponding spherical distribution by no more than 3%. Why is the g term ($l = 4$) in $\sigma(\mathbf{r})$ above not a physically good choice?

References

The numbers in *italics* after the entries refer to the page in the book where the reference occurs, enabling the list of references to be used as an author index as well.

Abarenkov, I. V. and Heine, V. (1965). *Phil. Mag.*, **12**, 529. (*72*) *See also* Heine and Abarenkov

Abrikosov, A. A., Gorkov, L. P. and Dzaloshinski, I. E. (1963). *Methods of Quantum Field Theory in Statistical Physics*, Prentice-Hall, Englewood Cliffs. (*590*)

Adams, E. N. (1953). *J. Chem. Phys.*, **21**, 2013. (*44*)

Allan, G., Lomer, W. M., Lowde, R. D. and Windsor, C. G. (1968). *Phys. Rev. Letters*, **20**, 933. (*507*)

Als-Nielsen, J. (1970). *Phys. Rev. Letters*, **25**, 730. (*500*)

Als-Nielsen, J., Dietrich, O. W., Marshall, W. and Lindgard, P. A. (1967). *Solid State Commun.*, **5**, 607. (*499*)

Altmann, S. L. (1958). *Proc. Roy. Soc.*, **A244**, 141, 153. (*49, 50*)

Ambegaokar, V., Conway, J. M. and Baym, G. (1964). *Lattice Dynamics*, Conference Proceedings (Ed. R. F. Wallis), Pergamon, Oxford, p. 261. (*457, 459, 640*)

Ambegaokar, V. and Kohn, W. (1960), *Phys. Rev.*, **117**, 423. (*210, 211, 282*)

Anderson, P. W. (1963). *Concepts in Solids*, Benjamin, New York. (*326*)

Animalu, A. O. E. (1965). *Phil. Mag.*, **11**, 379. (*73*)

Ascarelli, P. and Raccah, P. M. (1970). *Phys. Letters*, **31A**, 549. (*472*)

Ashcroft, N. W. (1963). *Phil. Mag.*, **8**, 2055. (*648*)

Austin, B. J., Heine, V. and Sham, L. (1962). *Phys. Rev.*, **127**, 276. (*73*)

Baltensberger, W. (1966). *Phys. Kondens. Mater.*, **5**, 341. (*397*)

Bardeen, J. (1936). *Phys. Rev.*, **50**, 1098. (*174*)

Bardeen, J. (1937). *Phys. Rev.*, **52**, 688. (*102, 258, 653*)

Bardeen, J. (1938). *J. Chem. Phys.*, **6**, 372. (*647*)

Bardeen, J. and Pines, D. (1955). *Phys. Rev.*, **99**, 1140. (*216*)

Barker, A. S. (1966). *Phys. Rev.*, **145**, 391. (*513*)

Barron, T. H. K. (1963). *Lattice Dynamics*, Conference Proceedings (Ed. R. F. Wallis), Pergamon, Oxford, p. 247. (*287*)

Batterman, B. W., Chipman, D. R. and de Marco, J. J. (1961). *Phys. Rev.*, **122**, 68. (*475*)

Baym, G. (1962). *Phys. Rev.*, **127**, 1391. (*421*)

Baym, G. and Kadanoff, L. P. (1961). *Phys. Rev.*, **124**, 287. (*421*)

Bennett, H. S. and Martin, P. C. (1965). *Phys. Rev.*, **138**, A608. (*377, 382, 499*)

Berk, N. F. and Schrieffer, J. R. (1966). *Phys. Rev. Letters*, **17**, 433. (*427*)

Bethe, H. A. (1929). *Ann. Phys.*, **2**, 133. (*472*)

Betts, D. D., Bhatia, A. B. and Wyman, M. (1956). *Phys. Rev.*, **104**, 37. (*246*)

Björkman, G., Lundqvist, B. I. and Sjölander, A. J. (1967). *Phys. Rev.*, **159**, 551. (*286*)

Bloch, F. (1929). *Z. Physik*, **57**, 545. (*654*)

Blount, E. I. (1962). *Solid State Phys.*, **13**, 305. (*44, 208*)

Boccara, N. and Sarma, G. (1965). *Physics*, **1**, 219. (*297*)

Bohm, D. and Pines, D. (1951). *Phys. Rev.*, **82**, 625. (*2, 7, 112*)

Bohm, D. and Pines, D. (1952). *Phys. Rev.*, **85**, 338. (*2, 7, 112*)

Bohm, D. and Pines, D. (1953). *Phys. Rev.*, **92**, 609. (*2, 7, 112*)

Bohm, D. and Staver, T. (1952). *Phys. Rev.*, **84**, 836. (*216, 220*)

Born, M. (1914). *Ann. Phys. Lpz.*, **44**, 605. (*272*)
Born, M. (1951). *Fest. Akad. d. Wiss. Göttingen, Mat.-phys. Klasse*, Springer-Verlag, Berlin.
Born, M. and Huang, K. (1954). *Dynamical Theory of Crystal Lattices*, University Press, Oxford. (*232, 280, 296*)
Bouckaert, L. P., Smoluchowski, R. and Wigner, E. P. (1936). *Phys. Rev.*, **50**, 58. (*25, 82*)
Brinkman, W. F., Platzman, P. M. and Rice, T. M. (1968). *Phys. Rev.*, **174**, 495. (*181*)
Brockhouse, B. N., Arase, T., Caglioti, G., Rao, K. R. and Woods, A. D. B. (1962). *Phys. Rev.*, **128**, 1099. (*258, 260, 640*)
Brockhouse, B. N. and Iyengar, P. K. (1958). *Phys. Rev.*, **111**, 747. (*275*)
Brockhouse, B. N. and Sinclair, R. N. (1960). *Phys. Rev.*, **120**, 1638. (*495*)
Brout, R. and Carruthers, P. (1963). *Lectures on the Many-electron Problem*, Interscience, New York. (*402*)

Cable, J. W., Lowde, R. D., Windsor, C. G. and Woods, A. D. B. (1967). *J. Appl. Phys.*, **38**, 1247. (*505*)
Callaway, J. (1958). *Solid State Phys.*, **7**, 199. (*647*)
Callaway, J. (1959). *Phys. Rev.*, **116**, 1368. (*212, 599*)
Carr, W. J. (1961). *Phys. Rev.*, **122**, 1437. (*194*)
Chell, G. G. (1970). *J. Phys.* C, **3**, 1861. (*298*)
Chen, S. H. (1967). *Phys. Rev.*, **163**, 532. (*609*)
Chester, G. V. and Houghton, A. (1959). *Proc. Phys. Soc.*, **73**, 609. (*612*)
Choquard, P. F. (1967). *The Anharmonic Crystal*, Benjamin, New York. (*297, 458*)
Claesson, A. (1969). *Solid State Commun.*, **8**, 283. (*198*)
Cochran, W. (1959a). *Phil. Mag.*, **4**, 1082. (*271*)
Cochran, W. (1959b). *Proc. Roy. Soc.*, **A253**, 260. (*271, 275*)
Cochran, W. (1963). *Proc. Roy. Soc.*, **A276**, 308. (*248, 464*)
Cochran, W. (1969). *Advan. Phys.*, **18**, 157. (*310, 508, 513*)
Coldwell-Horsfall, R. A. and Maradudin, A. (1960). *J. Math. Phys.*, **1**, 395. (*194*)
Collins, M. F., Pickart, S. J. and Windsor, C. G. (1960). *J. Appl. Phys.*, **37**, 1054. (*499*)
Cooper, M. (1971). *Advan. Phys.*, **20**, 453. (*484*)
Cooper, M., Williams, B. G., Borland, R. E. and Cooper, J. R. A. (1970). *Phil. Mag.*, **22**, 441. (*485*)
Coulson, C. A. (1954). *Proc. Phys. Soc.*, **67**, 608. (*648*)
Cowley, R. A. (1966). *Phonons in Perfect Lattices* (Ed. R. W. H. Stevenson), Oliver and Boyd, Edinburgh. (*625, 627*)

Dalton, N. W. (1970). *J. Phys.* C, **3**, 1912. (*83, 84*)
Daniel, E. and Vosko, S. H. (1960). *Phys. Rev.*, **120**, 2041. (*143*)
Daniels, W. B., Shirane, G., Frazer, B. C., Umebayashi, H. and Leake, J. A. (1967). *Phys. Rev. Letters*, **18**, 548. (*305*)
Davis, H. L. (1960). *Phys. Rev.*, **120**, 789. (*338, 340*)
Davis, H. L., Faulkner, J. S. and Joy, H. W. (1968). *Phys. Rev.*, **167**, 601. (*78*)
Davis, L. (1939). *Phys. Rev.*, **56**, 93.
de Gennes, P. G. (1958). *J. Phys. Chem. Solids*, **4**, 223. (*383, 499*)
de Launay, J. (1956). *Solid State Phys.*, **2**, 219. (*246, 247*)

de Leener, M. and Résibois, P. (1969). *Phys. Rev.*, **178**, 819. (*385*)
de Marco, J. J. and Weiss, R. J. (1965). *Phys. Letters*, **18**, 92. (*472*)
deSorbo, W. and Tyler, W. W. (1953). *J. Chem. Phys.*, **21**, 1660. (*248*)
Dembinski, S. T. (1964). *Physica*, **30**, 1217. (*336*)
Dick, B. G. and Overhauser, A. W. (1958). *Phys. Rev.*, **112**, 90. (*270*)
Dickey, J. M. and Paskin, A. (1969). *Phys. Rev.*, **188**, 1407. (*305*)
Dirac, P. A. M. (1930). *Proc. Camb. Phil. Soc.*, **26**, 376. (*6, 126, 200*)
Dirac, P. A. M. (1957). *The Principles of Quantum Mechanics*, University Press, Oxford. (*291*)
Donaghy, J. J. and Stewart, A. T. (1964). *Bull. Am. Phys. Soc.*, **9**, 238. (*450, 488*)
Doniach, S. (1967). *Proc. Phys. Soc.*, **91**, 86. (*399*)
Doniach, S. (1969). *Advan. Phys.*, **20**, 1. (*356*)
Doniach, S. and Engelsberg, S. (1966). *Phys. Rev. Letters*, **17**, 750. (*427*)
Donovan, B. and March, N. H. (1956). *Proc. Phys. Soc.*, **69**, 1249. (*488*)
du Mond, J. W. M. (1933). *Rev. Mod. Phys.*, **5**, 1. (*478*)
Dunifer, G., Schultz, S. and Schmidt, P. H. (1968). *J. Appl. Phys.*, **39**, 397. (*413*)
Dyson, F. J. (1956). *Phys. Rev.*, **102**, 1217, 1230. (*335, 336*)

Edwards, D. M. (1969). *J. Phys.* C, **2**, 84. (*413*)
Edwards, S. F. (1958). *Proc. Phys. Soc.*, **72**, 685. (*650*)
Ehrenreich, H. and Cohen, M. H. (1959). *Phys. Rev.*, **115**, 786. (*550*)
Elliott, R. J. and Wedgwood, F. A. (1963). *Proc. Phys. Soc.*, **81**, 846. (*432, 434*)
Elliott, R. J. and Wedgwood, F. A. (1964). *Proc. Phys. Soc.*, **84**, 63. (*432, 434*)

Faulkner, J. S., Davis, H. L. and Joy, H. W. (1967). *Phys. Rev.*, **161**, 656. (*58, 80*)
Fedders, P. A. and Martin, P. C. (1966). *Phys. Rev.*, **143**, 245. (*418, 421, 422*)
Feenberg, E. and Pake, G. (1959). *Notes on the Quantum Theory of Angular Momentum*, University Press, Stamford. (*330*)
Fischer, I. Z. (1964). *Statistical Theory of Liquids*, University of Chicago Press. (*458*)
Fisher, M. E. (1967). *Rep. Prog. Phys.*, **30**, 615. (*370, 371*)
Flower, M., March, N. H. and Murray, A. M. (1960). *Phys. Rev.*, **119**, 1885. (*54*)
Foreman, A. J. E. and Lomer, W. M. (1967). *Proc. Phys. Soc.*, **B70**, 1143. (*464*)
Fox, L. and Goodwin, E. T. (1949). *Proc. Camb. Phil. Soc.*, **45**, 373. (*657*)
Freeman, A. J., Furdyna, A. M. and Dimmock, J. O. (1966). *J. Appl. Phys.*, **37**, 1256. (*429*)
Friedel, J. (1954). *Advan. Phys.*, **3**, 446. (*213*)
Fuchs, K. (1935). *Proc. Roy. Soc.*, **A151**, 585. (*193*)
Fues, E. and Statz, H. (1952). *Z. Naturfurch*, **7a**, 2. (*69*)

Gaskell, T. (1962). *Proc. Phys. Soc.*, **80**, 1091. (*650*)
Gell-Mann, M. (1957). *Phys. Rev.*, **106**, 369. (*172*)
Gell-Mann, M. and Brueckner, K. A. (1957). *Phys. Rev.*, **106**, 364. (*111, 116, 172, 194, 574–5*)
Gilat, G. and Raubenheimer, L. J. (1966). *Phys. Rev.*, **144**, 390. (*238*)
Gillis, N. S., Werthamer, N. R. and Koehler, T. R. (1968). *Phys. Rev.*, **165**, 951. (*297*)
Green, H. S. (1952). *J. Chem. Phys.*, **20**, 1274. (*522*)
Griffiths, R. B. (1965). *Phys. Rev. Letters*, **14**, 623. (*371*)

Grindlay, J. and Howard, R. (1965). *Lattice Dynamics* (Ed. R. F. Wallis), Pergamon, Oxford. (*307*)
Grüneisen, E. (1933). *Ann. Physik*, **16**, 530. (*293, 294*)
Gutzwiller, M. C. (1963). *Phys. Rev. Letters*, **10**, 159. (*341*)
Gutzwiller, M. C. (1964). *Phys. Rev.*, **134**, A923. (*341*)

Halperin, B. I. and Hohenberg, P. (1969). *Phys. Rev.*, **177**, 952. (*385*)
Ham, F. S. (1955). *Solid State Phys.*, **1**, 127. (*79*)
Hamann, D. R. and Overhauser, A. W. (1966). *Phys. Rev.*, **143**, 183. (*403, 410, 412, 423*)
Hardy, J. R. (1961). *Phil. Mag.*, **6**, 27. (*270*)
Harrison, W. A. (1963). *Phys. Rev.*, **129**, 2512. (*73*)
Harrison, W. A. (1966). *Pseudopotentials in the Theory of Metals*, Benjamin, New York and Amsterdam. (*73*)
Hebborn, J. and March, N. H. (1970). *Advan. Phys.*, **19**, 175. (*393*)
Hebborn, J. and Sondheimer, E. H. (1960). *J. Phys. Chem. Solids*, **13**, 105. (*397*)
Hedin, L. and Lundqvist, S. (1969). *Solid State Phys.*, **23**, 1. (*213*)
Heine, V. (1957). *Proc. Roy. Soc.*, **A240**, 361. (*261*)
Heine, V. and Abarenkov, I. (1964). *Phil. Mag.*, **9**, 451. (*72, 450*)
Herman, F. (1955). *Proc. Inst. Radio Engrs*, **43**, 1703. (*86*)
Herman, F. (1958). *Rev. Mod. Phys.*, **30**, 102. (*86, 87*)
Herman, F. and Skillman, S. (1963). *Atomic Structure Calculations*, Prentice-Hall, Englewood Cliffs. (*657*)
Herman, F., van Dyke, J. P. and Ortenburger, I. B. (1969). *Phys. Rev. Letters*, **22**, 807. (*201, 202, 267*)
Herring, C. (1940). *Phys. Rev.*, **57**, 1169. (*48*)
Herring, C. (1966). "Exchange interactions among Itinerant Electrons", *Magnetism*, Vol. 4 (Ed. G. T. Rado and H. Suhl), Academic Press, New York. (*342, 350*)
Hohenberg, P. C. and Kohn, W. (1964). *Phys. Rev.*, **136**, B864. (*4, 5, 592, 593*)
Holstein, T. and Primakoff, H. (1940). *Phys. Rev.*, **58**, 177. (*331, 335*)
Hooton, D. J. (1955). *Phil. Mag.*, **46**, 485. (*297*)
Horton, G. K. and Schiff, H. (1956). *Phys. Rev.*, **104**, 32. (*248*)
Houston, W. V. (1948). *Rev. Mod. Phys.*, **20**, 161. (*244, 248*)
Hubbard, J. (1957). *Proc. Roy. Soc.*, **A243**, 336. (*193*)
Hubbard, J. (1963). *Proc. Roy. Soc.*, **A276**, 238. (*341*)
Hubbard, J. (1964a). *Proc. Roy. Soc.*, **A281**, 401. (*341*)
Hubbard, J. (1964b). *Proc. Roy. Soc.*, **A285**, 542. (*341, 355, 356*)
Hubbard, J. and Dalton, N. W. (1968). *J. Phys.* C, **1**, 1637. (*83*)
Hulthén, L. (1938). *Arkiv. Mat., Ast. Fys.*, **261**, 1011. (*629*)
Hwang, J. L. (1955). *Phys. Rev.*, **99**, 1098. (*244*)

Inglesfield, J. E. and Johnson, F. A. (1970). *Proc. Roy. Soc.*, **A317**, 293. (*316, 317*)
Izuyama, T., Kim, D. J. and Kubo, R. (1963). *J. Phys. Soc. Japan*, **18**, 1025. (*320, 414*)

Jaros, M. and Vinsome, P. K. W. (1969). *J. Phys.* C, **2**, 2373. (*603*) *See also* Vinsome and Jaros.
Johnson, F. A. (1969). *Proc. Roy. Soc.*, **A310**, 101. (*267, 316, 317*)
Johnson, L. E., Conklin, J. B. and Pratt, G. W. (1963). *Phys. Rev. Letters*, **11**, 538. (*94*)

Jones, W. and March, N. H. (1970). *Proc. Roy. Soc.*, **A317**, 359. (*265*)
Jones, W., March, N. H. and Tucker, J. W. (1965). *Proc. Roy. Soc.*, **A284**, 289. (*475*)
Josephson, B. D. (1969). *J. Phys.* C, **2**, 200. (*374*)

Kadanoff, L. P. (1966). *Physics*, **2**, 263. (*372, 373*)
Kadanoff, L. P. and Baym, G. (1962). *Quantum Statistical Mechanics*, Benjamin, New York and Amsterdam.
Kanamori, J. (1963). *Prog. Theor. Phys. Japan*, **30**, 275. (*341, 351, 357*)
Kane, E. O. (1956). *J. Phys. Chem. Solids*, **1**, 83.
Kastelejn, P. W. (1952). *Physica*, **18**, 104. (*370, 629*)
Kasteleyn, P. (1971). *Theory of Imperfect Crystals*, IAEA Vienna. (*370, 374*)
Kawasaki, K. (1967). *J. Phys. Chem. Solids*, **28**, 1277. (*385*)
Keffer, F. and Loudon, R. (1961). *J. Appl. Phys.* (Suppl.), **32**, 25. (*655*)
Kellermann, E. W. (1940). *Phil. Trans.*, **A238**, 513. (*613, 615*)
Kilby, G. E. (1963). *Proc. Phys. Soc.*, **82**, 900. (*657*)
Kilby, G. E. (1965). *Proc. Phys. Soc.*, **86**, 1037. (*479, 482*)
Kimball, G. (1935). *J. Chem. Phys.*, **3**, 560. (*9*)
Kirzhnits, D. A. (1957). *Soviet Phys.—JETP*, **5**, 64. (*201*)
Kittel, C. (1963). *Quantum Theory of Solids*, Wiley, New York. (*430, 503*)
Klein, M. L., Chell, G. G., Goldman, V. V. and Horton, G. K. (1970). *J. Phys.* C, **3**, 806. (*298*)
Kohn, W. (1958). *Phys. Rev.*, **110**, 857. (*209*)
Kohn, W. and Luttinger, J. M. (1957). *Phys. Rev.*, **108**, 590.
Kohn, W. and Luttinger, J. M. (1960). *Phys. Rev.*, **115**, 41. (*589*)
Kohn, W. and Rostoker, N. (1954). *Phys. Rev.*, **94**, 1111. (*58*)
Kohn, W. and Sham, L. J. (1965). *Phys. Rev.*, **140**, A1133. (*200*)
Korringa, J. (1947). *Physica*, **13**, 392. (*58, 62*)
Kubo, R. (1965). *Boulder Lectures on Theoretical Physics.*

Landau, L. D. (1956). *Soviet Phys.—JETP*, **3**, 920. (*100, 147*)
Landau, L. and Lifshitz, E. M. (1958). *Statistical Physics*, Pergamon, Oxford. (*621*)
Lang, G., DeBenedetti, S. and Smoluchowski, R. (1955). *Phys. Rev.*, **99**, 596. (*488*)
Lax, M. and Lebowitz, J. L. (1954). *Phys. Rev.*, **96**, 594. (*242*)
Leake, J. A., Daniels, W. B., Skalyo, J., Frazer, B. C. and Shirane, G. (1969). *Phys. Rev.*, **181**, 1251. (*305*)
Lee, E. W. and Wilding, M. D. (1965). *Proc. Phys. Soc.*, **85**, 955. (*434*)
Leibfried, G. and Ludwig, W. (1961). *Solid State Phys.*, **12**, 276. (*287, 293*)
Lidiard, A. B. (1970). *Comments in Solid State Physics*, **II**, 76. (*276*)
Lin, P. J. and Phillips, J. C. (1965). *Advan. Phys.*, **14**, 257. (*76*)
Lindhard, J. (1954). *Kgl. Danske Mat. Fys. Medd.*, **28**, 8. (*105*)
Lloyd, P. (1965). *Proc. Phys. Soc.*, **86**, 825. (*71*)
Lomer, W. M. (1962). *Proc. Phys. Soc.*, **80**, 489. (*418, 420, 507*)
Loucks, T. L. (1965). *Phys. Rev.*, **139**, A1333. (*95, 541, 547*)
Loucks, T. L. (1965). *Phys. Rev. Letters*, **14**, 693. (*98*)
Lowde, R. D. and Windsor, C. G. (1967). *Phys. Rev. Letters*, **18**, 1136. (*507*)
Luttinger, J. M. (1960). *Phys. Rev.*, **119**, 1153. (*144, 147, 349*)
Luttinger, J. M. *Phys. Rev.*, **121**, 942. (*349*)
Lyddane, R. H., Sachs, R. G. and Teller, E. (1941). *Phys. Rev.*, **59**, 673. (*279, 615*)

Ma, S. S. and Brueckner, K. A. (1968). *Phys. Rev.*, **165**, 18. (*201, 267, 594*)
Macke, W. (1950). *Z. Naturforsch.*, **5a**, 192. (*111*)
Maradudin, A. A., Montroll, E. W. and Weiss, G. M. (1963). *Solid State Phys.* Suppl., **3**, Academic Press, New York and London. (*238, 244, 247, 654*)
Maradudin, A. A. and Vosko, S. H. (1968). *Revs. Mod. Phys.*, **40**, 1. (*611*)
March, N. H. and Murray, A. M. (1960). *Phys. Rev.*, **120**, 830. (*3, 522*)
March, N. H. and Murray, A. M. (1961). *Proc. Roy. Soc.*, **A261**, 119. (*3, 104, 522*)
March, N. H., Young, W. H. and Sampanthar, S. (1968). *The Many Body Problem in Quantum Mechanics*, University Press, Cambridge. (*194, 560*)
Marshall, W. (1955a). *Proc. Roy. Soc.*, **A232**, 48. (*332*)
Marshall, W. (1955b). *Proc. Roy. Soc.*, **A232**, 69. (*338, 339, 629*)
Marshall, W. and Lovesey, S. W. (1971). *Theory of Thermal Neutron Scattering*, Clarendon Press, Oxford. (*455*)
Marshall, W. and Lowde, R. D. (1968). *Rep. Prog. Phys.*, **31**, 705. (*357*)
Marshall, W. and Murray, G. (1969). *J. Phys.* C, **2**, 539. (*496*)
Mattis, D. C. (1964). *Theories of Magnetism*, Harper, New York. (*340*)
Melvin, J. S., Pirie, J. D. and Smith, T. (1968). *Phys. Rev.*, **175**, 1082. (*277*)
Messiah, A. (1962). *Quantum Mechanics*, Vol. II, North-Holland, New York and Amsterdam, p. 339. (*51, 81, 457*)
Minkiewicz, V. J., Collins, M. F., Nathans, R. and Shirane, G. (1969). *Phys. Rev.*, **182**, 624. (*507*)
Misra, P. K. and Roth, L. M. (1969). *Phys. Rev.*, **177**, 1089. (*398*)
Mook, H. A. (1966). *Phys. Rev.*, **48**, 495. (*472*)
Morgan, G. J. (1966). *Proc. Phys. Soc.*, **89**, 365. (*68*)
Mori, H. and Kawasaki, K. (1962). *Prog. Theor. Phys. Japan*, **27**, 529. (*382, 383, 392, 499, 655*)
Mueller, F. M. (1967). *Phys. Rev.*, **153**, 659. (*83*)

Nakamura, T. (1950). *Prog. Theor. Phys.* (*Kyoto*), **5**, 213. (*244*)
Nozières, P. (1964). *Theory of Interacting Fermi Systems*, Benjamin, New York and Amsterdam. (*144, 147, 175, 579, 581*)

Oguchi, T. (1960). *Phys. Rev.*, **117**, 117. (*336*)
Onsager, L. (1952). *Phil. Mag.*, **43**, 1006. (*443, 446*)
Overhauser, A. W. (1962). *Phys. Rev.*, **128**, 1437. (*403, 418*)

Penn, D. (1962). *Phys. Rev.*, **128**, 2093. (*212, 598*)
Philipp, H. R. and Taft, E. A. (1958). *Phys. Rev.*, **113**, 1002. (*602*)
Philipp, H. R. and Taft, E. A. (1960). *Phys. Rev.*, **120**, 37. (*602*)
Phillips, J. C. (1956). *Phys. Rev.*, **104**, 1263. (*242, 540, 541*)
Phillips, J. C. and Kleinman, L. (1959). *Phys. Rev.*, **116**, 287. (*70, 73, 76*)
Pines, D. (1961). *The Many Body Problem*, Benjamin, New York. (*278*)
Pippard, A. B. (1957). *Phil. Trans. Roy. Soc. London* A, **250**, 325. (*78*)
Platzman, P. M. and Tzoar, N. (1965). *Phys. Rev.*, **139**, A410. (*479*)
Platzman, P. M. and Wolff, P. A. (1967). *Phys. Rev. Letters*, **18**, 280. (*413*)
Prokofjew, W. K. (1929). *Z. Physik*, **58**, 255. (*50, 261*)
Pratt, G. W. and Ferreira, L. G. (1964). *Physics of Semi-conductors* (Conference Proceedings; Ed. M. Hulin, Paris), Dunod, Paris, p. 69. (*92*)

Quinn, J. J. and Ferrell, R. A. (1958). *Phys. Rev.*, **112**, 812. (*598*)

Raimes, S. (1954). *Proc. Phys. Soc.*, **A67**, 52. (*50*)
Raimes, S. (1961). *Wave Mechanics of Electrons in Metals*, North-Holland, New York and Amsterdam.
Rajagopal, A. K. and Jain, K. P. (1972). *Phys. Rev.*, **A5**, 1475.
Résibois, P. and Piette, C. (1970). *Phys. Rev. Letters*, **24**, 515. (*385*)
Rice, T. M. (1965). *Ann. Phys. N.Y.*, **31**, 100.
Rose, M. E. (1957). *Elementary Theory of Angular Momentum*, Wiley, New York. (*96*, *97*, *330*)
Rushbrooke, G. S. (1963). *J. Chem. Phys.*, **39**, 842. (*371*)

Sawada, K. (1957). *Phys. Rev.*, **106**, 372. (*100*, *116*)
Schiff, L. I. (1949). *Quantum Mechanics*, McGraw-Hill, New York. (*66*, *432*)
Schlosser, H. and Marcus, P. M. (1963). *Phys. Rev.*, **131**, 2529. (*547*)
Schneider, T. (1970). *Solid State Commun.*, **8**, 279. (*504*)
Schofield, P. (1969). *Phys. Rev. Letters*, **22**, 606. (*374*)
Schröder, U. (1966). *Solid State Commun.*, **4**, 347. (*277*)
Schultz, S. and Dunifer, G. (1967). *Phys. Rev. Letters*, **18**, 283. (*413*)
Schultz, T. D. (1963). *J. Math. Phys.*, **4**, 666. (*340*)
Segall, B. (1962). *Phys. Rev.*, **125**, 109. (*82*)
Sham, L. J. (1965). *Proc. Roy. Soc.*, **A283**, 33. (*193*, *248*)
Sham, L. J. (1966). *Phys. Rev.*, **150**, 720. (*282*)
Sham, L. J. (1969). *Phys. Rev.*, **188**, 1431. (*279*)
Sham, L. J. and Kohn, W. (1966). *Phys. Rev.*, **145**, 561. (*4*, *590*)
Shaw, R. W. and Pynn, R. (1969). *J. Phys.* C, **2**, 2071. (*265*)
Shockley, W. (1950). *Phys. Rev.*, **78**, 173. (*87*)
Shull, C. G. and Yamada, Y. (1962). *J. Phys. Soc. Japan*, Suppl. Biii, **17**, 1. (*472*)
Silin, V. P. (1957). *J.E.T.P.*, **33**, 495. (*147*)
Silverstein, S. D. (1962). *Phys. Rev.*, **128**, 631. (*172*)
Singwi, K. S., Tosi, M. P., Land, R. M. and Sjölander, A. (1968). *Phys. Rev.*, **176**, 589. (*138*, *189*)
Singwi, K. S., Sjölander, A., Tosi, M. P. and Land, R. H. (1970). *Phys. Rev.*, **B1**, 1044. (*189*, *195*)
Sinha, S. K. (1969). *Phys. Rev.*, **177**, 1256. (*277*)
Skolt, G. (1969). *Acta phys. Acad. Scient. Hung.*, **26** (3), 261. (*270*)
Slater, J. C. (1930). *Phys. Rev.*, **35**, 509. (*629*)
Slater, J. C. (1937). *Phys. Rev.*, **51**, 846. (*49*, *62*)
Slater, J. C. (1951). *Phys. Rev.*, **81**, 385. (*6*)
Slater, J. C. (1953). *Phys. Rev.*, **92**, 603. (*62*)
Slater, J. C. (1965). *Quantum Theory of Molecules and Solids*, Vol. II, McGraw-Hill, New York.
Sondheimer, E. H. and Wilson, A. H. (1951). *Proc. Roy. Soc.*, **A210**, 173. (*440*)
Soven, P. (1965). *Phys. Rev.*, **137**, A1706. (*95*)
Stewart, A. T. and Roellig, L. O. (1967). *Positron Annihilation*, Academic Press, New York.
Stoddart, J. C., Beattie, A. M. and March, N. H. (1971). *Int. Journ. Quantum Chem.*, **4**, 35. (*201*, *590*, *593*, *594*)
Stoddart, J. C. and March, N. H. (1967). *Proc. Roy. Soc.*, **A299**, 279. (*201*, *202*, *525*, *526*, *590*)
Stoner, E. C. (1938). *Proc. Roy. Soc.* A, **165**, 372. (*427*, *656*)
Stoner, E. C. (1939). *Proc. Roy. Soc.* A, **169**, 339. (*427*, *656*)

Teaney, D. T., Moruzzi, V. L. and Argyle, B. E. (1966). *J. Appl. Phys.*, **37**, 1122. (*499*)
Thouless, D. J. (1967). *Proc. Phys. Soc.*, **90**, 243. (*338*)
Tolpygo, K. B. (1957). *Ukr. Fiz. Zh.*, **2**, 242. (*270*)
Toya, T. (1958). *J. Research Inst. Catalysis*, **6**, 161, 183. (*258*)
Tucker, J. W., Jones, W. and March, N. H. (1965). *Phys. Letters*, **19**, 366. (*475*)

van Hove, L. (1954). *Phys. Rev.*, **95**, 249. (*125, 641*)
van Leeuwen, J. H. (1919). Dissertation, Leiden. (*439*)
Vinsome, P. K. W. and Jaros, M. (1970). *J. Phys.*, C, **3**, 2140. (*603*)
von der Lage, F. C. and Bethe, H. A. (1947). *Phys. Rev.*, **71**, 612. (*50*)
von Weizsäcker, C. F. (1935). *Z. Phys.*, **96**, 431. (*200, 201*)
Vosko, S. H., Taylor, R. and Keech, G. H. (1965). *Can. J. Phys.*, **43**, 1187. (*248, 253, 259*)

Walker, C. B. (1956). *Phys. Rev.*, **103**, 558. (*242, 243*)
Walsh, W. M. and Grimes, C. C. (1964). *Phys. Rev. Letters*, **13**, 523. (*99*)
Wannier, G. (1937). *Phys. Rev.*, **52**, 153. (*446*)
Watson, R. E. (1960). *Phys. Rev.*, **119**, 1934.
Watson, R. E., Ehrenreich, H. and Hodges, L. (1970). *Phys. Rev. Letters*, **24**, 829. (*83, 84*)
Weiss, R. J. and de Marco, J. J. (1958). *Rev. Mod. Phys.*, **30**, 59. (*472*)
Wigner, E. P. (1934). *Phys. Rev.*, **46**, 1002. (*193*)
Wigner, E. P. (1938). *Trans. Faraday Soc.*, **34**, 678. (*193*)
Wigner, E. P. and Seitz, F. (1933). *Phys. Rev.*, **43**, 804. (*49, 52*)
Wilson, K. G. (1971). *Phys. Rev.*, **B4**, 3174, 3184. (*373*)
Windsor, C. G. (1967). *Proc. Phys. Soc.*, **91**, 353. (*499*)
Windsor, C. G., Briggs, G. A. and Kestigian, M. (1968). *J. Phys. Soc.*, **1**, 940. (*498*)
Windsor, C. G., Lowde, R. D. and Allan, G. (1969). *Phys. Rev. Letters*, **22**, 849. (*505, 507*)
Windsor, C. G. and Stevenson, R. W. H. (1966). *Proc. Phys. Soc.*, **87**, 501. (*498*)
Wohlfarth, E. P. (1953). *Rev. Mod. Phys.*, **25**, 211. (*427*)
Wolff, P. A. (1960). *Phys. Rev.*, **120**, 814. (*406, 427*)
Woll, E. J. and Kohn, W. (1962). *Phys. Rev.*, **126**, 1693. (*260*)
Wood, J. H. (1962). *Phys. Rev.*, **126**, 517. (*84*)
Woods, A. D. B., Cochran, W. and Brockhouse, B. N. (1960). *Phys. Rev.*, **119**, 980. (*271*)

Yarnell, J. L., Warren, J. L. and Koenig, S. H. (1965). *Lattice Dynamics*, Conference Proceedings (Ed. R. F. Wallis), Pergamon, Oxford, p. 57. (*465*)
Yosida, K. and Watabe, L. (1962). *Prog. Theor. Phys. Japan*, **28**, 361. (See also (1957). *Phys. Rev.*, **106**, 893.) (*435*)

Ziman, J. M. (1965). *Proc. Phys. Soc.*, **86**, 337. (*81, 545*)
Ziman, J. M. (1969). *The Physics of Metals, 1. Electrons*, Cambridge University Press (especially Heine's article, p. 1). (*83*)
Ziman, J. M. (1971). *Solid State Physics, Vol. 26* (Eds. H. Ehrenreich, F. Seitz and D. Turnbull), Academic Press, New York, p. 1.
Zubarev, D. N. (1960). *Sov. Phys. Usp.*, **3**, 320. (*346*)

Index